Methods in Enzymology

Volume 277

MACROMOLECULAR CRYSTALLOGRAPHY

Part B

METHODS IN ENZYMOLOGY

EDITORS-IN-CHIEF

John N. Abelson   Melvin I. Simon

DIVISION OF BIOLOGY
CALIFORNIA INSTITUTE OF TECHNOLOGY
PASADENA, CALIFORNIA

FOUNDING EDITORS

Sidney P. Colowick and Nathan O. Kaplan

Methods in Enzymology

Volume 277

Macromolecular Crystallography

Part B

EDITED BY

Charles W. Carter, Jr.

DEPARTMENT OF BIOCHEMISTRY AND BIOPHYSICS
THE UNIVERSITY OF NORTH CAROLINA AT CHAPEL HILL
CHAPEL HILL, NORTH CAROLINA

Robert M. Sweet

BIOLOGY DEPARTMENT
BROOKHAVEN NATIONAL LABORATORY
UPTON, NEW YORK

ACADEMIC PRESS
San Diego London Boston New York Sydney Tokyo Toronto

This book is printed on acid-free paper. ∞

Academic Press
15 East 26th Street, 15th Floor, New York, New York 10010, USA
http://www.apnet.com

Academic Press Limited
24-28 Oval Road, London NW1 7DX, UK
http://www.hbuk.co.uk/ap/

International Standard Book Number: 0-12-182178-1

PRINTED IN THE UNITED STATES OF AMERICA
97 98 99 00 01 02 MM 9 8 7 6 5 4 3 2 1

Table of Contents

Section I. Phases

A. Horizon Methods

B. Model-Independent Map Refinement

Section II. Models

A. Model Building

B. Refinement

C. Verification: Safe Crystallography

Section III. Dynamic Properties

A. From Static Diffraction Data

B. From Time-Resolved Studies: Laue Diffraction

Section IV. Presentation and Analysis

A. Illustrating Structures

B. Modeling Structures

C. Databases

D. Program Packages

Contributors to Volume 277

Article numbers are in parentheses following the names of contributors.
Affiliations listed are current.

ENRIQUE E. ABOLA (29), *Protein Data Bank, Department of Biology, Brookhaven National Laboratory, Upton, New York 11973*

DAVID J. BACON (26), *Department of Computer Science, New York University, New York, New York 10012*

JOHN BADGER (17), *Molecular Simulations Inc., San Diego, California 92121*

HELEN M. BERMAN (30), *Department of Chemistry, Rutgers University, Piscataway, New Jersey 08855*

PHILIP E. BOURNE (30), *San Diego Supercomputer Center, La Jolla, California 92037, and Department of Pharmacology, University of California, San Diego, San Diego, California 92093*

JAMES U. BOWIE (20), *Department of Chemistry and Biochemistry, University of California, Los Angeles, Los Angeles, California 90095*

GÉRARD BRICOGNE (2, 5), *MRC Laboratory of Molecular Biology, Cambridge CB2 2QH, United Kingdom, and LURE, Université Paris–Sud, 91405 Orsay, France*

AXEL T. BRÜNGER (13, 19), *Howard Hughes Medical Institute and Department of Molecular Biophysics and Biochemistry, Yale University, New Haven, Connecticut 06520*

MIKE CARSON (25), *Center for Macromolecular Crystallography, University of Alabama at Birmingham, Birmingham, Alabama 35294*

CHARLES W. CARTER, JR. (6), *Department of Biochemistry and Biophysics, University of North Carolina at Chapel Hill, Chapel Hill, North Carolina 27599*

ANTONY CHIVERTON (8), *Department of Chemistry, Queen's University, Kingston, Ontario K7L 3N6, Canada*

JAMES B. CLARAGE (21), *Department of Biochemistry and Cell Biology, Institute of Biosciences and Bioengineering, Rice University, Houston, Texas 77005*

I. J. CLIFTON (23), *Laboratory of Molecular Biophysics, Oxford University, Oxford OX1 3QU, United Kingdom*

KEVIN COWTAN (4), *Department of Chemistry, University of York, Heslington, York YO1 5DD, England*

ELEANOR J. DODSON (32), *Department of Chemistry, University of York, Heslington, York YO1 5DD, United Kingdom*

E. M. H. DUKE (23), *CCLRC Daresbury Laboratory, Warrington WA4 4AD, United Kingdom*

DAVID EISENBERG (20), *Laboratory of Structural Biology and Molecular Medicine, Molecular Biology Institute, University of California, Los Angeles, Los Angeles, California 90095*

BARRY C. FINZEL (12), *Pharmacia and Upjohn, Kalamazoo, Michigan 49007*

PAULA M. D. FITZGERALD (30), *Merck Research Laboratories, Rahway, New Jersey 07065*

SUZANNE FORTIER (8), *Departments of Chemistry and Computing and Information Science, Queen's University, Kingston, Ontario K7L 3N6, Canada*

WILLIAM F. FUREY (31), *Biocrystallography Laboratory, Veterans Administration Medical Center, Pittsburgh, Pennsylvania 15240, and Department of Crystallography, University of Pittsburgh, Pittsburgh, Pennsylvania 15260*

GARY L. GILLILAND (28), *Center for Advanced Research in Biotechnology of the Maryland Biotechnology Institute and National Institute of Standards and Technology, Rockville, Maryland 20850*

CHRISTOPHER J. GILMORE (5), *Department of Chemistry, University of Glasgow, Glasgow G12 8QQ, Scotland*

JANICE GLASGOW (8), *Department of Computing and Information Science, Queen's University, Kingston, Ontario K7L 3N6, Canada*

ROGER S. GOODY (24), *Abteilung Physikalische Biochemie, Max Planck Institut für Molekulare Physiologie, 44 139 Dortmund, Germany*

HERBERT A. HAUPTMAN (1), *Hauptman-Woodward Medical Research Institute, Inc., Buffalo, New York 14203*

LYLE H. JENSEN (18), *Departments of Biological Structure and Biochemistry, University of Washington, Seattle, Washington 98195*

T. ALWYN JONES (10, 11, 27), *Department of Molecular Biology, Biomedical Centre, Uppsala University, S-75124 Uppsala, Sweden*

M. KJELDGAARD (10), *Department of Chemistry, University of Aarhus, DK-8000 Aarhus, Denmark*

GERARD J. KLEYWEGT (11, 27), *Department of Molecular Biology, Biomedical Centre, Uppsala University, S-75124 Uppsala, Sweden*

VICTOR S. LAMZIN (14), *European Molecular Biology Laboratory (EMBL), DESY, 22603 Hamburg, Germany*

LAURENCE LEHERTE (8), *Laboratoire de Physico-Chimie Informatique, Facultés Universitaires Notre-Dame de la Paix, Namur, Belgium*

ROLAND LÜTHY (20), *Amgen, Thousand Oaks, California 91320*

PETER MAIN (4), *Department of Physics, University of York, Heslington, York YO1 5DD, England*

NANCY O. MANNING (29), *Protein Data Bank, Department of Biology, Brookhaven National Laboratory, Upton, New York 11973*

BRIAN MCMAHON (30), *International Union of Crystallography, Chester CH1 2HU, United Kingdom*

ETHAN A. MERRITT (26), *Department of Biological Structure, University of Washington, Seattle, Washington 98195*

KEITH MOFFAT (22), *Department of Biochemistry and Molecular Biology, University of Chicago, Chicago, Illinois 60637*

GEORGE N. PHILLIPS, JR. (21), *Keck Center for Computational Biology, Department of Biochemistry and Cell Biology, Rice University, Houston, Texas 77005*

JAIME PRILUSKY (29), *Bioinformatics Unit, Weizmann Institute of Science, Rehovot 76100, Israel*

FLORANTE A. QUIOCHO (9), *Howard Hughes Medical Institute, Baylor College of Medicine, Houston, Texas 77030*

ADAM RALPH (32), *CCLRC Daresbury Laboratory, Warrington WA4 4AD, United Kingdom*

RANDY J. READ (3, 7), *Department of Medical Microbiology and Immunology, University of Alberta, Edmonton, Alberta T6G 2H7, Canada*

Z. REN (23), *Department of Biochemistry and Molecular Biology, University of Chicago, Chicago, Illinois 60637*

LUKE M. RICE (13), *Department of Molecular Biophysics and Biochemistry, Yale University, New Haven, Connecticut 06520*

JOHN S. SACK (9), *Department of Macromolecular Crystallography, Bristol-Myers Squibb Pharmaceutical Research Institute, Princeton, New Jersey 08543*

ILME SCHLICHTING (24), *Abteilung Physikalische Biochemie, Max Planck Institut für Molekulare Physiologie, 44 139 Dortmund, Germany*

THOMAS R. SCHNEIDER (16), *Max Planck Institut für Molekulare Physiologie, 44 139 Dortmund, Germany*

GEORGE M. SHELDRICK (16), *Institut für Anorganische Chemie, Göttingen University, D37077 Göttingen, Germany*

JOEL L. SUSSMAN (29), *Department of Structural Biology, Weizmann Institute of Science, Rehovot 76100, Israel, and Protein Data Bank, Department of Biology, Brookhaven National Laboratory, Upton, New York 11973*

S. SWAMINATHAN (31), *Biocrystallography Laboratory, Veterans Administration Medical Center, Pittsburgh, Pennsylvania 15240, and Department of Crystallography, University of Pittsburgh, Pittsburgh, Pennsylvania 15260*

DALE E. TRONRUD (15), *Howard Hughes Medical Institute and Institute of Molecular Biology, University of Oregon, Eugene, Oregon 97403*

F. M. D. VELLIEUX (3), *Institut de Biologie Structurale, Jean-Pierre Ebel CEA CNRS, 38027 Grenoble Cedex 01, France*

S. WAKATSUKI (23), *European Synchrotron Radiation Facility, F-38043 Grenoble Cedex, France*

KEITH D. WATENPAUGH (30), *Physical and Analytical Chemistry, Pharmacia and Upjohn, Kalamazoo, Michigan 49001*

JOHN D. WESTBROOK (30), *Department of Chemistry, Rutgers University, Piscataway, New Jersey 08855*

KEITH S. WILSON (14), *European Molecular Biology Laboratory (EMBL), DESY, 22603 Hamburg, Germany*

MARTYN WINN (32), *CCLRC Daresbury Laboratory, Warrington WA4 4AD, United Kingdom*

SHIBIN XIANG (6), *Department of Biochemistry and Biophysics, University of North Carolina at Chapel Hill, Chapel Hill, North Carolina 27599*

KAM Y. J. ZHANG (4), *Division of Basic Sciences, Fred Hutchinson Cancer Research Center, Seattle, Washington 98109*

Preface

Macromolecular crystallography is an indispensable component of enzymology. Structural biology, with macromolecular crystallography as its central technique, makes fundamental contributions to enzymology: one can pose few enzymological questions without first checking to see what relevant structures may be in the Protein Data Bank. Only then can one go on to learn what it implies about the mechanism, what suggestions it makes for genetic variation, and so on. We present these volumes to provide both a reference for practitioners of macromolecular crystallography and as a portal to the field for those who want to learn.

Methods in Enzymology Volumes 114 and 115, edited by Wyckoff, Hirs, and Timasheff, were timely. They provided the basic outlines and many details of macromolecular crystallography to several scientific generations of structural biologists, and many chapters remain primary resources. Since the publication of these volumes in 1985, macromolecular crystallography has evolved from an immature science, when practitioners needed to evaluate and possibly redesign each step in the process, to a set of procedures that are increasingly likely to succeed. This trend toward automation had characterized small molecule crystallography during the previous two decades, and had begun for macromolecular crystallography at the time of the publication of the two Wyckoff *et al.* volumes. The trend has accelerated and doubtless will spawn a growth industry in "service" macromolecular crystallography. This is evidenced by the growing population of practitioners whose primary interest rests not with structure determination itself, but with what can be derived from subsequent analysis. Systematic studies and comparison of mutants, ligand complexes, and different structural polymorphs depend on the rapid determination of structures.

At the same time, fundamental experimental, theoretical, and computational underpinnings of the field have experienced a parallel explosion. These include improved crystal growth and handling to provide higher resolution data, synchrotron X-ray sources, better detectors, improved methods for solving the phase problem, fragment library-based algorithms for the interpretation of electron density maps, and new refinement methods that, on the one hand, increase the radius of convergence for marginal models and, on the other, provide sophisticated models to exploit high-resolution data. We are becoming more sensitive to the importance of avoiding errors in interpretation and in understanding the limitations placed on refined parameters by the data.

A consequence of these changes is that our volumes differ from the preceding set not only in content that has arisen 10 years later, but also in emphasis. We perceive that the original practitioners of the crystallographic art were physicists, who handed the tool to chemists. Many of those now solving macromolecular crystal structures have biology as their primary training and interest. The core personnel responsible for the continued development of the field have been diluted both by the dispersion into a broad variety of departments, professional disciplines, and industrial laboratories and by the increasing numbers of relatively naive "users." Moreover, the multitude of techniques available offer multiple solutions to most of the rate-limiting steps. Often the choice of which approach to take depends more on personal experience and taste than on respect for the underlying principles. Therefore, while emphasizing experimental methods, we have included many chapters that describe the fundamentals of recent advances that may spark further transformation of the field.

The chapters in these volumes present expert witness to the state-of-the-art for many individual aspects of the field. The two volumes provide the logical train of objectives necessary to solve, refine, and analyze a macromolecular crystal structure. Although these volumes may not serve as a simple textbook to introduce the field, individual chapters should provide a comprehensive background for each topic. Students and teachers alike will benefit from a careful reading of each chapter as it becomes relevant in the course of research.

Part A (Volume 276) deals with the three requisites for structure solution: crystals, data, and phases. The first section covers aspects of the theory and practice of growing high-quality crystals. Since exploiting intrinsic information from a crystal to its full extent depends on measuring the highest quality data, the second section provides information about radiation sources, instrumentation for recording, and software for processing these data. Finding the phases represents the final rate-limiting step in the solving of a structure. Therefore the third section includes a penetrating analysis of the statistical foundations of the phase problem and covers a broad range of experimental sources of phase information and the techniques for using them effectively. It ends with several "horizon" methods that may help transform phase determination in the coming decade.

Part B (Volume 277) continues the section on horizon methods for phase determination. It follows with various ways in which structures are built, refined, and analyzed. An important development since 1985 is in model-independent, real-space refinement. Construction of atomic models is the crucial step in converting electron density maps into structures. Chapters are included that present the increasing trend toward computer-assisted and/or automated map interpretation. Fragment libraries representing how proteins fold are already integral parts of some of the software described previously. Use of simulated annealing in model refinement has increased the radius of convergence; it has become integrated *de facto* into the process of solving structures. New tools for refinement of models to fit high-resolution data, when they can be measured, now permit the exploration of more detailed models. Procedures for cross-verification and *post hoc* analysis provide tools to help avoid unnecessary errors and possibly incorrect structures.

A long-term goal in structural biology is "molecular cinematography." The molecules we study undergo some kind of internal motion in carrying out its function. Some of these motions can be inferred experimentally by the analysis of the static diffraction patterns. Others require the use of multiple diffraction patterns recorded as a function of time. These topics are covered in the next section on dynamic properties.

The final sections sample widely used accessory software for manipulating, archiving, analyzing, and presenting structures. Databases, with tools for accessing specific information contained therein, are essential resources for those who study macromolecular structures, and even for those involved in crystal growth. Finally, we have documented some of the integrated packages of software which contain most of the tools needed for structure solution.

The ferocious march of technology places burdens on everyone concerned with the production of such a collection, and we are sincerely grateful to the authors for their cooperation and patience. The staff of Academic Press provided continuous and valuable support, which we both appreciate.

CHARLES W. CARTER, JR.
ROBERT M. SWEET

METHODS IN ENZYMOLOGY

VOLUME XVII. Metabolism of Amino Acids and Amines (Parts A and B)
Edited by HERBERT TABOR AND CELIA WHITE TABOR

VOLUME XVIII. Vitamins and Coenzymes (Parts A, B, and C)
Edited by DONALD B. MCCORMICK AND LEMUEL D. WRIGHT

VOLUME XIX. Proteolytic Enzymes
Edited by GERTRUDE E. PERLMANN AND LASZLO LORAND

VOLUME XX. Nucleic Acids and Protein Synthesis (Part C)
Edited by KIVIE MOLDAVE AND LAWRENCE GROSSMAN

VOLUME XXI. Nucleic Acids (Part D)
Edited by LAWRENCE GROSSMAN AND KIVIE MOLDAVE

VOLUME XXII. Enzyme Purification and Related Techniques
Edited by WILLIAM B. JAKOBY

VOLUME XXIII. Photosynthesis (Part A)
Edited by ANTHONY SAN PIETRO

VOLUME XXIV. Photosynthesis and Nitrogen Fixation (Part B)
Edited by ANTHONY SAN PIETRO

VOLUME XXV. Enzyme Structure (Part B)
Edited by C. H. W. HIRS AND SERGE N. TIMASHEFF

VOLUME XXVI. Enzyme Structure (Part C)
Edited by C. H. W. HIRS AND SERGE N. TIMASHEFF

VOLUME XXVII. Enzyme Structure (Part D)
Edited by C. H. W. HIRS AND SERGE N. TIMASHEFF

VOLUME XXVIII. Complex Carbohydrates (Part B)
Edited by VICTOR GINSBURG

VOLUME XXIX. Nucleic Acids and Protein Synthesis (Part E)
Edited by LAWRENCE GROSSMAN AND KIVIE MOLDAVE

VOLUME XXX. Nucleic Acids and Protein Synthesis (Part F)
Edited by KIVIE MOLDAVE AND LAWRENCE GROSSMAN

VOLUME XXXI. Biomembranes (Part A)
Edited by SIDNEY FLEISCHER AND LESTER PACKER

VOLUME XXXII. Biomembranes (Part B)
Edited by SIDNEY FLEISCHER AND LESTER PACKER

VOLUME XXXIII. Cumulative Subject Index Volumes I–XXX
Edited by MARTHA G. DENNIS AND EDWARD A. DENNIS

VOLUME XXXIV. Affinity Techniques (Enzyme Purification: Part B)
Edited by WILLIAM B. JAKOBY AND MEIR WILCHEK

VOLUME XXXV. Lipids (Part B)
Edited by JOHN M. LOWENSTEIN

VOLUME XXXVI. Hormone Action (Part A: Steroid Hormones)
Edited by BERT W. O'MALLEY AND JOEL G. HARDMAN

VOLUME XXXVII. Hormone Action (Part B: Peptide Hormones)
Edited by BERT W. O'MALLEY AND JOEL G. HARDMAN

VOLUME XXXVIII. Hormone Action (Part C: Cyclic Nucleotides)
Edited by JOEL G. HARDMAN AND BERT W. O'MALLEY

VOLUME XXXIX. Hormone Action (Part D: Isolated Cells, Tissues, and Organ Systems)
Edited by JOEL G. HARDMAN AND BERT W. O'MALLEY

VOLUME XL. Hormone Action (Part E: Nuclear Structure and Function)
Edited by BERT W. O'MALLEY AND JOEL G. HARDMAN

VOLUME XLI. Carbohydrate Metabolism (Part B)
Edited by W. A. WOOD

VOLUME XLII. Carbohydrate Metabolism (Part C)
Edited by W. A. WOOD

VOLUME XLIII. Antibiotics
Edited by JOHN H. HASH

VOLUME XLIV. Immobilized Enzymes
Edited by KLAUS MOSBACH

VOLUME XLV. Proteolytic Enzymes (Part B)
Edited by LASZLO LORAND

VOLUME XLVI. Affinity Labeling
Edited by WILLIAM B. JAKOBY AND MEIR WILCHEK

VOLUME XLVII. Enzyme Structure (Part E)
Edited by C. H. W. HIRS AND SERGE N. TIMASHEFF

VOLUME XLVIII. Enzyme Structure (Part F)
Edited by C. H. W. HIRS AND SERGE N. TIMASHEFF

VOLUME XLIX. Enzyme Structure (Part G)
Edited by C. H. W. HIRS AND SERGE N. TIMASHEFF

VOLUME L. Complex Carbohydrates (Part C)
Edited by VICTOR GINSBURG

VOLUME LI. Purine and Pyrimidine Nucleotide Metabolism
Edited by PATRICIA A. HOFFEE AND MARY ELLEN JONES

VOLUME LII. Biomembranes (Part C: Biological Oxidations)
Edited by SIDNEY FLEISCHER AND LESTER PACKER

VOLUME LIII. Biomembranes (Part D: Biological Oxidations)
Edited by SIDNEY FLEISCHER AND LESTER PACKER

VOLUME LIV. Biomembranes (Part E: Biological Oxidations)
Edited by SIDNEY FLEISCHER AND LESTER PACKER

VOLUME LV. Biomembranes (Part F: Bioenergetics)
Edited by SIDNEY FLEISCHER AND LESTER PACKER

VOLUME LVI. Biomembranes (Part G: Bioenergetics)
Edited by SIDNEY FLEISCHER AND LESTER PACKER

VOLUME LVII. Bioluminescence and Chemiluminescence
Edited by MARLENE A. DELUCA

VOLUME LVIII. Cell Culture
Edited by WILLIAM B. JAKOBY AND IRA PASTAN

VOLUME LIX. Nucleic Acids and Protein Synthesis (Part G)
Edited by KIVIE MOLDAVE AND LAWRENCE GROSSMAN

VOLUME LX. Nucleic Acids and Protein Synthesis (Part H)
Edited by KIVIE MOLDAVE AND LAWRENCE GROSSMAN

VOLUME 61. Enzyme Structure (Part H)
Edited by C. H. W. HIRS AND SERGE N. TIMASHEFF

VOLUME 62. Vitamins and Coenzymes (Part D)
Edited by DONALD B. MCCORMICK AND LEMUEL D. WRIGHT

VOLUME 63. Enzyme Kinetics and Mechanism (Part A: Initial Rate and Inhibitor Methods)
Edited by DANIEL L. PURICH

VOLUME 64. Enzyme Kinetics and Mechanism (Part B: Isotopic Probes and Complex Enzyme Systems)
Edited by DANIEL L. PURICH

VOLUME 65. Nucleic Acids (Part I)
Edited by LAWRENCE GROSSMAN AND KIVIE MOLDAVE

VOLUME 66. Vitamins and Coenzymes (Part E)
Edited by DONALD B. MCCORMICK AND LEMUEL D. WRIGHT

VOLUME 67. Vitamins and Coenzymes (Part F)
Edited by DONALD B. MCCORMICK AND LEMUEL D. WRIGHT

VOLUME 68. Recombinant DNA
Edited by RAY WU

VOLUME 69. Photosynthesis and Nitrogen Fixation (Part C)
Edited by ANTHONY SAN PIETRO

VOLUME 70. Immunochemical Techniques (Part A)
Edited by HELEN VAN VUNAKIS AND JOHN J. LANGONE

VOLUME 71. Lipids (Part C)
Edited by JOHN M. LOWENSTEIN

VOLUME 72. Lipids (Part D)
Edited by JOHN M. LOWENSTEIN

VOLUME 126. Biomembranes (Part N: Transport in Bacteria, Mitochondria, and Chloroplasts: Protonmotive Force)
Edited by SIDNEY FLEISCHER AND BECCA FLEISCHER

VOLUME 127. Biomembranes (Part O: Protons and Water: Structure and Translocation)
Edited by LESTER PACKER

VOLUME 128. Plasma Lipoproteins (Part A: Preparation, Structure, and Molecular Biology)
Edited by JERE P. SEGREST AND JOHN J. ALBERS

VOLUME 129. Plasma Lipoproteins (Part B: Characterization, Cell Biology, and Metabolism)
Edited by JOHN J. ALBERS AND JERE P. SEGREST

VOLUME 130. Enzyme Structure (Part K)
Edited by C. H. W. HIRS AND SERGE N. TIMASHEFF

VOLUME 131. Enzyme Structure (Part L)
Edited by C. H. W. HIRS AND SERGE N. TIMASHEFF

VOLUME 132. Immunochemical Techniques (Part J: Phagocytosis and Cell-Mediated Cytotoxicity)
Edited by GIOVANNI DI SABATO AND JOHANNES EVERSE

VOLUME 133. Bioluminescence and Chemiluminescence (Part B)
Edited by MARLENE DELUCA AND WILLIAM D. MCELROY

VOLUME 134. Structural and Contractile Proteins (Part C: The Contractile Apparatus and the Cytoskeleton)
Edited by RICHARD B. VALLEE

VOLUME 135. Immobilized Enzymes and Cells (Part B)
Edited by KLAUS MOSBACH

VOLUME 136. Immobilized Enzymes and Cells (Part C)
Edited by KLAUS MOSBACH

VOLUME 137. Immobilized Enzymes and Cells (Part D)
Edited by KLAUS MOSBACH

VOLUME 138. Complex Carbohydrates (Part E)
Edited by VICTOR GINSBURG

VOLUME 139. Cellular Regulators (Part A: Calcium- and Calmodulin-Binding Proteins)
Edited by ANTHONY R. MEANS AND P. MICHAEL CONN

VOLUME 140. Cumulative Subject Index Volumes 102–119, 121–134

VOLUME 141. Cellular Regulators (Part B: Calcium and Lipids)
Edited by P. MICHAEL CONN AND ANTHONY R. MEANS

VOLUME 142. Metabolism of Aromatic Amino Acids and Amines
Edited by SEYMOUR KAUFMAN

VOLUME 143. Sulfur and Sulfur Amino Acids
Edited by WILLIAM B. JAKOBY AND OWEN GRIFFITH

VOLUME 144. Structural and Contractile Proteins (Part D: Extracellular Matrix)
Edited by LEON W. CUNNINGHAM

VOLUME 145. Structural and Contractile Proteins (Part E: Extracellular Matrix)
Edited by LEON W. CUNNINGHAM

VOLUME 146. Peptide Growth Factors (Part A)
Edited by DAVID BARNES AND DAVID A. SIRBASKU

VOLUME 147. Peptide Growth Factors (Part B)
Edited by DAVID BARNES AND DAVID A. SIRBASKU

VOLUME 148. Plant Cell Membranes
Edited by LESTER PACKER AND ROLAND DOUCE

VOLUME 149. Drug and Enzyme Targeting (Part B)
Edited by RALPH GREEN AND KENNETH J. WIDDER

VOLUME 150. Immunochemical Techniques (Part K: *In Vitro* Models of B and T Cell Functions and Lymphoid Cell Receptors)
Edited by GIOVANNI DI SABATO

VOLUME 151. Molecular Genetics of Mammalian Cells
Edited by MICHAEL M. GOTTESMAN

VOLUME 152. Guide to Molecular Cloning Techniques
Edited by SHELBY L. BERGER AND ALAN R. KIMMEL

VOLUME 153. Recombinant DNA (Part D)
Edited by RAY WU AND LAWRENCE GROSSMAN

VOLUME 154. Recombinant DNA (Part E)
Edited by RAY WU AND LAWRENCE GROSSMAN

VOLUME 155. Recombinant DNA (Part F)
Edited by RAY WU

VOLUME 156. Biomembranes (Part P: ATP-Driven Pumps and Related Transport: The Na,K-Pump)
Edited by SIDNEY FLEISCHER AND BECCA FLEISCHER

VOLUME 157. Biomembranes (Part Q: ATP-Driven Pumps and Related Transport: Calcium, Proton, and Potassium Pumps)
Edited by SIDNEY FLEISCHER AND BECCA FLEISCHER

VOLUME 158. Metalloproteins (Part A)
Edited by JAMES F. RIORDAN AND BERT L. VALLEE

VOLUME 159. Initiation and Termination of Cyclic Nucleotide Action
Edited by JACKIE D. CORBIN AND ROGER A. JOHNSON

VOLUME 160. Biomass (Part A: Cellulose and Hemicellulose)
Edited by WILLIS A. WOOD AND SCOTT T. KELLOGG

Volume 161. Biomass (Part B: Lignin, Pectin, and Chitin)
Edited by Willis A. Wood and Scott T. Kellogg

Volume 162. Immunochemical Techniques (Part L: Chemotaxis and Inflammation)
Edited by Giovanni Di Sabato

Volume 163. Immunochemical Techniques (Part M: Chemotaxis and Inflammation)
Edited by Giovanni Di Sabato

Volume 164. Ribosomes
Edited by Harry F. Noller, Jr., and Kivie Moldave

Volume 165. Microbial Toxins: Tools for Enzymology
Edited by Sidney Harshman

Volume 166. Branched-Chain Amino Acids
Edited by Robert Harris and John R. Sokatch

Volume 167. Cyanobacteria
Edited by Lester Packer and Alexander N. Glazer

Volume 168. Hormone Action (Part K: Neuroendocrine Peptides)
Edited by P. Michael Conn

Volume 169. Platelets: Receptors, Adhesion, Secretion (Part A)
Edited by Jacek Hawiger

Volume 170. Nucleosomes
Edited by Paul M. Wassarman and Roger D. Kornberg

Volume 171. Biomembranes (Part R: Transport Theory: Cells and Model Membranes)
Edited by Sidney Fleischer and Becca Fleischer

Volume 172. Biomembranes (Part S: Transport: Membrane Isolation and Characterization)
Edited by Sidney Fleischer and Becca Fleischer

Volume 173. Biomembranes [Part T: Cellular and Subcellular Transport: Eukaryotic (Nonepithelial) Cells]
Edited by Sidney Fleischer and Becca Fleischer

Volume 174. Biomembranes [Part U: Cellular and Subcellular Transport: Eukaryotic (Nonepithelial) Cells]
Edited by Sidney Fleischer and Becca Fleischer

Volume 175. Cumulative Subject Index Volumes 135–139, 141–167

Volume 176. Nuclear Magnetic Resonance (Part A: Spectral Techniques and Dynamics)
Edited by Norman J. Oppenheimer and Thomas L. James

Volume 177. Nuclear Magnetic Resonance (Part B: Structure and Mechanism)
Edited by Norman J. Oppenheimer and Thomas L. James

VOLUME 213. Carotenoids (Part A: Chemistry, Separation, Quantitation, and Antioxidation)
Edited by LESTER PACKER

VOLUME 214. Carotenoids (Part B: Metabolism, Genetics, and Biosynthesis)
Edited by LESTER PACKER

VOLUME 215. Platelets: Receptors, Adhesion, Secretion (Part B)
Edited by JACEK J. HAWIGER

VOLUME 216. Recombinant DNA (Part G)
Edited by RAY WU

VOLUME 217. Recombinant DNA (Part H)
Edited by RAY WU

VOLUME 218. Recombinant DNA (Part I)
Edited by RAY WU

VOLUME 219. Reconstitution of Intracellular Transport
Edited by JAMES E. ROTHMAN

VOLUME 220. Membrane Fusion Techniques (Part A)
Edited by NEJAT DÜZGÜNEŞ

VOLUME 221. Membrane Fusion Techniques (Part B)
Edited by NEJAT DÜZGÜNEŞ

VOLUME 222. Proteolytic Enzymes in Coagulation, Fibrinolysis, and Complement Activation (Part A: Mammalian Blood Coagulation Factors and Inhibitors)
Edited by LASZLO LORAND AND KENNETH G. MANN

VOLUME 223. Proteolytic Enzymes in Coagulation, Fibrinolysis, and Complement Activation (Part B: Complement Activation, Fibrinolysis, and Nonmammalian Blood Coagulation Factors)
Edited by LASZLO LORAND AND KENNETH G. MANN

VOLUME 224. Molecular Evolution: Producing the Biochemical Data
Edited by ELIZABETH ANNE ZIMMER, THOMAS J. WHITE, REBECCA L. CANN, AND ALLAN C. WILSON

VOLUME 225. Guide to Techniques in Mouse Development
Edited by PAUL M. WASSARMAN AND MELVIN L. DEPAMPHILIS

VOLUME 226. Metallobiochemistry (Part C: Spectroscopic and Physical Methods for Probing Metal Ion Environments in Metalloenzymes and Metalloproteins)
Edited by JAMES F. RIORDAN AND BERT L. VALLEE

VOLUME 227. Metallobiochemistry (Part D: Physical and Spectroscopic Methods for Probing Metal Ion Environments in Metalloproteins)
Edited by JAMES F. RIORDAN AND BERT L. VALLEE

VOLUME 228. Aqueous Two-Phase Systems
Edited by HARRY WALTER AND GÖTE JOHANSSON

VOLUME 229. Cumulative Subject Index Volumes 195–198, 200–227

VOLUME 230. Guide to Techniques in Glycobiology
Edited by WILLIAM J. LENNARZ AND GERALD W. HART

VOLUME 231. Hemoglobins (Part B: Biochemical and Analytical Methods)
Edited by JOHANNES EVERSE, KIM D. VANDEGRIFF, AND ROBERT M. WINSLOW

VOLUME 232. Hemoglobins (Part C: Biophysical Methods)
Edited by JOHANNES EVERSE, KIM D. VANDEGRIFF, AND ROBERT M. WINSLOW

VOLUME 233. Oxygen Radicals in Biological Systems (Part C)
Edited by LESTER PACKER

VOLUME 234. Oxygen Radicals in Biological Systems (Part D)
Edited by LESTER PACKER

VOLUME 235. Bacterial Pathogenesis (Part A: Identification and Regulation of Virulence Factors)
Edited by VIRGINIA L. CLARK AND PATRIK M. BAVOIL

VOLUME 236. Bacterial Pathogenesis (Part B: Integration of Pathogenic Bacteria with Host Cells)
Edited by VIRGINIA L. CLARK AND PATRIK M. BAVOIL

VOLUME 237. Heterotrimeric G Proteins
Edited by RAVI IYENGAR

VOLUME 238. Heterotrimeric G-Protein Effectors
Edited by RAVI IYENGAR

VOLUME 239. Nuclear Magnetic Resonance (Part C)
Edited by THOMAS L. JAMES AND NORMAN J. OPPENHEIMER

VOLUME 240. Numerical Computer Methods (Part B)
Edited by MICHAEL L. JOHNSON AND LUDWIG BRAND

VOLUME 241. Retroviral Proteases
Edited by LAWRENCE C. KUO AND JULES A. SHAFER

VOLUME 242. Neoglycoconjugates (Part A)
Edited by Y. C. LEE AND REIKO T. LEE

VOLUME 243. Inorganic Microbial Sulfur Metabolism
Edited by HARRY D. PECK, JR., AND JEAN LEGALL

VOLUME 244. Proteolytic Enzymes: Serine and Cysteine Peptidases
Edited by ALAN J. BARRETT

VOLUME 245. Extracellular Matrix Components
Edited by E. RUOSLAHTI AND E. ENGVALL

VOLUME 246. Biochemical Spectroscopy
Edited by KENNETH SAUER

VOLUME 247. Neoglycoconjugates (Part B: Biomedical Applications)
Edited by Y. C. LEE AND REIKO T. LEE

Section I

Phases

A. Horizon Methods
Articles 1 and 2

B. Model-Independent Map Refinement
Articles 3 through 7

[1] *Shake-and-Bake:* An Algorithm for Automatic Solution *ab Initio* of Crystal Structures

By HERBERT A. HAUPTMAN

Introduction

The phase problem is formulated as a problem in constrained global minimization. Because the objective function has a myriad of local minima, the problem of locating the global minimum is presumed to be difficult. However, the ability to impose constraints, in the form of identities among the phases that must, of necessity, be satisfied, even if only incompletely and approximately, transforms the problem completely and enables one to avoid the countless (local) minima of the objective function and to arrive at the (essentially unique) constrained global minimum. Thus the computer program *Shake-and-Bake* (*SnB*), which implements the minimal principle formulated here, provides a routine and completely automatic solution of the phase problem when diffraction intensities to 1.2 Å or better are available. *Shake-and-Bake* yields initial values of the phases with an average error in the range of 20 to 30° in typical cases. Structures having as many as 600 independent nonhydrogen atoms have been solved routinely even when no heavy atom is present. Its ultimate potential is still unknown.

It is assumed for simplicity that the structure consists of N identical atoms per unit cell, although the method is easily generalized to the case in which unequal atoms are present. The relationship between the (complex) normalized structure factors $E_{\mathbf{H}}$ and the crystal structure is given by the pair of equations (1) and (2):

$$E_{\mathbf{H}} = |E_{\mathbf{H}}| \exp(i\phi_{\mathbf{H}}) = \frac{1}{N^{1/2}} \sum_{j=1}^{N} \exp(2\pi i H \cdot \mathbf{r}_j) \tag{1}$$

$$\begin{aligned} \langle E_{\mathbf{H}} \exp(-2\pi i H \cdot \mathbf{r}) \rangle_{\mathbf{H}} &= \frac{1}{N^{1/2}} \left\langle \sum_{j=1}^{N} \exp[2\pi i H \cdot (\mathbf{r}_j - \mathbf{r})] \right\rangle_{\mathbf{H}} \\ &= \frac{1}{N^{1/2}} \quad \text{if} \quad \mathbf{r} = \mathbf{r}_j \\ &= 0 \quad \text{if} \quad \mathbf{r} \neq \mathbf{r}_j \end{aligned} \tag{2}$$

where $\mathbf{r}_j$ is the position vector of the atom labeled j and $\mathbf{r}$ is an arbitrary position vector. Hence the positions of the maxima of the Fourier series [Eq. (2)] coincide with the atomic position vectors $\mathbf{r}_j$. To calculate the

0076-6879/97 $25.00

function of the position vector $\mathbf{r}$, Eq. (2), both the magnitudes $|E_{\mathbf{H}}|$ and phases $\phi_{\mathbf{H}}$ of the complex normalized structure factors $E_{\mathbf{H}}$ are needed.

In practice, the magnitudes $|E_{\mathbf{H}}|$ are obtainable (at least approximately) from the observed diffraction intensities but the phases $\phi_{\mathbf{H}}$, as defined by Eq. (1), cannot be determined experimentally. However, the system of equations [Eq. (1)] implies the existence of relationships among the normalized structure factors $E_{\mathbf{H}}$ because the (relatively few) unknown atomic position vectors $\mathbf{r}_j$ may, at least in principle, be eliminated. In this way one obtains a system of equations among the normalized structure factors $E_{\mathbf{H}}$ alone, i.e., among the phases $\phi_{\mathbf{H}}$ and magnitudes $|E_{\mathbf{H}}|$:

$$F(E) = G(\phi; |E|) = 0 \tag{3}$$

which depends on N but is independent of the atomic position vectors $\mathbf{r}_j$. For a specified crystal structure, the magnitudes $|E|$ are determined and the system of equations [Eq. (3)] leads directly to a system of identities among the phases ϕ alone:

$$G(\phi; |E|) \rightarrow H(\phi/|E|) = 0 \tag{4}$$

dependent now on the presumed known magnitudes $|E|$, which must of necessity be satisfied. By the term *direct methods* is meant that class of methods that exploits relationships among the phases to go directly from the observed magnitudes $|E|$ to the needed phases ϕ.

Structure Invariants

Equation (2) implies that the normalized structure factors $E_{\mathbf{H}}$ determine the crystal structure. However, Eq. (1) does not imply that, conversely, the crystal structure determines the values of the normalized structure factors $E_{\mathbf{H}}$ because the position vectors $\mathbf{r}_j$ depend not only on the structure but on the choice of origin as well. It turns out, nevertheless, that the magnitudes $|E_{\mathbf{H}}|$ of the normalized structure factors are in fact uniquely determined by the crystal structure and are independent of the choice of origin, but that the values of the phases $\phi_{\mathbf{H}}$ depend also on the choice of origin. Although the values of the individual phases depend on the structure and the choice of origin, there exist certain linear combinations of the phases, the so-called structure invariants, whose values are determined by the structure alone and are independent of the choice of origin. The most important class of structure invariants, and the only one to be considered here, consists of the three-phase structure invariants (triplets):

$$\phi_{\mathbf{HK}} = \phi_{\mathbf{H}} + \phi_{\mathbf{K}} + \phi_{-\mathbf{H}-\mathbf{K}} \tag{5}$$

where $\mathbf{H}$ and $\mathbf{K}$ are arbitrary reciprocal lattice vectors.

Probabilistic Background

Historical Background

The techniques of modern probability theory lead to the joint probability distributions of arbitrary collections of diffraction intensities and their corresponding phases. These distributions constitute the foundation on which direct methods are based. They have provided the unifying thread from the beginning (around 1950) until the present time. They have led, in particular, to the (first) minimal principle,[1–3] which has found expression in the *Shake-and-Bake* formalism,[4,5] a computer program that provides a completely automatic solution to the phase problem, *ab initio,* provided that diffraction data to at least 1.2 Å are available. Our experience shows that structures having as many as 600 independent nonhydrogen atoms are routinely accessible by this approach and suggests that its ultimate potential is greater still.

It should perhaps be pointed out that the minimal principle, the theoretical basis of *Shake-and-Bake,* replaces the phase problem with one of constrained global minimization, in sharp contrast to an earlier formulation by Debaerdemaeker and Woolfson[6] in which the (unconstrained) global minimum was sought. The distinction is crucial, not only on the theoretical level, but in the application as well: Not only does the constrained global minimum yield the correct values of the individual phases *ab initio,* but this formulation suggests how the minimum is to be reached (via the *Shake-and-Bake* program) and how to identify it. The failure to impose the constraints, as had been done in the earlier work, greatly limited the usefulness of that approach, in part because it was not at all clear how to reach or identify the unconstrained global minimum, and, even more important, because the unconstrained global minimum does not, in fact, solve the phase problem.

[1] H. Hauptman, *in* "Proceedings of the International School of Crystallographic Computing" (D. Moras, A. D. Podjarny, and J. C. Thierry, eds.), pp. 324–332. Published for the International Union of Crystallography by Oxford University Press, Oxford, 1991.

[2] H. Hauptman, D. Velmurugan, and H. Fusen, *in* "Direct Methods of Solving Crystal Structures" (H. Schenk, ed.), pp. 403–406. Plenum, New York, 1991.

[3] G. DeTitta, C. Weeks, P. Thuman, R. Miller, and H. Hauptman, *Acta Crystallogr.* **A50,** 203 (1994).

[4] C. M. Weeks, G. T. DeTitta, H. A. Hauptman, P. Thuman, and R. Miller, *Acta Crystallogr.* **A50,** 210 (1994).

[5] R. Miller, S. M. Gallo, H. G. Khalak, and C. M. Weeks, *J. Appl. Crystallogr.* **27,** 613 (1994).

[6] T. Debaerdemaeker and M. M. Woolfson, *Acta Crystallogr.* **A39,** 193 (1983).

Probabilistic Background

It is assumed that (1) a crystal structure is specified, (2) three nonnegative numbers R_1, R_2, R_3 are also specified, and (3) the reciprocal lattice vectors **H** and **K** are the primitive random variables, which are assumed to be uniformly and independently distributed in the subset of reciprocal space defined by

$$|E_{\mathbf{H}}| = R_1, \qquad |E_{\mathbf{K}}| = R_2, \qquad |E_{\mathbf{H+K}}| = R_3 \tag{6}$$

where the magnitudes $|E|$ are defined by Eq. (1). Then the structure invariant $\phi_{\mathbf{HK}}$ [Eq. (5)], as a function of the primitive random variables **H** and **K** [Eq. (1)], is itself a random variable.

Conditional Probability Distribution of $\phi_{\mathbf{HK}}$, Given $|E_{\mathbf{H}}|$, $|E_{\mathbf{K}}|$, $|E_{\mathbf{H+K}}|$

Under the three assumptions of the preceding section, the conditional probability distribution of the triplet $\phi_{\mathbf{HK}}$ [Eq. (5)], where $|E_{\mathbf{H}}|$, $|E_{\mathbf{K}}|$, and $|E_{\mathbf{H+K}}|$ are given by Eq. (6), is known to be

$$P(\Phi/R_1, R_2, R_3) = \frac{1}{2\pi I_0(\kappa_{\mathbf{HK}})} \exp(\kappa_{\mathbf{HK}} \cos \Phi) \tag{7}$$

where Φ represents the triplet $\phi_{\mathbf{HK}}$,

$$\kappa_{\mathbf{HK}} = \frac{2}{N^{1/2}} R_1 R_2 R_3 = \frac{2}{N^{1/2}} |E_{\mathbf{H}} E_{\mathbf{K}} E_{\mathbf{H+K}}| \tag{8}$$

and I_0 is the modified Bessel function. Equation (7) implies that the mode of $\phi_{\mathbf{HK}}$ is zero, and the conditional expectation value (or average) of cos $\phi_{\mathbf{HK}}$, given $\kappa_{\mathbf{HK}}$, is

$$\varepsilon(\cos \phi_{\mathbf{HK}}) = \frac{I_1(\kappa_{\mathbf{HK}})}{I_0(\kappa_{\mathbf{HK}})} > 0 \tag{9}$$

where I_1 is the modified Bessel function. It is also readily confirmed that the larger the value of $\kappa_{\mathbf{HK}}$ the smaller is the conditional variance of cos $\phi_{\mathbf{HK}}$, given $\kappa_{\mathbf{HK}}$. It is to be stressed that the conditional expected value of the cosine, Eq. (9), is always positive because $\kappa_{\mathbf{HK}} > 0$.

Minimal Principle

Heuristic Background

It is assumed that a crystal structure consisting of N identical atoms in the unit cell is fixed, but unknown, that the magnitudes $|E|$ of the normalized structure factors E are known, and that a sufficiently large base of phases, corresponding to the largest magnitudes $|E|$, is specified.

The mode of the triplet distribution [Eq. (7)] is zero and the variance of the cosine is small if $\kappa_{\mathbf{HK}}$ [Eq. (8)] is large. In this way, the estimate for the triplet $\phi_{\mathbf{HK}}$ [Eq. 5)] is obtained:

$$\phi_{\mathbf{HK}} = \phi_{\mathbf{H}} + \phi_{\mathbf{K}} + \phi_{-\mathbf{H}-\mathbf{K}} \approx 0 \tag{10}$$

which is particularly good in the favorable case that $\kappa_{\mathbf{HK}}$ is large, i.e., that $|E_{\mathbf{H}}|$, $|E_{\mathbf{K}}|$, and $|E_{\mathbf{H+K}}|$ are all large. The estimate given by Eq. (10) is one of the cornerstones of the traditional techniques of direct methods. It is surprising how useful Eq. (10) has proven to be in the applications, especially as it yields only the zero estimate of the triplet, and only those estimates are reliable for which $|E_{\mathbf{H}}|$, $|E_{\mathbf{K}}|$, and $|E_{\mathbf{H+K}}|$ are all large. Clearly the coefficient $2/N^{1/2}$ in Eq. (8), and therefore $\kappa_{\mathbf{HK}}$ as well, both decrease with increasing N, i.e., with increasing structural complexity. Hence the relationship [Eq. (10)] becomes increasingly unreliable for larger structures, and the traditional step-by-step sequential direct methods procedures based on Eq. (10) eventually fail. Even the rantan method of Yao[7] and the program SAYTAN, devised by Woolfson and Yao,[8] which have to some extent overcome this limitation, also break down for the larger structures. Here a different approach is described that extends the range of complexity still further; its ultimate potential, however, is still unknown.

In view of Eq. (9) and the previous discussion one now replaces the zero estimate [Eq. (10)] of the triplet $\phi_{\mathbf{HK}}$ by the estimate

$$\cos \phi_{\mathbf{HK}} \approx \frac{I_1(\kappa_{\mathbf{HK}})}{I_0(\kappa_{\mathbf{HK}})} \tag{11}$$

and expects that the smaller the variance, that is the larger $\kappa_{\mathbf{HK}}$ becomes, the more reliable this estimate will be. Hence one is led to construct the function (the so-called minimal function), determined by the known magnitudes $|E|$,

$$R(\phi) = \frac{1}{\sum_{\mathbf{H,K}} \kappa_{\mathbf{HK}}} \sum_{\mathbf{H,K}} \kappa_{\mathbf{HK}} \left[\cos \phi_{\mathbf{HK}} - \frac{I_1(\kappa_{\mathbf{HK}})}{I_0(\kappa_{\mathbf{HK}})} \right]^2 \tag{12}$$

which is seen to be a function of all those triplets that are generated by a prescribed set of phases $\{\phi\}$.

[7] J.-X. Yao, *Acta Crystallogr.* **A37,** 642 (1981).
[8] M. M. Woolfson and J.-X. Yao, *Acta Crystallogr.* **A46,** 409 (1990).

Because the triplets $\phi_{\mathbf{HK}}$ are defined by Eq. (5) as functions of the individual phases ϕ, Eq. (12) defines $R(\phi)$ implicitly as a function of the individual phases. One therefore naturally anticipates, as had been done by Debaerdemaeker and Woolfson,[6] that the set of values for the individual phases is best that minimizes the residual $R(\phi)$; this conjecture, however, turns out not to be true.

Minimal Principle

If one denotes by R_T the value of $R(\phi)$ obtained when the phases are equal to their true values for some choice of origin and by R_R the value of $R(\phi)$ when the phases are chosen at random, it may then be shown that[3]

$$R_T < \tfrac{1}{2} < R_R$$

thus tending to confirm the conjecture that the true values of the phases minimize $R(\phi)$. The conjecture itself, however, turns out to be false: There exist values for the phases that yield values for $R(\phi)$ even less than R_T, so that R_T is not the global minimum of R. As already described earlier, it is necessary to constrain the phases to satisfy certain identities among them [Eq. (4)] that are known to exist because their number generally exceeds by far the number of unknown parameters needed to specify the crystal structure. In short, it is the constrained global minimum of $R(\phi)$ that yields the desired phases (the minimal principle). The next section describes how this minimum is reached in practice.

Computer Program *Shake-and-Bake*

The six-part *Shake-and-Bake* (*SnB*)[4] phase determination procedure, shown by the flow diagram in Fig. 1, combines minimal-function phase refinement and real-space filtering. It is an iterative process that is repeated until a solution is achieved or a designated number of cycles have been performed. With reference to Fig. 1, the major steps of the algorithm are described next.

A. Generation of Invariants

Normalized structure-factor magnitudes ($|E|$ values) are generated by standard scaling methods and the triplet invariants [Eq. (5)] that involve the largest corresponding $|E|$ values are generated. Parameter choices that must be made at this stage include the numbers of phases and triplets to be used. The total number of phases is ordinarily chosen to be about 10 times the number of atoms whose positions are to be determined and the

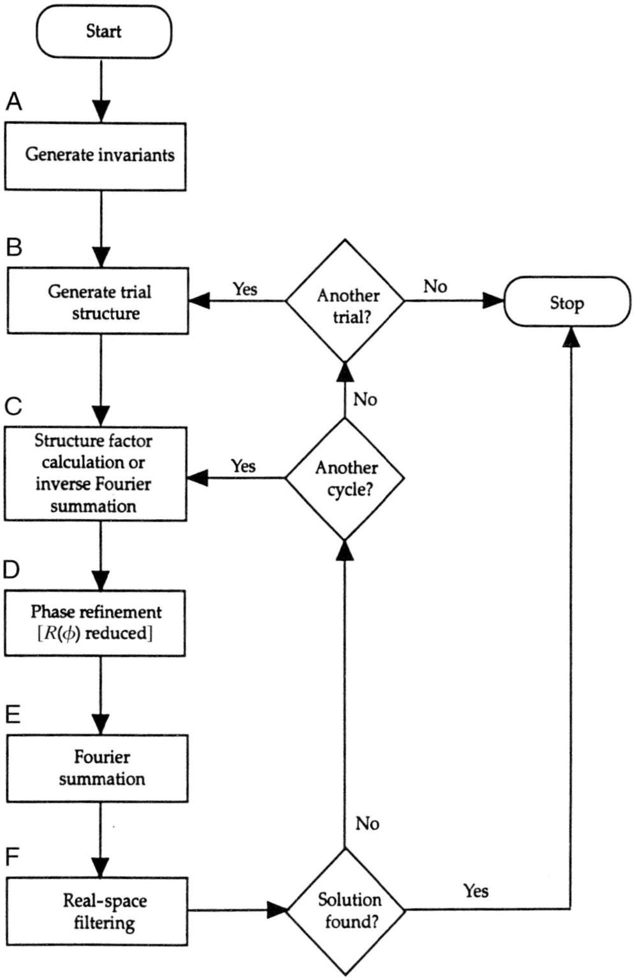

FIG. 1. Flow chart for *Shake-and-Bake*, the minimal-function phase refinement and real-space filtering procedure.

number of structure invariants is usually chosen to be at least 100 times the number of atoms.

B. Generation of Trial Structure

A trial structure or model is generated that is composed of a number of randomly positioned atoms equal to the number of atoms in the unit cell. The starting coordinate sets are subject to the restrictions that no two atoms are closer than a specified distance (normally 1.2 Å) and that no atom is within bonding distance of more than four other atoms.

C. Structure-Factor Calculation

A normalized structure-factor calculation [Eq. (1)] based on the trial coordinates is used to compute initial values for all the desired phases simultaneously. In subsequent cycles, peaks selected from the most recent Fourier series are used as atoms to generate new phase values.

D. Phase Refinement

The values of the phases are perturbed by a *parameter-shift* method in which $R(\phi)$, which measures the mean-square difference between estimated and calculated structure invariants, is reduced in value. $R(\phi)$ is initially computed on the basis of the set of phase values obtained from the structure-factor calculation in step C. The phase set is ordered in decreasing magnitude of the associated $|E|$ values. The value of the first phase is incremented by a preset amount and $R(\phi)$ is recalculated. If the new calculated value of $R(\phi)$ is lower than the previous one, the value of the first phase is incremented again by the preset amount. This is continued until $R(\phi)$ no longer decreases or until a predetermined number of increments has been applied to the first phase. A completely analogous course is taken if, on the initial incrementation, $R(\phi)$ increases, except that the value of the first phase is decremented until $R(\phi)$ no longer decreases or until the predetermined number of decrements has been applied. The remaining phase values are varied in sequence as just described. Note that, when the ith phase value is varied, the new values determined for the previous $i - 1$ phases are used immediately in the calculation of $R(\phi)$. The step size and number of steps are variables whose values must be chosen.

E. Fourier Summation

Fourier summation is used to transform phase information into an electron-density map [refer to Eq. (2)]. Normalized structure-factor amplitudes, $|E|$ values, have been used at this stage (rather than $|F|$ values) because phases are available for the largest $|E|$ values but not for all the largest $|F|$ values. The grid size must be specified.

F. Real-Space Filtering (Identities among Phases Imposed)

Image enhancement has been accomplished by a discrete electron-density modification consisting of the selection of a specified number of the largest peaks, made equal, on the Fourier map for use in the next structure-factor calculation. The simple choice, in each cycle, of a number of the largest peaks, corresponding to the number (n) of expected atoms (per asymmetric unit) has given satisfactory results. No minimum interpeak-distance criterion is applied at this stage. The rest of the map is set equal to zero, thus completing the first iteration. Successive iterations ordinarily converge in fewer than $n/2$ cycles and may or may not yield the constrained global minimum of $R(\phi)$, depending on the choice of the initial random structure. However, those trials that do lead to the constrained global minimum of $R(\phi)$, and therefore to the solution of the phase problem, are always readily identified by the final value of $R(\phi)$, which is considerably less than the final values of $R(\phi)$ corresponding to the nonsolutions. (See Figs. 2 and 3 for the solution of the 624-atom Toxin II structure.) It is important to stress that it is at this point that the identities among the phases [Eq. (4)] are imposed, albeit only in an approximate sense, because observed magnitudes $|E|$ have been used in the calculation of the Fourier series from which the strongest peaks have been extracted, thus preserving the dominant feature of the map. Hence the associated set of $|E|$ values approximates the observed $|E|$ values and the system of identities [Eq.

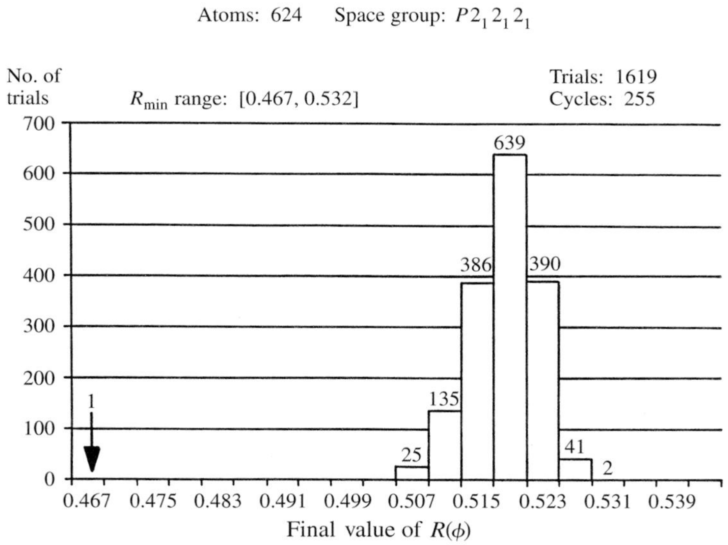

FIG. 2. Final values of the minimal function $R(\phi)$ after 255 cycles for 1619 Shake-and-Bake trials for the 624-atom Tox II structure, clearly showing the separation between the single solution and the 1618 nonsolutions.

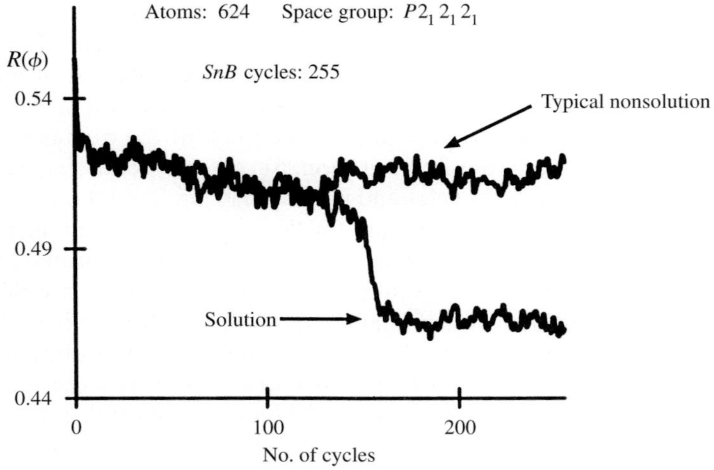

FIG. 3. The course of $R(\phi)$ for Tox II, as a function of cycle number, for the solution trial and for a typical nonsolution trial.

(4)] is approximately satisfied. Prior structural knowledge is also invoked because n peaks (the expected number) are selected.

Applications

Shake-and-Bake has been tested successfully, with no failure, using experimentally measured atomic resolution data for some 30 structures, many of which had been difficult to solve with existing techniques or had defeated all previous attempts. These structures range in size from 25 to 600 independent nonhydrogen atoms, and 6 of these are in the 300- to 600-atom range. Although a number of these structures had been previously known, this fact was not used in these applications. In all cases those trials that led to solutions were readily identified by the behavior of the minimal function $R(\phi)$ (see Figs. 2 and 3).

Notable Application: Determination by Shake-and-Bake of Previously Known Crystal Structure of Toxin II from Scorpion Androctorius australis Hector

The structure of Toxin II,[9] consisting of 64 amino acid residues (512 protein atoms, the heaviest comprising four disulfide bridges) and 112 water

[9] D. Housset, C. Habersetzer-Rochat, J. P. Astier, and J. C. Fontecilla-Camps, *J. Mol. Biol.* **238,** 88 (1994).

molecules (a total of 624 atoms), crystallizes in the space group $P2_12_12_1$ and diffracts to a resolution of 0.96 Å. A total of 50,000 triplets having the largest values of $\kappa_{\mathbf{HK}}$ were generated from the 5000 phases with the largest values of $|E|$ and used in the definition of the minimal function $R(\phi)$. A total of 1619 *Shake-and-Bake* trials were run, each for 255 cycles. The final value of $R(\phi)$ for the trial that led to the solution was 0.467, the value of the constrained global minimum of $R(\phi)$. The range of final values of $R(\phi)$ for the remaining 1618 trials was [0.507, 0.532] (see histogram, Fig. 2), clearly nonsolutions.

Figure 3 shows the course of the minimal function $R(\phi)$, as a function of cycle number, for the trial that led to the solution and for a typical nonsolution trial. Both trials show almost identical behavior for some 130 cycles when $R(\phi)$ for the solution trial drops precipitously from a value of about 0.50 to 0.467 and remains at about that level for all remaining cycles. For the nonsolution trial, however, $R(\phi)$ oscillates between 0.51 and 0.52 for all remaining cycles.

The structure was refined to a final R value of 0.160; and 418 hydrogen atoms were located. *Shake-and-Bake* had delivered the values of all 5000 phases with an initial average phase error, as compared with the refined values, of 19.1°.

Acknowledgments

I am indebted to Drs. George DeTitta, Russ Miller, and Charles Weeks for their essential contribution to the development of the *Shake-and-Bake* algorithm and to Dr. David Smith, who used X-PLOR to refine the Tox II structure without prior knowledge of the amino acid sequence. Research supported by NIH Grant No. GM-46733.

[2] *Ab Initio* Macromolecular Phasing: Blueprint for an Expert System Based on Structure Factor Statistics with Built-in Stereochemistry

By GÉRARD BRICOGNE

That which cannot be conceived through anything else must be conceived through itself.

Spinoza, *Ethics,* Axiom II

Introduction

So far the main efforts in formulating and implementing the Bayesian approach to structure determination[1] have been directed toward the design of more powerful methods for evaluating joint probability distributions of structure factors, and of a systematic protocol for forming hypotheses, for sampling them efficiently,[2] and for testing them against the available data.

In spite of these elaborations, the initial assumption on which the Bayesian statistical machinery is set to work remains as simplistic as that of standard direct methods: All atoms are still assumed to be independently distributed, so that chemical bonding rules are ignored.

Overcoming this embarrassing inadequacy, i.e., finding a way of incorporating *a priori* stereochemical knowledge into structure-factor statistics, has proved one of the most elusive dilemmas in theoretical crystallography. Classic papers by Harker,[3] Hauptman,[4] and Main[5] showed how some anomalies in intensity statistics could be predicted on the basis of such knowledge and taken into account at the stage of data normalization, but this was accompanied by loss of information and could have adverse effects on the structure determination process.

It is shown briefly here that the key concepts of saddlepoint approximation and maximum-entropy distributions can be applied to this problem to yield joint probability distributions of structure factors with built-in stereochemistry, i.e., *a priori* statistical criteria of stereochemical validity, opening up the possibility of constructing an expert system for knowledge-based macromolecular structure determination.

[1] G. Bricogne, *Methods Enzymol.* **276,** 361 (1997).
[2] G. Bricogne, *Methods Enzymol.* **276,** 424 (1997).
[3] D. Harker, *Acta Crystallogr.* **6,** 731 (1953).
[4] H. Hauptman, *Z. Krist.* **121,** 1 (1965).
[5] P. Main, *in* "Crystallographic Computing Techniques" (F. R. Ahmed, K. Huml, and B. Sedlacek, eds.), pp. 239–247. Munksgaard, Copenhagen, 1976.

0076-6879/97 $25.00

This chapter extends material presented in Bricogne[1,2] (in the sense that definitions and context found there are not repeated here) and is an expanded version of Ref. 6.

Molecular Placements

Consider a known molecular fragment (e.g., an amino acid side chain, or a stretch of polypeptide main chain in a given conformation) described in a reference position and orientation by a density ρ^{M} with transform Ξ^{M}. If ρ^{M} is rotated by $\mathbf{R}$ and translated by $\mathbf{t}$ [i.e., acted on by the placement operator $(\mathbf{R}, \mathbf{t})$] to give the copy of the fragment lying in the reference asymmetrical unit, then the density for the known partial structure in the crystal may be written (see Bricogne[1] for the notation) as

$$\rho^{\text{par}} = \sum_{g \in G} \tau_{S_g(\mathbf{t})} (\mathbf{R}_g \mathbf{R})^{\#} \rho^{M} \tag{1}$$

where the translation of a function by a vector $\mathbf{a}$ and its image under a rotation $\mathbf{R}$ are defined as usual by

$$(\tau_{\mathbf{a}} f)(\mathbf{x}) = f(\mathbf{x} - \mathbf{a}) \quad \text{and} \quad (\mathbf{R}^{\#} f)(\mathbf{x}) = f(\mathbf{R}^{-1}\mathbf{x}) \tag{2}$$

The parametrized partial structure factors, which play the same role here as the atomic structure factors $f(\mathbf{h})\Xi_{\mathbf{h}}(\mathbf{x})$ (see Bricogne[1]), are therefore

$$\Xi_{\mathbf{h}}^{\text{par}} (\mathbf{R}, \mathbf{t}) = \sum_{g \in G} e^{2\pi i \mathbf{h} \cdot S_g(\mathbf{t})}\, \Xi^{M} [(\mathbf{R}_g \mathbf{R})^{T} \mathbf{h}] \tag{3}$$

Random Molecular Placements and Structure-Factor Statistics

We now assume that the unknown structure under consideration is made up of N copies of the known fragment, with placements $(\mathbf{R}, \mathbf{t})$ randomly and independently distributed according to a probability distribution $q(\mathbf{R}, \mathbf{t})$ of such placements. This turns the structure factors [Eq. (3)] into random variables, just as the random position $\mathbf{x}$ does for the ordinary structure factor (see Bricogne[1]). A prior estimate $m(\mathbf{R}, \mathbf{t})$ of that distribution can, for instance, be obtained from a packing function.[7] If the unknown structure is made up of several types of known fragments, a "multichannel" treatment can be adopted, as previously described for isotropic scatterers.[8]

[6] G. Bricogne, *in* "ECCC Computational Chemistry" (F. Bernardi and J. L. Rivail, eds.), Vol. 330, pp. 449–470. American Institute of Physics, Woodbury, New York, 1995.

[7] M. T. Stubbs and R. Huber, *Acta Crystallogr.* **A47,** 521 (1991).

[8] G. Bricogne, *Acta Crystallogr.* **A44,** 517 (1988).

We may then use harmonic analysis over the whole group of Euclidean placements[9] (and not solely on the subgroup of translations, which gives rise to ordinary Fourier analysis) in order to calculate the moments of random structure factors required for a full probabilistic treatment of this "random fragment" model. Expanding the reference structure factor contributions $\Xi_{\mathbf{h}}^{\mathrm{M}}$ into spherical harmonics:

$$\Xi_{\mathbf{h}}^{\mathrm{M}} = \sum_{l=0} i^{l} \sum_{\mathrm{m}=-l}^{+l} b_{\mathrm{m}}^{l}(d_{\mathbf{h}}^{*})\, Y_{\mathrm{m}}^{l}(\hat{\mathbf{h}}) \tag{4}$$

(where $\hat{\mathbf{h}}$ denotes the unit vector along $\mathbf{h}$) we may now let the rotation $\mathbf{R}$ act on the spherical harmonics through the Wigner matrices[9,10] $\mathbf{D}^{l}(\mathbf{R})$, while the translation $\mathbf{t}$ acts through the usual scalar phase shifts. This yields an explicit formula for the dependence of $\Xi_{\mathbf{h}}^{\mathrm{par}}(\mathbf{R}, \mathbf{t})$ on the placement $(\mathbf{R}, \mathbf{t})$ in the form

$$\Xi_{\mathbf{h}}^{\mathrm{par}}(\mathbf{R}, \mathbf{t}) = \sum_{g \in G} e^{2\pi i \mathbf{h} \cdot S_{\mathrm{g}}(\mathbf{t})} \sum_{l=0} i^{l} \sum_{m=-l}^{+l} \sum_{m'=-l'}^{+l} b_{\mathrm{m}'}^{l}(d_{\mathbf{h}}^{*})\, Y_{\mathrm{m}}^{l}(\widehat{\mathbf{R}_{\mathrm{g}}^{\mathrm{T}}\mathbf{h}})\, D_{\mathrm{mm}'}^{l}(\mathbf{R}) \tag{5}$$

[the Wigner matrices of the point-group rotations $\mathbf{R}_{\mathrm{g}}$ could be used to write this expression in terms of the $Y_{\mathrm{m}}^{l}(\hat{\mathbf{h}})$ alone, but the final expression is rather cumbersome]. The moments of the real components of such structure-factor contributions can be calculated under any assumed distribution $q(\mathbf{R}, \mathbf{t})$ of random placements, using a linearization procedure similar to that of Bricogne[1]: products of complex exponentials are linearized in the usual way, whereas products of Wigner matrix elements are linearized by means of the Clebsch–Gordan expansion [Ref. 11, Eq. (69), p. 921]:

$$D_{\mathrm{m}_1\mathrm{m}_1'}^{l_1}(\mathbf{R})\, D_{\mathrm{m}_2\mathrm{m}_2'}^{l_2}(\mathbf{R}) = \sum_{l=|l_1-l_2|}^{l_1+l_2} \sum_{m=-l}^{+l} \sum_{m'=-l}^{+l} C_{l_1 m_1 l_2 m_2}^{l\mathrm{m}}\, C_{l_1\mathrm{m}_1' l_2\mathrm{m}_2'}^{l\mathrm{m}'}\, D_{\mathrm{mm}'}^{l}(\mathbf{R}) \tag{6}$$

which thus appears as the rotational equivalent of the Bertaut linearization formula (Bricogne[1]).

Maximum-Entropy Distributions of Random Molecular Placements

Given the structure factor vector $\mathbf{F}^{(\nu)} = (F_{\mathbf{h}_1}^{(\nu)}, F_{\mathbf{h}_2}^{(\nu)}, \ldots, F_{\mathbf{h}_{\mathrm{m}}}^{(\nu)})$ attached to a node ν, (see Bricogne[1]) we may build the maximum-entropy distribu-

[9] M. Hamermesh, "Group Theory and Its Application to Physical Problems." Addison-Wesley, Reading, Massachusetts, 1962.

[10] R. A. Crowther, *in* "The Molecular Replacement Method" (M. G. Rossmann, ed.), pp. 173–178. Gordon and Breach, New York, 1972.

[11] A. Messiah, "Mécanique Quantique," Vol. 2: Appendix C. Dunod, Paris, 1959.

tion of random placements compatible with these hypothetical structure factors as

$$q_{\nu}^{\mathrm{ME}}(\mathbf{R},\mathbf{t}) = \frac{m(\mathbf{R},\mathbf{t})}{Z(\lambda^{(\nu)})} \exp\left\{\sum_{j=1}^{n} \lambda_j^{(\nu)} X_j^{\mathrm{par}}(\mathbf{R},\mathbf{t})\right\} \tag{7}$$

with

$$Z(\lambda^{(\nu)}) = \int_{[0,1]^3} \int_{\mathrm{SO}_3/G} m(\mathbf{R},\mathbf{t}) \exp\left\{\sum_{j=1}^{n} \lambda_j^{(\nu)} X_j^{\mathrm{par}}(\mathbf{R},\mathbf{t})\right\} d^3\mathbf{R}\, d^3\mathbf{t} \tag{8}$$

where the $X_j^{\mathrm{par}}(\mathbf{R},\mathbf{t})$ are the real components of the various $\Xi_{\mathbf{h}}^{\mathrm{par}}(\mathbf{R},\mathbf{t})$ in the same order as the components of $\mathbf{F}^{(\nu)}$. The $\lambda_j^{(\nu)}$ are Lagrange multipliers determined by solving the maximum-entropy equations:

$$\nabla_\lambda(\log Z) = \mathbf{F}^{(\nu)} \quad \text{at} \quad \lambda = \lambda^{(\nu)} \tag{9}$$

which yields as by-products the maximal entropy value

$$S(\nu) = \log Z(\lambda^{(\nu)}) - \lambda^{(\nu)} \cdot \mathbf{F}^{(\nu)} \tag{10}$$

and the covariance matrix

$$\mathbf{Q}^{(\nu)} = \nabla_\lambda \nabla_\lambda^{\mathrm{T}}(\log Z) \quad \text{at} \quad \lambda = \lambda^{(\nu)} \tag{11}$$

completing the list of ingredients required to form the saddlepoint approximation to the joint probability of structure factors in $\mathbf{F}^{(\nu)}$ (see Bricogne[1]).

We have therefore succeeded in extending the "random atom model" of classic direct methods to a "random fragment model" capable of accommodating arbitrary amounts of *a priori* stereochemical information in the structure-factor statistics to which it gives rise.

For numerical implementation, the efficient sampling methods for molecular orientations described in Bricogne[2] may be invoked to keep calculations to the minimum required size.

Extension of Bayesian Strategy of Macromolecular Structure Determination

In the macromolecular field, where large structures are indeed built from a small number of basic building blocks, this is an extremely valuable extension. Until now "LEGO" fragments,[12] i.e., fragments small enough to be precisely conserved between many structures, were too small to be located in the conventional sense of the molecular replacement method.

[12] T. A. Jones and S. Thirup, *EMBO J.* **5,** 819 (1986).

The treatment just given uses probabilistic model building (as discussed in Bricogne[1]) through the device of maximum-entropy placement distributions for these fragments. Besides the phase hypotheses previously used in conjunction with the random atom model (and which will now modulate these placement distributions, causing much more intense extrapolation and thus increasing the sensitivity of the LLG), we may now use a hierarchy of structural hypotheses that vary the types and numbers of known fragments to be included in the "soup" of random ingredients according to the knowledge accumulated in the field of macromolecular architecture (rotamers, building blocks, tertiary fold classifications, etc.).

Sequential Bayesian inference can be conducted as described in Bricogne,[1] using these new statistical criteria with built-in stereochemistry in place of the old ones, testing increasingly complex structural hypotheses (prompted by expert knowledge) against simpler ones in a quantitative and fully automatic way, and merging phase determination and map interpretation into a unique sequential decision process.

Outlook

The supremely Bayesian viewpoint just reached provides the natural conceptual foundation, as well as all the detailed numerical procedures, on which to build a genuine expert system for knowledge-based structure determination, capable of consulting all relevant structural information to compensate for the relative paucity of diffraction data that distinguishes the macromolecular from the "small moiety" situation. Structure-factor statistics with built-in stereochemistry constitute the "missing link" that had so far precluded the possibility of even contemplating *ab initio* phasing at typical macromolecular (i.e., well below atomic) resolutions.

[3] Noncrystallographic Symmetry Averaging in Phase Refinement and Extension

By F. M. D. VELLIEUX and RANDY J. READ

Introduction

A well-known method of noise reduction in image processing is the addition of multiple, independent images of the same object to generate an improved, averaged image.[1] Using this method, one can increase the

[1] W. O. Saxton, "Computer Techniques for Image Processing in Electron Microscopy," p. 230. Academic Press, New York, 1978.

0076-6879/97 $25.00

signal-to-noise (S/N) ratio by a factor of $N^{1/2}$, when N-independent images are being averaged.[2] The technique has been used with great success in the processing of transmission electron microscopy images, where object recognition and alignment are performed before the averaging operation.[3]

A similar situation is encountered frequently in macromolecular crystallography, where the presence of multiple copies of the same object in the asymmetrical unit is known as "noncrystallographic symmetry." The potential uses of such redundancy in helping to generate improved electron-density maps have long been known.[4]

Background and Definitions

Crystallographic symmetry is global and applies to the whole contents of the crystal. By contrast, noncrystallographic symmetry (n.c.s.) is local, and a given n.c.s. operation applies only to a single set of objects within the entire crystal. Thus, n.c.s. operations, which relate objects to each other, can be applied only locally in a coordinate system used to describe such objects, i.e., molecular models. Hence, a Cartesian system is typically used for this purpose, and most conveniently, Cartesian angstroms.

Two separate types of n.c.s. exist (Fig. 1): (1) proper symmetry, in which the set of n.c.s. operations forms a closed group so that they apply equally to all objects in the assembly. For example, with an n-fold rotation axis, the n.c.s. operation that superimposes object 1 onto object 2 is identical to that superimposing object i onto object $i + 1$ or object n onto object 1; and (2) improper "symmetry," in which the set of n.c.s. operations applying to object 1 (i.e., superimposing object 1 onto object 2, etc.) is different from that applying to object 2, to object 3, etc. Note that both types of n.c.s. may be encountered simultaneously within an asymmetrical unit, e.g., a dimer with proper twofold symmetry plus a monomer. The overall redundancy is threefold, but in practical terms it is a case of improper symmetry.

A n.c.s. operation may be expressed, in matrix notation, as

$$\mathbf{x}_n = [R_n]\mathbf{x}_1 + \mathbf{t}_n$$

where $[R_n]$ is a rotation matrix, and $\mathbf{t}_n$ is the associated translation component.

[2] I. A. M. Kuo and R. M. Glaeser, *Ultramicroscopy* **1,** 53 (1975).

[3] M. van Heel, *in* "Crystallography in Molecular Biology" (D. Moras, J. Drenth, B. Strandberg, D. Suck, and K. Wilson, eds.), p. 89. NATO ASI Series. Plenum, New York, 1987.

[4] H. Muirhead, J. M. Cox, L. Mazzarella, and M. F. Perutz, *J. Mol. Biol.* **28,** 117 (1967).

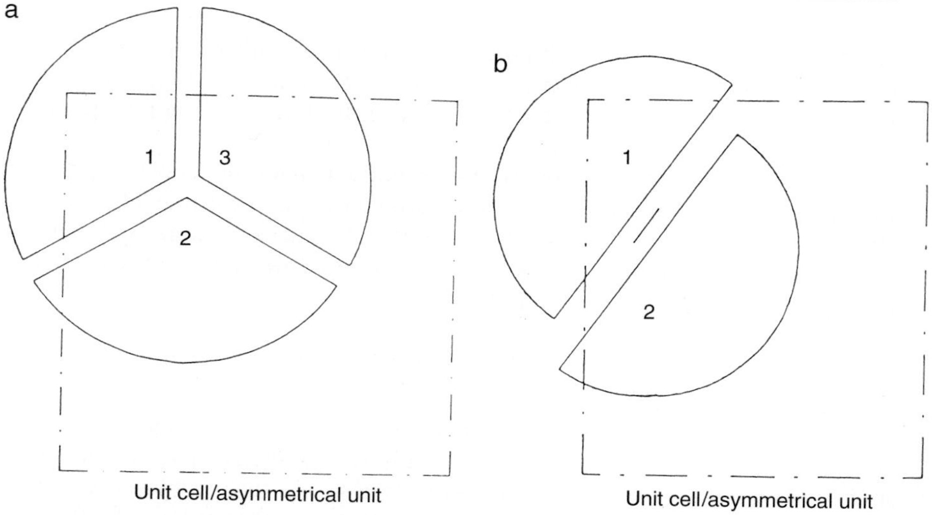

FIG. 1. The two types of noncrystallographic symmetry. (a) Proper local symmetry: example of a threefold local symmetry. (b) Improper local symmetry: example of an improper twofold screw symmetry.

The sort of n.c.s. that can be useful for symmetry averaging may arise from (1) the fortuitous presence in the asymmetrical unit of multiple copies of the same object, these being identical molecules, subunits, or even domains, and (2) the existence of several crystal forms of the same molecule (polymorphism).

Reciprocal Space Analysis of Electron-Density Modification Procedures

Density modification originates from the notion that one might use prior knowledge about electron-density distributions as a source of phase information to improve current estimates of phases. This idea was introduced, in terms of noncrystallographic redundancy, by Rossmann and Blow,[5] who addressed the problem of the detection of identical subunits in the asymmetrical unit. Later, Main and Rossmann[6] expressed the constraints due to the presence of n.c.s. as equations relating structure factors in reciprocal space (these being referred to hereafter as the "molecular-

[5] M. G. Rossmann and D. M. Blow, *Acta Crystallogr.* **15,** 24 (1962).
[6] P. Main and M. G. Rossmann, *Acta Crystallogr.* **21,** 67 (1966).

replacement equations"). Main[7] then incorporated into these equations the additional constraint of having featureless solvent regions. Crowther[8,9] reformulated the molecular replacement equations in a linear form. Bricogne proved the equivalence between the molecular-replacement equations and the procedure of real-space averaging of electron densities of the objects related by n.c.s.,[10] and described some algorithms and computer programs required for this purpose.[11] After this, n.c.s. averaging rapidly became a useful tool not only to improve current phase estimates, but also to generate new phase estimates to a higher resolution than the current resolution limit (phase extension).

By their simplicity, the molecular-replacement equations have the advantage of providing insight into the mechanism of phase improvement, and are therefore discussed below.

If a crystal is composed of N identical objects related by noncrystallographic symmetry, plus featureless solvent regions, its electron-density map will be unchanged when averaging and solvent flattening are carried out. Because of errors in both the initial experimental phases and the observed amplitudes, a map computed with these generally will not satisfy this constraint. However, an iterative procedure of modifying the density, back-transforming it, and computing a new map with the back-transformed phases will produce a set of phases that satisfies this constraint reasonably well. By setting $\rho(\mathbf{x}) = \rho_{\text{avg}}(\mathbf{x})$, and taking the Fourier transform of each side, one obtains the molecular replacement equations in reciprocal space. These equations express the n.c.s. constraint in reciprocal space and explain the propagation of phase information that leads to phase improvement and extension. For clarity, the derivation that follows is given for space group $P1$.

The equations are simpler if the (presumably constant) density in the solvent regions is zero. This can be arranged by subtracting ρ_s from all density values in the entire map. The contrast between protein and solvent regions is unchanged, and the only change in the corresponding structure factors is in the F_{000} term, which is reduced by $V\rho_s$. In the following, we use

$$\begin{aligned} \rho'(\mathbf{x}) &= \rho(\mathbf{x}) - \rho_s \\ &= \rho'_{\text{avg}}(\mathbf{x}) = 1/\text{N} \sum_{m=1}^{N} \sum_{n=1}^{N} \rho'(\mathbf{x}_{mn}) M_m(\mathbf{x}) \end{aligned} \tag{1}$$

[7] P. Main, *Acta Crystallogr.* **23,** 50 (1967).
[8] R. A. Crowther, *Acta Crystallogr.* **22,** 758 (1967).
[9] R. A. Crowther, *Acta Crystallogr.* **B25,** 2571 (1969).
[10] G. Bricogne, *Acta Crystallogr.* **A30,** 395 (1974).
[11] G. Bricogne, *Acta Crystallogr.* **A32,** 832 (1976).

where $M_m(\mathbf{x})$ is the mask function for object m, defined as 1 within the object and 0 outside, and $\mathbf{x}_{mn}$ is $\mathbf{x}$ transformed by the n.c.s. operation that superimposes object m onto object n, i.e.,

$$\mathbf{x}_{mn} = [R_{mn}]\mathbf{x} + \mathbf{t}_{mn} \tag{2}$$

When both sides of Eq. (2) are Fourier transformed, the product inside the double summation becomes a convolution in reciprocal space, as is seen in the following. The Fourier transform of $M_m(\mathbf{x})$ is defined as $U\mathbf{G}_m(\mathbf{s})$, so that $\mathbf{G}_m$ is the interference function; it is continuous because it does not obey crystallographic symmetry, hence the use of the reciprocal space vector $\mathbf{s}$ instead of a reciprocal lattice vector (h, k, l).

$$\mathbf{Fp} = \int_{\text{cell}} \rho'(\mathbf{x}) \exp(2\pi i\mathbf{px})\, d\mathbf{x} = \int_{\text{cell}} \rho'_{\text{avg}}(\mathbf{x}) \exp(2\pi i\mathbf{px})\, d\mathbf{x} \tag{3}$$

The expression for $\rho'_{\text{avg}}(\mathbf{x})$ is substituted from Eq. (1), while replacing $\rho'(\mathbf{x}_{mn})$ and $M_m(\mathbf{x})$ with their Fourier transforms.

$$\mathbf{Fp} = \int_{\text{cell}} 1/N \sum_{m=1}^{N} \sum_{n=1}^{N} 1/V \sum_{\mathbf{h}} \mathbf{Fh} \exp(-2\pi i\mathbf{hx}_{mn}) U/V \int_{\mathbf{s}} \mathbf{G}_m(\mathbf{s}) \tag{4}$$
$$\exp(-2\pi i\mathbf{sx})\, d\mathbf{s} \exp(2\pi i\mathbf{p}\cdot\mathbf{x})\, d\mathbf{x}$$

Now we substitute for $\mathbf{x}_{mn}$ from Eq. (2):

$$\mathbf{Fp} = \int 1/N \sum_{m=1}^{N} \sum_{n=1}^{N} 1/V \sum_{\mathbf{h}} \mathbf{Fh} \exp(-2\pi i(\mathbf{h}[R_{mn}])\cdot\mathbf{x})$$
$$\exp(-2\pi i\mathbf{h}\cdot\mathbf{t}_{mn})\, U/V \int_{\mathbf{s}} \mathbf{G}_m(\mathbf{s})$$
$$\exp(-2\pi i\mathbf{s}\cdot\mathbf{x})\, d\mathbf{s} \exp(2\pi i\mathbf{p}\cdot\mathbf{x})\, d\mathbf{x} \tag{5}$$

The terms in $\mathbf{x}$ are collected into the integral over the unit cell volume:

$$\mathbf{Fp} = U/NV \sum_{m=1}^{N} \sum_{n=1}^{N} 1/V \sum_{\mathbf{h}} \mathbf{Fh} \exp(-2\pi i(\mathbf{h}\cdot\mathbf{t}_{mn}) \int_{\mathbf{s}} \mathbf{G}_m(\mathbf{s})$$
$$\exp(-2\pi i\mathbf{s}\cdot\mathbf{x})\, 1/V \int_{\text{cell}}$$
$$\exp(2\pi i(\mathbf{p} - \mathbf{h}[R_{mn}] - \mathbf{s})\cdot\mathbf{x})\, d\mathbf{x}\, d\mathbf{s} \tag{6}$$

The integral over the unit cell disappears unless $\mathbf{s} = \mathbf{p} - \mathbf{h}[R_{mn}]$, in which case it is equal to V:

$$\mathbf{Fp} = U/NV \sum_{\mathbf{h}} \mathbf{Fh} \sum_{m=1}^{N} \sum_{n=1}^{N} \mathbf{G}_m(\mathbf{p} - \mathbf{h}[R_{mn}]) \exp(-2\pi i\mathbf{h}\cdot\mathbf{t}_{mn}) \tag{7}$$

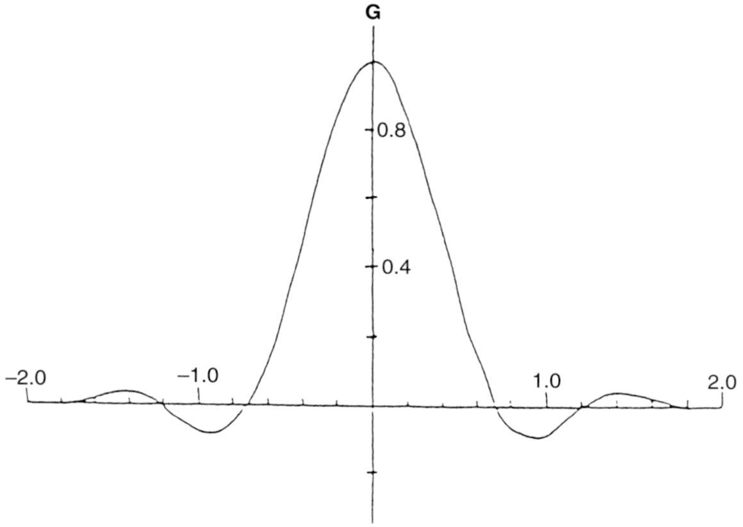

FIG. 2. The **G** function for a spherical molecular envelope. (Adapted from Rossmann and Blow.[5]) The abscissa is the dimensionless quantity $|[R_n]\mathbf{h} - \mathbf{p}|^*r$. The first quantity is the distance in reciprocal space by which the transformed point misses its target; the second is the radius of the molecular envelope.

Equation (7), also previously given by Arnold and Rossmann,[12] expresses the value of each structure factor **Fp** as the weighted sum, over the whole reciprocal space, of all structure factors including itself. The value taken by the weighting or interference function **G** will depend on the position and shape of the envelope of a single object, as well as on the n.c.s. operators $[R_{mn}]$ and $\mathbf{t}_{mn}$. Note that a similar equation may be derived when the only redundancy of information in the asymmetrical unit is the presence of large featureless regions of solvent. In this case of density modification by solvent flattening alone, the integral given in Eq. (6) is evaluated only at position $\mathbf{p} - \mathbf{h}$. Thus the conclusions drawn from this analysis are also applicable, with adaptation, to density modification by solvent flattening alone.

The shape of the **G** function, shown for a spherical envelope in Fig. 2, suggests that significant interference between structure factors will occur only in the first positive peak of this function, i.e., at or near the origin; the argument of the function must be small. For density modification by n.c.s. averaging, we therefore must have $\mathbf{h}[R_{mn}] - \mathbf{p} \approx 0$, i.e., $\mathbf{h} \approx [R_{mn}]^T\mathbf{p}$,

[12] E. Arnold and M. G. Rossmann, *Proc. Natl. Acad. Sci. U.S.A.* **83,** 5489 (1986).

where $[R_{mn}]^T$ is the transpose (and also the inverse) of the rotation matrix $[R_{mn}]$. The implication of these molecular-replacement equations in the case of n.c.s. averaging can be visualized in Fig. 3.

This has implications for the uses that may be made of n.c.s. averaging: the relationships among structure factors caused by the presence of n.c.s. do not extend over the whole of the reciprocal space, but are limited to the regions, related to each other by the rotation components of the n.c.s. operators, where significant interference among structure factors takes place. Thus, the presence of n.c.s. relates distant regions of reciprocal space, with the value of **Fp** being given by a small number of structure factors, including itself. Hence, n.c.s. averaging may be used not only to improve experimental electron-density maps at constant resolution, but also to assign

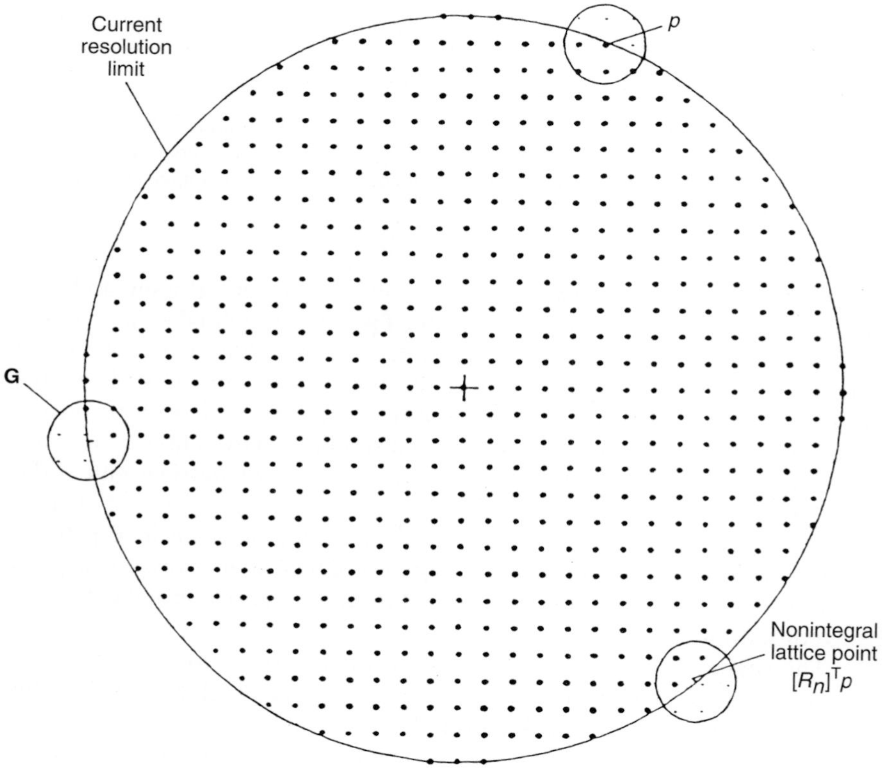

FIG. 3. Relationships among structure factors caused by the presence of a noncrystallographic proper threefold axis of symmetry. **G** is the volume in reciprocal space where the **G** function takes significant values.

map-inversion phases to unphased $|F_0|$ values in a procedure of phase completion or of phase extension. Furthermore, a consequence of these molecular-replacement equations is that some of the estimates for the missing reflections calculated by Fourier inversion of an averaged electron-density map have nonrandom values. Therefore, these can be used in a procedure of data completion.

The range over which map-inversion data computed from an averaged electron-density map can be used may be derived by considering the volume over which the **G** function takes significant values, i.e., within its first positive peak. As seen previously, **h** must transform close to **p** by the n.c.s. rotation operation for significant interference to take place between **Fh** and **Fp**, i.e., $\mathbf{h} \approx [R_{mn}]^T\mathbf{p}$. To understand roughly the propagation of phase information, one may assume a spherical object of radius r. Significant values of the **G** function are achieved when $|[R_{mn}]\mathbf{h} - \mathbf{p}|r < 0.7$, i.e., when $d < 0.7/r$, where d is the distance between $[R_{mn}]\mathbf{h}$ and $\mathbf{p}$. If we assume reciprocal-space parameters $1/a$, the number of reciprocal lattice points to be found within the volume of interest in any one direction is given by $n < 0.7a/r$. Such calculations may be used after map-inversion data have been obtained from the averaged map to decide, *a posteriori,* the validity of the estimates of structure factors provided by the averaging procedure.[13] The resolution of the **G** function (hence the degree to which phase information propagates) depends on the level of detail in the specification of the envelope. Because the envelope will not, in general, be spherical, the **G** function also will not be spherically symmetrical.

Practical Considerations

Detecting Noncrystallographic Symmetry

Noncrystallographic symmetry may arise in several ways and can be suspected or detected according to these criteria.

1. More than one crystal form of the same molecule has been grown.
2. Biochemical data indicate that a crystallized macromolecule is composed of several identical subunits.
3. Solvent-content calculations[14] indicate the presence of more than one molecule or subunit in the asymmetrical unit. This may be verified by measuring the density of the crystals.[15]

[13] F. M. D. A. P. Vellieux, J. F. Hunt, S. Roy, and R. J. Read, *J. Appl. Crystallogr.* **28,** 347.
[14] B. W. Matthews, *J. Mol. Biol.* **33,** 491 (1968).
[15] E. M. Westbrook, *J. Mol. Biol.* **103,** 659 (1976).

4. Self-rotation function calculations may indicate the presence of multiple copies of the same object in the asymmetrical unit. Note, however, that such Patterson superimposition calculations provide information only on the rotation component of a local axis of symmetry. Any information about possible translation components (e.g., local screw axes) is unavailable in the absence of a molecular-search model. Also, the only information that can be derived concerns the orientation of the local axis of symmetry, but not, in general, its location.

5. It frequently is found that an n.c.s. axis runs parallel to a crystallographic axis of symmetry of the same, or higher, symmetry.[16] In this case, the rotation peak in a self-rotation function is hidden under an origin peak due to the crystallographic symmetry. Nonetheless, the n.c.s. is betrayed by the presence of a nonorigin peak in the native Patterson function, which arises because the combination of crystallographic symmetry and n.c.s. generates translational pseudosymmetry. From the position of this peak one can deduce the position of the n.c.s. rotation axis relative to a crystallographic rotation axis, as well as its screw component.

6. Cross-rotation function calculations followed by translation function calculations in the case of multiple copies of the same object in the asymmetrical unit will reveal, when successful, both the orientation and location of the local axes of symmetry. These can be derived by superimposing the properly positioned objects onto each other.

7. When the presence of n.c.s. is suspected and the initial phasing is carried out using a heavy-atom method, information about the orientation and position of the local axis of symmetry often can be derived from the known location of the heavy-atom scatterers. Note, however, that (a) the presence of improper n.c.s. (e.g., a translation component along the rotation axis) will confuse the interpretation of heavy-atom positions in terms of a local axis of symmetry; (b) local proper twofold symmetry will require the presence of at least two pairs of heavy-atom sites to define both the orientation and position of the local symmetry axis (Fig. 4); and (c) all heavy-atom sites will not necessarily obey the local symmetry axis, perhaps because of crystal packing effects: some of the equivalent sites may be inaccessible to the heavy-atom reagent. Also, a heavy-atom site may be located at a special position, i.e., on an N-fold n.c.s. rotation axis.

8. Direct observation of multiple copies of the same object (molecule, subunit, or domain) in an experimental electron-density map: Even when the resemblance of regions of density cannot be detected visually, the n.c.s. operations can be determined by the use of domain rotation[17] and phased

[16] X. Wang and J. Janin, *Acta Crystallogr.* **D49,** 505 (1993).

[17] P. M. Colman, H. Fehlhammer, and K. Bartels, *in* "Crystallographic Computing Techniques" (F. R. Ahmed, K. Huml, and B. Sedlacek, eds.), p. 248. Munksgaard, Copenhagen, 1976.

translation functions,[17,18] as applied in the structure determination of pertussis toxin.[19]

Refinement of Noncrystallographic Symmetry Operators

To apply the technique of noncrystallographic symmetry averaging successfully in real space, accurate n.c.s. operators must be available. This usually means that one must refine the initial estimates for these operators. Such refinement may be performed in the following manner.

When atomic coordinates are available (either in the form of a molecular-replacement solution to the phase problem, or as a result of partial map interpretation), refinement may be accomplished readily by rigid-body refinement[20–22] of the independent models. Otherwise (in the case of an initial electron-density map obtained from heavy-atom phases), such refinement may be carried out by locating the site, in terms of rotation and translation parameters, of maximum correlation among regions related by the n.c.s.[11]

Note that the initial estimates of n.c.s. operators provided by molecular-replacement calculations usually still will require refinement: molecular replacement proceeds by Patterson superimposition operations. A characteristic of Patterson maps is that they contain broader features than the corresponding electron-density maps[23]; a lesser precision is obtained by superimposing Patterson maps than by superimposing electron densities.

Envelope Requirements

Because n.c.s. is, by definition, only local and applies to a single set of objects within the entire crystal, the operation of averaging electron density over objects related by n.c.s. requires important information about the position and shape of the objects whose density must be averaged. This information, known as an envelope (Fig. 5), is essential for n.c.s. averaging: Without this information, averaging using the n.c.s. operators may average density belonging to objects to which the operators do not apply. This will lead to deterioration of the electron density in some areas of the map.

[18] R. J. Read and A. J. Schierbeek, *J. Appl. Crystallogr.* **21,** 490 (1988).

[19] P. E. Stein, A. Boodhoo, G. D. Armstrong, S. A. Cockle, M. H. Klein, and R. J. Read, *Structure* **2,** 45 (1994).

[20] J. L. Sussman, S. R. Holbrook, G. M. Church, and S. H. Kim, *Acta Crystallogr.* **A33,** 800 (1977).

[21] A. T. Brünger, *Acta Crystallogr.* **A46,** 46 (1990).

[22] E. E. Castellano, G. Oliva, and J. Navaza, *J. Appl. Crystallogr.* **25,** 281 (1992).

[23] G. H. Stout and L. H. Jensen, "X-Ray Structure Determination: A Practical Guide," p. 280. John Wiley & Sons, New York, 1989.

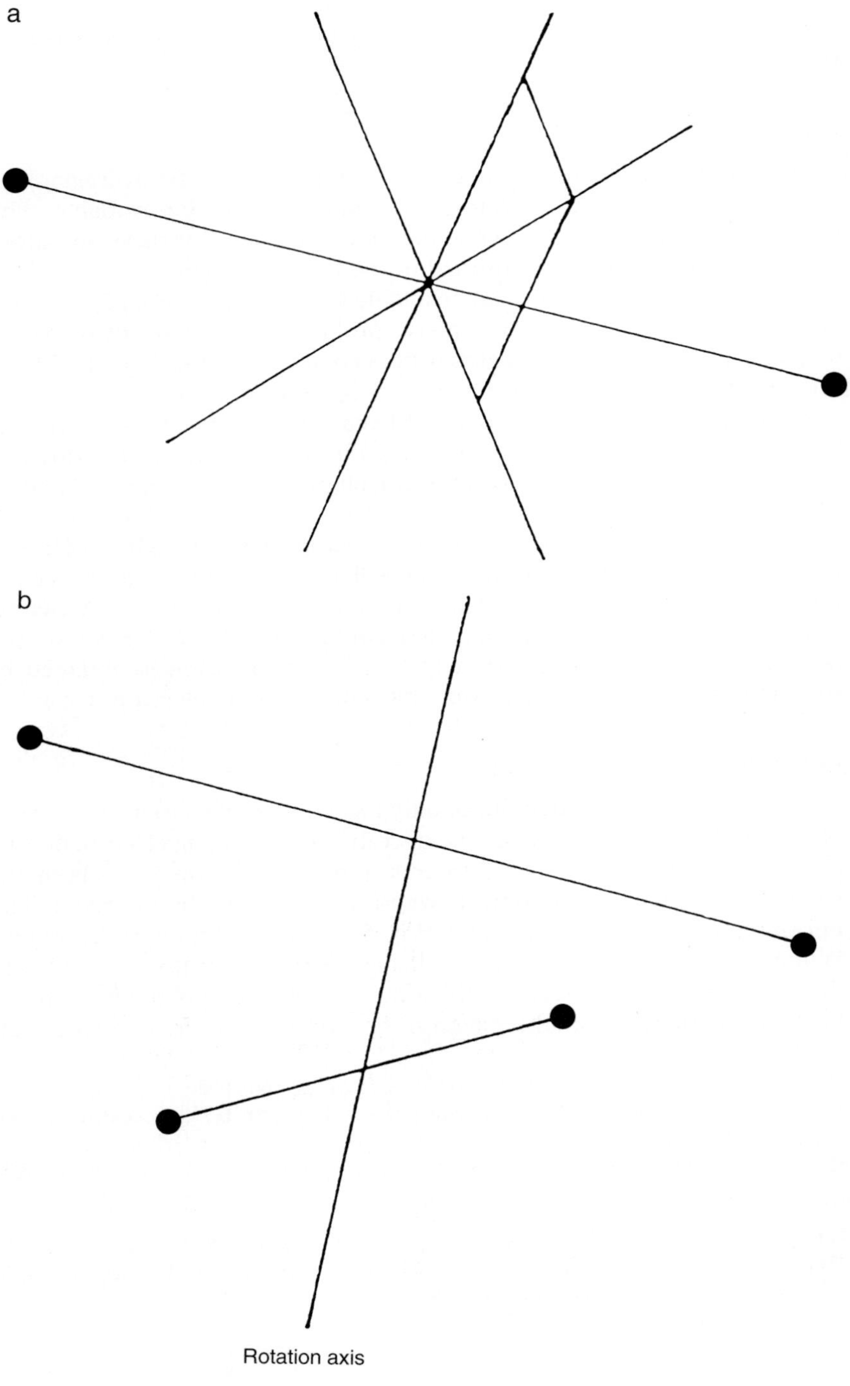

FIG. 4.

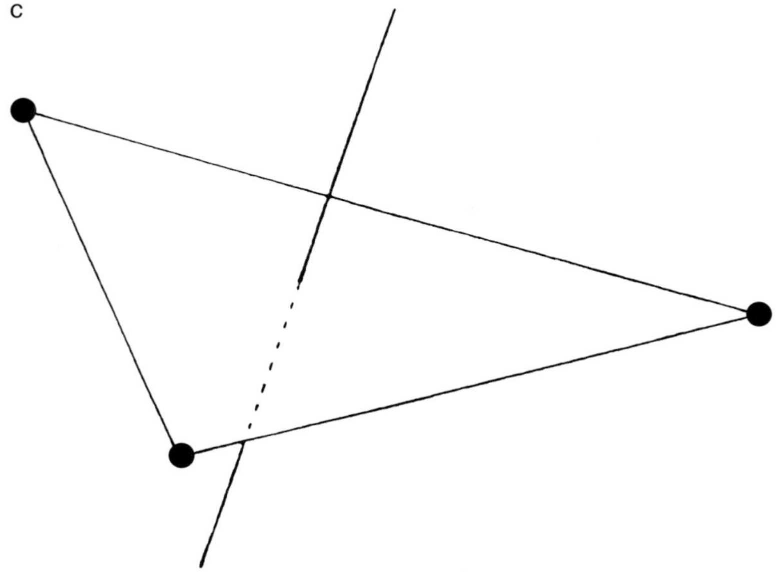

FIG. 4. Determination of n.c.s. operators from heavy-atom positions. (a) Proper local twofold symmetry, with one pair of heavy-atom positions: the location and orientation of the local axis of symmetry are not defined. (b) Proper local twofold symmetry, with two pairs of heavy-atom sites: the location and orientation of the local symmetry axis are defined uniquely. (c) Proper local threefold symmetry, with three heavy-atom sites: both location and orientation of the local axis of symmetry are defined by the heavy-atom positions.

Because electron density is computed as discrete points on a regular grid, the envelope also is generated on a similar grid.

The envelope may be obtained in several ways. When an atomic model is available, one can assign all points near to atomic positions to be part of the envelope. When an initial model is not available, one can trace the envelope with a mouse,[11] although this operation is time consuming and tedious. Alternatively, applying "envelope-free" averaging (i.e., without an envelope) to an experimental map may be used to define the envelope. The lack of envelope information will lead to application of the "wrong" n.c.s. operations to parts of the density. The density in these regions will deteriorate, whereas it will improve in the regions where the known n.c.s. operations apply. This procedure requires some knowledge of the volume where the n.c.s. operations apply, because the map to be averaged should cover all points within this volume. An alternative to the simple averaging of electron density to define an envelope is the computation of a "local

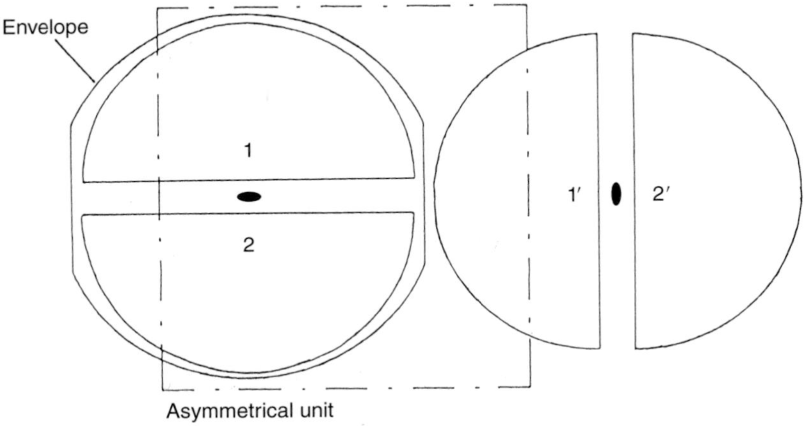

FIG. 5. Example of an envelope encompassing two subunits in the case of a proper twofold axis of symmetry.

correlation map," where the resulting map contains correlation coefficients between regions related by n.c.s.[13,19,24] The envelope then can be obtained readily by the procedure of automatic envelope definition developed by Wang[25] and Leslie.[26]

A perfect envelope can be obtained only when a complete atomic model has been obtained, i.e., when the structure has been solved. Hence, during structure determination, only imperfect envelopes can be generated at first. These should encompass the objects completely so as not to exclude regions of the macromolecules from the averaging process. Thus these envelopes must be larger than absolutely necessary. A consequence of this is that the envelope may extend into a region corresponding to another object, related to the first object either by crystal lattice translations, space group symmetry, or improper n.c.s. If these regions of overlap between envelopes are not taken care of properly, then some areas of the map will be averaged using the incorrect n.c.s. operator, leading to map deterioration. Thus, one algorithm used to take care of these regions of overlap between envelopes is simply to assign each point in the overlapping regions as belonging to the nearest, nonoverlapped object, as is shown in Fig. 6.

For the creation of the envelope, one must distinguish the cases of proper and improper noncrystallographic symmetries. In the case of proper n.c.s. with N identical objects, the set of n.c.s. operations relating these N

[24] B. Rees, A. Bilwes, J.-P. Samama, and D. Moras, *J. Mol. Biol.* **214,** 281 (1990).
[25] B. C. Wang, *Methods Enzymol.* **115,** 90 (1985).
[26] A. G. W. Leslie, *Acta Crystallogr.* **A43,** 134 (1987).

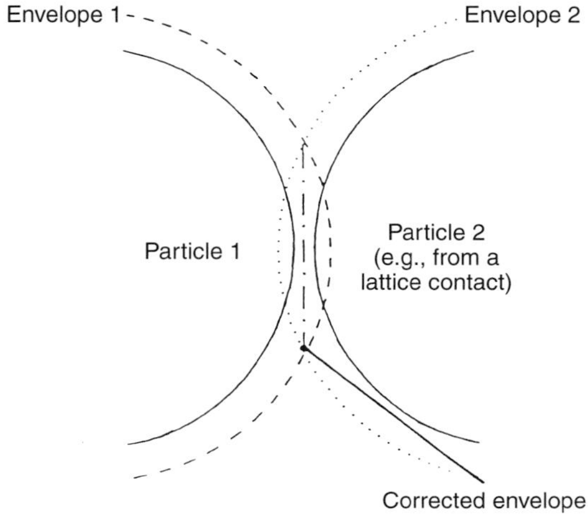

FIG. 6. The envelope overlap problem: case of an overlap region between two objects. The correction to be applied to the envelope assigns points in the overlap region to the nearest nonoverlapped grid points.

objects forms a closed group; it is not necessary to distinguish among the N objects to generate the envelope. A single envelope encompassing all objects is sufficient, and will carry information as to whether a point belongs to one of the N objects, or is part of the solvent region. For improper noncrystallographic symmetry, in which a different n.c.s. operator applies to each of the N objects, one must distinguish the envelope for each object. The envelope corresponding to each individual object must be created separately, then merged together, with all overlap regions dealt with properly. The resulting envelope then will convey information about a point belonging to the solvent region, to object 1, to object 2, etc. In all cases, envelope overlaps due to space group symmetry and crystal lattice translations must be taken care of.

Iterative Noncrystallographic Symmetry Averaging

To perform iterative noncrystallographic symmetry averaging, one must execute the following steps sequentially:

1. Map calculation [preferably by fast Fourier transform (FFT) methods]
2. Reduction of the map to the crystallographic asymmetrical unit
3. Symmetry averaging and solvent flattening
4. Generation of the unit cell from the averaged map

5. Computation of **Fc** values from the averaged map (preferably by FFT methods)
6. Scaling of the $|F_c|$ values to the $|F_0|$ values
7. Computation of weights, phase combination, and data extension

During iterative map improvement, possibly including phase extension, the overall strategy can be summarized as follows: the initial map to be averaged must be the best possible map that may be computed. In the case of a structure determination by heavy-atom methods, this means a figure-of-merit weighted $m|F_0| \exp(i\alpha_{best})$ map. When a molecular model is available for the computation of a starting set of phases, two types of map coefficients may be used to compute the initial map: either Sigmaa[27] map coefficients of the form $(2m|F_0| - D|F_c|) \exp(i\alpha_{model})$ for acentric data and $m|F_0| \exp(i\alpha_{model})$ for centric reflections, or as unweighted $(3|F_0| - 2|F_c|) \exp(i\alpha_{model})$ coefficients for acentric reflections and $(2|F_0| - |F_c|) \exp(i\alpha_{model})$ for centric data. Averaging is carried out then, starting with a few cycles during which only the reflections carrying an initial phase are used. After this, data completion, phase and amplitude extension, or extension of the resolution may be performed. Customarily one refines the phases with a few cycles of the averaging procedure in-between extension or completion steps. Because the phase or amplitude estimates provided by this method are initially nonrandom but not yet optimal, they are improved to better estimates by such iterations.

The parts of this cycle dealing with map calculation and map inversion by FFT procedures need not be discussed here, apart from mention of the amplitude coefficients to be used. These can be simply $m^*|F_0| \exp(i\alpha_{calc})$, or probably better coefficients of the form $m^*(2|F_0| - |F_c|) \exp(i\alpha_{calc})$ for acentric reflections and $m^*|F_0| \exp(i\alpha_{calc})$ for centric data, where m is usually computed using the Sim weighting scheme[28,29] (although other weighting schemes are also being used; see below).

Several algorithms are available to perform the averaging of densities and solvent flattening. Perhaps the best known of these, for historical reasons, is the double-sorting technique developed both by Bricogne[11] and Johnson[30] at a time when computers were not as large or powerful. Being cumbersome and difficult to implement, and not making full use of the capacity of present-day computers, the double-sorting technique is not discussed further.

[27] R. J. Read, *Acta Crystallogr.* **A42,** 140 (1986).
[28] G. A. Sim, *Acta Crystallogr.* **12,** 813 (1959).
[29] G. A. Sim, *Acta Crystallogr.* **13,** 511 (1960).
[30] J. E. Johnson, *Acta Crystallogr.* **B34,** 576 (1978).

Two other algorithms used for this step keep both the whole map to be averaged and the envelope in computer memory. In the first of these, all density to be averaged can be placed first in an artificial box, in which the averaging procedure is carried out before replacing the averaged density into its original crystal form.[31] We describe the second of these as it is implemented in the program suite DEMON/ANGEL.[13] Both the unaveraged map, corresponding to one complete crystallographic asymmetrical unit, and the envelope are kept in memory. The averaged electron-density map also will correspond to one asymmetrical unit and typically is generated on a grid corresponding to one-third of the high-resolution limit. The program sequentially scans each point of this output map. Each of these points is mapped to the envelope. When belonging to the solvent region, the density may be kept as its original value, or replaced by a fixed density, possibly corresponding to the previously computed average solvent density (solvent flattening). For a point belonging to an object, its Cartesian angstrom coordinates within the envelope are computed. The n.c.s. operators then are applied in turn to these Cartesian coordinates. Each n.c.s. equivalent point (also expressed in Cartesian angstroms) is mapped to the unit cell in fractional coordinates, then folded to the asymmetrical unit. The resulting fractional coordinates then are transformed into nonintegral grid points of the input map, and interpolation is performed to obtain the unaveraged electron density at this, possibly nonintegral, grid point of the input map. Several types of interpolation can be used. An eight-point linear interpolation requires an input map computed using a grid corresponding to at most one-fifth to one-sixth of the high-resolution limit. Alternatively, an 11-point nonlinear interpolation, or a 64-point interpolation by means of cubic splines may be used, using an input map computed with a grid size of ca. one-third to one-fifth of the resolution limit. The average density over the N n.c.s. equivalents is then computed, and this value is written to the output map.

This procedure also has been adapted for the averaging of densities between crystal forms, where the input and output maps do not correspond to the same crystal forms. The program still keeps only one input map plus one envelope in memory, but an additional input map, already containing density corresponding to a second crystal form, also is provided to the program. This second input map is on the same grid as the output map, which will contain, after the averaging operation, the averaged density between the two crystal forms.

[31] M. G. Rossmann, R. McKenna, L. Tong, D. Xia, J.-B. Dai, H. Wu, H.-K. Choi, and R. E. Lynch, *J. Appl. Crystallogr.* **25,** 166 (1992).

One must realize that, in the case of averaging between crystal forms, an additional step of scaling the protein densities within the program regions of the envelope should be carried out, especially if the X-ray data have been left on arbitrary scales. It is preferable to place all data sets used to compute the different maps on an absolute scale, but this is difficult to achieve accurately because of the limited resolution. In addition, electron-density maps for different crystal forms will have different nominal or effective resolution limits, either because the data have been collected or phased to different resolutions or because the figures of merit fall off differently. Furthermore, the FFT programs used to compute the electron-density maps usually generate maps on an arbitrary scale.

Two types of scaling of electron-density maps can be performed, the simplest being linear scaling. If the original densities in the protein regions of crystal forms 1 and 2 are called ρ_1 and ρ_2, respectively, then the linear scaling algorithm will determine the values of two parameters A and B that will minimize $\Sigma(\rho_2' - \rho_1)^2$ with $\rho_2' = A\rho_2 + B$. The second type of scaling, which may be performed subsequently, is by means of histogram mapping,[13,32,33] in which the histogram of densities of ρ_2 is made equal to that of ρ_1 (or both are matched to a theoretical histogram). This can be used either if the data sets are not on the same absolute scale, or if the two densities to be averaged do not have the same effective resolution (Fig. 7).

After FFT inversion of the averaged electron density, proper treatment of the map-inversion amplitudes involves the computation and application of a scale and isotropic temperature factor to bring the $|F_c|$ values onto the same scale as the $|F_0|$ values. There is one major reason for this requirement: the electron-density map from which the density values are fetched to calculate the averaged map usually is computed from figure-of-merit weighted structure-factor amplitues. These normally fall off with increasing resolution, so that a direct inversion of this map would lead to $|F_{c'}'|$ values having additional fall-off. This fall-off can be corrected by the application of an artificial isotropic temperature factor. In addition, Bricogne[11] suggested that the use of interpolation during the averaging process causes a loss of spectral power that amplifies at high resolution, also causing the need for correction with an artificial temperature factor. However, the improvement in density obtained by averaging can be sufficient to mask this effect, especially when a single interpolation step is performed as opposed to algorithms that use multiple interpolation steps.

After scaling of the map-inversion structure-factor amplitudes, the next

[32] K. Y. J. Zhang and P. Main, *Acta Crystallogr.* **A46,** 41 (1990).

[33] V. Y. Lunin, *Acta Crystallogr.* **D49,** 90 (1993).

step is computation of weights for the next iteration to downweight the map-inversion phases as a function of their probability. Three types of weighting scheme are used, none of which is entirely satisfactory. This is because these statistical weighting schemes derive phase probabilities by comparing the scaled $|F_c|$ values to the $|F_0|$ values. Because density modification procedures always rapidly bring good agreement between calculated and observed structure-factor amplitudes, the resulting weights are usually overestimated. This can be judged by test calculations, when errors are deliberately introduced into the averaging process. The figure-of-merit values still will increase (not shown). The three weighting schemes that are being used are as follows.

1. Sim weighting,[28,29] in which the averaged map-inversion amplitudes are assumed to be derived from a partial structure: The weights are calculated as $w = I_1(X)/I_0(X)$, where I_1 and I_0 are the modified Bessel functions of order 1 and 0, respectively, and $X = 2|F_0||F_c|/\langle||F_0| - |F_c||^2\rangle$.

2. Rayment weighting,[34] in which the probability of correctness of a phase is estimated from the pairwise comparison of structure-factor amplitudes, the weight being computed as

$$w = \exp[-(|F_c| - |F_0|)/|F_0|]$$

3. Sigmaa weighting,[27] where map-inversion amplitudes are assumed to be derived from a partial structure containing errors.

When data extension is also performed by replacing missing reflections as map-inversion amplitudes and phases, the weight given to the newly introduced reflections may be given as a fraction of the average weight in the resolution bin of the reflection being introduced.

At this stage, one may want to combine the phases obtained from the averaged map with the initial, experimental phases. This step is simply carried out by summing the Hendrickson–Lattman coefficients[35,36] defining the two phase-probability distributions. This may be required, e.g., if the envelope is suspected to truncate part of the protein so that, if solvent flattening is also carried out during averaging, the truncated protein density will reappear in the next map. One word of caution must be professed against this possible phase-combination step—probability distributions can be combined in this manner only if they refer to independent events. Obviously, the probability distribution for the phases obtained by Fourier transform of an averaged electron-density map is not independent from

[34] I. Rayment, *Acta Crystallogr.* **A39,** 102 (1983).
[35] W. A. Hendrickson and E. E. Lattman, *Acta Crystallogr.* **B26,** 136 (1970).
[36] W. A. Hendrickson, *Acta Crystallogr.* **B27,** 1472 (1971).

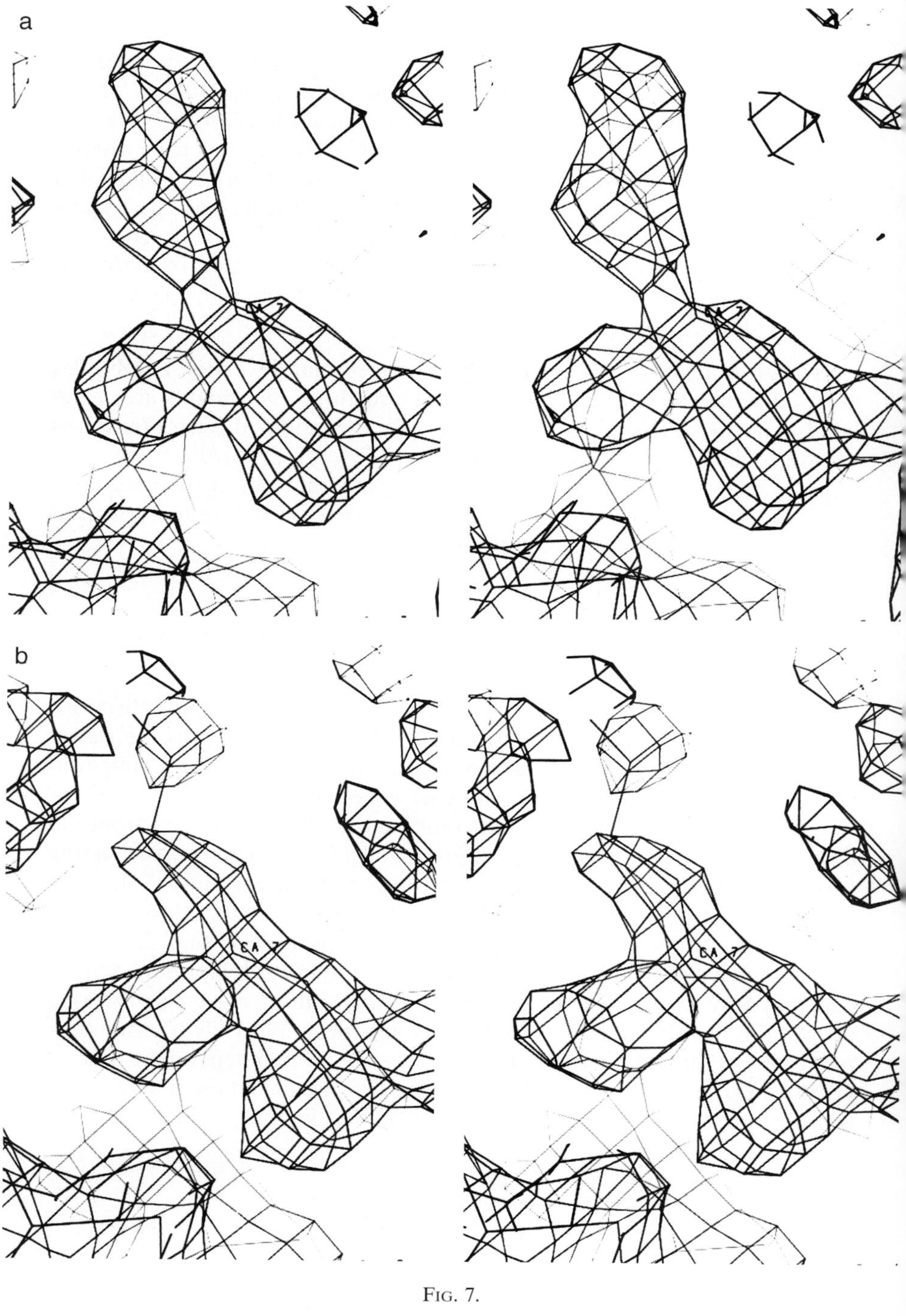

FIG. 7.

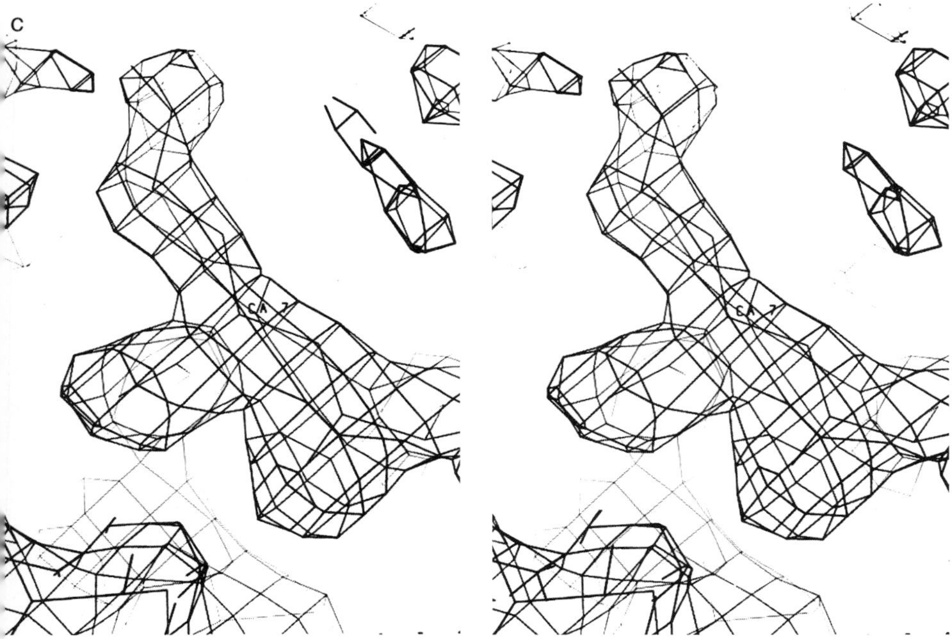

FIG. 7. Effects of map scaling by histogram-mapping techniques. (a) Twofold averaged electron density for residue Glu-7 of a monoclinic crystal form of hen egg white lysozyme (Protein Data Bank access code: 1LYM), computed to a nominal resolution of 2.5 Å. (b) Threefold averaged electron density obtained starting from the averaged map shown in (a). The map that has been added is computed using data from the triclinic crystal form of hen egg white lysozyme (PDB access code: 1LZT) to a nominal resolution of 2.0 Å, and linear scaling has been carried out to minimize the r.m.s. difference in the protein regions of the two maps. (c) Same as (b), except that a step of scaling by histogram mapping has been performed to match the histogram of densities in the protein regions of the 2.5-Å resolution map to that of the higher resolution map. Electron densities are contoured at the 1.0σ level.

that of the phases used to generate the initial electron-density distribution. Phase combination therefore will result in the overestimation of the quality of the resulting, combined phases. This effect can be alleviated by using an "inspired relative weighting" scheme,[11] in which damping factors are applied to both of the phase-probability distributions, with $p_{\text{comb}}(\alpha) = p(\alpha_{\text{initial}})^{\text{wi}} * p(\alpha_{\text{averaging}})^{\text{wa}}$.

Alternatively, an ad-hoc weighting function[13] can be derived if we assume that both phase sets should contribute equally to the resulting combined phase set: an initial phase-combination step is carried out, and the average cosines of the phase difference between the combined phase and each of the two starting phases are computed in resolution bins. From these, difference cosine values are obtained in resolution bins as $\Delta = \langle\cos(\alpha_{\text{combined}} - \alpha_{\text{initial}})\rangle - \langle\cos(\alpha_{\text{combined}} - \alpha_{\text{averaged}})\rangle$. The scale factor to be applied to the Hendrickson–Lattman coefficients of the initial phase set is then computed as $S = 1.0 - \Delta^{1/2}$ if Δ is positive, or as $S = 1.0/(1.0 - \Delta^{1/2})$ if Δ is negative. Such scale factors obviously can be applied in a bin-to-bin smoothly varying fashion, and should be applied only once to the set of coefficients corresponding to the initial phase set, during the initial phase-combination step.

Detection and Effect of Errors

As in any crystallographic procedure, errors may occur. These can have disastrous consequences. Therefore it is important to understand how these errors may arise, and to know what their effect will be and how one may attempt to correct them. A problem should be suspected when one fails to improve the electron density or to extend the resolution using n.c.s. averaging. The following are potential source of errors.

Insufficient Quality of Measured Amplitudes. Because these amplitudes constrain the density modification procedure, results will be better if measurements are accurate.

Inaccurate Definition of Noncrystallographic Symmetry Operators. The general rule is, use only well-refined n.c.s. operators. Two types of error can be made, either in the rotation or translation component of the operator. These will have different consequences: an error in the rotation matrix $[R]$ will lead to a progressive deterioration of map quality from the axis of rotation toward the surface of the object (Fig. 8). By contrast, an error made in the location of the rotation axis will lead to an overall deterioration of the map (Fig. 8).

Use of Erroneous Envelopes. The envelopes are usually larger than absolutely necessary. The use of far too large envelopes will limit the power of the method because the phase improvement from the flattening of solvent regions will be reduced. Envelopes that truncate part of the protein will lead to disappearance of protein density if solvent is being flattened. It is similarly bad if the envelope is in the wrong place (Fig. 9). Manual envelope editing may improve such erroneous envelopes.

Use of Invalid Estimates of Structure Factors. The relationships among structure factors caused by n.c.s. do not extend over the whole reciprocal space, but in practice concern only reciprocal lattice points located in small regions related by the rotation components of the n.c.s. operators. Hence, one should not increase resolution too fast during phase extension. Errors also would arise if map-inversion structure factors were used in place of missing data when large volumes of reciprocal space are missing systemati-

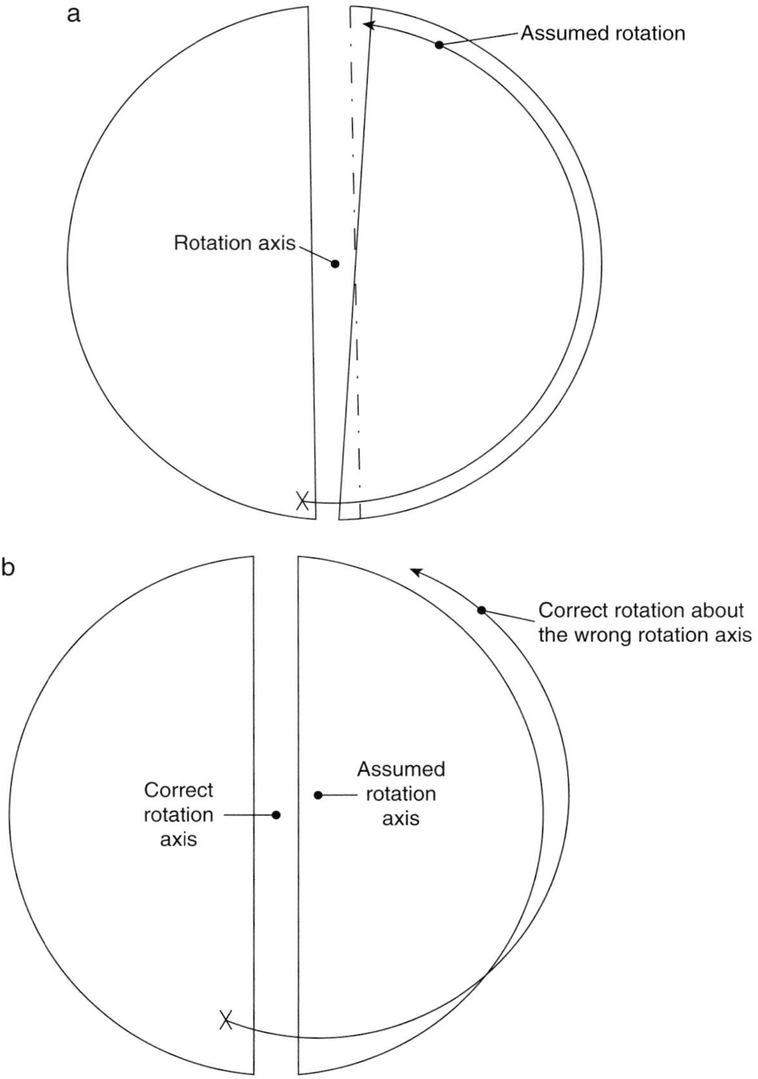

FIG. 8. The effect of errors made in the rotation matrix $[R]$ or in the translation component $[t]$. (a) The assumed rotation is 180°, whereas the correct rotation is 175°/185°; (b) error in the location of a proper local twofold axis of symmetry.

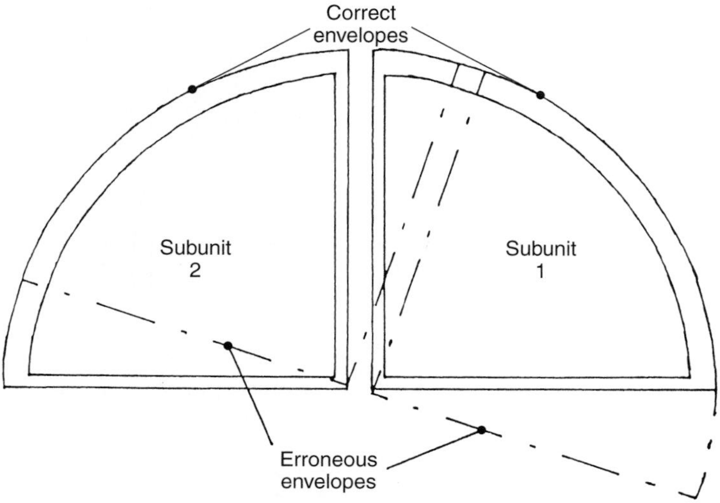

FIG. 9. Effect of placing the envelope in the incorrect location. Case of an improper twofold operation.

cally. Here (Fig. 10), only valid estimates should be used to replace the missing data; data should be completed stepwise. Also remember that in this work four sets of phases are compatible with the measured structure-factor amplitudes: those corresponding to the positive structure, with phases α; the phase set of the Babinet opposite of the structure, i.e., its negative image, with phases $\alpha + 180°$; the phase set of the enantiomeric structure, with phases $-\alpha$; and the phase set of the Babinet opposite of the enantiomer, with phases $-\alpha + 180°$. If one proceeds incautiously when different regions of reciprocal space are poorly tied together by the n.c.s. equations, a mixture of these different phase sets may arise (see also below).

Selected Applications of Method

From an educational point of view, only applications that have met with problems are discussed in this section.

Improvement of Heavy-Atom Maps

A difficult structure determination is that of aldose reductase, described by Tête-Favier *et al.*[37] The procedure that eventually succeeded is depicted

[37] F. Tête-Favier, J.-M. Rondeau, A. Podjarny, and D. Moras, *Acta Crystallogr.* **D49,** 246 (1993).

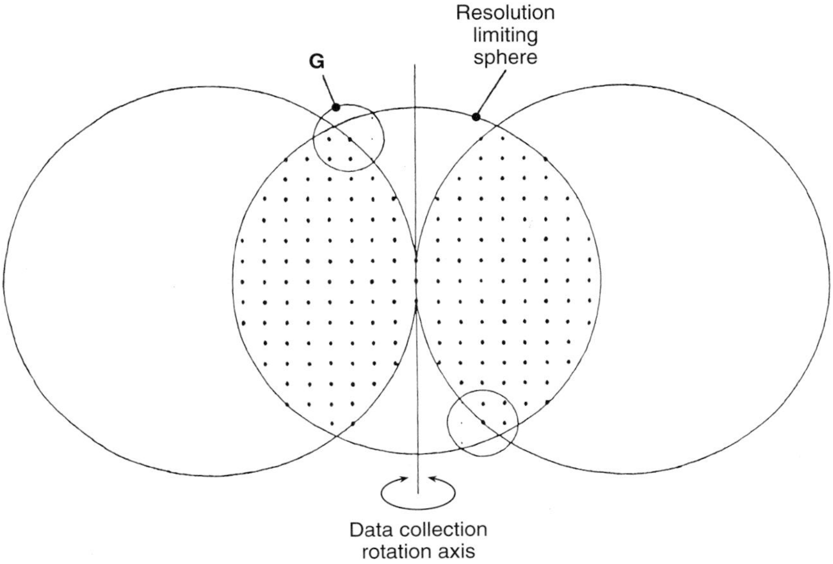

FIG. 10. Example of a nonrandom distribution of missing data: case of a missing cusp. In such a case, the introduction of missing data should be carried out stepwise, taking into account the volume of reciprocal space where the **G** function takes nonnegligible values.

in Fig. 11, and can be summarized as follows: the monomeric enzyme aldose reductase crystallizes in space group *P*1 with four molecules in the asymmetric unit, related to each other by pseudo 222 local symmetry. Initial SIR phases were computed using data from a single mercury derivative, and had a mean phase error of 70°. Self-rotation function calculations showed the presence of local 222 symmetry, although, as seen previously, these computations do not allow a test for translation components along the local twofold axes. The local symmetry was assumed to be 222 at this stage, and this was supported by the locations of the heavy-atom sites. The molecular envelope was generated from the SIR map by computing a correlation function-based map,[24] leading to an inflated solvent content of 45%. The initial SIR map, computed with data from 12- to 3.5-Å resolution, was subjected to averaging using the proper 222 n.c.s. operators, which included phase combination with the initial phases. This led to a phase set having a phase error of 90°, and gave an averaged map of lower quality than the initial map. In a second attempt, the envelope was judged as having been too small and therefore was enlarged by adding a 2-Å-thick layer at the surface of the previous envelope, giving a solvent content reduced to 17%. The averaging procedure was repeated with the assumed 222 opera-

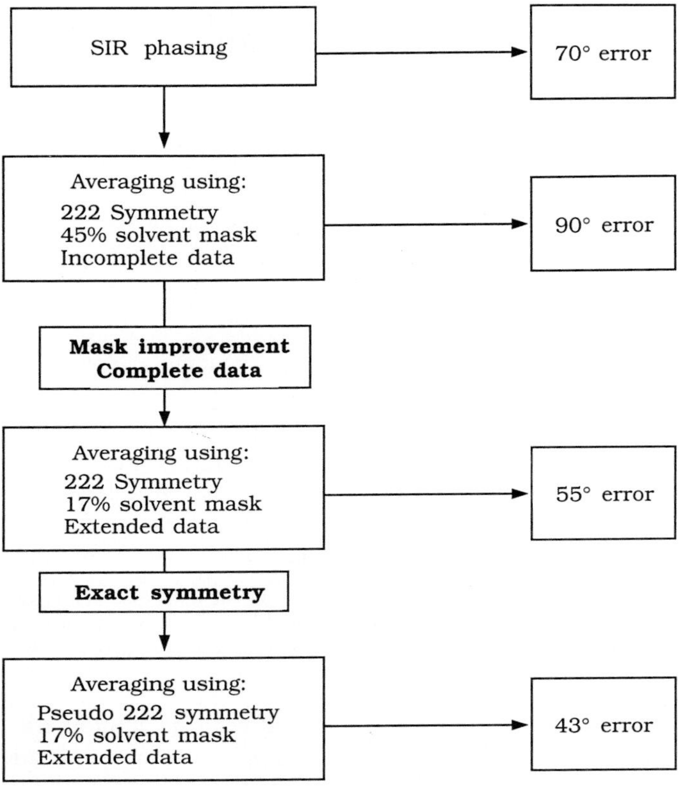

FIG. 11. Flow chart describing the three different attempts at real-space refinement in the structure determination of aldose reductase. (Reproduced with permission from Tête-Favier *et al.*[37])

tors, this time with inclusion of all unphased or unmeasured reflections between 83- and 10-Å resolution. The resulting 3.5-Å resolution phase set had a phase error of 55° and the resulting averaged map showed convincing density for 80% of the main-chain atoms and for 60% of the side-chain atoms. This allowed construction of an initial molecular model. Attempts to extend the phases further to 3.07-Å resolution failed, giving an average phase error of 74° between 3.5 and 3.07 Å. Next the n.c.s. operators were optimized by rigid-body refinement of the four independent monomers. This led to an improved definition of the n.c.s. as pseudo $2t2t2t$ in which the three twofold axes do not intersect. The envelope was recomputed on a monomer-by-monomer basis, and averaging was carried out as described previously. This led to a phase set having a mean phase error of 43°. Heavy-atom rerefinement versus the phases obtained from the averaging procedure led to improved SIR phases, which were used for a final procedure of map averaging in which the resolution was extended successfully to 3.07 Å. The average phase error to 3.5 Å had dropped to 38°, and that between 3.5- and 3.07-Å resolution was 44°, giving a map in which the complete model could be interpreted.

Removing Initial Model Bias

In the structure determination of the avidin–biotin complex,[38,39] Livnah *et al.* used a partially refined avidin model (provided by A. Pähler and W. A. Hendrickson) as a search model for molecular replacement. The molecular-replacement solution revealed two copies of the molecule in the asymmetrical unit, and gave an initial R factor of 46% in the resolution range 15–3.0 Å. After initial model rebuilding, the 3.0-Å resolution $2|F_0| - |F_c|$ electron-density map showed unclear electron density for several of the loop regions connecting β strands in the structure. Phase extension by twofold averaging improved the density in these regions. The molecular envelope was defined from the molecular model by assigning all points within 4.0 Å from any atomic coordinates to the protein region. The averaging procedure was initiated from a Sim weighted[28,29] $2|F_0| - |F_c|$ electron-density map computed using model phases in the resolution range 15–4.0 Å. During the real-space averaging, no phase combination was carried out, and 15 phase-extension steps extended the resolution from 4.0 to 3.0 Å. The resulting 3.0-Å resolution averaged map showed convincing density for the loop regions (Fig. 12) and allowed construction of the correct molecular model. Phase extension by twofold real-space averaging successfully removed the bias from the initial phasing model.

Data Completion

As seen previously, n.c.s. averaging may be used to generate meaningful estimates for missing reflections. An example is the structure determination of glycosomal glyceraldehyde-phosphate dehydrogenase from *Trypanosoma brucei.*[40,41] Very few crystals of this enzyme were grown using the little protein material purified from the glycosomes. Unfortunately these

[38] O. Livnah, E. A. Bayer, M. Wilcheck, and J. L. Sussman, *Proc. Natl. Acad. Sci. U.S.A.* **90,** 5076 (1993).

[39] O. Livnah, Ph.D. thesis. The Weizmann Institute of Science, Rehovot, Israel, 1992.

[40] F. M. D. Vellieux, J. Hajdu, C. L. M. J. Verlinde, H. Groendijk, R. J. Read, T. J. Greenhough, J. W. Campbell, K. H. Kalk, J. A. Littlechild, H. C. Watson, and W. G. J. Hol, *Proc. Natl. Acad. Sci. U.S.A.* **90,** 2355 (1993).

[41] F. M. D. Vellieux, J. Hajdu, and W. G. J. Hol, *Acta Crystallogr.* **D51,** 575.

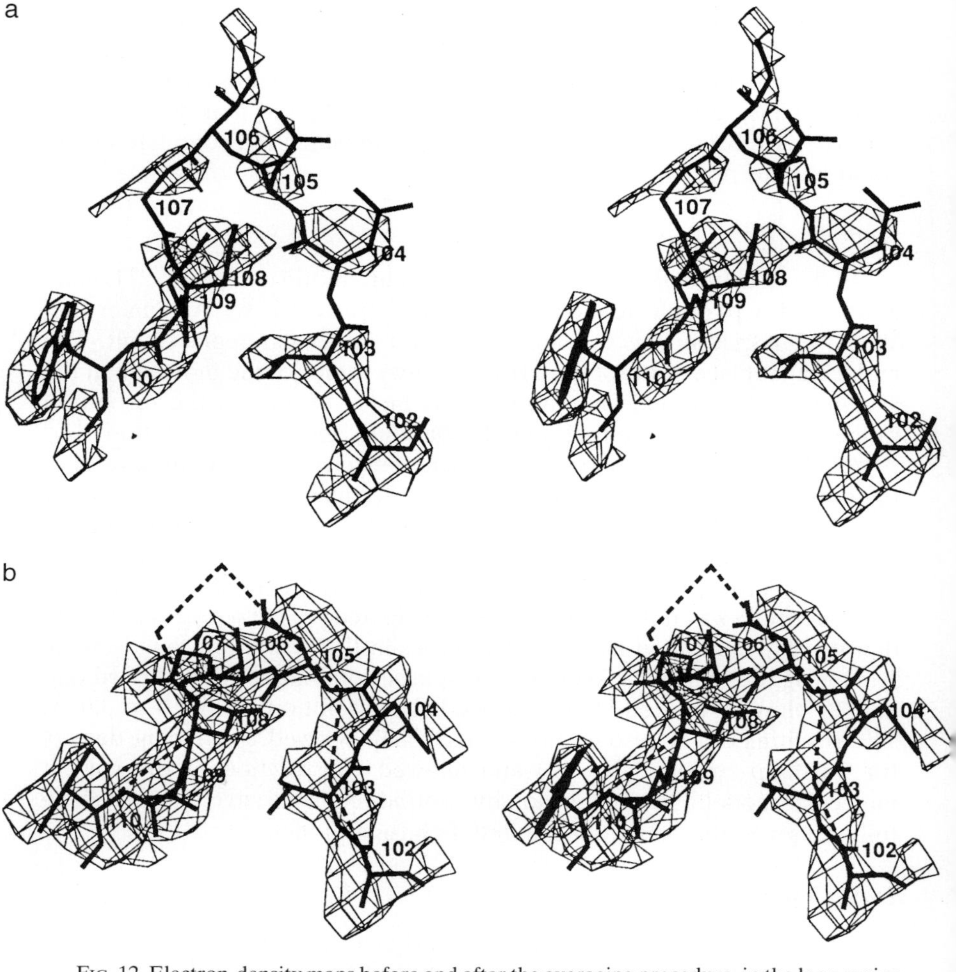

FIG. 12. Electron-density maps before and after the averaging procedure, in the loop region between residues 102 and 110. (Reproduced with permission from Livnah.[39]) Electron densities are contoured at the 1.0σ level. (a) The initial 8- to 3-Å resolution $2F_0 - F_c$ map, together with the initial, incorrect tracing of the loop. (b) The map obtained at completion of the averaging and phase-extension procedure, together with the correct model. Also shown in dashed lines is the $C\alpha$ tracing of the initial, incorrect model for this loop.

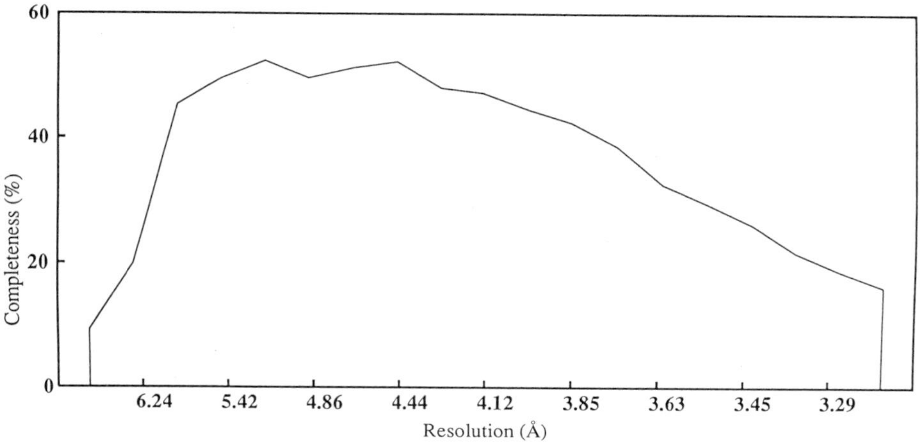

FIG. 13. Completeness of the Laue data set, used in the procedure of data completion by sixfold averaging, as a function of the resolution.

crystals also were extremely sensitive to X-ray radiation[42]: although the initial diffraction pattern recorded using monochromatic X-ray radiation at the Daresbury synchrotron source extends to ca. 2.8 Å, it rapidly decays beyond 5 Å. With the scarcity of available crystals, the crystal-efficient Laue method was used to collect a 3.2-Å native data set with only two crystals, whereas a monochromatic data collection strategy would have required ca. 20 crystals. The data obtained after processing of the four Laue film packs were only 37% complete in the resolution range of 7.0 to 3.2 Å (Fig. 13). The structure was solved by molecular replacement, using as a search model the homologous enzyme from *Bacillus stearothermophilus,*[43] which has 52% sequence identity to the glycosomal enzyme. This revealed the presence, in the asymmetrical unit, of one and a half tetrameric molecules, thereby indicating sixfold redundancy. Noncrystallographic symmetry operators were refined by maximizing the correlation coefficient between $|F_c|^2$ values and $|F_0|^2$ values, using the program BRUTE.[44] A model was built into the initial electron density to generate the envelope used during density modification. This was necessary owing to the presence of large surface insertions in the *T. brucei* enzyme subunit with respect to the bacterial enzyme. The initial map then was improved in an iterative proce-

[42] R. J. Read, R. K. Wierenga, H. Groendijk, A. Lambeir, F. R. Opperdoes, and W. G. J. Hol, *J. Mol. Biol.* **194,** 573 (1987).

[43] T. Skarżyński, P. C. E. Moody, and A. J. Wonacott, *J. Mol. Biol.* **193,** 171 (1987).

[44] M. Fujinaga and R. J. Read, *J. Appl. Crystallogr.* **20,** 517 (1987).

dure of 36 cycles of 6-fold averaging, in which the 63% missing reflections were introduced gradually as map-inversion amplitudes and phases. This procedure succeeded, largely because the missing data were randomly distributed throughout reciprocal space. The resulting averaged electron-density map, which had been computed with the 37% amplitudes from the Laue data set, plus the 63% missing reflections estimated during the averaging procedure, was of good quality. A complete molecular model was built that has been refined to an R factor of 17.6% using noncrystallographic symmetry restraints. Using the refined model phases as a reference set, one can analyze the results of the averaging procedure both for the observed and for the unobserved, missing data (Fig. 14). These scatter plots indicate, for the observed data, that the phases obtained from molecular replacement form a cloud along the main diagonal, thereby indicating their nonrandomness. This cloud shrinks along the diagonal during the averaging procedure, thereby indicating an improvement in the quality of the phases. For the unobserved reflections, a much more diffuse but similar cloud is also observed for the phases computed from the molecular replacement model (these phases were obviously not used for the computation of the initial map). Similarly, the cloud has collapsed along the main diagonal at completion of the averaging procedure. The phase estimates provided by the procedure for the missing reflections also are nonrandom. Similarly, using additional Laue data collected after completion of the averaging procedure, one can demonstrate that the structure-factor amplitudes estimated by this procedure also are nonrandom (not shown). Hence, all missing data were gradually replaced by nonrandom estimated values, thereby improving map quality.

Parts of Structure Not Obeying Assumed Symmetry

In their study of the three-dimensional structure of the homotetrameric enzyme flavocytochrome b_2, Xia *et al.*[45] have described an interesting situation of twofold n.c.s., in which one of the enzyme domains does not obey the n.c.s. Each monomer is made up of two domains, a 411-residue flavin-binding domain and an external, 100-residue cytochrome b_2 domain. The crystals of space group $P3_221$ contain a proper molecular fourfold axis coincident with the crystallographic twofold axis, thus the asymmetrical unit contains half a tetramer. In the solvent-flattened multiple isomorphous replacement (MIR) map, one of the two cytochrome b_2 domains shows no interpretable density and is disordered, whereas both flavin-binding do-

[45] Z.-X. Xia, N. Shamala, P. H. Bethge, L. W. Lim, H. D. Bellamy, N. H. Xuong, F. Lederer, and F. S. Mathews, *Proc. Natl. Acad. Sci. U.S.A.* **84,** 2629 (1987).

mains obey the twofold n.c.s. (Fig. 15). Although density modification by means of twofold averaging was not necessary to solve the structure, one could have improved the electron density by this procedure. This would have required an envelope showing the location of the two flavin-binding domains (where twofold averaging could have been performed), the ordered cytochrome b_2 domain (where the density should have been left unmodified), and the remaining regions (where solvent flattening could have been applied).

Refinement of Phases to Phase Sets Other Than That of Positive Structure

The most spectacular examples of the use of n.c.s. averaging are probably observed in the structure determinations of icosahedral viruses; these determinations make use of the high redundancy of information due to the high-order n.c.s. of the viral capsid. In procedures of phase extension applied to viral particle crystal data, several groups have observed that the technique can converge on phase sets other than those of the correct, positive structure.

In the case of the canine parvovirus structure determination, Tsao *et al.*[46] initiated phasing from a shell of uniform density, of inner radius 85 Å and outer radius 128 Å, which had been positioned according to the results of self-rotation function calculations. Phasing was initiated at 20-Å resolution, with 10 cycles of 60-fold averaging using $|F_0|$ values and calculated phases, after which the unobserved reflections were replaced by map inversion-calculated data in an additional 10 cycles of averaging at constant resolution. Phase extension was then carried out to 11-Å resolution in steps of one reciprocal lattice point at a time, then further to 9 Å, but the agreement between observed and map-inversion amplitudes dropped steadily as the resolution was increased and the procedure was judged not to be converging. A difference Fourier map from a partial K_2PtBr_6 derivative data set was computed and averaged over the 60 equivalent noncrystallographic asymmetrical units. This showed a single negative peak (Fig. 16), indicating that the set of phases generated by the averaging procedure had provided a Babinet opposite of the correct solution. This had been caused by the incorrect estimation both of the diameter of the hollow phasing shell and of the particle center. The SIR phasing at 8-Å resolution, followed by gradual phase extension to 3.25 Å, interspersed with refinement of the particle center, provided an interpretable 3.4-Å resolution electron-density map corresponding to the enantiomeric structure.

[46] J. Tsao, M. S. Chapman, H. Wu, M. Agbandje, W. Keller, and M. G. Rossmann, *Acta Crystallogr.* **B48,** 75 (1992).

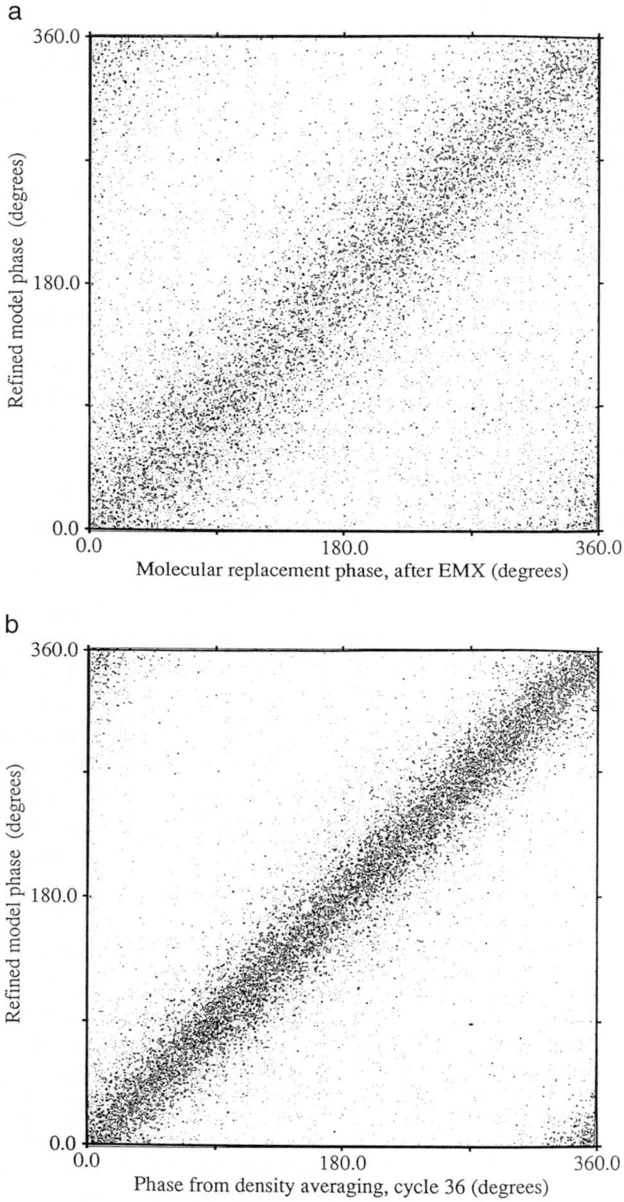

FIG. 14. Phase scatter plots comparing several phase sets to the phases computed from the refined molecular model. (Reproduced with permission from Vellieux *et al.*[41]) (a) Phase set computed from the initial molecular-replacement model, for the observed reflections only. (b) Phase set obtained at completion of the iterative averaging procedure, for the observed reflections only. (c) As in (a), for the unobserved, missing reflections. (d) As in (b), for the unobserved, missing reflections.

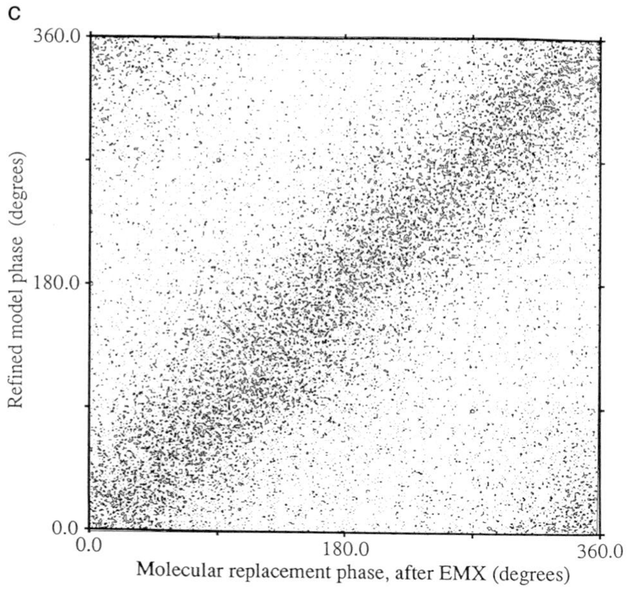

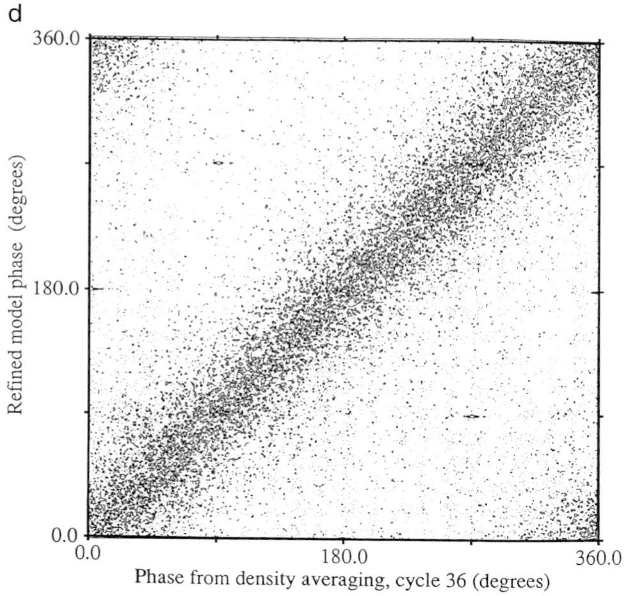

FIG. 14. (*continued*)

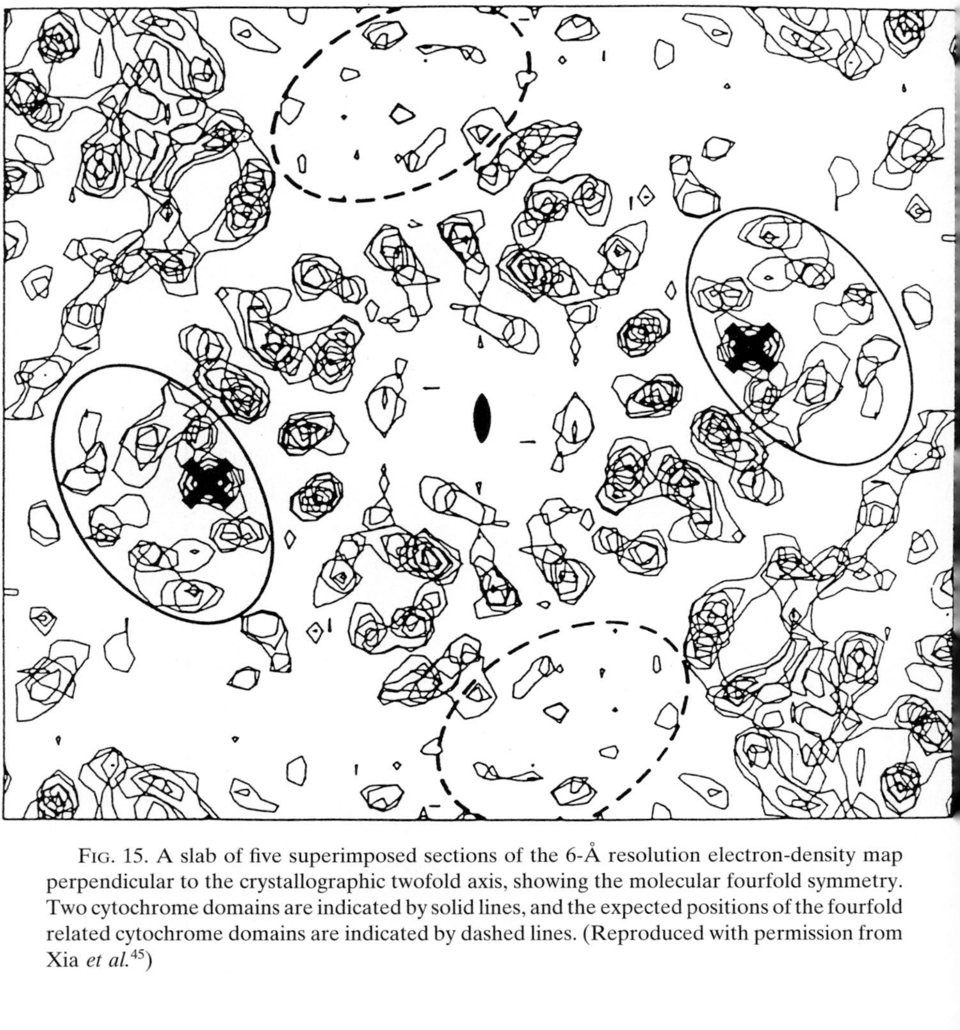

FIG. 15. A slab of five superimposed sections of the 6-Å resolution electron-density map perpendicular to the crystallographic twofold axis, showing the molecular fourfold symmetry. Two cytochrome domains are indicated by solid lines, and the expected positions of the fourfold related cytochrome domains are indicated by dashed lines. (Reproduced with permission from Xia *et al.*[45])

Another, similar example is the structure detemination of bacteriophage MS2,[47] in which the fivefold axes of icosahedral symmetry are aligned with the rhombohedral axes of the *R*32 space group of the crystals. Failure to interpret heavy-atom data prompted the authors to solve the structure by the molecular-replacement method, using a low-resolution search model made up of a truncated soybean mosaic virus (SBMV) structure.[48] In view

[47] K. Valegård, L. Liljas, K. Fridborg, and T. Unge, *Acta Crystallogr.* **B47,** 949 (1991).

[48] C. Abad-Zapatero, S. S. Abdel-Meguid, J. E. Johnson, A. G. W. Leslie, I. Rayment, M. G. Rossmann, D. Suck, and T. Tsukihara, *Nature* (*London*) **286,** 33 (1980).

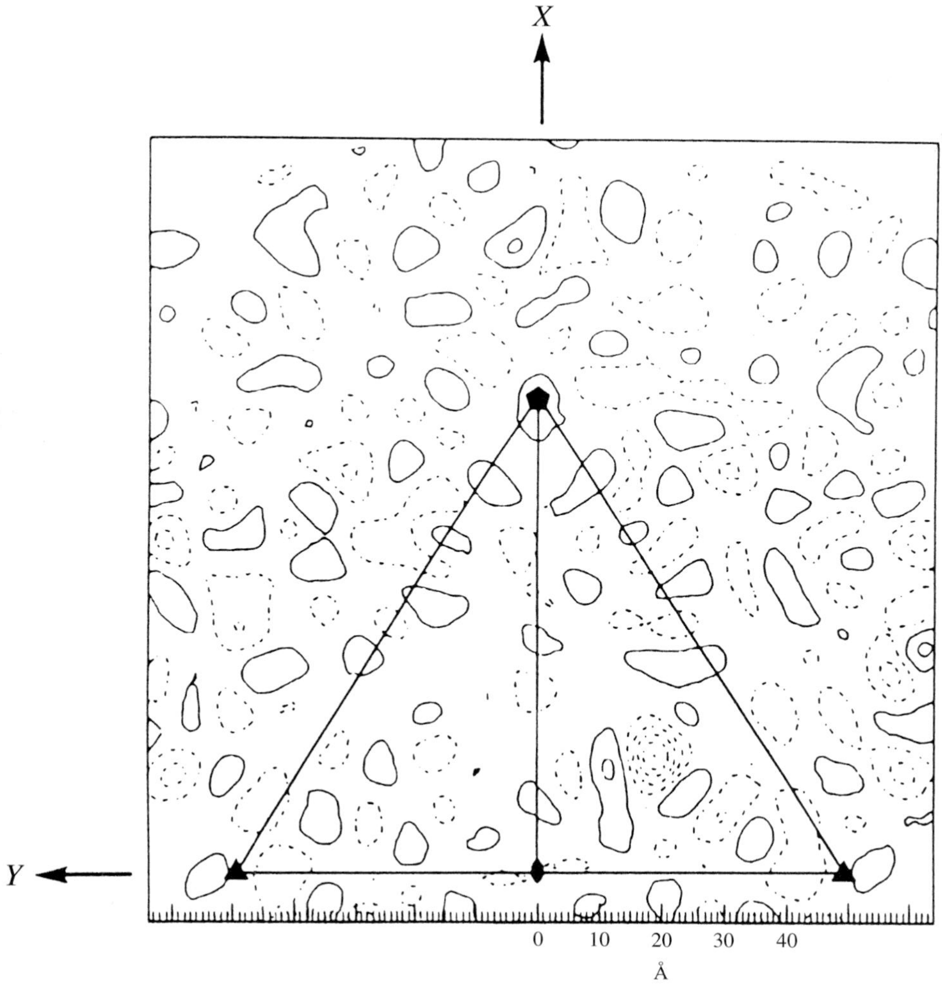

FIG. 16. A section of the skew-averaged difference-Fourier map calculated for the K_2PtBr_6 derivative, using the phases extended to 9-Å resolution from the hollow shell model. The large negative peak corresponds to the Pt site. (Reproduced with permission from Tsao *et al.*[46])

of the failure of an *R* factor search to provide the correct location of the model, the model was placed in the cell in a position compatible with the observed symmetries and where model clashes were prevented. A starting electron-density map was computed with measured data from 35 to 13 Å and with calculated structure factors used in place of all missing reflections to a lower resolution limit of 300 Å. It was subjected to phase extension to 3.4 Å by 30-fold n.c.s. averaging. Although the statistics obtained from the averaging procedure were good, giving a map inversion R factor of 20.4%, the resulting map was not interpretable. Reanalysis of the heavy-atom data using the phases generated by the averaging procedure showed heavy-atom sites as negative peaks in difference Fourier maps, thus indicating that phases had, again, refined to the wrong sign. Starting from isomorphous replacement phases, phase extension from 6 to 3.3 Å then provided an interpretable map. Comparison of the phase set computed from the refined molecular model to that generated from the initial averaging procedure (Fig. 17) shows that, while most of the phases had refined to the Babinet hand of the correct solution, some had refined to the enantiomorphic phases

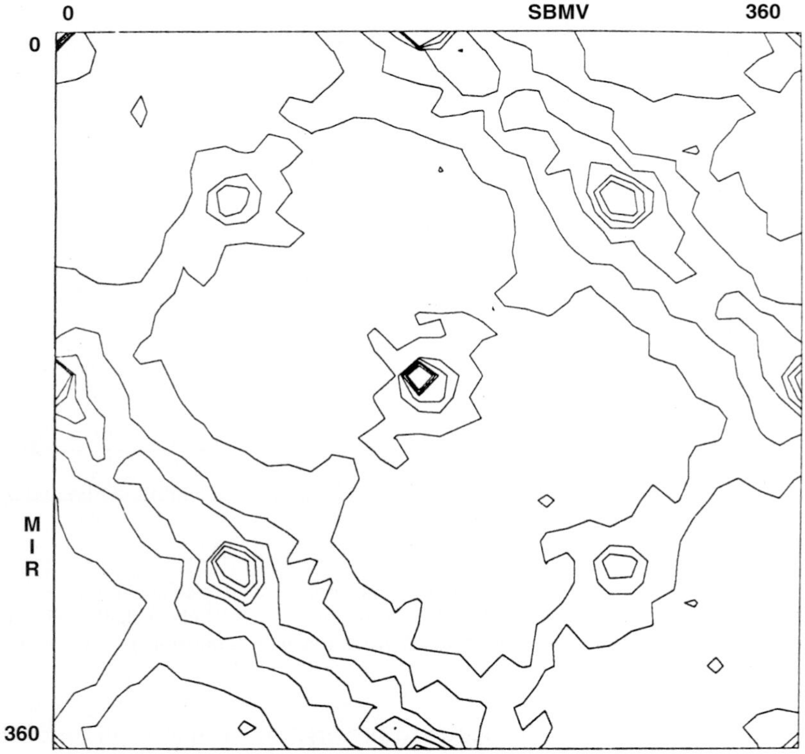

FIG. 17. A contoured histogram of phase correspondence gives a comparison of phase angles between the phase set after phase extension from the SBMV–MS2 model and the final phase set based on isomorphous replacement, in the resolution range of 15 to 9 Å. (Reproduced with permission from Valegård *et al.*[47])

whereas others had refined to its Babinet opposite, thus leading to the generation of a totally uninterpretable map.

Acknowledgments

We thank Dr. Fontecilla-Camps for critical reading of the manuscript. This is publication number 256 of the Institut de Biologie Structurale Jean-Pierre Ebel.

[4] Combining Constraints for Electron-Density Modification

By KAM Y. J. ZHANG, KEVIN COWTAN, and PETER MAIN

Introduction

The crystallographic phase problem is indeterminate given only the structure-factor amplitudes. It is only through the knowledge of the chemical or physical properties of the electron density that the phases can be retrieved. The characteristic features of electron density can often be expressed as mathematical constraints on the phases. The phasing power of a constraint depends on the number of density points affected and the magnitudes of changes imposed on the electron density. It also depends on the physical nature and accuracy of the constraint and on how rigorously the constraint is applied. The phasing power also increases with the number of independent constraints employed.

The phase problem can be solved *ab initio* under favorable circumstances by exploiting one or several constraints, such as positivity and atomicity used in direct methods for small molecules, and noncrystallographic symmetry averaging for spherical viruses. Some of the constraints successfully used in the *ab initio* phasing for small molecules are tabulated in Table I (atomic resolution). The available constraints at our disposal are, however, generally not enough to suffice for an *ab initio* solution to the phase problem for macromolecules. Nevertheless, these constraints could be used to refine the experimental phases or to extend them to the full resolution of the native data. Table I (nonatomic resolution) lists those constraints and how they are used for macromolecular phase improvement.

The density modification methods that are discussed in this chapter seek to improve the experimental phases by exploiting and combining constraints derived from some characteristic features of the electron-density map. The more constraints that are employed, the lower the resolution

0076-6879/97 $25.00

TABLE I
CONSTRAINTS USED IN PHASE REFINEMENT

Constraints	Use
In reciprocal space: For atomic resolution	
Positivity $\rho(\mathbf{r}) \geq 0$	Karle–Hauptman determinant
$\int \rho^3(\mathbf{r})\, dv \Rightarrow \max$	Tangent formula
Equal atom	Sayre's equation
$\int \rho(\mathbf{r}) \ln[\rho(\mathbf{r})]\, dv \Rightarrow \max$	Maximum entropy method
In real space: For nonatomic resolution	
Positivity $\rho(\mathbf{r}) \geq 0$	Attenuation of negative densities
Solvent flatness	Solvent flattening
Equal molecules	Molecular averaging
Map continuity	Skeletonization

from which the refinement and extension can be initiated. We will have a method with which to solve the macromolecular phase problem *ab initio* when we can start refinement and extension from randomly generated phases. Our goal is to find new characteristic features of electron density and to combine them with existing constraints to create more powerful methods for phase improvement.

An integrated procedure, known as SQUASH, which combines the constraints from the correct electron-density distribution, solvent flatness, correct local density shape, and equal molecules, has been demonstrated to be a powerful method for macromolecular phase refinement and extension.[1–5] In this chapter, we summarize previous results and describe progress in density modification using SQUASH.

Constraining Electron Density

The chemical or physical information of the underlying structure that the electron density represents serves as constraints on the electron density or structure factors. Because the structure-factor amplitudes are known, these constraints restrict the value of phases. For small molecules, the constraints of positivity and atomicity are sufficient to solve the phase problem, because crystals of small molecules generally diffract to atomic resolution. Because macromolecular crystals rarely diffract to atomic reso-

[1] K. Y. J. Zhang and P. Main, *Acta Crystallogr.* **A46,** 41 (1990).
[2] K. Y. J. Zhang and P. Main, *Acta Crystallogr.* **A46,** 377 (1990).
[3] P. Main, *Acta Crystallogr.* **A46,** 507 (1990).
[4] K. D. Cowtan and P. Main, *Acta Crystallogr.* **D49,** 148 (1993).
[5] K. Y. J. Zhang, *Acta Crystallogr.* **D49,** 213 (1993).

lution, no single constraint at our disposal is powerful enough to render the phase problem determinable. Because each constraint represents different characteristic features in an electron density, there is independence between these constraints. We believe that the combination of all available constraints would yield a larger constraint space and reduce the ambiguity of phase values. The optimum way to combine different constraints is to apply them simultaneously, so that the phases satisfy all of the constraints at the same time.

The constraints used in our work can be broadly divided into three categories. They are linear constraints, such as solvent flatness, density histograms, and equal molecules; nonlinear constraints, such as Sayre's equation; and structure-factor amplitude constraints. We describe them in the following section.

Linear Constraints: Density Modifications

The modification to the density value at a grid point from a linear constraint system is independent of the values at other grid points. These include solvent flattening, histogram matching, and molecular averaging. These density modification methods construct an improved electron density directly from an initial density map, as expressed by Eq. (1):

$$\rho(\mathbf{x}) = H(\mathbf{x}) \tag{1}$$

where $H(\mathbf{x})$ is the modified density.

This density modification can be expressed as a residual equation in real space that becomes zero when the constraints are satisfied. Different weights or confidence values could be applied to electron-density points to allow different density-modification techniques, which affect different regions of the map, to be weighted differently.

$$r_1(\mathbf{x}) = w_1(\mathbf{x})[H(\mathbf{x}) - \rho(\mathbf{x})] \tag{2}$$

where r_1 represents the residual and w_1 is the weight.

Solvent Flattening. Solvent flattening exploits the fact that the solvent region of an electron density is flat at medium resolution owing to the high thermal motion of atoms and disorder of the solvent. The existence of a flat solvent region in a crystal places strong constraints on the structure-factor phases.

The constraint of solvent flatness is implemented by identifying the molecular boundaries and replacing the densities in the solvent region by their mean density value. The molecular boundary that partitions protein from solvent is located by an automated convolution procedure proposed by Wang[6] and modified by Leslie.[7]

Histogram Matching. Histogram matching seeks to bring the distribution of electron-density values of a map to that of an ideal map.

The density histogram of a map is the probability distribution of electron-density values. The comparison of the histogram for a given map with that expected for a good map can serve as a measure of quality. Furthermore, the initial map can be improved by adjusting density values in a systematic way to make its histogram match the ideal histogram.

The ideal histograms for protein structures depend on resolution and overall temperature factors, but they are independent of the particular structures.[8] The ideal histogram can be taken from a known structure[9] or predicted by an analytical formula.[10,11]

The matching of histogram $P'(\rho)$ to $P(\rho)$ is achieved by finding the upper bound ρ_i' corresponding to ρ_i in Eq. (3),

$$\int_{\rho_0}^{\rho_i} P(\rho)\, d\rho = \int_{\rho_0'}^{\rho_i'} P'(\rho)\, d\rho \tag{3}$$

where ρ_0 and ρ_0' are the minimum densities in the histograms $P(\rho)$ and $P'(\rho)$, respectively. A linear transformation between ρ_i and ρ_i' can be used, provided that the sampling i is sufficiently fine,

$$\rho_i' = a_i \rho_i + b_i \tag{4}$$

where a_i and b_i are the scale and shift, respectively.

Note that histogram matching applies a minimum and a maximum value to the electron density, imposes the correct mean and variance, and defines the entropy of the new map. The order of electron-density values also remains unchanged after histogram matching.

Molecular Averaging. The averaging method enforces the equivalence of electron-density values between grid points related by the noncrystallographic symmetry (NCS).

When the same molecule occurs more than once in the asymmetric unit, the distinct molecules usually adopt a similar three-dimensional structure. It can be assumed that the corresponding density values between those related molecules are approximately equal. This places a strong con-

[6] B. C. Wang, *Methods Enzymol.* **115,** 90 (1985).
[7] A. G. W. Leslie, *Acta Crystallogr.* **A43,** 134 (1987).
[8] K. Y. J. Zhang and P. Main, *Acta Crystallogr.* **A46,** 41 (1990).
[9] K. Y. J. Zhang and P. Main, *Acta Crystallogr.* **A46,** 41 (1990).
[10] P. Main, *Acta Crystallogr.* **A46,** 507 (1990).
[11] V. Yu. Lunin and T. P. Skovoroda, *Acta Crystallogr.* **A47,** 45 (1991).

straint on the density values at equivalent positions among those related molecules.

The initial NCS operation obtained from rotation and translation functions[12] or heavy-atom positions can be refined by a density space R factor search method or a least-squares refinement method.[13]

Noncrystallographic symmetry-related regions of the density are averaged together to enhance the density features and suppress noises. Different domains of the molecule may be masked and averaged separately with different NCS operators.

The interpolation of density values between grid points is achieved by the use of linear or quadratic *b* splines, combined with a spectral correction in reciprocal space.[14] This approach gives a more accurate interpolant than conventional methods of the same order, and lends itself to the simultaneous calculation of density gradients, used in the refinement of NCS operators.

Iterative Skeletonization. The iterative skeletonization method enhances connectivity in the map. This is achieved by locating ridges of density, constructing a graph of linked peaks, and then building a new map using cylinders of density around the graph links.[15]

The connectivity graph, or "skeleton," may be efficiently calculated by the core-tracing algorithm[16]; this algorithm has the advantage that the skeleton can be analyzed as it is built to determine optimum connectivity parameters for a particular map.

Nonlinear Constraints: Sayre's Equation

Sayre's equation constrains the local shape of electron density. For equal and resolved atoms, the density distribution is equal to the squared density convoluted with an atomic shape function, as pointed out by Sayre.[17] This is normally expressed in terms of structure factors as

$$F(\mathbf{h}) = \frac{\theta(\mathbf{h})}{V} \sum_{\mathbf{k}} F(\mathbf{k})F(\mathbf{h} - \mathbf{k}) \tag{5}$$

where $\theta(\mathbf{h}) = f(\mathbf{h})/g(\mathbf{h})$ is the ratio of scattering factors of real and "squared" atoms and V is the unit cell volume.

[12] M. G. Rossmann and D. M. Blow, *Acta Crystallogr.* **15,** 24 (1962).
[13] K. Y. J. Zhang, *Acta Crystallogr.* **D49,** 213 (1993).
[14] K. D. Cowtan, in preparation (1997).
[15] D. Baker, C. Bystroff, R. J. Fletterick, and D. A. Agard, *Acta Crystallogr.* **D49,** 429 (1993).
[16] S. Swanson, *Acta Crystallogr.* **D50,** 695 (1994).
[17] D. Sayre, *Acta Crystallogr.* **5,** 60 (1952).

Sayre's equation can also be expressed in terms of the electron density,

$$\rho(\mathbf{x}) = \frac{V}{N} \sum_{\mathbf{y}} \rho^2(\mathbf{y}) \psi(\mathbf{x} - \mathbf{y}) \tag{6}$$

where

$$\psi(\mathbf{x} - \mathbf{y}) = \frac{1}{V} \sum_{\mathbf{h}} \theta(\mathbf{h}) \exp[-2\pi i \mathbf{h}(\mathbf{x} - \mathbf{y})] \tag{7}$$

which states that the convolution of squared electron density with a shape function produces the original electron density. It can be seen from Eq. (6) that Sayre's equation puts constraints on the local shape of electron density. The shape function is represented by $\psi(\mathbf{x} - \mathbf{y})$ in Eq. (7), which is the Fourier transformation of $\theta(\mathbf{h})$.

The real-space residual for Sayre's equation can be expressed as

$$r_2(\mathbf{x}) = \frac{V}{N} \sum_{\mathbf{y}} \rho^2(\mathbf{y}) \psi(\mathbf{x} - \mathbf{y}) - \rho(\mathbf{x}) \tag{8}$$

The scale factor θ, or its real-space equivalent ψ, can be predicted at high resolution from the known atomic shape. In the macromolecular case, atomic features are not visible; therefore, a spherical symmetric function is determined empirically to give the best agreement between the left- and right-hand sides of Sayre's equation.

Structure-Factor Constraints: Phase Combination

The observed structure-factor amplitudes serve as another constraint that the modified electron density must satisfy. Moreover, if experimental phases are known, it is also desirable to tether the phase values of the modified map to the observed ones. This is achieved by a phase combination procedure. The constraints of the observed structure-factor amplitudes are used to assess the accuracy of each modified phases and weigh them accordingly in the phase combination with observed phases.

The multiple isomorphous replacement (MIR) phase probability distribution is given by Blow and Crick.[18] The probability distribution for phases calculated from the modified map is determined by the Sim weighting scheme[19] as adapted by Bricogne.[20] The phases are combined by multiplying

[18] D. M. Blow and F. H. C. Crick, *Acta Crystallogr.* **12,** 794 (1959).

[19] G. A. Sim, *Acta Crystallogr.* **12,** 813 (1959).

[20] G. Bricogne, *Acta Crystallogr.* **A30,** 395 (1974).

their respective phase probabilities.[21] This multiplication of phase probabilities is simplified by adding the coefficients that code for phase probabilities.[22]

Combining Constraints and Solution to System of Constraint Equations

For a modified electron-density map to satisfy the preceding constraints at the same time, we must solve a system of simultaneous equations that includes all the residual equations,

$$\begin{cases} r_1(\mathbf{x}) = w_1[H(\mathbf{x}) - \rho(\mathbf{x})] \\ r_2(\mathbf{x}) = \dfrac{V}{N}\sum_{\mathbf{y}} \rho^2(\mathbf{y})\psi(\mathbf{x}-\mathbf{y}) - \rho(\mathbf{x}) \end{cases} \tag{9}$$

where w_1 is the relative weight between Sayre's equation [Eq. (8)] and the constraints from density modification [Eq. (2)].

The equations in Eq. (9) represent a system of nonlinear simultaneous equations with as many unknowns as the number of grid points in the asymmetric unit of the map and twice as many equations as unknowns. The functions $H(\mathbf{x})$ and $\psi(\mathbf{x})$ are both known. The least-squares solution, using either the full matrix or the diagonal approximation, is obtained using the Newton–Raphson technique with fast Fourier transforms (FFTs) as described by Main.[23]

Results

To demonstrate the effect on phase improvement, various density modification techniques are applied to an MIR data set for which the refined coordinates are available. The test structure is 5-carboxymethyl-2-hydroxymuconate Δ-isomerase (CHMI).[24] The protein was crystallized in space group $P4_12_12$ with a unit cell of $a = b = 90.4$ Å, $c = 130.1$ Å. There are three molecules in the asymmetric unit, related by a threefold noncrystallographic rotation axis. Native data were collected to 2.1 Å. Phasing is from one platinum and one gold derivative, with anomalous scattering data collected for both derivatives. These gave good phases to 3.7 Å, with the gold derivative providing weak phases to 2.7 Å. The initial MIR map is quite poor, although the structure was eventually solved by building frag-

[21] G. Bricogne, *Acta Crystallogr.* **A32,** 832 (1976).
[22] W. A. Hendrickson and E. E. Lattman, *Acta Crystallogr.* **B26,** 136 (1970).
[23] P. Main, *Acta Crystallogr.* **A46,** 372 (1990).
[24] D. B. Wigley, D. I. Roper, and R. A. Cooper, *J. Mol. Biol.* **210,** 881 (1989).

ments into each molecule in the asymmetric unit and averaging them using the NCS relationships.

The following density modification tests were applied to the MIR data:

1. Phase extension by solvent flattening
2. Phase extension by solvent flattening and histogram matching
3. Phase extension by solvent flattening and averaging
4. Phase extension by solvent flattening, histogram matching, and averaging
5. Phase extension by solvent flattening, histogram matching, and averaging followed by phase refinement using solvent flattening, histogram matching, and Sayre's equation
6. Phase extension by solvent flattening, histogram matching, and averaging followed by phase refinement using solvent flattening, histogram matching, and skeletonization

Phase extension was performed using the 3.7-Å data as a starting set, followed by introducing new reflections by resolution shell as the calculation progresses. Twenty cycles of phase extension were used to reach the resolution limit of 2.1 Å.

Sayre's equation and skeletonization are more dependent on the quality of the input map than the other techniques and cannot be applied to the initial map in this case; thus they have been applied to refinement of the final phases from method 4.

The MIR and density-modified maps, sharpened by the removal of the overall temperature factor, are compared with a map calculated from the refined model by means of the correlation between the two maps, calculated as

$$\frac{\overline{\rho_{\text{est}}\rho_{\text{calc}}} - \overline{\rho_{\text{est}}}\,\overline{\rho_{\text{calc}}}}{[(\overline{\rho_{\text{est}}^2} - \overline{\rho_{\text{est}}}^2)(\overline{\rho_{\text{calc}}^2} - \overline{\rho_{\text{calc}}}^2)]^{1/2}}.$$

The mean phase error between the modified phases and that calculated from the refined model in the regions ∞–3.7 Å and 3.7–2.1 Å is given in Table II.

Note that solvent flattening alone has led to a deterioration in the electron density. This is probably due to the weakness of solvent flattening as a phase extension technique. Of the other methods, histogram matching and averaging seem to contribute most strongly to the phase improvement, with their combination being particularly powerful.

To examine what is happening in more detail, it is useful to plot the quantity

$$\frac{\overline{w_{\text{est}}|F|^2 \cos(\phi_{\text{est}} - \phi_{\text{calc}})}}{(\overline{w_{\text{est}}^2|F|^2}\,\overline{|F|^2})^{1/2}}$$

TABLE II
5-CARBOXYMETHYL-2-HYDROXYMUCONATE Δ-ISOMERASE PHASE REFINEMENT RESULTS[a]

Data set	Correlation	Mean phase error	
		∞–3.7 Å	3.7–2.1 Å
0. MIR	0.143	72.90	88.59
1. sf	0.132	69.52	87.78
2. sf/hm	0.176	68.63	85.39
3. sf/av	0.200	63.11	86.21
4. sf/hm/av	0.371	61.78	72.49
5. (4) + sf/hm/sq	0.382	60.70	70.78
6. (4) + sf/hm/sk	0.399	61.76	69.91

[a] sf, Solvent flattening; hm, histogram matching; av, NCS averaging; sq, Sayre's equation; sk, skeletonization.

as a function of resolution for each data set, where ϕ_{est} is the estimated phase and w_{est} is the estimated figure of merit for that phase. When calculated over the whole data set this function is identical to the map correlation described earlier; however, by forming the correlation in reciprocal space we can

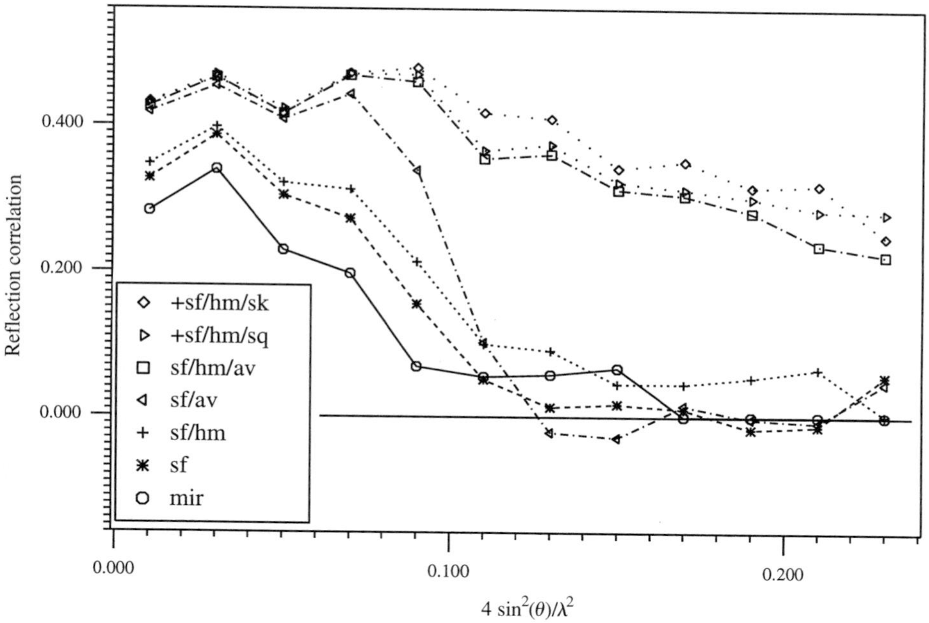

FIG. 1. CHMI density modification results: structure-factor correlation with resolution.

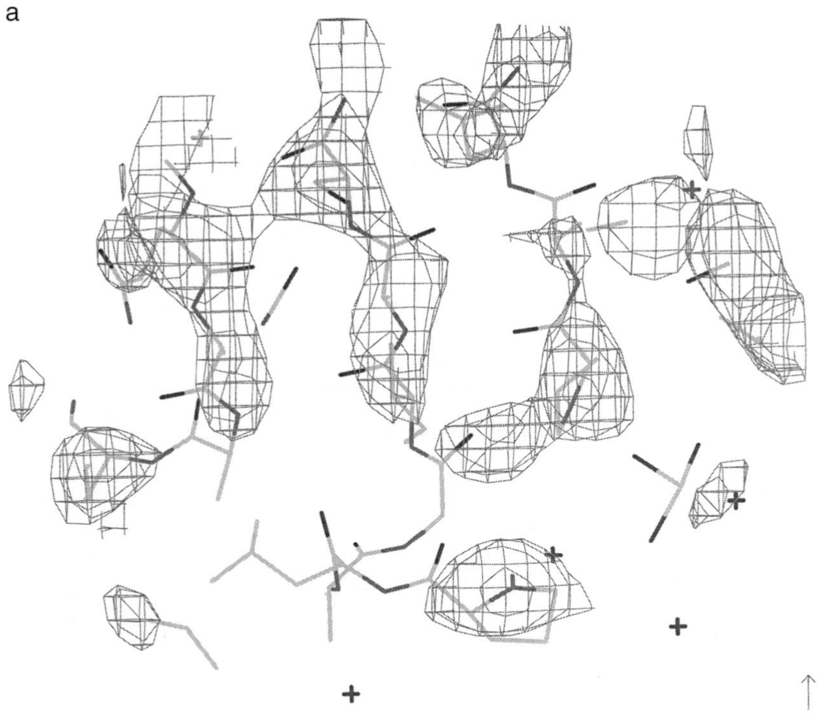

FIG. 2. CHMI density maps, before and after averaging; pictures produced using QUANTA. (a) MIR map. (b) Density-modified (sf/hm/av) map.

examine how the agreement with the calculated map varies for different regions of reciprocal space. This function is shown in Fig. 1.

From the graph in Fig. 1 it is clear the solvent flattening alone has improved the low-resolution data; however, the data in the range from 3.7 to 2.7 Å has gotten worse. There is no significant phase extension past this point. Histogram matching has further improved the low-resolution data, and in addition produced significant phasing to the limit of the native data.

Solvent flattening and averaging have further improved the low-resolution data, but again there is no significant information in the intermediate- or high-resolution data. Solvent flattening, histogram matching, and averaging give comparable results at low resolution, but now the phase extension has become very strong indeed, to the limit of the native data.

Sayre's equation and skeletonization both lead to a slight further improvement in the data, especially at high resolution; however, both tech-

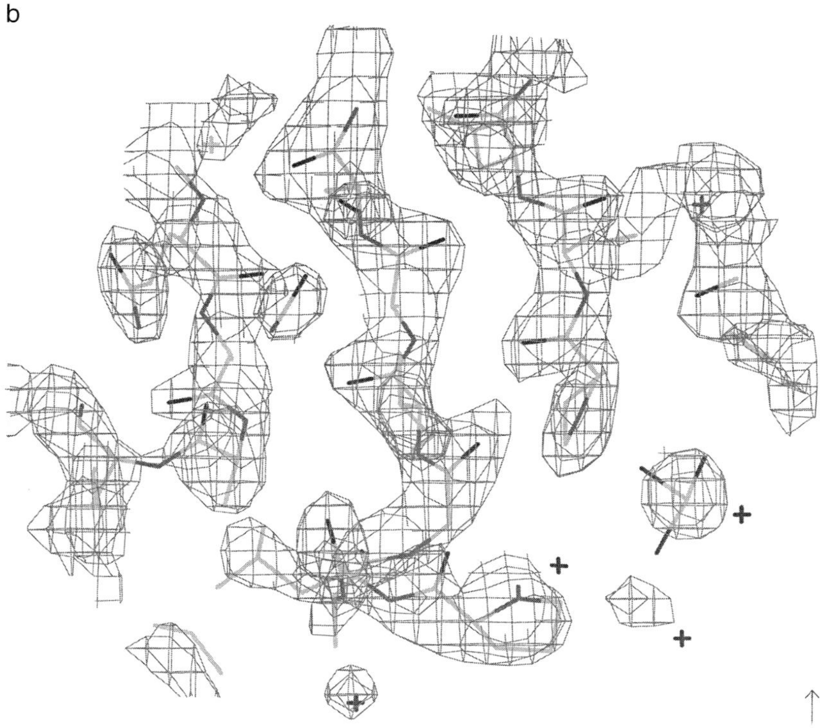

FIG. 2. (*continued*)

niques function better when they can be applied to the initial map rather than after a previous phase extension calculation.

The region of density around the β sheet that links the trimer together is shown in Fig. 2 for the MIR map and the modified map from method 4; this is a typical density-modification protocol, whereas the combined procedures of methods 5 and 6 are still relatively untried. The density is contoured at the 2σ level. The MIR map shows both breaks in the β strands and links across from one strand to another. In the density-modified map, the strands are complete and resolved from one another, and most of the side chains are clearly visible.

Discussion

Density-modification methods have become sufficiently powerful that it is possible to solve structures from comparatively poor initial maps, reducing the amount of effort required in finding further derivatives and collecting additional data sets. In the case of high-order noncrystallographic symmetry, density modification can even be used as part of an *ab initio* solution procedure,[25,26] and even when strong phase information is available it may simplify the process of map interpretation.

There are a number of software packages known to the authors for performing various density modifications. These include

1. SQUASH: An automated program to perform solvent flattening, histogram matching, averaging, and Sayre refinement[27]
2. DM: An automatic program to perform solvent flattening, histogram matching, skeletonization, averaging, and Sayre refinement; available as part of the CCP4 suite[28,29]
3. DEMON: A set of programs including facilites for solvent flattening, histogram matching, and averaging; written as part of the BIOMOL package[30]
4. RAVE: A set of programs for averaging, including averaging between crystal forms[31]
5. PRISM: A set of programs for density modification by improving the connectivity of electron-density maps using the iterative skeletonization method[32]

Acknowledgments

K. Cowtan is funded by the CCP4 project, Daresbury, England. Graphics were produced using QUANTA.[33]

[25] J. Tsao, M. S. Chapman, and M. G. Rossmann, *Acta Crystallogr.* **A48,** 293 (1992).
[26] K. Braig, Z. Otwinowski, R. Hegde, D. C. Boisvert, A. Joachimiak, A. L. Horwich, and P. B. Sigler, *Nature (London)* **371,** 578 (1994).
[27] K. Y. J. Zhang and P. Main, *Acta Crystallogr.* **A46,** 377 (1990).
[28] The CCP4 suite: Programs for protein crystallography. *Acta Crystallogr.* **D50,** 760 (1994).
[29] K. D. Cowtan, DM program. *Joint CCP4 ESF-EACBM Newslett.* **31,** 34 (1994).
[30] F. M. D. Vellieux, J. F. Hunt, S. Roy, and R. J. Read, *J. Appl. Cryst.* **28,** 347 (1995).
[31] G. J. Kleijwegt and A. T. Jones, "RAVE Suite for Symmetry Averaging." Uppsala, Sweden.
[32] D. Baker, C. Bystroff, R. J. Fletterick, and D. A. Agard, *Acta Crystallogr.* **D49,** 429 (1993).
[33] "QUANTA." MSI, Burlington, Massachusetts (1995).

[5] MICE Computer Program

By CHRISTOPHER J. GILMORE and GÉRARD BRICOGNE

Introduction

The MICE (maximum entropy in a crystallographic environment) computer program was begun as an experimental computer program designed to use maximum entropy (ME) and likelihood to solve crystal structures of small molecules,[1–4] especially those in which data quality was a problem, with specific applications to problems in powder and electron diffraction. The underlying theory is based on the methodology of ME coupled with likelihood estimation proposed by Bricogne,[5–11] and is discussed in detail in Volume 276 in this series.[11a] Quite early on, however, it was realized that the method exhibited behavior that was not found in standard direct method phasing programs: (1) It was stable regardless of data resolution or completeness; (2) it was robust with respect to observational errors, which can even be incorporated into the calculations; and (3) it had an ability to utilize many different constraints, not just those of amplitude and phase. In macromolecular crystallography the most obvious additional constraint is that of the envelope or mask in conjunction with solvent flattening, noncrystallographic symmetry (NCS), and partial structural information. Such properties led to applications of MICE to problems in

[1] G. Bricogne and C. J. Gilmore, *Acta Crystallogr.* **A46,** 284 (1990).

[2] C. J. Gilmore, G. Bricogne, and C. Bannister, *Acta Crystallogr.* **A46,** 297 (1990).

[3] C. J. Gilmore and G. Bricogne, *in* "Crystallographic Computing 5: From Chemistry to Biology" (D. Moras, A. D. Podjarny, and J. C. Thierry, eds.), p. 298. Oxford University Press, Oxford, 1991.

[4] C. J. Gilmore, *in* "Crystallographic Computing 6: A Window on Modern Crystallography" (H. D. Flack, L. Parkănyi, and K. Simon, eds.), p. 25. Oxford University Press, Oxford, 1993.

[5] G. Bricogne, *Acta Crystallogr.* **A40,** 410 (1984).

[6] G. Bricogne, *Acta Crystallogr.* **A44,** 517 (1988).

[7] G. Bricogne, *in* "Crystallographic Computing 4: Techniques and New Technologies" (N. W. Isaacs and M. R. Taylor, eds.), p. 60. Clarendon Press, Oxford, 1988.

[8] G. Bricogne, *in* "Maximum Entropy in Action" (B. Buck and V. A. Macaulay, eds.), p. 187. Oxford University Press, Oxford, 1991.

[9] G. Bricogne, *in* "Crystallographic Computing 5: From Chemistry to Biology" (D. Moras, A. D. Podjarny, and J. C. Thierry, eds.), p. 257. Oxford University Press, Oxford, 1991.

[10] G. Bricogne, *in* "The Molecular Replacement Method. Proceedings of the Study Weekend Held at Daresbury 1992" (W. Wolf, E. J. Dodson, and S. Glover, eds.), p. 60. Daresbury Laboratory, Warrington, UK, 1992.

[11] G. Bricogne, *Acta Crystallogr.* **D49,** 37 (1993).

[11a] G. Bricogne, *Methods Enzymol.* **276,** 361 (1997).

0076-6879/97 $25.00

macromolecular crystallography. However, compared to most of the methods described in this volume, ME is not yet a fully established technique in crystallography, and is still very much an experimental methodology.

In what follows, we shall be concerned with X-ray and electron diffraction. Electron microscopy is proving to be an exciting tool for studying systems that are inaccessible to X-ray crystallography, and in particular two-dimensional crystals, especially those of membrane proteins. In this case we often have a diffraction pattern plus phase information at a lower resolution derived from the Fourier transform of the image, and we are concerned with phasing the diffraction intensities.

Most of this chapter is concerned with the underlying methodology of MICE; however, there is a brief outline of its successful applications at the end, some of which is further amplified in [6] in this volume.[11b]

Outline of Method

1. The diffraction intensities are first processed using the RALF (recycling and likelihood estimation from fragments) computer program, which normalizes them to give unitary structure factors, $|U_{\mathbf{h}}|^{\text{obs}}$, and their associated standard deviations $\sigma_{\mathbf{h}}$. For macromolecular data sets this can be a problematic procedure: There is, for example, a tendency to produce an unrealistic overall temperature factor, and often one must impose a suitable one. The root of this problem lies partly with the use of Wilson statistics,[12] which assumes a random atom model in which the atoms are uniformly distributed. This is, of course, inappropriate for proteins, for which there is a need to use a multichannel formulation in which solvent and protein are treated independently,[6] but this has not yet been done in RALF. In the electron diffraction case, often using two-dimensional data, these problems are further exacerbated by radiation damage and dynamical effects giving rise to systematic errors in the intensities themselves. Structural information, in the form of more or less correctly placed fragments, if available, is incorporated into the normalization, and this provides $|U_{\mathbf{h}\in H}|^{\text{frag}}$ and $\phi^{\text{frag}}_{\mathbf{h}\in H}$ (where ϕ is a phase angle) for later use. The program optionally reads data from CCP4 MTZ files[13] and coordinates from Protein Data Bank (PDB) entries if required. Several data filters are provided. The

[11b] C. W. Carter, Jr. and S. Xiang, *Methods Enzymol.* **277,** [6], 1997 (this volume).

[12] A. J. C. Wilson, *Acta Crystallogr.* **2,** 318 (1949).

[13] "The CCP4 Suite—Computer Programs for Protein Crystallography." Daresbury Laboratory, Warrington, UK, 1993.

normalized data, and fragment U magnitudes and phases, are passed to the MICE program (see Fig. 1).

2. If envelope information is to be used, MICE provides options for its preparation: a suitable map is read from the CCP4 suite or from a previous run of MICE. It is smoothed using pixel smoothing, and normalized to give a nonuniform prior $m(\mathbf{x})$. It is also Fourier transformed to give coefficients $g_{\mathbf{h}}$. When no envelope knowledge is available, as in the *ab initio* determination of heavy atom positions, a uniform prior of the form $m(\mathbf{x}) = 1/\text{vol}(V)$ is used, where V is the unit cell of volume vol(V). It is possible to generate a fragment prior: a known fragment correctly placed in the cell has an exclusion zone placed around it to prevent the building of any further electron density in this region. Envelopes can be used in conjunction with fragment priors. Solvent-flattening options are also provided.

3. The observed U magnitudes are partitioned into two sets: the basis set $H = \{\mathbf{h}_1, \mathbf{h}_2, \ldots, \mathbf{h}_n\}$ comprises the reflections for which reliable phase information $\Phi = \{\phi_1, \phi_2, \ldots, \phi_n\}$ for n reflections is available. In phase-extension studies this set consists of all reflections with phases derived from isomorphous replacement techniques; in electron microscopy this set can be phases obtained from the Fourier transform of the image, whereas in the *ab initio* case this set comprises only the origin (and enantiomorph, if relevant) defining reflections with associated phases (n in this case is usually ≤ 4). The disjoint set $\{K\}$ comprises the remaining unphased reflections. Filters are provided to allow the user to select MIR phases on the basis of figure of merit, resolution, or minimum amplitude.

4. The basis set reflections are used as constraints in an entropy maximization. The distribution $q^{\text{ME}}(\mathbf{x})$ having maximum entropy

$$S_m(q) = -\int_V q(\mathbf{x}) \log[q(\mathbf{x})/m(\mathbf{x})]\, d^3\mathbf{x} \tag{1}$$

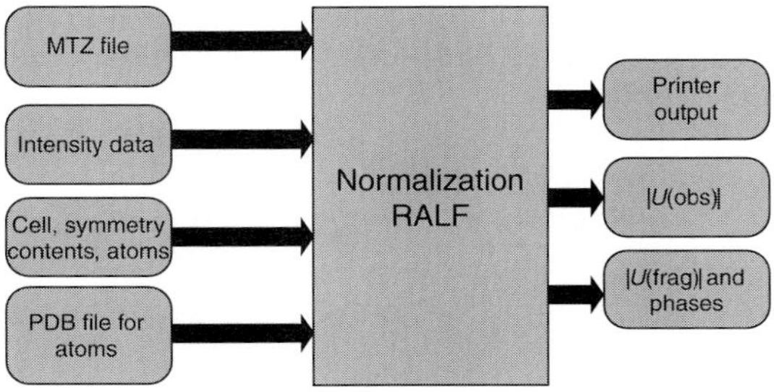

FIG. 1. The RALF program.

is generated subject to the constraints

$$\int_V q(\mathbf{x}) \exp(2\pi i h \cdot \mathbf{x})\, d^3\mathbf{x} = |U_{\mathbf{h}}^{\text{obs}}| \exp(i\varphi_{\mathbf{h}}) \quad \forall \mathbf{h} \in H \tag{2}$$

This means that $q^{\text{ME}}(\mathbf{x})$ must reproduce the intensities and phases of the basis set, but it will also have Fourier coefficients $U_{\mathbf{k}}^{\text{ME}}$ with nonnegligible amplitude for many nonbasis set reflections $\mathbf{k} \in K$, especially when $\mathbf{k}$ is in the second neighborhood of **H**. This is shown diagrammatically in Fig. 2, and is the process of maximum entropy extrapolation. It causes the conditional distribution of $|U_{\mathbf{k}}|$ to become a Rice distribution (which has the standard Gaussian multiplied by the appropriate Bessel function, which acts as an offset) rather than the simpler Gaussian distribution of Wilson statistics. The second neighborhood is defined by reflections $\mathbf{h}_1 \pm {}^t\mathbf{R}_g \cdot \mathbf{h}_2$ for $\mathbf{h}_1$, $\mathbf{h}_2 \in$ H, where ${}^t\mathbf{R}_g$ is the transpose of a rotation matrix obtained from the crystal space group.

If fragments are present, the fragment U magnitude, suitably scaled, is subtracted from the observed U before it is employed as a constraint. The $g_{\mathbf{h}}$ coefficient, if present, is similarly used. If noncrystallographic symmetry (NCS) is present, then this information is also passed to the program, and the $q^{\text{ME}}(\mathbf{x})$ maps generated are suitably symmetry averaged. Solvent-flattening conditions are also imposed using nonuniform priors.

The entropy maximization method used is an iterative exponential modeling algorithm,[1,2] a variant and approximation of the Newton–Raphson method, in which the optimization process is controlled by the use of line and plane searches. It has the benefit of stability under most circumstances, the ability to use additional constraints readily, and to incorporate nonuniform priors in a natural way. It works for basis sets as small as one reflection and extending to thousands of reflections. Because it is a Fourier method, the computer time it uses is proportional to $N \log N$, where N is the number of Fourier points. The penalty of this generality is a lack of efficiency for small basis sets, but this is far outweighed by the benefits when large sets are employed.

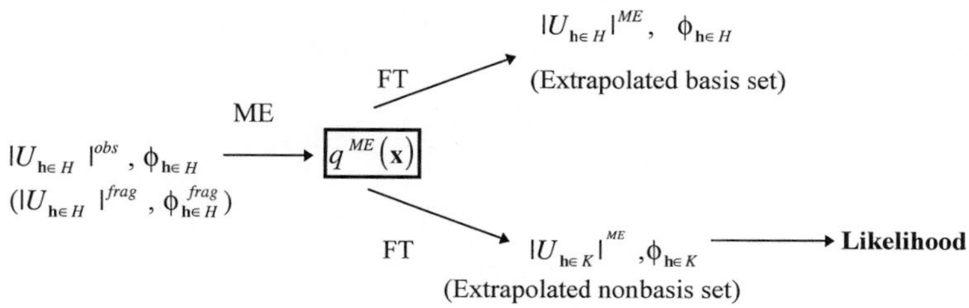

FIG. 2. The maximum entropy process.

The degree of fit between $|U_{\mathbf{h}\in H}|^{\text{obs}}$ and the extrapolate $|U_{\mathbf{h}}^{\text{ME}}|$ is very important. Too close a fitting results in artifacts appearing, whereas the reverse fails to achieve the optimum strength of extrapolation. A χ^2 statistic is used as a measure of fit, where

$$\chi^2 = \frac{1}{n} \sum_{\mathbf{h}\in H} \nu_{\mathbf{h}} ||U_{\mathbf{h}}|^{\text{obs}} - |U_{\mathbf{h}}^{\text{ME}}||^2 \tag{3}$$

and

$$\nu_{\mathbf{h}} = 1/(p\varepsilon_{\mathbf{h}}\Sigma + \sigma_{\mathbf{h}}^2) \tag{4}$$

n is the total number of degrees of freedom in the basis set reflections, which is given by $n = 2n_{\text{a}} + n_{\text{c}}$, where n_{a} is the number of acentric reflection and n_{c} the number of centrics. Σ is approximately the reciprocal of the number of nonhydrogen atoms in the unit cell (N), but is a refineable parameter via likelihood, ε is the standard crystallographic ε factor derived from point group symmetry, and p is an empirical factor usually set to 1. An appropriate value of χ^2 is unity.

5. At this point, however, the extrapolation is often too weak to be of any value. This is certainly the case for *ab initio* studies, and usually so for phasing electron diffraction data from image data, but it may not always be so when multiple isomorphous replacement (MIR) data are used. In the latter case, it is often, however, the situation that many strong reflections are still not yet reliably phased. Thus new phase information needs to be incorporated into the basis set. This is carried out by adding new strong reflections, which are hitherto inconclusively extrapolated and that optimally enlarge the second neighborhood of the current basis set, by permuting their phases. This gives rise to a series of phase choices, and each choice is represented as a node on the second level of the phasing tree in which the first, or root, node was defined by the original basis set. This is shown in Fig. 3.

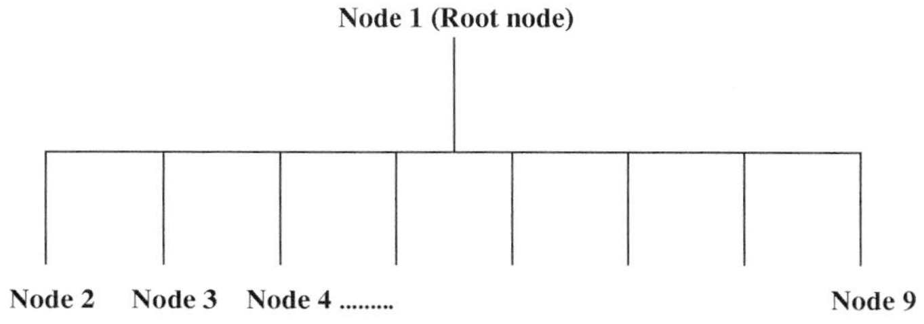

FIG. 3. A two-level phasing tree.

The number of strong reflections that we want to add to {H} can often be large; if there are n_c centric phases and n_a acentric phases to be added to the basis set this will generate $2^{n_c}4^{n_a}$ nodes in a full factorial design in which each centric phsae is given its two possible values, and each acentric is quadrant fixed, i.e., it takes on the four possible angles ±45°, ±135°. This rapidly becomes a combinatorial explosion, but fortunately techniques based on error-correcting codes,[11] and incomplete factorial designs[14] (IFDS), can alleviate the problem. This subject is too large to be discussed in any detail here. Error-correcting codes find their way into group theory, sphere packing in high-dimensional geometries, and information theory,[15–17] but here they are used as a source of optimal experimental designs. In particular they are used to sample phase space in a highly efficient way, and are so constructed that the full space can be partially recovered by relatively simple means.

Not all codes are a useful source of experimental designs; the codes and IFDS offered in MICE are as follows: (a) a Hadamard [7, 4, 3] code, which for 7 or 8 degrees of freedom (df) gives 16 nodes instead of $2^7 = 128$ or $2^8 = 256$. In the former case, one of these nodes will have at most one wrong phase choice; in the latter, two. (One could, of course, be lucky and have no incorrect phase choices!); (b) the Nordstrom–Robinson (16, 256, 6) or (15, 256, 5) code produces 256 nodes instead of $2^{16} = 65{,}536$ or $2^{15} = 32{,}768$ for 15 or 16 df. One of these nodes will have a maximum of 4 (for 16 df) or 3 (for 15 df) incorrect phase choices; (c) the Golay [24, 12, 8] or [23, 12, 7] code produces 4096 nodes instead of $2^{24} = 16{,}777{,}216$ or $2^{23} = 8{,}388{,}608$. One of these nodes will have a maximum of four incorrect phase choices for 24 df, or three for 23 df. These three codes come from the Bricogne BUSTER program.[11] For macromolecular work, the Golay and sometimes the Nordstrom–Robinson codes are not always practicable from the computational viewpoint, so as an alternative the following incomplete factorial designs are available from a database derived by C. Carter: (d) 10 df giving 48 nodes, (e) 12 df giving 64 nodes, and (f) 14 df giving 100 nodes. Although IFDS may not have all the properties of codes, here they provide optimal designs in the cases that are intermediate between the Hadamard and Nordstrom–Robinson codes.

6. Each node on the tree is now subjected to constrained entropy maximization just as before, to produce a revised nonuniform distribution for

[14] C. W. Carter, Jr. and C. W. Carter, *J. Biol. Chem.* **254,** 12219 (1979).

[15] R. Hill, "A First Course in Coding Theory." Clarendon Press, Oxford, 1991.

[16] F. J. MacWilliams and N. J. A. Sloane, "The Theory of Error Correcting Codes." North-Holland, Amsterdam, 1977.

[17] J. H. Conway and N. J. A. Sloane, "Sphere Packings, Lattices and Groups." Springer-Verlag, Berlin, 1993.

the random atomic positions. To rank the nodes, a Rice-type likelihood function is used that evaluates the agreement between the extrapolated structure-factor magnitudes from the relevant maximum entropy distribution and the experimentally measured ones. This criterion measures the extent to which the observed pattern of the unphased intensities has been rendered more likely by the phase choices made for the reflections in the basis set than they were under the null hypothesis of uniform distribution of the random positions. This distortion depends on assumptions concerning basis set phases, and can be detected by calculating the log-likelihood gain (LLG). The log-likelihood gain is defined as a sum of logarithms of probability ratios calculated under the two hypotheses for a sample of observed values of structure-factor amplitudes in the second neighborhood of the basis set. For each extrapolated reflection $\mathbf{h} \in H$, let $r = |U_{\mathbf{h}}|^{\text{obs}}$ and $R = |U_{\mathbf{h}}^{\text{ME}}|$. Then, summing separately over the centric and acentric reflections,

$$\begin{aligned} L(H_1) = &\sum_{h_{\text{acentric}}} \{\log I_0[(2N/\varepsilon_{\mathbf{h}})Rr] - N/\varepsilon_{\mathbf{h}}R^2\} \\ &+ \sum_{h_{\text{centric}}} \{\log\cosh[(N/\varepsilon_{\mathbf{h}})Rr] - N/2\varepsilon_{\mathbf{h}}R^2\} \end{aligned} \tag{5}$$

As is standard in tests of significance, a null hypothesis, $L(H_0)$, is constructed by setting $R = 0$. The LLG is then

$$\text{LLG} = L(H_1) - L(H_0) \tag{6}$$

The LLG will be largest when the phase assumptions for the basis set lead to predictions of deviations from the Wilson distribution in the unphased reflections, and in this context it is used as a powerful figure of merit.

7. The LLGs are analyzed for phase indications using the Student t test, which defines the significance level of the contrast between two means. The simplest t test involves the detection of the main effect associated with the sign of a single centric phase. The LLG average, μ^+, and its associated V^+ is computed for those nodes in which the sign of the phase under test is positive. The calculation is then repeated for those nodes in which the same sign is negative, to give the corresponding μ^-, and variance V^-. The t statistic is then

$$t = \frac{|\mu^+ - \mu^-|}{(V^+ + V^-)^{1/2}} \tag{7}$$

The use of the t test enables a sign choice to be derived with an associated significance level. This calculation is repeated for all the single-phase indications, and is then extended to combinations of two and three phases as justified by the absence of aliasing errors or confounding in the code. An extension to acentric phases is straightforward by employing two signs to define the phase quadrant both in permutation and in the subsequent analysis.[18] In general, only relationships with associated significance levels <2% are used, but this is sometimes relaxed with sparse data sets.

8. Only those nodes that are consistent with the t test results are kept, pruned further if necessary to 8–16 in a given level. Further reflections are then permuted and a new level of nodes generated.

If codes have been used, then an attempt is made to reconstruct the full dimensionality of the sampled phase space. The optimum method for this is a high-dimensional Fourier transform as used in BUSTER, which exploits the periodicity of phase angles and their phase relationships.[11] In MICE a simpler, but effective, algorithm is employed: (a) The 1-, 2-, and 3-phase relationships are derived from the t test, and given an associated figure of merit equal to the significance level with which they are determined; (b) all the possible phase combinations of the current set of trial phases in a full factorial design are generated, not just those sampled by the code or IFD. No entropy maximization is performed on them; (c) each set so generated is assigned a score related to the measure of agreement between the phase relationships weighted by their associated significance levels from (a) and the phases in the set itself; (d) those sets (usually ca. 100) that are highly consistent with the phase relationships are added to the tree and subjected to constrained entropy maximization; and (e) the 8–16 sets from (d) having the highest LLG are kept.

9. The tree-building and pruning procedure is continued until most large unitary structure factors have significant phase indications. This may not always be obvious, so maps should be consulted visually for each preferred node in each level of the tree. It must be remembered that the ME distributions, $q^{\mathrm{ME}}(\mathbf{x})$, associated with the various nodes are not electron-density maps in the traditional sense; rather, they are distributions of atomic positions. Conventional maps are therefore generated as centroid maps by means of a Sim-type filter in which each reflection is assigned a weight $w_{\mathbf{h}}$ computed as follows[1,2]:

For **h** centric,

$$w_h = \tanh(N/\varepsilon_{\mathbf{h}} Rr) \tag{8}$$

For **h** acentric,

$$w_h = I_1(X_{\mathbf{h}})/I_0(X_{\mathbf{h}}) \tag{9}$$

[18] K. Shankland, C. J. Gilmore, G. Bricogne, and H. Hashizume, *Acta Crystallogr.* **A49,** 493 (1993).

where

$$X_{\mathbf{h}} = (N/\varepsilon_{\mathbf{h}})Rr \tag{10}$$

The maps so produced are optionally compatible with the CCP4 format. Figure 4 outlines the structure and options of the MICE program.

Inherent Parallelism of Maximum Entropy-Likelihood Formalism and Its Exploitation

Entropy maximization for each node on a phasing tree is, from a computational point of view, an independent calculation: each node can be treated separately. Most crystallography laboratories now run networks of scientific workstations, and MICE has been adapted to exploit this. A phasing tree is split by a central server into the individual nodes, which are passed sequentially to all the workstations on a given net or subnet. When an individual client computer has finished a node, it is returned over the net and another sent. Workstations with more than one processor can be sent more than one node to process at a given time. A small file on the server contains a list of available computers, their addresses, and the number of available processors; not all those machines that are available need necessar-

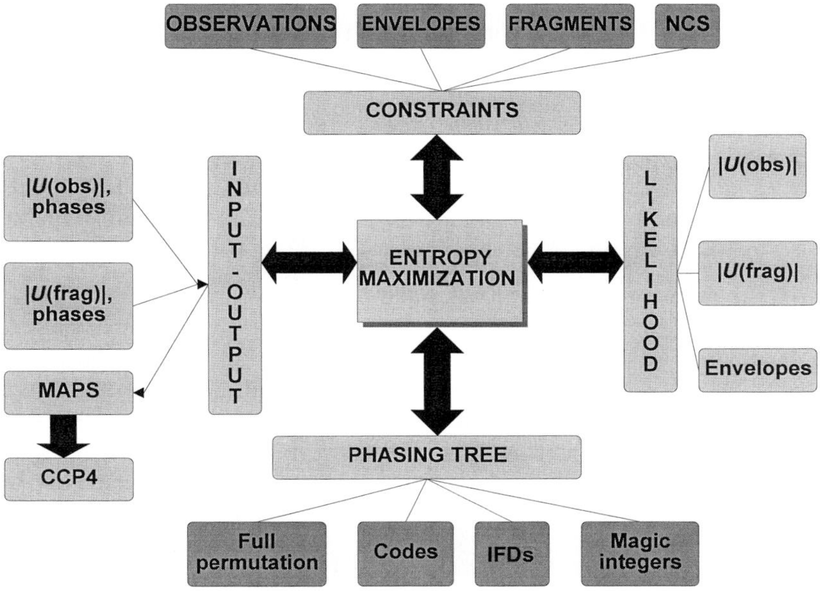

FIG. 4. The MICE computer program.

ily be used. The network need not be homogeneous, but all the computers used must share a common Fortran binary file structure. The use of such a computing environment makes it feasible to explore hundreds of nodes, even for macromolecules.

Applications to Macromolecules: X-Ray Diffraction Data

Ab Initio Studies

Avian pancreatic polypeptide (App) is a small protein made somewhat unusual by the availability of good 1-Å data,[19] and, because of this, it is used as a test structure by those interested in developing protein phasing techniques. The first use of the MICE program in a macromolecular environment was to identify the best phase sets for App from those generated by the SAYTAN direct methods computer program. In this study, sets of phases from SAYTAN using 1 Å data could not be ranked by any conventional figure of merit, although it was known that there were solutions present with mean phase errors of only 45°. These phase sets were passed to the MICE program using only the 2 Å data (to save computer time), and each solution was subjected to constrained entropy maximization coupled with likelihood evaluation. The sets with minimum phase error (around 45°) were clearly indicated by the LLG estimate.[20] No other figure of merit currently available could work so well in a crystallographic environment at this resolution.

Avian pancreatic protein contains one zinc atom per asymmetric unit, the position of which is not particularly difficult to locate using Patterson methods at 1 Å, but MICE can locate it to within 0.5 Å of its refined position *ab initio* using a combination of entropy maximization, likelihood estimation, and error-correcting codes at a resolution of only 3 Å. To do this the following procedure was used.

1. The data were normalized using RALF.
2. Two appropriate reflections were used to define the origin: App crystallizes in space group *C*2, so it is not possible to define the enantiomorph at this stage without the use of guesswork and/or luck, so the enantiomorph was left undefined. These reflections define the basis set for the root node.
3. Reflections totaling 24 degrees of freedom were given permuted phases using the Golay code. The latter generates only 4096 nodes compared

[19] I. Glover, I. Haneef, J. E. Pitts, S. P. Wood, D. Moss, I. J. Tickle, and T. Blundell, *Biopolymers* **22,** 293 (1983).

[20] C. J. Gilmore, A. N. Henderson, and G. Bricogne, *Acta Crystallogr.* **A47,** 832 (1991).

to $2^{24} = 16{,}777{,}216$ for the full factorial design. Each of these nodes was subjected to an entropy maximization.

4. The resulting nodes were analyzed via the t test and the 1-, 2-, and 3-phase interactions between basis set phases below the 2% confidence level were derived. Ideally this should be followed by Fourier analysis [9], but here the reconstruction procedure described in step 7 (see Outline of Method, above) was employed, and a subset of 16 phase sets that best satisfied the phase interactions was chosen. At this point the enantiomorph becomes defined by default.

5. These 16 sets were then further extended using the Nordstrom–Robinson code. This uses 16 df to generate 256 nodes (instead of 65,536). Thus $16 \times 256 = 4096$ nodes were generated, giving the third level of the phasing tree. Each of these nodes was subjected to entropy maximization and analyzed as in step 4. The preferred node with the highest likelihood produced a centroid map that clearly revealed the zinc atom as the most prominent feature, at less than 0.5 Å from its correct position.

The calculations took less than 24 hr on a heterogeneous network of six SUN and Silicon Graphics workstations.

Phase Extension from Multiple Isomorphous Replacement Phases

We present here a survey of successful applications of MICE to phase extension problems in macromolecular X-ray crystallography: MICE was adapted for use with macromolecules and used by Xiang *et al.*[21] to apply envelope and solvent flattening to bring about substantial model-free map improvement at 3.1 Å resolution for cytidine deaminase, in which the envelope was included as an active constraint in the entropy maximization from the very beginning. Even without the use of phasing trees, the conclusion was that the use of ME in this way was a superior technique for improving isomorphous replacement electron-density maps before model building. The fact that this is a model-free method is important: it minimizes the accumulation of error in the whole phasing process.

Doublié *et al.*[22,23] used the adapted version of MICE to solve the unknown tryptophanyl-tRNA synthetase (TrpRS) structure. In this case there were serious problems with a lack of isomorphism, and a poorly known molecular envelope exacerbated all the usual problems. These were over-

[21] S. Xiang, C. W. C. Carter, Jr., G. Bricogne, and C. J. Gilmore, *Acta Crystallogr.* **D49,** 193 (1993).

[22] S. Doublié, S. Xiang, C. J. Gilmore, G. Bricogne, and C. W. C. Carter, Jr., *Acta Crystallogr.* **A50,** 164 (1994).

[23] S. Doublié, G. Bricogne, C. J. Gilmore, and C. W. C. Carter, Jr., *Structure* **3,** 17 (1995).

come in a novel way: (1) Strong reflections for which there were only weak phase indications were given permuted phases; (2) different envelope characteristics were permuted, and analysis of the LLG was used to decide on the correct mask; and (3) phase permutation was used to phase strong reflections that were inconclusively indicated by MIR techniques. Incomplete factorial designs (IFDS) were employed to reduce the number of nodes needed for a given permutation experiment, but still allowing the selection of the correct phases by an appropriate use of significance testing. The result was a successful structure solution which is discussed in detail in [6] in this volume.[11b]

Lapthorn *et al.*[24] also used MICE in a similar non-tree-building mode, and with the inclusion of the envelope prior to generating centroid maps for the unknown human chorionic gonadotropin (hCG) structure. These methods were supplemented by phases determined directly for seven reflections by permutation and likelihood scoring. The improvement produced over standard maps made a significant contribution to the modeling of some of the loops in the protein.

Applications to Macromolecules: Electron Diffraction and High-Resolution Image Data

Electron diffraction techniques can be applied to biological macromolecules. In this case an *ab initio* phase determination is not practical because of the structural complexity of the molecules. However, instead of proceeding from a basis set that defines only the origin and enantiomorph, or from MIR phases, the phases of the basis set reflections can be obtained in this case from the Fourier transform of a suitable image. Such phase information is usually of relatively low resolution, and the practical problems of obtaining such image data for macromolecules are considerable. The electron diffraction data are, however, more easily obtained, and have a much higher resolution than the images. The problem then arises as to how to phase the diffraction data starting from lower resolution image transform phases when processing macromolecule data, and for this the ME method is ideally suited.

The first application of phase extension in a macromolecular environment involved a trial with the two-dimensional purple membrane data from *Halobacterium halobium* from Henderson *et al.*[25,26] This showed that it is

[24] A. J. Lapthorn, D. C. Harris, A. Littlejohn, J. W. Lustbader, R. E. Canfield, K. J. Machin, F. J. Morgan, and N. W. Isaacs, *Nature* (*London*) **369,** 455 (1994).

[25] R. Henderson, J. M. Baldwin, K. H. Downing, J. Lepault, and F. Zemlin, *Ultramicroscopy* **19,** 147 (1986).

[26] J. M. Baldwin, R. Henderson, E. Beckman, and F. Zemlin, *J. Mol. Biol.* **202,** 585 (1988).

possible to produce good phase extrapolation starting with basis sets having a resolution as low as 15 Å. The centroid maps produced using a 15 Å basis set extrapolating to intensities at 10 Å produced a level of quality and resolution similar to that generated from a conventional 10 Å data set.[27]

Cholera toxin is the protein responsible for the clinical symptoms of cholera. It consists of five B subunits (each with a molecular weight of 10,600) and one A subunit composed of two polypeptides (A_1 and A_2, with molecular weights of 23,500 and 5,500, respectively). The five B subunits are arranged in a pentameric ring around the central A core, exhibiting fivefold noncrystallographic symmetry. We have been working with a data set from a sample crystallized in two dimensions on lipid layers in which 56 unique image-derived phases were available at 8.8 Å resolution plus 1417 diffraction intensities extending to 4 Å in projection.[28] The problem is to phase the 4 Å data from the 56 known phases, imposing 5-fold noncrystallographic symmetry and solvent flattening, while working wholly with this projection.

Hitherto we have been using likelihood as an indicator of phase choices, but it is of course a general technique of statistical inference. Accordingly, in the cholera toxin case we have successfully tested the likelihood criterion as an accurate and reliable predictor of the following.

1. The effective number of atoms in the unit cell: At 4 Å resolution we do not have atomic resolution. The question then poses itself as to what is the effective number of scatterers, N_{eff}, at this resolution? This is not, in general, the number of atoms in the cell. Xiang *et al.*[21] proposed a procedure for using the maximum entropy-likelihood formalism to define this, and the method has been adapted for the cholera toxin case as follows.
   a. The data are normalized with an $N_{\text{eff}} = 2400$.
   b. The image-derived phases are used to define a basis set that is then subjected to entropy maximization. The LLG is calculated in each cycle, using all the nonbasis set diffraction intensities, and terminates when the LLG reaches a maximum (LLG_{max}).
   c. N_{eff} is reduced by 200 and procedures a and b are repeated. This continues until $N_{\text{eff}} = 1000$.
   d. The LLG_{max} is plotted against N_{eff}; the maximum value of the LLG defines the effective number of scatters in the cell.

[27] C. J. Gilmore, K. Shankland, and J. R. Fryer, *Ultramicroscopy* **49,** 132 (1993).

[28] A. Brisson and G. Moser, *J. Electron Microsc. Tech.* **18,** 387 (1991).

In this case there is a clear maximum at $N_{\text{eff}} = 2000$, which, perhaps coincidentally, corresponds to the number of amino acids in the unit cell. This value was used throughout the subsequent calculations.

2. In principle, the coordinates of the NCS rotation axis can be determined by autocorrelation Patterson methods, but the LLG should also be capable of determining them. Accordingly, starting with an approximate centroid derived from the initial 8.8 Å image-derived map, a grid search was used with 0.1 Å increments in the plane of projection. At each point the basis set was subjected to entropy maximization with fivefold symmetry imposed. The LLG was calculated using the complete diffraction data set. A plot of LLG against centroid coordinate showed a clear, unambiguous maximum that accords well with other estimates.
3. The final parameter we need is the radius of the cholera toxin. A one-dimensional grid search was used in which the radius was varied from 20 to 45 Å in 1 Å increments, and the LLG evaluated at each point. Again there is a clearly defined maximum at ca. 32 Å that agrees well with X-ray results.[29]
4. There is a space group ambiguity with electron diffraction data, and the LLG can be used to decide on the two possible choices.

We have used these results to carry out phasing extension on cholera toxin. The first results using a three-level tree and the Nordstrom—Robinson code are very encouraging. Preferred nodes show a good agreement in projection with the X-ray structure.[29] Work is now continuing with three-dimensional data derived from tilted specimens.

[29] E. A. Merritt, S. Sarfatym, F. van der Akker, C. L'Hoir, J. A. Martial, and W. G. J. Hol, *Protein Sci.* **3,** 166 (1994).

[6] Phase Improvement Using Conditional Probability Methods: Maximum Entropy Solvent Flattening and Phase Permutation

By CHARLES W. CARTER, JR., and SHIBIN XIANG

The work of Bricogne on statistical direct methods introduced a radically new approach to all aspects of the macromolecular phasing problem.[1–6] Such innovation brings with it numerous difficulties: Direct methods are unfamiliar to most macromolecular crystallographers, the proposed algorithms themselves are not intuitive, and computer programs written to implement them are not yet readily available. Nonetheless, the algorithms issuing from Bricogne's Bayesian phase determination paradigm, described elsewhere in this series,[7,8] have made significant contributions to the solution of several challenging protein crystal structures,[9–11] and promise to make further inroads as they become more fully implemented. For these reasons, it seems appropriate to summarize the status of practical work based on the new concepts.

Conditional probability methods from the Bayesian paradigm have been implemented in the computer programs MICE (see Refs. 12, 13, and [5] in this volume[14]) and BUSTER.[15] A specialized version of MICE, MICE_mx_8, has been adapted to accept and impose molecular-envelope constraints. These programs offer a progressive, direct approach to phase

[1] G. Bricogne, *Acta Crystallogr.* **A30,** 395 (1974).
[2] G. Bricogne, *Acta Crystallogr.* **A32,** 832 (1976).
[3] G. Bricogne, *in* "Computational Crystallography" (D. Sayre, ed.), p. 258. Oxford University Press, New York, 1982.
[4] G. Bricogne, *Acta Crystallogr.* **A40,** 410 (1984).
[5] G. Bricogne, *Acta Crystallogr.* **A44,** 517 (1988).
[6] G. Bricogne, *Acta Crystallogr.* **D49,** 37 (1993).
[7] G. Bricogne, *Methods Enzymol.* **276,** 361 (1997).
[8] G. Bricogne, *Methods Enzymol.* **276,** 424 (1997).
[9] S. Doublié, S. Xiang, C. J. Gilmore, G. Bricogne, and C. W. J. Carter, *Acta Crystallogr.* **A50,** 164 (1994).
[10] S. Doublié, G. Bricogne, C. J. Gilmore, and C. W. Carter, Jr., *Structure* **3,** 17 (1995).
[11] A. J. Lapthorn, D. C. Harris, A. Littlejohn, J. W. Lustbader, R. E. Canfield, K. J. Machin, F. J. Morgan, and N. W. Isaacs, *Nature* (*London*) **369,** 455 (1994).
[12] C. J. Gilmore, K. Henderson, and G. Bricogne, *Acta Crystallogr.* **A47,** 830 (1991).
[13] C. J. Gilmore, A. N. Henderson, and G. Bricogne, *Acta Crystallogr.* **A47,** 842 (1991).
[14] C. Gilmore and G. Bricogne, *Methods Enzymol.* **277,** [5], 1997 (this volume).
[15] G. Bricogne, *in* "International Tables for Crystallography" (U. Shmueli, ed.), p. 22. Kluwer Academic Publishers, Dordrecht, 1993.

0076-6879/97 $25.00

improvement. All sources of phase information are explicitly monitored and controlled, and the quality of the contributing information and the phase improvement are assessed by comprehensive cross-validation. Thus, while still at an early stage of implementation, the algorithms offer considerable advantages. These include accurately assessing the quality of experimental phases by a process that leads naturally to the identification of the most important missing information. Accurate choices between competing hypotheses regarding new phases, features of the mask for solvent flattening, and noncrystallographic symmetry operations[5] then can be made by sampling combinations of such choices, likelihood scoring based on conditional probability expressions, and statistical analysis.

The Bayesian paradigm considerably strengthened the foundations of direct methods[4] by deriving a real-space implementation of the conditional probability relations between structure factors in a basis set of phased structure factors, {**H**}, and demonstrating its equivalence to constrained entropy maximization of the distribution of atomic positions. Fourier transformation of the resulting probability distribution gives rise to extrapolated values for both amplitudes and phases of the remaining unphased structure factors, {**K**}. This maximum entropy extrapolation captures the statistical implications of the basis set reflections both more accurately and more completely than earlier, reciprocal-space conditional probability distributions, such as the Cochran distribution,[16] that underlie conventional direct methods.[17–20]

The importance of the Bayesian paradigm to the solution of macromolecular structures can be appreciated from the fact that the phases themselves embody the only explicit information about atomic coordinates. Thus, the "phase problem" subsumes the entire structure-solution process, from creation of the first electron-density map to the final atomic parameter refinement. Read illustrates the pernicious effect of model bias in [7] in this volume.[21] Model bias potentially afflicts all structure determinations, particularly those based on molecular-replacement phases. Conditional probability methods limit the effects of such bias through "model-independent" phase improvement, so that model building can be postponed until the best possible phases have been found.

For readers unfamiliar with maximum entropy methods we begin with an intuitive guide to procedures and results described in subsequent sections.

[16] W. Cochran, *Acta Crystallogr.* **5,** 65 (1952).

[17] H. Hauptman and J. Karle, "The Solution of the Phase Problem: I. The Centrosymmetric Crystal." Polycrystal Book Service, Pittsburgh, 1953.

[18] J. Karle, *Acta Crystallogr.* **A45,** 765 (1989).

[19] M. M. Woolfson, *in* "Computational Crystallography" (D. Sayre, ed.), p. 110. Clarendon Press, Oxford, 1980.

[20] M. M. Woolfson, *Acta Crystallogr.* **A43,** 593 (1987).

[21] R. J. Read, *Methods Enzymol.* **277,** [7], 1997 (this volume).

Recentered Conditional Probability Distributions, Extrapolation, and Likelihood

Conditional probability methods, and in particular entropy maximization, address the seemingly disingenuous question: "How are the atoms distributed in the crystal if they are in accordance with the measured amplitudes?" The answer to the question is provided by a joint, conditional probability distribution (JPD) of the atomic positions. Accurate approximations to the JPDs will differ for each distinct "basis set" $\{\mathbf{H}\}$ of phased reflections, because strong Bragg reflections introduce different nonuniformities in the probable distribution of atomic positions, depending on their phases. So it is crucial, particularly for macromolecules, that any JPD be recentered around the selected reciprocal-space constraints given by the phased basis set reflections. Bricogne[4] showed that the most conservative, and hence most reliable, representation of this recentering is the maximum entropy distribution of atomic positions compatible with $\{\mathbf{H}\}$. As noted by Bricogne,[5,6] this process can be applied iteratively to structure solution if, to seed the process, initial phases are assumed for some number of strong reflections.

The necessary recentering can be achieved by finding a set of Fourier coefficients (also called "exponential modeling parameters") for a map, the omega map $\omega(\mathbf{x})$, that, when exponentiated to give a probability distribution $q^{\mathrm{ME}}(\mathbf{x})$, has the basis set constraints in its Fourier spectrum. Thus, the recentering is carried out in real space, by constrained entropy maximization.

The modulation in $q^{\mathrm{ME}}(\mathbf{x})$ induced by $\{\mathbf{H}\}$ has an impact on the expected distribution of the structure factors for the remaining reflections, in $\{\mathbf{K}\}$, which can be expressed in closed form using the measured amplitudes and the structure factors obtained by Fourier transformation of the maximum entropy model. This conditional distribution of structure factors is used in two ways: (1) Centroid structure factors in $\{\mathbf{K}\}$ combine the observed amplitudes with the maximum entropy phases and Sim-like weights, which, together with the phased structure factors in $\{\mathbf{H}\}$, enable a centroid map to be calculated for display purposes and possibly for interpretation; (2) after integration over the phases, it provides a robust cross-validation criterion known as the log likelihood gain (LLG), which expresses the marginal probability of the observed amplitudes in $\{\mathbf{K}\}$. Details can be found in the literature cited above and in reviews.[22–25]

[22] G. Bricogne, *in* "Crystallographic Computing 4" (N. W. Isaacs and M. R. Taylor, eds.), p. 60. IUCR and Oxford University Press, Oxford, 1988.

[23] G. Bricogne, *in* "Direct Methods of Solving Crystal Structures" (H. Schenk, ed.), p. 215. Plenum, New York, 1991.

Distinct phase choices for the strong reflections can be ranked by comparing predicted and observed distributions for the amplitudes of the other reflections using the LLG. In essence, one can make successive changes to the trial phases of a "basis set" $\{\mathbf{H}\}$ of strong reflections and then use the observed amplitude data outside the basis set to gauge the validity of each choice of trial phases.

Experimental phase determination invariably produces for each reflection both a phase estimate and a probability distribution with a "centroid" figure of merit to indicate how reliable that estimate may be. From these figures of merit a subset of phases can be chosen whose average phase errors are expected to be less than a threshold value. These subsets form natural basis sets for Bayesian phase improvement.

The illustration in Fig. 1 may help to introduce unfamiliar terminology in Table I and to clarify relationships between entropy maximization and the recentering of joint conditional probability distributions. This example supplements the one-term example first suggested by Bricogne,[4] with enough specific detail to give tangible expression to two key concepts underlying the approach: exponentiation in real space and extrapolation in reciprocal space. A one-dimensional crystal structure constructed for a linear triatomic molecule such as $HgCl_2$ allows both real and reciprocal spaces to be represented graphically and accurately in their entirety. The light gray map and X-ray amplitude data represent the structure solution. The remaining "maps" and structure factors illustrate stages in the construction of the conditional probability distribution of atomic positions obtained by assigning a phase of 90° to the $h = 1$ reflection.

The one-term Fourier map $\omega^0(\mathbf{x})$ is a poor model for the distribution of atomic positions. Exponentiating it to give $q^0_{[1]}(\mathbf{x})$ is already an improvement. This "exponential model" is a proper conditional probability distribution, positive in the interval $0 \rightarrow 1.0$, for the atomic positions in the unit cell, with a maximum value close to the position of the mercury atom. However, it no longer has the value $|U_{\mathrm{obs},h=1}| = 0.61$ in its (unitary) Fourier spectrum, reflecting the fact that the probability is centered on the wrong reciprocal-space vector, i.e., it is centered around the value $|U_{h=1}| = 0.35$. Recentering the distribution proceeds by the now obvious path of adjusting the $[\zeta_1]$ coefficient of the $\omega^0(\mathbf{x})$ map so as to fit this constraint.

---

[24] G. Bricogne, *in* "Maximum Entropy in Action" (B. Buck and V. A. Macaulay, eds.), p. 187. Clarendon Press, Oxford, 1992.

[25] C. W. Carter, Jr., *in* "From First Map to Final Model" (R. Hubbard, D. Waller, and S. Bailey, eds.), p. 41. Science and Engineering Research Council, Daresbury Laboratory, Daresbury, UK, 1994.

Ensuring that the probability distribution includes the amplitude $|U_1| = 0.61$ in its Fourier spectrum makes the distribution conditional on that basis set reflection. It yields the desired constrained maximum entropy distribution, $q_{[1]}^{\mathrm{ME}}(\mathbf{x})$. One algorithm for finding appropriate Fourier coefficients, $\{\zeta_i\}$, for the omega map is called exponential modeling, and involves an iterative fitting process, imposing the amplitude and phase constraints in reciprocal space after each exponentiation by comparing observed and calculated values to derive shifts, $\{\Delta\zeta_i\}$ (see [5] in this volume[14]).

Two aspects of this formalism already have achieved useful map improvement for macromolecules. First is the combination of maximizing the conditional probability for a given basis set of reliably phased amplitudes under the joint constraint of solvent flatness[26]; the second involves hypothesis permutation, likelihood scoring, and regression analysis.[9] The former is a passive density modification procedure, in the sense that it improves a map solely on the basis of a reliable set of phases. The second involves using the same computational engine actively, to seek out critical new unphased or incorrectly phased reflections and/or other missing information, for example about the molecular envelope, and augment the basis set constraints with statistically reasonable choices from among the possible hypotheses. These two procedures are now introduced using the framework of the one-dimensional example in Fig. 1.

Maximum Entropy Solvent Flattening

Maximum entropy solvent flattening, or MESF,[26] is essentially a density modification procedure providing a superior alternative to the conventional solvent-flattening procedure introduced by Wang.[27] Increased effectiveness arises from combining the solvent flatness constraint with maximum conditional probability for the current basis set. Figure 1 suggests how this improvement occurs.

"Recentering" the probability distribution by fitting the Fourier coefficients $\{\zeta_i\}$ to the constraint(s) in $\{\mathbf{H}\}$ introduces new information in real space and strengthens the extrapolation in reciprocal space, as represented by the respective cross-hatched regions in both halves of Fig. 1a. Extrapolation from the recentered distribution, $q_{[1]}^{\mathrm{ME}}(\mathbf{x})$, centered at $|U_{h=1}| = 0.61$, is more robust than that from the original distribution, $q_{[1]}^{0}(\mathbf{x})$, which was centered at $|U_{h=1}| = 0.32$. Thus, extrapolation resulting from exponentiating the omega map is highly sensitive to where the conditional probability

[26] S. Xiang, C. W. Carter, Jr., G. Bricogne, and C. J. Gilmore, *Acta Crystallogr.* **D49,** 193 (1993).
[27] B. C. Wang, *Methods Enzymol.* **115,** 90 (1985).

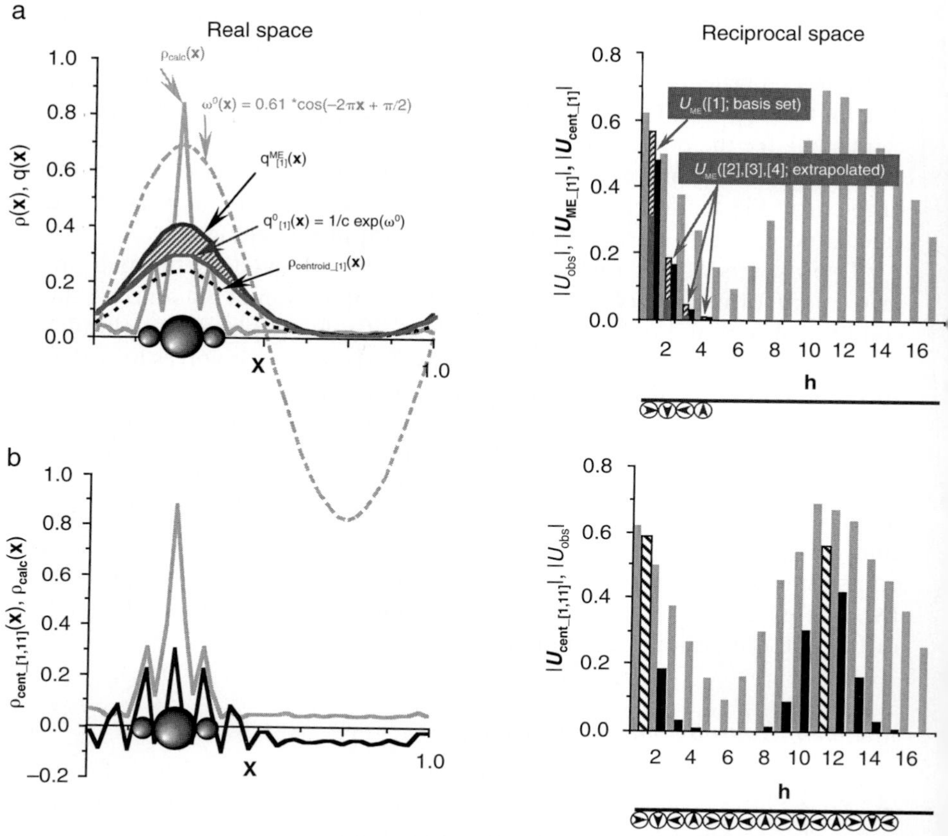

FIG. 1. The intuitive content of constructing and using conditional probability distributions for phase determination: construction of a "maximum entropy" distribution of atomic positions constrained by a single amplitude and phase ($h = 1$) for a one-dimensional crystal of a single linear triatomic molecule ($HgCl_2$) in the asymmetric unit. (a) One-term basis set. Successive real-space approximations include the one-term Fourier map $\omega^0(\mathbf{x})$ based on the basis set reflection $q^0_{[1]}(\mathbf{x})$, the exponentiation of $\omega^0(\mathbf{x})$, which is the initial approximation to the desired distribution, $q^{ME}_{[1]}(\mathbf{x})$. "Recentering" the distribution involves adjusting the parameter ζ_i to satisfy the constraint that its transform produce the basis set reflection in reciprocal space. It leads to a modest sharpening of the probability distribution (cross-hatched region between $q^0_{[1]}(\mathbf{x})$ and $q^{ME}_{[1]}(\mathbf{x})$. The resulting distribution is a good low-resolution representation of the electron density $\rho_{calc}(\mathbf{x})$. Exponentiated forms are normalized. They are all superior to the one-term Fourier map, which has a large negative region. Exponentiation in real space leads to extrapolation in reciprocal space, shown by the significant amplitudes $[U_{ME}]$ for reflections $h = 2$, 3, and 4. Recentering of the probability by imposing the basis set constraint leads to increased extrapolation of these amplitudes, which assume their correct phases (dials below the **h** axis). Improved agreement between observed and extrapolated amplitudes increases the log likelihood statistic (not shown). Nevertheless, because the extrapolation is not fully

distribution (CPD) is centered, i.e., to the phases and amplitudes of the basis set.

The new information represented by the cross-hatched regions in Fig. 1a arises from the statistical "coupling" between amplitudes and phases induced by positivity alone, but it illustrates the source of map improvement obtained for macromolecular data sets when the coupling from that constraint is enforced jointly with that arising from solvent flatness. This new information comes in two forms; the phases extrapolated for the $h = 2, 3$, and 4 reflections, indicated by the arrowheads below the **h** axis (Fig. 1a, right), are correct. Thus, the extrapolation brings with it much essentially correct new phase information. However, even at low resolution the agreement between extrapolated and observed amplitudes is relatively poor, and the resulting Sim-like weighting to give a "centroid" electron-density map broadens the shape of the molecule still further, as shown by the black dashed curve, $\rho_{\text{centroid\_[1]}}(\mathbf{x})$ (Fig. 1a, left). This figure-of-merit weighting effect provides additional information about which of the extrapolated phases are likely to be correct. The corresponding individual phase probability distributions can also be combined with any prior experimental phase probability information both within and outside the basis set.[21,28–31]

Phase Permutation

Because it is unconstrained by any high-resolution reflections, the exponential model $q^{\text{ME}}_{[1]}(\mathbf{x})$ and its centroid electron-density $\rho_{\text{centroid\_[1]}}(\mathbf{x})$ in Fig. 1 remain noncommittal regarding how atoms are distributed within the low-resolution envelope. The model cannot extrapolate beyond the corresponding low-resolution "envelope" in reciprocal space, and so is fundamentally incomplete. However, supplementing the basis set with one strong, correctly phased, high-resolution reflection ($h = 11$, $\phi = 90°$) leads to

[28] W. A. Hendrickson and E. E. Lattman, *Acta Crystallogr.* **B26,** 136 (1970).
[29] W. A. Hendrickson, *Acta Crystallogr.* **B27,** 1472 (1971).
[30] R. Read, *Acta Crystallogr.* **A42,** 140 (1986).
[31] K. Cowtan and P. Main, *Acta Crystallogr.* **D42,** 43 (1996).

expressed, the Sim-like weighting of the extrapolated reflections (black bars) reflects uncertainties of the extrapolated phases. This uncertainty broadens the "centroid" electron density, $\rho_{\text{centroid\_[1]}}(\mathbf{x})$ (black dashed curve) in real space. (b) Two-term basis set. The centroid map in real space (boldface curve) is based on the maximum entropy distribution $q^{\text{ME}}_{[1],[11]}(\mathbf{x})$ (not shown) based on $h = 1$ and $h = 11$ (cross-hatched bars). It closely resembles the calculated map. The pattern of extrapolation in reciprocal space extends to many more reflections, in particular those at high resolution.

TABLE I
TERMINOLOGY

Abbreviation	Meaning
JPD	Joint probability distribution, either of atomic positions (real space) or structure factors (reciprocal space)
CPD	Conditional probability distribution; a joint distribution subject to a set of imposed constraints or conditions, usually amplitudes and phases of structure factors, molecular envelopes, etc.
$\omega^0(\mathbf{x}) = \sum_{h=1}^{N} \zeta_i \exp(2\pi i \mathbf{h} \cdot \mathbf{x})$	The "omega map." A Fourier synthesis with one complex coefficient for each acentric and one real coefficient for each centric constraint reflection in the basis set $\{\mathbf{H}_i\}$.
$q_{\{\mathbf{H}\}}^{\mathrm{ME}}(\mathbf{x}) = \frac{\mathrm{m}(\mathbf{x})}{Z} \exp[\omega(\mathbf{x})]$	Probability distribution for the atomic positions, conditioned by the basis set $\{\mathbf{H}\}$. Also, the maximum entropy distribution subject to the constraints given by $\{\mathbf{H}\}$. $\mathrm{m}(\mathbf{x})$ is a prior distribution, which can have a value of unity, and Z is a normalization constant.
LLG	Log likelihood gain: $$\mathrm{LLG}\,(\{U_{\mathbf{H}}\}) = \sum_{h\in\{\mathbf{K}\}} \left[\log I_0 \left\{2\frac{N}{\varepsilon_h} \lvert U_{h\,\mathrm{obs}}\rvert \lvert U_{h\,\mathrm{ME}}\rvert\right\} - \frac{N}{\varepsilon_h}\lvert U_{h\,\mathrm{ME}}\rvert^2\right]$$
$\{\zeta_i\}$	The set of (complex) exponential modeling parameters adjusted to recenter the CPD on the reciprocal lattice vector defined by the basis set reflections
$\{\Delta\zeta_i\}$	Shifts for the exponential modeling parameters, calculated during each cycle of entropy maximization
$\lvert U_{\mathrm{obs},h=1}\rvert$	Observed unitary structure factor amplitude for reflection $h = 1$

considerably richer maximum entropy extrapolation, producing a centroid map (Fig. 1b) that resolves all three atoms.[25,32]

The correct phase for the $h = 11$ reflection can be found by permuting its phase to generate a new set of exponential models constrained by two reflections, one at low and one at high resolution. Each model is generated independently by constrained entropy maximization, and displays a different pattern of extrapolated amplitudes throughout the entire resolution range. Quantitative comparison of this extrapolation for competing phase

[32] C. W. Carter, Jr., S. Xiang, G. Bricogne, and C. Gilmore, *in* "Likelihood, Bayesian Inference, and Their Application to the Solution of New Structures" (G. Bricogne and C. W. Carter, Jr., eds.), p. 41 American Crystallographic Association, Buffalo, New York, 1997.

TABLE I (*continued*)

Abbreviation	Meaning
$\lvert U_{\mathrm{ME},h=1}\rvert$	Maximum entropy unitary structure factor amplitude for reflection $h = 1$
$\{\mathbf{H}\}$	The set of basis set reflections
$\{\mathbf{K}\}$	The set of reflections extrapolated from the basis set during entropy maximization. Occasionally this set is restricted to the second neighborhood of $\{\mathbf{H}\}$. Usually the complement of $\{\mathbf{H}\}$ for large basis sets
$\lvert U_{\mathrm{renorm}}\rvert$	Renormalized structure factor amplitude, obtained from the cosine law approximation, $\lvert U_{\mathrm{renorm}}\rvert = \lvert U_{\mathrm{obs}}\rvert^2 + \lvert U_{\mathrm{ME}}\rvert^2 - 2\lvert U_{\mathrm{obs}}\rvert\lvert U_{\mathrm{ME}}\rvert \mathrm{Wt}_{\mathrm{Sim}}$ (see Fig. 2).
$\mathscr{C}$	A reduced chi-squared-like statistic used in MICE to impose structure factor constraints, $\mathscr{C} = \frac{1}{2} \sum_{h_{\mathrm{acentric}} \in \{\mathbf{H}\}} \frac{(\lvert U_{\mathrm{obs}} - U_{\mathrm{calc}}\rvert^2)}{\varepsilon_h P \Sigma_h + \sigma_h^2} + \sum_{h_{\mathrm{centric}} \in \{\mathbf{H}\}} \frac{(\lvert U_{\mathrm{obs}} - U_{\mathrm{calc}}\rvert^2)}{\varepsilon_h P \Sigma_h + \sigma_h^2}$
Constraint	A condition to be met by a particular exponential model or node. This can be a phased basis set reflection, a molecular envelope, or a noncrystallographic symmetry equivalence
Node	An exponential model based on a particular set of constraints, including basis set reflections and other assumptions regarding, for example, the molecular envelope. A node may be one of a number of daughter nodes in a permutation design aimed at determining phases for some number of new reflections whose phases have been permuted

hypotheses is obtained by consulting the measured data outside the basis set, using the statistical criterion known as the LLG. With important qualifications that are met in the macromolecular applications described here, extrapolation beyond the basis set reflections matches the $\{\lvert U_{\mathrm{obs}}\rvert\}$ most closely and hence the LLG is greatest when the phases and amplitudes of the basis set are correct.[4,22,33] Thus, this structure actually can be "solved" by constructing exponential models from five different basis sets, one with only a single reflection and four with only two reflections.

For more complicated situations encountered with macromolecular problems, suitable choices must be made of those new reflections most likely to complement and enhance the basis set. Key to this choice is a cosine law approximation to the distance between extrapolated and observed

[33] G. Bricogne and C. J. Gilmore, *Acta Crystallogr.* **A46,** 248 (1990).

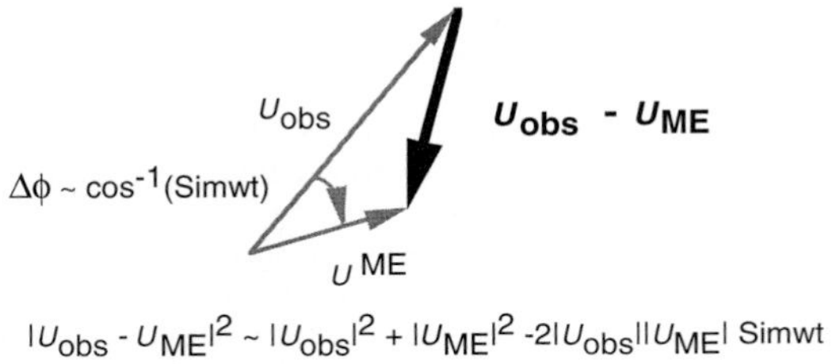

FIG. 2. Weak or incorrect extrapolation gives rise to large "renormalized" structure-factor amplitudes outside the basis set. Comparing renormalized amplitudes reveals their relative importance. (Reprinted from Ref. 32, with permission.)

structure-factor vectors $\langle|(U_{\text{obs}} - U_{\text{calc}})|\rangle$, called the renormalized structure-factor amplitude $|U_{\text{renorm}}|$ (Fig. 2).[34] Large renormalized amplitudes typically occur for intense reflections outside the basis set for which there is little or no maximum entropy extrapolation (Fig. 1a). Because such reflections are not strongly coupled to the current basis set, focusing on them tends to recruit the most useful new information. Complementary selection criteria based on the actual coupling patterns, based in turn on the neighborhood principle,[35,36] have been developed by Bricogne.[6]

Macromolecular Data Sets

The one-dimensional example in Fig. 1 captures many of the essential features found in macromolecular applications of entropy maximization and hypothesis permutation with likelihood scoring. The intensity data are markedly non-Wilsonian and are divided into low- and high-resolution domains. Maximum entropy extrapolation is strongest for reflections in the second neighborhood of the basis set, here defined by $h = (h_i \pm h_j)$, $i, j \in \{\mathbf{H}\}$. For the one-reflection basis set this includes only $h = 2$; for the two-reflection basis set it includes $h = 2$, 10, and 11. In this way, phasing the strongest reflections recruits successively more of the weaker reflections to which they are coupled in the right directions. Thus, the extrapolation strengthens as the number of basis set reflections increases.

In the simple example, only a modest amount of new information is introduced by maximizing conditional probability, because the extrapolation extends only to reflections that are strongly coupled to the one basis

[34] G. Bricogne, *in* "Molecular Replacement" (E. J. Dodson, S. Gover, and W. Wolf, eds.), p. 62. Science and Engineering Research Council, Daresbury Laboratory, Daresbury, Warrington, UK, 1992.

[35] H. A. Hauptman, *Acta Crystallogr.* **A33,** 553 (1977).

[36] H. A. Hauptman, *Acta Crystallogr.* **A34,** 525 (1978).

set reflection. In macromolecular applications involving many basis set reflections, the quality of the extrapolation may be significantly more extensive, leading to substantially correct phases for many more extrapolated reflections and resolving ambiguities in bimodal experimental phase probability distributions, via phase recombination. These effects lead to considerable improvement in the resulting centroid maps.[9,26]

Similarly, our experience has been that the extrapolation from the most reliable 30% or so of a set of marginal experimental phases at 3-Å resolution reaches all but a handful of strong reflections in the entire data set. Often, the latter can be determined quite accurately by phase permutation.

Phase permutation is a quintessential factorial design problem: the space representing all possible phase choices for the permutation must be sampled efficiently, but reliably enough to support decisions on the basis of likelihood scoring. Incomplete factorial designs[37,38] have excellent sampling properties, providing coverage of the important main effects using $\sim N^{1/2}/2$ tests from the full factorial. Scores from a sampled permutation experiment are then compared using the Student t test to obtain the significant phase choices.[6,9] More elegant sampling and analysis solutions exploiting the periodicity of phases arise from the mathematics of error-correcting codes.[6,8]

The LLG score repeatedly has proved accurate to a fraction of a percentage in real statistical tests. In keeping with that observation, we have also found that 10–20 strong reflections can have a substantial impact on the maps. Consequently, permutations involving 10–12 bits of phase information (three to five acentric phases sampled in the four quadrants with the remainder composed of centric reflections sampled at two levels) can lead to both accurate phase determination and meaningful map improvements.

Successful applications to macromolecular structure determinations[9,11,39] have involved data sets at moderate resolution of ~3.0 Å, far below the resolution limit of ~1.2 Å required by even the best of the direct methods based on the Cochran distribution (see Refs. 20, 40, 41, and [1] in this volume).[42] These examples required substantial amounts of experimental phase information provided by isomorphous or molecular replace-

[37] C. W. Carter, Jr., and C. W. Carter, *J. Biol. Chem.* **254,** 12219 (1979).

[38] C. W. Carter, Jr., "METHODS: A Companion to *Methods in Enzymology*," Vol. 1, p. 12. Academic Press, San Diego, California, 1990.

[39] V. Ilyin and C. W. Carter, Jr., *Acta Crystallogr.* in preparation (1997).

[40] G. T. DeTitta, C. M. Weeks, P. Thuman, R. Miller, and H. A. Hauptman, *Acta Crystallogr.* **A50,** 203 (1994).

[41] C. M. Weeks, G. T. DeTitta, H. A. Hauptman, P. Thuman, and R. Miller, *Acta Crystallogr.* **A50,** 210 (1994).

[42] H. A. Hauptman, *Methods Enzymol.* **277,** [1], 1997 (this volume).

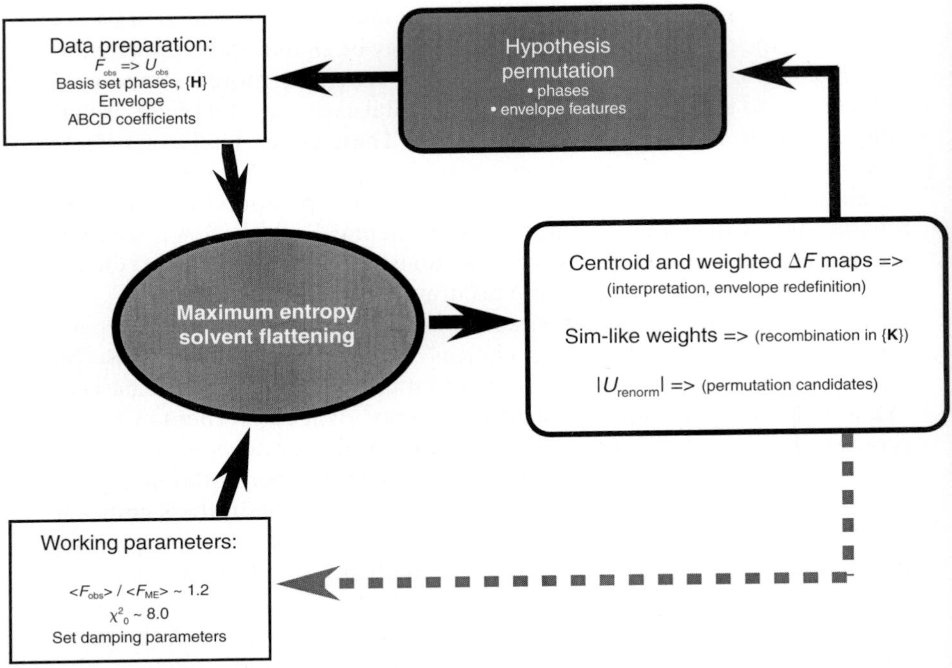

FIG. 3. Summary of map improvement procedures using the available entropy maximization algorithms.

ment. Nevertheless, conditional probability methods made decisive contributions to their structure solutions. The available experimental phasing power was insufficient to produce interpretable electron-density maps, and the complementary phasing power afforded by the Bayesian paradigm led directly to satisfactory maps. In all cases, it was also necessary to determine 7–30 unknown phases directly, via phase permutation, MESF, and likelihood scoring.[9]

Procedures

The map improvement procedures described here are summarized schematically in Fig. 3. Although similar in intent and form to other map improvement methods (see Refs. 43, 44, [4][45] and [3][46] in this volume, and

[43] K. Cowtan, *in* "Direct Methods of Solving Crystal Structures" (H. Schenk, ed.), p. 421. Plenum, New York, 1991.

[44] K. Y. J. Zhang and P. Main, *Acta Crystallogr.* **A46,** 377 (1990).

[45] K. Zhang, K. Cowtan, and P. Main, *Methods Enzymol.* **277,** [4], 1997 (this volume).

[46] F. M. D. Vellieux and R. J. Read, *Methods Enzymol.* **277,** [3], 1997 (this volume).

Refs. 47–53), they differ in both underlying motivation and implementation. Bayesian statistical inference proceeds along a metaphorical "phasing tree," whose branches connect successive "nodes" defined by increasing numbers of "basis set" reflections {**H**} with assumed phases. The computational engine for evaluating each node is the MESF process, and one proceeds from node to node toward a structure solution by the accompanying processes of hypothesis generation, scoring, and statistical analysis. These steps are now described briefly.

Data Preparation

Most of the steps in preparing data for the MICE_mx_8 program are straightforward reformatting procedures. Specific areas deserve some comment. Normalization of X-ray amplitude/intensity data is a standard procedure in direct phase determination. In practical terms, there are three separate modifications of the data. Data are placed on absolute scale, "sharpened" by a scale constant that increases exponentially with increasing resolution, and divided by an expectation value, which can be as crude as an average amplitide within a given resolution shell. The latter process is the normalization step; it identifies reflections that are conspicuously stronger or weaker than the average, or expected, value. The limited resolution of our studies, ~3.0 Å, imposes some restrictions on the effectiveness of these procedures. We have not used either sharpening or normalization in preparing macromolecular data sets for entropy maximization calculations. Rather, we have empirically "normalized" the data in a crude way, such that the largest amplitude is 5.0–8.0. The reason for this is simply that data are input to MICE in the format of a MITHRIL data file,[54] for which the largest stored value is 8.0. A related process, "renormalization," is applied later in the process, after exponential modeling has produced estimates for the amplitudes of reflections outside the basis set. The goal of such renormalization is to identify strongly observed reflections that are not given significant maximum entropy estimates.

In practice, absolute scaling of the amplitude data is more important than sharpening or normalization. The structure factors calculated inter-

[47] V. Y. Lunin, *Acta Crystallogr.* **A44,** 144 (1988).
[48] V. Y. Lunin and T. P. Skovoroda, *Acta Crystallogr.* **A47,** 45 (1991).
[49] V. Y. Lunin and E. A. Vernoslova, *Acta Crystallogr.* **A47,** 238 (1991).
[50] V. Y. Lunin, *Acta Crystallogr.* **D49,** 90 (1993).
[51] P. Main, *Acta Crystallogr.* **A46,** 507 (1990).
[52] S. Xiang and C. W. Carter, Jr., *Acta Crystallogr.* **D52,** 49 (1996).
[53] J. P. Abrahams and A. G. W. Leslie, *Acta Crystallogr.* **D42,** 30 (1996).
[54] C. J. Gilmore, *J. Appl. Crystallogr.* **17,** 42 (1984).

nally by MICE are presumed to be on absolute scale, and our experience has been that the average observed and calculated structure factors should be in the ratio of about 1.2:1 on the first cycle of exponential modeling by MICE.[26] Because Wilson scaling is less accurate without data beyond 3 Å, this criterion is achieved manually, by adjusting an empirical scaling factor that is applied by the program.

A choice must be made regarding the basis set phases. Our experience has been that it is satisfactory to use a figure of merit (FOM) threshold to select reflections with the most reliable phases. There is no infallible guide for the setting of this threshold. A useful rule of thumb is that roughly one-third of the data should be used in the initial basis sets. If the phases are reliable enough that this third of the data has $\langle FOM \rangle > 0.7$ or a maximum phase error less than ~45°, then MESF can be used to produce a map quite close in appearance to that obtained from refined coordinates. For *Bacillus stearothermophilus* tryptophanyl-tRNA synthetase,[9] where this fraction had $\langle FOM \rangle > 0.5$, or a mean phase error of 60°, the more elaborate Bayesian paradigm of supplementing MESF with phase and envelope permutation led to an excellent structure.

Envelope definition for imposing solvent flatness should be done carefully, as the envelope exerts a profound effect on the course of the density modification imposed by MESF. For much of our work, we had to make use of envelope editors[55] to adjust envelopes created using the Wang–Leslie algorithm.[56] We have used a standalone program called MORPHO, written by D. Hurwitz (University of North Carolina, Chapel Hill, unpublished). This program uses a morphological image-processing algorithm that maintains binarized copies of the input map, on one of which it carries out a sequence of openings and closings.[57] These operations connect peaks inside the molecule and regions of low density outside the molecule until a boundary is produced. The result is generally suitable as a mask without further editing. Executable versions of the MORPHO program are available from the authors for many UNIX workstations (see the concluding section).

Exponential Modeling

Maximum entropy solvent flattening is contrasted in Fig. 4 with the conventional solvent-flattening procedure, introduced by B. C. Wang.[27,56]

[55] W. Minor, "MAP_CCP4: A Program for Editing Molecular Envelopes." Purdue University, Lafayette, Indiana, 1992.

[56] A. W. G. Leslie, *Acta. Cryst.* **A43,** 134 (1987).

[57] R. C. Gonzales and R. E. Woods, "Digital Image Processing." Addison Wesley, Reading, Massachusetts, 1993.

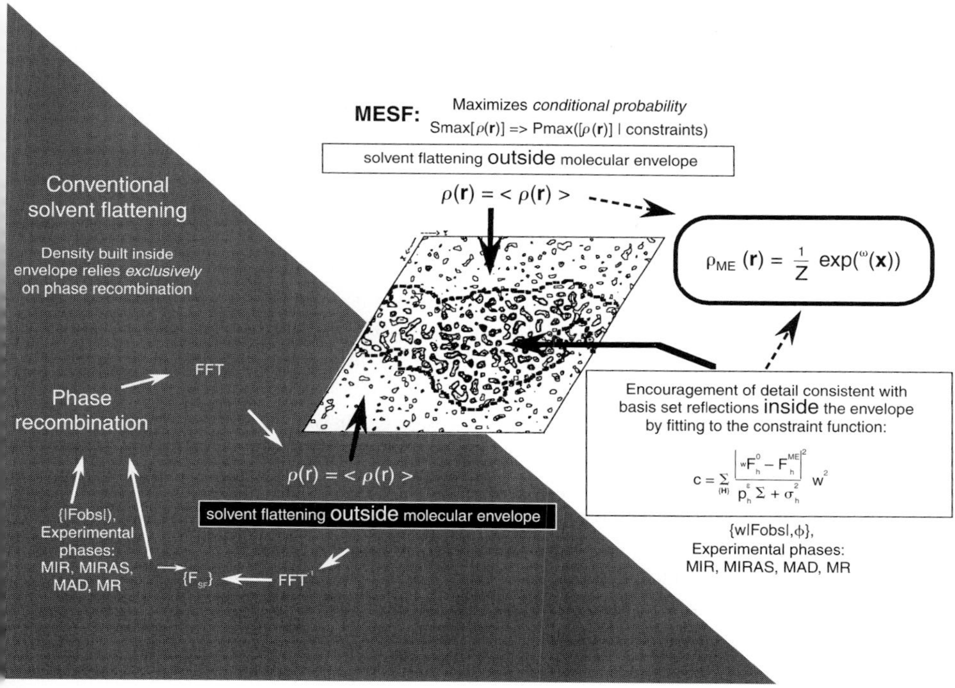

FIG. 4. Comparison between MESF (upper triangle) and conventional solvent flattening (lower, shaded triangle). Both methods involve solvent averaging. The difference between them is that in MESF, solvent averaging is enforced jointly with the maximum entropy requirement. Thus, the density modification is encouraged to take place only within the molecular envelope, and only in ways consistent with the basis set reflections. In contrast, the only way to communicate the solvent flatness constraint in the conventional algorithm is via the phase recombination step. (Reprinted from Ref. 32, with permission.)

Averaging the solvent while at the same time forcing an exponential model to agree with the most confidently known phases affords a more efficient coupling between the probabilistic and solvent flatness constraints. The envelope constraint imposes additional coupling between high-resolution reflections that form triplet phase relations with strong low-order reflections in the spectrum of the envelope, as outlined by Bricogne.[5,6] Enforcing this coupling during entropy maximization under the constraint of solvent flatness ensures that density features built up inside the envelope are consistent with the basis set structure factor constraints. If there is sufficient information contained in the basis set and/or envelope, this extrapolation is essentially correct and complete.[4] Thus, although MESF is only a first approximation to the procedure outlined by Bricogne,[5] it has proved to be a superior algorithm relative to the conventional procedure.[9,11,26,58]

Exponential modeling in MICE depends on having several internal parameters properly adjusted, in addition to the "absolute" scaling constant mentioned above. After each cycle, the extent to which the model fits the basis set constraints is evaluated using a reduced chi-squared-like statistic, $\mathscr{C}$ (Table I). The initial $\mathscr{C}$ can be adjusted by setting a parameter, *P*, which sets a balance between different contributors to the variance assumed for the structure-factor amplitudes in the denominator of the $\mathscr{C}$ expression. This value should be set such that $\mathscr{C}$ is initially equal to about 8.0, because the variances of the observed data are generally unreliable. A lower initial value leads to premature cessation of the fitting, whereas a higher initial value can lead to overfitting. For protein data sets, $\mathscr{C} = 1$ is often a less reliable stopping criterion than is achieving a maximum LLG.

Damping parameters invoked in each cycle to adjust the application of parameter shifts can be set by the user. This flexibility can be useful, and we generally use it empirically, so that the maximum LLG is achieved in ~30 cycles of fitting.

MICE_mx_8 accepts input *ABCD* coefficients for experimental phase probability distributions. These are used in calculating "phased likelihoods,"[5] whose function is similar to that of external phase recombination. Phased likelihoods offer increased sensitivity over unphased likelihoods in scoring phase permutation experiments.

Hypothesis Generation, Permutation, and Sampling

Renormalized structure-factor amplitudes (Fig. 2) are calculated by MICE_mx_8 whenever an output map is requested, and these are written to a formatted output file containing the observed amplitudes, the extrapolated (maximum entropy) amplitudes and phases, and the combined phase probability coefficients. Renormalized amplitudes are estimates for the error in maximum entropy extrapolation. As noted by Bricogne,[7] they also represent the contribution of a reflection to the root-mean-square (RMS) noise in the electron-density map. After sorting, a histogram of the largest values provides a rapid way to evaluate the need, and to select candidates for, phase permutation. Recruiting reflections with the largest renormalized amplitude will effect the most rapid increase in the signal to noise of the maps.

[58] S. Doublié, "2.9 Å Crystal Structure of *Bacillus stearothermophilus* Tryptophanyl-tRNA Synthetase Complexed to Its Adenylate, Tryptophanyl-5′AMP." University of North Carolina at Chapel Hill, Chapel Hill, North Carolina, 1993.

Once the decision is made to permute choices for missing constraint information, factorial designs should be used to set up new nodes, one for each distinct set of choices. We used incomplete factorial designs,[37,38,59] constructed according to two criteria. Randomized sampling helps to assure an even distribution of tests throughout the full factorial space. The randomization is limited, however, by a balance criterion, which assures that an equal number of tests is performed for each level of every factor, and that all two-factor interactions are also sampled as uniformly as possible.

A library of such designs for 24 nodes (10 bits of phase information), 60 nodes (12 bits), and 100 nodes (14 bits) has been compiled for this purpose. Utility programs are available for setting up constraint files with the appropriate phases, given a previous phase set and the appropriate design matrix, and for formatting the design matrix and scores once all nodes have been taken to maximum LLG. These tools are available from the authors (see the concluding section).

Sampling designs based on error-correcting codes[6] have been described elsewhere in this series.[8] These designs are similar in spirit to incomplete factorial designs, but because they are based on lattices, the LLG scores can be analyzed by Fourier transformation, which provides even more efficient coverage of higher-order interactions between phases. All but the smallest of these designs, however, require simultaneous development of between 250 and 4096 nodes, which is excessive with current computing resources. Clearly, however, the advent of an additional order of magnitude in computing power will bring them into range, even for problems like tryptophanyl-tRNA synthetase (TrpRS).

Multiple Regression and Analysis of Variance

Once the nodes of a permutation design have been expanded to maximum LLG, the phase bits that are significant "predictors" of the LLG must be identified and their statistical significance evaluated. The procedures we use for this process are exactly those described elsewhere in this series.[60]

Incomplete factorial designs and, to an even greater extent, error-correcting codes, exploit the power of averaging to enhance the signal-to-noise ratio in the effect any given phase has on the LLG. By simultaneously varying the phase choices of multiple reflections, the effect of changing a phase bit and the intrinsic fluctuation induced in the LLG scoring by the

[59] C. W. Carter, Jr., *in* "Crystallization of Proteins and Nucleic Acids: A Practical Approach" (A. Ducruix and R. Giegé, eds.), p. 47. IRL Press, Oxford, 1992.

[60] C. W. Carter, Jr., *Methods Enzymol.* **276,** 74 (1997).

overall permutation of new reflections is thereby averaged over as large a variety as practical of different contexts, with respect to the choices made for the other permuted reflections.

Log-likelihood gain scores from an 11-bit, 24-node design are all represented in all four histograms of Fig. 5, but the nodes are grouped differently in each case according to the sign given to each phase bit. For the binary

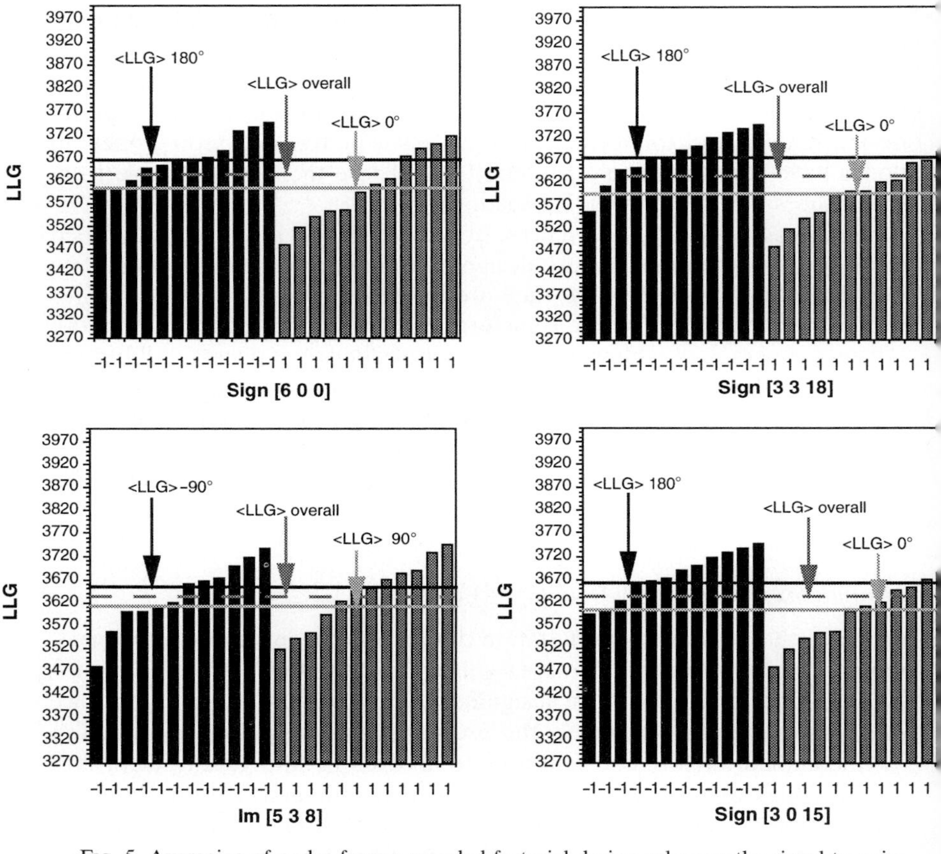

FIG. 5. Averaging of nodes from a sampled factorial design enhances the signal-to-noise in LLG scoring. Four phase bits from a 24-node permutation are shown. For each one, those experiments treated at the minus sign of the phase bit are shaded darkly; those treated with the plus sign are shaded lightly. The same 24 nodes are illustrated for each reflection; they are simply sorted differently, according to the phase assignments in the incomplete factorial design. The average LLG values for the two phase bits are compared to the overall ⟨LLG⟩, using the differently shaded horizontal lines. In each case the differences between the averages are highly statistically significant.

TABLE II
SIGNIFICANCE-TESTING LOG LIKELIHOOD GAIN SCORES FOR 9 PHASE BITS

Variable	Average effect	Standard error	Standard coefficient	Student's t	p (two-tail)	ϕ_{Ind} (°)	ϕ_{calc} (°)	$\Delta\phi$ (°)
Constant	2098.612	2.131	0.000	984.97	0.10×10^{-14}			
Re 6 4 48	26.199	2.485	0.547	10.542	0.25×10^{-7}	0	27	27
Re 7 4 41	10.374	2.268	0.217	4.575	0.36×10^{-3}	0	−1	1
Re 10 9 26	3.050	2.631	0.064	1.159	0.264			
Im 10 9 26	−7.061	2.850	−0.147	−2.477	0.026	−66	−101	35
Re 9 8 28	17.351	2.284	0.362	7.597	0.16×10^{-5}			
Im 9 8 28	17.824	2.488	0.372	7.164	0.33×10^{-5}	46	86	40
Sign 2 0 54	22.313	2.337	0.466	9.550	0.91×10^{-7}	90	90	0
Sign 0 0 48	−5.346	2.571	−0.112	−2.079	0.055	180	180	0
Sign 9 0 35	21.243	2.799	0.443	7.589	0.16×10^{-5}	135	135	0

choice accorded to each phase bit, factorial designs provide a comparison between averages of half of the nodes against the other half. This effect can be appreciated graphically by sorting the LLG score histograms according to the specific treatment of each phase bit (Fig. 5). The average score for those reflections with the same sign is, in general, different from the average of all 24 experiments. Averages for nodes given plus signs are either higher or lower than the overall average of 3630 by varying amounts, depending on the reflection.

These average differences are quite small, on the order of 1% of the average LLG, and it remains to assess their significance. Our analysis involved a least-squares regression model in which the LLG is expressed as a linear function of contributions from the individual phase bits. Stepwise regression and other manual adjustments of the model were used to select from the total list of phase bits the most satisfactory subset, using the program SYSTAT.[61] The best subset, from which phase indications were obtained, was taken to be that with the highest squared-multiple correlation coefficient and the best F ratio test from the analysis of variance, as shown for a different example in Table II. The former criterion assesses how much of the variation in LLG scores can actually be predicted by the model. A squared multiple correlation coefficient of 98% for the final model means that all but 2% of the variation in LLG score among the nodes can be attributed to the phase choices made for each node in the respective basis sets by the design matrix.

We conclude that the sign of each phase bit is likely to be positive or

[61] L. Wilkinson, M. Hill, and E. Vang, "Systat: Statistics," version 5.2. Systat, Inc., Evanston, Illinois, 1992.

negative, depending on which sign leads to a higher average score. The individual Student t test indicates which of the phase bits have significant impacts on the score. When coefficients for both real and imaginary bits of an acentric reflection have significant t tests, an estimate of its phase angle is obtained from the inverse tangent of their ratio.

Using Maximum Entropy Extrapolates for Missing Reflections

Maximum entropy extrapolation provides optimal estimates for amplitudes and phases for all extrapolated reflections, including those that are missing from the observed data set. Supplementing the data set, particularly missing low-order reflections that can be quite strong, with these estimates can be effective. A separate input file of normalized structure factors including these estimates should be prepared for use when calculating output maps.

Molecular Replacement Phases and Noncrystallographic Symmetry Averaging

One of the most powerful potential applications of the Bayesian paradigm involves recycling of information regarding known molecular fragments. Indeed, imposing a solvent mask is actually a special case in which the envelope is the known fragment. At present, there is no working implementation of fragment recycling in MICE. Nor is it yet possible to average redundant structures in the crystallographic asymmetric unit. Program development is underway to provide these capacities. However, we have found that MICE_mx_8 can be useful to help eliminate the model bias in a molecular-replacement structure with sixfold noncrystallographic symmetry if combined with several other programs to carry out the averaging (RAVE) and phase recombination (SigmaA) as shown schematically in Fig. 6.

Results

The Bayesian strategy of hypothesis permutation and LLG scoring played a crucial role in solving the tryptophanyl tRNA synthetase structure.[9] Early on it became clear that overcoming a serious lattice-preserving lack of isomorphism in two heavy-atom derivatives would be key to solving the structure. TrpRS is intrinsically capable of conformational polymorphism consistent with the tetragonal lattice. We have subsequently solved three different tetragonal structures from essentially identical crystals treated in two different ways for data collection. The structural variation between them is on the order of 4–6 Å for these loops and on the order of 1–2 Å over most of the molecule.

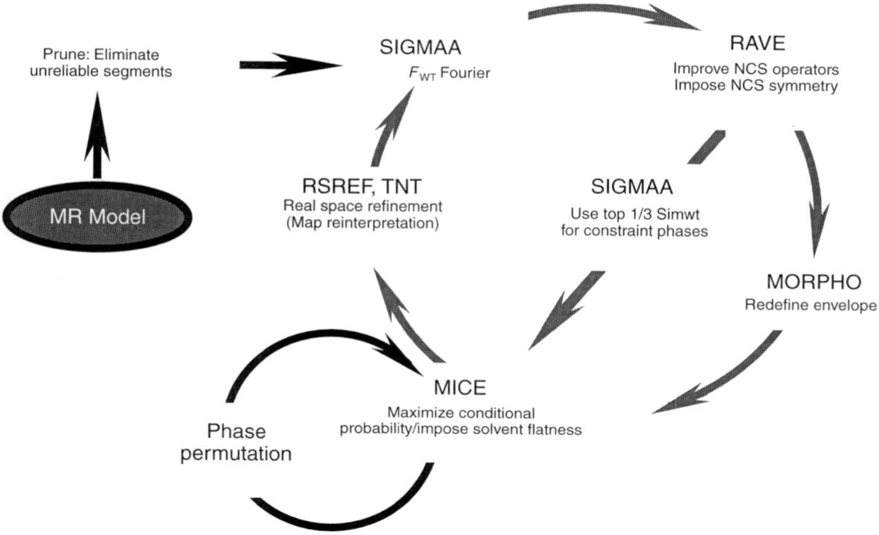

FIG. 6. Incorporation of MESF into the map improvement for a problem phased initially by molecular replacement, and for which there were six independent monomers in the crystallographic asymmetric unit. This algorithm provided an effective phase improvement algorithm that resulted in correcting substantial bias in the molecular-replacement model.

Our experience in solving the tetragonal form of tryptophanyl-tRNA synthetase (TrpRS) using a combination of experimental and conditional probability methods supports the following general conclusions.

Conditional Probability Methods: Bringing New Sources of Phasing Power to Bear on a Structure Solution

Figure 7 traces the progressive improvement, from our first map to the one that was finally interpreted, in the correlation coefficient between the current map and the final F_c map. The phase determination was discontinuous; during the process it became possible to incorporate new experimental phase information arising from isomorphous [selenium–sulfur] differences from data from selenomethionylated TrpRS as well as data from an isomorphous, but weakly substituted, trimethyllead acetate derivative. Interestingly, map improvement proceeded with roughly the same slope from both starting points, and showed no sign of leveling off.

The map improvement procedures described here use the conditional probability model constrained by the most reliable phases to generate more accurate phases directly for the remaining reflections with less reliable

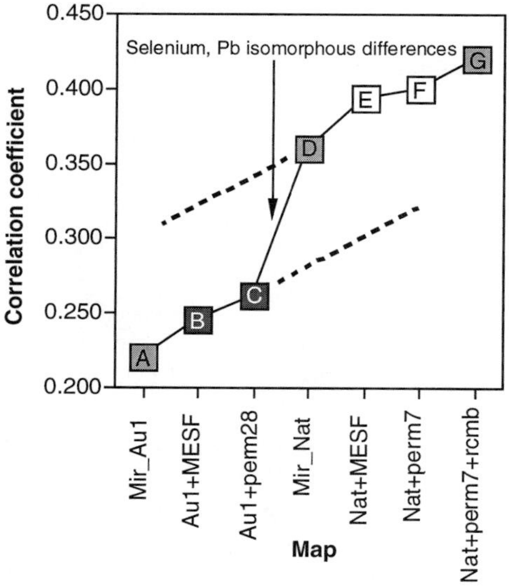

FIG. 7. Progressive improvement in the correlation coefficient between the model-independent TrpRS maps obtained at successive stages of direct-phase determination and the final F_{calc} map. The correlation coefficients were calculated only over the pixel values within the molecular envelope obtained from the refined structure. Maps are designated with capital letters in the sequence in which they were obtained, to facilitate direct comparisons of the maps in Figs. 8 and 9. Maps B, C, E, F, and G were obtained using MESF constrained by the same envelope, namely that obtained after phasing 28 reflections directly and permuting the envelope. Trend lines are drawn to indicate the similar rate of map improvement observed with the two different sets of starting phases, and the fact that there is no indication that either phase determination process was approaching convergence. Maps in lightly shaded boxes are illustrated in Fig. 8, maps in darkly shaded boxes are shown in Fig. 9. (Reprinted from Ref. 32, with permission.)

phases. Phases were improved in two ways. First, maximum entropy extrapolation outside the basis set yielded Sim-like probability distributions for the extrapolated reflections, and these were used to phase these reflections either directly or by recombination with existing phase probability distributions. Second, the pattern of extrapolation was used to identify rogue reflections (either within or outside the basis set) with unreliable phases and the greatest impact on the maps. Then, phases were obtained for these reflections by sampled permutation experiments.

The overall map improvement achieved with sustained application of the Bayesian paradigm can be substantial, even when compared with that obtained by supplementing the experimental phases with the contribution from additional derivatives. Figures 8 and 9 compare the major improvements in the most troublesome region of the TrpRS electron density, the

β sheet of the Rossmann fold. This region could not be interpreted until all sources of phase information were combined in a framework of MESF using a mask that was itself determined by a Bayesian process of hypothesis generation and permutation. That final map (Fig. 8c) afforded an unambiguous interpretation for all but six or eight residues in a disordered surface loop.

Figure 8a shows the initial map calculated with experimental MIRAS phases obtained from two highly nonisomorphous heavy-atom derivatives. As frequently happens, the difficulty of tracing the chain through this region was that the connectivity of the map ran predominantly across the sheet, rather than along the strands. The decisive contribution came from MESF, represented in Fig. 8c by the contributions of phase and envelope permutation, together with phase recombination.

Three additional sources of information were included as the structure was solved: (1) phase permutation, likelihood scoring, and regression analysis provided accurate phases for 28 reflections whose observed amplitudes were strong, but which had very small extrapolated amplitudes; (2) permutation of three structural features together with other binary characteristics of the molecular boundary led to an essentially correct mask for MESF; and (3) the experimental MIRAS phases were supplemented by isomorphous {selenium–sulfur} differences from selenomethionine-substituted TrpRS and from a lead derivative. None of these increments changed the appearance of the parallel β sheet significantly by itself. The similarity of the maps in Fig. 8b (the best experimental phases without MESF), Fig. 9b (MESF with the correct envelope, but without the 28 strong reflections), and Fig. 9c (MESF with the original envelope but with the 28 strong reflections) illustrates the strong interdependence of the different sources of phase information. Each increment of phase information improved both the definition of the molecular envelope for MESF, and the Sim probabilities and hence the recombined phases for the extrapolated reflections. This progressive improvement continued even after the selenomethionine and lead data sets were used to calculate MIRAS phases, and the map in Fig. 8c was obtained only after phasing seven additional phases by permutation and after MESF phase probabilities were used in phase recombination with the experimental phase probability distributions.

Bayesian map improvement was critical even for the location of the 10 selenium atoms, which could not be located in difference Fourier maps calculated using the heavy-atom MIR phases. After permuting 28 reflections and correcting the envelope, the eight top peaks in the difference map were selenium atom positions.

The effect of the starting experimental phases was significant, as illustrated by the sharp jump in the correlation coefficient between maps C

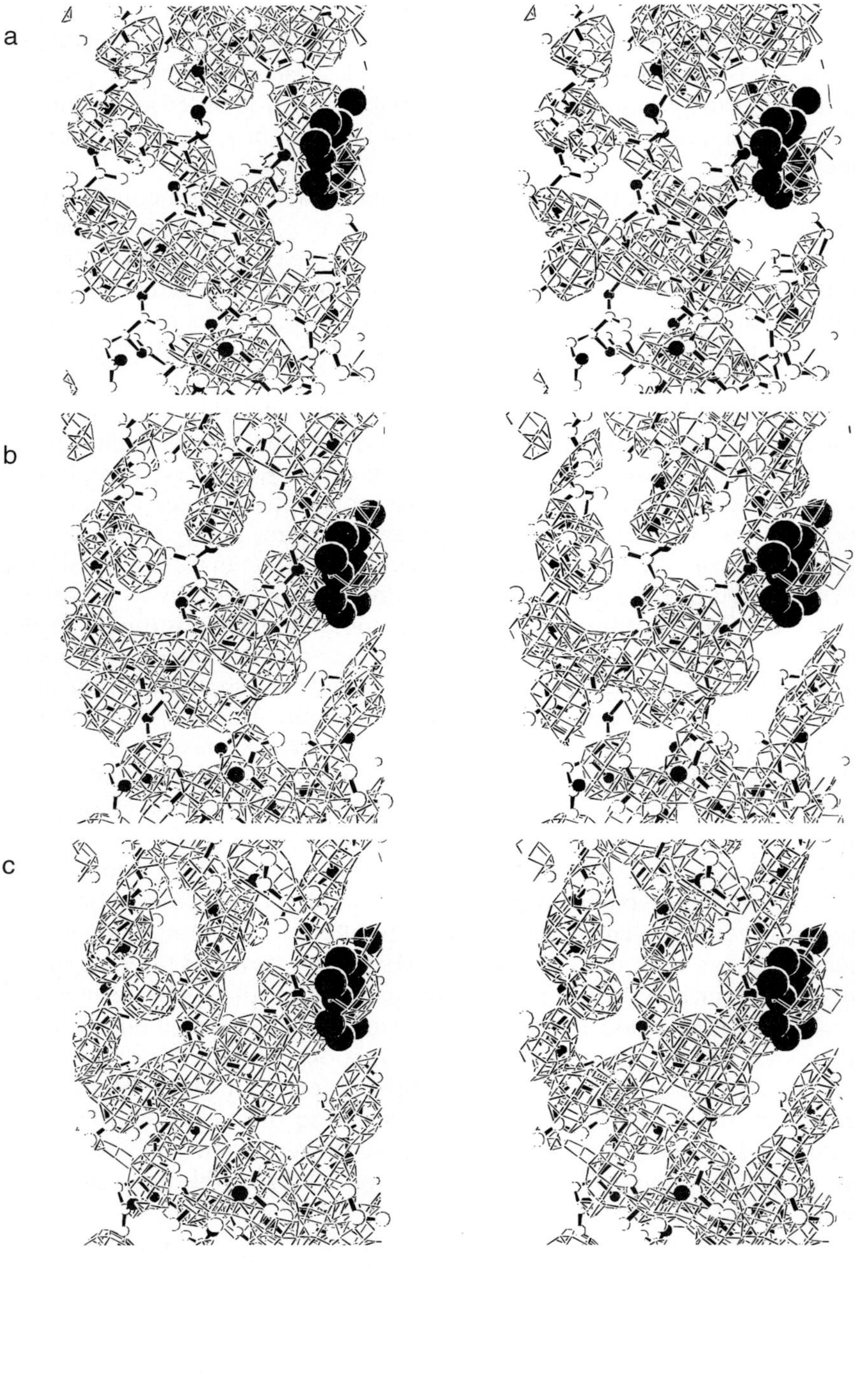
a
b
c

and D (Fig. 7). However, in the difficult-to-interpret region containing the parallel β sheet, the maps were almost equally confusing, irrespective of whether the additional isomorphous replacement information from selenomethionine and lead derivatives was included (compare Fig. 8a and b). The density in both maps is dominated by connectedness across the β sheet, rather than along the strands.

Maximum entropy solvent-flattening density modification brought about very little improvement in the starting map, consistent with the notion that it was expressing accurately the quality of the available information. It did improve the quality of the starting map somewhat (compare Fig. 9a and b), but it did not greatly improve the interpretability in this region. Neither the use of the correct envelope (Fig. 9b), nor the 28 permuted reflections (Fig. 9c), brought about decisive improvement. In contrast, when 28 additional phases and a correct envelope were used as constraints, the improvement obtained with MESF was readily apparent (Fig. 9d). We might have interpreted that map had additional selenomethionyl and lead derivative phasing information not become available. This shows that the efficacy of MESF for map improvement interacts strongly with the quality of the starting phases, with the correctness of the mask used to define the solvent boundary, and with supplemental phases determined directly by permutation. The progressive improvement in the maps in Fig. 9 demonstrate that even with marginal starting phases the full Bayesian paradigm can lead to an interpretable map.

Identifying Errors in Molecular Replacement Solutions

While working with a monoclinic crystal form of unliganded TrpRS, which we were attempting to solve using coordinates obtained from a tetragonal crystal form from which the ligand had been removed as a molecular replacement model, it became clear that the structure was quite different from that of the phasing model. After the fact, we found that the RMS deviation between the starting and final models was around 4 Å. Moreover, because there were six ~37-kDa monomers in the asymmetric

Fig. 8. Bayesian, model-independent phase improvement can contribute an amount comparable to that achieved with additional derivatives. Maps A, D, and G in Fig. 7, showing the parallel β sheet in TrpRS, are presented here. (a) The initial MIRAS map (A in Fig. 7) obtained from two nonisomorphous derivatives. (b) The initial MIRAS map (D in Fig. 7) obtained using phases from selenomethionyl TrpRS and a lead derivative, together with the derivatives in (a). (c) The final MESF map (G in Fig. 7) based on constraint phases with FOM > 0.61 from (c), after envelope and phase permutation and phase recombination with the MIRAS phase probabilities. All maps here and in Fig. 9 are stereopairs. (Reprinted from Ref. 32, with permission.)

a

b

c

d

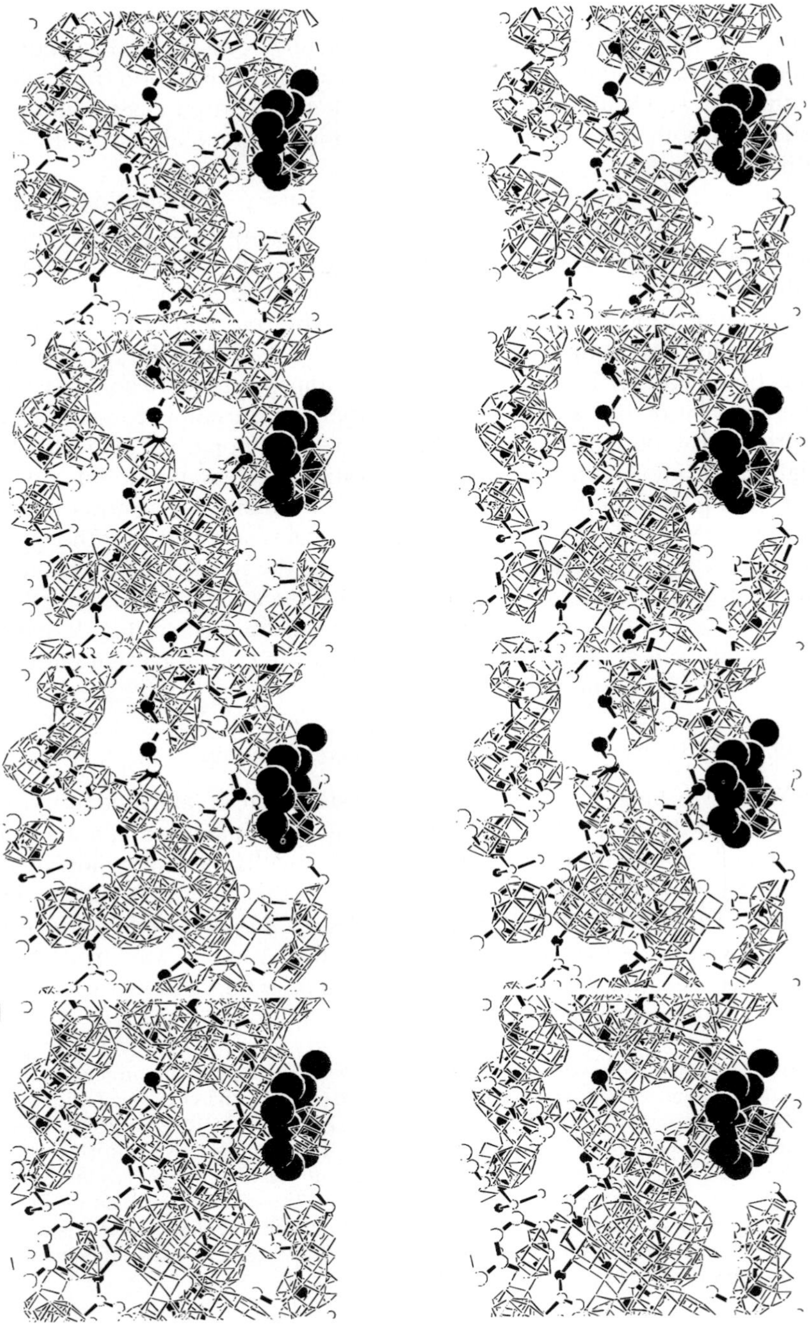

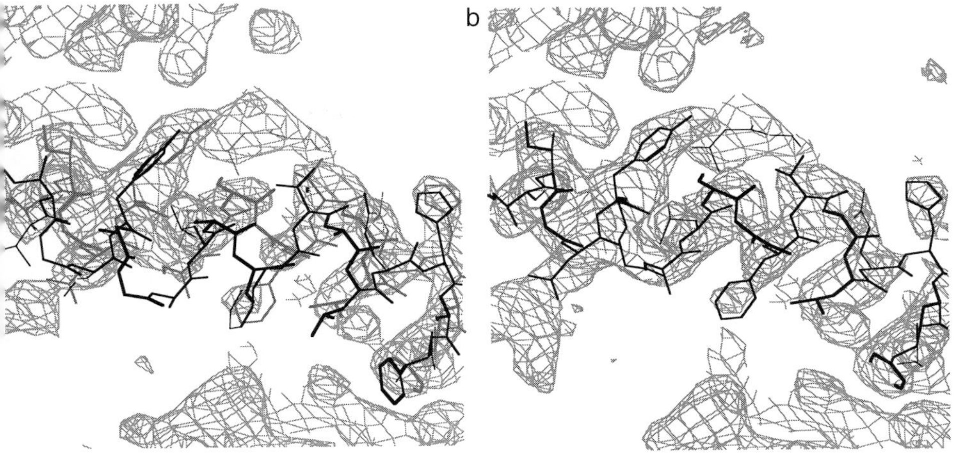

FIG. 10. Phase permutation for 16 reflections with very strong renormalized structure factor amplitudes after several rounds of map improvement using the algorithm shown schematically in Fig. 6. (a) An active-site helix was incorrectly placed in the initial model, and the density for that helix was discontinuous, even after sixfold averaging. (b) Bayesian phase determination led to phases for 10 of the 16 reflections with an average phase error of 38°, and confirmed the location of the helix.

unit, we were using noncrystallographic symmetry averaging to improve the molecular replacement phases. The phasing model was broken into domains and refined as rigid bodies in a manner suggested by the relative domain motion previously observed in several different polymorphic tetragonal crystal forms we had solved. Neither noncrystallographic symmetry (NCS) averaging nor rigid body refinement, however, provided unbroken density for the amino-terminal helix, which comprises a crucial part of the enzyme active site (Fig. 10a). Moreover, continued application of NCS averaging led to a deterioration of this region of the map.

We began a Bayesian process that made use of the stand-alone programs shown in Fig. 6 together with MICE and MESF. Examination of the renormalized structure-factor amplitudes after the first such cycle revealed ap-

---

FIG. 9. The effect of MESF depends jointly on the quality of the envelope and the content of the basis set. (a) Map calculated with entire set of starting phases (map A in Fig. 7) without further modification. (b) The effect of MESF using the final envelope after envelope permutation (map B in Fig. 7). (c) The effect of enhancing the basis set for MESF with 28 reflections phased by permutation but constrained with the original envelope calculated according to Leslie[56] (map not shown in Fig. 7). (d) The effect of MESF with the enhanced basis set (map C in Fig. 7). This map compares favorably with map G (compare with Fig. 8c). (Reprinted from Ref. 32, with permission.)

proximately 20 strong reflections that clearly were not reached either by the noncrystallographic symmetry or by the maximum entropy extrapolation. Phases were permuted for 16 of these reflections and significant indications were obtained for 1 centric and 9 acentric reflections. The mean phase error for the nine acentric reflections was 38°, and the resulting centroid map (Fig. 10b) confirmed the density for the helix by filling in across a significant discontinuity.

Summary

Solvent Flattening under the Constraint of Maximum Conditional Probability: Both More Conservative and More Effective Than Conventional Solvent Flattening

Maximization of the conditional probability in real space expresses optimally the phase information available from the statistical phase relations between phased basis set structure factors, which are commonly used in small-molecule direct methods.[4,5] Combining this information with additional constraints on the electron density, such as solvent flatness and noncrystallographic symmetry, therefore should approximate optimal implementations of those density modification techniques. Our experience in imposing solvent flatness is consistent with that conclusion.

Comparing maps for cytidine deaminase[26] revealed that the maximum entropy algorithm produced superior maps that were at the same time in better agreement both with the starting MIRAS map (correlation 0.68) and with the final $\{2F_{obs} - F_{calc}\}$ map (correlation 0.69) than were maps produced by conventional solvent flattening constrained by the same envelope (correlations 0.44 and 0.59, to the starting and target maps, respectively). Thus, the MESF map lay nearly on a direct path from the starting map to the target map, while the conventionally solvent-flattened map diverged from both starting and target maps. This phenomenon reflects the highly desirable "maximally noncommital" property of constrained maximum entropy distributions, which introduce the minimal new detail necessary to fit the constraints.

Because the traditional constraints imposed on the electron density are complementary,[62] we believe that the ultimate coordination of other sources of phase information into the Bayesian framework as envisioned by Bricogne[5] will yield a superior phase improvement algorithm.

[62] K. Zhang, *Acta Crystallogr.* **D49,** 213 (1993).

Hypothesis Permutation: Face to Face with Phase Problem

Although hypothesis permutation may strike one as making an unnecessarily long excursion, it has the eminently satisfying advantage that it brings the investigator face to face with the phase problem. The most important missing phase information at any stage of the phase determination can be identified with considerable confidence, and accurate values can be obtained with a modest investment of computation.

Accuracy of Permutation with Likelihood Scoring and Statistical Analysis: $\Delta\phi \sim 30°$

We have phased a total of only 45 reflections by permutation, likelihood scoring, and regression analysis. For 17 of these, we can evaluate the phase error from the phases of a refined model. Each of 4 centric reflections was given the correct sign, and for 13 acentric reflections the mean phase error was 33°.

Large Impact of Relatively Small Number of Reflections on Maps

Perhaps the most surprising aspect of the work we have done using the Bayesian paradigm is how significant an improvement can be achieved by directly phasing a handful of strong unphased reflections. Small numbers of reflections phased in this manner can supplement an insufficient basis set sufficiently to produce an interpretable map. Compare the maps in Fig. 9b and d (the effect of 28 reflections) and Fig. 10a and b (the effect of 10 reflections). Many marginal data sets can be improved substantially by MESF and phase permutation by taking advantage of this disproportionate influence of strong reflections.

Fallacy of Pursuing Maximum Entropy Extrapolation without Phase Permutation

Maximum entropy extrapolation represents a powerful mechanism to compensate for missing data in a manner that is most consistent with the basis set constraints. It cannot, however, build new information from thin air when the basis set itself is deficient. The strong extrapolation from insufficiently constraining data often can have nearly correct phases. However, selecting in this way for strong extrapolation from an incomplete basis set only reinforces the errors or prejudices resulting from the absence of information in the initial constraints. It constitutes a potent risk. We initially tried to iterate the MESF process as carefully as possible without adding any new information. The result was a disastrous decline in the LLG, as the basis set became increasingly infiltrated by extrapolated reflections at the expense of others that fit less well to the exponential model.[9]

Strong extrapolation, even if essentially correct, means only that structural information it expresses is already encoded in the basis set; one cannot recruit much helpful new information by reinforcing it. The only alternative in moving through the temporary impasse of an incomplete basis set is to select strongly observed reflections outside the current basis set from among those that are least well extrapolated. Such reflections can be identified from a sorted list of "renormalized" structure-factor amplitudes. These are also the reflections for which the least is known about the phases; phasing them requires permutation.

Management of Combinatorial Explosion by Hierarchical and Sampling Methods

Experimental design plays an all-important role in screening the very large number of possible combinations of phase (and envelope) choices. For the initial TrpRS structure, 49 phase bits and 5 envelope choices were screened, or 1.8×10^{16} possible nodes. We actually inferred correct choices for 40 of the 54 bits tested from a sample of only 136 nodes. This extraordinary economy was achieved in two stages.

First, the 54 bits were sampled hierarchically, in the fashion suggested by the phasing-tree metaphor, treating the most important reflections first. Sacrificing information about possible interactions between reflections permuted in different steps allowed the benefits of each new increment to the basis set to be expressed in all subsequent experiments. Early in the process, the selected reflections were very strong, bringing considerable new structural information with them into the basis set. Among the benefits was the ability to define better the molecular envelope from the improved centroid maps obtained after phase recombination.

The second source of economy came from examining only a small sample of the possible branches issuing from a particular node. As Bricogne has noted elsewhere,[6,8] the phase problem is a quintessential factorial problem, but if the factorial design is sampled effectively, the combinatorial explosion need grow only by roughly the square root of the growth in the number of new phase bits. Because the branches examined represent only a small proportion of the total number of branches, the correct branches are statistically unlikely to be members of this original sample, and the best branches in the sample will not, in general, have correct phase choices for all phase bits. Nevertheless, the most promising branches in the entire set can be identified by statistical analysis and t testing of the effects of each phase bit on the sample of LLG scores. From these choices one can construct new nodes outside the sample and along those branches that could lead to the structure.

The most effective strategy involves permuting the largest practical number of bits, which usually meant that 10–11 phase bits at a time could be permuted in 24-node designs within a 24-hr period using a network of four DEC alpha workstations. By first identifying the most critical missing information for permutation, we could use the Bayesian paradigm to achieve significant improvements within realistic amounts of computing time.

Branching

The phenomenon of branching represents one of the most significant hurdles remaining to be overcome before the Bayesian paradigm can begin to supplant experimental sources of phase determination on a more ambitious scale than we have demonstrated here. Branching is the quintessence of the phasing-tree metaphor. It is indicated whenever the effects of changing a phase bit are not statistically significant, as proved to be the case for 14 of the 54 bits we permuted during the TrpRS phase determination. For such reflections, the phasing tree still presents a number of branches along which the final structure may still lie. This multiplicity of viable branches, even at such an advanced stage of phase development, is a sobering reminder of the remaining challenges associated with the phase problem.

Note on Availability of MICE Program

The MICE_mx_8 program was written by C. Gilmore and G. Bricogne and modified to carry out MESF on protein data sets by S. Xiang. It is administered by Glasgow University, to which formal application must be made for the executable code. Information about how to obtain it should first be obtained from C. Gilmore (Chemistry Department, Glasgow University, Glasgow, Scotland, UK; chris@chem.gla.ac.uk) or the authors of this chapter (Biochemistry and Biophysics Department, University of North Carolina, Chapel Hill, NC 27599-7260; carter@med.unc.edu).

[7] Model Phases: Probabilities and Bias

By RANDY J. READ

Introduction

The intensities of X-ray diffraction spots measured from a crystal give us only the amplitudes of the diffracted waves. To reconstruct a map of the electron density in the crystal, the unmeasured phase information is also required. This is usually referred to as the phase problem of crystallography. In fact, the unknown phases are actually much more important to the appearance of the electron-density map than the measured amplitudes. When the phases are supplied by an atomic model, therefore, some degree of model bias is inevitable.

The optimal use of model phase information requires an estimate of its reliability, specifically the probability that various values of the phase angle are true. Such a probability distribution can be derived starting first with the relationship between the structure factor (amplitude and phase) of the model and that of the true crystal structure. The phase probability distribution can then be obtained from this and used, for instance, to provide a figure-of-merit weighting that minimizes the root mean square (rms) error from the true electron density.

Even with figure-of-merit weighting, model phased electron density is biased toward the model. The systematic bias component of model phased map coefficients can be predicted, allowing the derivation of map coefficients that give electron-density maps with reduced model bias. With the help of a few simple assumptions, a correction for bias can also be made when different sources of phase information are combined. The computation of map coefficients to reduce model bias is performed by the program SIGMAA, which also computes model phase probabilities and carries out phase combination.

Finally, the refinement of a model against the observed amplitudes allows a certain amount of overfitting of the data, which leads to an extra "refinement bias." Fortunately, the use of appropriate refinement strategies can reduce the severity of this problem.

Model Bias: Importance of Phase

Dramatic illustrations of the importance of the phase have been published. For instance, Ramachandran and Srinivasan[1] calculated an electron-

[1] G. N. Ramachandran and R. Srinivasan, *Nature* (*London*) **190,** 159 (1961).

0076-6879/97 $25.00

density map using phases from one structure and amplitudes from another. In this map there are peaks at the positions of the atoms in the structure that contributed the phase information, but not in the structure that contributed the amplitudes. Similar calculations with two-dimensional Fourier transforms of photographs (inspired by a similar example by Oppenheim and Lim[2]) show that the phases of one completely overwhelm the amplitudes of the other (Fig. 1).

These examples, although dramatic, are not completely representative of the normal situation, in which the structure contributing the phases is partially or even nearly correct. Nonetheless, model phases always contribute bias, so that the resulting map tends to bear too close a resemblance to the model.

Parseval's Theorem

The importance of the phase can be understood most easily in terms of Parseval's theorem, a result that is important to the understanding of many aspects of the Fourier transform and its use in crystallography. Parseval's theorem states that the mean square value of the variable on one side of a Fourier transform is proportional to the mean square value of the variable on the other side. Because the Fourier transform is additive, Parseval's theorem also applies to sums or differences.

If ρ_1 and ρ_2 are, for instance, the true electron density and the electron density of the model, Parseval's theorem tells us that the rms error in the electron density is proportional to the rms error in the structure factor, considered as a vector in the complex plane (Fig. 2).

$$\langle \rho^2 \rangle = \frac{1}{V^2} \sum_{\text{all } \mathbf{h}} |\mathbf{F}(\mathbf{h})|^2$$

$$\langle (\rho_1 - \rho_2)^2 \rangle = \frac{1}{V^2} \sum_{\text{all } \mathbf{h}} |\mathbf{F}_1(\mathbf{h}) - \mathbf{F}_2(\mathbf{h})|^2$$

This understanding of error in electron-density maps explains why the phase is much more important than the amplitude in determining the appearance of an electron-density map. As illustrated in Fig. 3, a random choice of phase (from a uniform distribution of all possible phases) will generally give a larger error in the complex plane than a random choice of amplitude (from a Wilson[3] distribution of amplitudes).

[2] A. V. Oppenheim and J. S. Lim, *Proc. IEEE* **69,** 529 (1981).
[3] A. J. C. Wilson, *Acta Crystallogr.* **2,** 318 (1949).

FIG. 1. Photographs of two key figures in phase determination—J. Karle (top left) and H. Hauptman (top right)—are used to illustrate the importance of the phase. In the image at bottom left, the amplitudes from the Fourier transform of the photograph of Karle have been combined with the phases from the photograph of Hauptman. The complementary combination is shown at bottom right. (The original photograph of H. Hauptman is courtesy of W. L. Duax, and J. Karle kindly supplied a photograph of himself.)

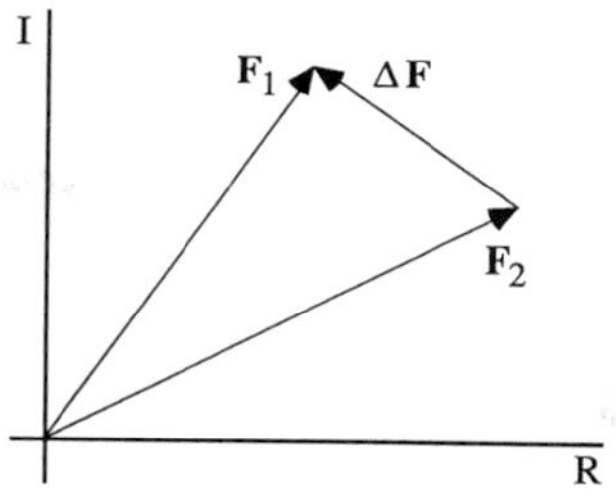

FIG. 2. Vector difference in complex plane.

Structure-Factor Probability Relationships

To use model phase information optimally, one needs to know the probability distribution for the true phase (or, equivalently, the distribution of the error in the model phase). Such a distribution can be derived first by working out the probability distribution for the true structure factor (or the distribution of the vector difference between the model and true structure factors). Then, by fixing the known value of the structure-factor amplitude and renormalizing, the phase probability distribution is obtained.

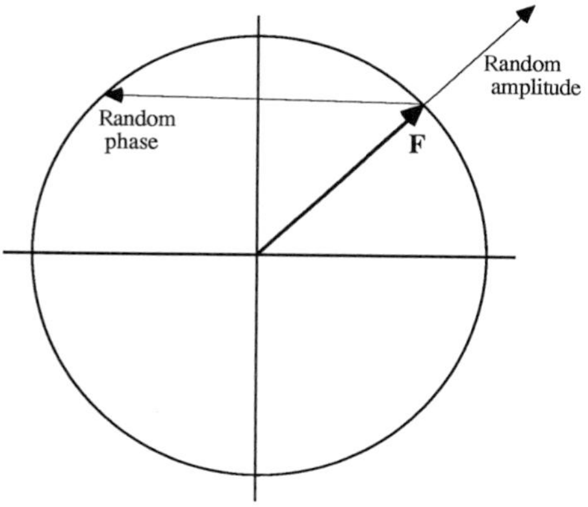

FIG. 3. Schematic illustrations of the relative errors introduced by a random choice of phase or a random choice of amplitude. The example has been constructed to represent the rms errors introduced by randomization (computed by averages over the Wilson distribution). Phase randomization will introduce rms errors of $2^{1/2}$ ($\approx$1.41) times the rms structure factor amplitude $|\mathbf{F}|$. By comparison, map coefficients weighted by figures of merit of zero would have rms errors equal to the rms $|\mathbf{F}|$, so a featureless map would be more accurate than a random phase map. Amplitude randomization will introduce rms errors of $[(4 - \pi)/2]^{1/2}$ ($\approx$ 0.66) times the rms $|\mathbf{F}|$, so a map computed with random amplitudes will be closer to the true map than a featureless map.

We begin with a brief discussion of the central limit theorem, because it will be invoked to justify the use of Gaussian distributions for structure factors. A number of related Gaussian structure-factor distributions have been derived, differing in the amount of information available about the structure and in the assumed form of errors in the model. These range from the Wilson distribution, which applies when none of the atomic positions is known, through to a distribution that applies when there are a variety of sources of error in an atomic model.

An implicit assumption in this approach is that all structure factors at different points in reciprocal space (with different Miller indices *hkl*) are independent of one another, even though we know from the success of direct methods and density modification that they are not. While there is certainly much valuable information to be obtained by considering higher order collections of structure factors,[4] the phase probabilities for single *hkl*'s turn out to be very useful in practice, and are also inexpensive to compute.

Central Limit Theorem

When a number of sources of error are added together, it is often found that the overall error follows a Gaussian distribution, regardless of the individual distributions of the various sources of error. The central limit theorem lays out the very general conditions under which the Gaussian distribution provides a good approximation. When these conditions are satisfied, it is possible to sidestep the laborious manipulations required to work out the probability distribution of a random variable.

For the central limit theorem to apply, two important conditions must be satisfied. First, there must be a sufficient number of independent random variables in the sum. Second, none of the variables may dominate the errors. If these conditions are satisfied, regardless (within some very general limits) of the form of the probability distributions of the individual random variables, the probability distribution of the sum will closely approach a Gaussian distribution. Two parameters, the centroid (mean) and the variance, are required to specify a Gaussian distribution. The centroid of the distribution of the sum is the sum of the centroids of the individual distributions, and the variance is the sum of the individual variances.

$$S = \sum_j x_j$$

[4] G. Bricogne, *Acta Crystallogr.* **D49,** 37 (1993).

where the x_j values are independent random variables with centroids (expected values) $\langle x_j \rangle$ and variances $\sigma^2(x_j) = \langle (x_j - \langle x_j \rangle)^2 \rangle$,

$$p(S) \approx \frac{1}{[2\pi\sigma^2(S)]^{1/2}} \exp\left(-\frac{(S - \langle S \rangle)^2}{2\sigma^2(S)}\right)$$

where $\langle S \rangle = \Sigma_j \langle x_j \rangle$, and $\sigma^2(S) = \Sigma_j\, \sigma^2(x_j)$.

Wilson and Sim Structure-Factor Distributions in P1

For the Wilson distribution,[3] it is assumed that the atoms in a crystal structure in space group $P1$ are scattered randomly and independently through the unit cell. In fact, it is sufficient to make the much less restrictive assumption that the atoms are placed randomly with respect to the Bragg planes defined by the Miller indices. The assumption of independence is somewhat more problematic because there are restrictions on the distances between atoms, large volumes of protein crystals are occupied by disordered solvent, and many protein crystals display noncrystallographic symmetry; as discussed in [3] in this volume,[5] the resulting relationships among structure factors are exploited implicitly in averaging and solvent-flattening procedures. The higher order relationships among structure factors are used explicitly in direct methods for solving small-molecule structures and are being developed for use in protein structures.[4] For the purposes of simpler relationships between the calculated and true structure factors for a single *hkl*, however, the lack of complete independence does not seem to create serious problems.

When atoms are placed randomly relative to the Bragg planes, the contribution of each atom to the structure factor will have a phase varying randomly from 0 to 2π. The overall structure factor can then be considered to be the result of a random walk in the complex plane (Fig. 4). In terms of the central limit theorem, the structure factor is the sum of the independent atomic scattering contributions, each of which has a probability distribution defined as a circle in the complex plane centered on the origin, with a radius of f_j. The centroid of this atomic distribution is at the origin, and the variance for each of the real and imaginary parts is $\frac{1}{2}f_j^2$. The probability distribution of the structure factor that is the sum of these contributions is thus a two-dimensional Gaussian, the product of the one-dimensional Gaussians for the real and imaginary parts. It is referred to as a Wilson distribution.[3] Because the variances are equal in the real and imaginary directions, it can be simplified as shown below and expressed in terms of a single distribution parameter Σ_N.

[5] F. M. D. Vellieux and R. J. Read, *Methods Enzymol.* **277,** [3], 1997 (this volume).

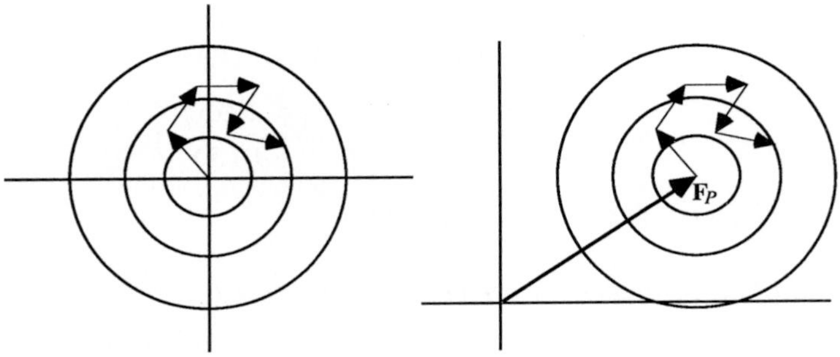

FIG. 4. Schematic illustrations of the Wilson (left) and Sim (right) structure-factor probability distributions for space group $P1$, as they arise from random walks in the complex plane.

$$\mathbf{F} = \sum_{j=1}^{N} f_j \exp(2\pi i \mathbf{h} \cdot \mathbf{x}_j) = A + iB$$

$$\langle A \rangle = \langle B \rangle = 0 \qquad \sigma^2(A) = \sigma^2(B) = \frac{1}{2}\sum_{j=1}^{N} f_j^2 = \frac{1}{2}\Sigma_N$$

so

$$p(A) = \frac{1}{(\pi\Sigma_N)^{1/2}} \exp\left(-\frac{A^2}{\Sigma_N}\right) \quad \text{and} \quad p(B) = \frac{1}{(\pi\Sigma_N)^{1/2}} \exp\left(-\frac{B^2}{\Sigma_N}\right)$$

$$p(\mathbf{F}) = p(A, B) = \frac{1}{\pi\Sigma_N} \exp\left(-\frac{|\mathbf{F}|^2}{\Sigma_N}\right)$$

The Sim distribution,[6] which is relevant when the positions of some of the atoms are known, has a very similar basis, except that the structure factor is now considered to arise from a random walk starting from the position of the structure factor corresponding to the known part $\mathbf{F}_P$ (Fig. 4). Atoms with known positions do not contribute to the variance, while the atoms with unknown positions (the "Q" atoms) each contribute $\frac{1}{2}f_j^2$ to each of the real and imaginary parts, as in the Wilson distribution. The distribution parameter in this case is referred to as Σ_Q. The Sim distribution is a conditional probability distribution, depending on the value of $\mathbf{F}_P$.

$$p(\mathbf{F}; \mathbf{F}_P) = \frac{1}{\pi\Sigma_Q} \exp\left(-\frac{|\mathbf{F} - \mathbf{F}_P|^2}{\Sigma_Q}\right)$$

[6] G. A. Sim, *Acta Crystallogr.* **12,** 813 (1959).

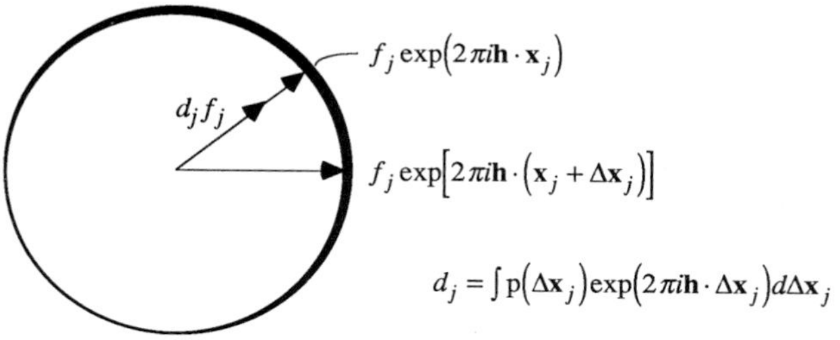

FIG. 5. Centroid of the structure-factor contribution from a single atom. The probability of a phase for the contribution is indicated by the thickness of the line.

The Wilson[3] and Woolfson[7] distributions for space group $P\bar{1}$ are obtained similarly, except that the random walks are along a line, and the resultant Gaussian distributions are one dimensional. (The Woolfson distribution is the centric equivalent to the Sim distribution.) For more complicated space groups, it is a reasonable approximation to treat acentric reflections as following the $P1$ distribution, and centric reflections as following the $P\bar{1}$ distribution. However, for any zone of the reciprocal lattice in which symmetry-related atoms are constrained to scatter in phase, the variances must be multiplied by the expected intensity factor ε for the zone, because the symmetry-related contributions are no longer independent.

Probability Distributions for Variable Coordinate Errors

In the Sim distribution, an atom is considered to be either exactly known or completely unknown in its position. These are extreme cases, because there will normally be varying degrees of uncertainty in the positions of various atoms in a model. The treatment can be generalized by allowing a probability distribution of coordinate errors for each atom. In this case, the centroid for the individual atomic contribution to the structure factor will no longer be obtained by multiplying by either zero or one. Averaged over the circle corresponding to possible phase errors, the centroid will generally be reduced in magnitude, as illustrated in Fig. 5. In fact, averaging to obtain the centroid is equivalent to weighting the atomic scattering contribution by the Fourier transform of the coordinate error probability distribution, d_j. By the convolution theorem, this in turn is equivalent to convoluting the atomic density with the coordinate error distribution. Intuitively, the atom is smeared over all of its possible positions. The weighting

[7] M. M. Woolfson, *Acta Crystallogr.* **9,** 804 (1956).

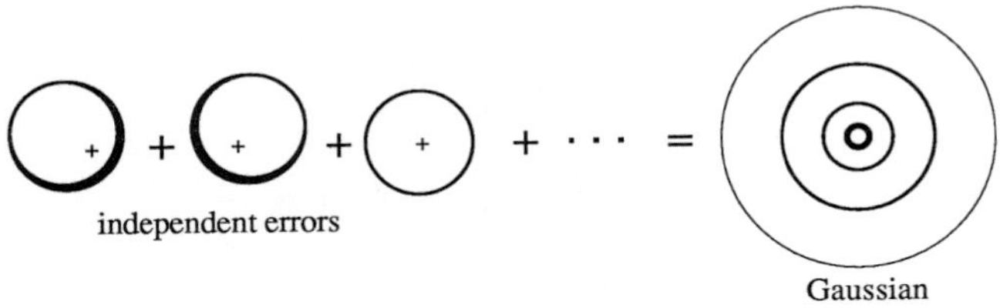

FIG. 6. Schematic illustration of the combination of errors in the calculated structure factor. Each small cross indicates the position of the centroid of an atomic scattering contribution.

factor d_j is thus analogous to the thermal motion term in the structure-factor expression.

The variances for the individual atomic contributions will vary in magnitude but, if there are a sufficient number of independent sources of error, we can invoke the central limit theorem again and assume that the probability distribution for the structure factor will be a Gaussian centered on $\Sigma d_j f_j \exp(2\pi i \mathbf{h} \cdot \mathbf{x}_j)$ (Fig. 6). If the coordinate error distribution is Gaussian, and if each atom in the model is subject to the same errors, the resulting structure-factor probability distribution is the Luzzati[8] distribution. In this special case, $d_j = D$ for all atoms, and D is the Fourier transform of a Gaussian, which behaves like the application of an overall B factor.

General Treatment of the Structure-Factor Distribution

The Wilson, Sim, Luzzati, and variable error distributions have very similar forms, because they are all Gaussians arising from the application of the central limit theorem. The central limit theorem is valid under many circumstances, and even when there are errors in position, scattering factor, and B factor, as well as missing atoms, a similar distribution still applies. As long as these sources of error are independent, the true structure factor will have a Gaussian distribution centered on $D\mathbf{F}_C$ (Fig. 7), where D now includes effects of all sources of error, as well as compensating for errors in the overall scale and B factor.[9]

$$p(\mathbf{F}; \mathbf{F}_C) = \frac{1}{\pi \varepsilon \sigma_\Delta^2} \exp\left(-\frac{|\mathbf{F} - D\mathbf{F}_C|^2}{\varepsilon \sigma_\Delta^2}\right)$$

in the acentric case, where $\sigma_\Delta^2 = \Sigma_N - D^2 \Sigma_P$, ε is the expected intensity factor, and Σ_P is the Wilson distribution parameter for the model.

[8] V. Luzzati, *Acta Crystallogr.* **5,** 802 (1952).
[9] R. J. Read, *Acta Crystallogr.* **A46,** 900 (1990).

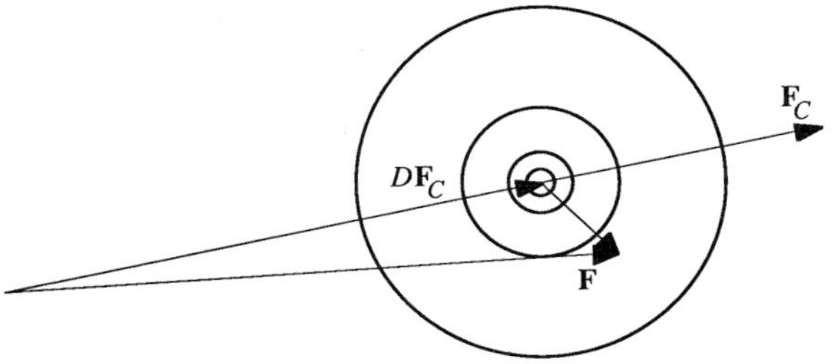

FIG. 7. Schematic illustration of the general structure-factor distribution relevant in the case of any set of independent random errors in the atomic model.

For centric reflections, the scattering differences are distributed along a line, so the probability distribution is a one-dimensional Gaussian.

$$p(\mathbf{F}; \mathbf{F}_C) = \frac{1}{(2\pi\varepsilon\sigma_\Delta^2)^{1/2}} \exp\left(-\frac{|\mathbf{F} - D\mathbf{F}_C|^2}{2\varepsilon\sigma_\Delta^2}\right)$$

Estimating σ_A

Srinivasan showed that the Sim and Luzzati distributions could be combined into a single distribution that had a particularly elegant form when expressed in terms of normalized structure factors, or E values.[10] This functional form still applies to the general distribution reflecting a variety of sources of error; the only difference is the interpretation placed on the parameters.[9] If **F** and $\mathbf{F}_C$ are replaced by the corresponding E values, a parameter σ_A plays the role of D, and σ_Δ^2 reduces to $(1 - \sigma_A^2)$. [The parameter σ_A is D corrected for model completeness; $\sigma_A = D(\Sigma_P/\Sigma_N)^{1/2}$]. By normalizing the structure factors one also eliminates overall scale and B-factor effects. The parameter σ_A that characterizes this probability distribution varies as a function of resolution. It must be deduced from the amplitudes $|\mathbf{F}_O|$ and $|\mathbf{F}_C|$, because the phase (thus the phase difference) is unknown.

A general approach to estimating parameters for probability distributions is to maximize a likelihood function. The likelihood function is the overall joint probability of making the entire set of observations, as a function of the desired parameters. The parameters that maximize the probability of making the set of observations are the most consistent with the data. The idea of using maximum likelihood to estimate model phase errors was introduced by Lunin and Urzhumtsev,[11] who gave a treatment

[10] R. Srinivasan, *Acta Crystallogr.* **20,** 143 (1966).

[11] V. Yu. Lunin and A. G. Urzhumtsev, *Acta Crystallogr.* **A40,** 269 (1984).

valid for space group $P1$. In a more general treatment that applies to higher symmetry space groups, allowance is made for the statistical effects of crystal symmetry (centric zones and differing expected intensity factors).[12]

The σ_A values are estimated by maximizing the joint probability of making the set of observations of $|\mathbf{F}_O|$. If the structure factors are all assumed to be independent, the joint probability distribution is the product of all the individual distributions. As discussed above, the assumption of independence is not completely justified in theory, but the results are fairly accurate in practice.

$$L = \prod_{\mathbf{h}} p(|\mathbf{F}_O|; |\mathbf{F}_C|)$$

The required probability distribution $p(|\mathbf{F}_O|; |\mathbf{F}_C|)$ is derived from $p(\mathbf{F}; \mathbf{F}_C)$ by integrating over all possible phase differences and neglecting the errors in $|\mathbf{F}_O|$ as a measure of $|\mathbf{F}|$. The form of this distribution, which is given in other publications,[9,12] differs for centric and acentric reflections. (It is important to note that, although the distributions for structure factors are Gaussian, the distributions for amplitudes obtained by integrating out the phase are not.) It is more convenient to deal with a sum than a product, so the log likelihood function is maximized instead. In the program SIGMAA, reciprocal space is divided into spherical shells, and a value of the parameter σ_A is refined for each resolution shell. Details of the algorithm are described elsewhere.[12]

A simpler, but somewhat less accurate, method for estimating σ_A is based on work by Hauptman,[13] who defined a functionally equivalent probability distribution. The equivalent parameter in the Hauptman distribution is estimated as the square root of the standard linear correlation coefficient between $|\mathbf{E}_O|^2$ and $|\mathbf{E}_C|^2$. In fact, this method is used to provide the initial estimate of σ_A for each resolution shell in the program SIGMAA. For both methods, the resolution shells must be thick enough to contain several hundred to a thousand reflections each to provide σ_A estimates with a sufficiently small statistical error. A larger number of shells (fewer reflections per shell) can be used for refined structures, because estimates of σ_A become more precise as the true value approaches one. If there are sufficient reflections per shell, the estimates will vary smoothly with resolution. As discussed below, the smooth variation with resolution can also be exploited through a restraint that allows σ_A values to be estimated from fewer reflections.

[12] R. J. Read, *Acta Crystallogr.* **A42,** 140 (1986).
[13] H. Hauptman, *Acta Crystallogr.* **A38,** 289 (1982).

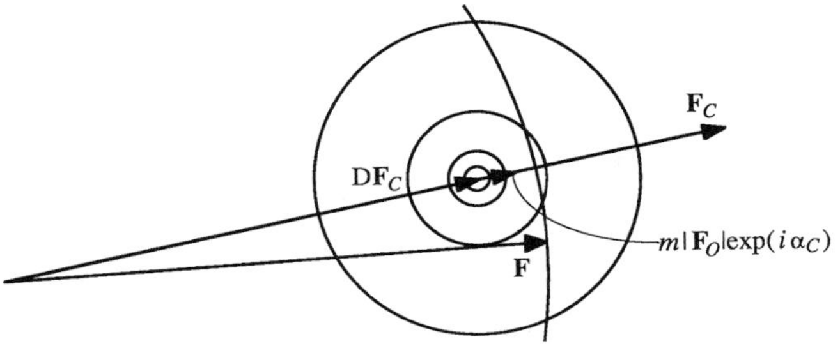

FIG. 8. Figure-of-merit weighted model phased structure factor, obtained as the probability-weighted average over all possible phases.

Figure-of-Merit Weighting for Model Phases

Blow and Crick[14] and Sim[6] showed that the electron-density map with the least rms error is calculated from centroid structure factors. This conclusion follows from Parseval's theorem, because the centroid structure factor (its probability-weighted average value, or expected value) minimizes the rms error of the structure factor. Because the structure factor distribution $p(\mathbf{F};\ \mathbf{F}_C)$ is symmetrical about $\mathbf{F}_C$, the expected value of $\mathbf{F}$ will have the same phase as $\mathbf{F}_C$, but the averaging around the phase circle will reduce its magnitude if there is any uncertainty in the phase value (Fig. 8). We treat the reduction in magnitude by applying a weighting factor called the figure of merit, m, which is equivalent to the expected value of the cosine of the phase error.

Map Coefficients to Reduce Model Bias

Model Bias in Figure-of-Merit Weighted Maps

A figure-of-merit weighted map, calculated with coefficients $m|\mathbf{F}_O|\exp(i\alpha_C)$, has the least rms error from the true map. According to the normal statistical (minimum variance) criteria, then, it is the best map. However, such a map will suffer from model bias; if its purpose is to allow us to detect and repair errors in the model, that is a serious qualitative defect. Fortunately, it is possible to predict the systematic errors leading to model bias, and to make some correction for them.

Main[15] dealt with this problem in the case of a perfect partial structure. Since the relationships among structure factors are the same in the general

[14] D. M. Blow and F. H. C. Crick, *Acta Crystallogr.* **12,** 794 (1959).
[15] P. Main, *Acta Crystallogr.* **A35,** 779 (1979).

a

b

c

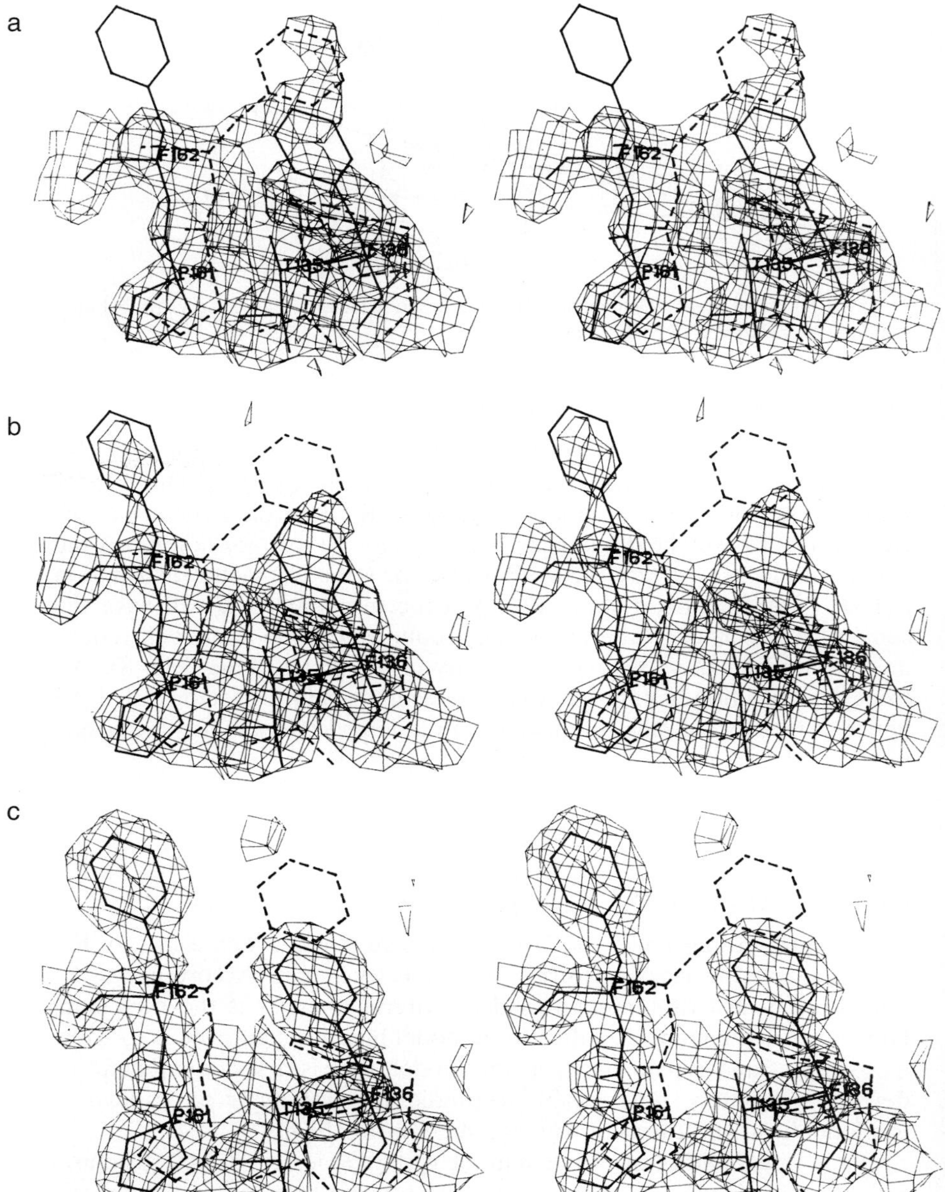

case of a partial structure with various errors, once $D\mathbf{F}_C$ is substituted for $\mathbf{F}_C$, all that is required to apply the results of Main more generally is a change of variables.[9,12]

In the approach taken by Main, the cosine law is used to introduce the cosine of the phase error, which is converted into a figure of merit by taking expected values. Some manipulations allow us to solve for the figure-of-merit weighted map coefficient, which is approximated as a linear combination of the true structure factor and the model structure factor.[12,15] Finally, we can solve for an approximation to the true structure factor, giving map coefficients from which the systematic model bias component has been removed.

$$m|\mathbf{F}_O| \exp(i\alpha_C) = \frac{\mathbf{F}}{2} + \frac{D\mathbf{F}_C}{2} + \text{noise terms}$$

$$\mathbf{F} \approx (2m|\mathbf{F}_O| - D|\mathbf{F}_C|) \exp(i\alpha_C)$$

A similar analysis for centric structure factors shows that there is no systematic model bias in figure-of-merit weighted map coefficients, so no bias correction is needed in the centric case.

Sample density maps in Fig. 9 (reproduced from an earlier publication[12]) illustrate the qualitative effect on interpretability.

Model Bias in Combined Phase Maps

When model phase information is combined with, for instance, multiple isomorphous replacement (MIR) phase information, there will still be model bias in the acentric map coefficients, to the extent that the model influences the final phases. However, it is not appropriate to continue to use the same map coefficients to reduce model bias, because some phases could be almost completely determined by the MIR phase information. It makes much more sense to have map coefficients that reduce to the coefficients appropriate for either model or MIR phases, in the extreme cases where there is only one source of phase information, and that vary smoothly between those extremes.

FIG. 9. Electron-density maps illustrating the reduction in bias from the use of SIGMAA map coefficients. The correct structure is depicted with solid lines, while the phasing model is shown with dashed lines. Each map is contoured at 1.25 times the rms electron density. (a) Figure-of-merit weighted map calculated with coefficients $m|\mathbf{F}_O| \exp(i\alpha_C)$. (b) SIGMAA map calculated with acentric coefficients $(2m|\mathbf{F}_O| - D|\mathbf{F}_C|) \exp(i\alpha_C)$ and centric coefficients $m|\mathbf{F}_O| \exp(i\alpha_C)$. (c) True map.

Map coefficients that satisfy these criteria (even if they are not rigorously derived) are implemented in the program SIGMAA. The resulting maps are reasonably successful in reducing model bias. Two assumptions are made: (1) the model bias component in the figure-of-merit weighted map coefficient, $m_{\mathrm{com}}|\mathbf{F}_{\mathrm{O}}| \exp(i\alpha_{\mathrm{com}})$, is proportional to the influence that the model phase has had on the combined phase; and (2) the relative influence of a source of phase information can be measured by the information content H[16] of the phase probability distribution. The first assumption corresponds to saying that the figure-of-merit weighted map coefficient is a linear combination of the MIR and model phase cases.

$$\text{MIR:} \qquad m_{\mathrm{MIR}}|\mathbf{F}_{\mathrm{O}}| \exp(i\alpha_{\mathrm{MIR}}) \approx \mathbf{F}$$

$$\text{Model:} \qquad m_{\mathrm{C}}|\mathbf{F}_{\mathrm{O}}| \exp(i\alpha_{\mathrm{C}}) \approx \frac{\mathbf{F}}{2} + \frac{D\mathbf{F}_{\mathrm{C}}}{2}$$

$$\text{Combined:} \quad m_{\mathrm{com}}|\mathbf{F}_{\mathrm{O}}| \exp(i\alpha_{\mathrm{com}}) \approx \left(1 - \frac{w}{2}\right)\mathbf{F} + \frac{w}{2} D\mathbf{F}_{\mathrm{C}}$$

$$\text{where } w = \frac{H_{\mathrm{C}}}{H_{\mathrm{C}} + H_{\mathrm{MIR}}}$$

$$\text{and } H = \int_0^{2\pi} p(\alpha) \ln \frac{p(\alpha)}{p_0(\alpha)}\, d\alpha; \quad p_0(\alpha) = \frac{1}{2\pi}$$

Solving for an approximation to the true **F** gives the following expression, which can be seen to reduce appropriately when w is 0 (no model influence) or 1 (no MIR influence).

$$\mathbf{F} \approx \frac{2m|\mathbf{F}_{\mathrm{O}}| \exp(i\alpha_{\mathrm{com}}) - wD\mathbf{F}_{\mathrm{C}}}{2 - w}$$

Estimation of Overall Coordinate Error

In principle, because the distribution of observed and calculated amplitudes is determined largely by the coordinate errors of the model, one can determine whether a particular coordinate error distribution is consistent with the amplitudes. Unfortunately, it turns out that the coordinate errors cannot be deduced unambiguously, because many distributions of coordinate errors are consistent with a particular distribution of amplitudes.[9]

If the simplifying assumption is made that all the atoms are subject to a single error distribution, then the parameter D (and thus the related

[16] S. Guiasu, "Information Theory with Applications." McGraw-Hill, London, 1977.

parameter σ_A) varies with resolution as the Fourier transform of the error distribution, as discussed above. Two related methods to estimate overall coordinate error are based on the even more specific assumption that the coordinate error distribution is Gaussian: the Luzzati plot[8] and the σ_A plot.[12] Unfortunately, the central assumption is not justified; atoms that scatter more strongly (heavier atoms or atoms with lower B factors) tend to have smaller coordinate errors than weakly scattering atoms. The proportion of the structure factor contributed by well-ordered atoms increases at high resolution, so that the structure factors agree better at high resolution than they would if there were a single error distribution.

It is often stated, optimistically, that the Luzzati plot provides an upper bound to the coordinate error, because the observation errors in $|\mathbf{F}_O|$ have been ignored. This is misleading, because there are other effects that cause the Luzzati and σ_A plots to give underestimates.[9] Chief among these are the correlation of errors and scattering power, and the overfitting of the amplitudes in structure refinement (discussed below). These estimates of overall coordinate error should not be interpreted too literally; at best, they provide a comparative measure.

Features of Program SIGMAA

The computer program SIGMAA[12] has been developed to implement the results described here. Model phase probabilities can optionally be combined with phase information from other sources such as MIR, using Hendrickson–Lattman coefficients,[17] and four types of map coefficient can be produced. Apart from the two types of coefficient discussed above, there are also two types of difference map coefficient:

Model phased difference map: $(m|\mathbf{F}_O| - D|\mathbf{F}_C|)\exp(i\alpha_C)$

General difference map: $m_{com}|\mathbf{F}_O|\exp(i\alpha_{com}) - D\mathbf{F}_C$

The general difference map, it should be noted, uses a vector difference between the figure-of-merit weighted combined phase coefficient (the "best" estimate of the true structure factor) and the calculated structure factor. When additional phase information is available, it should provide a clearer picture of the errors in the model.

The program SIGMAA also produces a σ_A plot to estimate overall coordinate error although, as discussed above, the estimate should not be interpreted too literally.

Several versions of SIGMAA are available. It has been implemented within the CCP4 [SERC (UK) Collaborative Computer Project 4, Daresbury Laboratory, UK, 1979] and Biomol (Laboratory of Chemical

[17] W. A. Hendrickson and E. E. Lattman, *Acta Crystallogr.* **B26,** 136 (1970).

Physics, University of Groningen) program systems, and the algorithms have been implemented in a release of X-PLOR[18] that should be currently available. Finally, a standalone version is available from the author or by anonymous ftp from the site *mycroft.mmid.ualberta.ca,* in the directory *pub/sigmaa.*

Refinement Bias

The structure-factor probabilities discussed above depend on the atoms having independent errors (or at least a sufficient number of groups of atoms having independent errors). Unfortunately, this assumption breaks down when a structure is refined against the observed diffraction data. Few protein crystals diffract to sufficiently high resolution to provide a large number of observations for every refinable parameter. The refinement problem is, therefore, not sufficiently overdetermined, so it is possible to overfit the data. If there is an error in the model that is outside the range of convergence of the refinement method, it is possible to introduce compensating errors in the rest of the structure to give a better, and misleading, agreement in the amplitudes. As a result, the phase accuracy (hence the weighting factors m and D) are overestimated, and model bias is poorly removed. Because simulated annealing is a more effective minimizer than gradient methods,[18] it is also more effective at locating local minima, so structures refined by simulated annealing probably tend to suffer more severely from refinement bias.

There is another interpretation to the problem of refinement bias. As Silva and Rossmann[19] point out, minimizing the rms difference between the amplitudes $|\mathbf{F}_O|$ and $|\mathbf{F}_C|$ is equivalent (by Parseval's theorem) to minimizing the difference between the model electron density and the density corresponding to the map coefficients $|\mathbf{F}_O| \exp(i\alpha_C)$; a lower residual is obtained either by making the model look more like the true structure, or by making the model phased map look more like the model through the introduction of systematic phase errors.

A number of strategies are available to reduce the degree or impact of refinement bias. The overestimation of phase accuracy has been overcome in a new version of SIGMAA that is under development (R. J. Read, unpublished, 1997). Cross-validation data, which are normally used to compute R_{free} as an unbiased indicator of refinement progress,[20] are used to obtain unbiased σ_A estimates. Because of the high statistical error of σ_A

[18] A. T. Brünger, J. Kuriyan, and M. Karplus, *Science* **235,** 458 (1987).
[19] A. M. Silva and M. G. Rossmann, *Acta Crystallogr.* **B41,** 147 (1985).
[20] A. T. Brünger, *Nature* (*London*) **355,** 472 (1992).

estimates computed from small numbers of reflections, reliable values can be obtained only by exploiting the smoothness of the σ_A curve as a function of resolution. This can be achieved either by fitting a functional form or by adding a penalty to points that deviate from the line connecting their neighbors. Lunin and Skovoroda[21] have independently proposed the use of cross-validation data for this purpose but, as their algorithm is equivalent to the conventional SIGMAA algorithm, it will suffer severely from statistical error.

The degree of refinement bias can be reduced by placing less weight on the agreement of structure-factor amplitudes. Anecdotal evidence suggests that the problem is less serious, in structures refined using X-PLOR,[18] when the Engh and Huber[22] parameter set is used for the energy terms. In this new parameter set the deviations from standard geometry are much more strictly restrained, so in effect the pressure on the agreement of structure-factor amplitudes is reduced.

If errors are suspected in certain parts of the structure, "omit refinement," in which the questionable parts are omitted from the model, can be a very effective way to eliminate refinement bias in those regions.[23,24]

If MIR or multiple-wavelength anomalous dispersion (MAD) phases are available, combined phase maps tend to suffer less from refinement bias, depending on the extent to which the experimental phases influence the combined phases. Finally, it is always a good idea to refer occasionally to the original MIR or MAD map, which cannot suffer at all from model bias or refinement bias.

Maximum Likelihood Structure Refinement

Conventional structure refinement is based on a least-squares target, which would be justified if the observed and calculated structure factor amplitudes were related by a Gaussian probability distribution. Unfortunately, the relationship between $|\mathbf{F}_O|$ and $|\mathbf{F}_C|$ is not Gaussian, and the distribution for $|\mathbf{F}_O|$ is not even centered on $|\mathbf{F}_C|$. Because of this, the author[9] and Bricogne[25] have suggested that a maximum likelihood target should be used instead, and that it should be based on probability distributions such as those described above. One of the expected advantages of maximum likelihood refinement is a decrease in refinement bias, as the

[21] V. Yu. Lunin and T. P. Skovoroda, *Acta Crystallogr.* **A51,** 880 (1995).

[22] R. A. Engh and R. Huber, *Acta Crystallogr.* **A47,** 392 (1991).

[23] M. N. G. James, A. R. Sielecki, G. D. Brayer, L. T. J. Delbaere, and C.-A. Bauer, *J. Mol. Biol.* **144,** 43 (1980).

[24] A. Hodel, S.-H. Kim, and A. T. Brünger, *Acta Crystallogr.* **A48,** 851 (1992).

[25] G. Bricogne, *Acta Crystallogr.* **A47,** 803 (1991).

calculated structure-factor amplitudes will not be forced to be equal to the observed amplitudes.

We have devised appropriate likelihood targets for structure refinement by incorporating the effect of measurement error.[26] A key ingredient in the implementation is the use of cross-validation data to estimate σ_A values, as discussed above. Our tests of the new refinement method show that, compared to least-squares refinement, maximum likelihood can be more than twice as effective in improving an atomic model.[26]

Acknowledgments

Bart Hazes, Allan Sharp, and Penny Stein helped greatly by pointing out where the presentation could be clarified. Bart Hazes also helped to develop the program used in the calculations for Fig. 1. This work was supported by the Medical Research Council of Canada, a scholarship from the Alberta Heritage Foundation for Medical Research and, in part, by an International Research Scholar award from the Howard Hughes Medical Institute. This chapter is a revised and expanded version of a contribution to the 1994 CCP4 proceedings.[27]

[26] N. S. Pannu and R. J. Read, *Acta Crystallogr.* **A52,** 659 (1996).

[27] R. J. Read, *in* "From First Map to Final Model: Proceedings of the CCP4 Study Weekend, 6–7 January 1994" (S. Bailey, R. Hubbard, and D. Waller, eds.), p. 31. Engineering and Physical Sciences Research Council, Daresbury, UK, 1994.

Section II

Models

A. Model Building
Articles 8 through 12

B. Refinement
Articles 13 through 18

C. Verification: Safe Crystallography
Articles 19 and 20

[8] Critical-Point Analysis in Protein Electron-Density Map Interpretation

By SUZANNE FORTIER, ANTONY CHIVERTON, JANICE GLASGOW, and LAURENCE LEHERTE

Introduction

Need for Automated Approaches to Map Interpretation

The interpretation of electron-density maps from protein crystals has been facilitated greatly by the advent of powerful graphics stations coupled with the implementation of computer programs, notably FRODO[1] and O,[2] designed specifically for the visual analysis of these maps. Most map interpretation tasks still rely, however, on significant human intervention and require considerable time investments. It would be useful therefore to have available automated approaches that could assist the user in model building and, in particular, in the detailed interpretation of maps at several levels of resolution and refinement.

While automated approaches can be helpful, irrespective of the method used for structure determination, our own work on the design and implementation of computer approaches for the automated interpretation of protein maps has been motivated by an effort to adapt direct methods of structure determination to macromolecules. Traditional direct methods explore the phase space and evaluate phasing paths by using only general chemical constraints—nonnegativity of the electron-density distribution and atomicity—and constraints imposed by the amplitude data. While these constraints have proven sufficiently limiting for application to small molecules, they are not sufficient for protein structures, which have a much broader search space containing many local minima. To date, successful applications of direct method approaches to proteins all have required the introduction of additional constraints, particularly in the form of partial structure information. Direct method procedures intended for macromolecular applications are thus usually updated to include some form of map interpretation and density modification algorithms in the phasing cycles.

The major breakthroughs made by Hauptman's group[3] in the successful

[1] T. A. Jones, *J. Appl. Crystallogr.* **11,** 268 (1978).
[2] T. A. Jones, J. Y. Zou, S. W. Cowan, and M. Kjeldgaard, *Acta Crystallogr.* **A47,** 110 (1991).
[3] R. Miller, G. T. DeTitta, R. Jones, D. A. Langs, C. M. Weeks, and H. A. Hauptman, *Science* **259,** 1430 (1993).

0076-6879/97 $25.00

ab initio determination of several small proteins, using their computer program *Shake-and-Bake,*[4] now have demonstrated convincingly the applicability of the direct methods approach to protein structures. However, a major hurdle remains: All of the successful *ab initio* applications to date have required high-resolution data (1.2 Å or higher), a resolution level that is seldom available. Previous calculations[5] and more recent applications[6] indicate, however, that the barrier is not intrinsic to the direct method tools. The problem may arise in the identification and integration of partial structure information when the diffraction data are at less than atomic resolution. These results clearly point to the need for map interpretation and density modification approaches that can be used at several levels of resolution. Furthermore, because direct method procedures usually involve the iterative exploration of several hundred possible phasing sets, it becomes essential that these approaches be fully automated and highly efficient.

Map Interpretation as an Exercise in Molecular Scene Analysis

Interpreting an electron-density map is a complex scene analysis[7] exercise. To gain an understanding of the molecular scene represented in the map, one needs to locate the features of interest and furthermore to recognize what these features are. In other words, one needs to segment the map into its meaningful parts, which then must be classified and identified. For example, at low resolution, i.e., >5 Å, one may want only to segment the map into two regions and to classify these regions as protein or solvent. At medium resolution, i.e., around 3 Å, one may want to attend to the protein region, segmenting it into main and side chains and classifying main-chain segments in terms of secondary structure motifs. Segmentation of the main chain into individual residues, followed by their classification into residue classes, and finally their identification can then proceed as resolution increases.

Although map interpretation can be divided into two tasks: segmentation and pattern recognition. One must realize that often these tasks are

[4] G. T. DeTitta, C. M. Weeks, P. Thuman, R. Miller, and H. A. Hauptman, *Acta Crystallogr.* **A50,** 203 (1994); C. M. Weeks, G. T. DeTitta, H. A. Hauptman, P. Thuman, and R. Miller, *Acta Crystallogr.* **A50,** 210 (1994); R. Miller, S. M. Gallo, H. G. Khalak, and C. M. Weeks, *J. Appl. Crystallogr.* **27,** 613 (1994).

[5] C. M. Weeks, personal communication (1994); S. Fortier, C. M. Weeks, and H. Hauptman, *Acta Crystallogr.* **A40,** 646 (1984); H. Hauptman, S. Potter, and C. M. Weeks, *Acta Crystallogr.* **A38,** 294 (1982).

[6] B. Sha, S. Liu, Y. Gu, H. Fan, H. Ke, J. Yao, and M. M. Woolfson, *Acta Crystallogr.* **D51,** 342 (1995).

[7] M. A. Fischler and O. Firschein, "Intelligence: The Eye, the Brain, and the Computer." Addison-Wesley, Reading, Massachusetts, 1987.

not independent of one anohter or applied sequentially. For example, analysis of the spatial relationships among various segmented parts may be essential to their identification. Similarly, partial identification can help guide further segmentation.

In Search of Representation for Automated Interpretation of Protein Maps

One of the most important questions to address in designing an automated approach to map interpretation is how to represent the electron-density distribution. This is equivalent to asking which properties of the electron-density distribution help us most in performing the tasks of segmentation and pattern recognition, and how these properties can be made explicit. Both the segmentation and pattern recognition tasks draw primarily from an analysis of shape properties and spatial relationships in the electron-density map. Indeed, it is from the local and global topology of the map that structural features such as protein and solvent regions, main chains, helices, individual residues, etc., can be located, segmented, and identified. While shape and spatial relationship information is contained implicitly in the three-dimensional grid representation of electron-density maps, this is not a form that is suitable for efficient computation. Alternative representations that synthesize and make explicit selected shape and spatial properties of the electron-density distribution thus have been investigated and used. For example, the skeletonization approach introduced by Hildith,[8] and adapted to protein crystallography by Greer,[9] has been used extensively to capture connectivity in protein maps. Our own work[10] has focused on critical-point mapping, which has been used as a way of representing the topological properties of the electron-density distribution, $\rho(\mathbf{r})$. This approach is well known among the theoretical chemistry and accurate electron-density distribution research communities, where it is applied to high-resolution theoretical and experimental maps of small molecules.[11,12] It has not been used extensively yet in mainstream crystallography despite the fact that its potential usefulness for lower resolution maps of proteins was

[8] C. J. Hildith, *Machine Intell.* **4,** 403 (1969).

[9] J. Greer, *J. Mol. Biol.* **82,** 279 (1974).

[10] L. Leherte, S. Fortier, J. Glasgow, and F. H. Allen, *Acta Crystallogr.* **D50,** 155 (1994).

[11] R. W. Bader, "Atoms in Molecules—a Quantum Theory." Oxford Clarendon Press, Oxford, 1990.

[12] S. T. Howard, M. B. Hursthouse, C. W. Lehman, P. R. Mallinson, and C. S. Frampton, *J. Chem. Phys.* **97,** 5616 (1992).

recognized early on by Johnson,[13,14] who implemented it in the computer program ORCRIT.[15] The core tracing method, which also draws from the topological properties of electron-density maps, has been proposed by Swanson.[16]

It is important to appreciate that a good representation model should not only capture the features of interest but also discard unnecessary and distracting details. The critical-point mapping approach, for instance, summarizes the shape and spatial properties of the electron-density distribution through its critical points while discarding unnecessary detailed background information. A good test for a representation model is to check it against the criteria established in the context of machine vision by Marr and Nishihara[17] for general shape representation:

accessibility: The representation should be derivable from the initial image at a reasonable computing cost

scope and uniqueness: The representation should provide a unique description of all possible shapes in the domain of interest

stability and sensitivity: The representation should capture both the more general (less variant) and the subtle (of finer distinction) properties of shape

Theoretical Background and Terminology

The procedure we have adopted for the critical-point analysis of protein maps consists of three main steps: (1) location of critical points, (2) connection and identification of the critical points, and (3) recognition of secondary structure motifs in critical-point segments.

Locating Critical Points

The critical points of the three-dimensional electron-density distribution, $\rho(\mathbf{r})$, are its points of inflection. They are found at locations for which the first derivatives of $\rho(\mathbf{r})$ vanish and are characterized as maxima, minima, or saddle points by the second derivatives of $\rho(\mathbf{r})$. Critical-point analysis thus requires the computation of the gradients (∇), Hessians ($\mathbf{H}$), and

[13] C. K. Johnson, "Proceedings of the American Crystallography Association Meeting," Evanston, Illinois. Abstract B1. 1976.

[14] C. K. Johnson, "Proceedings of the American Crystallography Association Meeting," Asilomar, California. Abstract JQ6. 1977.

[15] C. K. Johnson, "ORCRIT." The Oak Ridge Critical Point Network Program, Chemistry Division, Oak Ridge National Laboratory, Oak Ridge, Tennessee, 1977.

[16] S. M. Swanson, *Acta Crystallogr.* **D50,** 695 (1994).

[17] D. Marr and H. K. Nishihara, *Proc. R. Soc. London Ser. B.* **200,** 269 (1978).

Laplacians (∇^2) of the electron-density function.[11,18] The Hessian matrix of a continuous three-dimensional function such as the electron density, $\rho(\mathbf{r})$, is built from its second derivatives:

$$\mathbf{H}(\mathbf{r}) = \begin{vmatrix} \partial^2\rho/\partial x^2 & \partial^2\rho/\partial x\partial y & \partial^2\rho/\partial x\partial z \\ \partial^2\rho/\partial y\partial x & \partial^2\rho/\partial y^2 & \partial^2\rho/\partial y\partial z \\ \partial^2\rho/\partial z\partial x & \partial^2\rho/\partial z\partial y & \partial^2\rho/\partial z^2 \end{vmatrix} \quad (1)$$

For each critical point, this real and symmetric matrix can be put in a diagonal form, which corresponds to finding a rotation of the original coordinate system that aligns the new coordinate system with the principal axes of the critical point, resulting in:

$$\mathbf{H}'(\mathbf{r}) = \begin{vmatrix} \partial^2\rho/\partial x'^2 & 0 & 0 \\ 0 & \partial^2\rho/\partial y'^2 & 0 \\ 0 & 0 & \partial^2\rho/\partial z'^2 \end{vmatrix} \quad (2)$$

The three diagonal elements of $\mathbf{H}'(\mathbf{r})$ are the eigenvalues of $\mathbf{H}$. They correspond to the three main curvatures of the electron-density function at the critical point $\mathbf{r}$ while the eigenvectors, resulting from the diagonalization of $\mathbf{H}(\mathbf{r})$, correspond to the three principal axes of curvature. The Laplacian is the trace of $\mathbf{H}'(\mathbf{r})$, that is, the sum of the diagonal elements of the Hessian matrix. It gives details about the local concentration (sign <0) or depletion (sign >0) of electron density. The rank of $\mathbf{H}'$ is the number of nonzero eigenvalues, and the signature, s, is the algebraic sum of their signs. When the rank of $\mathbf{H}'$ is 3, as is normally the case for critical points in three-dimensional electron-density distributions, then four possible cases arise:

$s = -3$ corresponds to a local maximum or *peak*, i.e., the electron-density function adopts maximum values along each of the three principal directions x', y', and z'

$s = -1$ corresponds to a saddle point or *pass* (with peaks on both sides) and there are only two negative eigenvalues

$s = +3$ corresponds to a local minimum or *pit*, i.e., the electron-density function adopts minimum values along each of the three principal directions

$s = +1$ corresponds to a saddle point or *pale* (with pits on both sides) characterized by only one negative eigenvalue

[18] V. H. Smith, Jr., P. F. Price, and I. Absar, *Isr. J. Chem.* **16,** 187 (1977).

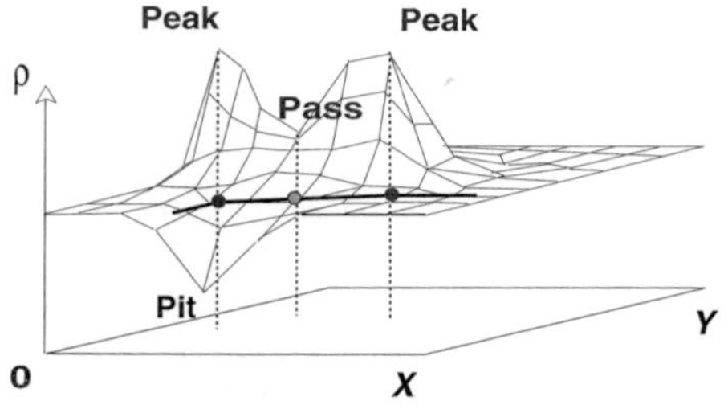

FIG. 1. Critical points in an electron-density distribution.

Peaks, pits, and saddle points are depicted in Fig. 1. Thus, from computing the gradient, Hessian, and Laplacian of the electron-density distribution, one obtains information not only on the location and types of critical points in the function, but also on their local environment. In particular, retaining information on their heights, curvatures, and principal directions of curvature allows a reconstruction of local electron-density distributions and, in turn, of the global distribution. This is important in macromolecular direct methods procedures, where the purpose of map interpretation is to achieve density modification, and in turn phase refinement, through the Fourier transform of direct-space information into reciprocal space.

Connecting and Identifying Critical Points

While rich information is contained in individual critical points, it is primarily their interrelationships that help locate and identify structural features in the map. An important step, therefore, is the construction of a graph that usefully connects the set of critical points. A few terms from graph theory, which are used below, are introduced at this point.

A *graph* is a representation of a set of points and their connectivity. In particular, a *connected graph* consists of a set of points (vertices) in which any pair may be connected by a path formed by one or several edges in the graph. A connected graph containing no *circuit* or *cycle* is called a *tree,* and a graph whose components are trees is called a *forest.* A subgraph of a connected graph G, that is, a tree containing all the vertices of G, is called a *spanning tree.* A spanning tree thus can be obtained from a connecting graph G by removing edges from G in such a way that all circuits are eliminated while all vertices remain connected. In general, a graph may have several possible spanning trees. A *minimum-weight spanning tree* is obtained when graph edges are assigned weights and the edge-removal algorithm minimizes the sum of the edge weights. For example, in ORCRIT minimum spanning trees are obtained through the use of weights derived from attributes of adjacent critical points, such as distances, density heights, types of critical points, and eigenvector projections.

Recognizing Secondary Structure Motifs in Critical-Point Segments

Once a critical-point graph is constructed, it can be analyzed so as to classify substructures in the protein. In particular, a geometric analysis of segments of connected critical points can be used for the recognition of secondary structure motifs. In general, torsion angles among four sequential peak critical points and distances between the peaks at the extremities of the four-member segment are the most useful parameters for the identification of β sheets and α helices. Length of a secondary structure segment and existence of parallel segments with similar geometry also are useful criteria for motif recognition. Specific rules or templates to be used in recognition procedures are usually resolution dependent, because the critical points catch different atomic moieties at different resolution levels.

Because the geometric parameters may have value ranges that match more than one type of secondary structure and because different parameters may yield conflicting results for a given segment, a method must be used to establish a *confidence level* in the results obtained. One possible approach is through the calculation of *certainty factors,*[19] as implemented in the diagnosis system MYCIN.

Applications to Maps at 3-Å Resolution

To develop and test a methodology that could be used at medium resolution, initial testing was made on calculated (error-free) maps of the following proteins: phospholipase A_2 (1BP2),[20] ribonuclease T1 complex (1RNT),[21] and trypsin inhibitor (4PTI).[22] Structure factors were calculated for these three test proteins (Table I), using protein and solvent atomic coordinates and thermal parameters as stored in the Protein Data Bank (PDB).[23] From these calculated structure factors, 3-Å resolution electron-density maps were generated. The approach also has been tested on calculated and experimental maps of penicillopepsin[24] and crambin[25] (Table I).

[19] E. Rich and K. Knight, "Artificial Intelligence," 2nd Ed. McGraw-Hill, New York, 1991.

[20] B. W. Dujkstra, K. H. Kalk, W. G. J. Hol, and J. Drenth, *J. Mol. Biol.* **147,** 97 (1981).

[21] R. Arni, U. Heinemann, M. Maslowska, R. Tokuoka, and W. Saenger, *Acta Crystallogr.* **B43,** 549 (1987).

[22] M. Marquart, J. Walter, J. Deisenhofer, W. Bode, and R. Huber, *Acta Crystallogr.* **B39,** 480 (1983).

[23] F. C. Bernstein, T. F. Koetzle, G. J. B. Williams, E. F .Meyer, M. D. Brice, J. R. Rodgers, O. Kennard, T. Shimanouchi, and M. Tasumi, *J. Mol. Biol.* **112,** 535 (1977).

[24] M. N. G. James and A. R. Sielecki, *J. Mol. Biol.* **163,** 299 (1983).

[25] M. M. Teeter, *Proc. Natl. Acad. Sci. U.S.A.* **81,** 6014 (1984).

TABLE I
CHEMICAL AND STRUCTURAL PARAMETERS OF TEST STRUCTURES

Source and parameter	Value
Phospholipase A_2 (bovine pancreas)	
Code	1BP2
Resolution (Å)	1.7
R factor	0.170
Cell parameters (Å and degrees)	47.070, 64.450, 38.158, 90, 90, 90
Space group	$P2_12_12_1$
Solvent/heteroatoms	$109H_2O$, $2C_6H_{14}O_2$, Ca^{2+}
Number of amino acids	123
Percentage α	54
Percentage β	8
Number of turns	10
Number of disulfide bonds	7
Ribonuclease T1 complex (*Aspergillus oryzae*)	
Code	1RNT
Resolution (Å)	1.9
R factor	0.180
Cell parameters (Å and degrees)	46.810, 50.110, 40.440, 90, 90, 90
Space group	$P2_12_12_1$
Solvent/heteroatoms	$91H_2O$, $C_{10}H_{14}N_5O_8P$
Number of amino acids	104
Percentage α	16
Percentage β	35
Number of turns	10
Number of disulfide bonds	2
Trypsin inhibitor (bovine pancreas)	
Code	4PTI
Resolution (Å)	1.5
R factor	0.162
Cell parameters (Å and degrees)	43.100, 22.900, 48.600, 90, 90, 90

The experimental map of penicillopepsin was generated using all reflections in the resolution range of 10 to 3 Å. Observed structure-factor amplitudes and multiple isomorphous replacement phases with their figures of merit were used in the calculation of the map. The phases were derived from eight heavy-atom derivatives and had an overall figure of merit of 0.9 for all observed data to 2.8-Å resolution. The experimental maps of crambin were calculated using all reflections to 3-Å resolution. The observed amplitudes were combined with phases obtained from the application of *Shake-and-Bake* to a high-resolution data set. As mentioned above, it is not possible yet to apply *Shake-and-Bake* to low-resolution data. Therefore,

TABLE I (*continued*)

Source and parameter	Value
Space group	$P2_12_12_1$
Solvent/heteroatoms	$60H_2O$
Number of amino acids	58
Percentage α	17
Percentage β	33
Number of turns	3
Number of disulfide bonds	3
Penicillopepsin (fungus)	
Code	3APP
Resolution (Å)	1.8
R factor	0.136
Cell parameters (Å and degrees)	97.370, 46.640, 65.470, 90, 115.4, 90
Space group	*C*2
Solvent/heteroatoms	$320H_2O$
Number of amino acids	323
Percentage α	15
Percentage β	52
Number of turns	2
Number of disulfide bonds	1
Crambin (*Crambe abyssinica*)	
Code	1CRN
Resolution (Å)	1.5
R factor	0.114
Cell parameters (Å and degrees)	40.960, 18.650, 22.520, 90, 90.77, 90
Space group	$P2_1$
Solvent/heteroatoms	
Number of amino acids	46
Percentage α	47
Percentage β	17
Number of turns	1
Number of disulfide bonds	3

we used phase solutions obtained from a high-resolution application of the program, which then were truncated to 3 Å. Critical-point analyses of both the calculated and experimental maps were carried out using the program ORCRIT. Following are some observations derived from these studies.

Locating Critical Points

Electron-density distributions of proteins are rich in inflection points. While many of the critical points truly represent structural features in the distribution, others may not; that is, they may be the result of Fourier series termination effects or noise in experimental maps. Without the use of some selection criterion, collecting and analyzing all of the critical points easily could become a formidable task. One possible criterion is the electron-density height. We have found that a $\rho_{\min}$ value corresponding to one-third of the maximum value in the map was a useful cutoff value. With this cutoff value, however, most of the pale and pit critical points are eliminated from the analysis. It has also been found that in experimental maps, high-density critical points often are surrounded by a constellation of lower density points. Pruning of the critical points thus may be required before proceeding to the graph generation and analysis stages.

Locating critical points should correspond ideally to locating important chemical features within the protein. For convenience, protein backbones are commonly represented by a C_α tracing and comparison of models to structures is often done using C_α atoms. It is important to note, however, that at medium resolution, C_α atoms do not correspond to "important" chemical features, that is, they do not correspond to peaks in a 3-Å electron-density function. Indeed, as shown in Table II, the peak critical points in the test proteins are located closer to the center of mass of $C_\alpha CON$ groups than to the C_α atoms themselves. Critical points also are found in the vicinity

TABLE II
AVERAGE DISTANCE AND ROOT-MEAN-SQUARE DEVIATION OF PEAK CRITICAL POINTS[a]

Protein	Model	Distance (Å)	RMS (Å)
1BP2	C_α	1.29	0.37
	$C_\alpha CON_\omega$	0.52	0.41
	$C_\alpha CO_\omega$	0.59	0.33
1RNT	C_α	1.38	0.39
	$C_\alpha CON_\omega$	0.49	0.39
	$C_\alpha CO_\omega$	0.55	0.36
4PTI	C_α	1.27	0.48
	$C_\alpha CON_\omega$	0.66	0.45
	$C_\alpha CO_\omega$	0.68	0.35
3APP			
Calculated map	C_α	1.05	0.48
	$C_\alpha CON_\omega$	0.41	0.24
	$C_\alpha CO_\omega$	0.50	0.28
Experimental map	C_α	1.34	0.48
	$C_\alpha CON_\omega$	1.25	0.49
	$C_\alpha CO_\omega$	1.23	0.47

[a] From various atomic group centers of mass and from C_α atoms.

of side-chain atoms or atomic groups. Their occurrence and location depend on the size and thermal vibration of the side chains as well as on bonding effects. For example, some disulfide bridges may be represented by two peak critical points, associated with each sulfur atom with a pass in between, while others will have a single peak critical point located near the center of the bond, as shown in Fig. 2. Finally, critical points associated with solvent molecules also are found in the maps.

As mentioned earlier, sufficient information on the critical points is retained to reconstruct their local electron-density functions, and from these the global electron-density distribution, as demonstrated in a study by Leherte and Allen on DNA–drug systems.[26] Using the density, eigenvalues, and eigenvectors associated with the critical points, each maximum of the electron-density function, i.e., each peak, can be considered as a center of expansion of a Gaussian function, which then can be fitted to define a volume around each peak. This allows for an ellipsoidal and therefore anisotropic representation of local electron-density distributions, which may be important in direct method procedures where map interpretation is used principally for the purpose of density modification and phase refinement. At 3-Å resolution, the use of an isotropic electron-density distribution around the peaks of the function would be expected to lead to significant errors.

Connecting and Identifying Critical Points

From work on calculated maps at around 2-Å resolution, Johnson[15] previously had observed that the minimal spanning trees constructed from the collection of critical points showed . . . peak–pass–peak–pass–peak . . . sequences of adjacent points, which traced out ridge lines following the polymeric protein molecule. Similarly, . . . pit–pale–pit–pale–pit . . . sequences should be found to trace out valley lines separating adjacent molecules, although these lines may not be so clearly discernible in practice because the pits and pales normally have much smaller signal-to-noise ratios. In our applications to calculated maps at 3 Å, because a cutoff $\rho_{\min}$ value was used, only peak and pass critical points were considered in the construction of the minimum weight spanning trees. We also observed that adjacent critical points tended to outline . . . peak–pass–peak–pass–peak . . . motifs. Furthermore, each test structure had a principal spanning tree with a long . . . peak–pass–peak–pass–peak . . . "branch," with smaller branches jutting out of it, corresponding to the protein main chain and side chains, respectively. This pattern of critical points is depicted in Fig. 3. For

[26] L. Leherte and F. H. Allen, *J. Comput. Aided Mol. Design* **8,** 257 (1994).

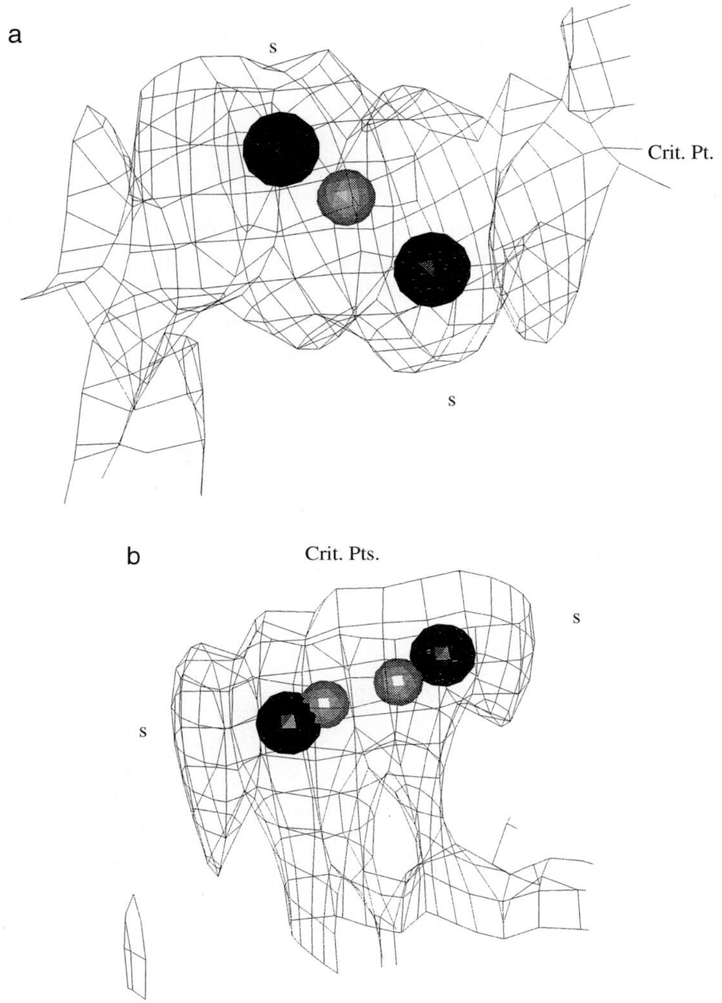

FIG. 2. Electron-density distributions of disulfide bridges depicting (a) the presence of a single peak in between the two sulfur atoms and (b) the presence of two peaks, each being associated with a sulfur atom. Dark and light spheres are used to represent the sulfur atoms and peaks, respectively.

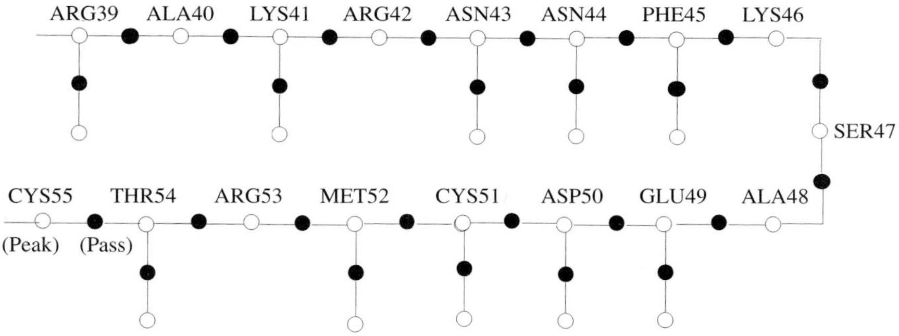

FIG. 3. Planar schematization of a typical critical-point motif in a 3-Å resolution protein map.

the calculated maps of IBP2, 1RNT, and 4PT1, we found that the main branch of the longest spanning tree captures the whole amino acid sequence. Of particular importance is the observation that successive peaks in the main branch of the longest spanning tree correspond to successive residues in the sequence, and that, in general, each residue is associated with one and only one peak in the main branch of the tree. Thus at 3 Å, the electron-density distribution of the protein along its main chain can be represented as a sequence of *superatoms,* which correspond to the residues. These results are summarized in Table III. It also was found that the critical points capture accurately the conformation of the protein, with the proviso that (as mentioned before) they are located close to the center of mass of

TABLE III
LOCATION AND CONNECTIVITY OF CRITICAL POINTS FOR CALCULATED ELECTRON-DENSITY MAPS OF 1BP2, 1RNT, AND 4PTI[a]

Parameter	1BP2	1RNT	4PTI
Number of critical points	532	365	196
Total number of spanning trees	25	19	10
Total number of peaks in main branch of longest spanning tree	134	109	60
Number of residues represented in main chain of longest spanning tree	123	104	58
Number of residues in main branch of longest spanning tree represented by one and only one peak	114	99	56

[a] At 3-Å resolution.

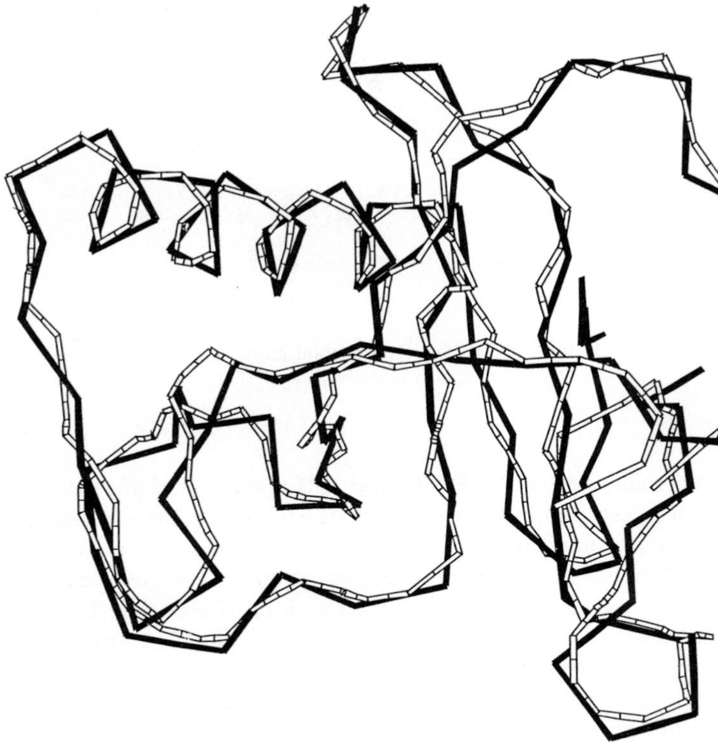

FIG. 4. Perspective view depicting the superposition of the C_α (dark) chain and the peaks of the main branch of the longest spanning tree (light) for the protein 1RNT.

C_αCON rather than to the C_α atoms. Figure 4 shows a superposition of a tracing through the C_α atoms and a tracing through the peaks in the main branch of the principal tree for 1RNT.

Insufficient testing has been done to date to draw general conclusions on the critical-point representation of the side chains of the residues. A preliminary analysis of the 1BP2, 1RNT, and 4PT1 test structures shows that a broad distinction can be made between large and small residues. That is, the occurrence of a small branch jutting out of the main branch of a tree usually is associated with the presence of a large residue.

In the application to the penicillopepsin experimental map, we found that the critical points captured 316 or 98% of the residues, as depicted in Fig. 5. Table IV shows that the trees also were found to trace significant portions of the protein main chain with 278 or 86% of the residues repre-

```
Seq. 1-50     A-A-S-G-V-A-T-N-T-P-  T-A-N-D-E-E-Y-I-T-P-  V-T-I-G-G-T-T-L-N-L-  N-F-D-T-G-S-A-D-L-W-  V-F-S-T-E-L-P-A-S-Q-
Theor.        Tree01......................01  * Tree01.......01 Tree01.............................................................
Exper.        ***** Tree01.........01 * ***** * * *** *   Tree02......................02 *** Tree02.........02 Tree02....

Seq. 51-100   Q-S-G-H-S-V-Y-N-P-S-  A-T-G-K-E-L-S-G-Y-T-  W-S-I-S-Y-G-D-G-S-S-  A-S-G-N-V-F-T-D-S-V-  T-V-G-G-V-T-A-H-G-Q-
Theor.        .....................................................................tree01 Tree01.........................
Exper.        ...02 Tree02...............................02 * ******* Tree02......  ..................*..............

Seq. 101-150  A-V-Q-A-A-Q-Q-I-S-A-  Q-F-Q-Q-D-T-N-N-D-G-  L-L-G-L-A-F-S-S-I-N-  T-V-Q-P-Q-S-Q-T-T-F-  F-D-T-V-K-S-S-L-A-Q-
Theor.        .............01 Tre01................01 Tree01..................01 Tree01..............................
Exper.        .02 Tree.02 Tree02***** * * *** Tree02......................02   Tree02.........................02 Tree03

Seq. 151-200  P-L-F-A-V-A-L-K-H-Q-  Q-P-G-V-Y-D-F-G-F-I-  D-S-S-K-Y-T-G-S-L-T-  Y-T-G-V-D-N-S-Q-G-F-  W-S-F-N-V-D-S-Y-T-A-
Theor.        .......................................................................................01 Tre01 Tree01..
Exper.        .................................................................03 Tree02.....................

Seq. 201-250  G-S-Q-S-G-D-G-F-S-G-  I-A-D-T-G-T-T-L-L-L-  L-D-D-S-V-V-S-Q-Y-Y-  S-Q-V-S-G-A-Q-Q-D-S-  N-A-G-G-Y-V-F-D-C-S-
Theor.        ..........................01 Tree01...............................01 Tree01..........................
Exper.        ..............................02 Tree04...............04 T04   Tree05..........05   Tree...06 Tree.07

Seq. 251-300  T-N-L-P-D-F-S-V-S-I-  S-G-Y-T-A-T-V-P-G-S-  L-I-N-Y-G-P-S-G-D-G-  S-T-C-L-G-G-I-Q-S-N-  S-G-I-G-F-S-I-F-G-D-
Theor.        .............01 Tree01...............................01 Tree01..........................................
Exper.        Tree04...04   Tree02...............02        *******   ***     *   * * Tre07 Tree04.......................

Seq. 301-323  I-F-L-K-S-Q-Y-V-V-F-  D-S-D-G-P-Q-L-G-F-A-  P-Q-A
Theor.        ..................................................01
Exper.         04 Tree02..........................................02
```

FIG. 5. Description of the critical-point networks of the calculated and experimental electron-density maps of penicillopepsin at 3-Å resolution. Principal spanning trees and smaller spanning trees are represented by dotted line and asterisks, respectively. Jumps between nonconnected residues are represented by spaces.

TABLE IV
LOCATION AND CONNECTIVITY OF CRITICAL POINTS FOR CALCULATED AND EXPERIMENTAL ELECTRON-DENSITY MAPS OF PENICILLOPEPSIN[a]

Parameter	3APP Calculated	3APP Experimental
Number of critical points	1576	1544
Total number of spanning trees	37	223
Number of spanning trees considered as principal[b]	1	7
Total number of peaks in main branch of principal spanning tree(s)	353	325
Number of residues represented in main branch of principal spanning tree(s)	321	278
Number of residues in main branch of principal spanning tree(s) represented by one and only one peak	267	176

[a] At 3-Å resolution.
[b] To be considered a principal spanning tree, a sequence of connected critical points must represent at least five residues.

sented in their main branches. However, we found that only 55% of the residues were captured by one and only one peak along the main branches of the trees. This is because, as mentioned above, high-density peaks often are surrounded by a constellation of lower density peaks. Better results are obtained when the set of critical points is subjected to a pruning procedure before one attempts to connect them. In addition, the graph generation algorithm did not yield a single spanning tree capturing the whole protein. Rather, the backbone of the protein mapped onto several trees of varying lengths. Improvement in the connectivity results was achieved by enforcing a peak–pass–peak motif, that is, by enforcing the presence of a pass between two connected peaks. The construction of the tree(s) was carried out from the collection of peak–pass–peak building blocks, rather than the collection of single critical points.

However, problems remained with some of the trees not strictly following the main-chain path. Jumps were observed, in particular, at the disulfide bridge and in regions where two side chains make close contacts. Ideally, one would like to see a single connected graph capturing the whole protein backbone. This may not be found, however, when a simple tree construction algorithm is chosen for the graph representation, because trees do not allow for closed circuits. Therefore, when disulfide bridges or close side-chain contacts are encountered, the tree construction algorithm must choose between following a main-chain or side-chain path, the latter resulting in the main chain being broken into several trees. This is illustrated in Fig.

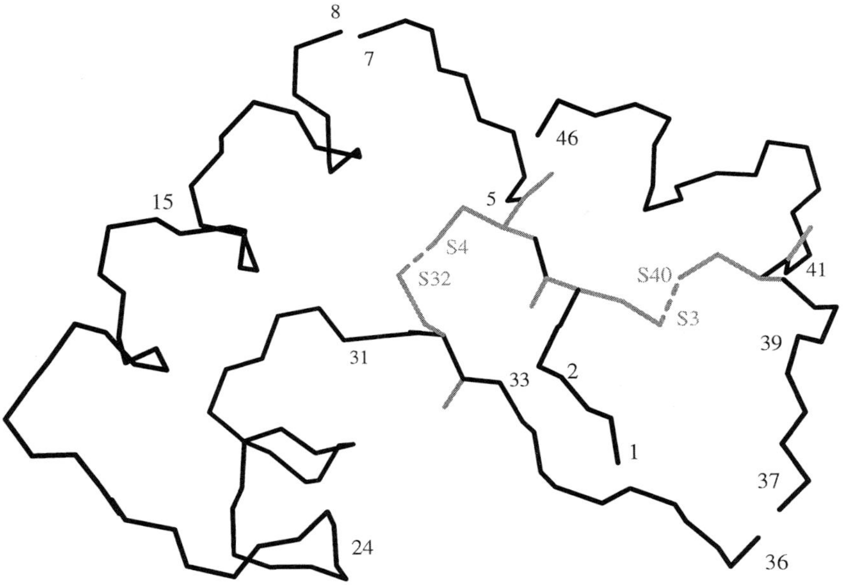

FIG. 6. Description of minimum spanning tree bridge crossing in crambin. The longest tree connects the residues as follows:

```
8-9- ······· -31-32 - 4 - 3-40-41- ····· -45-46
              |     |   | |
     36-35-34-33    5   2 39-38-37
                    |   |
                    6   1
                    |
                    7
```

6, which shows the trees obtained when the procedure is applied to crambin, a protein containing three disulfide bridges.

Several of the mishaps in the tracing of the penicillopepsin experimental map are due to the restrictions of the tree construction algorithm. For example, a jump is found between residues 229 and 254, whose side chains are sufficiently close to allow the building of a path from the five-membered ring of Pro-254 to the phenol ring of Tyr-229. Clearly, a problem with the current tree construction algorithm is that it is based solely on local attributes and parameters. One might anticipate that adding some global parameters, for example, relative density of the critical points and sequence, and increasing the size of the critical-point neighborhoods, should yield improved results.

Identification of Sulfurs and Heteroatoms

It is possible to detect the presence of sulfurs and heteroatoms among the critical points by looking at their density values. The distribution of the critical-point density values for the calculated map of 1BP2 is shown in Fig. 7. Figure 8 shows similar distributions for the calculated and experimental maps of penicillopepsin (3APP). Critical points associated with sulfurs or heteroatoms are found in regions of high-density values.

Recognizing Secondary Structure Motifs in Critical-Point Segments

Because the critical-point representation is able to capture accurately the conformation of protein backbones at 3-Å resolution, a geometric analysis of the critical-point networks can be used for the recognition of secondary structure elements. Shown in Fig. 9 are C_α and peak critical-point tracings through a helical region in the protein crambin. Starting from a graph sequence of alternating passes and peaks, it is found that the most useful parameters are the torsion angles t_{1234} among four consecutive peaks and the distances d_{14} between the peaks at the ends of a four peak segment. The angles a_{123} among three consecutive peak may also be used, but they

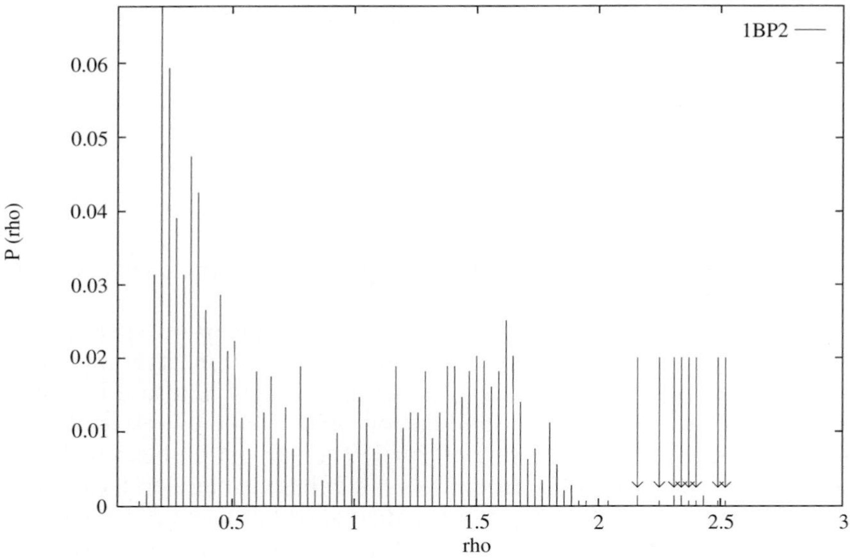

FIG. 7. Distribution of critical-point density values for the calculated map of 1BP2. Arrows point to sulfur atom peaks.

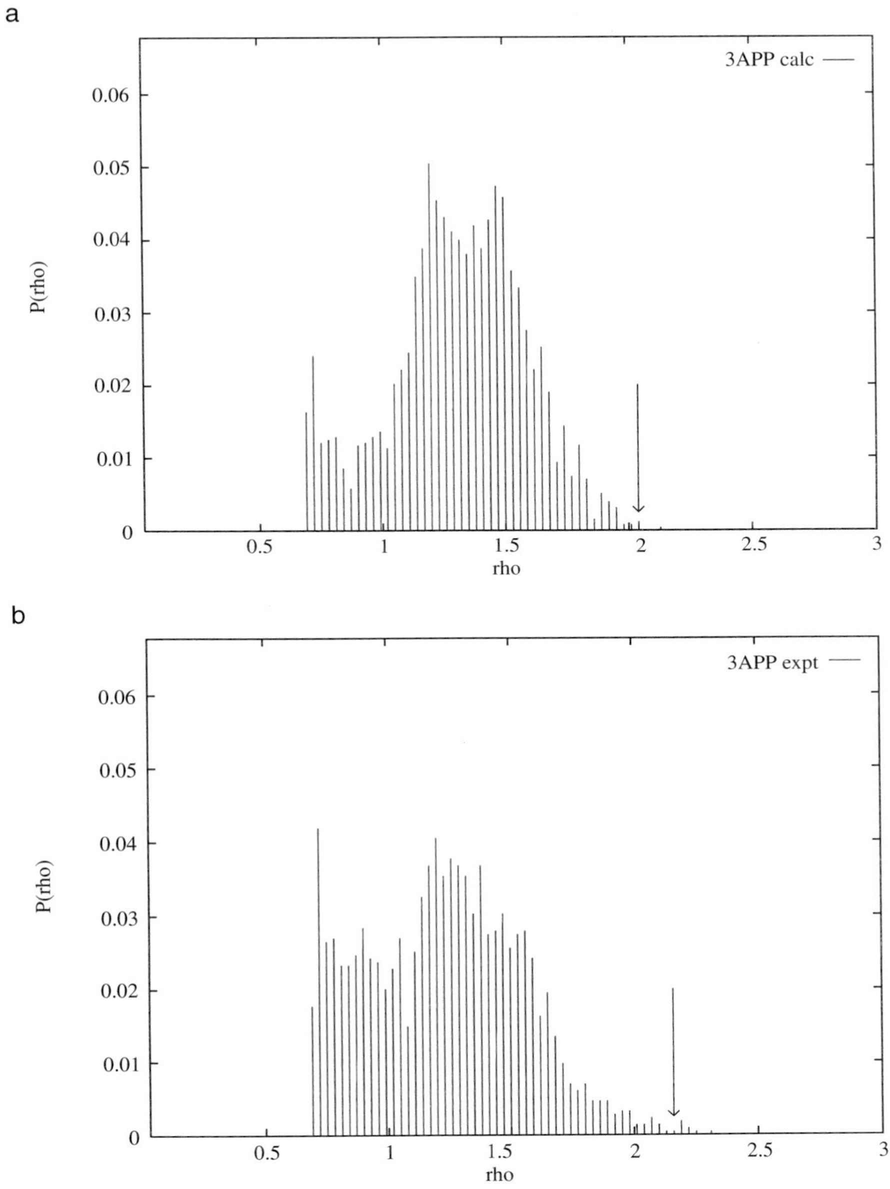

FIG. 8. Distribution of critical-point density values for the (a) calculated and (b) experimental maps of 3APP. Arrows point to sulfur atom peaks.

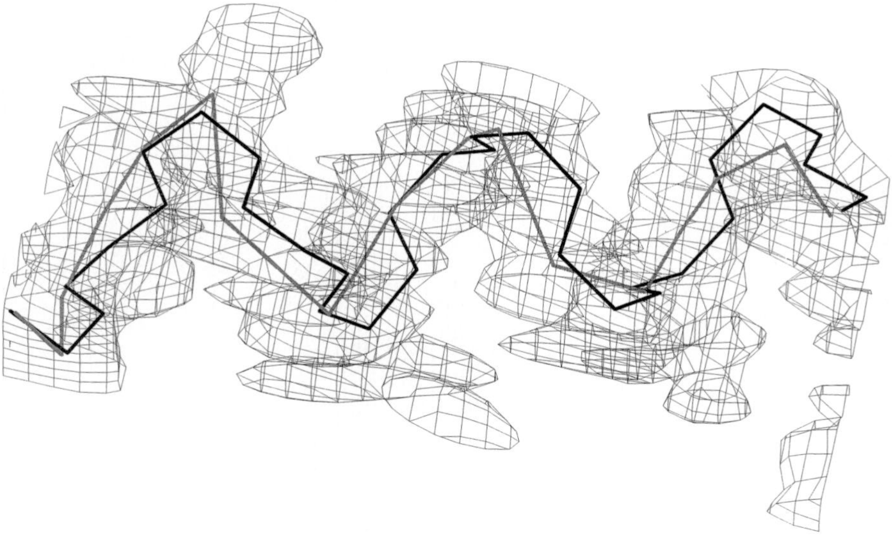

FIG. 9. C_α (dark) and critical-point tracing (light) through a helical region of crambin (1CRN).

do not show as clear a partitioning of their values for different secondary structure motifs.

Because it is found in calculated maps at 3 Å that each peak in the main branch of the trees corresponds to one and only one residue, ideal geometric parameters for α helices and β sheets can be derived using the parameters for prominent linear groups in polypeptide chains reported by Schulz and Schirmer.[27] These are given in Table V. For the purpose of recognition, it is more useful to have available for each parameter of interest frequency distributions, $f(\text{ssm}, g)$, and $f(\text{not\_ssm}, g)$, given a secondary structure motif (ssm) and a geometric constraint g. These were computed, using 63 nonhomologous protein structures retrieved from the PDB, for α helices and β sheets and for geometric constraints based on the torsion angle t_{1234}, distance d_{14}, and angle a_{123} and are presented in Figs. 10 to 12. The frequency distributions then can be used to establish ranges of values for a given geometric parameter for which a given secondary structure motif can be assigned to a peak under consideration. Alternatively, they can be used to calculate measures of belief based on $f(\text{ssm}, g)$ and disbelief

[27] G. E. Schulz and R. H. Schirmer, "Principles of Protein Structure." Springer-Verlag, New York, 1988.

TABLE V
IDEAL GEOMETRIC PARAMETERS FOR α HELICES AND β STRANDS[a]

Characteristic	α helix	β sheet
Parameters		
Number of residues (C_α) per turn	3.6	2.0
Rise between two residues along the helix axis (Å)	1.5	3.3
Helix radius (Å)	2.3	1.0
Results		
Distance between residues i and $i + 1$ (Å)	3.8	3.9
Distance between residues i and $i + 3$ (Å)	5.1	10.1
"Bond" angle (degrees)	90.2	117.6
"Torsion" angle (degrees)	50.1	180.0

[a] At 3-Å resolution.

based on f(not_ssmn, g), in the assignment of a given secondary structure motif to critical points. These in turn are used to calculate certainty factors. With the first approach, a secondary structure hypothesis will be considered to be either true or false, while in the second approach no hypothesis is rejected. Rather it is given a certainty factor, which can range from -100 to $+100$.

Regardless of the approach selected for secondary structure assignment, it has been found useful to do some preprocessing and pruning before proceeding to the assignment steps so as to eliminate and/or merge peaks that are within too short a distance of one another and caused by ripple effects. Table VI presents the results obtained from applying the rule-based recognition procedure to the theoretical and experimental maps of 3APP. As can be seen, more than 80% of the peaks for which secondary structure assignment is sought are classified correctly. Of those that are not, many are peaks located at the extremities of the secondary structures. Incorrect assignments also are found when short jumps between nonconnected residues occur, and in regions of the trees where a given residue is represented by more than one peak along the main chain. Figure 13 maps certainty factors for both α-helix and β-sheet hypotheses onto the sequence, as obtained for a calculated map of 4PTI. As can be seen, the approach correctly identifies helices and sheets in the structure. It should be noted that the high certainty factor associated with a helical motif in a turn region is not unexpected. Both motifs have similar geometry and are differentiated by their length.

Preliminary testing also has been done on experimental maps of crambin calculated using phase information derived from the application of *Shake-and-Bake*. Thirty phase sets truncated to 3-Å resolution were provided to

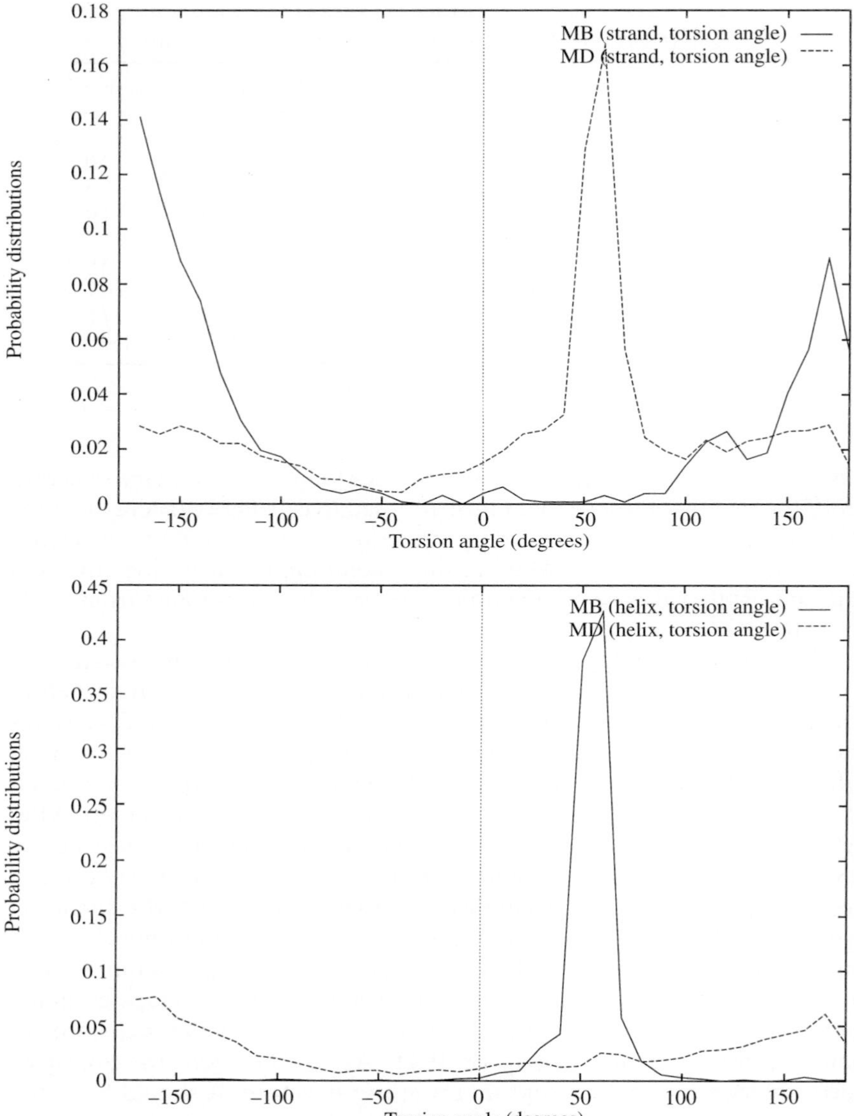

FIG. 10. Probability distributions computed at 3-Å resolution for measures of belief (MB) and disbelief (MD) of helix and strand structures for peak torsion angles.

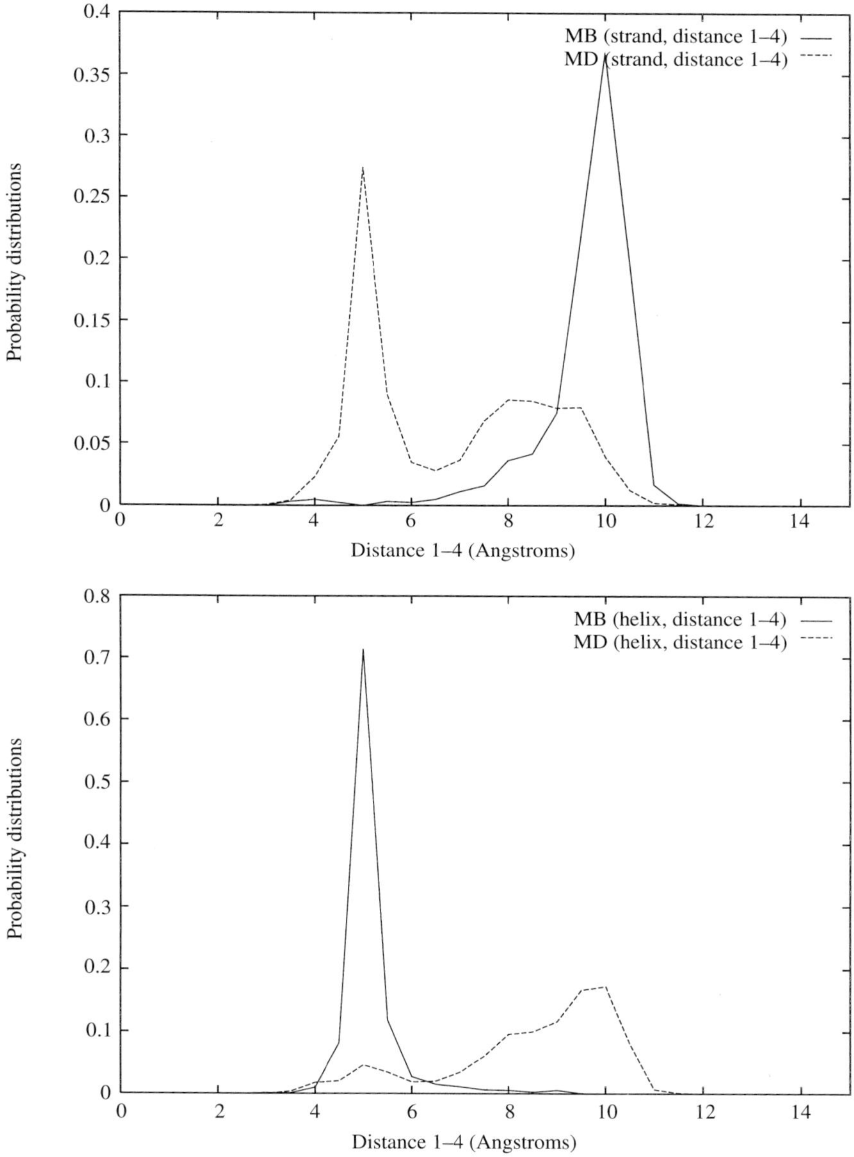

FIG. 11. Probability distributions computed at 3-Å resolution for measures of belief (MB) and disbelief (MD) of helix and strand structures for peak d_{14} distances.

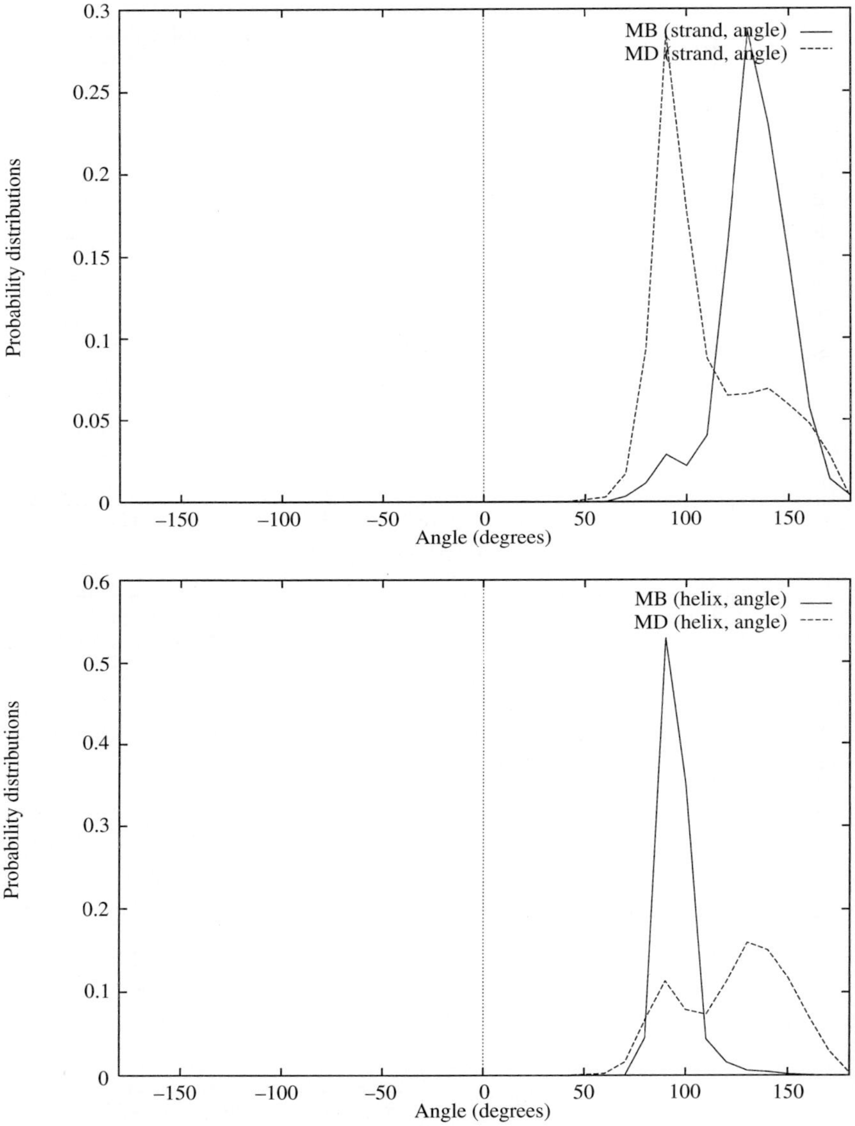

FIG. 12. Probability distributions computed at 3-Å resolution for measures of belief (MB) and disbelief (MD) of helix and strand structures for peak angles.

TABLE VI
APPLICATION OF RULE-BASED SECONDARY STRUCTURE RECOGNITION PROCEDURE TO CALCULATED AND EXPERIMENTAL MAPS OF PENICILLOPEPSIN[a]

Penicillopepsin map	Calculated	Experimental
Number of peaks used for identification procedure	353	325
Percentage of identified peaks	**58**	**55**
Percentage of correctly identified peaks	86	82
Percentage of incorrectly identified peaks	14	18
Percentage of unidentified peaks	**42**	**45**
Percentage of unidentified peaks for which identification is provided in PDB	31	38
Percentage of unidentified peaks for which identification is not provided in PDB	69	62

[a] At 3-Å resolution.

us to calculate 30 electron-density maps. We knew at the outset that at least one of the phase sets corresponded to a solution, as identified by the minimal function.[4] The purpose of our exercise was to see if the critical-point mapping approach could help recognize "good" and "bad" maps and consequently good and bad phase sets. Density profiles of the critical points, tree lengths, and identification of secondary structure motifs were used as discriminating criteria. On the basis of these criteria, two maps were identified correctly as good maps. In the best phase set, 43 (or 93%) of the residues were represented by critical points. There were no ghost peaks among the top 87 peaks above the ρ_{min} cutoff value.

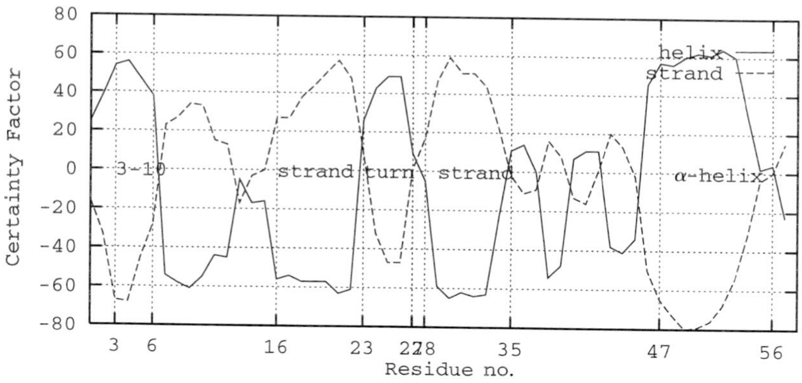

FIG. 13. Certainty factors obtained for helix and strand hypotheses for an ideal critical-point representation of protein 4PTI at 3-Å resolution. Also denoted is the correct interpretation for residues 16–23 (strand), 23–27 (turn), 28–35 (strand), and 47–56 (helix).

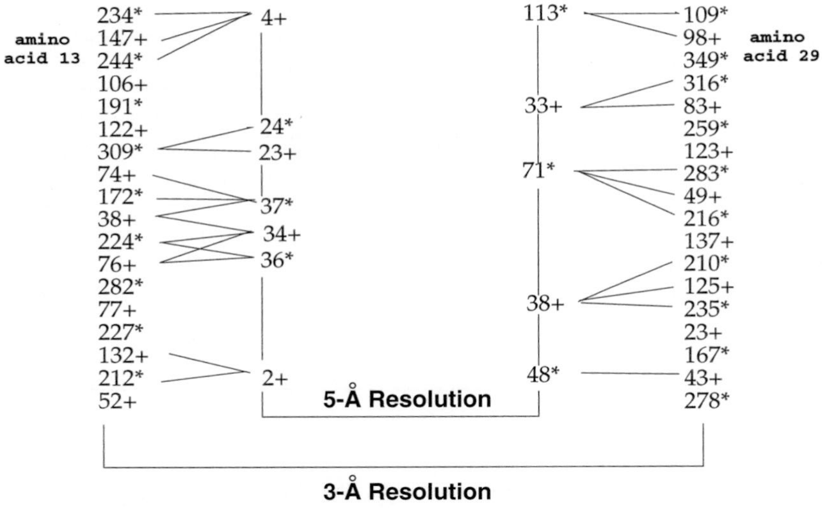

FIG. 14. Relationship between critical points at 3- and 5-Å resolution for residues 13 to 29 in 1RNT; plus symbols (+) and asterisks (*) denote, respectively, peaks and passes in the main chain. Critical points are assigned numerical values corresponding to their rank in a decreasing density value list.

Applications at Other Resolution

The critical-point mapping approach has also been tested at high (1-Å) and low (5-Å) resolution. Initial testing was made on calculated electron-density maps of a hexadecapeptide[28] whose secondary structure consists of an α helix followed by a 3_{10} helix followed by three β bends. As expected, at high resolution we find that every peak is associated with an atom in the structure. Again, . . . peak–pass–peak–pass–peak . . . motifs are observed, with passes corresponding to chemical bonds. At low resolution, again as expected, fewer critical points are found, and connectivity among sequential residues is found only in the case of helices. Connectivity of the β-sheet peaks no longer follows the chain direction. These results correspond to the observations reported by Lesk.[29] In 5-Å calculated maps, it is found that solvent regions are characterized by the presence of low-density critical points, primarily passes, pales, and pits while the protein

[28] I. J. Karle, J. Flippen-Anderson, M. Sukumar, and P. Balaram, *Proc. Natl. Acad. Sci. U.S.A.* **84,** 5087 (1987).

[29] A. M. Lesk, "Protein Architecture" (D. Rickwood and B. D. Hames, eds.). IRL Press, Oxford, 1991.

regions are populated with higher density points. The topological features thus can be used for envelope definition.

Test calculations with the protein structure 1RNT at 3 and 5 Å suggest the presence of a hierarchy between the peaks and passes at 5 Å and those at 3 Å. As illustrated in Fig. 14, individual critical points at 5 Å are found to be associated with individual or multiple points at medium resolution.

Concluding Remarks

Critical-point mapping has the potential to become an effective tool for the segmentation of protein electron-density maps and the recognition of structural motifs within the maps. Being built on a well-defined mathematical framework, it is well suited for implementation as a fully automated computer approach to map interpretation. Current efforts are focused on reimplementing the program to gain speed, and to allow for the incorporation of intelligence and information. In particular, the algorithm developed by Lathrop and Smith[30] to thread sequence onto a set of possible structures for the purpose of structure prediction is being modified and integrated to the critical-point mapping procedure.[31] As described above, critical-point mapping takes as input an electron-density map and produces a provisional interpretation consisting of hypothesized three-dimensional structures. The threading algorithm in turn attempts to find the most plausible ways to superimpose a given sequence onto hypothesized structures through the use of a scoring function. Threading thus can be used for the evaluation and validation of map interpretation results. Furthermore, it can be used for the classification and identification of individual residues in medium-resolution electron-density maps, providing a potentially very useful tool for model building and refinement.

Acknowledgments

We thank Carroll K. Johnson for providing the ORCRIT program and for many helpful discussions, and Marie Fraser for making available to us the penicillopepsin experimental map. Financial assistance from the Natural Sciences and Engineering Research Council of Canada (NSERC) and the Belgian National Council for Scientific Research (FNRS) is gratefully acknowledged.

[30] R. H. Lathrop and T. F. Smith, "Proceedings of the 27th Hawaii International Conference on System Sciences," p. 365. 1994.

[31] K. Baxter, E. Steeg, R. Lathrop, J. Glasgow, and S. Fortier, "Proceedings on Intelligent Systems for Molecular Biology." 1996.

[9] CHAIN: A Crystallographic Modeling Program

By JOHN S. SACK and FLORANTE A. QUIOCHO

Introduction

Computer graphics have played a critical role in the remarkable advances made in the field of structural biology. The ability to examine and manipulate macromolecular structures easily has led to a new understanding of protein structure and function. Nowhere have the developments in computer graphics been more noticeable, and more swiftly embraced, than in the field of protein crystallography. Interactive computer graphics are used routinely in the fitting of atomic coordinates to the experimental electron density.[1,2]

The program CHAIN was developed to assist the crystallographer in building atomic models from an electron-density map. By using a set of simple commands, the user easily can build, display, and manipulate atomic models and electron-density maps on a computer graphics device. The user first prepares data files containing the desired coordinate and density information. CHAIN then displays the electron-density contours, superimposed with atomic coordinates, on the screen. One can then make modifications to the coordinates interactively.

A number of additional features are included with the program to regularize or refine the coordinates automatically and to display them in a variety of ways. The program also provides a number of interfaces to read in coordinates from and write out coordinates to non-CHAIN programs.

An attempt has been made to allow a great deal of flexibility in the program while keeping it simple and easy to use, particularly for the novice user. Command files, display menus, function keys, and interactive dials allow for easy use without the need to remember a large number of commands. In addition, both the terminal and the interactive display have HELP commands to obtain information on any option.

Design Criteria

The main design criterion used during the development of CHAIN was that the program should be modular. This was implemented in three ways.

[1] T. A. Jones, *Methods Enzymol.* **115,** 252 (1985).

[2] J. W. Pflugrath, M. A. Saper, and F. A. Quiocho, New generation graphics system for molecular modeling. *In* "Methods and Applications in Crystallographic Computing" (S. Hall and T. Ashiaka, eds.), p. 407. Clarendon Press, Oxford, 1984.

0076-6879/97 $25.00

First, the command-line (nongraphical) portion of the program was designed as a series of tighly coupled subroutines connected by a main parser. As keyword commands are given by typing the command on the terminal keyboard, touching the command on the display menu, or by using a command file, the parser interprets the command and invokes the required routine to perform the desired function.

Second, all the graphics-dependent portions of the program are segregated in specific subroutines. This allows for the introduction of different graphics devices without modification to the entire program. At present, the program runs on the Evans & Sutherland ESV and Silicon Graphics IRIX systems.

Finally, by standardizing the calls to the data files, the graphics unit, and to the terminal, new subprograms can be added easily to CHAIN, provided they are modified to read in the CHAIN files. Additional entries can be added to the keyword list along with the appropriate calling routines.

CHAIN Parser

The CHAIN parser, originally developed by J. Pflugrath (Molecular Structure Corporation, The Woodlands, Texas), is used to set all the parameters before entering the graphics mode. These terminal commands are invoked by typing the command, followed by the new parameter value(s). For most parameters, the user needs only to enter the desired command and the new value for the parameter. For example, RADIUS 6.0 will reset the display radius to 6.0 Å. If no value is given with the command, the program will prompt for it.

The parser has a number of special features. For example, one needs only to type the unambiguous portion of a command; the first three letters define a unique CHAIN command. If a carriage return is entered as the response to any terminal command, the old value of the parameter is kept. If one gives an incorrect command, entering a carriage return will exit the command without changing any values. Commands may be typed using either capital or small letters.

The parser also has a series of special symbols to be used when interpreting commands. For example, the @ command is used to read and parse commands found in a command file. The $ command is used to invoke any valid UNIX command. The ? (question mark) command displays a list of all the legal commands available at this level of the program. If the program is currently requesting a value, the command displays the current value(s) of the parameter.

Command File

Many special or complicated tasks can be implemented by the use of the indirect command files. The command files store terminal commands to be read by the parser. It is equivalent to typing the commands manually while in terminal control. The command file is a standard formatted file in which each record contains a keyword followed by the appropriate values and an optional comment. The keyword is the same as the CHAIN command that changes this variable. The value field contains the current value(s) for this parameter. There are three command files invoked at the beginning of the program.

SYSTEM file: Stores system site-specific commands for the graphics unit in use. This file is read and parsed each time any user executes the program

USER file: Stores user-specific commands. Each user of the system may have a USER file that is parsed only when that specific user executes the program. One also can have a file that is parsed at the end of that user's session

SESSION file: Stores the parameters for a particular working session. The file is used to save and restore various parameters between CHAIN modeling sessions. It reflects the current state of the parameters that can be updated or changed. The user can have a separate SESSION file for each structure, etc. This allows the user to end a session (using the command END) and then to resume the session at a later time (with the command BEGIN), with all of the parameters set to the same values as before

The user can also invoke a command file at any time using the @ command.

CHAIN Rules

In addition to the special symbols, CHAIN also has a number of special rules and conventions used throughout the program. This simplifies many of the commands.

Color types: Whenever the program requests a COLOR, the user may respond with any of the available color type parameters. One can specify either an absolute color (WHITE, RED, ORANGE, YELLOW, GREEN, BLUE, CYAN, VIOLET, PURPLE, or GRAY), coloring based on the setting of the appropriate color dial (DIAL), or, when appropriate, coloring based on information contained in the coordinate file such as the atom type, the acid type, the residue secondary structure code, or the atom's temperature factor

Atom names: The program allows up to six characters to designate atom names, although normally only four characters are used. When atom names are requested by the program, it is not necessary to include any preceding blank spaces, because the program left-justifies atom names before doing any comparisons

Atom types: When a set of atom types is requested, the user can specify either the exact name of the atoms or type "*" to indicate any matching character. Thus, "C" indicates that CHAIN is to act only on carboxyl carbons whereas "C*****" indicates that CHAIN is to act on all carbon atoms, etc. A minus sign (−) as the first atom type designation requires that CHAIN select all *but* the specified atom types

Residue names: The program allows up to six characters to designate residue names (including the chain name and insertion character). When residue names are requested by the program, one should include the entire name, without any blank spaces. The program left-justifies names before doing any comparisons. Although residues are normally given integer names, the program stores the residue names as characters, and thus one can use any alphanumeric name except for some special words (END, FIRst, LASt, MOLecule, STArt)

Residue ranges: When a range of residues is requested by the program, the user may specify either a single residue, a range of residues (by specifying the first and last residue in the range), or a set of multiple ranges

There are several special words that can be used in place of the actual names of the residues, such as FIRST (the first residue in a file), LAST (the last residue in the file), ALL (all residues in the file), or a series of one to four dots to indicate the last or next 5, 10, 15, or 20 residues. Thus "FIRST LAST" is equivalent to "ALL," and "101 ..." would indicate 15 residues starting at residue 101

ATOMIC POSITION or SCREEN CENTER: Whenever an atomic position or screen center is requested, the user can specify this value in orthogonal angstroms, fractional coordinates (by giving the letter "F" followed by the fractional coordinates), or as a residue–atom pair. If a residue name is given without a corresponding atom name, the α carbon is assumed. If no α carbon is present, the first atom in the coordinate list is used

Utility Commands

The utility commands are used to start or end a CHAIN session and to set or examine various system parameters. There are also utility commands to obtain help on any CHAIN command or to attach to a subprocess.

Command BEGIN starts a new modeling session. It is equivalent to ending the current active session and beginning a new session. This command is automatically given at the start of the program. The routine requests the name of the new session. The program searches for the SESSION file owned by that user. If no session file exists with this name, a new SESSION file is created.

Command END terminates a CHAIN modeling session. All files currently open are closed, the current parameters are written to the SESSION file, and the program is terminated. Alternatively, the command QUIT ends a CHAIN session but the new parameters are *not* written to the session file.

Command CURRENT displays all of the parameters that are set for the current session.

GRAPHICS and SET Commands

The GRAPHICS and SET commands allow one to change the parameters that define the graphics system and display. The GRAPHICS commands specify the graphics terminal and peripherals in use. The SET commands set a number of user preferences while running CHAIN. The general form of the command is SET ⟨parameter⟩ ⟨new_value⟩. Many of the SET commands define additional atom or acid types, or determine which subdisplays are shown on the screen. Others set the default colors or values for various parameters.

ANCHOR	Set anchor atoms
ATOMTYPES	Set atom type codes
BELL	Turn "no-pick" bell on and off
CONTACTS	Set the contact distance
DIALS	Define the dial sensitivity factor
DICTIONARY	Change the name of the dictionary file in use
DISTANCES	Update distances and angles automatically
FKEYS	Turn function key display on or off
HIGHLIGHT	Control highlighting of selected menu items
HYDROGENS	Control the use of hydrogen atoms
IDENTS	Change the default identifications
IDSIZE	Change the default identification size
MAINCHAIN	Set main-chain atoms
MAPS	Master electron-density display switch
MESSAGES	Determine which types of messages are to be printed
MODEL	Name temporary model file
ORTHO	Default orthogonalization matrix
PEPTIDE	Set peptide bond atoms
PICK	Set atom picking area
REFINE	Select refinement parameters

RESIDUE	Check for duplicate residue names
ROCK	Select default rocking speed and angle
SCALEBAR	Turn scalebar display on or off
SECONDARY	Define secondary structure
SESSION	Update session file
STEREO	Default stereo width separation and angle
SYMMETRY	Determine the coloring of the symmetry atoms
TEMPER	Default atom temperature factor
TITLE	Define a title for the display
UPDATE	Continuous update control
WEIGHT	Default atom weight

Atomic Coordinates and Electron Density

The main purpose of program CHAIN is to display atomic coordinates and electron densities, and then to make modifications to the coordinates. In the terminal mode of the program one has commands to permit a wide range of options, such as to define an atomic coordinate or electron-density file, read in data from an external source, and make modifications or additions to the coordinate file. Within the program, the atomic coordinates are stored in a binary random-access file. Modifications to the atomic coordinates are continuously updated in this file, and can be saved between CHAIN sessions or written out in an external format.

Defining Atomic Coordinates

One must define the atomic coordinate data file using the commands DISPLAY and FILENAME before being able to display atomic coordinates. DISPLAY defines which of multiple files one is about to define (or modify) and FILENAME gives a name to this coordinate set.

The program allows for four active coordinate displays. In a typical session only one display is used, and the DISPLAY command is not needed. One can, however, define additional displays by giving the command DISPLAY 2 or DISPLAY 3. The program will operate from these alternate displays until one returns, using the DISPLAY 1 command.

The FILENAME is the name of a coordinate file on the disk that will hold the atomic coordinates and other information about the structure. This file is in a special format for program CHAIN, so that commands must be used to transfer data into the file, list the information in the file, or to write out data for use with other programs. This file usually has the extension name ".atm"; included in the distribution is a sample atom file "test.atm." To define this file as the coordinate file, one would give the command: DISPLAY 1 FILE test.atm.

Multiple commands can be placed on a single line provided there is no confusion. If, for example, one were giving a list of atom types, the program would not know where the list of atom types ended and the next command began. In this case, however, the DISPLAY and FILENAME commands each have only one value associated with them, so there is no problem with putting them on the same line.

Each command can be abbreviated to three letters. Thus we could have given the command "DIS 1 FIL test" to do the same thing. Also note that there are certain default values for each parameter. The default display is 1; thus if it had not been changed, one would not have to give the display number. One can always determine the current value of a parameter by giving the command followed by a question mark. If one had given the command "DIS ?," the program would have responded with the number 1. Furthermore, because the default atomic coordinate file name is ".atm," we did not have to give the complete file name.

There are commands to examine the current atomic coordinate data file. A complete description of each command is found in the reference section of the manual. Some of the more common ones are as follows.

LIST lists out the residues or atoms in the coordinate file. The format of this command is "LIST ⟨atom/residue⟩ ⟨residue_range⟩." One selects whether to list each atom or each residue in the coordinate file, and then the range of residues to list.

CHECK examines the coordinates for undefined atoms and for incomplete residues. The format of the command is "CHECK ⟨residue_range⟩," thus the command CHECK ALL would give a complete listing of all undefined/incomplete residues in the file.

CONTACT lists all the inter- and intramolecular contacts in the file. The routine asks a number of questions such as the name of a listing file, the contact distance range to report, the residue range to search, and whether one wants intermolecular or intramolecular contacts.

Reading and Writing Atomic Coordinates

When using coordinates from an external source, one must first read them into CHAIN using the READATOM command. This is required because CHAIN uses a special random-access disk file to retrieve atomic coordinates.

The program uses the Protein Data Bank as the default external format. However, it also allows one to read in the coordinates in a number of other formats such as the R. Diamond BUILDR program (DIAMOND), the Hendrickson PROTIN program, the expanded UPDATE format, the Fitzgerald MERLOT format (MERLOT), or the Brunger XPLOR program.

After modification, the coordinates can be written out in any of these formats.

There are a number of special parameters that can be set, for example, to require that only certain amino or nucleic acid types or atom names should be read in from the external file, to set the temperature factors or occupancies, and to control how the atomic coordinates are added to the file. The new coordinates can replace, overwrite, or be appended to the current atom list. The input can be fractional or angstrom coordinates. Once the READ command is given, any of these parameters can be set or changed until the GO command is given. Similar commands can be used to write out the coordinates to external formats.

In addition, the atomic coordinate file allows for the inclusion of secondary structural information. Much of this information is not required by CHAIN but can be useful in keeping track of the progress of the structure determination. The atomic coordinate file contains all the information used by CHAIN for a particular molecular structure. It is an unformatted, random-access file with 64 bytes per record.

The file contains both a residue section and an atom section. The residue section records each of the amino or nucleic acid residues in the structure. For each residue, there is one record containing the residue name and acid type for the residue and the center and radius of the pseudosphere that encloses this residue. This information is used in the RADIUS mode to select residues quickly for display. If the pseudosphere is wrong, then the display generated in RADIUS mode may not be correct. Also included are status and structure codes, and pointers to the atom records. The ATOM RECORDS contain information on each of the atoms in the structure, including the atom name, its position in orthogonal angstrom coordinates, and its temperature and occupancy factors.

At this point, one may need to make some modifications to the new coordinate file. In particular, if the input file did not contain the space group or the unit cell dimensions, one will want to define them in the file in order to display symmetry-related molecules. One also may want to make modifications to the file, such as inserting or replacing coordinates. Some of the more common modification commands are as follows.

BRIDGE	Define disulfide bridges in the coordinate file
CELL	Specify the real-space cell parameters
CODE	Add residue code information
DELETE	Remove a residue from the file
IMPOSE	Apply a rotation/translation operation
INSERT	Add a new residue to the atomic coordinate data file
MOVE	Move atoms to new positions
NUMBER	Renumber the residues in the atom file

RENAME	Change the residue names in the coordinate file
MUTATE	Replace a residue of one amino or nucleic acid type
SECOND	Add secondary structure information
SPACE	Add the crystallographic space group information
STATUS	Change the atom status of the atomic coordinates
TITLE	Add a title to the coordinate file

There is also a regularization option (REGULARIZE) based on the REFINE program by Hermans and McQueen.[3] The routine regularizes the bond lengths and bond angles of existing amino or nucleic acids, using a dictionary file of ideal bond lengths and bond angles.

Defining Coordinate Display

The last step before displaying the coordinates is to set all of the parameters that determine how these atoms are to be displayed on the screen. One can select which residues or atoms will be displayed from the file. With the RANGE option, for example, the program will display residues in a region along the chain of the molecule. The RADIUS option displays all of the atoms within a given radius of the center of the display. One can also display a combination of both RANGE and RADIUS, either the INTERSECTION or the UNION of the two.

Other commands that affect how the atoms are displayed are CENTER (to defined the center point of the display), BONDING (to determine how bonds are to be made), SYMMETRY (to turn the symmetry atom-generating option on), and COLOR (to determine the coloring of the atomic bonds). There are also options to display only selected acid or atom types. Once the SHOW command is given, the program will prepare and display the atomic coordinates on the screen.

Electron Density

There are three main steps involved in displaying the electron density. First, the density map must be created, outside of CHAIN, using an appropriate density calculation program. The density map then is written in CHAIN format into a direct access file used internally to create the electron-density map displayed. The electron density is stored in small (8-cubic grid point) blocks of data in byte values. This minimizes the size of the map files, and allows for selective reading of only those density blocks that are needed to create the current display. The MAP command is used to display the electron density. Any combination of up to eight different electron-

[3] J. Hermans, Jr. and J. E. McQueen, *Acta Crystallogr.* **A30,** 730 (1974).

density maps or contour levels can be displayed at the same time. Each display contains a single contour level from an electron-density file. One begins by giving the MAP command followed by the number of the display to be created. Parameters of each of the eight electron-density displays consist of (1) the name of electron-density data file, (2) the contouring level to be displayed from the density map, (3) the radius or cubic volume of electron-density contours to be drawn, and (4) the color of the display. Additional options permit the selection of a cover distance to remove vectors from the electron-density display that are greater than a set distance from any atom in the display and to set the display line type.

The electron-density map can also be displayed as a ridge-line representation.[4] The display can be modified interactively by adding, removing, or coloring the ridge lines to make a pseudocoordinate display.

Preparation of Static Models

While CHAIN was not designed as a molecular display or presentation program, there is a primitive set of routines to define static MODEL displays containing molecular wire models, and dots to indicate van der Waals or interfacial surfaces. Each MODEL display can be made up of multiple commands to create a complex display. Each MODEL can be made up from a series of commands. This enables the user to view the entire protein structure for study or photography without limiting the number of atoms displayed. Because it is a static display, however, one cannot pick or move atoms in this display. The command MODEL begins the creation of a MODEL display. One then gives the number of the MODEL display to be defined. There are 16 MODEL displays. One can then give any of the MODEL commands and then will be prompted for any necessary input. When the MODEL has been completely defined, the command SHOW (or another MODEL command) terminates a MODEL display.

As modeling commands are given, they can be stored in the model file. One can clear (rewind) this file by giving the MODEL CLEAR command. To display an old model file, one first can define the file by the SET MODEL FILE command, and then display it using the MODEL ON command.

MSFILE	Dot surface display from MS style file[5]
BARRY	Barry surface around a given range
DOTS	Dot surface for a set of atoms
DRAW	Atomic model from the commands in the current display
HBONDS	Display of possible hydrogen bonds in the structure

[4] J. Greer, *J. Mol. Biol.* **82,** 279 (1976).

[5] M. L. Connolly, *J. Appl. Cryst.* **16,** 548–558 (1983).

IDENTS	Atom and residue labels
JOIN	Extra bonds between given atoms
OUTLINE	Outline of the unit cell on the screen
RIBBON	Stylized representation of the secondary structure
VECTOR	Display from vectors listed in an external file

Interactive Display

The interactive display can be entered from the terminal only by use of the SHOW command. When one enters the interactive display, there is a pause as the program determines which atoms will be displayed and the connectivity of these atoms, and then sends the atom and bond vector lists to the graphics unit. It then determines which, if any, electron-density contours need to be drawn, does the contouring, and sends these vectors to the graphics unit. Control is then sent to the graphics unit. The display menu is cleared, and the program then waits for new picks of either atoms or commands. Once in the interactive routine, the user has four ways to interact with the display: the keyboard, the dials, the function keys, and the MIDDLE button on the mouse. Most actions involve using the mouse to pick atoms to be labeled or moved and to select options from the display menu. The molecule can be rotated, and the colors of the various displays can be changed by means of the eight control knobs. The 12 function keys along the top of the keyboard are used to turn selected displays on or off.

Function Keys

The function keys along the top of the keyboard activate different parts of the display. For example, one selectively can turn the menu, the axes, the atoms, and the different electron-density maps on and off. Each of the keys controls a different function, depending on which function keyboard is active.

The VIEW set controls the display view rotations. The keys select the view matrix, and allow one to rotate 90° about any of the three axes or to set the view to be down a given axis.

The DISPLAY set controls the display of the menu and viewport objects. Each key can turn one part of the display (menu, axes, identifications, the terminal emulator, and contact display) on or off.

The ATOMS set controls the display of the four atom and four symmetry-atom displays. Any of the eight displays can be turned on or off, without deleting them, by pressing the appropriate function key. The labels on the function keys change to indicate whether the displays are on or off.

The DENSITY set controls the display of the eight electron density maps created. Any of the eight displays can be turned on or off, without deleting them, by pressing the appropriate function key. The labels on the function keys change to indicate whether the displays are on or off.

The MODEL sets controls the display of the first MODEL objects created in the MODEL routine. Any of the displays can be turned on or off, without deleting them, by pressing the appropriate function key. The labels on the function keys change to indicate whether the displays are on or off.

The RESET key set resets different parts of the display. The keys reset the rocking speed and angle, the stereo parameters, the default colors of the displays, the size and position of the identification, and the screen intensity to their default values.

Function Dials

The function dials can have different functions, depending on which dial set is active. The keys of the numeric keypad control the use of the eight function dials. When one of the numeric keys is pressed, the function of the dials will change. The ENTER key is used to return quickly to the terminal from the display mode. CHAIN automatically switches the dial function whenever function MOVE, FRAGMENT, or TORSION is activated. The user always can switch to any of the dial sets even if the functions are not active (such as going to the FRAGMENT dial set when there is no active fragment). There is also a scaling function for the dials. Using the ALT keys on either side of spacebar, one can change the sensitivity of the dials to one-tenth normal. Touching the ALT key again resets the normal dial sensitivity. The program also has options to use a spaceball device or to use lines on the screen as pseudodials for systems without dials.

Display Menu

The commands along the edge of the screen are known as the *menu*. These menu commands are activated by moving the cursor over the desired menu command and clicking on the middle button of the mouse. The selected menu command will light up.

Most of the menu commands work first by clicking on the desired atom or atoms and then by clicking on the menu item. For example, to determine the position of an atom, first pick the desired atom from the active display and then pick the POSITION menu command. The coordinates will appear on the terminal and on the lower left of the display. To obtain the distance between two atoms, first pick the two atoms, and then the DISTANCE menu command.

TERMINAL returns control of the program from the display menu to the keyboard. If any coordinates have been changed, the user is asked if the modified coordinates should be saved in the atomic coordinate data file when leaving the interactive mode.

CLEAR is the master switch to clear any active menu items on the display. All active displays (MOVE, FRAGMENT, TORSION, and ROTOMER) are REJECTED and all monitoring (PLANE, POSITION, DISTANCE, ANGLE, and CONTACTS) is turned off.

REJECT is used to reject the changes made to the structure or system. The modified coordinates are not saved in the coordinate file. All active displays (MOVE, FRAGMENT, TORSION, and ROTOMER) are turned off, and the coordinates are reset to their initial positions.

RESET resets changes made to the structure or system. All active displays (MOVE, FRAGMENT, and TORSION) are reset to their initial positions while leaving the function active. The view translation is reset to its last value sent from the host program.

SAVE/UNSAVE allows one either to save or to discard changes made to the atomic coordinates since the last SAVE command was given. If the SAVE command is given, the coordinates of all modified atoms are written to the atomic coordinate data file. With UNSAVE, any changes made to the coordinates are reversed.

IDENT clears the atom identifications on the display and/or the atom information in the text area. It is also used to change the format of the identifications on the screen.

REDRAW shifts the screen center to the location of the last atom touched on the screen. The atomic coordinates and electron density are redrawn based on the new screen center. It is equivalent to returning to terminal mode and giving new CENTER and SHOW commands.

CENTER shifts the screen center to the location of the last atom touched on the screen. The screen center is shifted without any new atoms or electron-density vectors being drawn.

BOND makes or breaks a bond between the last two atoms picked on the screen. If the initial screen connectivity was not determined by the dictionary, the correct bonding pattern does not appear on the display. These bonds can be formed or removed using this command. It can also be used to isolate atomic fragments prior to a FRAGMENT, TORSION, or REFINE command. Atomic bonds are not saved between sessions in the coordinate file.

MOVE allows an atom to be moved without affecting the rest of the molecule. The dial settings change automatically to those controlling the movement of the atom. Bonds involving this atom are "rubber banded" as the atom moves. The MOVE command also can be used to add atoms to the display.

FRAGMENT moves a group of connected atoms independent of the rest of the display. All the atoms connected to the last atom touched move as a rigid group. The picked atom acts as the central pivot point.

TORSION interactively modifies the torsion angles of a structure. Up to six consecutive torsion angles can be defined and rotated independently.

REFINE initiates the real-space refinement procedure, using the electron-density map file. All the atoms connected to the last atom touched move as a rigid group in an attempt to minimize the difference between the observed and calculated density.

ROTOMER displays the most likely conformations of the side chains for the selected residue based on the work of Ponder and Richards.[6] The routine allows the user to select a side chain rotomer.

EXTEND adds the next residue in the sequence to the display. The coordinates of the new residues in the display range are read in from the coordinate file. For proteins, the action is to move along the sequence, residue by residue, from the N terminal to the C terminal.

STATUS changes the fix/free atom-status words of all subsequently picked atoms. This is useful prior to regularization to fix or recalculate the position of certain atoms before the regularization.

POSITION displays the current atomic coordinates of the last atom picked on the screen. A terminal listing also is produced, giving all the information that exists concerning the atom (including the atomic temperature factor, weight, atom type, and atom-status word).

DISTANCE displays the distance between the last two atoms picked on the screen. This distance is updated on the screen if either atom is moved.

ANGLE displays the angle defined by the last three atoms picked on the screen. Every triplet of atoms picked produces one angle calculation. This angle is updated on the screen if any of the three atoms moves.

PHIPSI displays the ϕ/ψ angles for a selected residue on a small Ramachandran plot, which appears on the screen.

CONTACTS creates a display on the screen showing all atoms that are within a certain distance of the last atom touched. Contacts between this atom and atoms in the same residue are not displayed.

PLANE makes a least-squares plane calculation for a set of picked atoms. The deviation from the plane of each atom is given.

[6] J. W. Ponder and R. M. Richards, *J. Mol. Biol.* **193,** 775 (1987).

PLOT writes the current display (atoms, bonds, identifications, and electron-density contours) to a plotter, laser printer, or a plot file for making plots of the screen display on an external plotter. The program provides PLOT and SPOOL options for creating hard-copy plots of the display on an HP plotter or a PostScript laser printer.

STORE/RECALL saves and subsequently recalls a view in memory. These viewport memories hold the current view (rotation, translation, clip, and scale). The command also can report back the current view to the user.

STEREO switches the stereo viewing of the display on and off. The program provides an interface for a StereoGraphics stereo viewer. The command GRAPHICS STEREO determines if a Stereographics stereo viewer is present. If a stereo viewer is available, this variable should be set to ON, otherwise it should be set to OFF.

ROCK switches the rocking of the display on and off. The rocking speed and maximum angle are set using the SET ROCK command.

IMAGES writes an rgb screen image of the current screen for later display. It also has the option of writing a file of the current atoms being displayed for use with the RASTER3D program.

RIDGE activates the ridgeline display and editing routines.

Program Modification

One of the major design features of program CHAIN was to allow the user to make modifications to the program. The simplest of these factors is the size limit for various parameters. Most program limits are set in the common block, and can be changed easily by the user. For example, if one is working with a large number of different metal ions, and wishes to add entries to the ATOMTYPE table, the ATMAX limit of 10 atom types could be increased in the common block and the program recompiled. While some parameters, such as the maximum number of ATOM or MAP files, cannot be increased without making major modifications to the program, most limits can be changed simply by making a change in the common block file.

Another common modification is in the input and output formats of coordinate and electron-density maps. While the program allows for the use of a number of standard formats, the user may wish to have additional formats available. There is a user-defined USER format that can be modified to read alternate coordinate files or electron-density maps.

A more ambitious modification would be to add a new routine to the program. While the program contains "hooks" in the parser to add new functions, the programmer must have some knowledge of the internal subroutines used to parse commands, read and write coordinates, and send messages to the terminal. The program also allows for additional menu commands. This requires a careful examination of both the host program and the function networks. One can use an existing menu routine as a template. The program documentation contains several examples of this type of modification.

Acknowledgments

We thank Phil Jeffrey, Timothy Reynolds, Lynn Rodseth, Steven Sheriff, John Spurlino, Mina Vyas, Nand Vyas, Connie Wallace, Herb Klei, and many others for assistance in the preparation and testing of the program. We also thank those people who have provided additional source material or assistance in the development of program CHAIN.

[10] Electron-Density Map Interpretation

By T. A. JONES and M. KJELDGAARD

Introduction

The life of a macromolecular crystallographer is sometimes exciting. We would list the high points of a crystallographic project as follows:

$$\text{Material} \rightarrow \text{crystals} \rightarrow \text{phases} \rightarrow \text{model} \rightarrow \text{publication}$$

Each step is crucial to the successful completion of the project and most certainly worth celebrating in some way, as it is likely to be followed by a period in which little progress is made. Each step can be rate determining. Fortunately, significant progress has been made in arriving at each step, except perhaps for the last. Interpreting a new electron-density map as a detailed molecular model is perhaps the most exciting step in a crystallographic project. Seeing the culmination of the efforts of many people appearing before one's eyes is a moment that is rarely forgotten. For a short time, you may be the only person in the world who knows what the structure actually looks like, or at least you think that you know.

Any errors that occur in a crystallographic project usually will be found before publication. Often, if an error has been made, the project will stall and there will be no publication. Introducing a serious error in a model can be different. In this chapter we discuss the kinds of error that might be made (with examples taken from our own work) and why these errors are made. We discuss some of the features of the crystallographic model-

0076-6879/97 $25.00

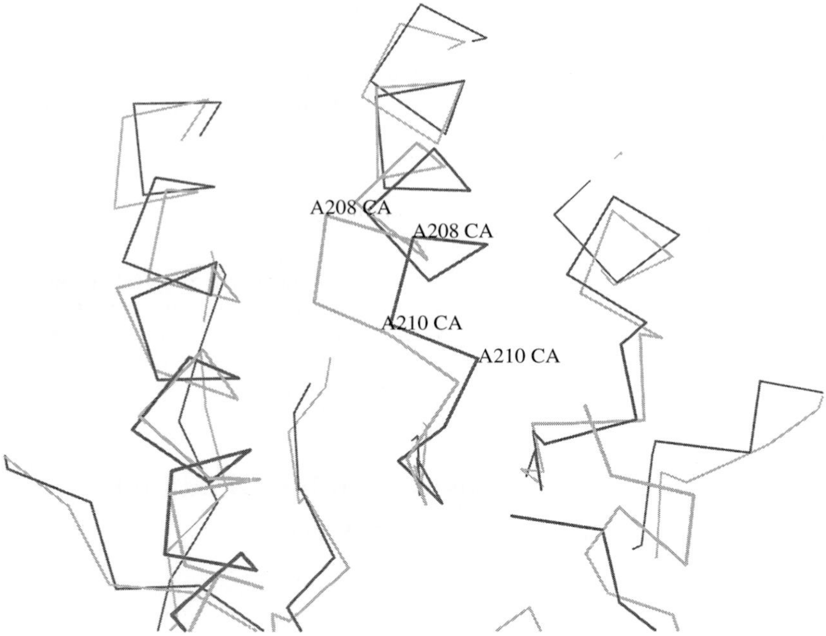

FIG. 1. Locally wrong structure in CBHII.

building program O[1] and then present an outline of how we believe maps should be interpreted.

Different Kinds of Model Error

Our primary concern is not with simple deviations from expected stereochemistry, and inconsistencies with database definitions. Real errors in models occur with frequencies that are, fortunately, inversely proportional to the seriousness of the error. They can include the following.

Totally wrong fold: If proper steps are taken, it is difficult to do and requires some determination. As yet, we do not have our own example of this kind of error. In an experiment in which we deliberately traced the chain backward through the map (i.e., we placed the correct N-terminal residue of the sequence at the C terminus of the structure and then followed the fold to the bitter end, placing the correct C-terminal residue at the N terminus), we could phrase a description of the usual quality criteria of the

[1] T. A. Jones, J. Y. Zou, S. W. Cowan, and M. Kjeldgaard, *Acta Crystallogr.* **A47,** 110 (1991).

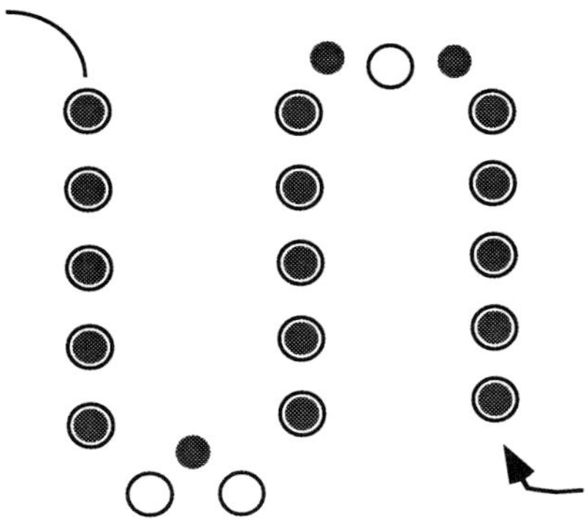

FIG. 2. Schematic out-of-register error.

structure so as not to cause alarm to a potential referee or reader.[2] This model had an R factor of 21.4% after refinement at 3-Å resolution. The free R factor[3] could not, however, be fooled, and had a value of 61.7%.

Locally wrong fold: In a multisubunit structure, for example, the folding of one domain could be totally incorrect. As another example, the connectivity between a subset of secondary structure elements may be wrong. If a large part of the structure is essentially correct, the situation may be hard to correct if the experimental map is ignored after the first model has been built. Provided that a sensible refinement strategy is used, it should be straightforward to recognize that a problem exists. Solving the problem may not be so easy.

Locally wrong structure: The main chain could, for example, be built through strong side-chain density or an unexpected metal ligand, resulting in a small region of structure that is totally incorrect. An example of this kind of error is shown in Fig. 1 and is taken from our work on cellobiohydrolase II (CBHII).[4] The lightly drawn C_α trace is the very first model built into the multiple isomorphous replacement (MIR) map, phased to 2.7 Å. The darker line is a trace of the structure after refinement to 1.8-Å resolution. There was a main-chain density break at the C terminus of this helix that prompted us to make a detour through strong density for the local side chains. This error was picked up and corrected in the third macrocycle of refinement and rebuilding.

Out-of-register errors: Although the local fold is correct, the placement of the sequence into the density may be out of register by a number of

[2] G. J. Kleywegt and T. A. Jones, *Structure* **3,** 535 (1995).
[3] A. T. Brünger, *Nature (London)* **355,** 472 (1992).
[4] J. Rouvinen, T. Bergfors, T. Teeri, J. K. C. Knowles, and T. A. Jones, *Science* **249,** 380 (1990).

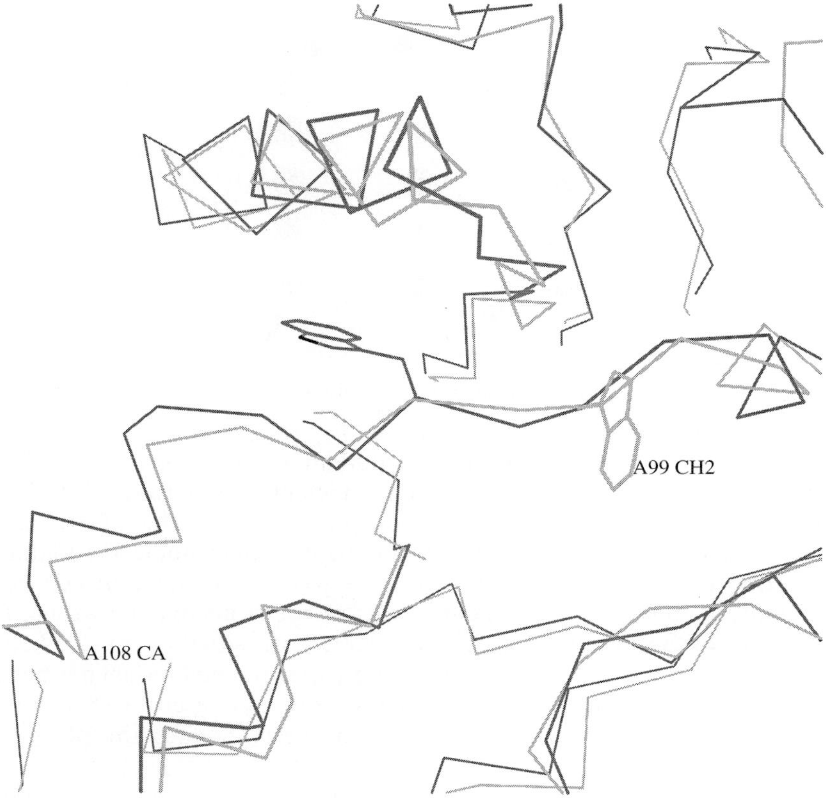

FIG. 3. Out-of-register error in CBHII.

residues. If two placement errors are made, the problem can become self-correcting so that only a zone of connected residues is wrong. Placement errors often occur in loop regions connecting secondary structural elements, as schematically illustrated in Fig. 2. Figure 3 shows a two-residue out-of-register error that occurred in the CBHII study. The first model was massaged to place the indole ring of residue 99 into very strong density. In the correct structure (darker line, Fig. 3), this density is occupied by an unexpected glycosylation of Thr-97. Forcing the model into the wrong density led to a distortion at the N terminus of the next helix, but the sequences came back into register in the middle of this helix at residue 108. Again, the error was corrected in the third macrocycle of refinement.

Wrong side-chain conformation: Wrong side-chain conformation errors are very frequent in the first model. The situation is exacerbated by changing

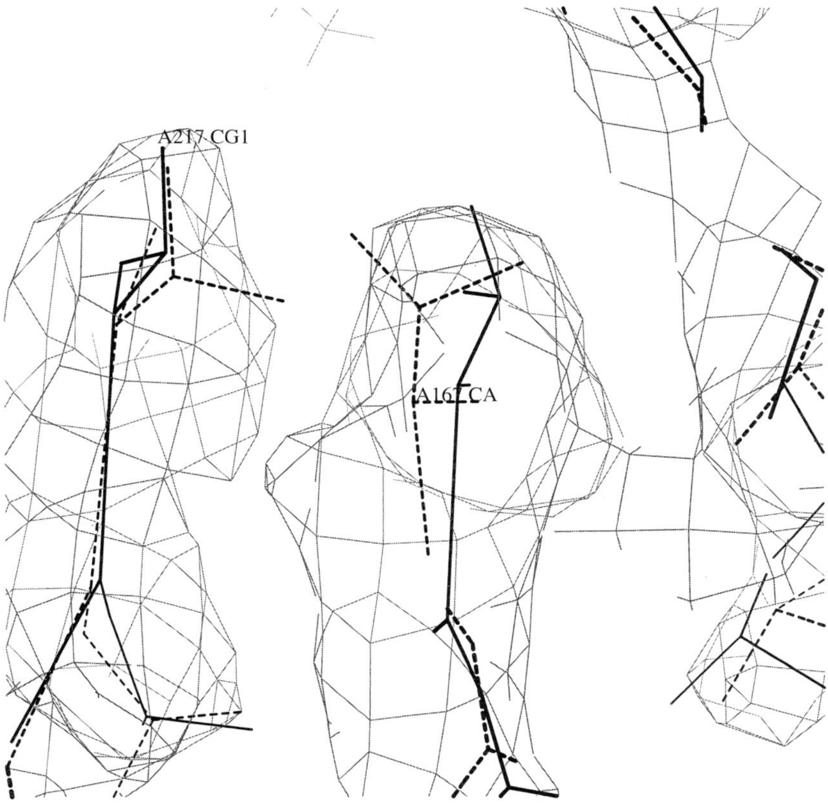

FIG. 4. Wrong side-chain conformation CBHII.

side-chain torsion angles in the model-building program instead of using rotamers. Figure 4 comes from the CBHII study and shows two valine residues on neighboring strands. The first model was traced with O, but because the program did not have full functionality at the time, the detailed side-chain placements were made with Frodo.[5] Both residues were built in energetically unfavorable conformations whereas in the final refined structure they both adopt the most frequent rotamer conformation. Many incorrect side-chain conformations will be corrected automatically by the refinement program, and those that are not usually can be identified during careful rebuilding.

[5] T. A. Jones, *J. Appl. Crystallogr.* **11,** 268 (1978).

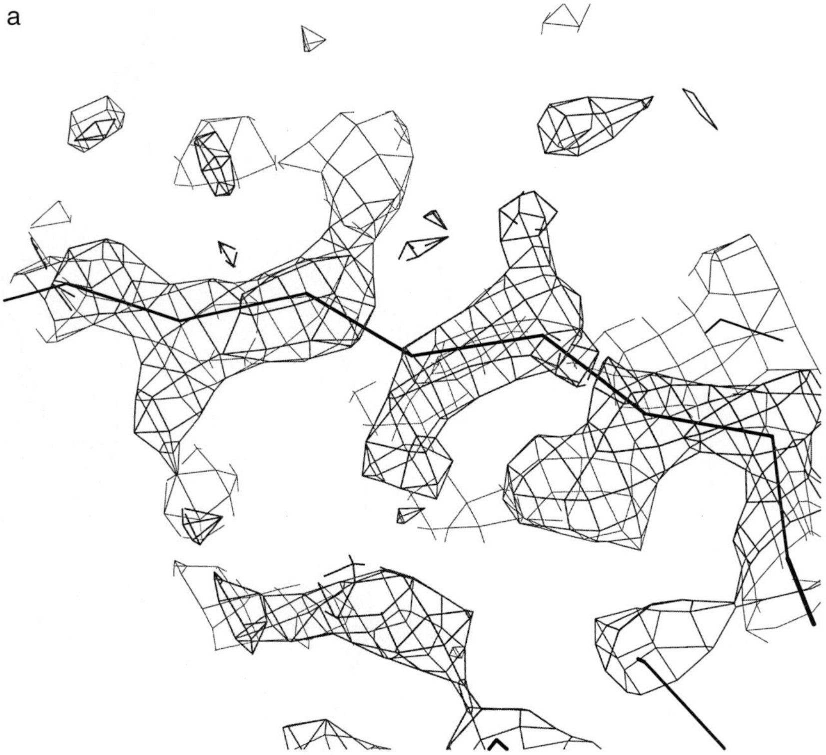

FIG. 5. (a) Experiment MIRAS map of P2 myelin protein. (b) Same region after cyclic threefold averaging.

Wrong main-chain conformation: The placement of the peptide linkage is a critical step in correctly defining the main-chain conformation. If the plane of the linkage is correctly oriented, the coordinate error of the carbonyl oxygen atom can still be more than 3 Å.

Most of the errors described above can be located and corrected, provided a careful refinement and rebuilding protocol is used (see [11] in this volume[6]).

Why There Are Errors in Models

The main causes and reasons why errors are made during map interpretation are noted in this section. Often a combination of events may lead to the error.

[6] G. J. Kleywegt and T. A. Jones, *Methods Enzymol.* **277,** [11], 1997 (this volume).

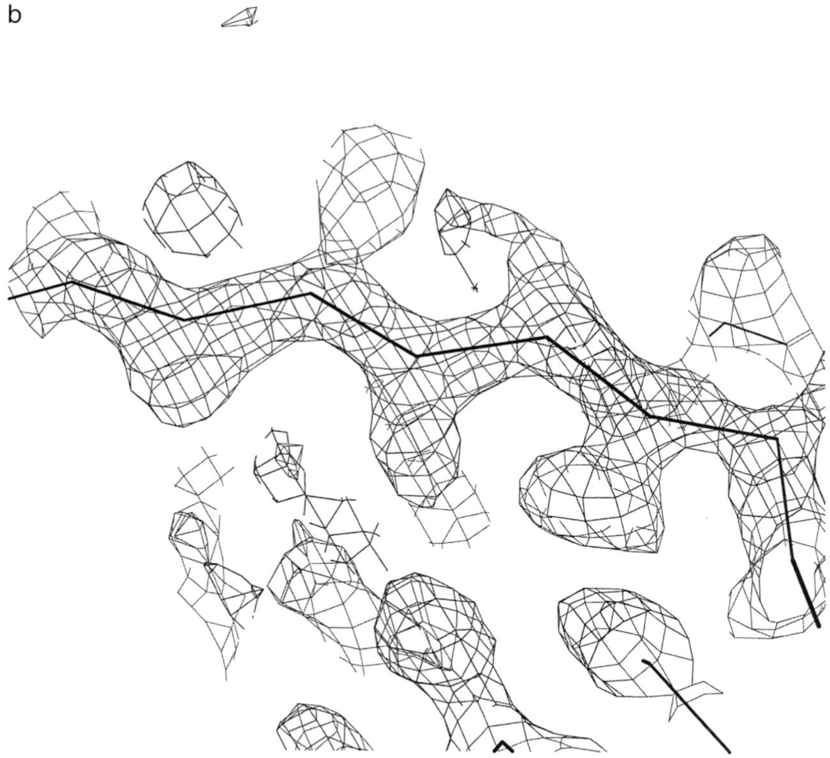

FIG. 5. (*continued*)

Phase Errors

The experimental techniques that are presently employed, be they MIR/MIRAS, multiple anomalous diffraction (MAD), or molecular replacement, cannot prevent phase errors in the diffraction data. Depending on the severity of these errors, the maps that are generated may be easy, difficult, or quite impossible to interpret. In our terminology, we call such maps "good, bad, and ugly" maps. In the last few years, we have seen substantial technical improvements in data collection (e.g., use of area detectors, synchrotron radiation, crystal cryocooling) and in phasing methods (e.g., MAD, site-directed heavy-atom incorporation, improved heavy-atom refinement algorithms, density modification techniques, and easier-to-use averaging programs). Together, this has meant that structures are now being solved where previously success would have been much more difficult.

Even good maps will have bad or ugly bits of density. This could be functional, a result of mobility or disorder, or due to phase error. Figure 5a shows a part of the map used to solve the structure of P2 myelin protein.[7] It corresponds to a β strand, and the up-and-down direction of the side chains can perhaps be recognized. Breaks in the density for the main chain are due to phase errors. This structure has three molecules in the asymmetric unit, and after cyclic averaging with A (a forerunner of the RAVE package[8]) the breaks disappear (Fig. 5b).

Resolution

The level of detail in an electron-density map will depend also on the quantity and quality of the diffraction data. With no phase error, a chain can be traced correctly at 4 to 4.5-Å resolution. Such examples are usually special cases, for example, viruses, for which phase refinement by cyclic averaging of a large number of noncrystallographically related units can give good-quality maps. In the work on satellite tobacco necrosis virus, the chain was traced at a nominal resolution of 3 Å,[9] but it turned out that the phasing was essentially perfect to around 3.8 Å and random between 3.8 and 3 Å.[10] The tracing of the first model was correct but with major local errors caused, in particular, by unexpected ion-binding sites. In the work on bacteriophage MS2, the first model was constructed at 4.2-Å resolution.[11]

After the first model has been built, the resolution of the diffraction data should be the single factor most likely to determine the accuracy of the final model. The diffraction data sets should be complete, with a high signal-to-noise ratio, and have high multiplicity of individual measurements.

Lack of Experience

The most experienced people in a research group usually do not carry out the model building. It is also true that students must learn the trade. In our opinion, the research leader/advisor must allow the student to make the chain trace and build the structure, but the work must be monitored carefully. We have often seen people start working on their own maps before learning the pitfalls by studying examples. Unless a determined effort is made, this problem will get worse as more molecular biologists

[7] T. A. Jones, T. Bergfors, J. Sedzik, and T. Unge, *EMBO J.* **7,** 1597 (1988).

[8] G. J. Kleywegt and T. A. Jones, *in* "From First Map to Final Model" (S. Bailey, R. Hubbard, and D. Waller, eds.), p. 59. SERC Daresbury Laboratory, Warrington, U.K. 1994.

[9] L. Liljas, T. Unge, T. A. Jones, K. Fridborg, S. Lövgren, U. Skoglund, and B. Strandberg, *J. Mol. Biol.* **159,** 93 (1982).

[10] T. A. Jones and L. Liljas, *J. Mol. Biol.* **177,** 735 (1984).

[11] K. Valegård, L. Liljas, K. Fridborg, and T. Unge, *Nature* (*London*) **344,** 36 (1990).

use crystallography, and as crystallographic software takes on the properties of a "black box."

Competition

In the rush to be first with a new exciting structure, the published structure may not have been refined as carefully as it should have been, or the study may be incomplete. For example, the crystals in question may actually diffract to better resolution than was collected.

Belief in a Number

Except at the highest resolution, the crystallographic *R* factor is not a sufficiently good indicator by itself of the correctness of a model.[12] With the efficient minimization algorithms available in refinement programs, it is not difficult to reduce the *R* factor, without actually improving the model. The reason for this is the large number of parameters in a refinement of a macromolecular structure, compared to the relatively small number of diffraction measurements. Therefore, after a certain stage, the refinement algorithm will begin to fine-tune the parameters to fit the noise in the data. Common bad practices that might be used in an attempt to reduce the crystallographic *R* factor include removal of diffraction data, manipulation of the resolution range, removal or reduction in the weight of stereochemical restraints, increasing the number of parameters being refined [for example, by removing noncrystallographic symmetry (NCS) constraints and/or restraints, or by the use of an inappropriate temperature factor model], and uncritical addition of water molecules. These types of mistakes can be reduced if the free *R* factor[3] is used to monitor the progress of the refinement, instead of the "classic" crystallographic *R* factor. For a further discussion of refinement practice, see Ref. 2 and [11] in this volume.[6]

Lack of Equipment

Computers and graphics workstations are now less expensive, but the number of projects has also increased. This means that getting access to the equipment remains a problem in many laboratories.

Bad Refereeing

Referees take some responsibility for the scientific content of what gets published. Unfortunately, the amount of time spent on each paper may

[12] C.-I. Brändén and T. A. Jones, *Nature (London)* **343,** 687 (1990).

vary enormously. The ultimate responsibility for the structure must remain with the authors.

Bad Journals

The most prestigious journals aim to sell to a general audience and, therefore, do not want papers to be swamped with crystallographic detail. If a journal refuses to publish sufficient crystallographic detail, the referee still should insist on seeing it.

Data Deposition

Although most journals now insist on the deposition of coordinate data in the Protein Data Bank (see [29] in this volume[12a]), few journals insist on the deposition of diffraction data. Insisting on the immediate deposition of all experimental data would be the single most useful way to improve the quality of crystallographic models.

What Crystallographers Need to Do

Crystallographers should try to collect good-quality diffraction data. There can be no substitute for accurate, complete, multiply recorded, high-resolution diffraction data. The following should also be remembered:

1. Learn how to use the tools. Whatever model-building program is to be used, try to solve the example structure distributed with O. This includes the experimental MIRAS map and skeleton used to solve P2 myelin protein at 2.7-Å resolution.[7] This example has many of the errors that can be expected in an experimental map, but the protein is small enough to be traced in a few days, even by a beginner. This structure has three molecules in the asymmetric unit, so the map can be averaged. The resulting map is easy to interpret.
2. Every model should be treated as a hypothesis. Be aware of the tendency to believe too early in a particular trace.
3. The model should make chemical sense and satisfy all that is known about the macromolecule. Remember, however, that what has been published previously about the molecule could be wrong.
4. Keep the experimental map. It is not tainted by your model, nor beliefs. During the refinement process, this map can be used to check any major changes in the model.

[12a] E. E. Abola, J. L. Sassman, J. Prilusky, and N. O. Manning, *Methods Enzymol.* **277,** [29], 1977 (this volume).

5. Expect the unexpected.
6. Adopt good refinement practices.

What O Offers Crystallographers

It is outside the scope of this chapter to describe too many details about O, but the main functions of the program are shown in Table I.

TABLE I
MAIN FUNCTIONS AND KEYWORDS OF O, GROUPED ACCORDING TO CATEGORY

Function and keyword	Comment
Display functions	
Draw	Create and display molecular objects
Paint	Color atoms, residues according to properties. Color objects
Select	Select/display atoms/residues according to properties
Crystallographic tools	
Manip	Move atoms or groups of atoms interactively
Lego	Database modeling
Bones/Trace	Manipulate and modify skeletonized electron density
Baton	Build protein main chain using a dipeptide
Symmetry	Display symmetry and packing
Map/Qmap	Display electron-density maps
Slider	Determine where sequence matches density.
Mask	Display and manipulation of molecular envelopes
Patterson	Solve Patterson functions
RSR	Real-space refinement into electron density
Refi	Hermans and McQueens regularization
Sam	Coordinate input/output in common formats. Water adder
MolRep	Evaluate and modify MR solutions interactively
Merge	Merge coordinate data from different molecules
Structure analysis	
LSQ	Least-squares alignment of related molecules. Object transformations
Trig	Analyze distances, angles. Display H-bond pattern
Yasspa	Make assignments of secondary structure
RS_fit	Analyze real-space electron-density fit
Pep_flip	Analyze peptide plane orientation
RSC_fit	Analyze side-chain rotamers
Presentation of structures	
Sketch	Make stylized pictures of proteins/nucleic acids
Plot	Generate hardcopy output
Graph	Interactive graphing of database entries (including Ramachandran plots)
Miscellaneous	
OHeap	Read, write, copy, add, delete datablocks

Building a molecular model from electron density is a complicated process. During the interpretation of an electron-density map, the basic function of the molecular graphics program is to assist the scientist in imagining, and then remembering, the three-dimensional folding and features of the structure. Thus, it is important to be able to change the model quickly, and not to be interrupted by the details of operating a computer program. To facilitate the rapid building and rebuilding of molecular models, O incorporates autobuild options, allowing the user to create a molecular structure quickly from a rough three-dimensional sketch. This has the drawback of possibly making it even easier to build a wrong structure.

O provides a database that allows the crystallographer to work with any molecule-related data. The O database is the equivalent of a simple file system, and the "files" are called datablocks. Each datablock is a vector, and has an associated header that contains data such as name, size, read–write status, and data type. Four data types are allowed: real, integer, character, and text. Most of the datablocks in the database are data items created by the program itself, but the user can also enter and extract information.

In dealing with macromolecules, O adopts a naming convention so that structural information is stored in particular datablock entries. To access the orthogonal coordinates, for example, the program would look for a datablock composed of the user-defined molecule name appended with the string *_atom_xyz*. A datablock of this type is called an atomic property. Likewise, there are residue properties; an example of this is the *_residue_type,* an array of character variables describing the amino acid or nucleotide sequence of the macromolecule. Finally, there are molecule properties. An example of such a property is *_molecule_ca,* which defines the name of the central atom of a residue in the molecule, for example, "CA" for protein, or "P" for nucleic acid. A standard molecular data set consists of a number of datablocks describing the atom names and coordinates, residue names, etc. An external structural format, such as that used by the Protein Data Bank, is converted into a series of vector datablocks. There is no data dictionary in O, and the user is free to load anything into the database, either created using a text editor or generated by standalone programs. Some commands in O will generate new entries; for example, the YASSPA command generates a character variable array of secondary structure assignments. The atom, and residue-based properties, can be used to select what part of the structure is to be viewed and in what colors it should be seen.

The O database can be used for more than just molecule-related information. In Fig. 6, we show a number of useful macros for use with map interpretation. Macro *mc* generates the main-chain skeleton representation of a density trace within 50 Å of the current screen center. The macro *msc*

a. Macro *both* to generate a detailed view of the skeleton close to the screen centre, and an overview of the main-chain trace about the molecular centre.

```
! macro name:  both
@msc
copy save_centre         .active_centre
copy .active_centre    molecule_centre
@mc
copy .active_centre    save_centre
```

b. Macro *msc* to generate all main- and side-chain skeleton atoms within 15 Å of the current screen centre.

```
! Macro name: msc
bone_setup start msc  15. 1 2 3 ;
bone_draw
```

c. Macro *mc* to generate all main-chain skeleton atoms within 50 Å of the current screen centre.

```
! Macro name: mc
bone_setup start mc  50. 1   3 ;
bone_draw
```

FIG. 6. Some skeleton-related macros.

generates an object containing all skeleton atoms within a shorter distance from the center. The macro *both* generates both objects, but things get a bit more interesting. The detailed macro is called in line 2, generating an object containing all skeleton atoms within 15 Å of the current screen center. In O, the current screen center is saved in a vector *.active_centre*, so the third line of *both* saves this in a vector *save_centre*. In line 4, the vector *molecule_centre* is copied to *.active_centre*. The vector *molecule_centre* has been created earlier by the user to specify the center of the structure being built. On line 5, the *mc* macro is called to generate an object showing the main-chain trace around the center of the molecule. In line 6, the initial value for the screen center is restored. The result of this, perhaps obscure, macro is a detailed view of the skeleton at the current point of interest to the crystallographer, and another object showing the current state of the trace. There are many opportunities in O to carry out such manipulations.

O provides an overview of the map by displaying a skeleton representation of the electron density and a detailed view by the display of three-dimensional (3-D) line contours or semitransparent surfaces.

One design philosophy of O is to address one of the problems of building a correct structure by enforcing a systematic mode of model construction, one that makes use of bits and pieces of existing, well-refined structures. Users are encouraged to use side-chain rotamers instead of changing dihedral angles. Other facilities are available for spotting and analyzing potential trouble spots in the structure. This, in an effective yet unobtrusive way, improves on the crystallographer's knowledge of protein structure. It also reduces the number of degrees of freedom in the model. Access to a fold library is also useful during the initial interpretation stage.

O incorporates the use of residue-based checking criteria. Many useful verification criteria are available and some are described in more detail in [11] in this volume.[6] They can easily be added to a user's database, and used for atomic selection or coloring. The fit of the structure to the electron density is particularly interesting during the map interpretation and building stage. The residue-based residuals of Jones *et al.*[1] are explained in Fig. 7. The matrix ρ_{obs} is the experimental map with a particular set of grid spacings. The existing model is used to generate ρ_{calc} on an identical grid. In O this is done by assuming a Gaussian electron-density distribution for each atom

G1	G2	G3
ρ_{obs}	ρ_{calc}	ρ_{env}
experimental	from model	envelope from model, containing fragment of interest

For all nonzero ρ_{env}, calculate:

$$\text{RSRF} = \frac{\Sigma |\rho_{obs} - \rho_{calc}|}{\Sigma |\rho_{obs} + \rho_{calc}|}$$

or

RSCC = Correlation coeff. (ρ_{obs}, ρ_{calc})

FIG. 7. Quantitative fit of a model to a map (real-space *R* factor and real-space correlation coefficient). The electron-density functions G1, G2, G3 are calculated on identical grids.

that makes use of resolution-dependent parameters. Each atom is assumed to have the same temperature factor. The third density, ρ_{env}, is also a calculated electron density, but built from only a subset of the current model. For every nonzero value in ρ_{env}, it then becomes possible to evaluate how well the observed density fits the calculated density. Either an *R* factor-like formulation can be used or a correlation coefficient. The subset of atoms used to create the envelope ρ_{env} can be changed according to what the crystallographer wants to do.

O incorporates 3-D notes. One of the few advantages of a wire model over a computer graphics model is the ability to stick notes on the wireframe! This can be useful during map interpretation, when it may be desirable to jot down such things as chain directionality, or where one is in the sequence. In O, the 3-D notes are associated with a point in space, so that it is possible to display all notes within 10 Å of the current center, for example.

The creation of "any number" of graphic objects is allowed. An object consists of a collection of graphics items such as lines, text, or polygons. Two general object types are available: pickable and nonpickable. Although it is normally desirable to be able to identify atoms by clicking on them with the mouse, it is usually not so interesting to do the same with the vertices of the electron-density wireframe representation. For each object created, its name appears on the screen, and the visibility of the object can be toggled by clicking. Most objects are created explicitly by the user, whereas other objects are created by the program. Any number of objects can be made from a molecule stored in the O database. A molecular object, however, is built from only one molecule. An object description language allows the user to create objects by using a text editor or with a suitable standalone program (e.g., MAMA[8]).

Steps in Building a Model

We identify five key steps in building a model: (1) generating a main-chain trace, (2) determining where the sequence matches the density, (3) building a rough model, (4) optimizing the fit of the model to the density, and (5) evaluating the model. These steps may not lead to a model that completely matches the full sequence of the molecule being studied. Indeed, there are many cases in the literature in which the first model to be built was just a partial polyalanine chain. Frequently, part of the structure can be built that matches the sequence while the rest is polyalanine. Such structures will usually become more complete after another round of calculations that could include phase combination, for example.

Using Skeletons

Map interpretation requires both an overview and a detailed representation of the electron density. This can be achieved by simultaneous viewing of skeletonized and contoured density, respectively. Skeletonized density can be thought of as a piece of string passing through the map. This representation was introduced by Greer[13] in early attempts to automate map interpretation. We use this representation as a means of indicating where we have been in the map and to show our interpretation of the density in terms of a main-chain trace. When a skeleton is used instead of a contoured map, one experiences a great loss of detail. The benefit of this representation, however, is that one is able to view a much larger volume of space without clutter. The skeleton data structure also allows us to associate extra information with this representation, such as our current hypothesis for the fold. In O, the skeletonized density data structure is similar to that used to describe molecular structures.

A number of different skeletonization algorithms have been described.[14,15] The final O data structures are straightforward and easy for someone with another algorithm to adopt [K. Cowtan (University of York, U.K.), for example, can generate O-style skeletons in the density modification program *dm*]. The Greer algorithm requires the definition of both a base level and a density step value. In this algorithm, all points below the base level are removed. The density is then searched for points with values of base+step. They will be removed unless they are needed to preserve continuity or are end points. This is repeated at a level of base+2*step, and so on, until just a skeleton of connected points remains. In the program *bones,* we first calculate a skeleton with base and step values typically 1.25 and 1.0 times the root-mean-square (r.m.s.) deviation level of the map in the whole asymmetric unit. If there are too many connections when viewed in O, it is necessary to increase the base level and recalculate the skeleton. If we see that too few skeleton atoms are connected, the base level needs to be decreased and the skeleton recalculated. The step value is not a sensitive parameter. It should be noted that the value of the density at each skeleton point is not part of the data structure and therefore is not passed on to O. This was an early design decision that 10 years of use seems to have justified. Instead, the crystallographer worries about skeleton connectivity and status. The *bones* program calculates initial status codes that are assigned to skeleton atoms based on the length of the connected fragment of which the atom is a member. Within O, these codes can be used to view and paint the trace according to any scheme that may be

[13] J. Greer, *J. Mol. Biol.* **82,** 279 (1974).

[14] C. K. Johnson, *Acta Crystallogr.* **A34,** S-353 (1977).

[15] S. M. Swanson, *Acta Crystallogr.* **D50,** 695 (1994).

helpful. In the simplest case, one may merely change the codes to mark clearly where one has passed through the map, deciding what is main chain and what is not. In a more complicated scheme, one could use one color for part of the trace where one has placed the sequence in the density, another color where one is sure of the main-chain trace, a third color where a main-chain branch could be wrong, etc.

When the skeleton is being modified, it is vital that the crystallographer view the 3-D contour representation of the density to provide extra detail such as size and shape. This representation is sensitive to the density level; if necessary, therefore, the contoured map can be viewed at higher or lower values. Two sets of commands exist within O to view contoured maps. One works with density brick files (*map* commands), the other reads in the whole map file (*qmap* commands; various map formats are supported). The latter commands maintain a database of contoured brick objects so that if the level is left unchanged, the recontouring time is much reduced.

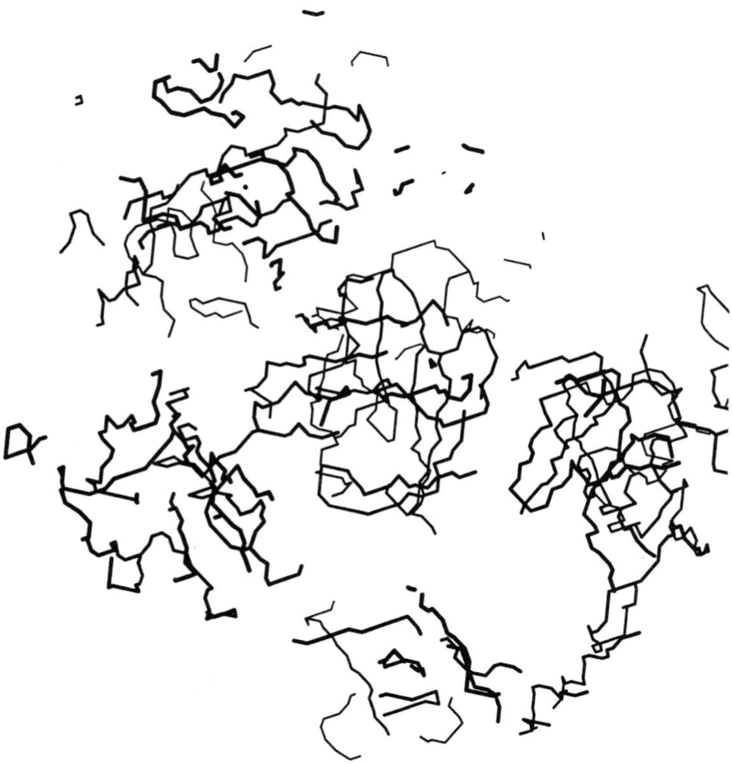

FIG. 8. Overview skeleton from P2 myelin MIRAS map. All "main-chain" bone atoms are drawn that are within 50 Å of the screen center.

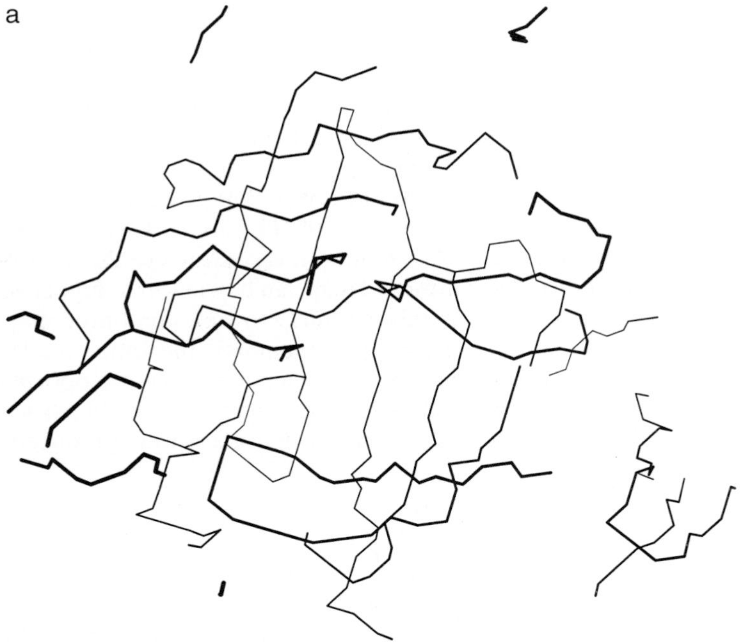

FIG. 9. (a) Starting a main-chain skeleton with the wrong connections. (b) Two connections have been broken (between strands, and between a helix and a strand), and a path has been defined through part of the skeleton (dashed line).

From the "bones molecule," the crystallographer should make a number of different objects. Normally, one object would represent the current main-chain trace and, therefore, occupy a large volume of space. Another would show all skeleton atoms that occur within the current region of interest. These objects could be generated by the macros described earlier in Fig. 6.

Generating a Main-Chain Trace

During the initial inspection process, the crystallographer attempts to determine the molecular boundary. Figure 8 shows the P2 myelin main-chain skeleton around one of the three molecules in the asymmetric unit. In this example, there are good solvent boundaries around the macromolecule although one connection between molecules is apparent. Defining the boundary can be difficult sometimes, especially if chains form tightly interacting dimers, for example. Care must be taken not to move into another molecule. It may, therefore, be useful to generate a symmetry object of

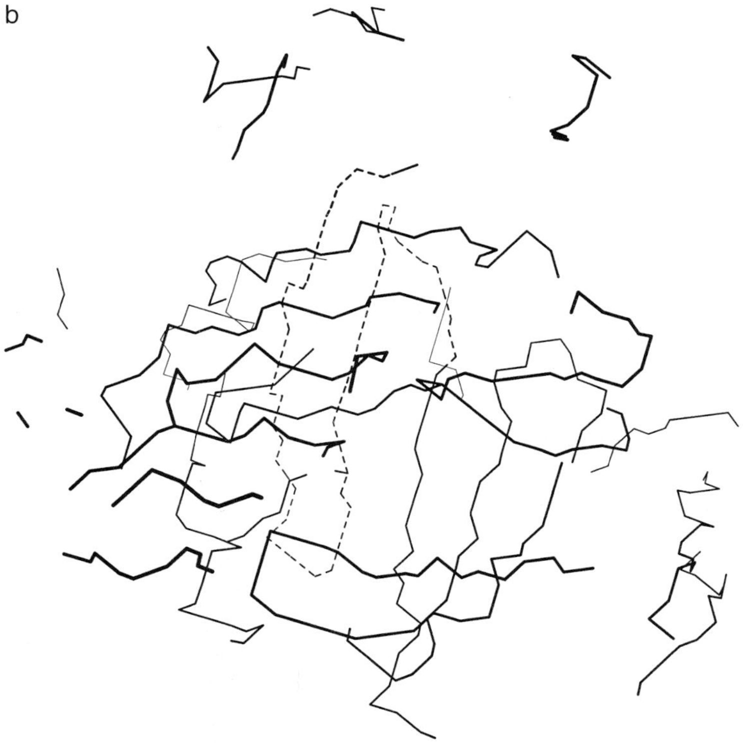

FIG. 9. (*continued*)

the current main-chain trace object. If noncrystallographic symmetry is present, one can also work at improving the mask defining a single copy. This is best done with a macro to include or exclude points close to an identified skeleton atom.

The most important job in the inspection process is to recognize and correct local errors in the skeleton. One should start by looking at the overview trace object (Fig. 9a). If a strand or helix can be recognized it should be followed (looking at the contoured density) until it is not clear what to do (the dashed line in Fig. 9b). One should then return to the overview object and try to recognize another strand or helix, and repeat the process. From the trace in Fig. 9a, it is possible to recognize a set of β strands immediately. Two of these strands appear connected, owing to a main-chain hydrogen bond (Fig. 10). Such connectivity errors in the skeleton should be corrected immediately, either by making or, in this case, by breaking bonds between the bone atoms (Fig. 9b). At the same time,

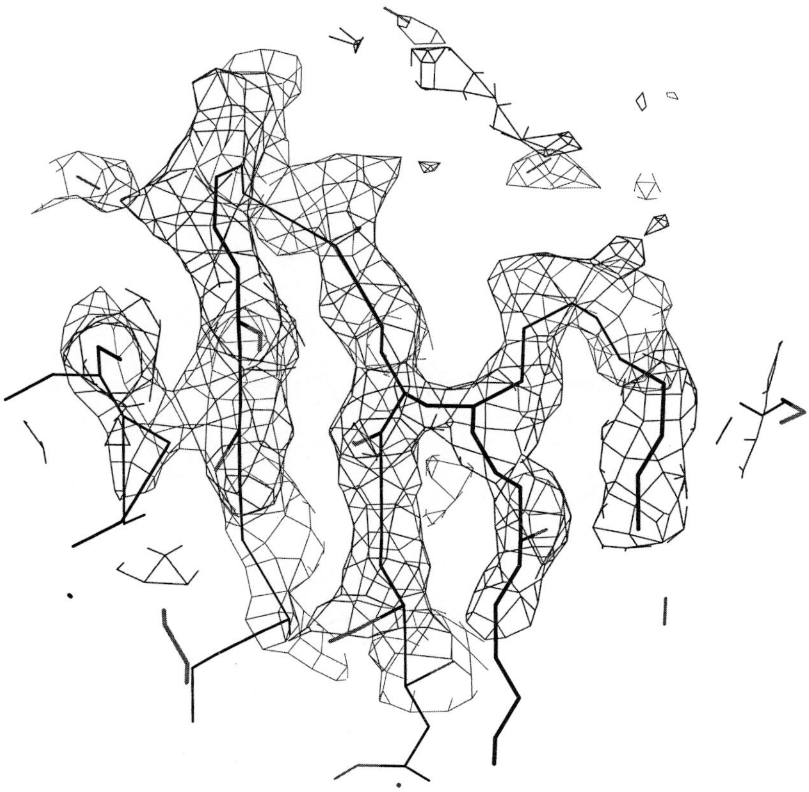

FIG. 10. Close-up view of the connection error between strands, caused by a main-chain hydrogen bond connection.

classification errors should also be corrected, because in many places the main chain may be incorrectly marked as a side chain and vice versa. This is illustrated in Fig. 11 where a side chain from a neighboring α helix interacts with a β-strand side chain. The path taken along the trace can be recognized in Fig. 11 because some of the skeleton atoms have been reclassified already as part of the likely main chain. Methods have been developed that assist in recognizing structural entities such as helices, strands, and other atomic groupings.[16] These are based on the real-space convolution of the experimental map with a structural template. The resulting map has density where the correctly oriented template fits the experimental density.

[16] G. J. Kleywegt and T. A. Jones, *Acta Crystallogr.* **D53,** 179–185 (1997).

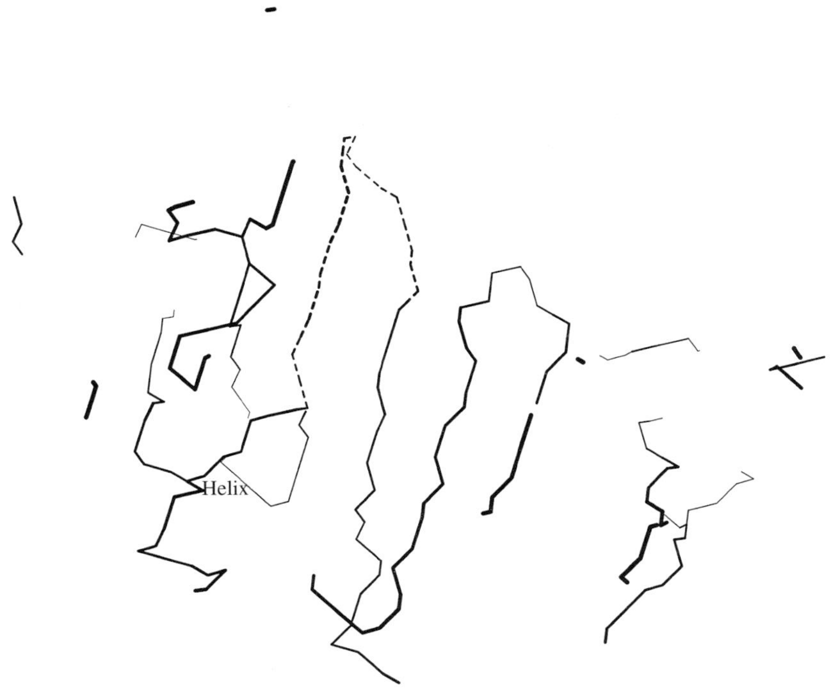

FIG. 11. Close-up of the connection error between a helix and a strand. The helix is to the left and the error is a series of links to the end of the dashed trace.

The level of care taken in repositioning the bone atoms depends on how one plans to build the structure. If the main chain is to be constructed by the placement of C_α guide points in the skeleton (see below), the skeleton atom at the intended position must be placed as accurately as possible. The r.m.s. error in placement is approximately the same as the error in the coordinates of the main-chain atoms that are built using databases,[1] and so it pays to be careful. If the main chain is to be built with the *baton* commands (see below), it is not necessary to position the guide atoms very accurately. With both methods it is a good idea to make the trace continuous and to have a branch point at the (possibly rough) C_α guide point. If the density is smooth at this position, there is a special command in O to add a skeleton branch point. In this case one should not be concerned with the direction of the branch. It is not necessary that the side-chain skeleton atoms accurately mimic the real side chain. We place a branch point even for glycine residues, because this helps later when constructing the main-chain trace.

In an earlier program, Frodo,[5] a number of different ways were introduced to build a structure in an electron density that did not depend on the use of skeletons.[17] In O, the skeleton plays an important role in the construction process. In one method, this can be a direct role wherein the crystallographer creates a so-called bones C_α trace that accurately mimicks a C_α trace in terms of the connectivity and the number of atoms. In this trace, each skeleton atom in a connected segment will be associated with the C_α coordinates of a particular residue in the structure under construction. By specifying a start and end point in a portion of skeleton, one can merge the coordinates of each bone atom into consecutive C_α coordinates in a portion of the protein. It is not necessary to edit the skeleton extensively to produce this bones C_α trace. Rather, once the skeleton has been edited to remove connectivity errors and branch points are added (or removed if in error), a connected section then can be processed to produce the bones C_α trace. The processing algorithm keeps all skeleton atoms that have a branch point and then investigates the separation between these atoms. If this is too long, an atom is added so that neighboring skeleton atoms are approximately separated by the interresidue C_α–C_α distance of 3.8 Å. The connection between the two identified atoms in Fig. 12a is made up of many short bonds between skeleton atoms positioned at grid points. After processing (Fig. 12b), only atoms that are roughly 3.8 Å apart are left. Note that when this trace is merged into the molecule of interest, a decisive step has been taken, namely a decision has been made as to where the electron density and sequence are to be matched. Tools to assist in this process are described below.

When editing the skeleton, one must pay attention to possible false connections such as hydrogen bonds between strands, interacting side chains that can make the density continuous, or disulfide bridges. The latter usually have rather strong density and can be mistaken for the main chain. One will frequently see breaks in the density. In some circumstances, one can make use of the local secondary structure to decide where the side chains are or should be pointing (Fig. 5a). When in doubt, one should proceed in the direction of the secondary structure element.

When moving along the skeleton, one should be attempting to decide on chain directionality and where the sequence of the molecule can be recognized in the density. It then becomes useful to jot down ideas on 3-D notes. In O, these are related to positions so that later one can display all notes within 10 Å of the current screen center, as in Fig. 11, for example, where we were able to recognize a helix. The local chain direction can be

[17] T. A. Jones, *in* "Computational Crystallography" (D. Sayre, ed.), p. 303. Clarendon Press, Oxford, 1982.

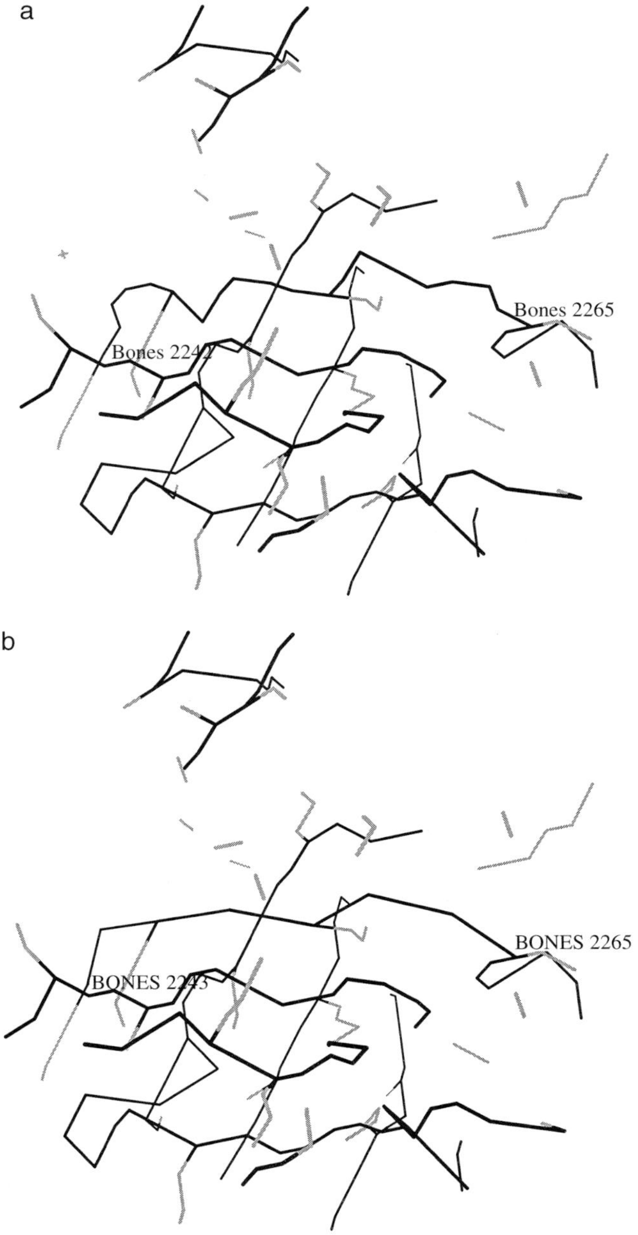

FIG. 12. Creation of a bones C_α trace. In (a) the skeleton is drawn showing the assignments that have been made for main and side chains. In (b) the connection between the identified atoms has been contracted to just those atoms at branch points or separated by ~3.8 Å.

determined in a number of ways. One way is based on the α-helix Christmas tree effect. This is a result of the direction of the C_α–C_β bond relative to the helix axis, so that side chains tend to point toward the N terminus of the helix. This can also be done quantitatively, by measuring the goodness of fit of a polyalanine segment to the electron density when built in either direction. A second, more general, approach requires good phasing and high resolution, ~2.5 Å or better. It then becomes possible to recognize peptide branching. The separation of branch-point pairs corresponding to atoms C_{α_i} and C_i is different from that of pairs C_{α_i} and C_{i-1}, and this allows one to determine the chain directionality.

When thinking about sequence placement, as well as when studying electron density (see below), one can sometimes make use of other information. For example, it may be easy to determine the active site by the location of a heavy atom if one is present, or by binding studies where the experimental phases may be good enough to locate ligands before solving the structure. In many cases, the identity of residues in the active site may be known already. In an MIR map, the known preference of some heavy-atom compounds for particular amino acid side chains may provide useful information. When using the MAD phasing method or when heavy-atom binding sites have been engineered by introducing cysteine residues, there is even more useful sequence information available.

As one moves through the map, one should try to recognize local protein-like features such as α helices, β strands, special side chains (for example, large aromatics), cofactors, ligands, or any metal-binding sites. After a while, one may recognize supersecondary structure motifs such as β–α–β units. Eventually one may recognize one of the more common domain folds such as the NAD-binding domain or a TIM barrel. With the avalanche of new structures, it may be useful to interrogate databases using programs such as DEJAVU (see [27] in this volume[18]). In general, time spent in the library learning about proteins will not be wasted. However, care must be taken for structural variations from the "classic" fold since this has led to tracing errors in the past. In one case, for example, photoactive yellow protein was first described[19] as having a β-clam structure (like P2 myelin protein), but has now been resolved and shown to have a fold more like that of the SH2 domain.[20]

A set of tools has been developed in an attempt to automate the production of a bones C_α trace. The tools aim to make a skeleton more like a C_α

[18] G. J. Kleywegt and T. A. Jones, *Methods Enzymol.* **277,** [27], 1997 (this volume).

[19] D. McRee, J. Tainer, T. Meyer, J. Van Beeuman, M. Cusanovich, and E. Getzoff, *Proc. Natl. Acad. Sci. U.S.A.* **86,** 6533 (1989).

[20] G. E. O. Borgstahl, D. R. Williams, and E. Getzoff, *Biochemistry* **34,** 6278 (1995).

trace by applying a set of filters to an existing skeleton to produce a new, "better" skeleton. These tools need either a good map, or require that the skeleton be edited to remove most of the serious connectivity errors. The pruning filter keeps the connectivity only between atoms with branch points. The fill filter places C_α atoms along the trace at suitable spacings (~3.8 Å). Applying just these commands to the skeleton obtained from the averaged P2 myelin map in Fig. 5b allowed us to create automatically a polyalanine model that matched 115 of 131 residues of P2 with an r.m.s. deviation on C_α atoms of 1.2 Å. The errors in this model result from defects in the skeletonization algorithm and because some side chains interact to produce continuous density. We have stopped the development of this approach for the moment, because the interactive method described above gives more control to the user and the *baton* method is even more popular with users.

Placing the Sequence in the Density

Placing the sequence in the density is the crucial step in building a model, and this is where the qualitative aspects of model building are most apparent. The main reason for this is that the quality of the density for an amino acid side chain depends on where it is in the structure. This means that an external, floppy, tryptophan residue could have just smooth main-chain density, thus looking very much like a glycine (note that a glycine residue cannot look like a tryptophan). Glycines can be useful markers during this important step. Unfortunately, glycines are often found in loops that are usually external and frequently have bad density. Glycine residues in the middle of some solid density are more useful.

Large aromatic residues and disulphide bridges are usually the most useful markers for locating the sequence in the electron density. One must, however, be on the lookout for the unexpected, including unexpected glycosylation sites such as the one that caused the error in Fig. 3, and unexpected ligands. The latter may often have strong density, such as the endogenous fatty acid in the barrel of P2 myelin,[7] and the mixed peptide population in the first HLA structure.[21]

Slider Commands

To help in deciding where the sequence fits the density, O can assist the crystallographer with the *slider* commands.[22] One group of commands is

[21] P. J. Bjorkman, M. A. Saper, B. Samraoui, W. S. Bennett, J. L. Strominger, and D. C. Wiley, *Nature (London)* **329,** 506 (1987).

[22] J. Y. Zou and T. A. Jones, *Acta Crystallogr.* **D52,** 833 (1996).

qualitative in nature, whereas another group is quantitative. The qualitative *slider* commands allow the user to enter a guess of the sequence on the basis of the shape of the density. This guess is compared with the real sequence, using a scoring matrix to decide where we are in the sequence. The matrix is a look-up table that is needed to evaluate how well the guess scores for each of the 20 amino acids. In a simple system we could use the letters *b*, *m*, and *s* to indicate big, medium, and small residues, for example. With such a system, a tryptophan in the sequence would have a high score if it were guessed as *b*, lower if *m*, and even lower if *s*. Similarly, a glycine in this system would score low, higher, and even higher for guesses *b*, *m*, and *s*, respectively. The standard matrix distributed with O is more complicated and the guesses appear at first sight like the one-letter amino acid code (the matrix can, of course, be modified by the user). If the user

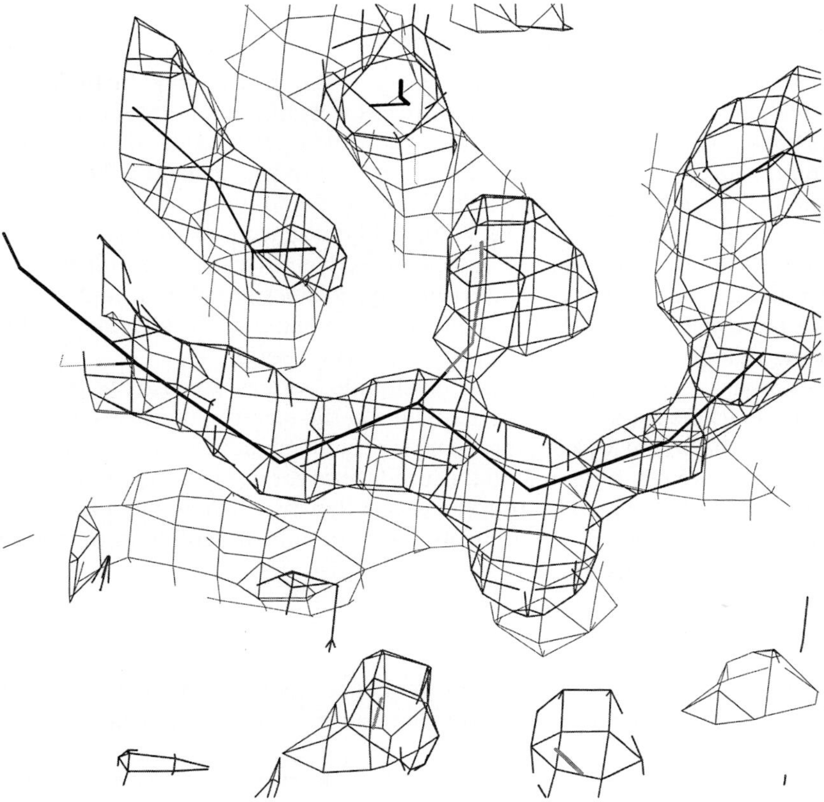

FIG. 13. Electron density and skeleton around a well-defined aromatic ring.

wishes, the result can be associated with a portion of a molecule and stored in the database. It may often be worthwhile to build a portion of a polyalanine chain in the density of interest to see how well the different amino acids fit the density. The *slider_rotamer* command then allows one to display each kind of amino acid side chain and its equivalent set of rotamers. Especially for beginners, this helps in judging the size of different side chains relative to the density of interest.

One of the problems with *slider* is that a long stretch of residues is needed to be sure that the correct result appears at the top of the scoring list. Unfortunately, a long stretch may have an insertion or deletion in it that has already been missed. Therefore, once a number of guesses have been made, they can be combined with the introduction of a variable-

```
Slid> Estimated sequence  : gvwa      Slid> Estimated sequence   : awvg
Slid> Average=0.44,rms=0.12           Slid> Average=0.44,rms=0.12
Slid>                       GVWA      Slid>                        AWVG
Slid>  Fit   1 0.875 A6     GTWK      Slid>  Fit   1 0.800 A96     KWNG
Slid>  Fit   2 0.725 A33    GNLA      Slid>  Fit   2 0.750 A3      KFLG
Slid>  Fit   3 0.700 A84    VTLA      Slid>  Fit   3 0.725 A33     GNLA
Slid>  Fit   4 0.675 A26    GLAT      Slid>  Fit   4 0.675 A112    KMVV
Slid>  Fit   5 0.675 A55    SPFK      Slid>  Fit   5 0.650 A82     STVT
Slid>  Fit   6 0.675 A122   VVCT      Slid>  Fit   6 0.625 A84     VTLA
Slid>  Fit   7 0.650 A99    GNET      Slid>  Fit   7 0.600 A23     LGVG
Slid>  Fit   8 0.650 A82    STVT      Slid>  Fit   8 0.600 A53     TESP
Slid>  Fit   9 0.650 A111   GKMV      Slid>  Fit   9 0.600 A40     VIIS
Slid>  Fit  10 0.625 A28    ATRK      Slid>  Fit  10 0.600 A25     VGLA
Slid>  Fit  11 0.625 A89    GSLN      Slid>  Fit  11 0.600 A80     TKST
Slid>  Fit  12 0.600 A62    ISFK      Slid>  Fit  12 0.600 A26     GLAT
Slid>  Fit  13 0.600 A40    VIIS      Slid>  Fit  13 0.600 A21     KALG
Slid>  Fit  14 0.600 A72    ETTA      Slid>  Fit  14 0.575 A64     FKLG
Slid>  Fit  15 0.600 A24    GVGL      Slid>  Fit  15 0.575 A111    GKMV
Slid>  Fit  16 0.600 A50    TIRT      Slid>  Fit  16 0.575 A81     KSTV
Slid>  Fit  17 0.600 A22    ALGV      Slid>  Fit  17 0.575 A60     TEIS
Slid>  Fit  18 0.575 A102   TTIK      Slid>  Fit  18 0.575 A123    VCTR
Slid>  Fit  19 0.550 A79    KTKS      Slid>  Fit  19 0.575 A9      KLVS
Slid>  Fit  20 0.550 A73    TTAD      Slid>  Fit  20 0.575 A72     ETTA
```

FIG. 14. *Slider* example from the tetrapeptide density shown in Fig. 13. The two sets of output search for the guess in both directions in the sequence. The output above is a pasting together of two runs with the *slider_guess* command.

length gap between them. One other drawback remains, and that concerns the chain directionality. If this is unknown, the reverse sequence must also be tried. For example, in the portion of density from P2 myelin shown in Fig. 13, one may guess that the sequence is GVWA after careful inspection and use of the *slider_rotamer* command. The comparison of both directions with the sequence gives the result shown in Fig. 14. In this case, the correct direction gives the best score.

A quantitative approach has been developed[22] that looks promising. Given a well-fitting polyalanine model, in this method we mutate each residue to each of the 20 different amino acids. For each amino acid, the fit of each rotamer is optimized to the density. Only a rotational search is carried out, pivoting the whole residue about its C_α atom. The real-space fit of the best-fitting rotamer is then used as an index of fit to determine how well that particular amino acid type fits the density. The resulting scoring matrix can then be used in the *slider* system. For the averaged P2 myelin map, this method works rather well. If one searches for a fragment of five residues, the correct answer is the top score 35% of the time. Increasing the length to 8 and 15 residues gives 67 and 86% correct scores, respectively. The main problem is caused by interacting side chains, often resulting in better fits for incorrect but longer side chains. As one develops the model, this problem should become less serious, because more and more density is occupied by fitting atoms. It should also be noted that an amino acid such as alanine will fit snugly into the density of a phenylalanine. This problem can be overcome by using envelopes for evaluating the goodness of fit that are larger than normal for the small side chains.

Generating the First Rough Model

The first rough model is usually built from a C_α trace, which is then "fleshed out" using databases. The C_α trace can be made either from a skeleton (or from multiple skeletons) as described above or by use of the *baton* commands. In its simplest mode, this command bears some resemblance to the methods used in Frodo[5] and to a prototype program developed by J. Pflugrath. In Frodo, one could "pop" along the sequence, adding the coordinates for the next residue in a standard conformation, and connected to the previous residue with standard bond lengths and angles. One could then rotate and translate the residue so that the main chain fitted the density and the side chain pointed off toward appropriate density. The *baton* method is similar but only a dipeptide is drawn and normally only the C_α atoms of the dipeptide are viewed. The rotation operation is now a pivot around the first C_α of the dipeptide and the position of the second residue is where the next residue to be built will be placed. As with Frodo,

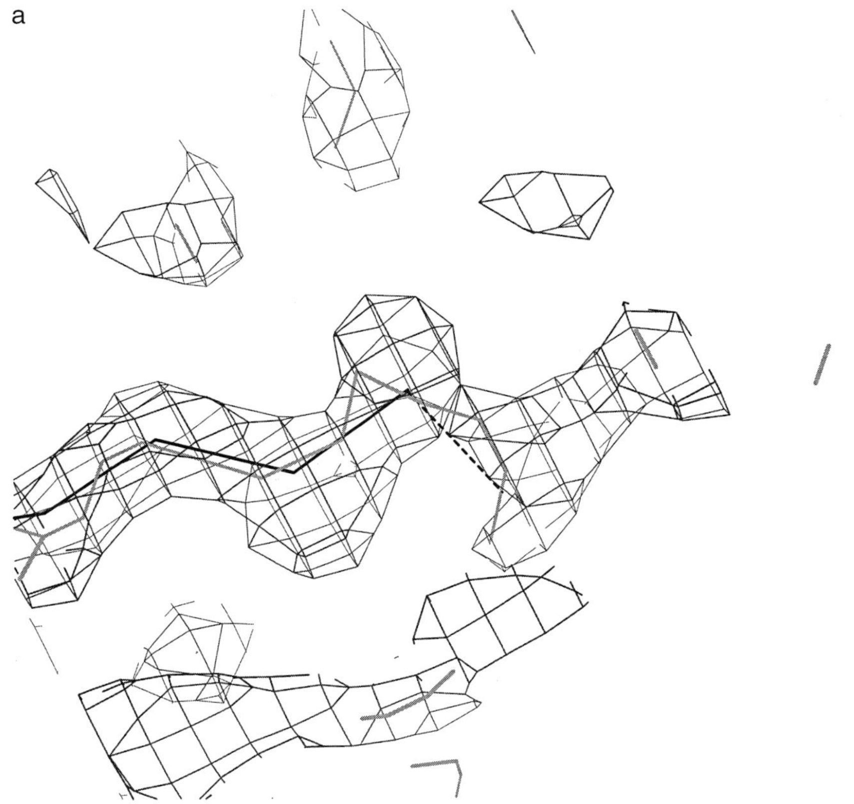

FIG. 15. Use of the baton to build the C_α trace. In (a) the dipeptide baton follows the skeleton, while in (b) it is positioned to continue building a β strand.

a buildup of errors occurs that requires an occasional translation of the dipeptide to maintain the fit to the density. In the present implementation, the user is not concerned with the placement of the side-chain atoms because this will be approximately correct after later database building. When accepted, the coordinates of the second residue of the dipeptide are written into the main chain of the structure being built. The dipeptide is repositioned so that it now pivots about the C_α position of this residue. The advantage of this method over the skeleton method described above is that one has better control in placing adjacent residues at ~3.8-Å intervals. As one gains experience with this method, the dipeptide indeed spins through the density much like a baton cast into the air.

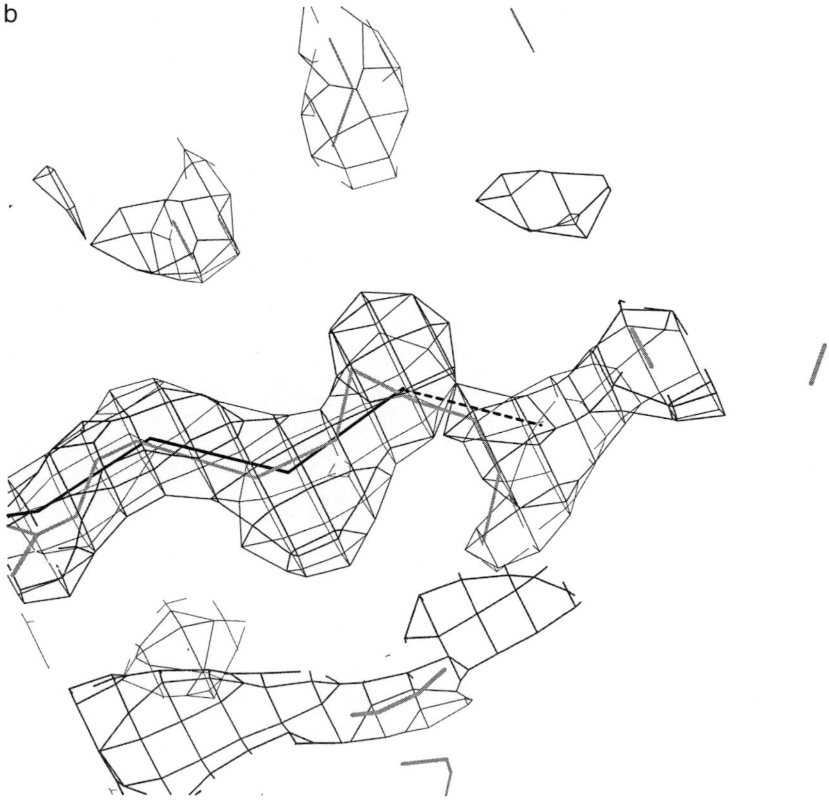

FIG. 15. (*continued*)

The next placement of the baton after the user has accepted the current position actually depends on a mode switch. In the simplest mode, it is always placed in a fixed orientation and the user then needs to change some dials to point it into the density. Alternatively, depending on the mode switch, it can be positioned to "follow" a particular skeleton or to build a standard piece of secondary structure (Fig. 15). When "skeleton sniffing," the baton may have a number of alternative possibilities due, for example, to branching. These can be cycled one at time via an algorithm that shows what are considered to be the most likely alternatives. In every alternative, the second C_α of the baton will be placed on a line joining connected skeleton atoms (Fig. 15a). The order of preference is based on the codes of these linked skeleton atoms, so that a placement between a pair of main-chain skeletons is taken as the most likely possibility, and a placement between linked side-chain atoms as the most unlikely. The baton tip will not be placed between skeleton atoms in the direction from which the structure has just been built. When building in secondary structure mode, the last four residues that have been built are used to decide where to place the dipeptide after making a least-squares comparison with short α-helix and β-strand units. Whichever mode is in use, the crystallographer is free to decide where the dipeptide is ultimately to go. In Fig. 15a, a break in the main-chain density results in an incorrect initial placement of the baton. On changing to secondary structure mode, the baton appears at roughly the correct place (Fig. 15b).

Whether the C_α trace has been made from skeletons or using the *baton* commands, the main chain is then (re-)made from a database of well-refined structures. Jones and Thirup[23] showed that the protein main chain could be generated from short fragments taken from a library of well-refined structures. For retinol-binding protein (RBP), on average, fragments of ~7 residues could be found that locally matched with r.m.s. deviations of <0.5 Å on C_α atoms. With such small deviations, the peptide planes of these fragments were well aligned with those in the RBP structure. Therefore, provided the correct fragment is located, one can be fairly confident that the carbonyl oxygens will be correctly oriented. In a refinement of the basic algorithm for use in O, Jones *et al.*[1] have discussed how well the method works when errors are introduced into the coordinates of the C_α guide atoms. Not surprisingly, the largest deviations were found for the carbonyl oxygen atoms. The reconstructed main chain had an r.m.s. error approximately equal to the r.m.s. error introduced into the guide coordinates. The benefits of using the database approach in map interpretation have been described by Zou and Mowbray,[24] who compared three structures of the periplasmic glucose/galactose-binding protein (GBP). One was built and refined with databases at 2.4 Å, and then further refined to 1.7 Å. The third structure was from a related GBP refined by more traditional methods. They demonstrated that the use of databases both speeded up the modeling process and improved the quality of the models.

The algorithm in O (referred to as autobuilding) searches for the best-fitting pentapeptide in the current window being built, but keeps only the main-chain coordinates of the central three residues. The window is then moved forward three residues (i.e., ensuring a two-residue overlap with the previous window) and the next pentapeptide is determined. Eventually, only the first and last residues in the region of interest remain to be built.

[23] T. A. Jones and S. Thirup, *EMBO J.* **5,** 819 (1986).
[24] J. Y. Zou and S. L. Mowbray, *Acta Crystallogr.* **D50,** 237 (1994).

This can be done with *baton* or by extending the region to be built artificially by one residue at both ends, prior to autobuilding. By choosing a five-residue comparison window, we ensure that a similar conformation will be found in the database. It is likely that the peptide orientations would be "correct" even if a somewhat bigger zone is used, but there is no need to take that risk. Nothing is gained with a longer window except some minor speed improvements.

At first sight, it might appear surprising that the fragment approach works so well. However, it has become apparent as more and more high-resolution structures are determined that protein molecules adopt energetically preferred conformations. This results in tight clustering in their Ramachandran plots and is a good indicator of the accuracy of a protein model.[25] If we restrict ourselves to just the center of the regions defining the α helix, β strand, and left-handed α helix, the number of possible fragments needed to define a pentapeptide from C_α atoms alone becomes $2 \times 3 \times 3 \times 3 = 54$ pentapeptides (note that the first peptide contributes just a ψ variation). This is a somewhat smaller set than that defined by Sussman and coworkers.[26]

The side-chain atoms are generated from a rotamer database. Initially they are added in the most likely rotamer conformation. The database used in O is based on an unpublished analysis made in 1996 by G. Kleywegt of all high-resolution structures, and includes all conformations that have a frequency greater than 5%. In some of the longer side chains, the torsional angles do not form strong rotamer groupings. Rotamers for arginine, for example, have only χ_1 and χ_2 restrictions.

Optimizing the Fit of the Model to the Density

Provided care has been taken in the placement of the C_α atoms making up the initial trace, the main chain of the rough model should have a good fit to the density. The peptide planes should be "correct" provided the C_α atoms have been placed with an accuracy of 0.5–1 Å. The side chain should be pointing in the right direction, but for many side chains it will be necessary to optimize their fit to the density.

The great success and widespread adoption of Frodo for the construction of a model was due in part to the strategy of allowing the crystallographer to rip apart the model by a combination of group or single atom movements, and/or localized or extended dihedral rotations; in essence an electronic

[25] G. J. Kleywegt and T. A. Jones, *Structure* **4,** 1395–1400 (1996).

[26] R. Unger, D. Harel, S. Wherland, and J. L. Sussman, *Proteins Struct. Funct. Genet.* **5,** 355 (1989).

"Richards box." After placement in the density, the model could then be regularized to adopt standard geometry. While giving great freedom to the crystallographer, this approach could also lead to models that had bad stereochemistry. O has a large set of tools for changing a model (Table II) but in the first rounds of building and refinement, we try to persuade users to restrict themselves to the stereochemically most reasonable conformations of main and side chains. This is accomplished for side chains by using the rotamer library described above. An improved fit to the density can often be made without ripping the model apart. This can be accomplished by moving the whole residue as a rigid body and then selecting the best-fitting rotamer. If one uses the *move_zone* command and double clicks on an atom, that atom becomes the pivot point. Frequently, one would pivot about the C_α atom to point the C_α–C_β bond correctly and then the correct side-chain conformation could be chosen from the rotamer library. This can also be done automatically with a real-space density-optimizing command (*RSR_rotamer*) that pivots around the C_α atom, trying each rotamer in turn. The usefulness of automated methods depends a great deal on the quality of the map.

Whether fitted manually or automatically, the resulting chain still becomes distorted and requires regularization. The deviation of the peptide units from "standard" values can be monitored at any time by determining the "pepflip value," an indication of how much each carbonyl oxygen

TABLE II
MODEL MANIPULATION COMMANDS IN O

Interactive	Real-space refinement
move_atom	*RSR_zone*
move_zone	*RSR_rigid*
move_fragment	*RSR_rotamer*
flip_peptide	*RSR_dgnl*
tor_residue	
tor_general	
grab_atom	
grab_group	
lego_side_chain	
lego_ca	
lego_auto_mc	
lego_auto_sc	
baton_build	
refi_zone	
bond_make	
bond_break	

deviates from similar conformations in the database.[1] Some of the longer side chains will still need to be fitted because they do not have preferred rotamer conformations. This is accomplished with a single click on an atom to define torsional adjustments (*tor_residue* command) and/or fragment moves (the *move_fragment* command). Groups of connected atoms can also be generated by breaking bonds; these then are either moved as a rigid unit (the *group_grab* command) or changed by dihedral rotations around some of the remaining bonds (the *tor_general* command). Once regularized, the *rsc_fit* command can be used to monitor how much each side-chain conformation deviates from one of the preferred rotamers.[1] By using the graphing features of O, the user can also plot one residue error indicator against another as suggested by Zou and Mowbray.[24]

Eventually a quantitative measure of how well each residue fits the density can be evaluated[1] (Fig. 7). In the original description, we used an *R* factor-like grid summation, but one can also calculate a correlation coefficient. The latter approach has the disadvantage that if a group of atoms fits snugly in weak density, it will still score well. Once calculated, the residue-based properties can, of course, be used to color atomic objects, making it easy to recognize the good or badly fitting portions of the molecule. The property can be evaluated for any set of atoms within each kind of residue. By choosing to evaluate the goodness of fit of just the main-chain atoms, one can see how well the main-chain trace follows through the density. By evaluating just the side-chain atoms, it may be possible to detect out-of-register errors by identifying clusters of poorly fitting adjacent residues.

Evaluate the Model Continuously

As more of the sequence is fitted to the density, the process should become easier if the interpretation is correct. More frequently, the crystallographer should see aspects of what is known about the molecule become clearly understood from the model under construction. Interacting side chains from residues that are far apart in the sequence should make chemical sense, e.g., an arginine residue should form an ion pair with a glutamic acid residue, instead of being buried in a patch of hydrophobic side chains. Interactions between noncrystallographically and/or crystallographically related molecules should also make chemical sense. It may be useful to evaluate the overall distribution of some residues, looking for buried charges, patches of exposed hydrophobic residues (they might be functional and not errors), glycine, and proline distributions. Any program that evaluates residue- or atom-based environments or properties could be modified easily to produce O data structures or macros.

At this stage, the crystallographer must begin checking if the model satisfies what is known about the molecule from biochemical data. If certain residues have been identified as being in the active site, are they close together in the model? If the molecule has disulfide bridges, are they formed in the model? However, the crystallographer must be aware that this kind of published data can also contain errors. In our work on CBHI,[27] for example, an aspartic acid that had been identified as being involved in catalysis, using epoxide labeling techniques, turned out to be ~25 Å away from the active site. Crystallographers must also be careful with their notes. In our attempts to fit the sequence of RBP[28] to the electron density, for example, it was impossible to fit the expected disulfide linkages until it became apparent that they had been written down incorrectly in the notebook (the structure was then retraced in a couple of hours). Similarly, in the structure determination of α-glutathione transferase (GST),[29] the wrong sequence was entered for one residue. The first refinement macrocycles were made with an aspartic acid side chain where there should have been a glycine. This could actually be a good way of evaluating a refinement because in the GST refinement, the side chain had a group temperature factor of more than 60 Å^2 when we detected the error, while the temperature factor for the main chain was 2 Å^2.

If the structure is related at the primary sequence level to a family of proteins, it will be necessary to evaluate the sequence similarities in light of the structure. A conserved hydrophobic core would be good to see; large deletions mapping to central β strands could indicate problems.

Learning to Use the Tools

The skeletons and maps described in this chapter are available from the authors. These were used to solve the structure of P2 myelin protein and are a good start for learning about map interpretation. It should be possible to make a successful interpretation after about 2 days work. A series of macros are also available that take one through the map, pointing out different kinds of errors in various regions of interest. Other introductions to the program (*O for Morons* and *O for the Structurally Challenged*)

[27] C. Divne, J. Ståhlberg, T. Reinikainen, L. Ruohonen, G. Pettersson, J. K. C. Knowles, T. T. Teeri, and T. A. Jones, *Science* **265,** 524 (1994).

[28] M. E. Newcomer, T. A. Jones, J. Åqvist, J. Sundelin, U. Eriksson, and P. A. Peterson, *EMBO J.* **7,** 1451 (1984).

[29] I. Sinning, G. J. Kleywegt, S. W. Cowan, P. Reinemer, H. W. Dirr, R. Huber, G. L. Gilliland, R. AN. Armstrong, X. Ji, P. G. Board, B. Oln, B. Mannervik, and T. A. Jones, *J. Mol. Biol.* **232,** 192 (1993).

have been written by G. Kleywegt, who also maintains a frequently asked questions (FAQ) list. A manual for the program is available as PostScript or via the web (http://www.imsb.au.dk/~mok/O) and we also run an information server. Enquiries should be addressed to *alwyn@xray.bmc.uu.se.*

[11] Model Building and Refinement Practice

By GERARD J. KLEYWEGT and T. ALWYN JONES

Introduction

An initial model built into an electron-density map calculated with experimental phases, or into a map poorly phased by molecular replacement, will usually contain many errors (see Refs. 1 and 2, and [10] in this volume[3]). To produce an accurate model, one must carry out crystallographic refinement as well as rebuilding at the graphics display in a cyclic process of gradual improvement of the model. Depending on the size of the structure, the automatic (refinement) or the manual (rebuilding) part may be rate limiting. Refinement programs change the model to improve the fit of observed and calculated structure-factor amplitudes. Many different refinement programs exist (see Kleywegt and Jones[1] for some history), but most contemporary programs use reciprocal-space methods. Because of the limited resolution typically obtained in biomacromolecular crystallography, the relatively scarce experimental data are augmented by chemical information, for instance on ideal bond lengths and angles. Rebuilding the model at an interactive graphics workstation is necessary to remove errors that cannot be remedied by the refinement program. Such errors often require a major new interpretation of parts of the model and, at present, can be done only by crystallographers at a three-dimensional workstation.

Model refinement has been a personalized affair for which laboratories have their preferred strategies, programs, etc. This has resulted in models with distinctive features of both the groups concerned and the software used. In this chapter we propose our views on how a macromolecule should be refined, and argue that the present practices in the community are often far from optimal, especially when only low-resolution data are available.[1]

[1] G. J. Kleywegt and T. A. Jones, *Structure* **3,** 535 (1995).
[2] C. I. Brändén and T. A. Jones, *Nature* (*London*) **343,** 687 (1990).
[3] T. A. Jones and M. Kjeldgaard, *Methods Enzymol.* **277,** [10], 1997 (this volume).

0076-6879/97 $25.00

All refinement programs nowadays use empirical restraints or constraints to ensure that a reasonable structure ensues during the refinement steps. This can result in a model with good stereochemical properties, but also in a model in which molecules related by noncrystallographic symmetry (NCS) are forced to have similar (restrained) or identical (constrained) conformations. Nevertheless, unless special precautions are taken, overfitting the data (i.e., adjusting the model in a manner that is not warranted by the quantity and/or quality of the experimental data) is almost guaranteed to take place. Because of the limited resolution of the diffraction data in a typical macromolecular crystallographic study, the number of parameters in the model is often similar to, or even greater than, the number of experimental observations. "Popular" methods to push the conventional R factor down are based on either increasing the number of model parameters or decreasing the number of reflections used in refinement. Examples of such methods include ignoring NCS[4] (otherwise the single most powerful method to reduce the number of degrees of freedom in the model), refining individual temperature factors and modeling alternative conformations at resolutions where this is not warranted,[5] removal of data[6] [by inappropriate resolution and $F/\sigma(F)$ cutoffs], and inclusion of spurious entities (such as solvent molecules). These methods invite the refinement program to fit error terms, and sometimes such overfitting can even mask gross errors in the model.[5,7]

When one is rebuilding a model, the original experimental map should always be kept (if it is available), and at each stage one should try to reinterpret it in the light of the current model, and with the help of the current $2F_o - F_c$, $F_o - F_c$, and other maps. One should keep in mind the kind of errors that might still be present in the model, and should try to locate places in the map that could be the result of such errors. In addition, the accumulated knowledge concerning macromolecular structures should be used to locate places in the current model that deviate from our expectations and previous experience (as pertaining to quality of the fit to the map, stereochemistry, preferred main-chain and side-chain conformations, and environments of residues). If such deviations are found, they could indicate local errors. Alternatively, they may occur in regions of potential biological interest, but only if the crystallographic data permit this should liberties be taken with the model.

[4] G. J. Kleywegt, *Acta Crystallogr.* **D52,** 842 (1996).

[5] G. J. Kleywegt, H. Hoier, and T. A. Jones, *Acta Crystallogr.* **D52,** 858 (1996).

[6] E. J. Dodson, G. J. Kleywegt, and K. S. Wilson, *Acta Crystallogr.* **D52,** 228 (1996).

[7] G. J. Kleywegt and T. A. Jones, *in* "Making the Most of Your Model" (W. N. Hunter, J. M. Thornton, and S. Bailey, eds.), p. 11. SERC Daresbury Laboratory, Daresbury, UK, 1995.

The aim of model building and refinement should be to construct a model that adequately explains the experimental observations, while making physical, chemical, and biological sense. It is a fact of life that low-resolution data can yield only low-resolution models. The refinement process, in particular, should therefore always be tailored for each problem individually, keeping in mind the amount, resolution, and quality of the data. At low resolution, it will often be impossible to produce a precise model, but that does not mean that one cannot produce an accurate model. The distinction between precision and accuracy is an important one, but unfortunately the two concepts are often confused. Precision is related to level of detail, accuracy to how close to the "truth" something is. For instance, the number 4.987453637 is a very precise, but not very accurate, approximation of the value of π; the number 3.14, however, is a not very precise, but much more accurate estimate. In the case of protein crystallography, a 3-Å structure with individual temperature factors, unrestrained NCS, and hundreds of water molecules built in may seem very precise, but it is doubtful whether the atoms on average are even within 1 Å of their actual positions. Similarly, a hydrogen-bonding distance reported as 2.83 Å for a 3-Å structure is quite precise, but not necessarily accurate. Even at low resolution one can build accurate models as well as the data allow, but only high-resolution data may yield a model that is both accurate and precise. Even high-resolution data alone are no guarantee of an accurate model[1,4,7]; in addition one must use sensible refinement and rebuilding procedures, and monitor the quality of the model throughout. When the final model is completely wrong (and this has happened even at 2.4 Å), clearly the data played a minor role. However, even at atomic resolution the final model still contains traces of the subjective decisions of the crystallographer (e.g., with respect to the inclusion of alternative conformations, hydrogen atoms, and solvent entities, and the choice between isotropic and anisotropic temperature-factor modeling for certain parts of the model).

Figure 1 shows a schematic view of the refinement and rebuilding process (with NCS present) as we tend to carry it out. The "loop" inside the boxed area we refer to as a macrocycle (i.e., one cycle of map calculations, quality control, rebuilding, and refinement). In the past one would typically go through dozens of such cycles to produce the "final" model. In our experience, good models now can be produced in a fraction of the time (typically 5 to 10 macrocycles).

In the following discussion, use of X-PLOR[8] and O[9] for refinement and

[8] A. T. Brünger, "X-PLOR: A System for Crystallography and NMR." Yale University, New Haven, Connecticut, 1990.

[9] T. A. Jones, J. Y. Zou, S. W. Cowan, and M. Kjeldgaard, *Acta Crystallogr.* **A47,** 110 (1991).

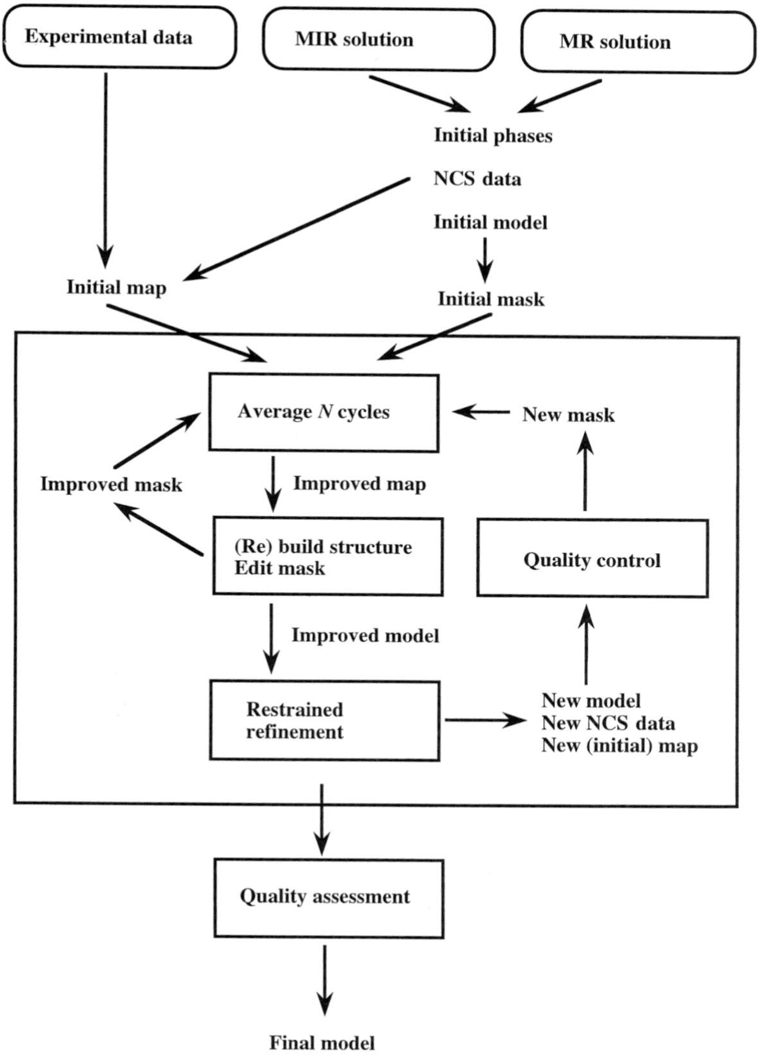

FIG. 1. Overview of the structure-building process for crystallographically determined macromolecular structures. The items inside the boxed area are collectively known as one macrocycle.

rebuilding, respectively, is assumed. Both programs provide powerful tools to help build good models.

The process of building an initial model into a multiple isomorphous replacement (MIR) map using the tools available in O is described in [10] in this volume.[3] These tools include the Trace and Bones commands to edit skeletonized electron density[10]; the Baton command to build the initial C_α trace[10]; and/or the Lego commands to generate the initial (partial) model, using a database of well-refined structures to generate the main-chain and side-chain atoms.[11]

In the case of a molecular replacement (MR) problem, one will often need to "mutate" some of the residues (with the Mutate commands in O), build loops with insertions or deletions (using the Baton and Lego commands), and select different side-chain conformations. Because this is mostly a rebuilding problem, most of the tools discussed in the section about this topic can also be used to generate an initial model for refinement.

Refinement

R_{free}

Brüger has introduced a cross-validation scheme based on the so-called free R factor, or R_{free} (see Refs. 12, 13, and 13a, and [19] in this volume[14]). The idea is to set aside a small fraction of the data (the "test set") that is not used in the refinement, but for which an R factor is nevertheless calculated all the time. Comparing the values of the conventional and free R factor tells something about the extent to which one has overfitted the data as well as about the quality of model and data.

A particularly striking demonstration of the power of the free R factor and the poor performance of the conventional R factor as an indicator of the correctness of a model was given by intentionally tracing the structure of cellular retinoic acid-binding protein (CRABP) type II (previously solved at 1.8-Å resolution[15]) backward, and refining this "model" using data to only 3-Å resolution.[1,7] Using an "established" refinement protocol, the conventional R factor came down as low as 0.214, and this model had what

[10] T. A. Jones and M. Kjeldgaard, *in* "From First Map to Final Model" (S. Bailey, R. Hubbard, and D. A. Waller, eds.), p. 1. SERC Daresbury Laboratory, Daresbury, UK, 1994.

[11] T. A. Jones and S. Thirup, *EMBO J.* **5,** 819 (1986).

[12] A. T. Brünger, *Nature (London)* **355,** 472 (1992).

[13] A. T. Brünger, *Acta Crystallogr.* **D49,** 24 (1993).

[13a] G. J. Kleywegt and A. T. Brünger, *Structure* **4,** 897 (1996).

[14] A. T. Brünger, *Methods Enzymol.* **277,** [19], 1997 (this volume).

[15] G. J. Kleywegt, T. Bergfors, H. Senn, P. Le Motte, B. Gsell, K. Shudo, and T. A. Jones, *Structure* **2,** 1241 (1994).

is usually termed "excellent stereochemistry." The free R factor, however, could not be fooled; it ended up at a value of 0.617, slightly worse than the value expected for a random set of scatterers. A consequence of the fact that the conventional R factor is not correlated with the accuracy of a model (unless the data-to-parameter ratio is high) is that coordinate error estimates derived from conventional Luzzati plots[16] are meaningless. We therefore proposed that this quantity be estimated from an R_{free} Luzzati plot[15] instead. In the case of the backward-traced CRABP model, the estimated coordinate error based on a Luzzati plot using the conventional R factor is ~0.35 Å, whereas that based on the free R factor is "infinite," which, at least in spirit, is more accurate. There are other indications that, at least at low resolution, an R_{free}-based Luzzati plot gives a more accurate estimate of coordinate error than one based on conventional R factors (see Ref. 13a and [19] in this volume[14]).

Besides enabling one to identify cases in which the model is seriously flawed, the free R factor can also be used to tune the refinement protocol for each individual case. For example,[13a]

1. To decide on a proper weight for the crystallographic pseudo-energy term[13]
2. To decide on the best temperature-factor model (one B per residue; two B's per residue; atomic B factors; weak or strong bonded restraints; weak, strong, or no NCS restraints; etc.)
3. To decide on the best way to model NCS (constraints, strong restraints, weak restraints, no restraints, a mixture of constraints and restraints, different restraint weights for different domains or for different sets of main-chain and side-chain atoms, etc.)
4. To find out if simulated annealing (SA) can be applied meaningfully. At low resolution it is not at all guaranteed that SA refinement will actually improve the model, but this will have to be investigated from case to case
5. To identify (sometimes) problems with a dataset. For instance, when the resolution of the data has been grossly overestimated, R_{free} will probably be extremely high in the outer shells. If something has gone wrong during data processing, e.g., all high-intensity reflections were overloaded on the detector, this also will often result in "stubborn" behavior of R_{free}

Some researchers do not like using R_{free}, or even claim that "it doesn't work with 3-Å data," because they are used to obtaining cosmetically pleasing low R factors. However, it is exactly at low resolution that R_{free} is most useful: At low resolution one has relatively few data (often also weak and incomplete) so that one is very close to a data-to-parameter ratio

[16] V. Luzzati, *Acta Crystallogr.* **5,** 802 (1952).

of one (and often below one). In these cases, the danger of overfitting is obviously at its greatest, and this may even lead to a masking of gross errors in the model.[5]

Some crystallographers include R_{free} calculations, but fail to "listen" to what R_{free} is telling them. This occurs when R and R_{free} values are reported that differ by 0.1–0.15 (this has even lead to the claim that "R_{free} is not a sensitive indicator of model quality"). Because the reflections used for the calculation of R_{free} are not used in the refinement, one always expects to obtain lower conventional than free R factors. If the data are good, and if the structure adequately models the data, then the structure factors used in refinement must be highly correlated to those not used, i.e., R and R_{free} must have similar values. In our experience, with good data sets we are able to obtain differences between R and R_{free} of ~0.02; if the data are of poorer quality or if the resolution is low, the difference may be as high as 0.05–0.08.

Analysis of the Protein Data Bank (PDB[17]) in May 1995, showed that only 62 (out of almost 3000) X-ray structures included a free R factor. The conventional and free R factors of these structures are shown as a function of resolution in Fig. 2. In the resolution range between 1.5 and ~3 Å, the conventional R factors are more or less identical for all structures (~0.20), whereas the free R factors (and, hence, the difference between the two) increase almost linearly with resolution.

There are still several limitations on and "undecided" issues with respect to the use of the free R factor [see also the discussions in [19] in this volume,[14] and in Refs. 6 and 13a].

1. Space group errors are unlikely to be detected by R_{free}, because most of the test set reflections will have a symmetry-related reflection in the work set.[5]

2. In the case of NCS, most of the reflections in the test set will be related to some of the reflections in the work set, unless special care is taken in the selection of the test set, e.g., by selecting them in thin resolution shells,[1] or as small spheres or cones related by the G function (P. Metcalf, personal communication, 1995; R. Read, personal communication, 1995).

3. Questions exist as to what constitutes a "significant drop" in R_{free}, i.e., the precision of R_{free}. Studies by Brünger (see [19] in this volume[14]) show that the precision depends mainly on the size of the test set (varies as $1/N^{1/2}$), and ranges from ~0.5% at high resolution to ~3% at low resolution. If an improvement of R_{free} is of the same magnitude as the precision,

[17] F. C. Bernstein, T. F. Koetzle, G. J. B. Williams, E. F. Meyer, M. D. Brice, J. R. Rodgers, O. Kennard, T. Shimanouchi, and M. Tasumi, *J. Mol. Biol.* **112,** 535 (1977).

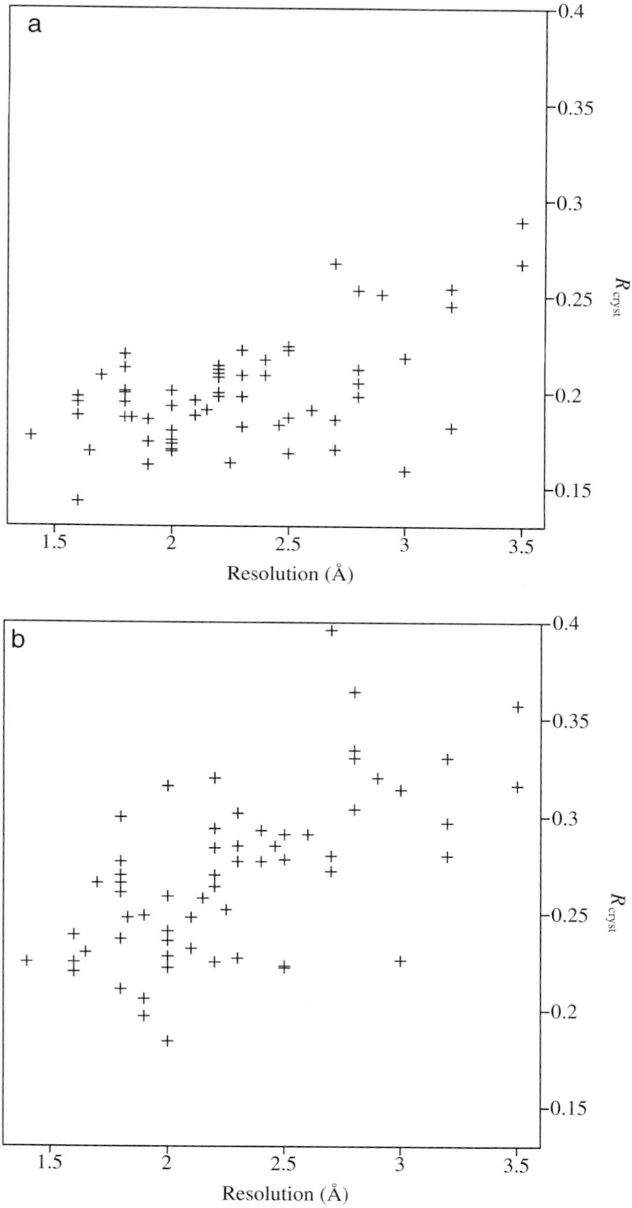

FIG. 2. Plot of the distribution of conventional (a) and free R factors (b) for the 62 X-ray structures in the May 1995 version of the PDB [F. C. Bernstein, T. F. Koetzle, G. J. B. Williams, E. F. Meyer, M. D. Brice, J. R. Rodgers, O. Kennard, T. Shimanouchi, and M. Tasumi, *J. Mol. Biol.* **112,** 535 (1977)] for which both values were reported.

one could carry out several calculations with different test sets to see if the reduction in R_{free} is a reproducible effect. For instance, when deciding on the most appropriate temperature-factor model, one could reset all temperature factors to an average value and then refine them in different ways and using different test sets. This will, of course, consume a great deal of computer time; if one wants to avoid that, err on the side of caution.

4. There is a lack of agreement on the necessary fraction, and minimum and maximum number of reflections, one should include in the test set. It appears that a minimum of ~500 reflections is required to attain meaningful statistics (see [19] in this volume[14]); however, there seems no point in having a test set of more than ~2000 reflections.[6]

5. Questions exist as to whether test set reflections should be used in map calculations. Strictly speaking, this would introduce bias toward the test set, but if one uses SA refinement in every cycle most of this bias is likely to be removed again. We prefer to use all reflections in map calculations for a pragmatic reason: we attempt to build the best possible model.

The free R factor is a global statistic, i.e., it relates to the mean absolute phase error.[13] It cannot be used for "micromanagement," for instance to decide if the addition of five water molecules improves the model. It also means that local errors in a model may well go undetected if only R_{free} is used to assess the correctness of a model. In particular, out-of-register errors will be hard to detect because, usually, only a fairly small number of scatterers is involved. In the process of refining the structure of *Trichoderma reesei* endoglucanase I (EGI),[18] we detected and corrected such an error, involving a stretch of 20 residues that were out of register with the density by 7 residues (an insertion had not been correctly accounted for). At that stage, R_{free} was already at 0.302 (R 0.245), i.e., well below our usual skepticism threshold of 0.35. In general, out-of-register errors can be detected only by a combination of experience, (SA-omit) maps, common sense (regarding the environments of residues), temperature factors, databases (to detect unusual main-chain arrangements), and alignment with homologous structures (if any are available). One should always be on the lookout for such errors. When doubting the correctness of a part of the trace, suspect residues can be omitted temporarily from the model, or cut back to alanines.

[18] G. J. Kleywegt, J. Y. Zou, C. Divne, G. J. Davies, I. Sinning, J. Ståhlberg, T. Reinikainen, M. Srisodsuk, T. T. Teeri, and T. A. Jones, *J. Mol. Biol.* in press (1997).

Simulated Annealing

Simulated annealing (SA) refinement (see [13] in this volume,[19] and Refs. 20 and 21) is a powerful method that we use in almost every refinement cycle. The major benefit of SA refinement lies in its large radius of convergence. If the model and the data are good, an SA calculation rarely does any harm to the structure. On the contrary, we find that if an SA calculation does lead to a poorer model (in terms of R_{free} or map quality), this is a strong indication that something is wrong with the input model, or that there is a problem with the data.

A good example of successful SA refinement is the structure of a complex of *Candida antarctica* lipase B with the nonionic detergent Tween 80[22] (a long, floppy, and chemically ill-characterized compound). The protein structure had been solved and refined at 1.55-Å resolution,[23] but for the Tween 80 complex only 2.5-Å data were available. The protein structure in the complex was solved by molecular replacement and refined using strict twofold NCS. Simulated annealing had not been used for fear of ruining a well-refined structure by exposing it to low-resolution data. This had yielded a final model with a low value for R_{free} of 0.209 (R 0.187). Unfortunately, there was relatively poor density in the presumed location of the Tween 80 molecule, even after electron-density averaging (see Fig. 3). We then carried out a slow cool from 4000 K, using weak harmonic restraints on all atomic positions. The rationale for this was to force the protein back into a structure similar to the highly refined starting model, while leaving some room for the structure to relax (i.e., to adjust its conformation to the new data), in the hope of obtaining better density for the Tween 80 molecule (which was, after all, the interesting part of the structure). The slow cool was successful and converged back to a model similar to the starting structure, but with slightly lower values of R_{free} and R. Most important, however, was the appearance of well-connected density in the difference map into which a crude partial model of the Tween 80 could be built easily. This model was then subjected to another slow cool from 4000 K, again using weak harmonic restraints for all atoms, except those of the Tween 80 molecule and nearby protein residues. Again, the refinement was successful, yielding an R_{free} of 0.188 (R 0.165). Density for the Tween

[19] A. T. Brünger and L. M. Rice, *Methods Enzymol.* **277,** [13], 1997 (this volume).

[20] A. T. Brünger, J. Kuriyan, and M. Karplus, *Science* **235,** 458 (1987).

[21] A. T. Brünger and A. Krukowski, *Acta Crystallogr.* **A46,** 585 (1990).

[22] J. Uppenberg, N. Öhrner, M. Norin, K. Hult, G. J. Kleywegt, S. Patkar, V. Waagen, T. Anthonsen, and T. A. Jones, *Biochemistry* **34,** 16838 (1995).

[23] J. Uppenberg, M. Trier Hansen, S. Patkar, and T. A. Jones, *Structure* **2,** 293 (1994).

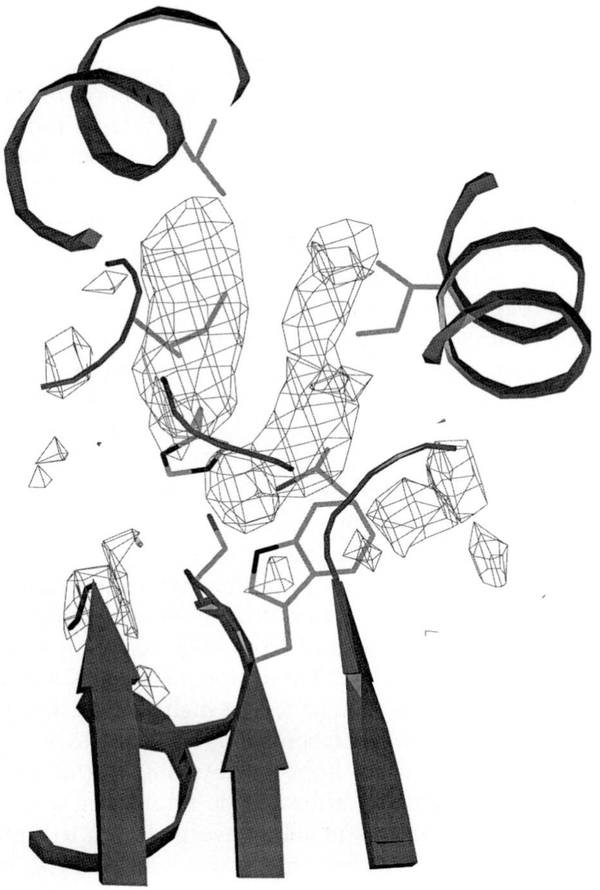

FIG. 3. Averaged difference density for *Candida antarctica* lipase B in complex with Tween 80 after refinement of the structure without SA [J. Uppenberg, N. Öhrner, M. Norin, K. Hult, G. J. Kleywegt, S. Patkar, V. Waagen, T. Anthonsen, and T. A. Jones, *Biochemistry* **34,** 16838 (1995)]. Note that the density for the Tween 80 molecule is virtually uninterpretable. Refer to text for details. (Figure kindly provided by J. Uppenberg, Uppsala.)

80 molecule in the $2F_o - F_c$ map after the slow cool is shown in Fig. 4. Note that this case is a splendid illustration of the fact that even low-resolution structures with conservative assumptions (strict NCS, grouped temperature factors) can yield low (free) R factors.

From release 4.0 onward, X-PLOR will contain a facility to use molecu-

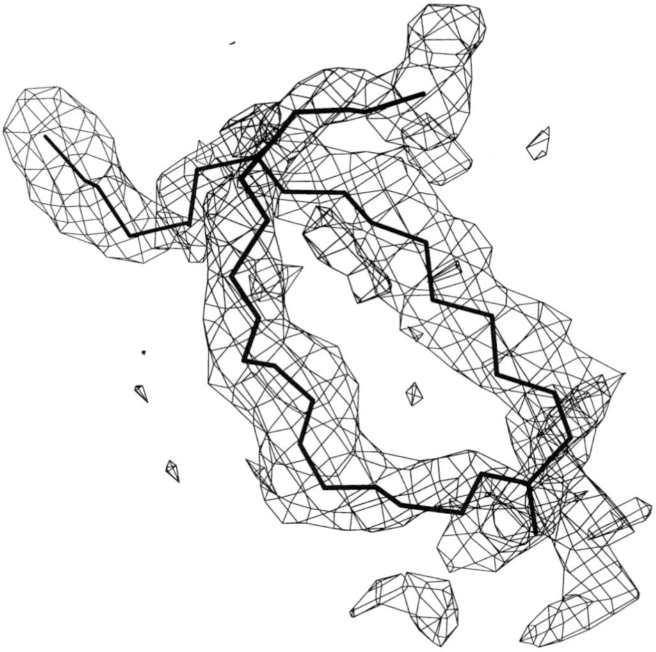

FIG. 4. Density for a Tween 80 model after refinement with SA [J. Uppenberg, N. Öhrner, M. Noren, K. Hult, G. J. Kleywegt, S. Patkar, V. Waagen, T. Anthonsen, and T. A. Jones, *Biochemistry* **34,** 16838 (1995)]. One refinement cycle yielded interpretable difference density for a part of the Tween 80 molecule. A crude model was built and refined in a second SA-refinement cycle. The density for the Tween 80 after this second cycle is shown (compare to Fig. 3). Refer to text for details.

lar dynamics (MD) refinement in torsion-angle space,[19,24] and we expect significant benefits from this. Carrying out SA refinement in torsion-angle space rather than in Cartesian coordinate space reduces the number of degrees of freedom from $3N$ (where N is the number of atoms) to something on the order of $N/3$.[24] Initial results indicate that this, plus the fact that SA refinement can now be carried out at considerably higher temperatures (up to 10,000 K), increases the radius of convergence from ~1.2 to ~1.7 Å.[24] In addition, one obtains near-ideal bond lengths and bond angles "for free."

Of course, there are still limitations to what SA refinement can do. In Cartesian coordinate space, the root-mean-square (RMS) radius of conver-

[24] L. M. Rice and A. T. Brünger, *Proteins Struct. Funct. Genet.* **19,** 277 (1994).

gence is ~1.2 Å,[20] although local changes of up to ~8 Å are possible.[20,25] Simulated annealing refinement will not make changes that require the breaking of covalent bonds (which is often done temporarily during manual rebuilding). Nor can SA refinement optimize the fit of individual residues. For example, the side-chain oxygen and nitrogen atoms of asparagine and glutamine residues, and the rings of histidine residues, often end up flipped by 180° (a careful human would take the whole hydrogen-bonding network into account to decide on the proper orientation). Also, SA refinement does not always produce rotamer-like side-chain conformations (although these could be enforced by applying strong restraints on dihedral angles). Finally, gross errors (in tracing, connectivity, and sequence) cannot be fixed by a refinement program. For these reasons, manual rebuilding of protein structures is still necessary at present.

Force Field and Dictionaries

Every refinement program presently uses geometric and other restraints to augment X-ray data. As for protein geometry, the best set of bond and angle parameters available today is probably the set developed by Engh and Huber[26] on the basis of an analysis of small-molecule crystal structures. These parameters are now available for the most widely used refinement and rebuilding programs.[27] When used in combination with a reduced weight for the crystallographic pseudo-energy term, protein models with good stereochemistry are virtually guaranteed (but note that good stereochemistry says absolutely nothing about how well the structure models the data). As more protein structures are solved at atomic resolution, better dictionaries can be expected.

However, problems remain in deriving "ideal" parameters for nonprotein entities. Writing an X-PLOR topology and parameter file for a ligand, for example, is a cumbersome, time-consuming, and error-prone process. What often is not realized is that, in fact, one must specify exactly what one would like the ligand, substrate, or cofactor to look like (with the exception of torsions around freely rotatable bonds) in the absence of any crystallographic information. This means that every sp^1 and sp^2 carbon gives rise to a "flatness" restraint, that the chirality of every chiral carbon must be restrained, etc. If one is lucky, the structure of a ligand has been solved separately and can be retrieved from a small-molecule database. In that case, we suggest that one derive all ideal values from this structure

[25] P. Gros, M. Fujinaga, B. W. Dijkstra, K. H. Kalk, and W. G. J. Hol, *Acta Crystallogr.* **B45,** 488 (1989).

[26] R. A. Engh and R. Huber, *Acta Crystallogr.* **A47,** 392 (1991).

[27] J. P. Priestle, *Structure* **2,** 911 (1994).

and use heavy weights for the bond, angle, and fixed-dihedral restraints. In other cases, one may be able to find the structure of a common cofactor or ligand in another structure in the PDB. If no structure is available, one will have to use "rule-of-thumb" values, or resort to quantum chemical or molecular mechanics calculations. In the case of low-resolution data, it may be best almost to constrain the ligand (by using very heavy weights for bond lengths and angles) so that it effectively has only a few degrees of freedom left (freely rotatable carbon–carbon bonds, for instance). Again we must emphasize that one must be extremely cautious when refining against low-resolution data. If the ligand dictionary allows ring puckering, for instance, the refinement program is invited to take liberties (in this fashion even aromatic rings can easily be "refined" into a nonplanar conformation).

Prior to including the ligand in crystallographic refinement, one always should apply energy minimization of only the ligand without the X-ray term. The result shows the structure that the refinement program will attempt to produce when the complete model is refined with inclusion of the experimental data.

Quality Control

Quality control is now an integral part of our model-building and refinement process, i.e., it is not something that is done only once, *a posteriori,* for the sole purpose of filling in some tables in the publication. Quality control entails the use of our knowledge with respect to the structure of macromolecules to find places in the model that need special attention and are possibly in error. Zou and Mowbray[28] have described the benefits that can be attained by the use of empirical knowledge (as embodied in databases) during protein rebuilding and refinement.

In judging the quality of a model produced by the refinement program we use the following criteria.

Stereochemistry: Deviations from ideality of bond lengths, bond angles, and violations of dihedral, flatness, and chirality restraints are conveniently analyzed with X-PLOR and ProCheck.[29]

NCS: RMS distances, RMS ΔB values, $\Delta\phi$, $\Delta\psi$ plots,[1,4,30] and multiple-model χ_1, χ_2 plots[4] for NCS-related molecules are assessed if no constraints are used.

[28] J. Y. Zou and S. L. Mowbray, *Acta Crystallogr.* **D50,** 237 (1994).

[29] R. A. Laskowski, M. W. MacArthur, D. S. Moss, and J. M. Thornton, *J. Appl. Crystallogr.* **26,** 283 (1993).

[30] A. P. Korn and D. R. Rose, *Prot. Eng.* **7,** 961 (1994).

Temperature factors: A plot of average B values versus residue number, RMS ΔB for bonded atoms, average B values for protein, ligand, substrate, waters, etc. are assessed. A radial temperature-factor plot [average temperature factor of atoms in shells as a function of their distance to the core of the molecule (E. J. Dodson, personal communication, 1995)] should be shaped roughly like a parabola for a globular protein, with a minimum for the residues in the core of the molecule.

Ramachandran plot[31,31a]: A Ramachandran plot is produced and compared to that of the previous model, as is a multiple-model Ramachandran plot in the case of (un-)restrained NCS.[4]

Preferred rotamer analysis: Side-chain conformations are analyzed with the RSC_fit command in O,[9,28] which calculates the RMS distance between the side-chain atoms of a residue and those of the most similar preferred rotamer conformation; a value greater than ~1.5 Å means that the side chain is not in a preferred rotamer conformation.[28]

Peptide orientation: The Pep_flip command in O,[9] is used to superimpose the C_α atoms of a residue and its two nearest neighbors at both sides with similar fragments found in a database of well-refined structures. After the superpositioning, the RMS distance between the carbonyl oxygen atom of the residue and the corresponding atom in the database fragments is calculated. A value greater than ~2.5 Å implies that the peptide plane is flipped compared to the orientation most frequently observed in the database.

Residue real-space electron-density fit[9]: This property is calculated for all atoms, main-chain atoms only, or side-chain atoms only. It can be expressed either as an R factor or as a correlation coefficient between density calculated from the atomic positions and that in a $2F_o - F_c$ or experimentally phased map; both can be computed with the RS_fit command in O. The correlation coefficient has the advantage that it is independent of the scales of the two densities. However, this also means that weak observed density may still correlate well with the calculated density, leading to the impression that the fit is good even though no density may be visible at a normal contour level. It is not possible to give absolute guidelines as to what cutoff to use between "good" and "bad" values. Usually, we use a cutoff of "the average minus one or two sigma" for the correlation coefficient, and "the average plus one sigma" for the R factor.

[31] C. Ramakrishnan and G. N. Ramachandran, *Biophys. J.* **5,** 909 (1965).

[31a] G. J. Kleywegt and T. A. Jones, *Structure* **4,** 1395 (1996).

Close contacts: Close contacts due to NCS and crystal packing should also be inspected. These are easily detected with X-PLOR.

Assessing all these criteria for each and every residue during each and every rebuilding session is cumbersome. Therefore, we use a program called OOPS[32] to aid in this process. The program makes use of the ability of O to generate and use so-called residue and atom properties (these are explained in more detail in [10] in this volume[3]). The idea behind OOPS is to calculate the values of some of the quality indicators in O before starting the rebuild, and to generate others on the fly from the coordinate file of the present model. OOPS gathers and integrates all this information on a per-residue basis. On the basis of this information, a set of O macros is generated that takes the crystallographer from one bad or suspect residue to the next and explains what may be wrong with each of them. In addition, OOPS produces plots of various properties as a function of residue number, statistics for all properties, and a residue-by-residue "critique" of the structure. The plots can be used to reveal areas where the structure is particularly bad, the statistics are useful to judge the overall quality and to decide which cutoff values to use, and the residue-by-residue listing is saved as an electronic notebook file that can be annotated during the rebuilding session.

It should be pointed out that there are two types of residue that give rise to violations: those that are wrong (errors), and those that actually do have an unusual conformation ("outliers for a reason"). The latter type is often found in the interesting places in a structure, for example, in a ligand- or substrate-binding site.[33] They can be recognized as such by convincing density, and a tendency to return to the same conformation, even after rebuilding and SA refinement. Residues that are wrong, however, almost always have poorly fitting or absent density and either will maintain their rebuilt conformation after SA refinement (indicating that the error has been fixed), or they will end up in yet another conformation (usually, this happens for surface and loop residues that are poorly defined by the data). As a rule of thumb, for a well-refined, high-resolution model, one would expect <2% outliers for a reason in the Ramachandran plot, ~1–2% residues with unusual peptide orientations, and ~5–10% residues with nonpreferred rotamer conformations.[4,7]

Maps

For rebuilding we tend to use $3F_o - 2F_c$, $2F_o - F_c$, and $F_o - F_c$ maps. $F_o - F_c$ difference maps should be contoured both at positive and negative

[32] G. J. Kleywegt and T. A. Jones, *Acta Crystallogr.* **D52,** 829 (1996).
[33] O. Herzberg and J. Moult, *Proteins Struct. Funct. Genet.* **11,** 223 (1991).

levels. In troublesome areas, SA-omit maps can be calculated.[34] Standard omit maps are not a good idea, since it may be impossible to tell whether the reappearance of density is real or due to model bias.[34] After an SA-omit run, we calculate both $2F_o - F_c$ and $F_o - F_c$ omit maps. The density in the omitted area should be very similar for both maps. During difficult refinements, we occasionally use systematic SA-omit maps. In this procedure, all residues are omitted in turn (using stretches of 5–10 residues at a time) and they are rebuilt in the resulting SA-omit map. Naturally, if an experimentally phased map is available, it should also be consulted during rebuilding.

In the case of NCS, we invariably use real-space electron-density averaging. Averaging is a well-established and powerful method for map improvement. Not only does it often improve the density in areas where the unaveraged map has no visible density at all, it also helps in the identification of regions where the NCS breaks down. In our experience, deviations from NCS are often much smaller than one would expect on the basis of published structures that were refined without making use of the NCS.[4] For example, in (NCS) disordered loops, there are often only one or two residues for which even after averaging no density is visible. Instead of (or in addition to) NCS, one sometimes has multiple nonisomorphous crystal forms, which means that cross-crystal averaging can be used to obtain better maps.

Rebuilding

Manual rebuilding of a structure is somewhat of a "black art," best learned by practising it on many different structures. Nevertheless, there are a number of simple questions that one must ask oneself all the time, for every residue in turn. The answers to these questions will determine what action must be taken.

If a residue violates a certain quality criterion, is there a special feature in the structure that may explain the violation? Can the model be rebuilt so as not to violate the criterion, and still satisfy the density? For example, proximity to a ligand, ion, or substrate may explain an unusual peptide orientation, or a "forbidden" combination of ϕ, ψ values for a nearby residue; formation of salt links, hydrogen bonds, or coordination to a metal ion may induce an unusual side-chain conformation, and the same is true for NCS and/or crystal packing interactions.

Even if a residue does not violate any criterion, one should still wonder: does the residue make sense in this place in the structure and in this conforma-

[34] A. Hodel, S. H. Kim, and A. T. Brünger, *Acta Crystallogr.* **A48,** 851 (1992).

tion? For example, if one encounters a single buried charged residue inside a hydrophobic pocket, there could be an error (either a sequence error or an out-of-register error). Also, at low resolution and in crude models one should be reluctant to "believe" in nonpreferred rotamer conformations. Often a preferred rotamer will fit the density equally well (perhaps after a small rotation and translation of the whole residue, or after minor adjustment of one or two torsion angles; see Fig. 5 for an example) and is therefore to be preferred given the high odds that it is closer to the real conformation.

If one decides to rebuild a residue, one should ask oneself afterward: did the rebuild improve things? After a residue has been rebuilt, the geometry of the residue and the nearby residues (e.g., two on each side) should be regularized to restore proper bond lengths, etc. For example, if one encounters a residue with a pep-flip value of 2.5 Å and poor main-chain density, rebuilds the structure and then obtains a pep-flip value of 3.2 Å, the rebuild was probably unwarranted. In such a case, the original structure should be restored.

In the following, we shall discuss violations of specific criteria, their possible causes, and possible remedies (using the tools in O). Again, all rebuilding should be followed by regularization to restore proper stereochemistry (with the Refi_zone command[35]).

Stereochemistry: If a residue has poor stereochemistry, usually this is because it is poorly defined by the data, owing to mobility, disorder, or breakdown of the NCS. The best way to fix this for amino acid side chains is to replace the residue by a preferred rotamer with the Lego_side_chain command (and perhaps adjust one or two torsion angles with the Tor_residue or Tor_general command). The stereochemistry of the main chain and of nonprotein entities can be improved with the Refi_zone command.

Peptide orientation: If a residue has a high pep-flip value (e.g., exceeding 2.5 Å), one should look for a structural reason. If none is obvious and the density is ambiguous, the peptide plane should be "flipped" (with the Flip_peptide command). If the original orientation was correct after all, this may show up after the next round of SA refinement.

Nonpreferred rotamers: One may safely assume 90–95% of the residues to have a side-chain conformation resembling that of a preferred rotamer.[4,7] Very often, preferred rotamers can be generated (with the Lego_side_chain command) and fitted to the density through rigid-body movements of the whole residue (with the Move_zone command, using the C_α atom as the pivot point) and regularization (Fig. 5 shows an example of this). Some residue types (glutamate, glutamine, lysine, arginine, and methionine) adhere less strictly to preferred rotamer conformations beyond the χ_2 torsion

[35] J. Hermans and J. E. McQueen, *Acta Crystallogr.* **A30,** 730 (1974).

a

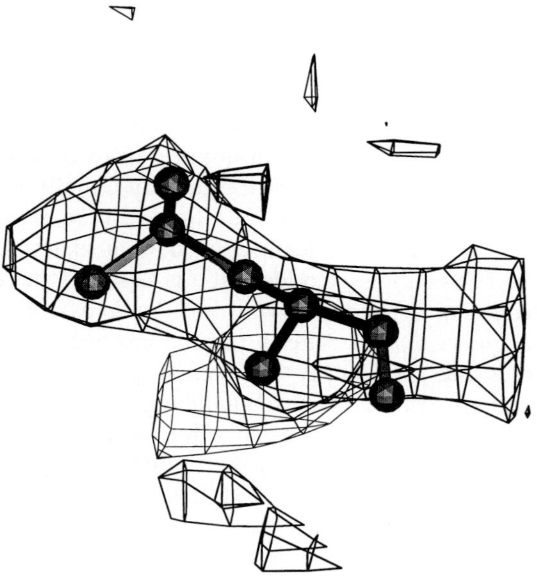

b

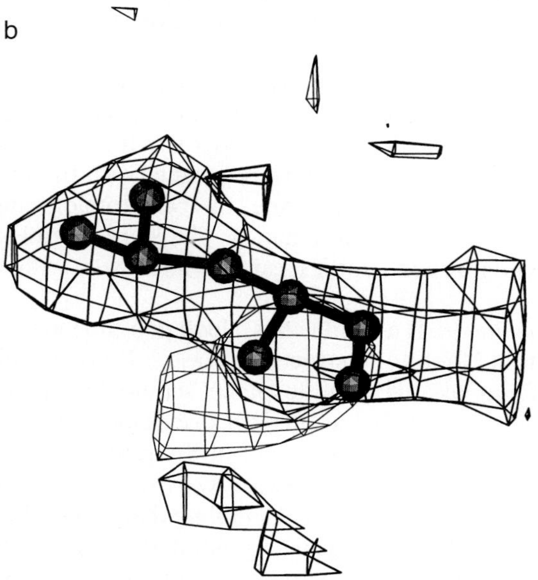

FIG. 5. Example of a case in which a nonrotamer side-chain conformation was built in a low-resolution (3.0-Å) map, which can easily be replaced by a rotamer conformation that fits the density equally well, if not better. (a) A nonrotamer conformation for a leucine residue. The RSC-fit value, as calculated by O, is 2.09 Å, with $\chi_1 = -70°$ and $\chi_2 = -25°$. (b) A better fitting rotamer for the same residue. The RSC-fit value is 0.63, with $\chi_1 = -38°$ and $\chi_2 = 175°$ (close to the values for the most common leucine rotamer, namely $\chi_1 = -60°$ and $\chi_2 = 180°$).

angle; they can then be fitted with rotations around torsion angles, but energetically favored conformations should be tried first. We strongly advise against moving individual atoms in amino acid residues, unless in a well-refined model all attempts to use common sense (i.e., database conformations) fail to yield a satisfactory fit to the density.

Poor density: There are many reasons why a residue may have poor density (and, usually, accompanying high temperature factors). Surface residues often have poor density due to mobility and should be replaced by a rotamer or stripped back to an alanine. Loops and the C and N termini sometimes have poor density due to disorder or mobility. In the case of NCS, poor averaged density may indicate areas where the NCS breaks down, especially if good density exists in each of the unaveraged maps (which can be viewed in O by applying suitable transformations to map objects). Water molecules with poor density are probably not tightly bound and should be removed from the model. Sometimes a single wrong peptide orientation puts strain on a short stretch of residues; this is an error in the structure and should be remedied by flipping the peptide and regularizing the zone afterward. Biologically interesting nonprotein entities with poor density should be treated with the greatest caution.

Bad regions: Sometimes one encounters a stretch of consecutive residues that all violate several quality criteria. One should then consider the possibility that there is a serious error in that part of the structure. An SA-omit map should be calculated with the suspect region omitted from the model. In the case of an MR structure, bad regions may indicate areas where the new structure is significantly different from the search model, but beyond the reach of the refinement protocol. Also in this case, an SA-omit map may provide hints as to how the structure should be rebuilt. Rebuilding loops, etc., can be done in O with the Baton commands or the Lego_loop command, followed by Lego_auto_sc and Lego_side_chain to put the side chains in as preferred rotamers. If one refines a model with strict NCS, poor regions are often characterized by high temperature factors and may indicate a local breakdown of the NCS. If one has sufficiently high-resolution data, one could then experiment with NCS restraints instead of constraints, using a lower force constant for the areas where the NCS may have broken down. However, one should be conservative, because such NCS breakdowns may easily become self-fulfilling prophecies.[1,4,5] A refinement program will gladly accept any extra degrees of freedom it is given, to "fit" errors in the data. In other words, if a zone of residues is subjected to only weak NCS restraints, it will definitely become different in each of the NCS-related molecules. Therefore, one should always inspect averaged maps: if the average density for a loop is good, with the exception of two residues, only those two residues should be restrained weakly, and not the entire loop. If one does not have high-resolution data, one should simply accept the locally poor structure as it is.

Nucleic acids: The current version of O lacks specific support for nucleic acids in the Lego and Baton commands. Otherwise they can be treated with the same tools as proteins.

Water molecules: One should postpone the addition of solvent entities until the protein model is essentially complete, and sufficiently well refined, to prevent placing waters in noise peaks or in places where side-chain atoms ought to go. In the case of NCS and low-resolution data, averaged difference maps should be used so as to include only waters that obey the NCS. Potential water molecules tend to show up as concomitant peaks in $F_o - F_c$ and $2F_o - F_c$ maps. They should have plausible hydrogen-bonding partners within ~3.5 Å. Waters that move around a great deal during refinement, those that obtain high temperature factors, and those that do not have good density after refinement should be omitted from the model. In O, water molecules can be moved into the peak of the density by hand (with the Move_atom command), or, preferably, automatically (with the RSR_rigid command). Tradition has it that most single-peak solvent entities are modeled as waters, even though there may have been several other isoelectronic entities in the crystallization solution (e.g., Na^+, NH_4^+, F^-). Electron density alone cannot be used to discern between these ions and water.

Small molecules: As explained in the section on refinement, it helps if the structure of a ligand, substrate, carbohydrate, inhibitor, or cofactor can be retrieved from a database. Inside O, several commands can be used to optimize the fit between model and map. The RSR_rigid command does a rigid-body optimization of (a fragment of) the molecule against a map. The Tor_general command can be used to adjust torsion angles, which is usually sufficient. If one creates appropriate dictionaries, the Tor_residue and Refi_zone commands can also be used. Again, moving individual atoms around should be avoided, unless all else fails, because it will distort bond lengths and bond angles, and may even introduce chirality errors. One of our utility programs contains options to generate some of the O dictionaries for heteroentities from a PDB file.

Final Model

When refinement and rebuilding of the structure have converged, a final refinement round can be carried out using all diffraction data and employing only energy minimization and temperature-factor refinement. After this, a final assessment of the quality of the structure must be carried out. Factors to be taken into account are similar to those that should be checked in every macrocycle. In addition, one may estimate the average coordinate error, for example from a Luzzati plot,[16] using R_{free} rather than the conventional R factor.[13a,15] If NCS is present and has not been constrained, differences between NCS-related molecules should be analyzed skeptically.[4] A particularly sensitive way of analyzing differences between NCS-related molecules is by comparing the main-chain and side-chain torsion angles of corresponding residues.[1,4,7,30]

How does one assess the quality of published structures? If no coordinates (and structure factors) are available, one is dependent on what is written in the publication.[7] The first thing to check is the quality of the data: What is the multiplicity, R_{merge}, completeness, and $I/\sigma(I)$ ratio, both overall and in the highest resolution shells? If complete tables of these quantities are included, one may roughly assess the effective resolution of the data (as opposed to the Bragg spacing of the single highest resolution reflection that was indexed). The second important task to perform is to assess the quality and strategy (if any) of the refinement. Even in the absence of details one usually can guess how the refinement was carried out. If R_{free} is not mentioned, it was probably not used; if it is mentioned, check if it was used throughout, and if the difference between R and R_{free} is small. If NCS is present, but no mention is made of NCS constraints or restraints, one may safely assume that the NCS was ignored during refinement. Temperature factors have been refined for individual atoms, unless specifically stated otherwise. If the Ramachandran plot is not shown or mentioned, it may have been poor (or never produced).

If coordinates are available, many of the quality criteria can be checked easily.[7,36] However, for a full evaluation of the structure, observed structure-factor amplitudes are needed. Only with those can one check if the structure is an adequate model for the data or not, if necessary by redoing the refinement. We therefore strongly recommend that structure factors be deposited together with coordinates (in fact, in an ideal world unmerged intensities would be deposited as well!).

Finally, when "validating" a model it is important to realize that any property that has been constrained or heavily restrained during refinement, and any property that has been closely monitored during rebuilding, cannot be used as an independent criterion to assess (or "prove") the quality of the model. For instance, most refinement programs operate by minimizing the difference between observed and calculated structure-factor amplitudes; therefore, the value of the conventional R factor is hardly an independent quality criterion (in particular when there were few observations per refined parameter). Similarly, most refinement programs tightly restrain bond

[36] M. W. MacArthur, R. A. Laskowski, and J. M. Thornton, *Curr. Opin. Struct. Biol.* **4,** 731 (1994).

lengths, bond angles, and certain (improper) torsion angles; therefore, low RMS deviations from ideal geometry cannot be presented as proof of the quality of the structure. Also, if side-chain conformations are monitored, and preferred rotamers are used in the rebuilding, a low fraction of residues with nonpreferred rotamers is not necessarily a hallmark of a correctly traced structure. With the widespread use of the program ProCheck and its pretty output, a standard phrase has begun to creep up in papers describing protein structures: "the model has a quality better than expected for structures at this resolution." Again, this is a rather meaningless statement if the criteria that ProCheck assesses have been restrained during the refinement. Moreover, the ranges of expected values that ProCheck uses (as a function of resolution) were derived using a set of structures that predate the Engh and Huber parameter set.[26] With a judicious choice of weights it is actually quite difficult to produce a model that has a lower quality than expected. In fact, apart from the Ramachandran plot, the backward-traced structure of CRABP type II[1] is of "better than average" quality according to ProCheck. For this reason, it is important to have one or two independent quality checks that are not applied in the refinement and rebuilding process, but only to assess the final model.

Acknowledgments

This work was supported by the Swedish Natural Science Research Council and Uppsala University. The many fruitful discussions with other crystallographers in Uppsala, with Dr. Eleanor Dodson (York) and the other members of the EU-funded Biotech group, the participants in the York meeting on statistical validators in protein crystallography,[6] and in particular Dr. Axel Brünger (Yale, New Haven, CT) are gratefully acknowledged. We also thank Dr. Jonas Uppenberg (Uppsala) for providing us with lipase B–Tween 80 data.

[12] LORE: Exploiting Database of Known Structures

By BARRY C. FINZEL

Introduction

As the collection of known protein structures has grown exponentially, crystallographers and modelers have begun to develop a variety of software tools that exploit the information in these structures to improve their own structure determination or prediction efforts. The work of Jones and

0076-6879/97 $25.00

Thirup,[1] who introduced a rapid algorithm for the identification of structurally homologous fragments of protein structure, has led to the widespread use of fragment databases. For electron-density interpretation, the ca_lego and lego_auto_mc commands of the *O* program provide easy access to a library of peptide pentamers that can be used to convert a set of primitive C_α positions extracted from density "bones" to a complete backbone model in semiautomated fashion.[2] Similar tools have been implemented more recently in *Xfit* of the *XtalView* package.[3] A host of "spare parts" algorithms exists to exploit the predictive potential of local conformational features in homology modeling[4,5] or just to expand on an existing C_α model.[6,7] In fact, the notion that all protein structures are composed of combinations of a limited set of well-known peptide fragments is so well established that comparisons against pentamer libraries now constitute a well-accepted structure validation technique (e.g., the Pep_flip command of *O*[2], or the validation tools of *WHATCHECK*[8]).

In the development of all of this software, emphasis has been placed on the convenient use of simple fragments of structure, because metrics used to evaluate these are more straightforward, and therefore more amenable to full or partial automation. In concentrating primarily on protein fragments of five (or fewer) residues, however, a great deal of information about larger elements of protein structure is being undervalued, or simply ignored. We would argue that a β sheet, for example, is always more easily recognized and constructed as a sheet, than as short individual strands, but no commonly used modeling software conveniently utilizes larger hierarchical assemblies from known structures.

We have developed more general protein substructure modeling tools within software we call *LORE*. Although *LORE* was written initially as an aid to electron-density interpretation in crystallography, the program has evolved to include a variety of features for the manipulation of the database of atomic coordinates. The result is a program that is a versatile molecular editor with potential usefulness not only in map fitting, but also in a broad range of model-building and structure analysis applications.

[1] T. A. Jones and S. Thirup, *EMBO J.* **5,** 819 (1986).
[2] T. A. Jones, J.-Y. Zou, and S. W. Cowan, *Acta Crystallogr.* **A47,** 110 (1991).
[3] D. E. McRee, "Practical Protein Crystallography." Academic Press, San Diego, California, 1993.
[4] M. Claessens, E. Van Cutsem, I. Lasters, and S. Wodak, *Protein Eng.* **2,** 335 (1989).
[5] M. Levitt, *J. Mol. Biol.* **226,** 507 (1992).
[6] L. Holm and C. Sander, *J. Mol. Biol.* **218,** 183 (1991).
[7] J. J. Wendoloski and F. R. Salemme, *J. Mol. Graphics* **10,** 124 (1992).
[8] G. Vriend, *J. Mol. Graphics* **8,** 52 (1990).

Program Design

LORE is primarily an atomic coordinate manipulation tool. It is implemented currently as a mixed Fortran and C langugage subprogram of the modeling software *CHAIN* (Sack[9]; and see [9] in this volume[10]). The interface of these programs is necessary for the display of structural fragments, but the symbiosis is also convenient because the map display and conventional model-building tools of *CHAIN* are desirable complements to the substructure manipulation capabilities of *LORE*. When a user invokes *LORE* from *CHAIN*, the current molecular model of *CHAIN* becomes the residue set on which *LORE* operates. These atomic coordinates serve as the reference for target selection and substructure superposition.

At any time, the user may choose to replace atomic coordinates of portions of this model with those identified from selected library substructures. Because substructures need to be evaluated in the context of a developing molecular model, the flexible superposition of coordinates is vital to the effective use of the search algorithm. *LORE* is designed to give the user ultimate control over the precise details of this overlay process.

The software was designed initially to take maximum advantage of a particular substructure search algorithm (described below). An easily updated protein structure database format was devised that is consistent with the efficient implementation of this search engine, and early versions of the program essentially comprised a flexible user interface to this algorithm. As the program has evolved, however, program improvements have been included to aid in the characterization and selection of fragments, and to improve coordinate-handling capabilities. Some of this design has been described in detail elsewhere.[11] Here, we elaborate further on those elements of the design that are important to the applications discussed below: those that set *LORE* apart from other database-modeling software.

Search Algorithm

The substructure search algorithm used in *LORE* is founded on the observation of Rossman and Liljas[12] that similar protein structures have characteristic patterns of distances between α carbons. Jones and Thirup[1]

[9] J. S. Sack, *J. Mol. Graphics* **6,** 224 (1988).

[10] J. S. Sack and F. A. Quiocho, *Methods Enzymol.* **277,** [9], 1997 (this volume).

[11] B. C. Finzel, *Acta Crystallogr.* **D51,** 450 (1995).

[12] M. G. Rossman and L. Liljas, *J. Mol. Biol.* **85,** 177 (1974).

a

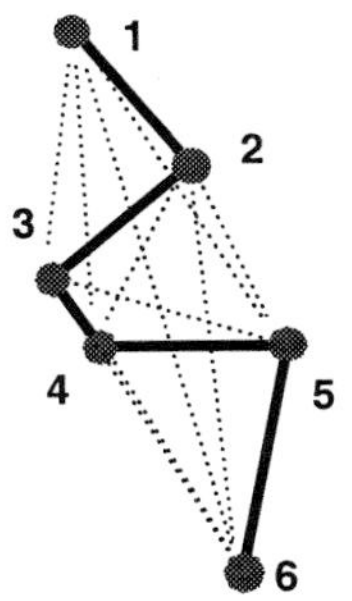

Cα	1	2	3	4	5	6
1	0	383	553	517	689	908
2	383	0	380	527	521	642
3	553	380	0	383	526	429
4	517	527	383	0	380	555
5	689	521	526	380	0	384
6	908	642	429	555	384	0

b

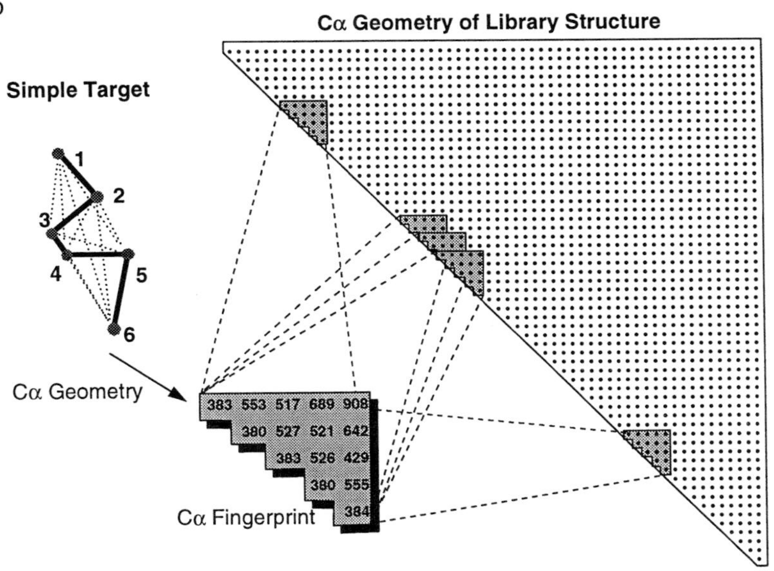

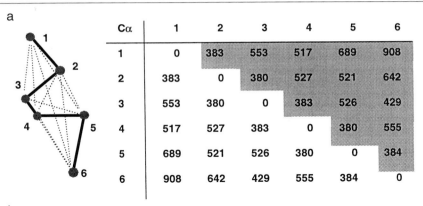

FIG. 1. (a) A simple *LORE* target geometry and corresponding C_α fingerprint, composed of all unique distances between α carbons. (b) A search of similar substructures in a larger library structure identifies portions of the larger fingerprint that are identical to the target. The search can result in any number of successes. Only the shaded half-matrix is stored. All distances are stored and manipulated as 2-byte integers.

later utilized these "C_α fingerprints" (Fig. 1a) as the basis of a substructure search algorithm. A C_α fingerprint derived from a target α-carbon geometry is compared to the appropriate elements of a triangular matrix of C_α–C_α distances from a library structure being searched for homologs (Fig. 1b).

A match of all elements of the fingerprint (within a tolerance) represents the match of a substructure conformation.

This algorithm has the advantages of being both fast (if library structure C_α fingerprints are precomputed) and flexible. It is easy to modify the target geometry specification to accommodate missing or poorly known α-carbon positions (Fig. 2a). This often is useful in fitting broken electron density, such as frequently occurs in β hairpins, or just to speed up the modeling process. It also represents a way to introduce conformational flexibility when substructure variety is required. We have extended the search procedure further to apply it to multiple disconnected segments of chain, such as the strands of a β sheet, by looking for the fingerprints of the individual strands, and then requiring that the off-diagonal matrix elements that define the relationship between C_α atoms in different segments match as well (Fig. 2b). The current *LORE* target specification is general. Targets of any size, and with up to 10 disconnected segments, may be specified. The implemented search algorithm also allows for specification of amino acid sequence restrictions on identified protein fragments. Some sample targets and search results are shown in Fig. 3.

Database

The database of known protein structures that we maintain for *LORE* is layered to enhance efficiency and flexibility in searching.[11] The lowest layer contains reformatted atomic coordinate data from selected Protein Data Bank (PDB) entries[13] in a binary format that simplifies access to selected residue ranges when required. Overlaying this is a database of "chain information," which consists of amino acid sequence and geometry data (in the form of precomputed C_α fingerprints) for individual polypeptide chains in the PDB. This layer also contains pointers to the complete atomic coordinate data in the lower layer. Chain information represents the primary data used in searching. The more detailed atomic coordinate data are referenced only after homology has been confirmed through examination of sequence and geometry data contained in the chain information, or by direct user specification.

The highest layer is a text-based index of chains that can be customized to fit the needs of particular users. The index points to chain information and includes a numerical chain ranking that can be helpful in the run-time selection of a specific subset of chains most appropriate to a particular application. In our laboratory, rankings based on properties such as se-

[13] F. C. Bernstein, T. F. Koetzle, G. J. B. Williams, E. F. Meyer, Jr., M. D. Brice, J. R. Rodgers, O. Kennard, T. Shimanouchi, and M. Tasumi, *J. Mol. Biol.* **112,** 535 (1977).

a

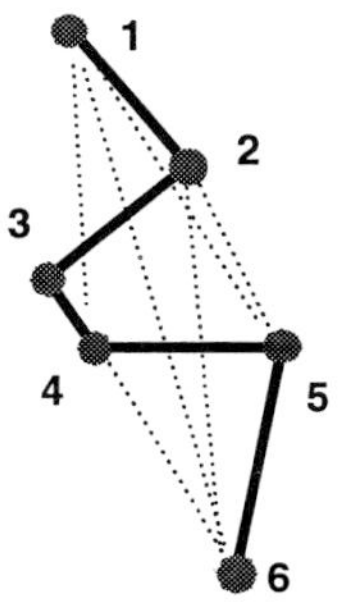

$C\alpha$	1	2	3	4	5	6
1	0	383			689	908
2	383	0			521	642
3	553	380	0			
4	517	527	383	0		
5	689	521	526	380	0	384
6	908	642	429	555	384	0

b

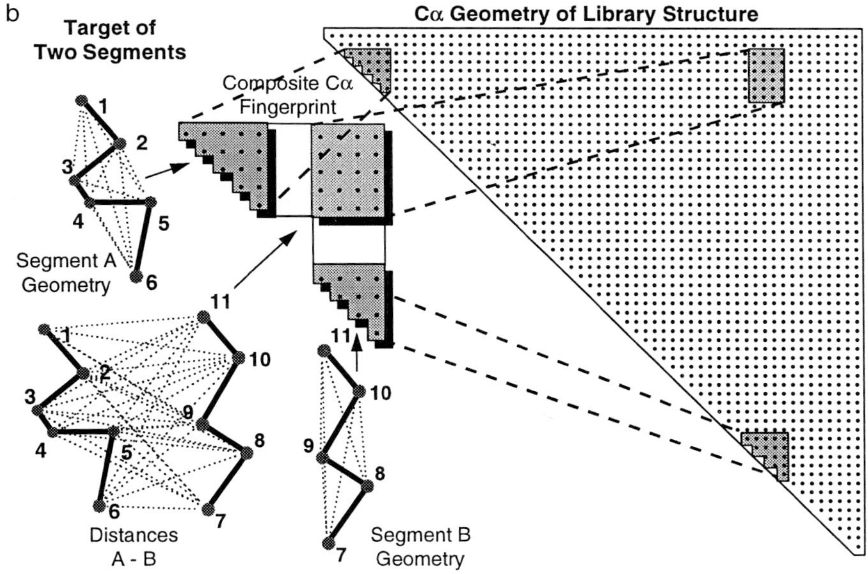

FIG. 2. (a) α-Carbon positions in the target can be flagged as "undefined." The corresponding elements of the fingerprint (in this example, residues 3 and 4) are then ignored while looking for homologous substructures. (b) C_α fingerprints for multiple segmented targets are built up from each component segment. Different tolerances can be applied to intrasegment distances (triangular block) and intersegment distances (the off-diagonal rectangular matrix) during the search. Segment A may or may not precede segment B in the library chain sequence.

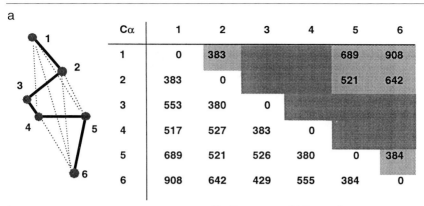

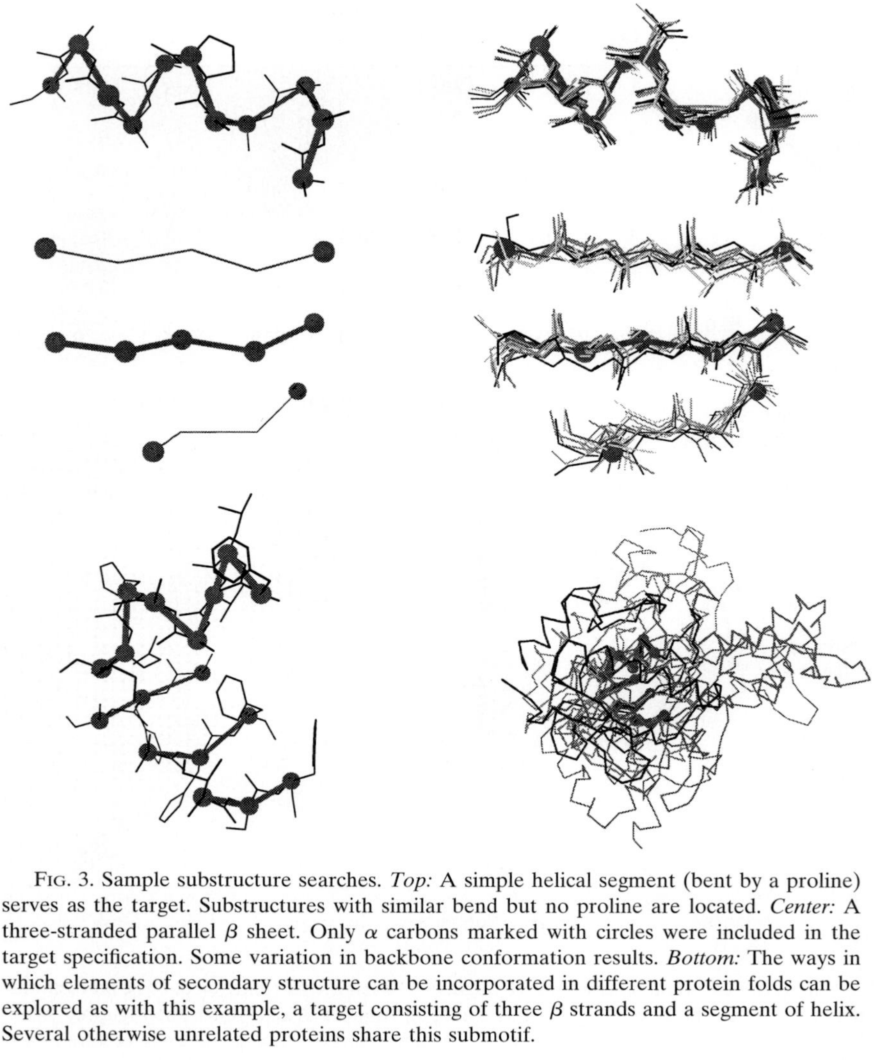

FIG. 3. Sample substructure searches. *Top:* A simple helical segment (bent by a proline) serves as the target. Substructures with similar bend but no proline are located. *Center:* A three-stranded parallel β sheet. Only α carbons marked with circles were included in the target specification. Some variation in backbone conformation results. *Bottom:* The ways in which elements of secondary structure can be incorporated in different protein folds can be explored as with this example, a target consisting of three β strands and a segment of helix. Several otherwise unrelated proteins share this submotif.

quence uniqueness, resolution of the structure determination, and *R* value are generated directly from PDB information. The "best" chains are selected for most applications unless a larger database subset is required. Other arbitrary ranking schemes could be implemented easily. Only chains that are unique at the 35% homology level[14] are included in our library, except when a specific need arises to include all members of a family of

[14] U. Hohorm, M. Scharf, R. Schneider, and C. Sander, *Protein Sci.* **1,** 409 (1992).

structures. Once added to the library, a user may restrict a search only to proteins from a given structural family simply by generating a chain index that includes only that subset of PDB entries.

Fragment Disposition

To make effective use of structural fragments from a library, these must be oriented to overlay the original target geometry. Most fragment-modeling programs assume that an overlay of corresponding α-carbon positions is sufficient and desirable. Often it is, but we have preferred to take a more general approach. The search and overlay procedures are conducted as separate steps in *LORE.* While the substructure search is confined to a comparison of α-carbon geometry, the superposition step need not suffer from this limitation. Often, other information about the desired structural fragments can be built into a target (such as a desired C_β position, or a main-chain carbonyl direction), and the goodness of fit of substructures to this original target geometry (as defined by all available atoms or some specified subset) can be an excellent metric by which to select the best fragment.

There is another advantage of separating search and superposition steps. Just as in the search phase, when it is convenient to assume that some α-carbon positions are simply unknown, it is convenient to ignore atoms from some residues during the superposition step; the ignored residues need not be the same if the steps are separated. As an example, suppose a third strand of a β sheet needs to be modeled, and strands one and two are already in place. A crude C_α model can be improvised to represent the third strand, and used as the basis for a three-segment substructure search, but the superposition is best accomplished using only the reliable part of the model (strands 1 and 2). In adopting the desired conformation for the third strand, the existing model for strands 1 and 2 can be retained or replaced as desired. This precise control is easy to achieve with *LORE* because of an interface feature known as the "residue mask,"[11] which allows the user to turn target residues on and off at any time, and thereby to influence the outcome of operations involving that residue position.

When searching for substructures with C_α fingerprints, the resulting alignment of residues in the target and library fragments is known explicitly. Each residue in the fragment is matched in a one-to-one correspondence to a residue in the target. This is a necessary prerequisite to superposition. But with *LORE,* a user also may declare an *ad hoc* alignment explicitly, so that the superposition software can be employed for general coordinate transformation tasks. Created fragments may contain any number of residues, including prosthetic groups, inhibitors, solvent molecules, or higher oligomeric assemblies. Only the orientation of a fragment need be tethered to the original target. General superposition of protein models is therefore a common application of *LORE* in our laboratory. Once overlaid, fragment residues can be interchanged with molecular editing operations comparable to the cutting and pasting that occurs with modern text editors.

Side Chains

When one is mixing and matching protein substructures with disparate amino acid sequences, one of the immediate problems to be dealt with is that of the amino acid side chains. Like many other modeling programs (*O*, *CHAIN*, *XtalView*), we have adopted provisions for looking up side-chain "rotamers," common conformations of amino acid side chains that represent geometries of particular stability.[15] The addition of a rotamer look-up capability to other *LORE* algorithms is beneficial, because the user then may assess the frequency of occurrence of different side-chain conformations in given main-chain conformational contexts. This functionality permits the interactive extension of generalizations about the relationship of side-chain conformations and secondary structure to specific substructural motifs. Frequency analyses of this type can be quite predictive.[16]

Applications

LORE is an indispensable component of our laboratory modeling software. This is primarily because questions concerning other protein structures are so often highly pertinent. *LORE* is designed to allow the user to pose questions of this sort in general terms, and few assumptions are made about what the identified atomic coordinates will be used for. Potential applications of such a tool arise unexpectedly. For example, the generalized model superposition tools are used frequently in our laboratory to examine simultaneously many different inhibitors of an enzyme or enzyme family following superposition of active site coordinates. This is not an application imagined in the design; the program just seems more convenient than other available software for this purpose. The open-ended fragment definition leaves the user free to gain new insights whenever structural information is manipulated.

To provide this flexibility, some ease of use has been sacrificed. Separate typed commands must be entered to define a target for substructures, to initiate a search, to load and overlay atomic coordinates, and to replace selected residues. With each step in this process, the user may elect just to

[15] J. W. Ponder and F. M. Richards, *J. Mol. Biol.* **195,** 775 (1987).

[16] B. C. Finzel, S. Kimatian, D. H. Ohlendorf, J. J. Wendoloski, M. Levitt, and F. R. Salemme, *in* "Crystallographic and Modeling Methods in Molecular Design" (C. E. Bugg and S. E. Ealick, eds.), p. 175. Springer-Verlag, New York, 1990.

continue on, to reconfigure something, or to begin again with a new approach. Sometimes the logic of coordinate manipulation can seem complicated, but flexibility has a price. The program does nothing automatically. *LORE* can be used to guide electron-density interpretation, or to create a full backbone from α carbons, or as a means to validate a structure. It may not be the most convenient tool for any one of these applications, but neither is it limited to any one. In using *LORE* for model building, the distinction between these as separate modeling activities blurs. Tentative α-carbon positions lead to the identification of substructures, which in turn suggest alternate or additional α-carbon positions. The number of such similar substructures serves to validate the original interpretation, and so on.

Homology Modeling

With *LORE,* homology modeling can be accomplished in a fashion similar to that described by Levitt (program *SeqMod*[5]). An α-carbon backbone model derived from a homolog structure can be expanded to include all backbone atoms by overlapping short substructures from the database. The length of fragments and extent of overlap are chosen at the discretion of the user. Long segments known to be conserved in homolog structures can be incorporated intact, thereby saving time and improving overall model integrity. Variable length loops can be modeled by identifying database substructures with the proper endpoint geometry.

LORE includes no energy minimization routines. For this reason it may not be the most convenient tool for this application, but we have used it successfully by employing simple geometric regularization routines in *CHAIN,* and following these with more detailed molecular dynamics with other programs after an initial model is completed. *LORE* can be applied easily to help validate and fine-tune a model against a population of homolog structures, because fragments of structure from the various homologs can be interchanged. Fairly reliable protein models [~1.0-Å root-mean-square (RMS) error] have been generated from C_α backbones using *LORE.*

Side-Chain Modeling

Amino acid side chains also can be placed with the aid of database information. Frequently, a rapid survey of occurrences of a particular amino acid within a protein substructure (e.g., an α helix, a β strand, a particular hairpin) can reveal a similarity among the side-chain conformations that appears strong enough to be predictive. *LORE* has tools to permit an interactive analysis of side-chain rotamer frequencies within any substructural motif.

Consider valine as a simple example. Valines typically occupy one of three distinctly favorable side-chain conformations: *trans* ($\chi_1 = 180°$), g^- ($\chi_1 = -60°$), and g^+ ($\chi_1 = 60°$).[15] In general, the *trans* conformation is observed most frequently (Table I). This preference becomes slightly more pronounced when valines resie in an extended β-like substructure, and almost exclusive in either parallel β structures, or helices. The overall frequency distribution, however, suggests that there must be some local main-chain conformations where other rotamers are preferred. In a search for valines in the central position (residue n) of pentamer substructures with (ϕ, ψ) angles for residue $n-1$ (−70, −20), n (−110, 10), and $n+1$ (−140, 170), the g^- conformation is most common (Table I). This is a backbone conformation occasionally found at the very end of an α helix, but is not otherwise well characterized. *LORE* is able to find such correlations because it looks only at the atomic coordinates, not inferred secondary structure assignments, to identify homologous substructures. It also can be noted that this distribution is not seen in populations of shorter (trimeric) substructures defined only by the (ϕ, ψ) angles of the central valine. A larger conformational context is required.

Because *LORE* can be applied to search for larger hierarchical assemblies, it sometimes can be used to resolve side-chain packing ambiguities at the interface between secondary structure elements, e.g., the point of closest approach between two crossing helices, the interior of a β barrel, or the packing interface between a β sheet and an overlaying helix. The same frequency analyses for side-chain conformations can be performed no matter how complicated the structural motif.

Density Interpretation

Electron-density interpretation has many similarities to homology modeling. Many of the same rules apply; the fit to the density is another im-

TABLE I
FREQUENCY OF VALINE SIDE-CHAIN CONFORMATIONS IN PROTEINS

	Rotamer		
Conformational subset	*trans* ($\chi_1 = 180°$)	g^- ($\chi_1 = -60°$)	g^+ ($\chi_1 = 60°$)
Overall	0.67	0.26	0.05
In extended β strand	0.73	0.12	0.09
Antiparallel β sheet	0.73	0.10	0.10
Parallel β sheet	0.87	0.06	0.06
α Helix	0.92	0.02	0.04
Helix terminator[a]	0.43	0.57	0.00

[a] (ϕ, ψ) angles for residue $n-1$ (−70, −20), n (−110, 10), and $n+1$ (−140, 170).

portant measure of success. The object is to devise a stereochemically reasonable model consistent with the amino acid sequence and the observed electron density. Any general knowledge of protein structure can help this process, but ready access to database structural information can be enormously useful in speeding the model construction, affirming an accurate interpretation, or suggesting viable alternatives.

In this application, the complementary functions of *LORE* and *CHAIN* are essential. *CHAIN* includes electron-density display features that reduce an electron-density map to a skeleton of ridge lines.[17] These may be edited (moved or reconnected), colored to reflect main-chain or side-chain assignments, or converted easily into α-carbon positions. Once C_α positions are created, they can serve as substructure targets for *LORE*.

The approach to building a model is incremental. By working with small portions of the structure, positioning some α carbons, searching for substructures that fit these positions, replacing the backbone, making adjustments to correct the fit of side chains and backbone carbonyls to density, and repeating the substructure search to optimize the fit to all well-known atom positions (not just the C_α positions), one can make the most of both the database and the developing interpretation of the density. The initial substructure search will present possible structures that include side-chain positions not yet modeled. A good fit of these to density serves to validate the initial interpretation, whereas bad correspondences necessitate reinterpretation. For small (pentameric or hexameric) segments, a large number of nearly identical substructures will signal success; if the C_α positions are unrealistic, no structures will be found.

The constant iteration of hypothesis testing and model improvement is a major strength of database modeling, but one that is largely underutilized in semiautomated approaches. Often, it is not the "best" fit to the initial α-carbon positions that is most revealing; it is the preponderance of similarity in a population that is distinctly different from this initial guess. It is easy to see such trends when structures are examined graphically, especially in the context of the electron density.

The flexible substructure definitions of *LORE* can be invaluable in density interpretation. Substructure searches can be contrived to help identify acceptable models that span gaps in the observed electron density, or modeling can be hurried along simply by placing a couple of C_α atoms at either end of a helix, and then performing a substructure search to fill in the missing residues. As pieces of a new structure are assembled, *LORE* can be used to identify homologous structural components arranged in a similar fashion in other proteins. Two modeled strands of a sheet can lead

[17] J. Greer, *J. Mol. Biol.* **82,** 279 (1974).

to identification of a third strand in the density, then a fourth. A helix position may be suggested by analogy to some other structure. The oriented pieces may correspond to a well-characterized fold that, once identified, can provide clues for connectivity that are helpful for completing the tracing of the polypeptide chain.

LORE is also widely applicable to resolve smaller stereochemical problems that crop up during structure refinement. Local distortions of geometry often occur and go uncorrected by refinement programs because of subtle failures in the interpretation of density; a side chain is modeled in the wrong configuration, or a peptide bond is flipped. These areas often can be corrected by simply examining the population of similar substructures, and selectively replacing atoms to correct the problem. The tools described above for the modeling of side chains are just as applicable to model building in the context of electron density. In refinement, it is as important not to move atoms that are already correct, so attention to detail is important, and *LORE* can provide this.

Summary

The collection of known protein structures contains a wealth of information, not only about small building blocks of proteins, but also about the way these elements assemble in real structures. Both types of information can be valuable to the modeler or crystallographer. With the software known as *LORE,* we provide substructure search and coordinate manipulation tools that place a wide variety of different types of structural information at the user's fingertips. Substructures, varying in complexity from small segments to complicated elements of a protein fold, can be identified and used in a wide range of applications, from density interpretation to protein engineering. To our knowledge, no other database modeling software provides such generalized coordinate manipulation capabilities. We have found the software to be a valuable addition to our molecular modeling tools.

[13] Crystallographic Refinement by Simulated Annealing: Methods and Applications

By AXEL T. BRÜNGER and LUKE M. RICE

Introduction

X-Ray crystallography contributes ever increasingly to an understanding of the structure, function, and control of biological macromolecules. Developments in molecular biology and X-ray diffraction data collection have allowed nearly exponential growth of macromolecular crystallographic studies. The analysis of diffraction data from these studies generally requires sophisticated computational procedures including methods of phasing, density modification, chain tracing, refinement, and structure validation. Many of these procedures can be formulated as chemically constrained or restrained nonlinear optimization of a target function, which usually measures the agreement between observed diffraction data and data computed from a model. This target function normally depends on several parameters such as structure factor phases, scale factors between structure factors, or atomic coordinates.

Here we focus on crystallographic refinement, a technique aimed at optimizing the agreement of an atomic model with both observed diffraction data and chemical restraints. Optimization problems in macromolecular crystallography generally suffer from there being multiple minima, which arise largely from the high dimensionality of the parameter space (typically at least three times the number of atoms in the model). The many local minima of the target function tend to defeat gradient descent optimization techniques such as conjugate gradient or least-squares methods.[1] These methods are simply not capable of shifting the atomic coordinates enough to correct errors in the initial model.

This limited radius of convergence arises not only from the high dimensionality of the parameter space, but also from what is known as the crystallographic "phase problem."[2] With monochromatic diffraction experiments on single crystals one can measure the amplitudes of the reflections, but not the phases. The phases, however, are required to compute electron-density maps, which are obtained by Fourier transformation of the structure factor described by a complex number of each reflection. Phases for new

[1] W. H. Press, B. P. Flannery, S. A. Teukolosky, and W. T. Vetterling, "Numerical Recipes." Cambridge University Press, Cambridge, 1986.
[2] H. A. Hauptman, *Physics Today* **42,** 24 (1989).

0076-6879/97 $25.00

crystal structures usually are obtained from experimental methods such as multiple isomorphous replacement.[3] However, electron-density maps computed from a combination of native crystal amplitudes and multiple isomorphous replacement phases sometimes are insufficiently accurate to allow a complete and unambiguous tracing of the macromolecule. Furthermore, electron-density maps for macromolecules usually are obtained at lower than atomic resolution and therefore are prone to human error on interpretation. A different problem arises when structures are solved by molecular replacement,[4,5] which uses a structure similar to that of a search model. In this case the resulting electron-density maps can be severely "model biased," that is, they seem to confirm the existence of the search model without providing clear evidence of actual differences between it and the true crystal structure. In either case, initial atomic models usually require extensive refinement.

This chapter addresses the common case in which experimental phases are either unavailable or inaccurate. In the unusual case that good experimental phases are available, refinement is much more straightforward.[6] Experimental phase information tends to increase the degree to which the global minimum of the target function can be distinguished from local minima. Its omission from the refinement process exacerbates the multiple minima problem to a point that gradient descent methods have little chance of finding the global minimum.[6a]

Simulated annealing[7–9] is an optimization technique particularly well suited to the multiple-minima characteristic of crystallographic refinement. Unlike gradient descent methods, simulated annealing can overcome barriers between minima, and thus can explore a greater volume of the parameter space to find "deeper" minima. Following its introduction in 1987,[10] crystallographic refinement by simulated annealing (often referred to as molecular dynamics refinement) was accepted quickly in the crystallographic community because it significantly reduced the amount of human labor required to determine a crystal structure. In fact, more than 75% of all crystal

[3] D. M. Blow and F. H. C. Crick, *Acta Crystallogr.* **12,** 794 (1959).
[4] W. Hoppe, *Acta Crystallogr.* **10,** 750 (1957).
[5] M. G. Rossmann and D. M. Blow, *Acta Crystallogr.* **A15,** 24 (1962).
[6] W. I. Weis, R. Kahn, R. Fourme, K. Drickamer, and W. A. Hendrickson, *Science* **254,** 1608 (1991).
[6a] L. M. Rice, Y. Shamoo, and A. T. Brünger, in preparation (1997).
[7] S. Kirkpatrick, C. D. Gelatt, and M. P. Vecchi, Jr., *Science* **220,** 671 (1983).
[8] P. J. M. Laarhoven and E. H. L. Aarts (eds.), "Simulated Annealing: Theory and Applications." D. Reidel, Dordrecht, 1987.
[9] M. E. Johnson, *Am. J. Math. Manage. Sci.* **8,** 205 (1988).
[10] A. T. Brünger, J. Kuriyan, and M. Karplus, *Science* **235,** 458 (1987).

structures published during the past 3 years were refined by this method.[11–13] This chapter summarizes the theory, applications, and developments of crystallographic refinement by simulated annealing.

Crystallographic Refinement

Before attempting to understand the chemistry of the crystallized macromolecule, one must correct any errors in the initial atomic model. Crystallographic refinement can correct some of the errors. Crystallographic refinement can be formulated as a search for the global minimum of the target function[14]

$$E = E_{\text{chem}} + w_{\text{xray}} E_{\text{xray}} \tag{1}$$

E_{chem} comprises empirical information about chemical interactions; it is a function of all atomic positions, describing covalent (bond lengths, bond angles, torsion angles, chiral centers, and planarity of aromatic rings) and noncovalent (van der Waals, hydrogen-bonding, and electrostatic) interactions. E_{xray} describes the difference between observed and calculated diffraction data, and w_{xray} is a weight chosen to balance the forces arising from each term. Several algorithms have been developed to minimize E, including least-squares optimization,[15–17] conjugate gradient minimization,[14,18] and simulated annealing refinement.[10]

Crystallographic Residual E_{xray}

The most common form of E_{xray} consists of the crystallographic residual, defined as the sum over the squared differences between the observed $[|\mathbf{F}_{\text{obs}}(\mathbf{h})|]$ and calculated $[|\mathbf{F}_{\text{calc}}(\mathbf{h})|]$ structure-factor amplitudes:

$$E_{\text{xray}} = \sum_{\mathbf{h}} [|\mathbf{F}_{\text{obs}}(\mathbf{h})| - k|\mathbf{F}_{\text{calc}}(\mathbf{h})|]^2 \tag{2}$$

[11] W. A. Hendrickson and K. Wüthrich, "Macromolecular Structures 1991: Atomic Structures of Biological Macromolecules Reported during 1990." Current Biology, London, 1991.

[12] W. A. Hendrickson and K. Wüthrich, "Macromolecular Structures 1992: Atomic Structures of Biological Macromolecules Reported during 1991." Current Biology, London, 1992.

[13] W. A. Hendrickson and K. Wüthrich, "Macromolecular Structures 1993: Atomic Structures of Biological Macromolecules Reported during 1992." Current Biology, London, 1993.

[14] A. Jack and M. Levitt, *Acta Crystallogr.* **A34,** 931 (1978).

[15] J. L. Sussman, S. R. Holbrook, G. M. Church, and S.-H. Kim, *Acta Crystallogr.* **A33,** 800 (1977).

[16] J. H. Konnert and W. A. Hendrickson, *Acta Crystallogr.* **A36,** 344 (1980).

[17] W. A. Hendrickson, *Methods Enzymol.* **115,** 252 (1985).

[18] D. E. Tronrud, L. F. Ten Eyck, and B. W. Matthews, *Acta Crystallogr.* **A43,** 489 (1987).

where $\mathbf{h} = (h, k, l)$ are the indices of the reciprocal lattice points of the crystal. The scale factor k usually is obtained by minimization of Eq. (2). This can be accomplished analytically by setting it to the value that makes the derivative of E_{xray} with respect to k equal to zero. The structure factor of the atomic model is given by

$$\mathbf{F}_{\text{calc}}(\mathbf{h}) = \sum_s \sum_i q_i f_i(\mathbf{h}) \exp[-B_i(\mathcal{F}^* \cdot \mathbf{h})^2/4] \exp[2\pi i \mathbf{h} \cdot (\mathcal{O}_s \cdot \mathcal{F} \cdot \mathbf{r}_i + \mathbf{t}_s)] \quad (3)$$

The first sum extends over all space group symmetry operators $(\mathcal{O}_s, \mathbf{t}_s)$ composed of a rotation matrix $\mathcal{O}_s$, and a translation vector $\mathbf{t}_s$. The second sum extends over all unique atoms i of the system. The quantity $\mathbf{r}_i$ denotes the orthogonal coordinates of atom i in angstroms. $\mathcal{F}$ is the 3×3 matrix that converts orthogonal (Å) coordinates into fractional coordinates; $\mathcal{F}^*$ is its transpose. B_i, q_i are, respectively, the atomic temperature factor and occupancy for atom i. The atomic form factors $f_i(\mathbf{h})$ typically are approximated by an expression consisting of several Gaussians and a constant[19]:

$$f_i(\mathbf{h}) = \sum_k a_{ki} \exp[-b_{ki}(\mathcal{F}^* \cdot \mathbf{h})^2/4] + a_{0i} \quad (4)$$

The structure-factor expression given by Eq. (3) is too computation intensive for practical purposes. Approximations are made usually to make crystallographic refinement feasible. One such approximation consists of computing $\mathbf{F}_{\text{calc}}(\mathbf{h})$ by numerical evaluation of the atomic electron density onto a finite grid, followed by fast Fourier transformation of the electron density. This speeds up the calculation by at least an order of magnitude.[20,21] Another approximation, applied to the minimization process itself, keeps the first derivatives of E_{xray} constant during the refinement process until any atom has moved by more than a specified small distance from the position at which the derivatives were last computed.[22]

The standard crystallographic residual [Eq. (2)] incorporates information about the amplitudes of the observed reflections only. However, a penalty term ("phase restraints"),[23] based on the difference between experimental phases and those calculated from the model, can be added to the residual:

$$E_{\text{xray}} = \sum_{\mathbf{h}} [|\mathbf{F}_{\text{obs}}(\mathbf{h})| - k|\mathbf{F}_{\text{calc}}(\mathbf{h})|]^2 + w_{\text{p}} \sum_{\mathbf{h}} f[\phi_{\text{obs}}(\mathbf{h}) - \phi_{\text{calc}}(\mathbf{h})] \quad (5)$$

[19] J. Ibers and W. C. Hamilton (eds.), "International Tables for X-Ray Crystallography." International Union of Crystallography, The Kynoch Press, Birmingham, 1974.

[20] L. F. Ten Eyck, *Acta Crystallogr.* **A29,** 183 (1973).

[21] A. T. Brünger, *Acta Crystallogr.* **A45,** 42 (1989).

[22] A. T. Brünger, M. Karplus, and G. A. Petsko, *Acta Crystallogr.* **A45,** 50 (1989).

[23] A. T. Brünger, *J. Mol. Biol.* **203,** 803 (1988).

Here w_p is the weight given to the phase restraint, and f is a square-well function with a width equal to the arccosine of the figure of merit $[m(\mathbf{h})]$ for each reflection. Another possible form of E_{xray}, which we call the "vector residual," does not use the amplitude residual at all but instead simultaneously restrains the real (A) and imaginary (B) parts of the structure factor.[24] It has the form

$$E_{xray} = \sum_{\mathbf{h}} m(\mathbf{h})\{[A_{obs}(\mathbf{h}) - kA_{calc}(\mathbf{h})]^2 + [B_{obs}(\mathbf{h}) - kB_{calc}(\mathbf{h})]^2\} \quad (6)$$

Chemical Term E_{chem}

A possible choice of E_{chem} is an empirical potential energy function[25–30]

$$\begin{aligned} E_{chem} = &\sum_{\text{bonds}} k_b(r - r_0)^2 + \sum_{\text{angles}} k_\theta(\theta - \theta_0)^2 \\ &+ \sum_{\text{dihedrals}} k_\phi \cos(n\phi - d) + \sum_{\text{chiral, planar}} k_\omega(\omega - \omega_0)^2 \\ &+ \sum_{\text{atom pairs}} (ar^{-12} + br^{-6} + cr^{-1}) \end{aligned} \quad (7)$$

Empirical energy functions were developed originally for energy-minimization and molecular dynamics studies of macromolecular structure and function (see Ref. 31 for an introduction). The parameters of the empirical potential energy E_{chem} are inferred from experimental as well as theoretical investigations, in particular vibrational spectroscopy and small-molecule crystallography.[25–30]

Because these energy functions were designed for another purpose, it is not surprising that they require some modification for use in crystallographic refinement. For example, empirical energy functions must be extended to simulate contacts between molecules related by crystallographic or noncrystallographic symmetry.[22,32] Empirical energy functions also behave poorly at the high simulation temperatures characteristic of simulated annealing.

[24] E. Arnold and M. G. Rossmann, *Acta Crystallogr.* **A44,** 270 (1988).

[25] S. Lifson and P. Stern, *J. Chem. Phys.* **77,** 4542 (1982).

[26] B. R. Brooks, R. E. Bruccoleri, B. D. Olafson, D. J. States, S. Swaminathan, and M. Karplus, *J. Comput. Chem.* **4,** 187 (1983).

[27] G. Némethy, M. S. Pottie, and H. A. Scheraga, *J. Phys. Chem.* **87,** 1883 (1983).

[28] J. Hermans, H. J. C. Berendsen, W. F. van Gunsteren, and J. P. M. Postma, *Biopolymers* **23,** 1513 (1984).

[29] L. Nilsson and M. Karplus, *J. Comput. Chem.* **7,** 591 (1986).

[30] S. J. Weiner, P. A. Kollman, D. T. Nguyen, and D. A. Case, *J. Comput. Chem.* **7,** 230 (1986).

[31] M. Karplus and G. A. Petsko, *Nature (London)* **347,** 631 (1990).

[32] W. I. Weis, A. T. Brünger, J. J. Skehel, and D. C. Wiley, *J. Mol. Biol.* **212,** 737 (1989).

They also must be modified to cope with the addition of experimental restraints (E_{xray}). To prevent distortions of aromatic rings, peptide bonds, and chiral centers, certain energy constants k_ϕ, k_ω in Eq. (7) often need to be increased.[23] Furthermore, the energy constant k_ϕ for the proline ω angle can be decreased to enable *cis*-to-*trans* isomerizations. (However, experience has shown[32] that this constant should be set to its original value during the final stages of refinement to obtain acceptable geometry about these peptide bonds.) Finally, since bulk solvent usually is omitted from refinement, the charged groups of Asp, Glu, Arg, and Lys residues have to be screened to avoid formation of artificial interactions with backbone atoms. This can be accomplished by setting the charges to zero.[33,34]

Apart from the modifications discussed above, crystallographic refinement is not sensitive to the accuracy of the empirical energy function. Thus, the electrostatic term in Eq. (7) is sometimes purposely omitted to avoid possible bias. Furthermore, one can use a "geometric" energy function consisting of terms for covalent bonds, bond angles, chirality, planarity, and nonbonded repulsion, where the corresponding parameters are derived from equilibrium geometry and root-mean-square (r.m.s.) deviations of bond lengths and angles observed in a small-molecule database.[35] The differences between a geometric energy function and an empirical energy function mainly affect regions that are not well determined by the experimental information. Little difference is observed for well-defined structures. For instance, the r.m.s. difference for backbone atoms between a structure of crambin refined at 2-Å resolution by PROLSQ,[16] a program that effectively uses a geometric energy function, and the same structure refined by conjugate gradient minimization using X-PLOR[36] was only 0.05 Å.[22] Comparison of DNA structures refined by different programs (NUCLSQ, TNT, and X-PLOR) and different parameter sets showed no significant differences within the estimated error of the atomic positions.[37]

Additional Restraints and Constraints

Additional constraints or restraints may be used to improve the ratio of observables to parameters. For example, atoms can be grouped so that they move as rigid bodies during refinement, or bond lengths and bond

[33] A. T. Brünger, A. Krukowski, and J. Erickson, *Acta Crystallogr.* **A46,** 585 (1990).

[34] M. Fujinaga, P. Gros, and W. F. van Gunsteren, *J. Appl. Crystallogr.* **22,** 1 (1989).

[35] R. A. Engh and R. Huber, *Acta Crystallogr.* **A47,** 392 (1991).

[36] A. T. Brünger, "X-PLOR: A System for X-Ray Crystallography and NMR," version 3.1. Yale University Press, New Haven, Connecticut, 1992.

[37] M. Hahn and U. Heinemann, *Acta Crystallogr.* **D5,** 468 (1993).

angles can be kept fixed.[15,38,39] The existence of noncrystallographic symmetry in a crystal can be used to average over equivalent molecules and thereby to reduce noise in the data. This is especially useful for virus structures: Noncrystallographic symmetry can be used to "overdetermine" the problem, assisting the primary phasing and the subsequent refinement.[32,40,41]

Weighting

The weight w_{xray} [Eq. (1)] balances the forces arising from E_{xray} and E_{chem}. The choice of w_{xray} can be critical: if w_{xray} is too large, the refined structure will show unphysical deviations from ideal geometry; if w_{xray} is too small, the refined structure will not satisfy the diffraction data. Jack and Levitt[14] proposed that w_{xray} be chosen so that the gradients of E_{chem} and E_{xray} have the same magnitude for the current structure. This approach implies that w_{xray} must be readjusted frequently during the course of the refinement. Brünger *et al.*[22] developed an empirical procedure for obtaining a value for w_{xray} that can be kept constant throughout the refinement. It consists of first performing a short molecular dynamics simulation with w_{xray} set to zero, then calculating the final r.m.s. gradient due to the empirical energy term E_{chem} alone. Next one calculates the gradient due to the experimental restraints E_{xray} alone, and chooses w_{xray} to balance the two. A correction to this procedure is to divide the resulting w_{xray} by two; this produces optimal phase accuracy as judged by the free R value[42,43,44] (A. T. Brünger, unpublished data).

Simulated Annealing Refinement

Annealing denotes a physical process wherein a solid is heated until all particles randomly arrange themselves in a viscous liquid phase, and then it is cooled slowly so that all particles arrange themselves in the lowest energy state. By formally defining the target E [Eq. (1)] to be the equivalent of the potential energy of the system, one can simulate the annealing

[38] R. Diamond, *Acta Crystallogr.* **A27,** 436 (1971).

[39] L. M. Rice and A. T. Brünger, *Proteins Struct. Funct. Genet.* **19,** 277 (1994).

[40] J. N. Champness, A. C. Bloomer, G. Bricogne, P. J. G. Butler, and A. Klug, *Nature (London)* **259,** 20 (1976).

[41] M. G. Rossmann, E. Arnold, J. W. Erickson, E. A. Frankenberger, J. P. Griffith, H.-J. Hecht, J. E. Johnson, G. Kamer, M. Luo, A. G. Mosser, R. R. Rueckert, B. Sherry, and G. Vriend, *Nature (London)* **317,** 145 (1985).

[42] A. T. Brünger, *Nature (London)* **355,** 472 (1992).

[43] A. T. Brünger, *Acta Crystallogr.* **D49,** 24 (1993).

[44] A. T. Brünger, *Methods Enzymol.* **277,** [19], 1997 (this volume).

process.[7] Simulated annealing is an approximation algorithm: There is no guarantee that it will find the global minimum (except in the asymptotic limit of an infinite search).[8] Compared to gradient descent methods where search directions must follow the gradient, simulated annealing achieves more optimal solutions by allowing motion against the gradient.[7] The likelihood of countergradient motion is determined by a control parameter referred to as "temperature": the higher the temperature the more likely the optimization will overcome barriers. It should be noted that the simulated annealing temperature normally has no physical meaning and merely determines the likelihood of overcoming barriers of the target function.

The simulated annealing algorithm requires a generation mechanism to create a Boltzmann distribution at a given temperature T,

$$B(q_1, \ldots, q_i) = \exp\left[\frac{-E(q_1, \ldots, q_i)}{k_b T}\right] \tag{8}$$

where E is given by Eq. (1), k_b is the Boltzmann constant, and $q_1, \ldots, q_i$ are adjustable parameters, such as the coordinates of the atoms. Simulated annealing also requires an annealing schedule, that is, a sequence of temperatures $T_1 \geq T_2 \geq \cdots \geq T_l$ at which the Boltzmann distribution is computed. Implementations of the generation mechanism differ in the way they generate a transition or "move" from one set of parameters to another that is consistent with the Boltzmann distribution at given temperature. The two most widely used generation mechanisms are Metropolis Monte Carlo[45] and molecular dynamics[46] simulations. Metropolis Monte Carlo can be applied to both discrete and continuous optimization problems, but molecular dynamics is restricted to continuous problems.

Monte Carlo

The Metropolis Monte Carlo algorithm[45] simulates the evolution to thermal equilibrium of a solid for a fixed value of the temperature T. Given the current state of system, characterized by the parameters q_i of the system, a "move" is applied by a shift of a randomly chosen parameter q_i. If the energy after the move is less than the energy before, i.e., if $\Delta E < 0$, the move is accepted and the process continues from the new state. If, on the other hand, $\Delta E \geq 0$, then the move may still be accepted with probability

$$P = \exp\left(-\frac{\Delta E}{k_b T}\right) \tag{9}$$

[45] N. Metropolis, M. Rosenbluth, A. Rosenbluth, A. Teller, and E. Teller, *J. Chem. Phys.* **21,** 1087 (1953).

[46] L. Verlet, *Phys. Rev.* **159,** 98 (1967).

where k_b is Boltzmann's constant. Specifically, if P is greater than a random number between 0 and 1 then the move is accepted. In the limiting case of $T = 0$, Monte Carlo is equivalent to a gradient descent method; the only moves allowed are the ones that lower the target function until a local minimum is reached. At a finite temperature, however, Monte Carlo allows uphill moves and hence allowing for crossing barriers between local minima.

The advantage of the Metropolis Monte Carlo algorithm is its simplicity. A particularly troublesome aspect concerns the efficient choice of the parameter shifts that define the Monte Carlo move. Ideally, this choice should in some way reflect the topology of the search space as characterized by the variables q_i.[8] In the case of a monoatomic liquid or gas, for example, the coordinates of the atoms of the gas are essentially uncoupled so that the coordinate shifts can be chosen in random directions. In the case of a covalently connected macromolecule, however, random shifts of atomic coordinates have a high rejection rate: they immediately violate geometric restrictions such as bond lengths and bond angles. This problem can be alleviated in principle by carrying out the Monte Carlo simulation in a suitably chosen set of internal coordinates such as torsions about bonds, or normal modes of vibration, or by relaxing the strained coordinates through minimization.[47–49]

Cartesian Molecular Dynamics

A suitably chosen set of continuous (smoothly varying) parameters q_i can be viewed as generalized coordinates that are propagated in time by the Hamilton equations of motion[50]

$$\frac{\partial H(p,q)}{\partial p_i} = \frac{dq_i}{dt}$$
$$\frac{\partial H(p,q)}{\partial q_i} = \frac{dp_i}{dt} \tag{10}$$

Here $H(p, q)$ is the Hamiltonian (the sum of the potential and kinetic energy) of the system and p_i is the generalized momentum conjugate to q_i. If the generalized coordinates represent the atomic coordinates of a molecular system, this approach is referred to as molecular dynamics.[46] If one makes the assumption that the resulting trajectories cover phase space

[47] M. Saunders, *J. Am. Chem. Soc.* **109,** 3150 (1987).
[48] Z. Li and H. A. Scheraga, *Proc. Natl. Acad. Sci. U.S.A.* **84,** 6611 (1987).
[49] R. Abagyan and P. Argos, *J. Mol. Biol.* **225,** 519 (1992).
[50] H. Goldstein, "Classical Mechanics," 2nd Ed. Addison-Wesley, Reading, Massachusetts, 1980.

(or more specifically, are ergodic) then they generate a statistical mechanical ensemble.[51]

Molecular dynamics can be coupled to a heat bath (see below) so that the resulting ensemble asymptotically approaches that generated by the Metropolis Monte Carlo acceptance criterion [Eq. (9)]. Thus, molecular dynamics and Monte Carlo are equivalent for the purpose of simulated annealing, although in practice one implementation may be more efficient than the other. Comparative work[51a] has shown the molecular dynamics implementation of crystallographic refinement by simulated annealing to be more efficient than the Monte Carlo implementation.

In the special case that the generalized coordinates q_i represent the Cartesian coordinates of n point masses and, furthermore, that momenta can be separated from coordinates in the Hamiltonian H, the Hamilton equations of motion reduce to the more familiar Newton's second law:

$$m_i \frac{\partial^2 \mathbf{r}_i}{\partial t^2} = -\nabla_i E = \frac{\mathbf{F}_i(\mathbf{r})}{m_i} \tag{11}$$

The quantities m_i and $\mathbf{r}_i$ are, respectively, the mass and coordinates of atom i, $\mathbf{F}_i$ is the force acting on atom i, and E is the potential energy. In the context of simulated annealing, E denotes the target function being optimized [Eq. (1)], which contains "physical" energies such as covalent and nonbonded energy terms as well as "nonphysical" energies that correlate observed and calculated diffraction data. The solution of the partial differential equations [Eq. (11)] is normally achieved numerically using finite-difference methods.[46] Initial velocities are usually assigned from a Maxwell distribution at the appropriate temperature.

Torsion Angle Molecular Dynamics

Although Cartesian molecular dynamics places restraints on bond lengths and bond angles, one might want to implement these restrictions as holonomic constraints, i.e., fixed bond lengths and bond angles. This is supported by the observation that the deviations from ideal bond lengths and bond angles are usually small in X-ray crystal structures. There are essentially two possible approaches to solve Newton's equations [Eq. (11)] with holonomic constraints. The first involves a switch from Cartesian coordinates $\mathbf{r}_i$ to generalized internal coordinates $\mathbf{q}_i$. Having thus redefined the system, one would solve equations of motion for the generalized coordi-

[51] D. A. McQuarrie, "Statistical Mechanics." Harper & Row, New York, 1976.

[51a] P. D. Adams, L. A. Rice, and A. T. Brünger, unpublished.

nates analogous to the Cartesian ones. This formulation has the disadvantage that it is difficult (but not impossible) to calculate the generalized gradients. Because the gradients are functions of the generalized coordinates only, however, conventional finite-difference integration schemes[46] can be used. A second possible approach is to retain the Cartesian coordinate formulation so that the gradient calculation remains relatively straightforward and topology independent.[39] In this formulation, however, the expression for the acceleration becomes a more complicated function of positions and velocities:

$$\mathbf{a}(\mathbf{r}, \dot{\mathbf{r}}) = \mathbf{M}^{-1}(\mathbf{r})\mathbf{Q}(\mathbf{r}, \dot{\mathbf{r}}) \tag{12}$$

where $\mathbf{a}$ represents the system acceleration vector, and $\mathbf{M}$ and $\mathbf{Q}$ denote the (6×6) system inertia matrix and (6×1) generalized force vector, respectively. This does not present insurmountable difficulties, but instead requires different integration schemes such as a fourth order Runge Kutta integration scheme.[52]

The equations of motion for constrained dynamics in this formulation are derived in complete generality by Bae and Haug.[53,54] We have also produced a slightly simpler derivation specific for fixed bond lengths and bond angles.[39] What follows is a simple sketch of this one particular implementation of molecular dynamics with holonomic constraints.

Consider two bodies, i and j, connected by a bond of fixed length $|\mathbf{h}_{ij}|$. Assuming that the only allowable relative motion between the two bodies is a rotation about $\mathbf{h}_{ij}$, let $\mathbf{r}_i$ and $\mathbf{r}_j$ locate (with respect to an arbitrary inertial frame) the center of mass of body i and j, respectively. Let $\mathbf{s}_{ij}$ ($\mathbf{s}_{ji}$) locate the end point of $\mathbf{h}_{ij}$ on body i (j) with respect to its center of mass. Thus, $\mathbf{s}_{ij}$ is a vector from the center of mass of body i to the end of $\mathbf{h}_{ij}$. The position of the center of mass of body j with respect to that of body i is simply $\mathbf{r}_{ij} = \mathbf{r}_j - \mathbf{r}_i$. Finally, the scalar q_{ij} measures the relative angle of rotation about the bond $\mathbf{h}_{ij}$ (see Ref. 39 for more details).

The assumption that the only allowable relative motion between the two bodies is a rotation about the bond connecting them implies a relationship between the angular velocity $\mathbf{w}$ of their respective centers of mass measured in an inertial ("lab") frame:

$$\mathbf{w}_j = \mathbf{w}_i + \hat{\mathbf{h}}_{ij}\dot{q}_{ij} \tag{13}$$

[52] M. Abramowitz and I. Stegun, "Handbook of Mathematical Functions," Vol. 55. Applied Mathematics Series. Dover, New York, 1968.

[53] D.-S. Bae and E. J. Haug, *Mech. Struct. Mach.* **15,** 359 (1987).

[54] D.-S. Bae and E. J. Haug, *Mech. Struct. Mach.* **15,** 481 (1988).

Here $\dot{q}_{ij}$ denotes the time derivative of the relative angle between the two bodies and $\hat{\mathbf{h}}_{ij} = \mathbf{h}_{ij}/|\mathbf{h}_{ij}|$ is the unit vector along the bond connecting them. The expression for $\mathbf{r}_j$ can be rewritten

$$\mathbf{r}_j = \mathbf{r}_i + \mathbf{r}_{ij} = \mathbf{r}_i + \mathbf{s}_{ij} + |\mathbf{h}_{ij}|\hat{\mathbf{h}}_{ij} - \mathbf{s}_{ji} \tag{14}$$

This expression can be differentiated and then rearranged, resulting in an expression for the center of mass velocity of body j in terms of that of body i:

$$\begin{aligned} \dot{\mathbf{r}}_j &= \dot{\mathbf{r}}_i + \dot{\mathbf{s}}_{ij} + |\mathbf{h}_{ij}|\dot{\hat{\mathbf{h}}}_{ij} - \dot{\mathbf{s}}_{ji} \\ &= \dot{\mathbf{r}}_i + \mathbf{w}_i \times \mathbf{s}_{ij} + |\mathbf{h}_{ij}|\mathbf{w}_i \times \hat{\mathbf{h}}_{ij} - \mathbf{w}_j \times \mathbf{s}_{ji} \\ &= \dot{\mathbf{r}}_i - \mathbf{s}_{ij} \times \mathbf{w}_i - |\mathbf{h}_{ij}|\hat{\mathbf{h}}_{ij} \times \mathbf{w}_i + \mathbf{s}_{ji} \times \mathbf{w}_i - \dot{q}_{ij}\hat{\mathbf{h}}_{ij} \times \mathbf{s}_{ji} \\ &= \dot{\mathbf{r}}_i - \mathbf{r}_{ij} \times \mathbf{w}_i - (\hat{\mathbf{h}}_{ij} \times \mathbf{s}_{ji})\dot{q}_{ij} \end{aligned} \tag{15}$$

Thus, assuming certain constraints act between atoms or groups of atoms, one can obtain an expression for the velocity of one group in terms of another. This relationship can be differentiated to give a relationship between accelerations, and integrated to give a relationship between positions.

The algorithm is recursive, so the equations of motion for two bodies easily extend to many. In our implementation, atoms are grouped into rigid bodies, allowing only torsion-angle motions between bodies. The connectivity of these bodies defines a treelike topology for a macomolecule, with one arbitrarily chosen body identified as the base (or root). As with any molecular dynamics algorithm, the torsion angle one begins with required positions, velocities, and forces for all atoms at the first step. Center of mass positions, velocities, and forces are then computed for each rigid group. Starting at the outer ends of the tree topology, each chain is "reduced" one body at a time by solving for the relative acceleration between the tip and the directly inner body. The inertial properties of the tip are then mapped, or aggregated, into those of the inner body, resulting in a chain that is effectively one link shorter. This process continues until an expression for the acceleration of the base body is obtained. After solving for the acceleration of the base (which requires only inversion of a 6×6 matrix), the aggregation of bodies is reversed. The acceleration of the body "outboard" of the base is determined by the acceleration of the base and the relative acceleration between the bodies. This outward expansion is continued until the tree has been completely covered (see Ref. 39 for more details). A Runge Kutta integration step then updates positions and velocities. Finally, new forces are calculated, and the whole process begins anew. The formalism is a general one: Several treelike topologies can be handled, as can closed topological loops such as those formed by disulfide bonds.

Temperature Control

Simulated annealing requires the control of the temperature during molecular dynamics. The three most commonly used methods are velocity scaling, Langevin dynamics, and temperature coupling. The current temperature T_{curr} is computed from the kinetic energy $[E_{\text{kin}} = \sum_i^n \frac{1}{2} m_i (\partial \mathbf{r}_i / \partial t)^2]$ of the molecular dynamics simulation

$$T_{\text{curr}} = \frac{2E_{\text{kin}}}{3nk_{\text{b}}} \tag{16}$$

Here n is the number of degrees of freedom and k_{b} is Boltzmann's constant.

Velocity scaling consists of periodic uniform scaling of the velocities $\mathbf{v}_i$, i.e.,

$$\mathbf{v}_i^{\text{new}} = \frac{\partial \mathbf{r}_i}{\partial t} \left(\frac{T}{T_{\text{curr}}} \right)^{1/2} \tag{17}$$

for all atoms i where T is the target temperature. The numerical integration of the equations of motion needs to be restarted using the new velocities $\mathbf{v}_i^{\text{new}}$ and current coordinates $\mathbf{r}_i$.

Langevin dynamics incorporates the influence of a heat bath into the classic equations of motion,

$$m_i \frac{\partial^2 \mathbf{r}_i}{\partial t^2} = -\nabla_i E_{\text{total}} - m_i \gamma_i \frac{\partial \mathbf{r}_i}{\partial t} + R(t) \tag{18}$$

where γ_i specifies the friction coefficient for atom i and $R(t)$ is a random force. $R(t)$ is assumed to be uncorrelated with the positions and velocities of the atoms. It is described by a Gaussian distribution with mean of zero and variance

$$\langle R(t) R(t') \rangle = 2 m_i \gamma_i k T \delta(t - t') \tag{19}$$

where k is Boltzmann's constant and $\delta(t - t')$ is the Dirac delta function.

The temperature-coupling method of Berendsen *et al.*[55] is related to Langevin dynamics except that it does not use random forces and it applies a temperature-dependent scale factor to the friction coefficient:

$$m_i \frac{\partial^2 \mathbf{r}_i}{\partial t^2} = -\nabla_i E^{\text{total}} - m_i \gamma_i \frac{\partial \mathbf{r}_i}{\partial t} \left(1 - \frac{T}{T_{\text{curr}}} \right) \tag{20}$$

[55] H. J. C. Berendsen, J. P. M. Postma, W. F. van Gunsteren, A. DiNola, and J. R. Haak, *J. Chem. Phys.* **81,** 3684 (1984).

The second term on the right-hand side of Eq. (20) represents positive fraction if $T_{curr} > T$, thus lowering the temperature; it represents negative "friction" if $T_{curr} < T$, thus increasing the temperature.

Annealing Schedule

The success and efficiency of simulated annealing depends on the choice of the annealing schedule,[56] that is, the sequence of numerical values $T_1 \geq T_2 \geq \cdots T_l$ for the temperature. Note that multiplication of the temperature T by a factor s is formally equivalent to scaling the target E by $1/s$. This applies to both the Monte Carlo as well as the molecular dynamics implementation of simulated annealing. This is immediately obvious on inspection of the Metropolis Monte Carlo acceptance criterion [Eq. (9)]. For molecular dynamics this can be seen as follows. Let E be scaled by a factor $1/s$ while maintaining a constant temperature during the simulation,

$$m_i \frac{\partial^2 \mathbf{r}_i}{\partial t^2} = -\nabla_i \frac{E}{s} \tag{21}$$

$$E_{\text{kin}} = \sum_i^n \frac{1}{2} m_i \left(\frac{\partial \mathbf{r}_i}{\partial t} \right)^2 = \text{constant} \tag{22}$$

This is equivalent to

$$m_i \frac{\partial^2 \mathbf{r}_i}{\partial t'^2} = -\nabla_i E \tag{23}$$

$$E'_{\text{kin}} = \sum_i^n \frac{1}{2} m_i \left(\frac{\partial \mathbf{r}_i}{\partial t'} \right)^2 = s E_{\text{kin}} \tag{24}$$

with $t' = t/s^{1/2}$, i.e., the kinetic energy and, thus, the temperature T_{curr} is scaled by s.

The equivalence between temperature control and scaling of E suggests a generalization of simulated annealing schedules where in addition to the overall scaling of E, relative scale factors between or modifications of the components of the target E are introduced, i.e., simulated annealing is carried out with an adjustable target function. In this case, the annealing schedule denotes the sequence of scale factors or modifications of components of E. A particular example of this type of generalized annealing schedule is the use of a "soft" van der Waals potential during high-temperature molecular dynamics followed by a normal van der Waals potential during the cooling stage.[39]

[56] D. G. Bounds, *Nature* (*London*) **329,** 215 (1987).

Annealing Control

The analogy of simulated annealing with the physical annealing of solids can be expressed more formally through a connection to statistical mechanics. Both Monte Carlo and molecular dynamics simulations can create statistical–mechanical ensembles.[51] Approximately, at least, one can use a statistical–mechanical language to describe the progress of simulated annealing. For example, changes in the degree of the order of the system can be viewed as phase transitions. They can be detected by finding large values of the specific heat c during the simulation,

$$c = \frac{\langle [E(t) - \langle E(t) \rangle]^2 \rangle}{k_b T^2} \tag{25}$$

where the brackets $\langle \rangle$ denote the mean computed over appropriate intervals of the simulation. It has been suggested[7] that the cooling rate be reduced at phase transitions because the system is in a critical state where fast cooling might trap the system in a metastable state. The observed fluctuations in E are relatively small during simulated annealing refinement, however, indicating local conformational changes rather than global phase transitions.[33] Thus, control of the annealing schedule by monitoring c has not yet been attempted and annealing schedules consisting of a predefined sequence of temperatures and modifications of E_{chem} are used.

Commonly Used Annealing Schedules

The two early implementations of simulated annealing refinement made use of the equivalent methods of temperature scaling[10] or energy scaling.[57] The influences of the temperature control method, energy term weighting, cooling rate, and duration of the heating stage were studied,[33] and it was found that the temperature-coupling method of Berendsen and co-workers[55] is preferable to velocity scaling because velocity scaling sometimes causes large temperature fluctuations at high temperatures. Temperature coupling also outperformed Langevin dynamics in the context of simulated annealing because the always positive friction of Langevin dynamics tends to slow atomic motions. Slow-cooling protocols (typically 25 K temperature decrements every 25 fsec) produced lower R values than faster cooling protocols. Constant-temperature dynamics represents the extreme limit of slow-cooling annealing with an infinitely long cooling rate. Thus, constant temperature protocols can outform slow-cooling ones.[39]

[57] P. Gros, M. Fujinaga, B. W. Dijkstra, K. H. Kalk, and W. G. J. Hol, *Acta Crystallogr.* **B45,** 488 (1989).

A typical constant-temperature protocol consists of a high-temperature molecular dynamics stage at 5000 K over a period of 2–4 psec, followed by a fast-cooling stage at 300 K over a period of 0.1 psec. The more robust torsion-angle molecular dynamics algorithm outlined above allows conformational sampling at much higher temperatures than are possible with conventional unconstrained molecular dynamics. For refinements of α-amylase inhibitor at 5- to 2-Å resolution, torsion-angle dynamics at 10,000 K produces better results than at 5000 K. However, more testing is required to establish the generality of this observation at different resolution ranges and data qualities. Thus, at the present stage, a certain amount of experimentation is required to find the optimal annealing schedule and temperature for each specific refinement problem. Choice of temperature is most influenced by model quality and resolution.

Radius of Convergence

A number of realistic tests on crambin,[22] aspartate aminotransferase,[23] myohemerythrin,[58] phospholipase A_2,[34] thermitase complexed with eglin c,[57]and immunoglobulin light chain dimers[59] have shown that simulated annealing refinement starting from initial models (obtained by standard crystallographic techniques) produces significantly improved overall R values and geometry compared to those produced by least-squares optimization or conjugate gradient minimization.

In tests,[39] arbitarily "scrambled" models were generated from an initial model of α-amylase inhibitor built using experimental phase information from multiple isomorphous replacement diffraction data.[60] Scrambling of this initial model was obtained by increasingly long molecular dynamics simulations at 600 K computed without reference to the X-ray data. Errors were thereby distributed throughout the structure and are probably typical of those found in molecular replacement models or in poorly built initial models. To compare the power of refinement techniques, a series of these models was refined using two standard methods: conjugate gradient minimization and slow-cooling simulated annealing.

Results are presented in Fig. 1, which depicts the backbone atom r.m.s. coordinate deviations before and after refinement for a number of different refinement methods. A similar graph for a perfect refinement technique would be a straight line along the horizontal axis: regardless of the initial

[58] J. Kuriyan, A. T. Brünger, M. Karplus, and W. A. Hendrickson, *Acta Crystallogr.* **A45,** 396 (1989).

[59] Z.-B. Xu, C.-H. Chang, and M. Schiffer, *Protein Eng.* **3,** 583 (1990).

[60] J. W. Pflugrath, G. Wiegand, R. Huber, and L. Vértesey, *J. Mol. Biol.* **189,** 383 (1986).

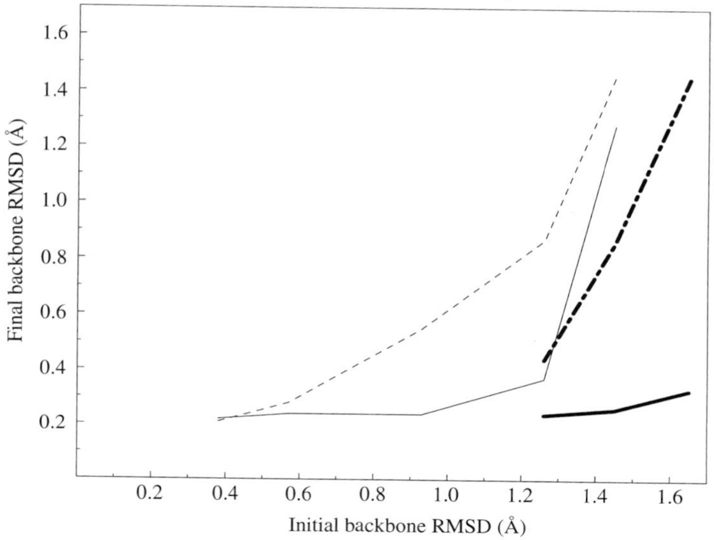

FIG. 1. Radius of convergence of conjugate gradient minimization, Cartesian slow cooling, and torsion-angle simulated annealing. Convergence is measured by the final backbone atom r.m.s. coordinate deviation to the crystal structure. Thin lines show the result from one conjugate gradient minimization (dashed) or one slow-cooling simulated annealing refinement (solid). The thick dot-dashed line shows the average backbone atom r.m.s. coordinate deviation obtained from 10 high-temperature torsion-angle refinements at 5000 K, and the thick solid line shows the backbone atom r.m.s. coordinate deviation achieved by the torsion-angle refinement with the lowest free R value.

errors, the final model would be in good agreement with the crystal structure. Clearly this is not the case for conjugate gradient minimization, or even for Cartesian simulated annealing, although Cartesian simulated annealing is a more powerful refinement technique than conjugate gradient minimization. For refinements carried out between 5 and 2 Å, slow-cooling simulated annealing can correct backbone atom r.m.s. coordinate deviations of around 1.3 Å.

Constant-temperature torsion-angle refinements (Fig. 1) outperform the slow-cooling Cartesian protocol on average, dramatically so if one considers only the best model from each series. The torsion-angle refinements are able to correct backbone atom r.m.s. coordinate deviations of at least 1.65 Å. Clearly, the backbone atom r.m.s. coordinate deviation is available only if one knows the crystal structure in advance. Figure 2, however, shows the strong correlation between R_{free} (see Refs. 42 and 43, and [19] in this volume[44]) and backbone r.m.s. coordinate deviations. Thus in practice R_{free}

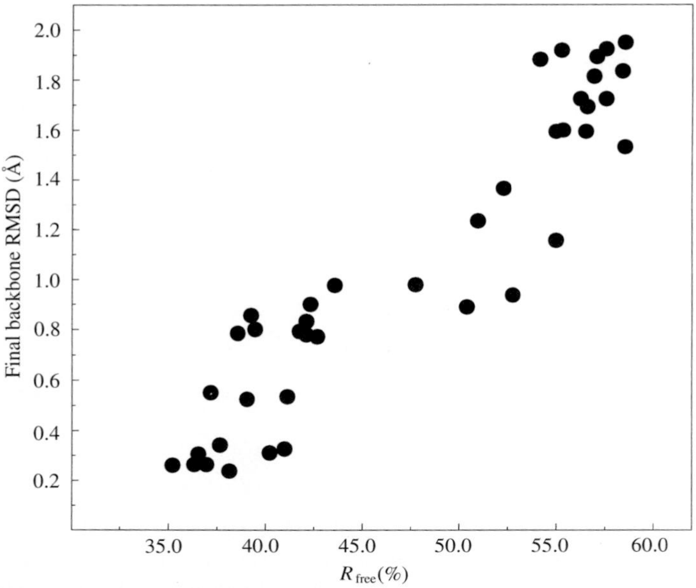

FIG. 2. Free R value vs backbone atom r.m.s. coordinate deviation for torsion-angle constant-temperature refinements using α-amylase inhibitor[60] as a test case.

can be used to identify the best models from a series of refinements (see [19] in this volume[44]).

Simulated annealing has made crystallographic refinement more efficient by automatically moving side-chain atoms by more than 2 Å, by changing backbone conformations, or by flipping peptide bonds without direct human intervention. Figure 3 shows a representative case in which simulated annealing refinement has essentially converged to a manually refined structure of the enzyme aspartate aminotransferase.[23] The imidazole ring of the histidine side chain has undergone a 90° rotation around the χ_1 bond during simulated annealing refinement. This rotation was accompanied by significant structural changes of the backbone atoms. This resulted in convergence of the refined structure to the manually refined structure. Conjugate gradient minimization could not arrive at an equally good model without rebuilding. Large rigid body-like corrections of up to 10° resulting from simulated annealing refinement were observed by Gros *et al.*[57]

Simulated annealing refinement is most useful when the initial model is relatively crude. Given a well-refined model, it offers little advantage over conventional methods, with the possible exception of providing information

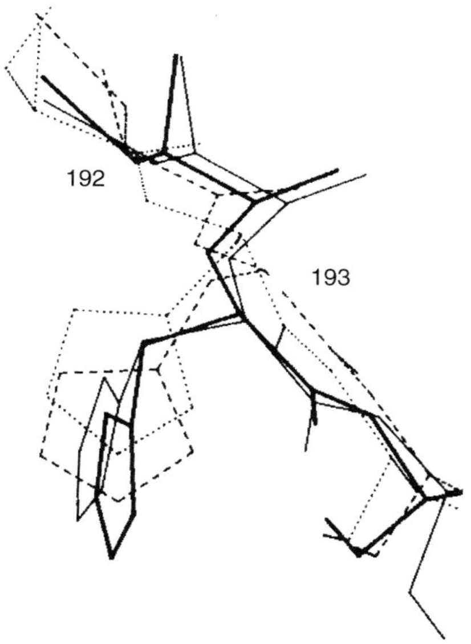

FIG. 3. The segment consisting of residues Cys-192 and His-193 of the 2.8-Å resolution structure of a single site mutant of aspartate aminotransferase.[23] Superimposed are the initial structure (dotted lines) obtained by fitting the atomic model to a multiple isomorphous replacement map, the structure obtained after several cycles of rebuilding and restrained least-squares refinement (thick lines), the structure obtained after simulated annealing refinement (thin lines), and the structure obtained after conjugate gradient minimization (dashed lines).

about the accuracy and conformational variability of the refined structure.[61] However, when only a crude model is available, simulated annealing refinement is able to reduce significantly the amount of human intervention required. The initial model can be as crude as one that is obtained by automatic building based on C_α positions alone.[62]

In spite of the success of simulated annealing refinement, the importance of manual inspection of the electron-density maps after simulated annealing refinement cannot be overemphasized. Manual inspection is essential for the placement of surface side chains and solvent molecules, for example, and for checking regions of the protein where large deviations from idealized

[61] F. T. Burling and A. T. Brünger, *Israel J. Chem.* **34,** 165 (1994).

[62] T. A. Jones, J.-Y. Zou, S. W. Cowan, and M. Kjeldgaard, *Acta Crystallogr.* **A47,** 110 (1991).

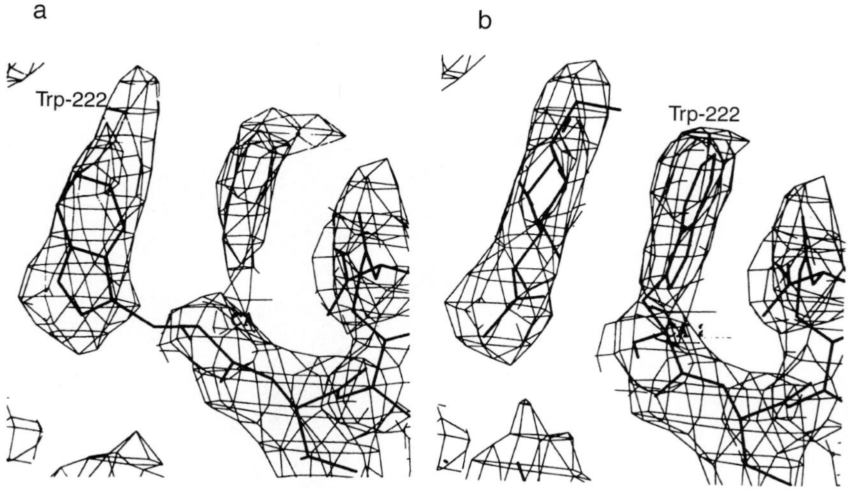

FIG. 4. Simulated annealing refinement can move atoms far from their initial positions to compensate for missing model atoms. Residue Trp-222 of influenza virus hemagglutinin, which has weak side-chain density, moved into strong N-linked carbohydrate density in the first round of simulated annealing refinement, before carbohydrate was added to the model. The electron-density maps were computed from $(2F_{obs} - F_{calc})$ amplitudes using F_{calc} phases corresponding to the atomic models at 3.0-Å resolution. It shows density corresponding to missing atoms as well as the current model. The electron-density maps are displayed as "chicken wire," which represents the density at a constant level of 1 standard deviation above the mean. The maps have been averaged about the threefold noncrystallographic symmetry axis. (a) Electron density and coordinates after round 1, showing missing density for Trp-222 C_β. (b) As in (a), showing properly built N-linked carbohydrate and Trp-222.

geometry occur. Figure 4 illustrates a problem that occurred during simulated annealing refinement of influenza virus hemagglutinin.[32] A poorly defined tryptophan side chain moved into strong density belonging to N-linked carbohydrate that was not included in the model used in the first round of simulated annealing refinement (Fig. 4a). Simulated annealing can move atoms far enough to compensate at least partially for missing parts of the model. The model was rebuilt manually in this region, and the missing carbohydrate was added. In subsequent rounds of simulated annealing, proper model geometry and fit to the electron-density map were maintained (Fig. 4b).

Simulated annealing refinement can produce *R* values in the twenties for partially incorrect structures. For example, after refinement of the protease from human immunodeficiency virus HIV-1 a partially incorrect structure[63] produced an R value of 0.25 whereas the correct structure produced an R value of 0.184[64] with comparable geometry.

Why Does It Work?

The goal of any optimization problem is to find the global minimum of a target function. In the case of crystallographic refinement, one searches for the conformation or conformations of the molecule that best fit the diffraction data at the same time that they maintain reasonable covalent and noncovalent interactions. As the preceding examples have shown, simulated annealing refinement has a much larger radius of convergence than do gradient descent methods. It must therefore be able to find a lower minimum of the target E [Eq. (1)] than the local minimum found simply by moving along the negative gradient of E. Paradoxically, the very reasons that make simulated annealing such a powerful refinement technique (the ability to overcome barriers in the target energy function) would seem to prevent it from working at all. If it crosses barriers so easily, what allows it to stay in the vicinity of the global minimum?

The answer lies in the temperature coupling. By specifying a fixed kinetic energy, the system essentially gains a certain inertia that allows it to cross energy barriers. The target temperature must be low enough, however, to ensure that the system will not "climb out" out from the global minimum if it manages to arrive there. While temperature itself is a global parameter of the system, temperature fluctuations arise principally from local conformational transitions—for example, from an amino acid side chain falling into the correct orientation. These local changes tend to lower the value of the target E, thus increasing the kinetic energy, and hence the temperature, of the system. Once the temperature coupling has removed this excess kinetic energy, the reverse transition is very unlikely, because it would require a localized increase in kinetic energy where the conformational change occurred in the first place. Temperature coupling maintains a sufficient amount of kinetic energy to allow local conformational corrections, but does not supply enough to allow escape from the global minimum. This explains the directionality of simulated annealing refinement, i.e., on average the agreement with the data will improve rather than worsen. It also explains the occurrence of small spikes in E during the simulated annealing process.[33]

[63] M. A. Navia, P. M. D. Fitzgerald, B. M. McKeever, C.-T. Leu, J. C. Heimbach, W. K. Herber, I. S. Sigal, P. L. Darke, and J.-P. Springer, *Nature* (*London*) **37,** 615 (1989).

[64] A. Wlodawer, M. Miller, M. Jaskólski, B. K. Sathyanarayana, E. Baldwin, I. T. Weber, J. M. Selk, L. Clawson, J. Schneider, and S. B. H. Kent, *Science* **245,** 616 (1989).

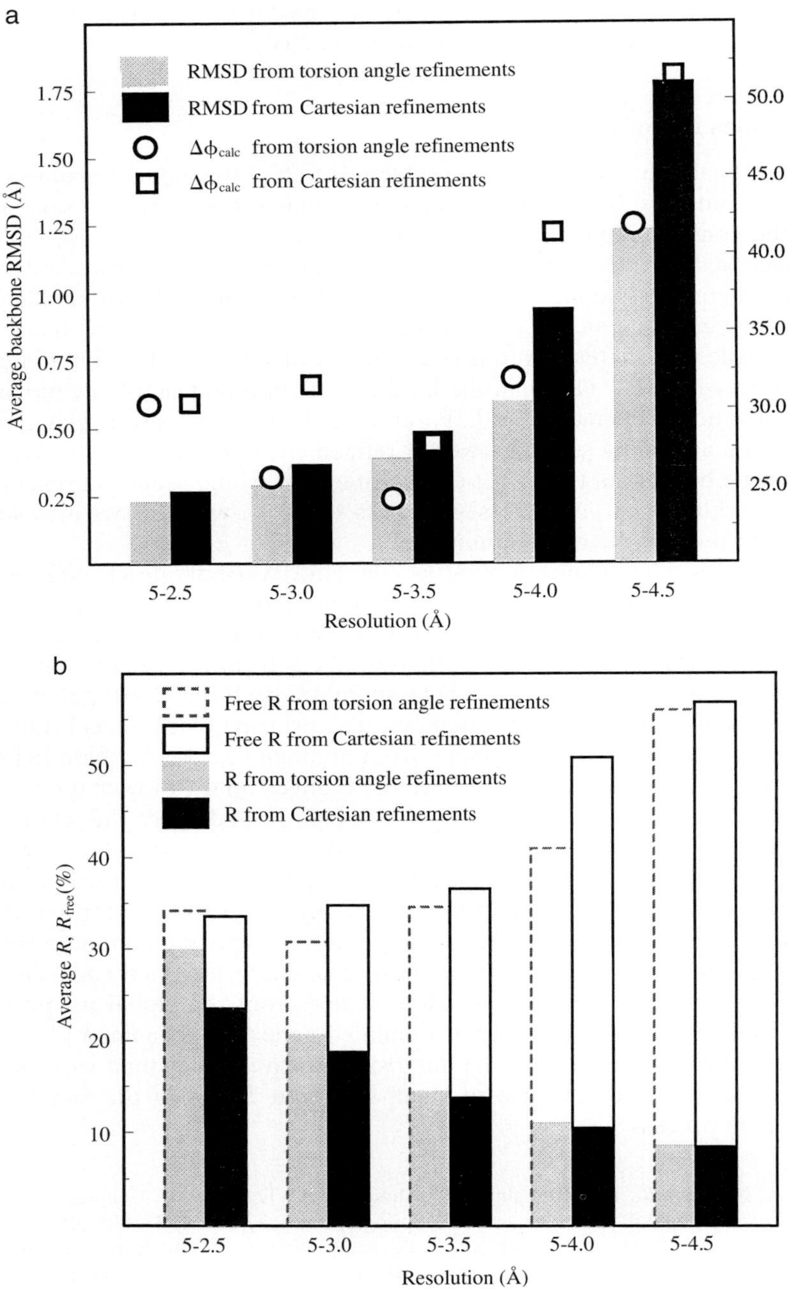
a
RMSD from torsion angle refinements
RMSD from Cartesian refinements
Δφcalc from torsion angle refinements
Δφcalc from Cartesian refinements
Average backbone RMSD (Å)
1.75
1.50
1.25
1.00
0.75
0.50
0.25
Average Δφcalc (°)
50.0
45.0
40.0
35.0
30.0
25.0
5-2.5
5-3.0
5-3.5
5-4.0
5-4.5
Resolution (Å)
b
Free R from torsion angle refinements
Free R from Cartesian refinements
R from torsion angle refinements
R from Cartesian refinements
Average R, Rfree (%)
50
40
30
20
10
5-2.5
5-3.0
5-3.5
5-4.0
5-4.5
Resolution (Å)

If the temperature of the simulated annealing refinement is too high, numerical instabilities can result in unreasonably large conformational changes. By suppressing high-frequency bond vibrations, torsion-angle dynamics has significantly reduced the potential for this to happen. In fact, one may now use much higher temperatures than previously possible with the Cartesian molecular dynamics implementation.[39]

Refinements at 3 to 4-Å Resolution

There are two related issues concerning initial models and crystallographic refinements: convergence and determinacy. For example, it is possible to have high-resolution data but a very poor initial model, in which case the refinement is well determined but may encounter problems of convergence, or searching. It is equally possible to have an excellent starting model but only low- to medium-resolution data. Refinement under these conditions does not have a severe search problem in that the initial model lies close to the correct one, but it is hampered by the fact that the correct answer may not be well determined by the limited experimental data. The question therefore arises: when is it appropriate to use torsion-angle refinement, and when do Cartesian methods suffice?

To address this question, a series of refinements were carried out at progressively lower resolution (Fig. 5). The crystal structure coordinates of the α-amylase inhibitor were subjected to 10 constant-temperature torsion-angle refinements and 10 constant-temperature Cartesian molecular dynamics refinements at resolutions of 5–2.5, 5–3.0, 5–3.5, 5–4.0, and 5–4.5 Å. For the refinements at 5- to 2.5-Å resolution, the two methods yield more or less equivalent results, but at resolutions below 3 Å the torsion-angle method shows considerable advantage. Thus for good initial models, Cartesian methods are sufficient if data better than 3-Å resolution are available. Torsion-angle dynamics should be used otherwise.

The free *R* value shows a high correlation with the accuracy of the model as assessed by the r.m.s. difference and phase difference between the low-resolution refined model and the 2-Å crystal structure (Fig. 5), except for the refinement at 5- to 4.5-Å resolution, where the differences in free *R* values are very small. However, in the latter case the free *R* value

FIG. 5. Cartesian and torsion-angle refinements at low to medium resolution. Averages over 10 refinements are shown. The crystal structure of the α-amylase inhibitor was refined against artificially truncated sets of the original diffraction data.[60] (a) Atomic r.m.s. differences and phase differences between the models refined at the specified resolution range and the deposited crystal structure. (b) Free *R* and *R* values.

has nearly reached the limit for a random distribution of atoms in noncentric space groups (57%), indicating divergence of refinement at 4.5-Å resolution.

The poorer performance of Cartesian molecular dynamics below 3-Å resolution is a consequence of the underdeterminacy of refinement at that resolution (the number of reflections is half the number of coordinates). Restricting refinement to torsion angles makes the search more efficient and it improves the representation of the model at low-to-medium resolution. For example, at 3-Å resolution, torsion-angle dynamics is superior to Cartesian dynamics. If the torsion dynamics is followed by a very brief Cartesian slow-cooling stage starting at low temperature, the performance is even better at resolutions around 3 Å (not shown). Thus, torsion-angle dynamics achieves a better search, but once the structure has reached the vicinity of the global minimum, adjustment of all parameters (including bond lengths and bond angles) may be beneficial, even at 3-Å resolution.

Simulated Annealing Omit Maps

Simulated annealing refinement is usually unable to correct very large errors in the atomic model or to correct for missing parts of the structure. The atomic model needs to be corrected by inspection of a difference electron-density map. To improve the quality and resolution of the difference electron-density map, the observed phases are often replaced or combined with calculated phases as soon as an initial atomic model has been built. These combined electron-density maps are then used to improve and refine the atomic model. The inclusion of calculated phase information brings with it the danger of biasing the refinement process toward the current atomic model. This model bias can obscure the detection of errors in atomic models if sufficient experimental phase information is unavailable. In fact, during the past decade several cases of incorrect or partly incorrect atomic models have been reported in which model bias may have played a role.[65]

Difference electron-density maps phased with simulated annealing refined structures often show more details of the correct chain trace.[23] However, the omission of some atoms from the computation of a difference electron-density map does not fully remove phase bias toward those atoms if they were included in the preceding refinement. More precisely, small rearrangements of the included atoms can bias the phases toward the omitted atoms.[66] Thus, the structure needs to be refined with the questionable region omitted before the difference electron-density map can be

[65] C. I. Brändén and A. Jones, *Nature* (*London*) **343,** 687 (1990).

[66] R. J. Read, *Acta Crystallogr.* **A42,** 140 (1986).

computed. Simulated annealing is a particularly powerful tool for removing model bias.[67] The improved quality of simulated annealing refined omit maps has been used to bootstrap about 50% of missing portions of an initial atomic model of a DNase–actin complex.[68] It should be noted that this is a rather extreme case for the amount of omitted atoms. Usually, omit maps are computed with about 10% of the atoms omitted.

In general, the improvement of the electron-density map achieved in simulated annealing refinement is a consequence of conformational changes distributed throughout the molecule. This is a reflection of the fact that the first derivatives of the crystallographic residual [Eq. (1)] with respect to the coordinates of a particular atom depend not only on the coordinates of that atom and its neighbors but also on the coordinates of all other atoms including solvent atoms in the crystal structure.

Refinement with Phase Restraints

As demonstrated throughout, simulated annealing has a large radius of convergence. The use of torsion-angle molecular dynamics combined with a repeated high-temperature annealing schedule significantly increased the radius of convergence compared to a slow-cooling Cartesian protocol at relatively high resolution. However, for refinements at lower resolution (~3 Å), no significant extension of the radius of convergence was observed when refining against structure-factor amplitudes alone.[39] Convergence at this resolution range can be sparse: the limited resolution drastically reduces the number of reflections (observables) and can result in a severely underdetermined search problem.

This adverse observable-to-parameter ratio can be improved using experimental phase information, for example, phases obtained from multiple isomorphous replacement diffraction data. Use of phase restraints [Eq. (5)] improves the radius of convergence somewhat, and the vector residual [Eq. (6)] shows a significantly increased radius of convergence.[39] Figure 6 summarizes convergence for refinements at 5- to 3-Å resolution, again using backbone atom r.m.s. coordinate deviations from the crystal structure as a measure of convergence. As for high-resolution refinements, torsion-angle refinements consistently outperform slow-cooling Cartesian refinement (Fig. 6). These refinements were performed without cross-validation because the phase accuracy could be assessed by direct comparison with the experimental phases.[44]

[67] A. Hodel, S.-H. Kim, and A. T. Brünger, *Acta Crystallogr.* **48,** 851 (1992).

[68] W. Kabsch, H. G. Mannherz, D. Suck, E. F. Pai, and K. C. Holmes, *Nature* (*London*) **347,** 37 (1990).

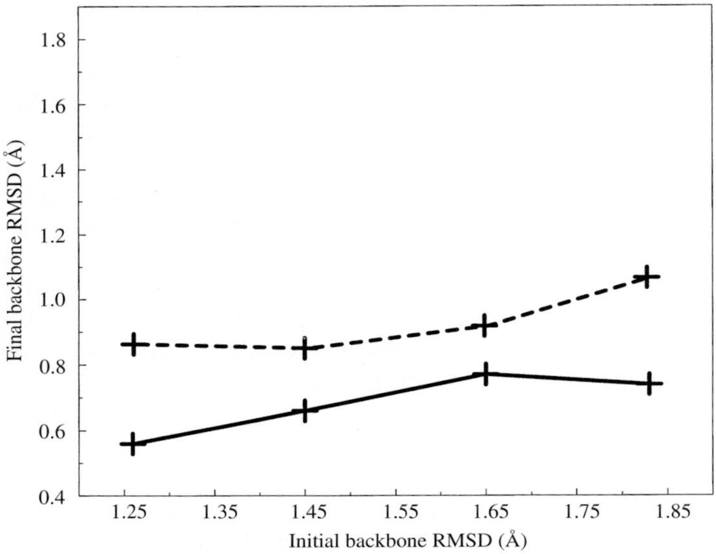

FIG. 6. Convergence of medium-resolution slow-cooling (starting at 5000 K, dashed line) and constant-temperature torsion-angle (10,000 K, solid line) refinements against the vector residual [Eq. (6)] at 5–3 Å of increasingly worse models for amylase inhibitor.[60] Convergence is measured by backbone atom r.m.s. coordinate deviations from the crystal structure, taking the best structure from a series of 10 independent refinement obtained by using different initial random velocities for simulated annealing. The best models can be identified by a low free R value.[39]

Conclusion

Simulated annealing has improved the efficiency of crystallographic refinement significantly. However, simulated annealing refinement alone is still insufficient to refine a crystal structure automatically without human intervention. Thus, crystallographic refinement of macromolecules proceeds in a series of steps, each of which consists of simulated annealing or minimization of E [Eq. (1)] followed by manual refitting of the model to difference electron-density maps, using interactive computer graphics.[62] During the final stages of refinement, solvent molecules are usually included, and alternate conformations for some atoms or residues in the protein may be introduced.

With currently available computing power, tedious manual adjustments using computer graphics to display and move positions of atoms of the model in the electron-density maps can represent the rate-limiting step in the refinement process. Further automation of refinement by judicious combination of reciprocal and real-space refinements should be possible.

However, all automation attempts ultimately will have to address the problem of pattern recognition of macromolecular features in noisy electron-density maps. The human brain appears to be highly efficient at solving this problem whereas computer algorithms have been rather slow in achieving this ability. Thus, for the near-term future, the main goal should be to eliminate tedious bookkeeping and computer graphics-bound intervention by automating all aspects of refinement that do not require significant pattern recognition.

Note Added in Proof

Recent developments of crystallographic refinement are reviewed in Ref. 70.

Acknowledgment

We are grateful to Erin Duffy for critical reading of the manuscript. L. M. R. is an HHMI predoctoral fellow. This work was funded in part by a grant from the National Science Foundation (DIR 9021975).

[70] A. T. Brünger, P. D. Adams, and L. N. Rice, *Structure* **15,** 325–336 (1997).

[14] Automated Refinement for Protein Crystallography

By VICTOR S. LAMZIN and KEITH S. WILSON

Introduction

Phase Problem

X-Ray crystallography is the most powerful tool for determining the three-dimensional structures of molecules. For calculation of the electron density it is necessary to have both the amplitude and phase of the structure factors. Only the amplitudes are provided by the X-ray diffraction experiment. The phases must then be evaluated. This is the so-called crystallographic phase problem.

For small and medium structures, the phase problem can be solved directly if diffraction data extend to atomic (about 1.2 Å) or higher resolution. To date, 1.2 Å seems to be the lowest resolution at which *ab initio* structure solution using direct statistical methods is possible.[1] The initial

[1] G. M. Sheldrick, *Acta Crystallogr.* **A46,** 467 (1990).

0076-6879/97 $25.00

set of phases is refined using, for example, tangent formula recycling. At present this generally reveals the complete set of ordered atoms. Peaks are assigned the appropriate atomic type on the basis of knowledge of the chemical structure. This atomic model is refined and updated in an automated way through iterative cycles of least-squares minimization alternating with Fourier syntheses. Stereochemical restraints are not required for well-ordered parts, but may be introduced for disordered regions. Such a scheme is realized in many programs used in small-molecule crystallography, e.g., SHELX (see [16] in this volume[2]).

In spite of developments in X-ray data collection techniques (use of synchrotron radiation, area detectors, and cryogenic temperatures) and the fact that atomic resolution data have been collected at the European Molecular Biology Laboratory (EMBL, Hamburg, Germany) alone for about 40 proteins, *ab initio* phase determination is still not possible for most of these. The success of Patterson superposition methods in the direct solution of rubredoxin[3] and cytochrome c_6[4] is dependent on the presence of metals in these proteins.

Structure solution for most proteins therefore requires experimental or model phases. These starting sets of phases play an analogous role to those obtained by direct or Patterson methods for small structures. Experimental phases can be obtained by multiple isomorphous replacement (MIR) and anomalous scattering. Model phases are provided by use of a known related model, using molecular replacement (MR). Initial phases need not necessarily be obtained for all the measured amplitudes. The starting phases can be improved in two senses. First, they should be extended to the full data set, typically by extending the resolution beyond that available from isomorphous derivatives. Second, refinement of the phase set can include various density-modification techniques such as solvent flattening, histogram matching, or noncrystallographic symmetry averaging (e.g., see Ref. 5). The partial atomic model obtained from interpretation of the MIR map or the MR solution is then refined using least-squares minimization between the observed and calculated X-ray amplitudes. However, for proteins at less than atomic resolution there is a paucity of X-ray terms, which must be supplemented by, for example, stereochemical restraints. Refinement does not proceed automatically, but requires adjustment of the model using

[2] G. M. Sheldrick, *Methods Enzymol.* **277,** [16], 1997 (this volume).

[3] G. M. Sheldrick, Z. Dauter, K. S. Wilson, H. Hope, and L. C. Sieker, *Acta Crystallogr.* **D49,** 18 (1993).

[4] C. Frazao, C. M. Soares, M. A. Carrondo, E. Pohl, Z. Dauter, K. S. Wilson, M. Hervas, J. A. Navarro, M. A. De la Rose, and G. M. Sheldrick, *Structure* **3,** 1159–1169 (1995).

[5] A. D. Podjarny, T. N. Bhat, and M. Zwick, *Annu. Rev. Biophys. Biophys. Chem.* **16,** 351 (1987).

computer graphics and the use of various Fourier syntheses, including omit maps.

Thus for proteins the solution of the phase problem does not proceed straightforwardly. There is no clear discrimination between the stages of phase improvement/extension and refinement of the model as is carried out for small molecules. This is because the model (hence the phase set) can become trapped in local minima, and least-squares minimization does not provide a means of escape.

Local Minima

Typical examples of problems found during least-squares refinement, with a side chain or whole loop out of density, are shown in Fig. 1. These often are referred to as local minima but this simply means that some atoms are in the wrong place! There are several reasons why least-squares minimization cannot move such wrongly placed atoms toward their correct positions. The directions of atomic displacements in crystallographic programs with conventional minimization are calculated using the first derivatives, and the length of the displacement is estimated from a parabolic approximation of the minimized function. This is approximately equivalent to the movement of an atom in the direction of the nearest positive peak in the difference Fourier synthesis. If the nearest peak is either too far distant, absent, or not well defined, appropriate positional gradients for atomic displacements cannot be computed.

To escape from "local minima," unrestrained refinement of positional and thermal atomic parameters seems to be the most straightforward. Atoms that can move to the nearest peak will do so. Others will remain out of density, and this will be compensated by an effective increase in

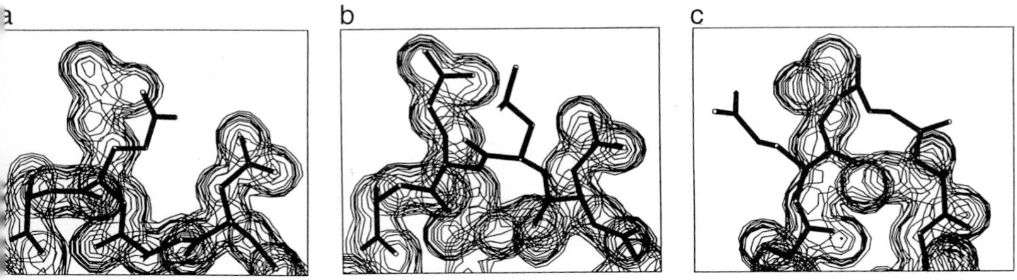

FIG. 1. Examples of a model trapped in a "local minimum." Wrongly positioned parts of the model and the ideal density are shown. (a) A side chain is in the wrong place. (b) A short loop is incorrectly positioned. (c) An "out-of-register" error, in which there is an incorrect insertion of a few residues.

their temperature factor. The latter actually makes the problem of defining gradients even worse, but substantially reduces the bias (memory) effect. For proteins, unrestrained refinement is, however, severely limited by the number of X-ray data. Consider diffraction to 2.0-Å resolution, about the mean resolution quoted for current structures in the Protein Data Bank.[6] According to Blundell and Johnson[7] there are about $2\pi/3d_{min}^3$ independent reflections in the primitive lattice for every 1 Å^3 volume of the unit cell. Assuming V_m between 2.0 and 2.8 Å^3/Da and an average atomic mass of 14 Da, at 2.0 Å resolution we expect 7 to 10 X-ray observations per atom. Provided the data are complete and three positional and one thermal atomic parameters are to be determined, this gives a ratio of observations to parameters of only two. The significant experimental errors in the data exacerbate this problem.

To increase effectively the number of observations, additional data are introduced as restraints based on *a priori* knowledge of protein stereochemistry. These can be a set of interatomic distances and angles, planarity, van der Waals radii, etc. Conventional restrained refinement (e.g., the PROLSQ package[8]) makes escape from "local minima" impossible in the early stages, during which many atoms are in the wrong place. For an atom in the vicinity of the correct position, the X-ray gradients will move the atom in the right direction as in unrestrained refinement. The geometric term will try to ensure the atom retains the specified connections with its neighbors. If some neighbors are wrongly located, they will not allow a well-positioned atom to improve its position. Wrongly located atoms also will introduce a memory effect.

Restrained refinement with molecular dynamics, e.g., the X-PLOR package,[9] sometimes may be advantageous in solving the local minimum problem. A side chain (Fig. 1a) can be rotated automatically or the peptide can flip to fit the electron density better. However, in the case of a wrongly positioned loop or an "out-of-register" error (Fig. 1b and c) molecular dynamics refinement is unlikely to improve the model. Thus for the phase/model improvement/extension stage for protein structures at resolution lower than atomic, the local minimum is a serious problem intrinsically related to the use of least-squares minimization alone.

[6] F. C. Bernstein, T. F. Koetzle, G. J. B. Williams, E. F. Meyer, Jr., M. D. Brice, J. R. Rodgers, O. Kennard, T. Shimanouchi, and M. Tasumi, *J. Mol. Biol.* **112,** 535 (1977).

[7] T. L. Blundell and L. N. Johnson, *in* "Protein Crystallography," pp. 248–249. Academic Press, New York, 1976.

[8] J. H. Konnert and W. A. Hendrickson, *Acta Crystallogr.* **A36,** 344 (1980).

[9] A. T. Brünger, J. Kuriyan, and M. Karplus, *Science* **235,** 458 (1987).

Solution

One way to solve the local minimum problem during protein structure determination is to extend the convergence radius of refinement by coupling least-squares minimization with some other approach. Any protein crystal structure solution is a combination of least-squares refinement with, e.g., density modification or manual rebuilding. Such a combination should be automated as implemented for small molecules.

We developed an automated refinement procedure (ARP) for proteins.[10] The basis of ARP is the iterative use of unrestrained least-squares minimization coupled with constant updating of the model. This is comparable to the iterative least-squares/Fourier synthesis approach for small molecules. It requires X-ray data to 2.0 Å or better to allow unrestrained refinement and improvement of the whole content of the unit cell. At lower resolutions, as a rule, only unrestrained parts of the model (e.g., solvent or ligand molecules) are expected to be improved. The quality of data and the initial phase set greatly influences the power of ARP.

ARP is still under development. Its present status is described in this chapter. The code is available on request. Many publications quoting use of ARP have already appeared. Only a limited number of applications are presented briefly here.

Method

General

Suppose that least-squares refinement is being used and we are faced with the situation shown in Fig. 1. Manual intervention consisting of moving atoms to a better place, using molecular graphics, density maps, and geometric assumptions, can solve the problem and allow refinement to proceed further.

It is possible to automate the movement of such atoms. The quickest way to change the position of an atom substantially is not actually to move it but to remove it from its current place and add a new atom at the new position. Such automatic correction of an atomic position requires two steps. First, the procedure should identify wrongly (or potentially wrongly) placed atoms and delete them from the model. Second, positions should be found for new atoms. This means that wrongly placed atoms no longer are required to move in a series of steps toward their true position. Indeed,

[10] V. S. Lamzin and K. S. Wilson, *Acta Crystallogr.* **D49,** 129 (1993).

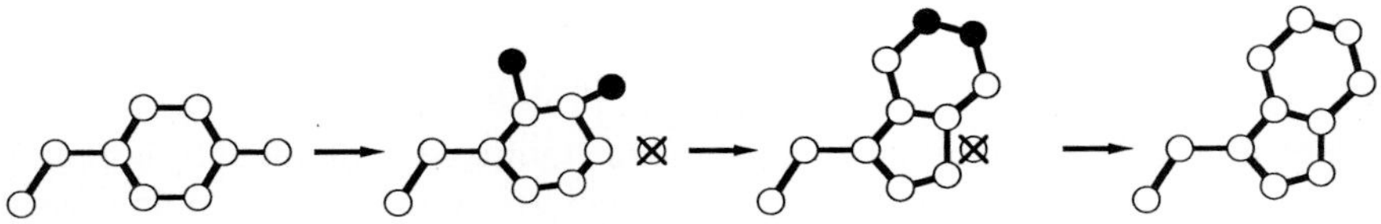

FIG. 2. Scheme of ARP. Successive automated updating interspaced with least-squares refinement can provide substantial changes in the model. Updating includes removal of some atoms (crossed out) and addition of new atoms (filled circles).

in an unrestrained ARP it does not matter if atoms migrate to a position that actually corresponds to a different, but correctly placed, atom. Such updating of the model does not imply that all rejected atoms are immediately repositioned somewhere else. The number of atoms to be added does not have to be equal to the number rejected.

After updating of the model, the new set of atoms is subjected to least-squares optimization of their parameters against the X-ray data and this is iterated (Fig. 2). There is no limitation, in principle, on the radius of convergence of the updating stage. The distance by which atoms can be "moved," the accuracy of their new positions, and the reliability of the model updating depend on the criteria used for atom selection. The current procedure requires density maps calculated using model phases and some *a priori* knowledge about protein structure. The actual convergence limit of the updating depends on the quality of phases.

Limits Imposed by X-Ray Data

Density-based atom selection for the whole structure is possible only if the X-ray data extend to a resolution at which atomic positions can be estimated from the Fourier syntheses with sufficient accuracy for them to refine to the correct position. The average interatomic distance in protein structures is about 1.4 Å. Features separated by this distance can be expected to be resolved if the nominal resolution is 2.0 Å or better.[11]

If the structural model is generally of reasonable quality, we may want to improve only part of it. For example, at 2.5 Å, although only part of the solvent structure can be located, ARP can be used to do this satisfactorily. ARP can also be used to model ligands, e.g., a short peptide.[12] As the

[11] R. W. James, "The Optical Principles of the Diffraction of X-Rays: The Crystalline State," Vol. II. Bell, London, 1957.

[12] P. M. D. Fitzgerald, A generalized approach to the fitting of non-peptide electron density. *In* "From First Map to the Final Model," pp. 125–132. Proceedings of the CCP4 study weekend, January 6–7, 1994. Daresbury Laboratory, Daresbury, Warrington, UK, 1994.

protein component of the initial model was essentially correct, ligand atoms were placed and refined satisfactorily without geometric restraints.

The X-ray data should be complete. If strong low-resolution data (e.g., 4 to 10 Å) are systematically incomplete, e.g., due to detector saturation, the electron density, even for good models, is often discontinuous. Because ARP involves updating on the basis of density maps, such discontinuity will lead to incorrect interpretation of the density and slow convergence or even noninterpretable output.

Limitations from Least-Squares Minimization

To improve the atomic positions, ARP requires unrestrained least-squares optimization of the atomic parameters. Apart from antibumping or van der Waals repulsion, no restraints are applied to solvent even if the protein part is highly restrained. Therefore use of ARP for solvent building does not depend on the refinement scheme and always can be used. This provides indirect improvement of the protein part. Some new solvent atoms will be placed within the protein, mimicking missing atoms and providing local improvement of these regions.

For new atoms in the protein region the problem arises of reassignment of atom type and interatomic connections. If unrestrained ARP is carried out, the output atoms can give improved and easily interpretable density while the atoms themselves do not make complete stereochemical sense unless the resolution is really atomic. To overcome this problem, unrestrained least-squares minimization should be coupled with the knowledge of new protein structure created by ARP. In other words, not only automated atom positioning but also automated atom-type assignment and model building are needed. Our first attempts to incorporate "protein image" recognition are presented below.

Automated Refinement Procedure Modes

ARP can be used in various modes as shown in Table I. If the initial model is more or less correct and only solvent needs to be built or improved,

TABLE I
AUTOMATED REFINEMENT PROCEDURE MODES

Mode	Input	Initial *R* factor (%)	LSQ	Resolution (Å)	To improve
1	"Good" model	Better than 30	Restrained	2.5 or better	Solvent
2	"Bad" model	30 or worse	Unrestrained	2.0 or better	Loops, side chains
3	Density map	—	Unrestrained	2.0 or better	Map

restrained refinement coupled with automatic adjustment of solvent using ARP is applicable (restrained ARP). Because ordered solvent comprises about 10% or more of the unit cell, improvement of solvent indirectly improves the density corresponding to the protein part. As a rough guide, an R factor of about 30% suggests the model is sufficiently good to be thus refined.

If the initial model is poor and needs to be improved substantially, then unrestrained least-squares minimization with updating of all atoms in the current model is necessary (unrestrained ARP). If no model but only experimental MIR phases are available, a "protein-like" dummy model can be built in the MIR map and refined with an unrestrained ARP.

The output of ARP is a set of atoms (ARP model), which for unrestrained ARP may not make complete geometric sense (unless automatic rebuilding is used; see below). A density map with coefficients ($3F_o - 2F_c$, α_c) must be calculated from the model and interpreted. For a restrained ARP both ($3F_o - 2F_c$, α_c) and ($F_o - F_c$, α_c) maps should be inspected.

Density Maps to Be Used

ARP selects atoms on the basis of maps calculated using model phases. The difference ($F_o - F_c$, α_c) map is the best indicator of missing parts of the structure and is used for addition of new atoms. This map in essence does not contain series termination errors.

Which density map should be used for identification of wrongly placed atoms? For a small protein refined at 2.0 Å resolution, two residues were omitted and the resulting features in several Fourier syntheses were inspected (Table II). If the atoms are merely omitted without subsequent least-squares minimization, the density for the missing atoms in the (F_o, α_c) synthesis is approximately 50% of the true height and in the ($2F_o - F_c$, α_c) synthesis is close to the correct height. This largely reflects the memory effect in the model: omitted atoms merely reappear at their old positions. In practice the model with the atoms omitted or missing is refined and then Fourier syntheses are used for analysis. The density in the (F_o, α_c) synthesis then falls to about 30%. This was described by Luzzati,[13] who showed that missing atoms do not appear with the same height as atoms involved in the phasing, and by Stout and Jensen,[14] who stated that the peak corresponding to a missing atom has a height of 50% for the centric and 30% for the acentric case.

[13] P. V. Luzzati, *Acta Crystallogr.* **6,** 142 (1953).

[14] G. H. Stout and L. H. Jensen, *in* "X-Ray Structure Determination: A Practical Guide," pp. 330–331. John Wiley & Sons, New York, 1989.

TABLE II
APPEARANCE OF OMITTED PARTS OF MODEL IN FOURIER SYNTHESES[a]

Map	Phe-10			Tyr-32		
	A	B	C	A	B	C
F_o, α_c	1.55 (100)	0.75 (48)	0.57 (37)	1.27 (100)	0.58 (46)	0.43 (34)
F_c, α_c	1.55 (100)	0.03 (2)	0.06 (4)	1.27 (100)	0.02 (2)	0.03 (2)
$F_o - F_c, \alpha_c$	0.01 (0)	0.72 (46)	0.51 (33)	0.00 (0)	0.57 (42)	0.40 (31)
$2F_o - F_c, \alpha_c$	1.56 (100)	1.46 (94)	1.08 (70)	1.27 (100)	1.15 (91)	0.84 (66)
$3F_o - 2F_c, \alpha_c$	1.57 (101)	2.18 (141)	1.59 (103)	1.26 (99)	1.71 (135)	1.24 (98)

[a] The refined 2.0 Å model of subtilisin proteinase inhibitor eglin c was used.[b] The average density interpolated at the atomic centers, using an F_{000}/V of 0.2 e Å^{-3} for all syntheses except ($F_o - F_c, \alpha_c$), was calculated for two well-defined side chains. This was done for A (the final model), B (the final model with these two side chains omitted), and C (the final model with these two side chains omitted, and 10 cycles of restrained least-squares minimization carried out). The density is given in e Å^{-3}. Its normalized values relative to the (F_c, α_c) synthesis A are given as a percentage in parentheses.

[b] C. Betzel, Z. Dauter, N. Genov, V. Lamzin, J. Navaza, H. P. Schnebli, M. Visanji, and K. Wilson, *FEBS Lett.* **317,** 185 (1993).

Because protein data can be considered as predominantly acentric, use of the ($3F_o - 2F_c, \alpha_c$) synthesis is most appropriate. For a missing atom, this synthesis will show a peak height of about 100%. If an atom is present, but should be rejected (as wrongly positioned), the ($3F_o - 2F_c, \alpha_c$) height will be zero. An atom that has not yet acquired its final B factor will also produce approximately correct ($3F_o - 2F_c, \alpha_c$) density. This is not true if the phases are random. The more reliable are the phases, the more valid are these assumptions. The widely quoted ($2F_o - F_c, \alpha_c$) synthesis is less useful and probably resulted from small-molecule crystallography, where space groups often are centric. The effect of "model bias" may to some extent be a property of use of the ($2F_o - F_c, \alpha_c$) density map for acentric structures.

Atom Rejection

A crucial question concerns which criteria allow classification of an atom as wrongly or correctly placed. As we will show, the criteria we can define are best applied according to stochastic rules and statistical procedures. Tests were performed on two enzymes: phosphoribosylamino-imidazolesuccinocarboxamide (SAICAR) synthase at 1.9 Å with MIR phases to 2.5 Å and xylanase to 1.5 Å resolution with MIR phases to 2.1 Å. The expected number of atoms were positioned in the MIR map. This dummy set of atoms was subjected to 30 cycles of unrestrained least-squares minimization using all data. No atoms were added or deleted during these refinements. For every atom in the refined dummy model the distance to the nearest atom from the finally refined protein structure was calculated. This serves as a measure of correctness of atomic position (CAP).

The $(3F_o - 2F_c, \alpha_c)$ map should neither be negative nor only marginally positive at an atomic center. If the density interpolated at the atomic center is low, the position of the atom is suspect. Figure 3a and b shows the $(3F_o - 2F_c, \alpha_c)$ density interpolated at the center of a dummy atom versus CAP. An atom with lower density is on average further from the correct position. The first criterion is that an atom is potentially wrongly placed if it has low density at its center.

The density at the atomic center is, to some extent, inversely related to the atomic temperature factor. The advantage of using the density rather than the B value as a criterion is that the better the phases the more reliably the density at the atomic center indicate the quality of atom positioning. The B factor is a less satisfactory parameter because it may not have yet refined to a satisfactory minimum. In contrast, use of the density allows the rejection of weak scatterers rather than strong ones if both happen to be in question.

The second criterion is related to density shape: atomic density distribution is spherical. For the "sphericity," an isotropic Gaussian function was used to describe the density shape within a 0.7-Å radius around every atom. Both the variance of the function (analogous to the atomic B factor) and its amplitude equivalent to the atomic number) were refined. A value of 1.0 minus the normalized root-mean-square (r.m.s.) fit was used to define the atomic sphericity. Figure 3c and d shows plots of the sphericity of dummy atoms. Most atoms with low sphericity (e.g., less than 0.6) correspond to a large (1 to 4 Å) deviation from a correct position. An atom with highly nonspherical density distribution is likely to be wrongly placed and is considered for rejection. However, a substantial number of well-shaped dummy atoms are also far from a true position.

The two density-based criteria do not provide straightforward and guaranteed identification of wrongly placed atoms. Considered together they suggest which atoms are most likely to be wrongly placed. The criteria are more effective at higher resolution but still useful at 1.9 Å. Current implementation allows rejection if both criteria indicate an atomic position is suspicious. A correctly located atom that happens to be rejected is expected to be selected again and put back into the model.

The number of atoms to be rejected/selected every ARP cycle has not yet been established objectively. If a small enough part of the model is updated, refinement proceeds smoothly. An empirical guide is that the number of atoms to update is approximately proportional to the $1/d^3_{\min}$ and is 8% at 1.0 Å, 3% at 1.5 Å, and 1% at 2.0 Å resolution. It may be advantageous at the stage where refinement is deemed to have converged to remove a large (10 to 30%) part of the model and carry on with addition of new atoms. According to Figure 3a and b all atoms that are further than 1 Å from a correct position have a density height of less than 2 e Å^{-3} at their centers. Removal of all such atoms would result in an incomplete model but with every remaining atom essentially correctly placed.

Atom Addition

Addition of atoms uses the difference ($F_o - F_c$, α_c) density. All grid points where the density exceeds a given threshold are analyzed. The grid point with highest density and satisfying the distance constraints is selected as a new atom. Other grid points within a defined radius around this atom are rejected. The next highest grid point is then selected and this is iterated until the desired number of new atoms is found. This is analogous to the peak selection algorithm described by Lunin and co-workers.[15] ARP works with grid points, not with peaks. This is important, because it increases the number of atoms found and provides more accurate positioning. The rationale is that peaks in the difference density are often poorly defined and overlap with neighboring peaks or existing atoms. This is especially true if the resolution and phases are poor. Consideration of every grid point partly overcomes this and inserts an atom not necessarily at the peak but near it. Least-squares minimization optimizes the new atomic parameters.

The number of atoms that can be found strongly depends on the selected density threshold. We compute an optimum threshold from a statistical consideration. The difference density histogram follows a Gaussian distribution,[16] provided phases are ideal, because for the refined model the residual $F_o - F_c$ is normally distributed. If the model is not fully refined, positive and negative peaks in the difference map provide a non-Gaussian contribution to the histogram. The resulting histogram will be a complicated convolution of a Gaussian and some unknown distribution and there is a problem in deconvolving it. We chose a simple approximation as follows. A Gaussian function is fitted to the difference density histogram and two integral probability distribution functions are then compared: a Gaussian and the actual one (Fig. 4). The threshold is selected so as to have not more than, for example, 10% of a Gaussian contribution above it.

[15] V. Yu. Lunin, A. G. Urzhumtzev, E. A. Vernoslova, Yu. N. Chirgadze, N. A. Nevskaya, and N. P. Fomenkova, *Acta Crystallogr.* **A41,** 166 (1985).

[16] P. Main, *Acta Crystallogr.* **A46,** 507 (1990).

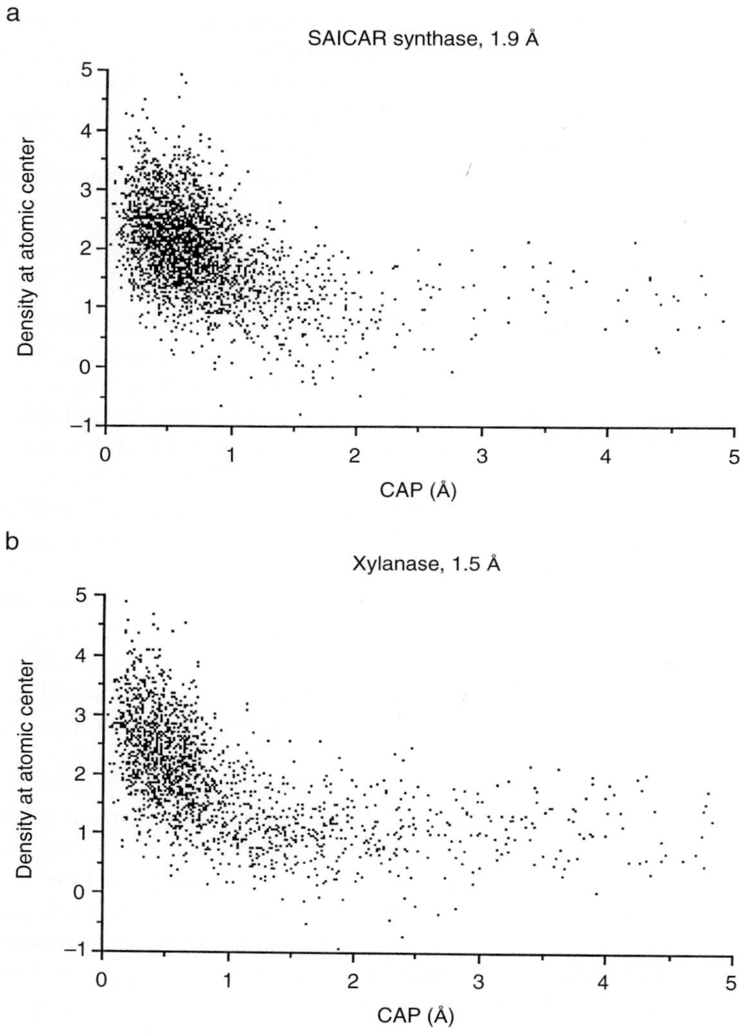

FIG. 3. Criteria for atom rejection. The expected number of equal dummy atoms was placed in MIR maps for SAICAR synthase and xylanase and then refined by unrestrained *xyzB* least-squares in the whole resolution range (see text). (a and b) $(3F_o - 2F_c, \alpha_c)$ density, $e/\text{Å}^3$, interpolated at the atomic center and (c and d) the atomic sphericity are plotted for each dummy atom as a function of CAP. CAP is the distance from a dummy atom to the nearest atom in the final model. An atom should be considered for removal if both the density at its center and its sphericity are low.

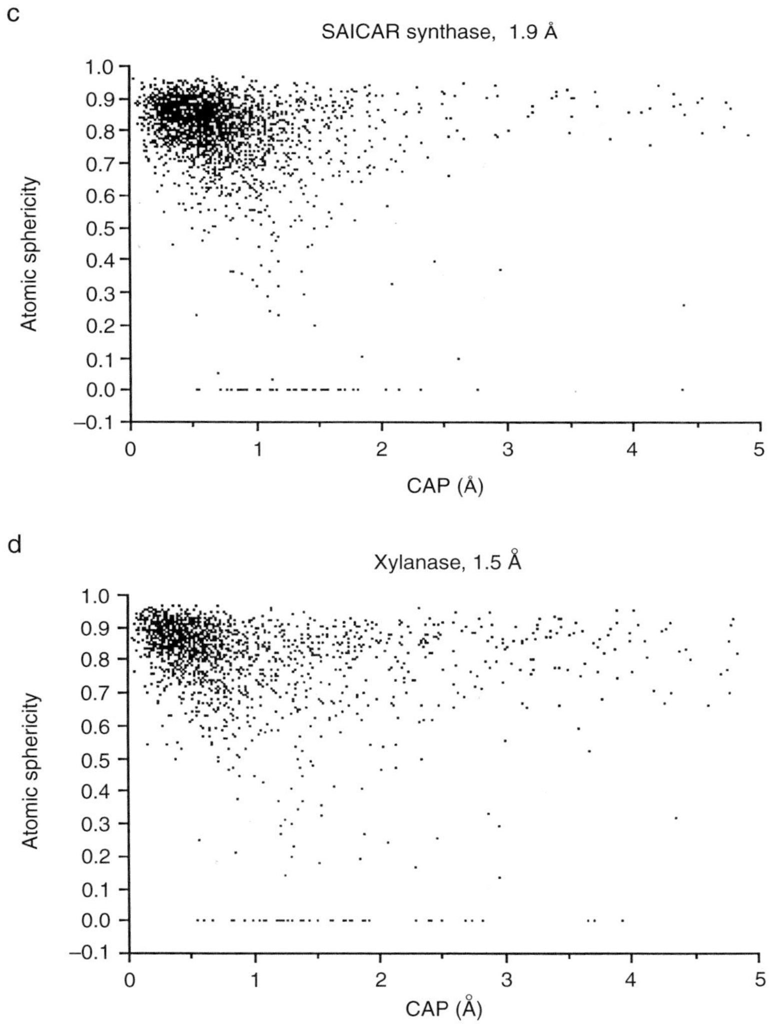

FIG. 3. (*continued*)

Use of automatic threshold determination was found to provide excellent results when phases are sufficiently good. When a restrained ARP is used (i.e., the protein part is refined in a standard way with stereochemical restraints and only solvent sites are allowed to be updated), the program stops finding new waters when convergence is achieved and there are only a few water molecules with B factors greater than 70 Å^2. This approach

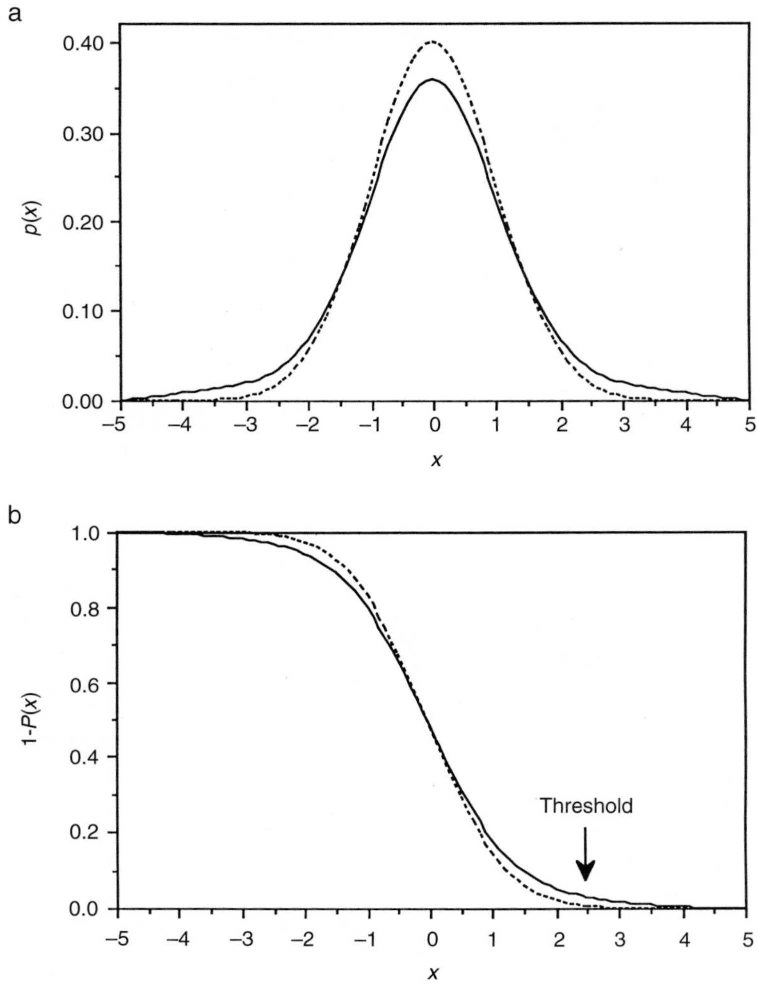

FIG. 4. Schematic representation of a difference electron-density histogram (—) and a Gaussian function (---) fitted in it. (a) Probability distribution functions. (b) Integral probability distribution functions. The threshold is selected so as to have only a small percentage of a Gaussian contribution above it.

works in the same manner at the late stage of refinement with unrestrained ARP. However, when phases are poor it seems not to work satisfactorily, and allows selection of only a few (if any) new sites in the beginning of unrestrained ARP. This probably reflects the fact that with poor phases there is a substantial (i.e., more than 10%) noise contribution at any threshold in the difference map. Therefore the optimum protocol for unrestrained ARP is to add a fixed number of new sites found above the 4σ threshold at every cycle. By the end of unrestrained ARP the number of atoms in the ARP model, taking into account both rejection and addition, should ideally be approximately equal to the expected number of atoms.

Real-Space Refinement

The $(3F_o - 2F_c, \alpha_c)$ density shape analysis around an atom can be used not only for identification of a potentially wrongly placed atom and its rejection, but also for the refinement of an atomic position. We noticed that weak solvent atoms, refined essentially without geometric limitations, tend to move to the edge of the density cloud. Even put manually at the center of the peak, such atoms move out after least-squares minimization. We cannot suggest a reasonable explanation for this effect. It may be related to the fact that these solvent atoms are not fully occupied and their occupancies are not refined. It also may reflect a weakness of the conjugate gradient minimization in the treatment of weak scatterers.

A practical solution to this problem is to move the atom back to the center of the peak using some property that differs from that minimized by least-squares refinement. The spherical shape of atomic density is one such function. The shifts applied to atoms in real space are restrained to be less than 0.2 Å and normally are less than 0.1 Å. They are substantially higher than shifts applied by least-squares minimization in reciprocal space. The sphericity-based real-space refinement keeps an atom in the center of the density cloud but influences the R factor and phase quality only slightly. In the current implementation this is applicable to well-separated atoms only, and therefore is used mostly in restrained ARP for solvent atoms. It was also useful in unrestrained ARP if the X-ray data extend to atomic resolution.

Geometric Constraints

ARP uses some geometric information from *a priori* knowledge of the structure being refined. This provides an improvement of the structure and acceleration of the refinement. Geometric constraints are applied in rejection and addition of atoms.

In protein structures the interatomic distance between carbon, nitrogen, or oxygen atoms is 1.2 to 1.6 Å. An H bond distance to solvent varies from about 2.2 to 3.3 Å. These limits are subject to an error reflecting the accuracy in the atomic position, and can be broader if atomic parameters are refined without geometric restraints. Atoms that, during refinement, approach each other too closely can be merged to one site with weighted average x, y, z, and B parameters. Typical ARP constraints used for merging are as follows. For updating of the whole structure two atoms are considered as one site if they are separated by a distance of 0.6 Å or less. For updating solvent structure two water molecules are merged if they are separated by a distance of 2.0 Å or less. A water that approaches protein atoms to within a 2.0 Å distance is rejected. An atom is also rejected if it has no neighbor within 3.5 Å. This mimicks the connectivity in the structure.

Similar constraints are applied for atom addition. For unrestrained ARP a distance between a new atom and an old atom should lie within 1.2 to 3.3 Å. The shortest distance is 2.2 Å in the case of updating solvent structure only. The shortest distance should be compatible with the resolving ability of the density map used for the atom search, and this is affected by the nominal resolution of the data.

It is not reasonable to apply these geometric limitations as formal restraints. The basis of ARP is that if, by chance, an atom is rejected wrongly, it will be picked up again and put back in the model. Restraining an atom to be at the edge of the allowed distance would introduce bias in the model.

Geometric Restraints and Model Building

Restraints on the interatomic distances during refinement can be applied only if the atoms have been assigned the correct atomic number (C, N, O, etc.). They should be applied to the protein only and should not introduce additional bias. This is especially important when using unrestrained ARP. The sets of atoms put as above in the MIR map and used for ARP of SAICAR synthase at 1.9 Å and xylanase at 1.5 Å were refined without geometric restraints (Table III). This revealed how such atoms could approach true peptide backbone positions. For most of the main-chain atomic positions a dummy model atom is within 0.6 Å, half of the shortest interatomic distance within a peptide. The success of correct recognition depends on the phase quality and resolution.

Several conclusions can be drawn. First, an atom refined without geometric restraints can approach the correct atomic position at a resolution typical for protein structures. Second, a position corresponding to a heavier atom can be found with higher confidence. Indeed, the percentage of matching for main-chain oxygen, nitrogen, and carbon atoms is essentially proportional to their atomic number. Third, the set of atoms refined without restraints can be used for pattern recognition and automatic model building.

As a first attempt to build the protein structure we tried to find sets of five atoms that would correspond to a C_α–C–O–N–C_α unit and satisfy

TABLE III
CORRESPONDENCE BETWEEN DUMMY MODEL REFINED WITHOUT RESTRAINTS AND MAIN-CHAIN ATOMS OF FINAL PROTEIN MODEL[a]

Atom	SAICAR synthase (1.9 Å)	Xylanase (1.5 Å)
CA	146 (49%)	116 (58%)
C	176 (59%)	88 (44%)
O	222 (74%)	157 (79%)
N	206 (69%)	128 (64%)

[a] The expected number of equal dummy atoms was put in two MIR density maps. They then were subjected to 30 cycles of unrestrained *xyzB* refinement using all data. The dummy atom positions were compared with those of the main-chain atoms from final models. The number of dummy atoms located within a 0.6 Å distance of a main-chain atom and its percentage of the total number of residues are given.

peptide geometry. A problem, however, arose from the fact that the probability of finding all five atoms within a peptide as a multiplication of individual atomic probabilities is very low. For the two test cases (Table III) it is about 10 to 15%.

To recognize a peptide unit, one must acknowledge that not all atoms are present in the ARP model. The atomic locations are also subject to positional error. The simplest assumption is that C_α atoms can be identified from a set of dummy atoms. Thus in the first step (Fig. 5) all pairs of atoms separated by a distance of 2.8 to 4.8 Å are selected as possible candidates for a C_α–C_α connection. All five atoms comprising an ideal peptide unit are then "superimposed" on these two supposedly C_α atoms, and the best orientation with respect to the fit to the electron density is obtained by rotation of the peptide around the C_α–C_α axis. Both directions of the chain are tested and can be accepted at this stage. The peptide is accepted if the $(3F_o - 2F_c, \alpha_c)$ density interpolated at each atomic center is higher than 1σ above the mean. Tentative peptides are then checked for possible connection, which is assumed to occur if the $C_\alpha - 1$ to $C_\alpha + 1$ distance is within 4.6 to 7.8 . Single peptides are skipped. Branch points are resolved such that the peptide in higher density has preference. The longest polypeptide is then considered. Additional checks for geometry of the peptide connections are performed. The chain is fitted to the density in real space, using geometric restraints. After the real-space fit, the orientation and type

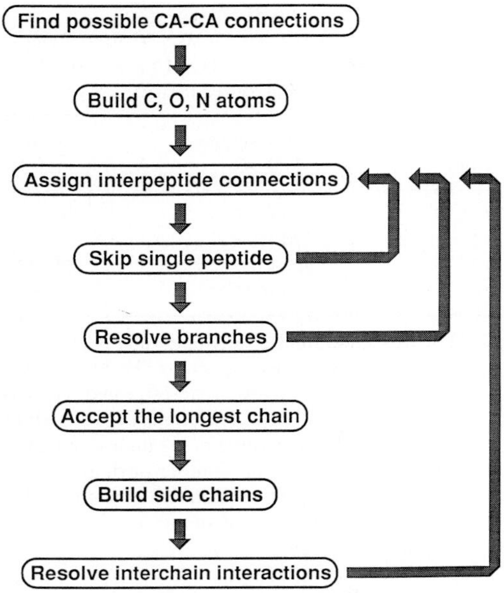

FIG. 5. The scheme of the automated model building on the basis of the ARP model and $(3F_o - 2F_c, \alpha_c)$ density.

of side chain are established on the basis of the density map. Tentative peptides lying too close to atoms from the accepted longest polypeptide are removed. This is iterated.

This procedure leads to a set of possible polypeptide fragments. The number of fragments, degree of connectivity, and correctness of atom identification depend on the quality of the set of atoms and the density map calculated using phases from these atoms. An advantage of the procedure is that no knowledge about the amino acid sequence and the number of peptides expected in the unit cell is required, although such extra information certainly would provide an improvement.

Building of the backbone is relatively straightforward because the connectivity constraint appears to be very powerful. The first problem we encountered is side-chain recognition. We assume that the side-chain type and, consequently, interatomic distances and a rotamer library are not known. Therefore, at present only C_β and C_γ atoms are identified. The second problem is implementation of this "density-driven" model building in the refinement. These problems will be addressed in future.

Software and Implementation

For least-squares minimization the SFALL (unrestrained), SFALL/PROLSQ (restrained), and for map calculation the FFT/EXTEND programs from the CCP4 suite[17] are used. Model updating/building is performed by the ARP program. It is in Fortran 77, does not use computer-specific codes, and is compatible with the CCP4 formats. We also implemented a scheme[18] wherein the ARP was used for solvent building and least-squares isotropic or anisotropic refinement was carried out using SHELXL-93.[19] The automated density map recognition and protein model building on the basis of ARP output is implemented at present as a standalone PEPTIDE program. It will be incorporated within ARP.

One cycle of refinement with the ARP and CCP4 programs using fast Fourier transform takes about 2.5 times longer than one cycle of unrestrained least-squares minimization. PEPTIDE takes about the same time as one cycle of the ARP.

Examples

We describe the use of ARP on several structures that differ in resolution for the native X-ray data and in terms of the quality of the initial phases. Unrestrained ARP was applied to four structures: phosphoribosylaminoimidazolesuccinocarboxamide (SAICAR) synthase (EC 6.3.2.6) at 1.9 Å resolution with MIR phases to 2.5 Å, space group $P2_12_12_1$, a = 62.6, b = 63.7, c = 81.2 Å[20]; xylanase at 1.5 Å with MIR phases to 2.1 Å, space group $C2$, a = 105.5, b = 47.9, c = 52.6 Å, β = 115.8°[21]; trypsin-like proteinase at 1.4 Å with phases from MR, space group $P2_12_12_1$, a = 40.4, b = 51.5, c = 69.3 Å[22]; and β-cyclodextrin complexed with a bromine-containing ligand at 1.03 Å with phases from one bromine atom identified from the Patterson synthesis, space group $P2_12_12_1$, a = 15.5, b = 17.0, c = 32.0 Å.[23] Restrained ARP was used to build solvent structure for dUTPase at 1.9 Å resolution,

[17] CCP4, *Acta Crystallogr.* **D50,** 760 (1994).

[18] J. Sevcik, Z. Dauter, V. S. Lamzin, and K. S. Wilson, *Acta Crystallogr.* **D51,** in press (1997).

[19] G. M. Sheldrick, "SHELXL-93: Program for Crystal Structure Refinement." University of Göttingen, Göttingen, Germany, 1993.

[20] V. M. Levdikov, V. V. Barynin, I. Albina, A. I. Grebenko, W. R. Melik-Adamyan, V. S. Lamzin, and K. S. Wilson, *Structure* in press (1997).

[21] V. S. Lamzin, Z. Dauter, M. Dauter, H. Bisgard-Frantzen, T. Halkier, and K. S. Wilson, in preparation (1997).

[22] V. S. Lamzin, Z. Dauter, H. Bisgard-Frantzen, T. Halkier, and K. S. Wilson, in preparation (1997).

[23] A. Perrakis, E. Antoniadou-Vyza, P. Tsitsa, V. S. Lamzin, K. S. Wilson, and S. J. Hamodrakas, *Carbohydrate Res.* in press (1997).

TABLE IV

APPLICATION OF AUTOMATED REFINEMENT PROCEDURE TO MULTIPLE ISOMORPHOUS REPLACEMENT PHASES AND MOLECULAR REPLACEMENT SOLUTION[a]

Characteristic	SAICAR synthase	Xylanase	Trypsin-like proteinase
Number of residues	306	200	196
Initial phases obtained by:	MIR	MIR	MR
X-Ray data			
Resolution (MIR) (Å)	2.5	2.1	—
MIR figure of merit	0.6	0.4	—
Resolution (native) (Å)	1.9	1.5	1.4
Number of atoms in ARP model	2800	1800	1800
Observations/atom ratio	9	20	15
Refinement			
MPE for MIR phases (°)	72	78	—
MPE for SQUASH phases (°)	70	—	—
MPE for initial atom set (°)	62	71	67
MPE for ARP model (°)	48	25	26
MCC for MIR map (%)	58	49	—
MCC for initial atom set (%)	55	44	47
MCC for ARP map (%)	72	91	88
R factor for ARP model (%)	16	17	17

[a] The final models are characterized by *R* factors of 14.4% at 1.9 Å for SAICAR synthase, 14.4% at 1.5 Å for xylanase, and 12.4% at 1.4 Å for trypsin-like proteinase.

space group *R*3, $a = 86.6$, $c = 62.3$ Å.[24] The X-ray data were collected at EMBL Hamburg using synchrotron radiation and an imaging plate scanner. The data are characterized by an overall *R* merge of less than 6%, three- to fourfold redundancy, more than 95% overall completeness, and about 99% completeness at low (10 to 4 Å) resolution with the exception of β-cyclodextrin.

The mean phase error to the final model in the whole resolution range (unless otherwise stated) is defined as MPE, the map correlation coefficient to the final $(3F_o - 2F_c, \alpha_c)$ density as MCC.

Example 1: Unrestrained Automated Refinement Procedure for Multiple Isomorphous Replacement

For SAICAR synthase and xylanase, MIR phases were available (Table IV, Fig. 6a and b). The MIR phases were of a reasonable quality with an MPE of about 51° for SAICAR synthase at 2.5 Å and 61° for xylanase at

[24] Z. Dauter, K. S. Wilson, G. Larsson, P. O. Nyman, and E. S. Cedergren-Zeppezauer, in preparation (1997).

2.1 Å. The MPEs in the whole resolution range, derived from an inverse Fourier transform of the MIR map, were 72 and 78°, respectively. Histogram matching and solvent flattening using SQUASH[25] have been applied to the SAICAR synthase MIR map at 2.5 Å. This reduced the MPE by 6° to 2.5 Å and from 72 to 70° for all data.

The expected number of equal dummy atoms (2800 for SAICAR synthase and 1800 for xylanase) were put into the MIR maps. These were estimated as number of residues times 8 plus 400 waters. The MIR maps were calculated with 0.3 Å grid separation. Starting from about 200 peaks the dummy atoms were iteratively added in the highest density within a distance of 1.2 to 1.6 Å from existing atoms. When the number of atoms in the model reached about three times the number expected (thus ensuring that even disconnected pieces of the MIR maps were filled by atoms), the one-third in the highest density were selected to make up the starting model.

Such placing of dummy atoms in the MIR map results in significant improvement of phases for resolution beyond that available from MIR, even before their *xyzB* parameters are refined (Table IV, Fig. 6a and b). Thus placing atoms in the density map is itself a quick phase extension method.

The course of the refinement for xylanase is shown in Fig. 6c and d. In each cycle 60 "worst" atoms in $(3F_o - 2F_c, \alpha_c)$ density less than 1.5σ above the mean were rejected and 60 new ones added. A merging distance of 0.6 Å was used throughout. The MPE went down steadily from 72° and reached 28° after 100 cycles. The drop in R factor correlated with the drop in the MPE. An alternative "shock" protocol has been tested. After 50 cycles the 500 worst atoms (about 30% of the model) were removed. This resulted in an increase in the R factor, but surprisingly the MPE remained the same. During the next 50 cycles more atoms were put in than removed, giving a model with the expected number of atoms. The shock protocol substantially improved convergence. The MPE for ARP model is 25°. Both MPE and R factor are better for the model obtained with the shock protocol. For comparison, unrestrained least-squares refinement alone (without updating the model) improved the MPE only to 52° (Fig. 6d).

A similar shock protocol was used for SAICAR synthase. Fifty atoms were updated during first 40 cycles when the ARP essentially converged. The MPE improved from 62 to 51°. Two shocks were applied with rejection of 800 worst atoms followed by 30 more cycles of refinement. This gave a small improvement. The resulting ARP model gives an R factor of 16% and an MPE of 48°.

[25] K. Y. J. Zhang and P. Main, *Acta Crystallogr.* **A46,** 41 (1990).

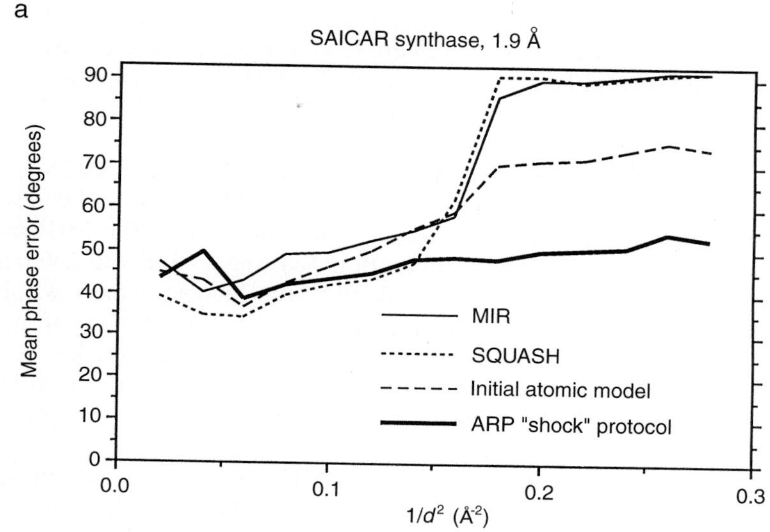

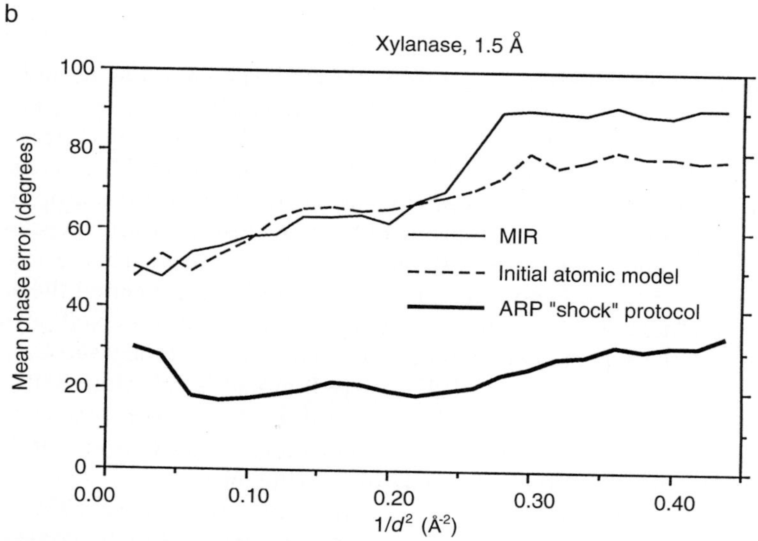

FIG. 6. The mean phase error to the final model phases as a function of resolution for (a) SAICAR synthase and (b) xylanase. (c) *R* factor and (d) the mean phase error for xylanase as a function of refinement cycle.

c

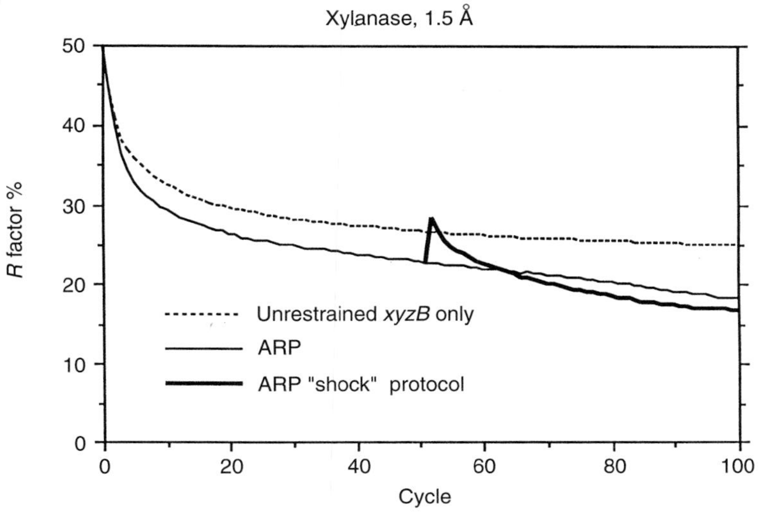

d

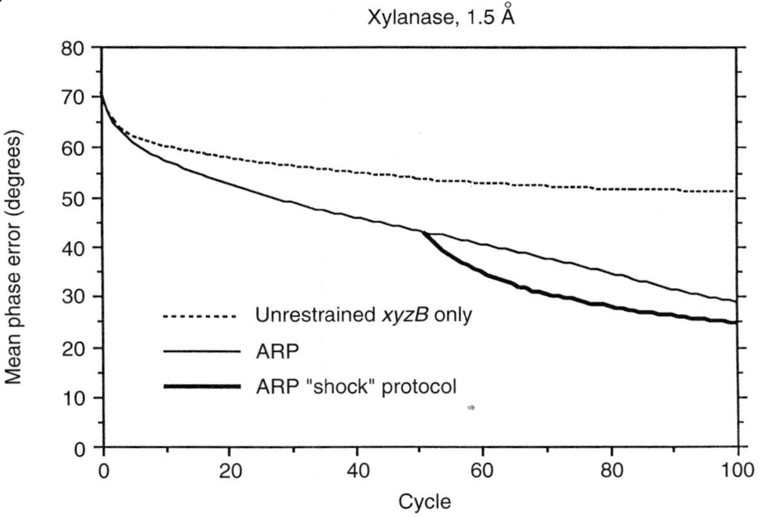

FIG. 6. (*continued*)

The MPE for the ARP models as a function of resolution is shown in Fig. 6a and b. It is approximately flat, i.e., the ARP phases are equally good in all resolution ranges. For SAICAR synthase, phases improved only for the resolution beyond that available for MIR. Combination of ARP and MIR phases may provide further improvement. For xylanase, phases dramatically improved in the whole range of resolution. The success of ARP is strongly correlated with resolution of the X-ray data. Density maps calculated with MIR and ARP phases are shown in Fig. 7.

The ARP model for xylanase consists of 1806 atoms, with 1528 of these (85%) lying within 0.6 Å of an atom from the final model. The ARP model for SAICAR synthase has 1336 (48%) such atoms.

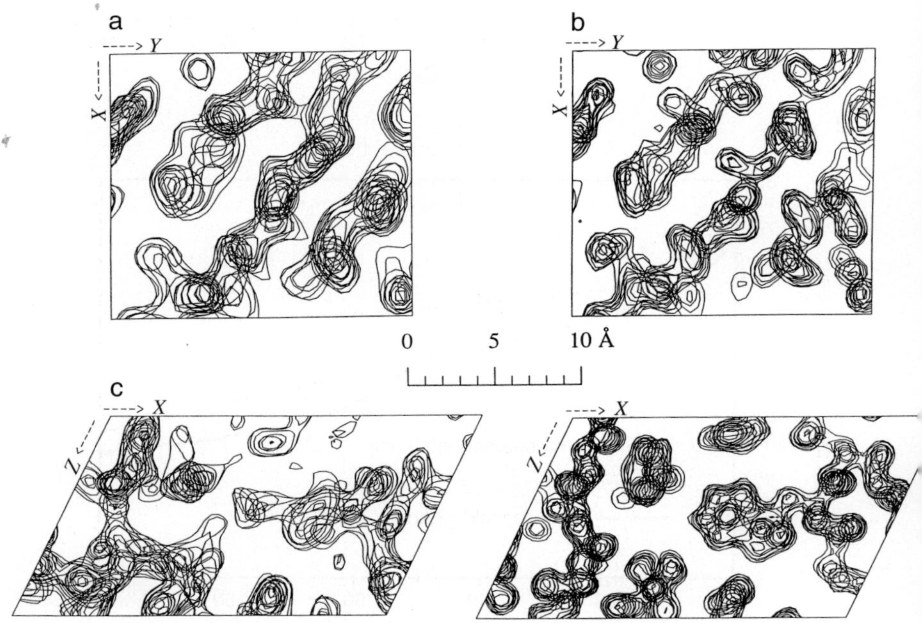

FIG. 7. Application of ARP for MIR phase improvement. (a) SAICAR synthase MIR (F_o, α_{squash}, FOM_{squash}) map, 2.5 Å; (b) SAICAR synthase ARP ($3F_o - 2F_c$, α_c) map, 1.9 Å; (c) xylanase MIR (F_o, α_{best}, FOM) map, 2.0 Å; (d) xylanase ARP ($3F_o - 2F_c$, α_c) map, 1.5 Å. Contour levels are 1, 2, and 3 σ above the mean. Map limits are as follows: x 98/132 to 132/132; y 28/132 to 66/132; z 1/168 to 8/168 for SAICAR synthase and x 28/280 to 88/280, y 10/128 to 19/128, z 22/144 to 60/144 for xylanase. Therefore all maps are plotted in approximately the same scale.

Example 2: Unrestrained Automated Refinement Procedure for Molecular Replacement

The initial phases for trypsin-like proteinase (Table IV) were obtained by molecular replacement using a related proteinase with 50% sequence identity and a slightly different number of residues. The initial model (waters were omitted) contained about 1400 atoms and gave an R factor of 53% at 1.4 Å resolution. The MPE of 67° and MCC of 47% are similar to those for the initial dummy models of SAICAR synthase and xylanase.

The ARP shock protocol was applied (Fig. 8a). During the first 40 cycles the number of atoms was increased steadily to 1800 by removing 50 and adding 60 atoms in each cycle. The R factor dropped to 22% and the MPE to 45°. The 500 worst atoms were removed and the number of atoms increased to 1800 during the next 30 cycles. This was repeated twice. The first shock reduced the R factor to 20% and the MPE to 36°, the second to 17% and 26°. The ARP map looked like the map for the final structure. Quick rejection of a large part of the model accelerated refinement to a greater extent than for SAICAR synthase but less than for xylanase. This is related to the observations/parameters ratio (Table IV).

For the ARP model the MPE is approximately the same in all resolution ranges (Fig. 8b) as for SAICAR synthase and xylanase. The ARP model consists of 1801 atoms, with 1374 of these (76%) within 0.6 Å of an atom in the final model. The density improvement is shown in Fig. 9.

By the end of refinement only 860 atoms of 1400 in the initial model (60%) had not been removed, remaining instead at approximately the same position. All others were "moved" in the ARP manner, i.e., they were taken out and new ones added. Of 6000 new atoms put in the model only 941 (16%) were finally accepted; others were subsequently rejected. This demonstrates that ARP works in a probabilistic manner. Only a fraction of the added atoms are correctly placed; the wrong atoms are eventually rejected.

Example 3: Unrestrained Automated Refinement Procedure as Automated Heavy-Atom Method for Medium-Size Structure

For a complex of β-cyclodextrin with a bromine-containing ligand, only one atom was used to provide starting phases: a bromine atom derived from the Patterson synthesis. (The bromine defined the correct enantiomorph.) The initial model gave an R factor of 56% at 1.03-Å resolution. The MPE to the final model (refined anisotropically to an R factor of 7.5%) was 71° and the MCC was 34%. About 120 nonhydrogen atoms were expected to be in the model, although the exact number was not known.

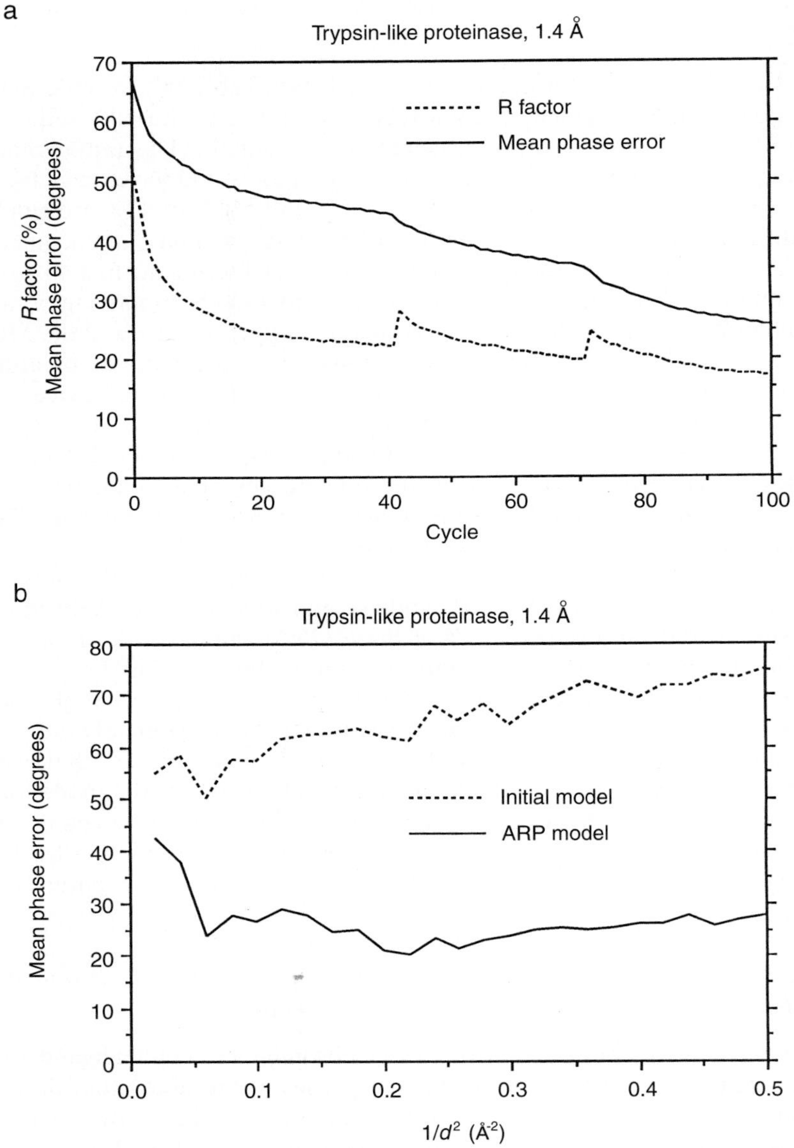

FIG. 8. The mean phase error to the final model phases as a function of (a) refinement cycle and (b) resolution for trypsin-like proteinase. *R* factor is also shown in (a).

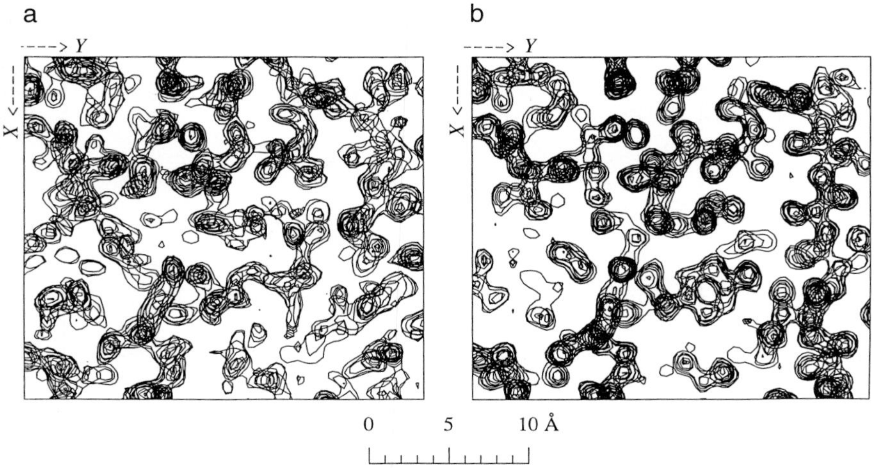

FIG. 9. Application of ARP to MR solution for trypsin-like proteinase at 1.4 Å. (a) ($3F_o - 2F_c$, α_c) map for the initial model; (b) ARP ($3F_o - 2F_c$, α_c) map. Contour levels are 1, 2, and 3 σ above the mean. Map limits are as follows: x 0/120 to 65/120, y 50/144 to 120/144, z 12/200 to 23/200.

The course of the refinement is shown in Fig. 10. During the first 25 cycles up to two atoms in ($3F_o - 2F_c$, α_c) density below 1.5σ above the mean were rejected and 6 atoms were added. The model consisted of 1 bromine atom and 110 new atoms. Ten more cycles were carried out with updating up to 6 atoms in each cycle and automated threshold determination for selection of new atoms. A shortest merging distance of 0.6 Å was used throughout and the sphere-based real-space fit was used during the last 10 cycles.

The drop in both R factor and MPE was relatively slow during the first 20 cycles. At that stage the ARP model became complete and accurate enough to allow rapid convergence. The resulting ARP model gave an R factor of 16%, an MPE of 15°, and an MCC of 94% even though all atoms except bromine were refined as oxygens, i.e., with the wrong scattering power. The model was composed of 1 bromine and 123 "water" atoms. One hundred and fifteen of these represented all nonhydrogen atoms in the final model with a r.m.s. deviation of 0.06 Å; 2 compensated for anisotropic motion of bromine and 3 corresponded to hydrogen atoms. Representative density for initial and ARP models is shown in Fig. 11. The observations/atom ratio is about 36.

The criteria for ARP atom selection were developed for protein structures at a resolution of 1.5 to 2.0 Å. It is encouraging that these criteria

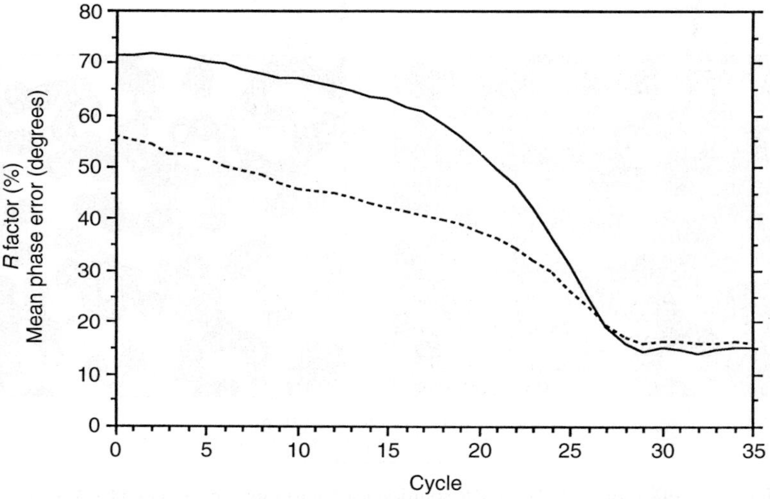

FIG. 10. R factor (···) and mean phase error (—) to the final model phases as a function of refinement cycle for the β-cyclodextrin complex (1.0 Å).

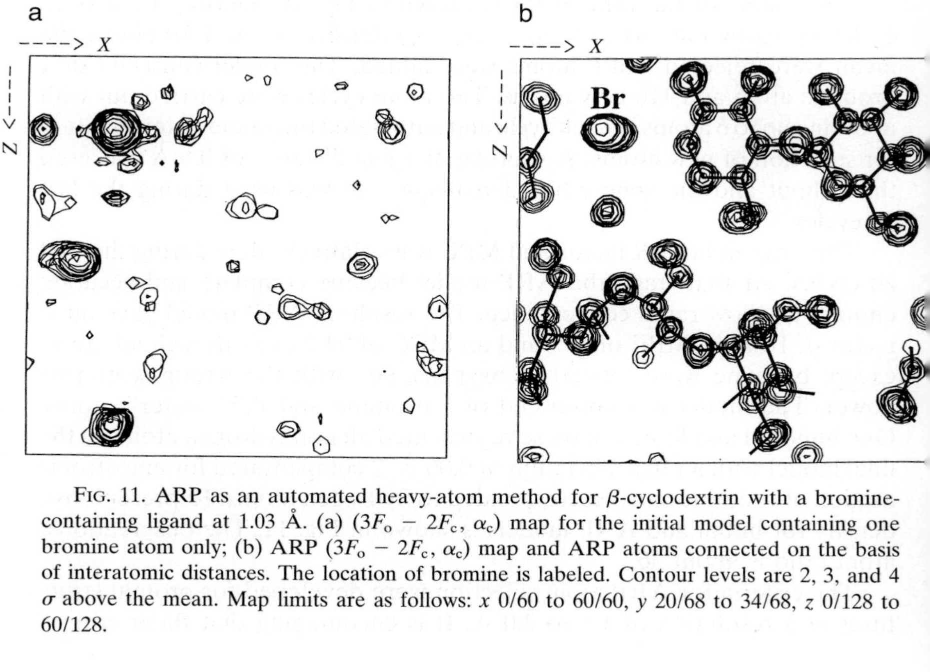

FIG. 11. ARP as an automated heavy-atom method for β-cyclodextrin with a bromine-containing ligand at 1.03 Å. (a) ($3F_o - 2F_c$, α_c) map for the initial model containing one bromine atom only; (b) ARP ($3F_o - 2F_c$, α_c) map and ARP atoms connected on the basis of interatomic distances. The location of bromine is labeled. Contour levels are 2, 3, and 4 σ above the mean. Map limits are as follows: x 0/60 to 60/60, y 20/68 to 34/68, z 0/128 to 60/128.

TABLE V
AUTOMATED MODEL BUILDING[a]

Characteristics	SAICAR synthase	Xylanase	Trypsin-like proteinase
Final model			
Number of residues	306	200	196
Number of CB atoms	279	178	153
Number of CG (OG, SG) atoms	319	204	180
Model built from ARP output			
Number of peptides found	162	193	173
Number of chains peptides comprise	58	5	12
Number of peptides in longest chain	9	70	36
Model-building quality			
Number of CA atoms	162 (90)	193 (192)	173 (173)
Number of C atoms	162 (92)	193 (192)	173 (173)
Number of O atoms	162 (91)	193 (192)	173 (173)
Number of N atoms	162 (79)	193 (192)	173 (173)
Number of CB atoms	37 (28)	159 (158)	125 (125)
Number of CG (OG, SG) atoms	46 (23)	184 (173)	153 (145)

[a] The protein structures were built automatically on the basis of the ARP model and density. Two numbers characterizing "model-building quality" are the overall number of atoms of a particular type and (in parentheses) the number of correctly identified atoms, i.e., lying within 0.6 Å of a corresponding atomic position in the final model.

are applicable to a medium-size nonprotein structure at atomic resolution. Here ARP behaves as an automated heavy-atom method. According to the "rule of thumb"[26] the atom was proposed to be heavy enough to allow "bootstrap" phasing in an iterative manner if the ratio $Z^2_{\text{heavy}}/\sum Z^2_{\text{light}}$ is about 0.2. For β-cyclodextrin the ratio is 0.2. Presently the required ratio has been found to be substantially lower (G. Sheldrick, private communication, 1997).

Example 4: Automated Protein Construction

Often, unrestrained ARP affords significant improvement in the electron density. With the exception of SAICAR synthase, the ARP set of atoms essentially represents the final atomic set. Complete models were built manually for examples 1–3 using graphics and were used for further restrained ARP.

We have developed an automated model-building protocol using the ARP model and $(3F_o - 2F_c, \alpha_c)$ density maps as an alternative to manual reconstruction (Table V). Of the main-chain atoms, 96% were correctly

[26] G. H. Stout and L. H. Jensen, *in* "X-Ray Structure Determination: A Practical Guide," pp. 292–295. John Wiley & Sons, New York, 1989.

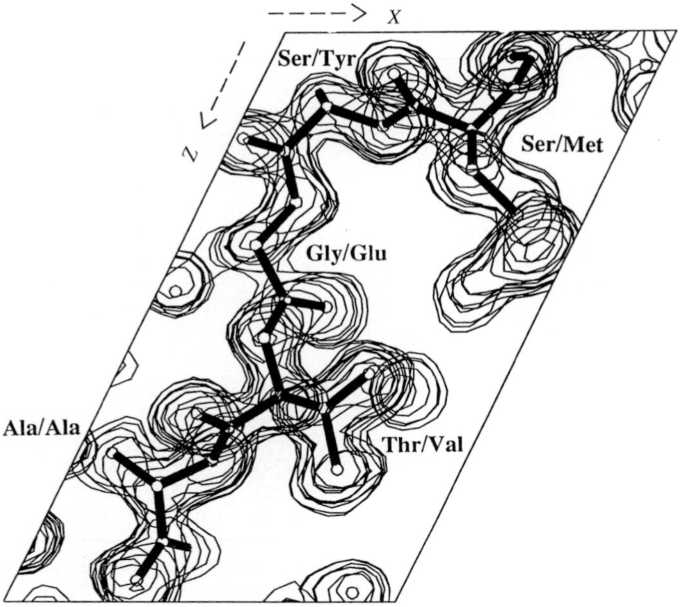

FIG. 12. Automated model building for the xylanase ARP model and ($3F_o - 2F_c$, α_c) density map at 1.5 Å. Main-chain atoms are correctly identified whereas side chains are built out to the C_γ atoms at best. Two side-chain types are given: that obtained by the model building and that in the final model. Contour levels are 1 and 2 σ above the mean. Map limits are as follows: x 21/280 to 46/280, y 23/128 to 32/128, z 4/144 to 47/144.

identified for xylanase and 88% were correctly identified for trypsin-like proteinase. Only one peptide was built incorrectly for xylanase. The side-chain building is slightly worse and 80 to 90% of all CB atoms and 80 to 85% of C_γ atoms were found with some misidentification (Fig. 12). For SAICAR synthase only about one-third of main-chain atoms were built correctly and the misidentification is substantial. Only 10% of the side chains were recognized.

The accuracy of the model building with respect to the final models is about 0.2 Å for the main-chain atoms and 0.3 Å for the side-chain atoms of xylanase and trypsin-like proteinase and about 0.4 Å for SAICAR synthase. These values agree to some extent with the estimated accuracy of coordinates derived from the σ_A plots.[27] The model-building procedure is highly dependent on the phase quality and needs further development. At present

[27] R. J. Read, *Acta Crystallogr.* **A42,** 140 (1986).

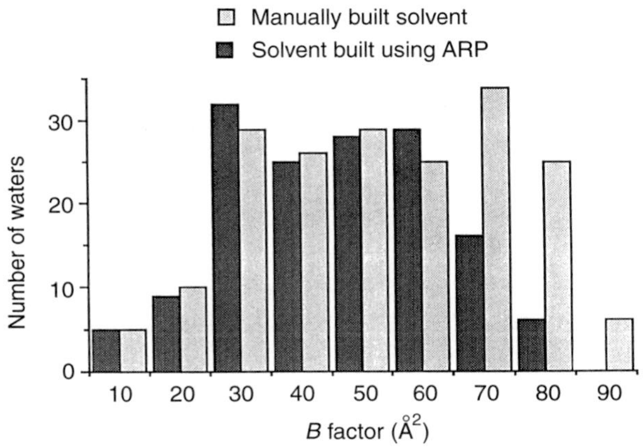

FIG. 13. Number of water molecules as a function of their *B* factor in dUTPase.

it can be helpful in retrieving a part of the model refined with an unrestrained ARP without manual intervention.

Example 5: Restrained Automated Refinement Procedure for Building Solvent

The structure of dUTPase was refined previously to an *R* factor of 14.6%.[28] The model of 136 residues contained 189 water molecules, which were introduced manually on the basis of the difference Fourier synthesis. This took a few days of graphics sessions. Restrained refinement now has been repeated with ARP and without manual intervention. All previously built waters were taken out of the model. Up to 10 water molecules were allowed to be added or rejected in each cycle. This number was reduced to four by the end of refinement. The threshold for new atom selection in the difference synthesis was determined automatically by the program. After 20 cycles the refinement converged. The model has the same *R* factor and geometric characteristics as that refined previously but the number of solvent sites has fallen to 141.

The solvent structure built by the ARP does not include as many water molecules with very high temperature factors as compared to the model with solvent built manually (Fig. 13). Most of the water molecules from the two models are essentially equivalent. One hundred and sixteen of 141 water molecules found by the ARP are located within 0.4 Å of waters

[28] E. S. Cedergren-Zeppezauer, G. Larsson, P. O. Nyman, Z. Dauter, and K. S. Wilson, *Nature* (*London*) **355,** 740 (1992).

built manually. The additional 48 high-temperature factor waters in the previously refined model did not improve the *R* factor and probably are not significant.

Convergence

Convergence Criteria

The mean phase error and the map correlation to the final model are good indications of the progress of refinement with unrestrained ARP for test cases but are not applicable in practice, because final phases are not known until refinement is complete. The crystallographic *R* factor appears to be a reasonable indicator. Its absolute value in the case of unrestrained ARP probably is meaningless, i.e., at convergence it is the same for all three of our examples, (Table IV). However, the relative value of the *R* factor seems to indicate the progress during refinement because it correlates with the mean phase error. Therefore unrestrained ARP can be continued as long as the *R* factor improves and the number of atoms (i.e., observations/parameter ratio) remains constant. Visual inspection of the density map is also useful. Various statistical tests, e.g., cross-validation where R_{free}[29] is monitored, can be applied. We found, however, that omitting even a small part of the data seriously affects refinement, especially at a resolution where the observations/parameter ratio is just enough to apply ARP.[29a]

Radius of Convergence of Unrestrained Automated Refinement Procedure as Function of Accuracy of Initial Atomic Set

The "radius of convergence" of ARP has been estimated roughly in terms of completeness of the initial model and accuracy of the atomic positions. At 1.8-Å resolution with a 75% complete model having C_α atoms deviating by about 1.4 Å from their correct positions, ARP provided substantial improvement.[10]

The r.m.s. deviation summed over all atoms also can serve as a characteristic of the power of unrestrained ARP. The correctness of atomic position (CAP) is the closest distance between an atom in the initial model and any atom in the final model. Initial atomic sets deviated from the final sets by 1.0 to 1.3 Å in the three sample structures. This should not be confused with the r.m.s. deviation between equivalent pairs of atoms. For comparison a random set of atoms would have an overall r.m.s. CAP of about 2.2 Å.

[29] A. T. Brünger, *Acta Crystallogr.* **D49,** 24 (1993).

[29a] E. J. van Asselt, A. Perrakis, K. H. Kalk, V. S. Lamzin, and B. W. Dijkstra, *Acta Crystallogr.* in press (1997).

If X-ray data extend to high enough resolution, the ARP atoms correspond satisfactorily to the correct structure even if no geometric restraints were applied during refinement (Fig. 14). The interatomic distances in the ARP model are not ideal. Some atoms may be missing although the ARP density with high quality represents a protein density. The CAP of the ARP set is related to the quality of the atomic positions in the initial model. Figure 15a shows such dependence and demonstrates that even if initial and final models deviate in some places up to 5 Å, an unrestrained ARP is able to refine it. The same dependence, but where every point is an average for about 100 atoms, is shown in Fig. 15b. There is an approximately linear correlation between a positional error for groups of atoms in initial and ARP models. If an atom in the initial model on average is 1 Å from the nearest atom, in the final model unrestrained ARP improves it to about 0.3 Å. If the error is about 3 Å, ARP is able to reduce it to about 0.7 Å. The slope of the straight line shown in Fig. 15b depends on two factors: the overall quality of the initial model/phases and the quality and resolution of the X-ray data.

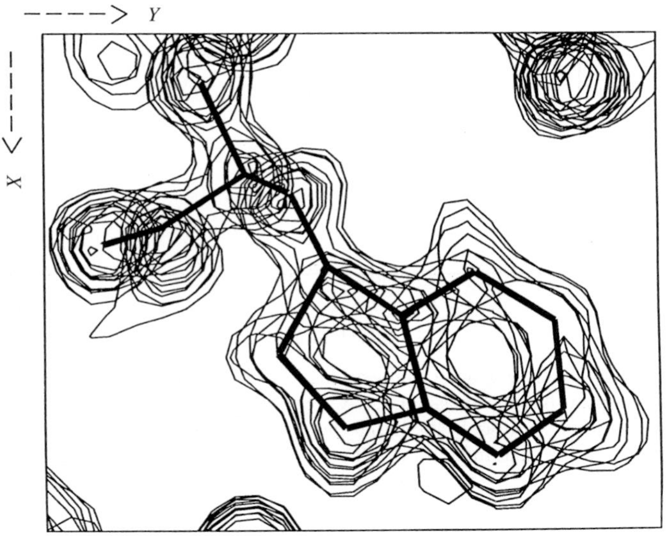

FIG. 14. A fragment of the ARP ($3F_o - 2F_c, \alpha_c$) map for trypsin-like proteinase at 1.4 Å with ARP atoms connected on the basis of interatomic distances. Molecular geometry has not been imposed and thus ARP atoms deviate from the final, correct positions. Contour levels are 1, 2, and 3 σ above the mean. Map limits are as follows: x 91/120 to 114/120, y 108/144 to 134/144, z 5/200 to 16/200.

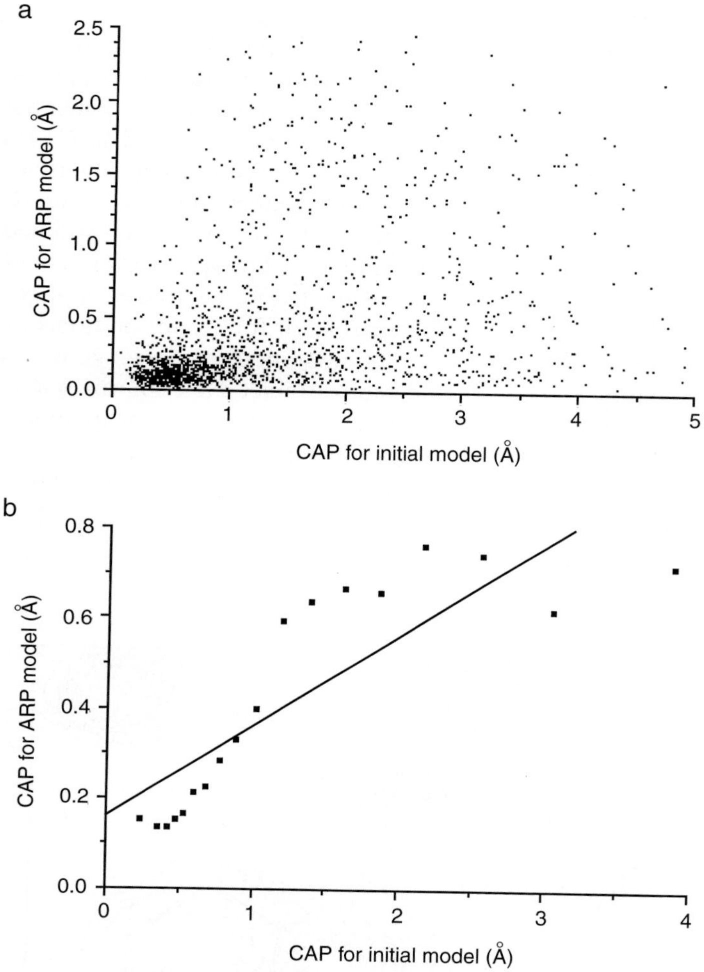

FIG. 15. Radius of convergence of unrestrained ARP as a function of accuracy of initial atomic set for trypsin-like proteinase at 1.4 Å. The "correctness of atomic position," CAP, is the distance between an atom in the initial (horizontal axis) or ARP (vertical axis) atomic set and any atom in the final model. (a) All atoms. (b) Every point is an average of about 100 atoms. An approximately linear correlation between the positional error for initial and ARP models is shown by the straight line.

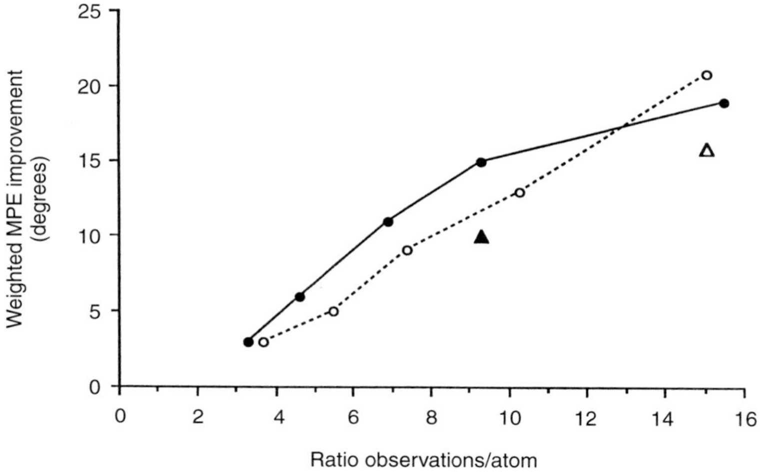

FIG. 16. Radius of convergence of unrestrained ARP as a function of the ratio of number of data to the number of atoms, which is related to resolution of the data. A weighted MPE is the difference in arccos(MCC) for the initial and final models. (●) Refinement against F_c values, SAICAR synthase; (○) refinement against F_c values, trypsin-like proteinase; (▲) refinement against F_o values, SAICAR synthase; (△) refinement against F_o values, trypsin-like proteinase.

Radius of Convergence of Unrestrained Automated Refinement Procedure as a Function of Resolution

The ARP radius of convergence is dependent on the ratio of number of data to the number of parameters. This ratio is related to the resolution of the data. Test calculations were performed as follows. Initial models for SAICAR synthase and trypsin-like proteinase were the same as described above (set of atoms built in MIR map and MR solution, respectively). Forty cycles of unrestrained ARP were carried out against structure factors calculated from the final models at different resolutions (varying between 1.4 and 2.7 Å). Forty cycles are enough to see a qualitative improvement but more cycles are required for full convergence.

The results are presented in Fig. 16. Because the MCC is the F^2-weighted cosine of the phase error,[30] arccos(MCC) can serve as a weighted MPE. The reduction of weighted MPE was chosen as a measure of improvement. For example, for trypsin-like proteinase at 1.4-Å resolution the MCC improved from 0.47 to 0.76. This corresponds to a reduction in the weighted MPE from 62 to 41°, i.e., 21° improvement.

[30] V. Y. Lunin and M. M. Woolfson, *Acta Crystallogr.* **D49,** 530 (1993).

In conclusion, first, the behavior of ARP is similar for the two different proteins in this test, and depends strongly on the observations-to-parameters ratio. The small differences reflect the different quality of starting models and crystal packing: a value of seven observations per atom corresponds to a resolution of 2.1 Å for SAICAR synthase and 1.9 Å for the trypsin-like proteinase. Second, it is very important to have as accurate data as possible. Indeed, refinement against F_o values gave less improvement compared to the use of F_c values calculated from the final models. Of course, nature is not so simple: she does not completely conform to our hopeful models. Third, an unrestrained ARP provides some improvement if there are as few as three observations per atom. For a crystal with a V_m of 2.4 this corresponds to about 2.8 Å resolution. An improvement at such low resolution is, however, marginal and mostly caused by unrestrained least-squares minimization because the density maps have essentially no atomic features and updating of the model cannot be performed satisfactorily.

Discussion

ARP, which constantly updates a model during refinement, resembles alternating cycles of least-squares and difference Fourier syntheses originally used in small-molecule crystallography when atomic resolution data were available. Applied to the refinement of a medium-size structure at 1.0 Å resolution and starting from one heavy-atom position, ARP determined the complete structure in a fully automated manner. ARP uses atomicity as the main property of the structure and differs completely from, e.g., direct methods, which are based on atomicity through statistical relationships between amplitudes of structure factors.[31]

Agarwal and Isaacs[32] placed dummy atoms in an insulin MIR map (1.9 Å) followed by further unrestrained refinement at high resolution (1.5 Å). They refined dummy atoms, but that gave only a marginal improvement and a mean phase error of 70° to the final phases. The limited success may be related to the quality of data and minimization procedures used at that time. In contrast, placing of dummy atoms in the xylanase MIR map followed by ARP at 1.5 Å resulted in a mean phase error of 25°.

Restrained refinement coupled with updating part of the model, resembling restrained ARP, has been proposed for finding missing protein parts[15] and for building solvent structure.[33] The authors paid attention to

[31] D. Sayre, *Acta Crystallogr.* **A28,** 210 (1972).

[32] R. C. Agarwal and N. W. Isaacs, *Proc. Natl. Acad. Sci. U.S.A.* **74,** 2835 (1977).

[33] H. Tong, A. M. Berghuis, J. Chen, and Y. Luo, *J. Appl. Crystallogr.* **A46,** 41 (1994).

removal of "bad" atoms as an important part of the procedure. However, no criterion other than an atomic *B* factor has been suggested for classification of an atom as potentially wrongly positioned.

Various procedures to reconstruct a protein model on the basis of a density map have been suggested (see, e.g., Ref. 34 for an overview). They include skeletonization of the density map, which provides details of connectivity and approximate C_α atom positions, followed by retrieving of the main- and side-chain atoms on the basis of density. The most widely used is an approach realized in the O program package.[35] Our first attempts to build a model on the basis of ARP and use of ARP atoms as candidates for C_α atoms allowed us to skip the skeletonization step. The set of polypeptides derived simplifies the manual rebuilding stage but is as yet too erroneous to be used iteratively in the refinement.

In conclusion, ARP can be used in two different ways, restrained and unrestrained, reflecting the presence or absence of additional geometric limitations used in the least-squares method. In both modes, updating of the model is applied to whichever atoms are atoms refined without restraints: all atoms in an unrestrained ARP and, e.g., solvent only in a restrained ARP. The procedure is strongly dependent on the initial model/phases, X-ray data, and resolution. An unrestrained ARP starts to give a small improvement provided there are three or more observations per atom, i.e., a resolution of about 2.8 Å. Such a low resolution cannot be recommended in practice. To obtain a significant improvement, there should be about eight observations per atom, roughly 2.0 Å resolution. If the number of observations per atom is about 15 (1.5 Å), not only the density but the ARP atoms themselves start to represent the true atomic structure. At the resolution typical for proteins, an unrestrained ARP can be successfully used if the initial atomic model gives a mean phase error of 70° or less and the map correlation to the final map is about 40%.

Acknowledgments

The authors thank Zbigniew Dauter and George Sheldrick for helpful discussion and Vladimir Levdikov and Anastasis Perrakis for providing data and models prior to publication. We thank the EC for partial support of this work through BIOTECH contract CT92-0524.

[34] S. M. Swanson, *Acta Crystallogr.* **D50,** 695 (1994).

[35] T. A. Jones, J.-Y. Zou, S. W. Cowan, and M. Kjeldgaard, *Acta Crystallogr.* **A47,** 110 (1991).

[15] TNT Refinement Package

By DALE E. TRONRUD

What Is TNT?

The TNT package of programs was designed by L. F. Ten Eyck in the late 1970s when protein refinement was just becoming practical. The program PROLSQ[1] was being used quite successfully in a number of projects. However, Ten Eyck wanted a refinement procedure that was quicker, easier to use, and easily changed to incorporate new methods and concepts. The programs he designed, which later were implemented and extensively modified by this author,[2–4] have met with considerable success in all these areas.[5]

The scope of TNT is to optimize atomic coordinates with respect to a series of observations, usually diffraction data and ideal stereochemistry. It also will provide information to aid in the examination and correction of the refined model. These include the difference maps to be displayed along with the model, the automatic location of peaks in these maps, and indicators of the problem spots in the refined model that should be examined directly by the crystallographer. The principal advantages of TNT are listed in Table I.

Although TNT is an excellent package for everyday refinement, its flexibility allows it to be used as a tool for the development of new refinement techniques. One can reorder TNT components, add new components to implement the refinement of the model against a novel set of observations, or make other kinds of changes without having to modify the existing programs. Many of the current features of TNT were tested first in this fashion, which reduced the time invested in an idea before our determining its usefulness.

The package has been used in this way several times. Chapman[6] has implemented novel methods for low-resolution real-space refinement, and

[1] W. A. Hendrickson and J. H. Konnert, *in* "Computing in Crystallography" R. Diamond, S. Ramaseshan, and K. Venkatesan, eds.), pp. 13.01–13.26. Indian Academy of Sciences, Bangalore, 1980.

[2] D. E. Tronrud, L. F. Ten Eyck, and B. W. Matthews, *Acta Crystallogr.* **A43,** 489 (1987).

[3] D. E. Tronrud, *Acta Crystallogr.* **A48,** 912 (1992).

[4] D. E. Tronrud, *J. Appl Crystallogr.* **29**(2), 100 (1996).

[5] B. Finzel, *Curr. Opin. Struct. Biol.* **3,** 741 (1993).

[6] M. S. Chapman, *Acta Crystallogr.* **A51,** 69 (1995).

0076-6879/97 $25.00

TABLE I
ADVANTAGES OF TNT REFINEMENT PACKAGE

Advantage	Comments
Easy to use	Create description of model and proceed
Flexible geometry definition	Can define any type of molecule
	All standard groups are predefined
	New groups can be defined from small-molecule data
	Chemical links to symmetry mates can be incorporated
Tools for model building	Provides lists of worst bond lengths, angles, etc.
	Produces map files for display
	Finds highest peaks in difference map
Preconditioned conjugate gradient method	Fewer cycles of refinement are required
	B factors are less biased toward their starting values
	All parameters can be varied in each cycle
Fast execution time	Usually 20 times faster than other programs
	Near real-time performance
Stereochemical restraints on *B* factors	Low-resolution models have *B* factors that are consistent with those of high-resolution models
Disordered solvent model	Solvent model is simple and accurate
	Allows use of data of lowest resolution
	Better estimate of average *B* factor
	Better maps at surface of molecule

Bricogne and Irwin[7] have used very sophisticated protocols to interface TNT and their maximum-likelihood refinement program BUSTER. Neither project, both of which implement ideas far from the thoughts of TNT's creators, required significant modifications to the TNT code.

Other investigators have implemented their ideas by modifying parts of TNT itself.[8,9] In these cases it was decided that the computational efficiency of adding new code directly to TNT justified the extra time required to learn the internal workings of TNT programs.

Philosophy of Refinement

To solve any computational problem, one establishes a protocol that is based on a collection of assumptions about the nature of the problem. Although some assumptions may be justified mathematically, with very

[7] G. Bricogne and J. J. Irwin, *in* "Proceedings of the Macromolecular Crystallographic Computing School" (P. Bourne and K. Watenpaugh, eds.), in preparation (1997).

[8] J. P. Abrahams, *in* "Macromolecular Refinement: Proceedings of the CCP4 Study Weekend, January 1996" (E. Dodson, M. Moore, A. Ralph, and S. Bailey, eds.). Daresbury Laboratory, Daresbury, Warrington, UK, 1996.

[9] N. S. Pannu and R. J. Read, personal communication (1996).

complicated problems, others cannot. Because the choice of basic assumptions will govern the success of a refinement package, it is important to examine them carefully.

Many of the assumptions made in TNT are shared by most of the available macromolecular refinement packages. These assumptions include the notions that anomalous scattering and anisotropic temperature factors can be ignored (the noteworthy exception being SHELXL, described in [16] in this volume[10]). In addition, the model is refined against the amplitude of the structure factor, rather than the more statistically correct intensity of the diffracted ray.[10a,11]

In some instances, when authors initially made differing assumptions, one particular assumption won out over time. Although the simultaneous refinement against diffraction data and stereochemical knowledge was quite radical when first introduced in PROLSQ,[1] it now has been incorporated within all packages as the best way to compensate for the lack of high-resolution data.

The convergence of assumptions continues as more experience in refinement is acquired. In the past there has been a strong dichotomy between the refinement packages that minimize a least-squares residual and those that minimize empirical energy functions. The most popular refinement program, X-PLOR,[12] uses the formalism of energy minimization to ensure that the model is consistent with ideal stereochemistry. However, on adopting the standard parameters of Engh and Huber,[13] Brünger abandoned the "force constants" of an energy function and began using the σ values of least squares. Now all of the major refinement packages are least-squares refinement packages.

The structure of TNT embodies several assumptions that differ from those found in other packages.

Assumption: Local Minima Are Not a Big Problem

One cannot determine mathematically the importance of local minima in the refinement function. The function is extremely complicated and exists

[10] G. Sheldrick and T. R. Schneider, *Methods Enzymol.* **277,** [16], 1997 (this volume).

[10a] Refining against amplitudes as opposed to intensities has three principal effects. First, the relative importance of the weak reflections versus the strong is altered. Second, some observations must be rejected arbitrarily because a negative intensity cannot be converted to a structure-factor amplitude. Third, the nonlinear transformation of F^2 to F will introduce a systematic error in F if there are significant, random measurement errors in F^2.[11]

[11] D. Schwarzenback, S. C. Abrahams, H. D. Flack, W. Gonschorek, T. Hahn, K. Huml, R. E. Marsh, E. Prince, B. E. Robertson, J. S. Rollett, and A. J. C. Wilson *Acta Crystallogr.* **A45,** 63 (1989).

[12] A. Brünger, K. Kuriyan, and M. Karplus, *Science* **235,** 458 (1987).

[13] R. A. Engh and R. Huber, *Acta Crystallogr.* **A47,** 392 (1991).

in a space of many thousands of dimensions. No one has performed an analysis of the distance between or the height of the barriers between the local minima. Objectively, one cannot rule out their importance, but if one assumes they are not a problem, the programming code becomes simpler to write and much quicker to execute.

Some support can be found for the assumption that entrapment in local minima is not significant. The minimization method used in TNT proceeds to the local minimum whereas that of X-PLOR has the capability to leap from one minimum to another. In spite of this difference, the final models from each program are quite similar.

Assumption: The Weak Power of Convergence of Minimization Techniques Is a Big Problem

The minimization methods used in crystallographic refinement—conjugate gradient[14–16] and preconditioned conjugate gradient (also known as conjugate direction)[3,17]—require, in theory, many thousands of cycles to reach the local minimum if that minimum is described by a perfectly quadratic function; it is not. There is no way to estimate how much better our models would be if we could run this many cycles of refinement. The minimizer in TNT has been constructed to be the most powerful (in terms of using more second derivatives of the function) of the packages available for refinement at nonatomic resolution. Because it uses more of the second derivatives of the function, it will approach the local minimum in fewer cycles.

Assumption: The Main Problem with Models Is Not in Their Parameters, but in Their Parameterization

Any refinement package is limited by the fact that it can change only the values of the parameters of the model; it can neither add nor remove parameters. The number of amino acids, amino acid sequence, and the common occurrence of unmodeled electron density are examples of properties of a model that cannot be changed automatically. Although TNT optimizes well the fit of atoms to their density, it does not attempt to tear the atoms out of density and place them in nearby unexplained density.

[14] R. Fletcher and C. Reeves, *Comput. J.* **7,** 81 (1964).
[15] M. J. D. Powell, *Math. Programming* **12,** 241 (1977).
[16] J. H. Konnert, *Acta Crystallogr.* **A32,** 614 (1976).
[17] O. Axelsson and V. Barker, "Finite Element Solution of Boundary Value Problems," pp. 1–63. Academic Press, Orlando, Florida, 1984.

Assumption: Computers (Algorithms) Are Not Good at ab Initio Interpretation of Electron-Density Maps

The programs in TNT do not replace the need to examine a model, using a molecular graphics program. TNT provides tools to ease the job of rebuilding, but it can never eliminate the need for this vital step in the refinement of a macromolecular model.

Assumption: Low-Resolution Data Are Important to Density Map Appearance, and Should Be Used

The low-resolution portion of the diffraction data includes some of the strongest reflections. Their omission will cause distortions in the appearance of any $2F_o - F_c$ map, and these distortions will appear principally on or near the surface of the molecule. The calculation of an appropriate map requires that these terms be included. However, the low-resolution reflections can be used only if the model includes compensation for the scattering of the disordered solvent in the crystal.

By default, TNT includes a model for this scattering, and the use of all low-resolution data is encouraged. The section Modeling a Solvent Continuum (below) describes the disordered-solvent model used in TNT.

Overview of Refinement

Macromolecular refinement is an iterative process. Each step consists of an automated optimization of the model, followed by the detailed examination of the remaining discrepancies between the model and the data: the electron-density map and ideal stereochemical restraints. TNT performs the optimization and provides information that the graphics program can display. After manual corrections have been applied to the model, it is returned to the optimizer.

The first time the model is "exposed" to refinement is the most difficult. The parameters of the model may fail to reach their optimum values because the errors in the model may be so large or of such a character that the automated procedures fail. To detect and correct these errors, one must anticipate the sort of error that might be found in a model. The following points are relevant to all refinement programs.

There are basically three sources for the sort of model one would use to begin refinement: multiple isomorphous replacement (MIR) phasing followed by model building, molecular replacement with the model of a similar molecule in another crystal, or molecular substitution (use of a model from an isomorphous crystal form with a small modification in the molecule, as with mutant or inhibitor structures). The nature and magnitude of starting errors in a model depend on which of these methods was employed.

A model built from an MIR map typically will contain some errors as large as 1 or 2 Å, many side chains may be built in the wrong rotamer, and there may be chain-tracing errors. While these are usually the worst starting models, the automated refinement programs will operate quite well. In spite of their inability to correct the latter two kinds of errors, positional errors of up to 3 Å can be corrected and the majority of smaller errors also will be reduced.

The refinement programs work well with these models because the errors of the models match the assumptions built into the programs. Most refinement programs presume that the error in one parameter is unrelated to the errors in the other parameters in the model (i.e., the off-diagonal elements of the normal matrix are ignored). In an MIR model the errors in the positions of two atoms are uncorrelated if the atoms are more than about 5 Å apart. A good MIR model can be subjected to individual-atom refinement using the highest resolution data available.

The errors in a molecular replacement model are correlated. One expects that significant rigid-body shifts will be required for the entire molecule, and then for individual domains with respect to each other. Often it will be difficult to see in a difference map that these shifts are needed, and because all refinement packages basically attempt to flatten the difference map, the automated programs will not identify the problem either. If individual-atom refinement is performed, the R value typically will drop to between 35 and 25% and fail to improve further.

Because there is no indicator of correlated errors, they must always be presumed to exist. For every molecular replacement model, rigid-body refinement should be performed. Each molecule in the model must be defined as a rigid group and refined. Each molecule must then be split into domains and refined again. If the molecule has parts that might be expected to be variable, then they should also be refined as rigid groups.

One performs rigid-body refinement to encourage large shifts of large objects; only the low-resolution diffraction data should be used. A resolution cutoff of 5 Å is recommended. Once the rigid-body refinement is complete, the model can be refined with individual atoms to the highest resolution data available.

Refinement of models derived from the third source, molecular substitution, can prove difficult simply because the seriousness of the errors is underestimated. One expects that the starting model will be quite accurate because the changes in the molecule are so slight. However, these models often are subject to the same invisible problems as those of molecular replacement models.

For example, cell constants may change from crystal to crystal. A subtle consequence of such a change is that if the model happens to be built into an asymmetric unit far from the origin of the unit cell, a significant translation of the entire molecule will be required. This error will be difficult to correct with individual atom refinement; rigid-body refinement is needed. Rigid-body refinement also is required when the domains of the molecule move in response to the structural modification.

Even though the *R* value of the starting model may be low and the errors are expected to be quite small, these models must be treated with the same skepticism as molecular replacement models.

Temperature-Factor Refinement

In TNT, all parameters of an atom can be varied in the same cycle; having this ability does not mean that one should use it. The occupancy parameters cannot be refined without quite high-resolution diffraction data. With 1.6-Å data one can refine group occupancies but individual occupancies are poorly behaved. In most cases, all atomic occupancy fractions in a model should be fixed equal to one.

Temperature factors present a larger problem. With 1.6-Å data one can refine individual temperature factors without resorting to restraints that depend on stereochemical relationships among atoms. At the other extreme, with only 3-Å data nothing is gained by refinement of individual temperature factors. Handling the ground in between is subject to judgment. The decisions depend on the resolution of the diffraction data and its quality. Table II lists some criteria for the application of restraints to the temperature factors of a model.

TABLE II
RESOLUTION LIMIT AND TEMPERATURE-FACTOR RESTRAINTS[a]

High-resolution limit	Comments
$\infty \geq$ limit $\geq$ 3 Å	Do not refine individual temperature factors
3 Å $\geq$ limit $\geq$ 2.5 Å	Refine temperature factors with restraints
2.5 Å $\geq$ limit $\geq$ 2 Å	Refine with restraints if required
2 Å $\geq$ limit	Refine without restraints

[a] Generally a model should be refined with the smallest possible number of nondiffraction restraints. This avoids the introduction of bias from inappropriate restraints. If the model agrees with the temperature-factor restraint library without its arbitrary imposition then the restraints are not required.

Modeling Solvent Continuum

Large portions of the unit cell of a macromolecular crystal are filled with disordered solvent. X-Rays scatter from this solvent, and the structure factors are affected. To reproduce the scattering of the crystal, both the ordered and the disordered portions must be modeled accurately.

Models of proteins often include no description of the disordered solvent, and as a consequence they are particularly poor at reproducing the low-resolution scattering. Usually the problem is simply ignored by deleting the low-resolution data. Although this solution avoids embarrassingly high *R* values it does not solve the problem.

Density maps calculated without the low-resolution data will be significantly distorted, particularly near the surface of the molecule. Models that contain no description of the disordered solvent will have systematic errors in many of their parameters when they are refined against the low-resolution data. In particular, the *B* factors all will be underestimated. The only way to avoid these problems is to include all of the data and to incorporate a model of the disordered solvent.

To be useful, a solvent model must be simple to incorporate, quick to calculate, and must contain a minimum number of parameters. The method used in TNT has all these properties. The review by Brünger[18] discusses a number of other solvent-modeling methods.

Solvent Scattering

Before one can understand the solvent model in TNT, one must examine the effect of the disordered solvent on the X-ray scattering (see also [17] in this volume[19]). A good place to start is to consider the difference between the observed structure factors, $F_o(\mathbf{s})$, and the structure factors calculated from the refined model of the ordered, atomic portion of the structure, $F_m(\mathbf{s})$. To avoid bias toward some assumed model of solvent scattering, the ordered model should have been refined with no attempt at modeling solvent.

Let us consider the average agreement between $F_o(\mathbf{s})$ and $F_m(\mathbf{s})$ as a function of resolution ($\sin\theta/\lambda$ or s). One can do this by splitting reciprocal space into thin shells, and determining a linear scale factor for each shell. We can determine the scale factor K for each shell by minimizing the function:

$$f = \sum^{\text{Shell}} [F_o(\mathbf{s}) - KF_m(\mathbf{s})]^2 \qquad (1)$$

[18] A. Brünger, *Acta Crystallogr.* **A43,** 489 (1994).

[19] J. Badger and D. L. Caspar, *Methods Enzymol.* **277,** [17], 1997 (this volume).

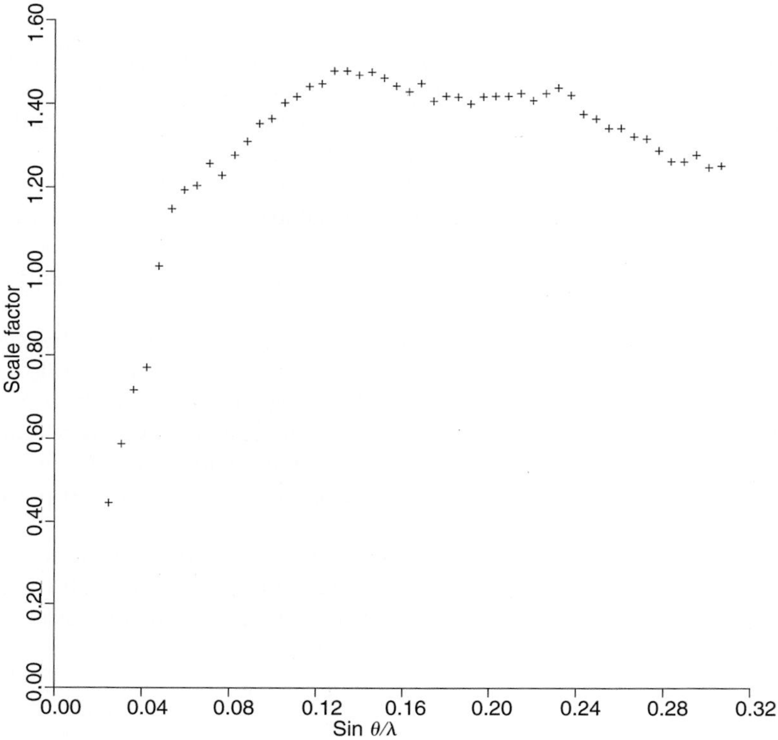

FIG. 1. Plot of K as a function of sin θ/λ for data from 20 to 1.6 Å. The data were from the thermolysin:inhibitor complex TLN:ZGPLL refined to the 10- to 1.6-Å data without a model for the disordered solvent.

One can rearrange this to show that

$$K = \sum^{\text{Shell}} F_o(\mathbf{s})F_m(\mathbf{s}) \Bigg/ \sum^{\text{Shell}} F_m^2(\mathbf{s}) \tag{2}$$

Figure 1 shows K as a function of resolution for the thermolysin:inhibitor complex TLN:ZGPLL[20] for data from 20 to 1.6 Å. One can see in Fig. 1 that K changes smoothly with resolution. There are two roughly straight sections with a break at about 4-Å resolution.

It is best to consider the two sections of the curve separately. In the high-resolution portion, K does not vary rapidly with resolution. The slight variation that is evident can be removed by addition of a temperature-

[20] D. E. Tronrud, H. M. Holden, and B. W. Matthews, *Science* **235,** 571 (1987).

factor correction to $F_m(\mathbf{s})$, equivalent to adding some constant to the B factor of each atom in the model. If the B values are properly refined, no correction will be required, and the K values for the high-resolution data will form a horizontal line. The line is not horizontal in Fig. 1 because the individual B factors were biased in refinement by the lack of a disordered solvent model. (The bias in the example is limited because only the data between 10 and 1.6 Å were used for this refinement.)

In the low-resolution portion of the curve, K changes rapidly. K at 20 Å is much smaller than K at 2.0 Å. This implies [by Eq. (2)] that $F_m(\mathbf{s})$ is systematically too large at low resolution.

One can understand the behavior of the low-resolution portion of the curve by examining the difference between the usual model of a macromolecule and the electron density that actually occupies the crystal. The electron density of a model without a disordered solvent contribution drops to zero outside the envelope of the molecule. Because the crystal contains significant amounts of electron density in these regions the contrast of the model is much greater than it should be. This high contrast results in the inappropriately large size of the low-resolution structure factors calculated from the model. It is important to note that the phases of the low-resolution coefficients from these two models will be similar in spite of the large difference in amplitude.

TNT Local Scaling Method

Moews and Kretsinger[21] proposed several methods of modeling solvent by scaling $F_m(\mathbf{s})$ to $F_o(\mathbf{s})$.

$$\text{Scale factor function} = 1 - k_{sol} \exp(-B_{sol}s^2) \tag{3}$$

The scale factor function has proved to be particularly useful. This function is not simply an empirical fit, but can be justified as an actual model of the scattering of the disordered solvent.

If we assume that every piece of density is either part of the molecule or part of the solvent, we can state that

$$\rho_c(\mathbf{r}) = \rho_m(\mathbf{r}) + \rho_{sol}(\mathbf{r}) \tag{4}$$

or, because the Fourier transform is a linear operator,

$$F_c(\mathbf{s}) = F_m(\mathbf{s}) + F_{sol}(\mathbf{s}) \tag{5}$$

[21] P. C. Moews and R. H. Kretsinger, *J. Mol. Biol.* **91,** 201 (1975).

When the contrast between $\rho_m(\mathbf{r})$ and $\rho_{sol}(\mathbf{r})$ is low, and assuming that there is little detail in either function, as at low resolution, Babinet's principle can be used to state

$$F_{sol}(\mathbf{s}) \simeq -F_m(\mathbf{s}) \tag{6}$$

To restrict this relationship to low resolution, $F_m(\mathbf{s})$ in Eq. (6) is multiplied by $\exp(-B_{sol}s^2)$. Also, because there really is a difference in contrast between the solvent and the molecule, $F_m(\mathbf{s})$ should be multiplied by $\overline{\rho_{sol}(\mathbf{r})}/\overline{\rho_m(\mathbf{r})}$, which will be renamed k_{sol}. With these modifications Eq. (6) becomes

$$F_{sol}(\mathbf{s}) \simeq -k_{sol}\exp(-B_{sol}s^2)F_m(\mathbf{s}) \tag{7}$$

and Eq. (5) becomes

$$F_c(\mathbf{s}) \simeq F_m(\mathbf{s}) - k_{sol}\exp(-B_{sol}s^2)F_m(\mathbf{s}) \tag{8}$$

or

$$F_c(\mathbf{s}) \simeq [1 - k_{sol}\exp(-B_{sol}s^2)]F_m(\mathbf{s}) \tag{9}$$

Parameter Determination. The parameters k_{sol} and B_{sol} can be determined empirically from the discrepancy between $F_o(\mathbf{s})$ and $F_m(\mathbf{s})$. The most convenient time for the calculation is when the calculated and observed data are scaled to each other. In TNT, the scaling parameters are found by minimizing

$$f = \sum \{F_o(\mathbf{s}) - k\exp(-Bs^2)[1 - k_{sol}\exp(-B_{sol}s^2)]F_m(\mathbf{s})\}^2 \tag{10}$$

This is a four-parameter scaling function. The parameter k converts the units of the calculated structure factors to the arbitrary units of the observed structure factors. B is the correction for the error in the mean temperature factor of the model. k_{sol} and B_{sol} have been discussed.

Minimizing Eq. (10) can be described as fitting a curve with the sum of two Gaussians. This type of problem is difficult because the exponential coefficients B and B_{sol} are highly correlated. The method chosen to fit these parameters is the Levenberg–Marquardt method.[22] This method combines the best of the full-matrix and steepest descent methods and usually will converge to a reasonable set of parameters within 10 cycles.

[22] W. Press, B. Flannery, S. Teukolsky, and W. Vetterling, "Numerical Recipes—the Art of Scientific Computing," pp. 523–528. Cambridge University Press, Cambridge, 1986.

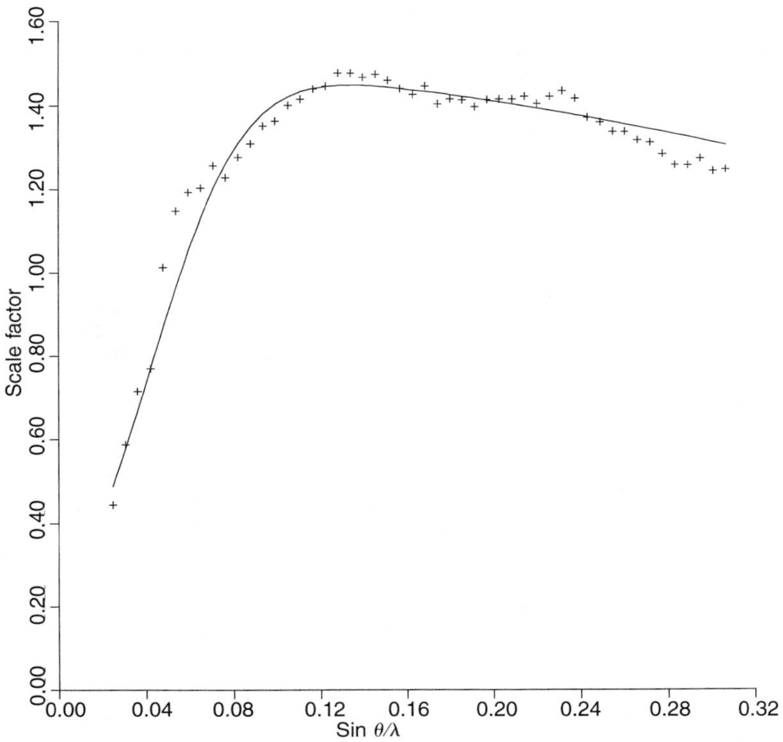

FIG. 2. Plot of K with the local scaling curve superimposed. The superimposed line is the scale factor function that results from the scaling parameters given in Table III. The data are from the thermolysin:inhibitor complex TLN:ZGPLL.

Figure 2 shows the individual shell scale factors from Fig. 1 with the solvent-corrected scale factor function superimposed. The curve fits the individual points in all major respects.

Efficacy of Solvent Model

Table III shows the solvent and scaling parameters for three structures. It shows that the addition of a solvent model can lower the R value of a refined structure by up to 2%.

Table III also shows that the solvent parameters for these three structures are remarkably similar. This similarity is consistent with the notion that "water is water." It seems reasonable that solvent will be excluded in a similar fashion from all protein molecules, and because B_{sol} describes the

TABLE III
SOLVENT PARAMETERS FOR SEVERAL PROTEIN STRUCTURES[a]

	TLN:ZGPLL (20–1.6 Å)		BCL (20–1.9 Å)		T4L (15–1.7 Å)	
Parameter	No solvent model	With solvent model	No solvent model	With solvent model	No solvent model	With solvent model
k_{sol}	—	0.805	—	0.845	—	0.889
B_{sol}	—	288	—	262	—	299
k	1.321	1.495	0.502	0.624	3.532	4.136
B	−1.379	1.375	−3.651	2.353	−3.385	0.550
R	18.8%	17.5%	21.7%	19.5%	19.2%	17.5%
R_{low}	36.2%	26.8%	39.3%	29.5%	43.3%	36.3%

[a] Solvent and scaling parameters for three structures, with the resulting R values. The first column for each shows the scale factors without any attempt to model solvent. The second column for each shows the parameters determined using the TNT method. R is the R value calculated using all data in the specified range. R_{low} includes only the data from the low-resolution limit to 6 Å. The first structure is the ZGPLL inhibitor complex of thermolysin. The second structure is model 5R of the bacteriochlorophyll *a*-containing protein [D. E. Tronrud, M. F. Schmid, and B. W. Matthews, *J. Mol. Biol.* **188,** 443 (1986)]. The third structure is model T4165RF of T4 lysozyme [J. A. Bell, K. P. Wilson X.-J. Zhang, H. R. Faber, H. Nicholson, and B. W. Matthews, *Proteins* **10,** 10 (1991)].

transition from solvent to protein it should be similar from one structure to another.

However, not all mother liquors have the same electron density. k_{sol} should vary from structure to structure. The structures in Table III are listed in order of increasing solvent electron density. Thermolysin crystals are stored in water with 7% (v/v) dimethyl sulfoxide (DMSO) and little else. T4 lysozyme crystals are immersed in 2 *M* phosphate buffer. Although mass density is not strictly proportional to electron density, the electron density of T4 lysozyme mother liquor should be higher than that of thermolysin.

Because the structures were refined without the low-resolution data and without solvent models, there is a correction to the average temperature factor in both sets of scale factors for each structure. In each case, the average temperature factor increases when the solvent model is added. An additional trend can be observed. The increase for the thermolysin inhibitor (which was refined to 1.6-Å resolution) was 2.8 Å^2, whereas the increase for the bacteriochlorophyll-containing protein (refined to 1.9 Å) was 6.0 Å^2. The lower the resolution of the data set, the more important is the contribution of the lowest resolution data.

It is true that the *R* value calculated with both low-resolution data and a solvent model will be higher than the *R* value for the same coordinate set calculated with the low-resolution data eliminated. As seen in Table III, the *R* value for the thermolysin–inhibitor complex using the 20- to 1.6-Å data is 17.5%. When the inner resolution limit is set of 5.5 Å the *R* value drops to 17.1%. Some investigators may consider the lower *R* value as justification for rejecting these data. This is incorrect reasoning. In a statistical analysis, the model never can be used to assess the data. It is clear that the lower *R* value does not represent a better model—the same model is used in both cases. The lower *R* value is simply an artifact of the method used in its calculation.

Acknowledgments

This work was supported in part by NIH Grant GM20066 to B. W. Matthews. The author also acknowledges the continuing support of B. W. Matthews.

[16] SHELXL: High-Resolution Refinement

By George M. Sheldrick and Thomas R. Schneider

Introduction

SHELXL-93 was originally written as a replacement for the refinement part of the small-molecule program SHELX-76. However, in the 10 years required to develop and test the new program, computers have become at least two orders of magnitude more powerful, so that it is now entirely feasible to use it for high-resolution (2.5 Å or better) refinement of small proteins and oligonucleotides. The program is designed to be easy to use and general for all space groups, and uses a conventional structure-factor calculation rather than a fast Fourier transform (FFT) summation; the latter would be faster, but in practice involves some small approximations and is not suitable for the treatment of anomalous dispersion or anisotropic thermal motion. The price to pay for the extra precision and generality is that SHELXL is much slower than programs written specifically for macromolecules. This is compensated for, to some extent, by the better convergence properties, reducing the amount of manual intervention required (and possibly also the *R* factor). A new version, SHELXL-97, was released in May 1997; this is the version described here. The changes are primarily designed to make the program easier to use for macromolecules.

0076-6879/97 $25.00

Excellent descriptions of crystallographic least-squares refinement strategies and of widely used macromolecular refinement programs have appeared in Vol. 115B of this series and elsewhere (see Refs. 1 to 3, and [15] and [19] in this volume[4,5]).The programs that are philosophically most similar to SHELXL are PROLSQ,[1] CORELS,[2] RESTRAIN,[3] and TNT (see [15] in this volume[4]). X-PLOR (see [19] in this volume[5]) is somewhat different because it makes extensive use of energy functions (which makes it applicable to lower resolution X-ray data) and also can be used to perform molecular dynamics. Watkin[6] has given an authoritative account of small-molecule refinement, including strategies implemented in the program CRYSTALS. SHELXL incorporates many ideas that were first implemented and tested in these programs; Refs. 1–6 should be consulted for mathematical details. This account also draws extensively on material presented at the SHELX Workshops in Göttingen (1993 and 1997), Newcastle (BCA Meeting, 1994), and Montreal (ACA Meeting, 1995), and at the Third Spanish Crystallography School in Oviedo, Spain (1994).

Advances in cryogenic techniques, area detectors, and the use of synchrotron radiation enable macromolecular data to be collected to higher resolution than was previously possible. In practice this tends to complicate the refinement because it is possible to resolve finer details of the structure; it is often necessary to model alternative conformations, and in a few cases even anisotropic refinement is justified. Although SHELXL provides a number of other features not found in many macromolecular refinement programs, e.g., treatment of twinned crystals, anomalous dispersion, extinction, variable wavelength (Laue) data, estimated standard deviations (ESDs) from the full covariance matrix, and CIF output, it is probably the flexible treatment of disorder and the facilities for restrained anisotropic refinement that are most likely to be of immediate interest to macromolecular crystallographers.

Program Organization

SHELXL is written in essentially machine-independent Fortran-77 and has been installed on a wide range of computers. To run the program only two input files are required (atoms/instructions and reflection data); because

[1] W. A. Hendrickson, *Methods Enzymol.* **115B,** 252 (1985).
[2] J. L. Sussman, *Methods Enzymol.* **115B,** 271 (1985).
[3] H. Driessen, M. I. J. Haneef, G. W. Harris, B. Howlin, G. Khan, and D. S. Moss, *J. Appl. Crystallogr.* **22,** 510 (1989).
[4] D. Tronrud, *Methods Enzymol.* **277,** [15], 1997 (this volume).
[5] A. Brünger, *Methods Enzymol.* **277,** [19], 1997 (this volume).
[6] D. J. Watkin, *Acta Crystallogr.* **A50,** 411 (1994).

both these files and the output files are pure ASCII text files it is easy to use the program on a heterogeneous network. The reflection data file (*name.hkl*) contains h, k, l, F^2, and $\sigma(F^2)$ in standard SHELX format: the program merges equivalents and eliminates systematic absences; the order of the reflections in this file is unimportant. Crystal data, refinement instructions, and atom coordinates are all stored in the file *name.ins;* further files may be specified as "include files" in the *.ins* file, e.g., for standard restraints, but this is not essential. Instructions appear in the *.ins* file as four-letter keywords followed by atom names, numbers, etc., in free format; examples are given in the following text. There are sensible default values for almost all numerical parameters. The more important keywords for macromolecular refinement are summarized in Table I. After each refinement cycle the results are output to a file *name.res*, which has the same format as the *.ins* file and may be edited and reinput for the next refinement job. The output files also include a listing file *name.lst* (which may be printed) and optionally a Protein Data Base (PDB) format atom coordinate file *name.pdb* and a CIF format file of phased structure factors (*name.fcf*).

An auxiliary interactive program SHELXPRO is provided as an interface between SHELXL and widely used macromolecular programs. SHELXPRO is able to reformat intensity data files and to generate an *.ins* file from a PDB format file, including the appropriate restraints. Usually a little editing by hand of the resulting *name.ins* file will still be required. SHELXPRO summarizes information from the refinement in the form of PostScript plots, prepares maps for graphic display programs, and generally serves as a user interface to prepare the *name.ins* file for the next SHELXL refinement job.

SHELXL always refines against F^2, even when F values are input. Experience with small molecules[6–8] indicates that refinement against all F^2 values is superior to refinement against F values greater than some threshold [say, $4\sigma(F)$]. More experimental information is incorporated (suitably weighted) and the weak reflections are particularly informative in pseudosymmetry cases (not unusual for macromolecules when noncrystallographic symmetry is present). It is difficult to refine against all F values because of the difficulty of estimating $\sigma(F)$ from $\sigma(F^2)$ when F^2 is zero or (as a result of experimental error) negative. Macromolecular data frequently include many weak reflections and the data-to-parameter ratio is almost always poorer than for small molecules, so it is particularly important to extract as much information as possible from the weak data in this way.

[7] F. L. Hirshfeld and D. Rabinovich, *Acta Crystallogr.* **A29,** 510 (1973).
[8] L. Arnberg, S. Hovmöller, and S. Westman, *Acta Crystallogr.* **A35,** 497 (1979).

TABLE I
SHELXL-97 KEYWORDS USEFUL FOR MACROMOLECULES

Keyword	Definition
DEFS	Set global restraint ESD defaults
DFIX	Restrain 1,2 distance to target (which may be a free variable)
DANG	Restrain 1,3 distance to target (which may be a free variable)
SADI	Restrain distances to be equal without specifying target
SAME	Generate SADI automatically for 1,2 and 1,3 distances using connectivity
CHIV	Restrain chiral volume to target (default zero; may be a free variable)
FLAT	Planarity restraint
DELU	Generate rigid bond U_{ij} restraints automatically using connectivity
SIMU	Generate similar U (or U_{ij}) restraints automatically using distances
ISOR	"Approximately isotropic" restraints
BUMP	Generate antibumping restraints automatically (including symmetry equivalents)
NCSY	Generate noncrystallographic symmetry restraints
FVAR	Starting values for overall scale factor and free variables
SUMP	Restrain linear combination of free variables
PART	Atoms with different nonzero PART numbers not connected by program
AFIX	Riding H, rigid groups, and other constraints on individual atoms
HFIX	Generate hydrogens and suitable AFIX instructions for their refinement
MERG	"MERG 4" averages equivalent reflections including Friedel opposites, sets all $\delta f''$ to 0
SHEL	Set maximum and minimum resolution (data ignored outside range)
STIR	Stepwise improvement of resolution during refinement
SWAT	Refine diffuse solvent parameter (Babinet's principle)
WGHT	Weighting scheme, probably best left at default "WGHT 0.1" throughout
CGLS	Number of cycles conjugate gradient least squares, select R_{free} reflections
BLOC, L.S.	Blocked-matrix least squares (for ESDs)
RTAB, MPLA, HTAB	Tables of bonds, angles, torsions, planes, H bonds, etc.
WPDB, ACTA, LIST	Output PDB and CIF files for archiving and data transfer

R indices based on F^2 are larger than (often more than double) those based on F. For comparison with older refinements based on F and a $\sigma(F)$ threshold, a conventional reliability index $R_1 = \Sigma|F_o - F_c|/\Sigma F_o$ for reflections with $F > 4\sigma(F)$ is also printed; it has the advantage that it is relatively insensitive to the weighting scheme used for the F^2 refinement (and so is difficult to manipulate).

Residues and Connectivity List

Macromolecular structures are conventionally divided into *residues,* for example, individual amino acids. In SHELXL residues may be referenced either individually, by "_" followed by the appropriate residue number, or as all residues of a particular class, by "_" followed by the class. For example, "DFIX 2.031 SG_9 SG_31" could be used to restrain a disulfide distance between two cystine residues, whereas "FLAT_PHE CB > CZ" would apply planarity restraints to all atoms between CB (C_β) and CZ (C_ζ) inclusive in all PHE (phenylalanine) residues. Plus and minus signs refer to the next and previous residue numbers respectively; for example, "DFIX_* 1.329 C_ − N" applies a bond length restraint to all peptide bonds [an asterisk (*) after the command name applies it to all residues]. This way of referring to atoms and residues is in no way restricted to proteins; it is equally suitable for oligonucleotides, polysaccharides, or structures containing a mixture of all three. It enables the necessary restraints and other instructions to be input in a concise and relatively self-explanatory manner. These instructions are checked by the program for consistency and appropriate warnings are printed.

A connectivity list is generated automatically by the program, and may be edited if necessary. It is used for setting up standard tables, defining idealized hydrogen atoms, and for the automatic generation of some restraints (SAME, DELU, CHIV). Different components of disordered groups should be given different PART numbers; bonds are never generated automatically between atoms with different PART numbers, unless one of them is zero (i.e., belongs to the undisordered part of the structure).

Constraints and Restraints

A *constraint* is an exact mathematical condition that leads to the elimination of one least-squares parameter because it can be expressed exactly in terms of other parameters. An example is the assignment of occupation factors of p and $1 - p$ to the two components of a disordered group; only one parameter p is refined, not two. In SHELXL p would be refined as a *free variable,* facilitating the refinement of common occupancies for the atoms of a disordered group. The free variable concept has been retained unchanged from SHELX-76, where it proved extremely useful; any atomic coordinate, occupancy, or displacement parameter, or any target distance or chiral volume, may be defined as a linear function of one of these least-squares variables. For example, a common isotropic displacement parameter ($U = 8\pi^2 B$) may be refined for a group of atoms by giving their U values symbolically as 21, which is interpreted as "1 times free variable number 2." The occupancies of two disorder components could be given as 31 [1 times fv(3)] and −31 {1 times [1 − fv(3)]} so that they add up to 1. In general, when the number given for atom coordinate, occupancy, or displacement parameter is greater than +15, it is interpreted as $10m + k$, and the parameter takes the value $k \cdot \mathrm{fv}(m)$, where fv(m) is the mth free variable. When the number is less than −15, it is interpreted as $-(10m + k)$, and the parameter takes the value $k[1 - \mathrm{fv}(m)]$. The starting values of these free variables are given on the FVAR instruction. The coordinates and anisotropic displacement parameters of atoms on special positions are handled invariably by means of constraints; in SHELX-76 the user had to input these constraints using free variables, but in SHELXL they are handled internally by the program for all possible special positions in all space groups. Further examples of constraints are riding hydrogen atoms and rigid groups. In addition, hydrogen atoms may be assigned U values equal to a specified constant times the (equivalent) isotropic U value of the atom to which they are attached.

A *restraint* is an additional condition that is not exact but is associated with a specified ESD (σ); restraints are applied in the form of additional "observational equations" and each contributes a quantity $w_r\Delta_r^2$ to the function to be minimized in the same way that reflection data do (where w_r is a weight proportional to $1/\sigma^2$ and Δ_r is the deviation of a function of the parameters from its target value). An example of a restraint is the condition that two chemically equivalent bond lengths should be equal within a specified ESD. If there are more than two components of a disorder, it is more convenient to restrain their sum of occupancies to 1 (using SUMP to restrain a linear sum of free variables) than to constrain it (as discussed above for the case of two components).

Rigid-group constraints enable a structure to be refined with few parameters, especially when the (thermal) displacement parameters are held fixed (BLOC 1). After a structure has been solved by molecular replacement using a rather approximate model for the whole protein or oligonucleotide, it may well be advisable to divide the structure into relatively rigid domains (using a few AFIX 6 and possibly AFIX 0 instructions) and to refine these as rigid groups, initially for a limited resolution shell (e.g., SHEL 8 3), then extending the resolution stepwise. Restraints still may be required to define flexible hinges and prevent the units from flying apart. In contrast to SHELX-76, SHELXL allows atoms in rigid groups to be involved in restraints as well.

The relative weighting of X-ray data and restraints is a difficult topic. In SHELXL an attempt is made to put both on an absolute scale. $1/\sigma^2$ (where σ is the standard deviation estimated by the user) is used directly as the weight for each restraint. The conventional X-ray weights $w_x = 1/$

$[\sigma^2(F^2) + gP^2]$ (where g is typically in the range 0.1 to 0.2 and $P = (F_o^2 + 2F_c^2)/3$; SHELXL also offers a choice of more complicated weighting schemes) are normalized by dividing them by the square root of the mean value of $w_x\Delta_x^2$, where $\Delta_x = F_o^2 - F_c^2$. This means that in the initial stages of refinement, when the model is inaccurate and the Δ_x values are relatively high, the model restraints are heavily weighted relative to the X-ray data. As the agreement with the diffraction data improves, the diffraction data automatically play a larger role. It is possible to use the Brünger R_{free} test (see below) to fine-tune the restraint ESDs. In practice, the optimal restraint ESDs vary little with the quality and resolution of the data, and the standard values (assumed by the program if no other value is specified) are entirely adequate for routine refinements. The program will suggest an X-ray weighting scheme designed to produce a flat analysis of variance, but in the case of macromolecules it is probably best to leave g at the standard value of 0.1 and to accept a "goodness of fit" greater than unity as a measure of the inadequacy of the model.

Geometric Restraints

SHELXL provides distance, planarity, and chiral volume restraints, but not torsion-angle restraints or specific hydrogen-bond restraints. Use of torsion-angle or H-bonding restraints probably would reduce the convergence radius substantially, and the fact that these quantities are not restrained makes them useful for checking the quality of a refined structure, for example, with the program PROCHECK.[9] For oligonucleotides, good distance restaints are available for the bases,[10] but for sugars and phosphates it is probably better to assume that chemically equivalent 1,2 and 1,3 distances are equal (using the SAME and SADI restraints) without the need to specify target values. In this way the effect of the pH on the protonation state of the phosphates and hence the P–O distances does not need to be predicted, but it is assumed the whole crystal is at the same pH. For proteins, because some amino acid residues occur only a small number of times in a given protein, it is probably better to use 1,2 and 1,3 target distances based on the study of Engh and Huber[11] (these are inserted into the *.ins* file by SHELXPRO).

Chiral volume restraints[1] are useful to prevent the inversion of α-carbon atoms and the β carbons of isoleucine and threonine, e.g., "CHIV_ILE 2.5

[9] R. A. Laskowski, M. W. MacArthur, D. S. Moss, and J. M. Thornton, *J. Appl. Crystallogr.* **26,** 283 (1993).

[10] R. Taylor and O. Kennard, *J. Mol. Struct.* **78,** 1 (1982).

[11] R. A. Engh and R. Huber, *Acta Crystallogr.* **A47,** 392 (1991).

CA CB." Although valine C_β and leucine C_γ atoms are not chiral, chiral volume restraints may be used to enforce the conventional numbering of the substituents on these atoms. As pointed out by Urzhumtsev,[12] conventional planarity restraints often either restrain the atoms to lie in a fixed least-squares plane through their current positions (thereby impeding convergence), or introduce a slight bias that would move the restrained atoms toward their centroid. SHELXL applies planarity restraints by restraining the volumes of a sufficient number of atomic tetrahedra to be zero, using the same algorithm as for the chiral volume restraints; this algorithm does not fix the plane in space and is robust because all derivatives are analytical. The planar restraint ESD thus should be given in units of cubic Ångströms, although the program also prints out the more familiar root-mean-square (RMS) deviation of the atoms from the planes (in Ångströms). A convenient (and equivalent) alternative to this FLAT restraint to restrain the three bonds to a carbonyl carbon atom to lie in the same plane is a chiral volume restraint (CHIV) with a target volume of zero (e.g., "CHIV_GLU 0 C CD" to restrain the carbonyl and carboxyl carbons in all glutamate residues to have planar environments).

Antibumping restraints[1] are distance restraints that are applied only if the two atoms are closer to each other than the target distance. They can be generated automatically by SHELXL, taking all symmetry-equivalent atoms into account. Because this step is relatively time consuming, in the 1993 release it was performed only before the first refinement cycle, and the antibumping restraints were generated automatically only for the solvent (water) atoms (however, they could be inserted by hand for any pairs of atoms). In practice this proved to be too limited, so in later releases the automatic generation of antibumping restraints was extended to all carbon, nitrogen, oxygen, and sulfur atoms (with an option to include H···H interactions) and was performed each refinement cycle. Antibumping restraints are not generated automatically for (1) atoms connected by a chain of three bonds or less in the connectivity array, (2) atoms with different nonzero PART numbers (i.e., in different disorder components), and (3) pairs of atoms of which the sum of occupancies is less than 1.1. Extra antibumping restraints may be added by hand if required. The target distances for the O..O and N..O distances are less than for the other atom pairs to allow for possible hydrogen bonds.

Restrained Anisotropic Refinement

There is no doubt that macromolecules are better described in terms of anisotropic displacements, but the data-to-parameter ratio is rarely

[12] A. G. Urzhumtsev, *Acta Crystallogr.* **A47,** 723 (1991).

adequate for a free anisotropic refinement. Such a refinement often results in "nonpositive definite" (NPD) displacement tensors, and at best will give probability ellipsoids that do not conform to the expected dynamic behavior of the macromolecule. Clearly constraints or restraints must be applied to obtain a chemically sensible model. It is possible to divide a macromolecule into relatively rigid domains, and to refine the 20 independent parameters of rigid-body motion for each domain.[5] This may be a good model for the bases in oligonucleotides and for the four aromatic side chains in proteins, but otherwise macromolecules are probably not sufficiently rigid for the application of such constraints, or they would have to be divided into such small units that too many parameters would be required. As with the refinement of atomic positions, restraints offer a more flexible approach.

The *rigid-bond restraint,* suggested by Rollett[13] and justified by the work of Hirshfeld[14] and Trueblood and Dunitz,[15] assumes that the components of the anisotropic displacement parameters (ADPs) along bonded 1,2 or 1,3 directions are zero within a given ESD. This restraint (DELU) should be applied with a low ESD, i.e., as a "hard" restraint. Didisheim and Schwarzenbach[16] showed that for many nonplanar groups of atoms, rigid-bond restraints effectively impose rigid-body motion. Although rigid-bond restraints involving 1,2 and 1,3 distances reduce the effective number of free ADPs per atom from six to fewer than four for typical organic structures, further restraints often are required for the successful anisotropic refinement of macromolecules.

The *similar ADP restraint* (SIMU) restrains the corresponding U_{ij} components to be approximately equal for atoms that are spatially close (but not necessarily bonded because they may be in different components of a disordered group). The isotropic version of this restraint has been employed frequently in protein refinements.[1] This restraint is consistent with the characteristic patterns of thermal ellipsoids in many organic molecules; on moving out along side chains, the ellipsoids become more extended and also change direction gradually.

Neither of these restraints is suitable for isolated solvent (water) molecules. A linear restraint (ISOR) restrains the ADPs to be approximately isotropic, but without specifying the magnitude of the corresponding equivalent isotropic displacement parameter. Both SIMU and ISOR restraints

[13] J. S. Rollett, *in* "Crystallographic Computing" (F. R. Ahmed, S. R. Hall, and C. P. Huber, eds.), p. 167. Munksgaard, Copenhagen, 1970.

[14] F. L. Hirshfeld, *Acta Crystallogr.* **A32,** 239 (1976).

[15] K. N. Trueblood and J. D. Dunitz, *Acta Crystallogr.* **B39,** 120 (1983).

[16] J. J. Didisheim and D. Schwarzenbach, *Acta Crystallogr.* **A43,** 226 (1987).

are clearly only approximations to the truth, and so should be applied as "soft" restraints with high ESDs. When all three restraints are applied, structures may be refined anisotropically with a much smaller data-to-parameter ratio, and still produce chemically sensible ADPs. Even when more data are available, these restraints are invaluable for handling disordered regions of the structure.

Geometric and ADP constraints and restraints greatly increase the radius and rate of convergence of crystallographic refinements, so they should be employed in the early stages of refinement whenever feasible. The difference electron-density syntheses calculated after such restrained refinements are often more revealing than those from free refinements. In larger small-molecule structures with poor data-to-parameter ratios, the last few atoms often cannot be located in a difference map until an anisotropic refinement has been performed with geometric and ADP restraints. Atoms with low displacement parameters that are well determined by the X-ray data will be affected relatively little by the restraints, but the latter may well be essential for the successful refinement of poorly defined regions of the structure.

The question of whether the restraints can be removed in the final refinement, or what the best values are for the corresponding ESDs, can be resolved elegantly by the use of R_{free} (Brünger[17]). To apply this test, the data are divided into a working set (about 95% of the reflections) and a reference set (about 5%). The reference set is used only for the purpose of calculating a conventional R factor, which is called R_{free}. It is important that the structural model not be based in any way on the reference set of reflections, so these are left out of all refinement and Fourier map calculations. If the original model was derived in any way from the same data, then many refinement cycles are required to eliminate memory effects. This ensures that the R factor for the reference set provides an objective guide as to whether the introduction of additional parameters or the weakening of restraints has actually improved the model, and not just reduced the R factor for the data employed in the refinement ("R-factor cosmetics").

R_{free} is invaluable in deciding whether a restrained anisotropic refinement is significantly better than an isotropic refinement. Experience indicates that both the resolution and the quality of the data are important factors, but that restrained anisotropic refinement is unlikely to be justified for crystals that do not diffract to better than 1.5 Å. An ensemble distribution created by molecular dynamics is an alternative to the harmonic de-

[17] A. Brünger, *Nature* (*London*) **355,** 472 (1992).

scription of anisotropic motion,[18,19] and may be more appropriate for structures with severe conformational disorder that do not diffract to high resolution.

Despite the excellent arguments for using R_{free} to monitor all macromolecular refinements, it is only a single number, and is itself subject to statistical uncertainty because it is based on a limited number of reflections. Thus R_{free} may be insensitive to small structural changes, and small differences in R_{free} should not be taken as the last word; one always should consider whether the resulting geometric and displacement parameters are chemically reasonable. The final refinement and maps always should be calculated with the full data, but without introducing additional parameters or changing the weights of the restraints.

Disorder

To obtain a chemically sensible refinement of a disordered group, one will probably need to constrain or restrain a sum of occupation factors to be unity, to restrain equivalent interatomic distances to be equal to each other or to standard values (or alternatively to apply rigid-group constraints), and to restrain the displacement parameters of overlapping atoms. In the case of a tight unimodal distribution of conformations, restrained anisotropic refinement may provide as good a description as a detailed manual interpretation of the disorder in terms of two or more components, and is simpler to perform. With high-resolution data it is advisable to make the atoms anisotropic before attempting to interpret borderline cases of side-chain disorder; it may well be found that no further interpretation is needed, and in any case the improved phases from the anisotropic refinement will enable higher quality difference maps to be examined. Typical warning signs for disorder are large (and pronounced anisotropic) apparent thermal motion (in such cases the program may suggest that an atom should be split and estimate the coordinates for the two new atoms) and residual features in the difference electron density. This information in summarized by the program on a residue-by-residue basis, separately for main-chain, side-chain, and solvent atoms.

Two or more different conformations should be assigned different PART numbers so that the connectivity array is set up correctly by the program; this enables the correct rigid-bond restraints on the anisotropic displacement parameters and idealized hydrogen atoms to be generated

[18] P. Gros, W. F. van Gunsteren, and W. G. Hol, *Science* **249,** 1149 (1990).

[19] J. B. Clarage and G.N. Phillips, *Acta Crystallogr.* **D50,** 24 (1994).

automatically even for disordered regions (it is advisable to model the disorder before adding the hydrogens).

Several different strategies may be used for modeling disorder with SHELXL, but for macromolecules the simplest is to use the same atom names for each component of the disorder, and to include all components in the same residue. Thus the corresponding atoms in different disorder components have the same names, residue classes, and residue numbers; they are distinguished only by their different PART numbers. This procedure enables the standard restraints dictionary to be used unchanged; the program uses the rules for interpreting the PART instructions to work out how the restraints, etc., are to be applied to the different disorder components. The Appendix illustrates this way of handling disorder, which was first implemented in the version SHELXL-96. With it, no special action is needed to add the disordered hydrogen atoms, provided that the disorder is traced back one atom further than it is visible, so that the hydrogen atoms on the PART 0 atoms bonded to the disordered components are also correct. In any case tracing a disorder one atom further back than it is visible improves the compatibility of the structure with the restraints.

Automatic Water Divining

It is relatively common practice in the refinement of macromolecular structures to insert water molecules with partial occupancies at the positions of difference electron-density map peaks to reduce the *R* factor (an example of "*R*-factor cosmetics"). Usually when two different determinations of the same protein structure are compared, only the most tightly bound waters, which usually have full occupancies and smaller displacement parameters, are the same in each structure. The refinement of partial occupancy factors for the solvent atoms (in addition to their displacement parameters) is rarely justified by R_{free}, but sometimes the best R_{free} value is obtained for a model involving water occupancies fixed at either 1.0 or 0.5.

Regions of diffuse solvent may be modeled using Babinet's principle.[3,20] This is implemented as the SWAT instruction and usually produces a significant but not dramatic improvement in the agreement of the low-angle data. Antibumping restraints may be input by hand or generated automatically by the program, taking symmetry equivalents into account. After each refinement job, the displacement parameters of the water molecules should be examined, and waters with high values (say, *U* greater than

[20] P. C. Moews and R. H. Kretsinger, *J. Mol. Biol.* **91,** 201 (1975).

0.8 Å^2, corresponding to a B value of 63) eliminated. The $F_o - F_c$ map is then analyzed automatically to find the highest peaks that involve no bad contacts and make at least one geometrically plausible hydrogen bond to an electronegative atom. These peaks then are included with full occupancies and oxygen scattering factors in the next refinement job. This procedure is repeated several times; in general R_{free} rapidly reaches its minimum value, although the conventional R index continues to fall as more waters are added. It should be noted that the automatic generation of antibumping restraints is less effective when the water occupancies are allowed to have values other than 1.0 or 0.5. This approach may be automated and provides an efficient way of building up a chemically reasonable (but not necessarily unique) network of waters that are prevented from diffusing into the protein, thus facilitating remodeling of disordered side chains, etc. The hydrogen bond requirement may lead occasionally to waters in hydrophobic pockets being ignored by the automatic water dividing, so the resulting structure should always be checked using interactive computer graphics. The occupancies of specific waters also may be tied (using free variables) to the occupancies of particular components of disordered side chains, where this makes chemical sense. A similar but more sophisticated approach (ARP: automated refinement procedure) described by Lamzin and Wilson (see [14] in this volume[21]) may also be used in conjunction with SHELXL.

Least-Squares Refinement and Estimated Standard Deviations

The conjugate gradient algorithm used in SHELXL for the solution of the least-squares normal equations is based closely on the procedure described by Hendrickson and Konnert.[22] The structure-factor derivatives contribute only to the diagonal elements of the least-squares matrix, but all restraints contribute fully to both diagonal and off-diagonal terms, although neither the Jacobian nor the least-squares matrix itself are ever generated or stored. The preconditioning recommended by Hendrickson and Konnert is used to speed up the convergence of the internal conjugate gradient iterations; it has the additional advantage of preventing the excessive damping of poorly determined parameters characteristic of other conjugate gradient algorithms.[23] If a rigid-group refinement with relatively few parameters is performed at the beginning of a structure refinement, it may be better

[21] V. S. Lamzin and K. S. Wilson, *Methods Enzymol.* **277,** [14], 1997 (this volume).

[22] W. A. Hendrickson and J. H. Konnert, *in* "Computing in Crystallography" (R. Diamond, S. Ramaseshan, and K. Venkatesan, eds.), p. 13.01. Indian Academy of Sciences, Bangalore, India, 1980.

[23] D. E. Tronrud, *Acta Crystallogr.* **A48,** 912 (1992).

to use the full-matrix rather than the conjugate gradient algorithm, because the rigid groups can give rise to large off-diagonal matrix elements.

A disadvantage of the conjugate gradient algorithm is its inability to estimate standard deviations. In the final stages of refinement, large full-matrix blocks may be employed instead; SHELXL allows these to be defined in a flexible manner. Thus because torsion angles are not restrained in the refinement, their ESDs may be estimated in this way. These ESDs take into account the information presented in the form of restraints, and in practice they are strongly correlated with the mean dispalcement parameters of the atoms involved.

A classic example in which ESDs are required is shown in Fig. 1. Because the heme group is not quite planar and the assumption of standard distances might introduce bias, it has been refined with chemically equivalent 1,2 and 1,3 distances restrained to be equal (SADI) and with the chiral volumes of the *sp*^{2}-hybridized carbon atoms restrained to be zero (CHIV). No overall planarity restraint was applied to the heme unit. Local fourfold

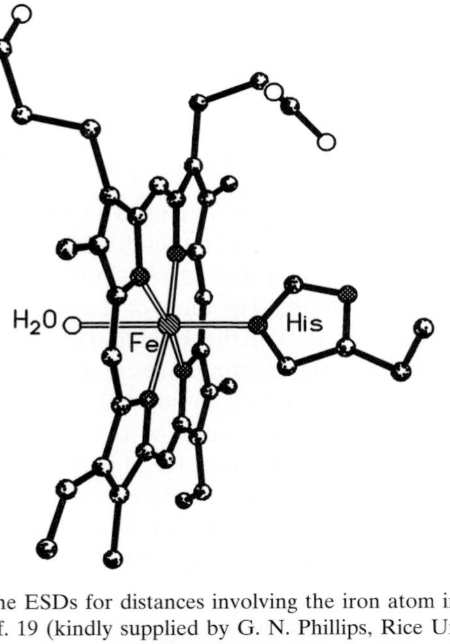

FIG. 1. Determining the ESDs for distances involving the iron atom in myoglobin, using the 1.75-Å data from Ref. 19 (kindly supplied by G. N. Phillips, Rice University, Houston, TX). The isotropic refinement was restrained as explained in text. The distance of the iron atom from the mean plane through the heme atoms (excluding the iron atom and side chains) was 0.245(11) Å (the iron atom lies on the histidine side of the plane), the Fe–O distance to the water molecule was 2.34(4) Å, and the Fe–N distance to the histidine was 2.11(3) Å.

symmetry was imposed by restraining the four iron–nitrogen distances to be equal (SADI); no other restraints were applied to the iron atom. After conjugate gradient refinement of all parameters had converged, a final cycle was performed with no damping and a shift multiplier of zero (DAMP 0 0) in which all the atoms of the heme group and the surrounding residues were included in one large full-matrix block. All the relevant correlation matrix elements were employed in determining the ESDs in the deviation of the iron atom from the mean heme plane and in the distances shown. These ESDs will, of course, be influenced by the restraints applied; provided that the ESDs input with the restraints are realistic estimates of their uncertainties, the ESDs of other parameters estimated from the restrained refinement also should be meaningful.

Refinement of Structures at Medium Resolution

Although the unique features of SHELXL are primarily useful for refinement against high-resolution data, tests indicated that the original release would require only small changes to extend its applicablity to medium-resolution data (say, to 2.5 Å). The most important of these changes were improved diagnostics and more sophisticated antibumping restraints (see above), and the addition of noncrystallographic symmetry (NCS) restraints. The use of NCS restraints considerably improves the effective data-to-parameter ratio, and the resulting Fourier maps often look as though they were calculated with higher resolution data than were actually used (because the phases are more accurate). Two types of NCS restraint may be generated automatically with the help of the NCSY instruction. The first type uses the connectivity table to define equivalent 1,4 distances, which are then restrained to be equal. The second restrains the isotropic U values of equivalent atoms to be equal. It is not normally necessary to restrain equivalent 1,2 and 1,3 distances to be equal because the DFIX and DANG restraints will have this effect anyway; but SAME may be used to add such restraints in the absence of DFIX and DANG. The use of restraints rather than applying NCS as an exact constraint (e.g., in the structure-factor calculation) is more flexible (but computationally slower) and does not require the specification of transformation matrices and real-space masks. Experience indicates that NCS restraints should be used wherever possible; it is not difficult to relax them later (e.g., for specific side chains involved in interactions with other non-NCS-related molecules) should this prove to be necessary. SHELXPRO includes facilities for analyzing NCS.

Radius of Convergence

A crucial aspect of any refinement program is the radius of convergence. A larger radius of convergence reduces the amount of time-consuming manual intervention using interactive graphics. Many cases of SHELXL giving *R* factors 1 or 2% lower than other programs have been tracked down either to subtle differences in the model or to not becoming trapped in local minima. The differences in the model include the treatment of solvent and hydrogen atoms, and the ability to refine common occupancies for disordered groups. This inclusion of dispersion terms and the use of a conventional rather than an FFT structure-factor summation is also more precise; the approximations in the FTT summation may become significant for high-resolution data and atoms with small displacement parameters. There are probably a number of factors contributing to the convergence typically observed with SHELXL, e.g., the inclusion of important off-diagonal terms in the least-squares algebra, the ability to refine all parameters at once (i.e., coordinates and displacement parameters in the same cycle), and the restriction to unimodal restraint functions; multimodal restraint functions such as torsion angles or hydrogen bonds tend to increase the number of spurious local minima. The errors in the FFT calculation of derivatives are larger than those in the structure factors (for the same grid intervals); this also would impede convergence.

In SHELXL-97, the STIR instruction (stepwise improvement of resolution) further increases the radius of convergence, and may be regarded as a primitive form of stimulated annealing. Typically data from infinity to 3.5 Å are used in the first cycle, and the resolution range is extended by 0.01 Å per cycle until the full resolution range of the experimental data has been reached. Large parameter shifts can occur in the early cycles, and the shifts are damped gradually (slow cooling) as more data are added, assisting the search for the global rather than a local minimum. At lower resolution there will be fewer local minima but the global minimum is broader. This technique is computationally efficient (relative to molecular dynamics) and is easy for the user to control; the only parameters needed are the initial high-resolution limit and the step size.

Typical SHELXL Refinement Using High-Resolution Data

An example of a typical SHELXL refinement against high-resolution data, for an inhibited form of serine protease, is summarized in Table II. Data were collected at 120 K on a synchrotron to 0.96 Å resolution with an overall mean I/σ of 15.2 and an R_{merge} (based on intensities) of 3.7%. Molecular dynamics refinement using X-PLOR (see [19] in this volume[5]) from initial *R* values of 42.5% produced the results shown as job 1. A reference set consisting of 10% of the reflections and a working set of the remaining 90% were held in separate files and used throughout the X-PLOR and SHELXL-93 refinement (in SHELXL-97 a single file could be

TABLE II
SHELXL REFINEMENT OF SERINE PROTEASE[a]

Job	Action taken	NP	NH	NW/$NW_{1/2}$/NX	N_{par}	R_1	R_{free}
1	Final X-PLOR, 1.1–8Å	1,337	0	176/0/19	6,129	19.47	21.14
2	Same atoms, SHELXL	1,337	0	176/0/19	6,129	17.15	18.96
3	SWAT added	1,337	0	176/0/19	6,130	17.07	18.95
4	All atoms anisotropic	1,337	0	176/0/19	13,790	12.96	16.10
5	Disorder, added solvent	1,376	0	207/0/34	14,565	11.46	14.20
6	More disorder and solvent	1,422	0	214/2/39	14,831	11.35	14.22
7	Disorder, half-occupied waters	1,447	0	213/20/37	15,478	11.13	14.10
8	Resolution: 0.96Å–∞	1,447	0	218/28/37	15,595	10.75	12.95
9	Riding hydrogens added	1,451	1,088	220/38/40	15,769	9.58	11.56
10	Minor adjustments	1,477	1,052	222/48/40	16,114	9.15	11.19
11	Minor adjustments	1,491	1,042	211/64/48	16,173	9.29	11.31
12	Weighting changed	1,491	1,029	222/84/38	16,357	8.74	10.85
13	Further refinement	1,499	1,025	212/96/38	16,353	8.76	10.79

[a] For a total of 188 residues. NP, Number of protein atoms (including partially occupied atoms); NH, number of hydrogens (all fully occupied); NW, number of fully occupied waters; $NW_{1/2}$, number of half-occupied waters; NX, number of other atoms (inhibitor, formate, gycerol, some of them partially occupied); and N_{par}, number of least-squares parameters. Further details are given in text.

used, with the reference set reflections flagged). The final X-PLOR and initial SHELXL refinements were performed with the resolution range restricted to 1.1 to 8 Å (48,495 working set reflections) to save computer time. Ten conjugate gradient cycles were performed in each of the SHELXL refinement jobs; where new atoms were introduced they were always refined isotropically for 2 cycles before making them anisotropic. The CPU times on a 150-MHz Silicon Graphics R4400 processor varied from 6.1 hr for job 2 to 21.7 hr for job 13; a 200-MHz Pentium Pro would be about twice as fast, and SHELXL-97 is also a little faster than the SHELXL-93 used in these tests. The weighting scheme was fixed at "WGHT 0.2" until jobs 12 and 13, where the two-parameter scheme with values suggested by the program was employed; an alternative strategy that we now prefer is to use fixed weights (WGHT 0.1) throughout. The restraints (DEFS 0.015 0.2 0.01 0.025) were made tighter than usual to make the refinements more comparable with X-PLOR; the mean distance deviation was 0.009 Å for X-PLOR and 0.014 Å for the final SHELXL job.

Introduction of the diffuse solvent parameter in job 3 (which started from the same parameters as job 2) was not significant, although usually it reduces R_{free} by about 0.5%; probably this was a consequence of leaving out the low-angle data at this stage. Making all atoms anisotropic resulted in almost a 3% drop in R_{free}, but from experience with similar structures we believe that the drop would have been larger if all the data had been used at this stage. This helps to explain the further drop in R_{free} on using all of the reflection data (job 8), and the fact that the difference between R_1 and R_{free} was about 3% for jobs 4 to 7 and about 2% for the remaining jobs. Particularly noteworthy is the drop in the R factors on introducing hydrogens (no extra parameters); a parallel job using exactly the same model but excluding hydrogens showed that 1.25% of the drop in R_{free} was contributed by the hydrogens. However, the drop in job 12 is caused almost entirely by the improvements to the model; the same job with the original weights gave an R_{free} of 10.90%. After using R_{free} to monitor the refinement as discussed here, a final refinement was performed against all 80,102 unique reflections without any further changes to the model; this converged to R_1 8.77%, essentially identical to the final R_1 for the working set. Figure 2 illustrates the anisotropic temperature factors from this refinement, and Fig. 3 shows some details of the final electron-density maps.

Appendix: Example of Instruction (*.ins*) File for SHELXL

The following extracts from the file *6rxn.ins* illustrate some of the points discussed above. This file is available, together with the file *6rxn.hkl* (reflection data), as a test job for SHELXL. This structure was determined by Stenkamp *et al.*,[24] who have kindly given permission for it to be used in this way. As usual in *.ins* files, comments may be included as REM (remark) instructions or after exclamation marks. The resolution of 1.5 Å does not quite justify refinement of all nonhydrogen atoms anisotropically ("ANIS" before the first atom would specify this), but the iron and sulfur atoms should be made anisotropic (ANIS_* FE SD SG).

```
TITL Rubredoxin in P1 (from 6RXN in the PDB)
CELL 1.54178 24.92 17.79 19.72 101.0 83.4 104.5 ! Lambda, cell
ZERR  1  0.04  0.03  0.03   0.1  0.1   0.1   ! Z and cell esds
LATT -1                     ! Space group P1
SFAC  C  H  N  O  S  FE     ! Scattering factor types and
UNIT  224 498 55 136 6 1    ! unit-cell contents

DEFS 0.02 0.2 0.01 0.04     ! Global default restraint esds

CGLS 20 -12  ! 20 Conjugate gradient cycles, every 12th refl.
SHEL 10 0.1  ! used for R(free), data with d > 10 A excluded
```

[24] R. E. Stenkamp, L. C. Sieker, and L. H. Jensen, *Proteins Struct. Funct. Genet.* **8,** 352 (1990).

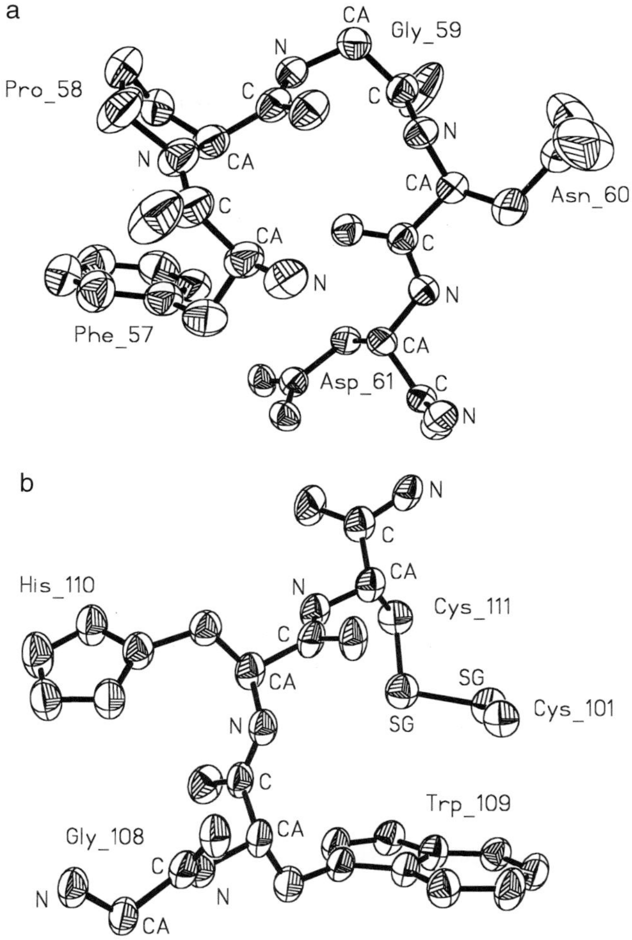

FIG. 2. ORTEP-style diagrams showing 50% probability ellipsoids for regions of the serine protease structure after anisotropic refinement with SHELXL. (a) The side chain of Asn-60 is exposed to the solvent and exhibits pronounced anisotropic motion. In contrast, both carboxylate oxygens of Asp-61 are tied down by well-defined intramolecular hydrogen bonds; the motion of this side chain is almost isotropic. The phenyl ring and most of the atoms in the peptide chain also show relatively small-amplitude isotropic motion, but there is evidence for puckering motion of the proline ring and the carbonyl oxygens of Phe-57 and Gly-59 show pronounced motion at right angles to the carbonyl planes. (b) Whereas the sulfur atoms of the disulfide bridge Cys-111–Cys-101 are relatively isotropic, the two aromatic residues Trp-109 and His-110 exhibit larger displacements perpendicular to their planes, consistent with rotations about their C_β–C_γ (CB–CG) bonds. Movement in the aromatic plane is more common, and is shown by other residues in this structure. The carbonyl oxygen of His-110 exhibits a larger displacement perpendicular to the plane of the carbonyl group.

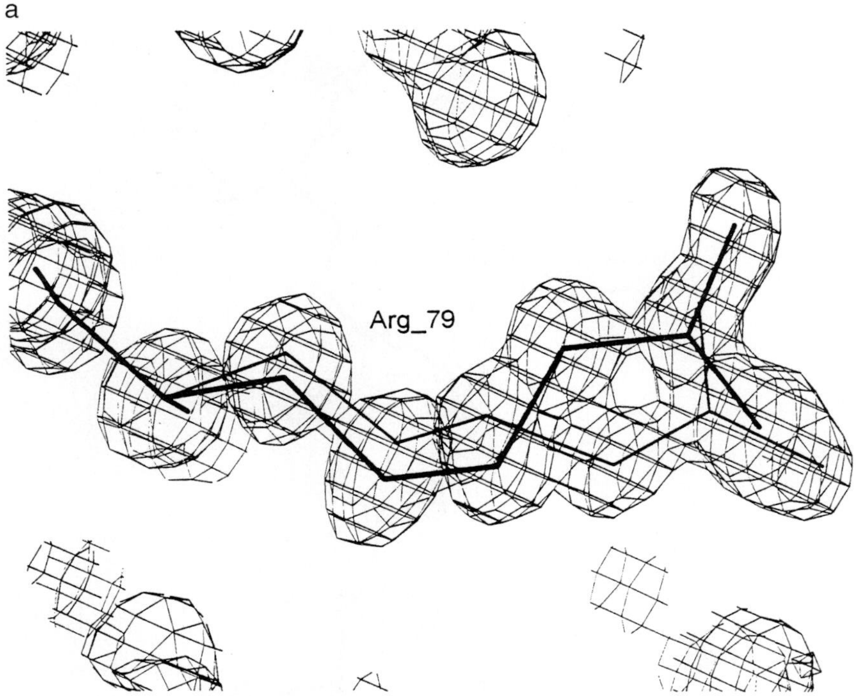

FIG. 3. $3F_o - 2F_c$ maps for the serine protease after SHELXL refinement, contoured at the 1σ level. (a) Isotropic refinement of this arginine side chain as a single conformation led to high U values. The improved quality of the difference maps after anisotropic refinement enabled two discrete conformations to be modeled. The occupancies were set to p (heavy lines) and $1 - p$ (light); the free variable p refined to 0.65. Note that the side chain has been split at C_β (CB); this gives fewer restraint violations than splitting at C_γ (CG), although the density around C_β appears to be spherical. (b) Coupled disorder of Tyr-147 and Thr-170, refined with a single free variable p. The occupancies of the components drawn with heavy lines were tied to p, the others to $1 - p$; p refined to 0.74. For the tyrosine alone a single conformation with anisotropic displacement parameters might have proved to be an adequate model, although this is more appropriate when the occupancies are equal. The major conformation is stabilized by a hydrogen bond (OG1_170..OH_147 2.67 Å, dashed), which is absent in the minor conformation. (c) Peaks A ($-2.3\ \sigma$) and B ($+2.3\ \sigma$) of the $2F_o - 2F_c$ map (the remaining contours are $3F_o - 2F_c$ at $1\ \sigma$) reveal that SHELXL has misplaced the hydrogen on OG1_32, causing it to come too close to the (correctly placed) hydrogen on OH_9. The AFIX 83 option (usually recommended for macromolecules) places hydrogens so that they make the "best" hydrogen bonds subject to standard bond O–H lengths and C–O–H angles, but here it has assigned two hydrogens to the same hydrogen bond (OG1_32..OH_9 2.91 Å) and none to the alternative (OG1_32..O_16 2.95 Å). The HFIX 143 option, which searches the difference density around the circle of geometrically possible hydrogen positions, probably would have placed the hydrogens correctly in this example.

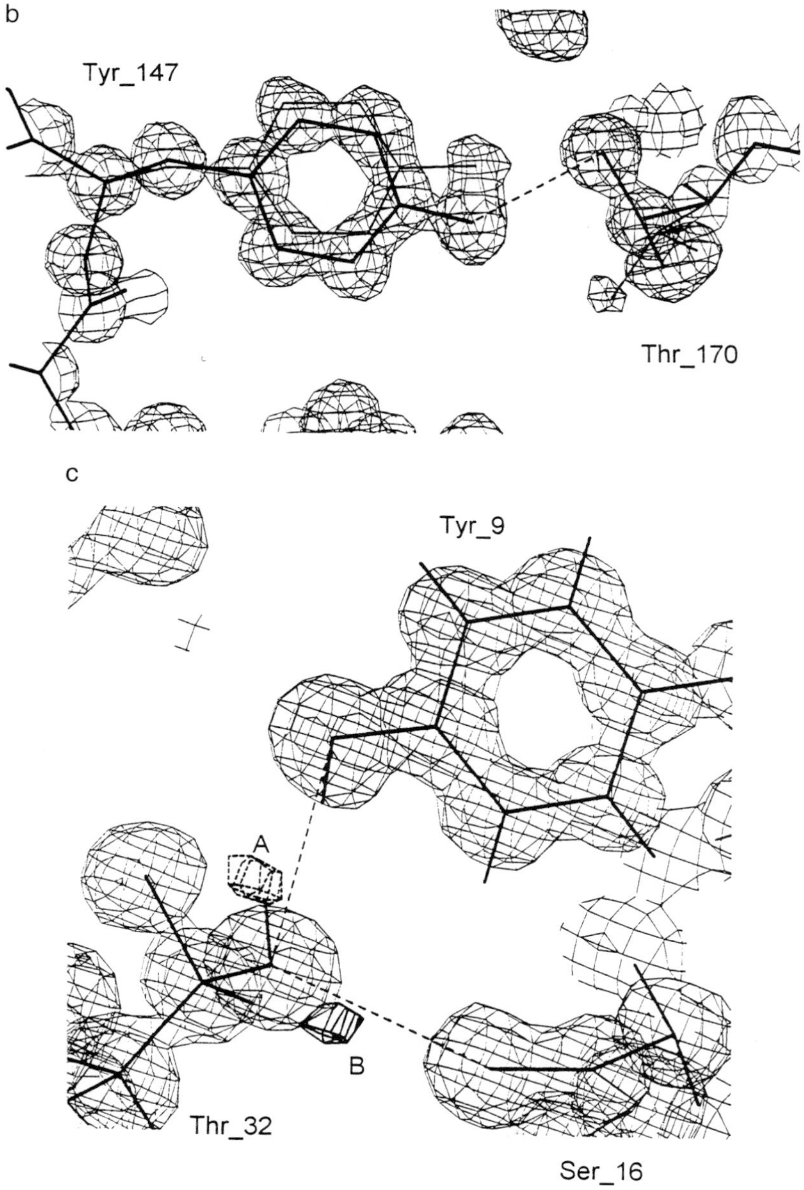

FIG. 3. (*continued*)

```
FMAP 2          ! Fo-Fc Fourier
PLAN 200 2.3 ! peaksearch and automatic water divining

LIST 6    ! Output phased reflection file (CIF) for maps etc.
WPDB      ! Write PDB format output file
HTAB      ! Output analysis of hydrogen bonds

DELU $C_* $N_* $O_* $S_*  ! Rigid bond restraints (anis only)
SIMU $C_* $N_* $O_* $S_*  ! Similar U restraints (iso/anis)
ISOR 0.1 O_201 > LAST    ! Approx. isotropic restraints (anis)

CONN 0 O_201 > LAST     ! Don't include water in connect. array
BUMP                    ! Automatic anti-bumping restraints

REM SWAT  ! Diffuse water not yet included

MERG 4    ! Remove if Friedel opposites should not be merged

MORE 1    ! MORE 0 for least, 2 or 3 for more verbose

REM Special restraints etc. specific to this structure follow:

REM HFIX 43 C1_1        ! Extra restraints etc. for O=C(H)-
DFIX C1_1 N_1 1.329     ! (formyl) at N-terminus
DFIX C1_1 O1_1 1.231    ! incorporated into residue 1
DANG N_1 O1_1 2.250     !
DANG C1_1 CA_1 2.453    !

DFIX_52 C OT1 C OT2 1.249      !
DANG_52 CA OT1 CA OT2 2.379    ! Ionized carboxyl at C-terminus
DANG_52 OT1 OT2 2.194          !

SADI_54 FE SG_6 FE SG_9 FE SG_39 FE SG_42  ! Equal Fe-S and
SADI_54 FE CB_6 FE CB_9 FE CB_39 FE CB_42  ! Fe..C distances

DFIX C_18 N_26 1.329      ! Patch break in numbering - residues
DANG O_18 N_26 2.250      ! 18 and 26 are bonded but there is a
DANG CA_18 N_26 2.425     ! gap in numbering for compatibility
DANG C_18 CA_26 2.435     ! with other rubredoxins, which
FLAT 0.3 O_18 CA_18 N_26 C_18 CA_26  ! contain an extra loop

REM Remove 'REM ' in following to activate H-atom generation

REM HFIX_ALA 43 N      ! Amide H
REM HFIX_ALA 13 CA     ! Tertiary CH
REM HFIX_ALA 33 CB     ! Methyl H (staggered)
```

```
REM HFIX_ASN 43 N
REM HFIX_ASN 13 CA
REM HFIX_ASN 23 CB      ! -CH2- hydrogens
REM HFIX_ASN 93 ND2     ! Amide -NH2 hydrogens

... etc. ...

REM Peptide standard torsion angles and restraints

RTAB_* Omeg CA C N_+ CA_+   ! Create tables of torsion angles
RTAB_* Phi C_- N CA C
RTAB_* Psi N CA C N_+
RTAB_* Cvol CA              ! Tabulate chiral volumes

DFIX_* 1.329 C_- N          ! Global restraints for peptide links.
DANG_* 2.425 CA_- N         ! Note the use of '_+' and '_-' for
DANG_* 2.250 O_- N          ! next and previous residues (in
DANG_* 2.435 C_- CA         ! numerical order).

FLAT_* 0.3 O_- CA_- N C_- CA

REM Standard amino-acid restraints etc.

CHIV_ALA C            ! Local planarity for carbonyl carbon
CHIV_ALA 2.477 CA     ! Chiral volume restraint

DFIX_ALA 1.231 C O
DFIX_ALA 1.525 C CA
DFIX_ALA 1.521 CA CB
DFIX_ALA 1.458 N CA
DANG_ALA 2.462 C N
DANG_ALA 2.401 O CA
DANG_ALA 2.503 C CB
DANG_ALA 2.446 CB N

... etc. ...

FVAR 1 .5 .5 .5 .5   ! OSF and Occupancies of 4 disordered
REM  side-chains refined as p and 1-p; (all p start at 0.5)
```

```
RESI    1  MET
C1      1  -0.01531    0.34721    0.44300   11.00000    0.12665
O1      4   0.00972    0.32433    0.48418   11.00000    0.16376
N       3   0.00836    0.35147    0.37770   11.00000    0.11842
CA      1   0.06045    0.32983    0.35232   11.00000    0.10259
C       1   0.10625    0.38660    0.39888   11.00000    0.10119
O       4   0.10422    0.45625    0.42001   11.00000    0.11221
CB      1   0.07538    0.34064    0.27785   11.00000    0.15363
CG      1   0.03097    0.29012    0.23051   11.00000    0.18022
SD      5   0.04911    0.31954    0.14408   11.00000    0.24621
CE      1   0.11104    0.28805    0.12036   11.00000    0.16148

RESI    2  GLN
N       3   0.14743    0.35596    0.40654   11.00000    0.10056
CA      1   0.19048    0.39787    0.45497   11.00000    0.08752
C       1   0.22208    0.47658    0.43518   11.00000    0.07612
O       4   0.25001    0.48505    0.37998   11.00000    0.08764
CB      1   0.23059    0.34657    0.45937   11.00000    0.11335
CG      1   0.27488    0.38810    0.51157   11.00000    0.09119
CD      1   0.24990    0.39218    0.58558   11.00000    0.08904
OE1     4   0.22830    0.33225    0.60733   11.00000    0.14198
NE2     3   0.24806    0.46366    0.62136   11.00000    0.08004

... etc. ...

RESI   12  GLU
N       3   0.41369    1.09134    0.48128   11.00000    0.05737
CA      1   0.37974    1.01164    0.48169   11.00000    0.05484
C       1   0.36179    0.97476    0.40893   11.00000    0.04661
O       4   0.34036    1.00559    0.37164   11.00000    0.07143
PART    1
CB      1   0.32730    1.01183    0.53146   21.00000    0.09562
CG      1   0.29991    0.92977    0.55047   21.00000    0.13982
CD      1   0.26259    0.93560    0.61660   21.00000    0.18618
OE1     4   0.24946    0.99654    0.64467   21.00000    0.19314
OE2     4   0.23951    0.87009    0.63348   21.00000    0.21341
PART    2
CB      1   0.32730    1.01183    0.53146  -21.00000    0.09562
CG      1   0.28412    0.93273    0.52861  -21.00000    0.16287
CD      1   0.23321    0.93266    0.57828  -21.00000    0.21467
OE1     4   0.22191    0.89782    0.62732  -21.00000    0.24634
OE2     4   0.19907    0.96700    0.56264  -21.00000    0.28319
PART 0

... etc. ...
```

```
RESI   52   ALA
N          3    0.33355    0.63065    0.69656  11.00000    0.07409
CA         1    0.30791    0.68723    0.74732  11.00000    0.09461
CB         1    0.33840    0.77331    0.74353  11.00000    0.13463
C          1    0.24600    0.67587    0.73680  11.00000    0.10360
OT1        4    0.22257    0.72488    0.77528  11.00000    0.08752
OT2        4    0.22619    0.61588    0.69459  11.00000    0.10233

RESI   54   FE
FE         6    0.72039    1.22308    0.43798  11.00000    0.08701

REM Only waters with high occupancies and low U's have been
REM retained, and all the occupancies have been reset to 1,
REM with a view to running the automatic water divining.
REM Water residue numbers have been changed to start at 201.

RESI 201   HOH
O          4    0.13784    0.53256    0.60976  11.00000    0.12412
RESI 202   HOH
O          4    0.84834    0.53945    0.69175  11.00000    0.13248

... etc. ...

HKLF 3 ! Read h,k,l,F and sigma(F) (F-squared would be better)
END
```

Acknowledgments

We are grateful to the large number of crystallographers, too many to name individually, who made valuable contributions to the development and testing of the program. G. M. S. is grateful to the DFG for support under the Leibniz Program, which gave him time to work on this project. T. R. S. thanks EMBL for a Predoctoral Fellowship and Novo Nordisk for providing the serine protease used in some of the examples. The programs O[25] and SHELXTL[26] were used to prepare Figs. 1–3. O also proved very useful in monitoring progress and rebuilding the model in the intermediate stages of the serine protease refinement.

[25] T. A. Jones, J.-Y. Zou, S. Cowan, and M. Kjeldgaard, *Acta Crystallogr.* **A47,** 110 (1991).
[26] Siemens Analytical Instruments, "SHELXTL," version 5. Siemens Analytical Instruments, Madison, Wisconsin, 1990.

[17] Modeling and Refinement of Water Molecules and Disordered Solvent

By JOHN BADGER

Introduction

The solvent content in protein crystals is typically ~50% of the total crystal volume.[1] Although most of the solvent is poorly ordered, and is frequently approximated by a structureless electron-density continuum, some solvent molecules are tightly localized by multiple hydrogen bonds to the protein surface, and have *B* factors comparable to the protein atoms. This chapter reviews techniques in X-ray crystallography that may be used for modeling both the disordered and the well-ordered solvent in protein crystals.

Bulk Solvent Scattering

Effect of Bulk Solvent Scattering

It is standard practice for most X-ray crystallographers to exclude low-resolution ($d > 10$ Å) diffraction data from the refinements of their protein atomic models. Although it may be argued that these data provide little extra information on the details of the protein structure, it obviously would be better if it were possible to include all experimental measurements in the refinement. One might expect that these data would be easy to fit because the structure factors calculated at low resolution should not be sensitive to small errors in atomic positions. In fact, measured and calculated structure-factor amplitudes show large differences at low resolution unless the scattering from the bulk solvent is taken into account.

The primary effect of the bulk solvent scattering is to reduce the observed structure factor amplitudes relative to the predictions from the protein atomic model. This reduction in scattering occurs because a large component of the scattered intensity at low resolution is due to the scattering difference between the average protein and average solvent densities (see Ref. 2 for an account of this effect and how it might be exploited). The true difference between the mean protein and solvent scattering densities is

[1] B. W. Matthews, *J. Mol. Biol.* **33,** 491 (1968).

[2] C. W. Carter, K. V. Crumley, D. E. Coleman, F. Hage, and G. Bricogne, *Acta Crystallogr.* **A46,** 57 (1990).

0076-6879/97 $25.00

less than $0.1e/\text{Å}^3$ in most crystals, whereas the calculated scattering density for a protein model in the absence of solvent is $\sim 0.43e/\text{Å}^3$. At higher resolution the diffraction pattern is dominated by density variations in the protein part of the crystal (i.e., the details of the protein structure) and the scattering contrast between average protein and solvent densities has little effect. If low-resolution data are included in the refinement and no account is taken of bulk solvent scattering the data will be badly misfit at low resolution and the overall scale factor between observed and calculated structure factors will be distorted. Two common approximations used to compensate for bulk solvent scattering, one applied in real space and one in reciprocal space, are discussed in the next two sections.

Real-Space Models of Solvent Continuum

In protein crystallography the usual method for the calculation of structure factors from an atomic model is to use the fast Fourier transform, in which a density model of the protein is first built on a large three-dimensional grid. The fast Fourier transform of this density grid[3,4] gives calculated structure factors and phases for the crystal. The most obvious way to treat the bulk solvent is simply to set the density value of grid points outside the protein volume to the expected value for the solvent density. Two important issues with this method are (1) deciding exactly what volume the solvent should fill and (2) finding an efficient method to smooth the sharp and unphysical edge where the flat bulk solvent density meets the edge of the protein density model. The former problem may be dealt with geometrically, by demarking a solvent-accessible volume outside the van der Waals exclusion zone of the protein. The latter problem is usually handled by carrying out separate calculations to obtain the partial structure factors for the protein model in the absence of solvent, $\mathbf{F}_c(\text{protein})$, and the flat solvent density alone, $\mathbf{F}_c(\text{solvent})$. Applying a large artificial temperature factor, B_{sol}, to the transform of the solvent density continuum is equivalent to rounding edges of the solvent density at the protein–solvent surface. Thus, the complete calculated structure factor is

$$\mathbf{F}_c(\text{total}) = \mathbf{F}_c(\text{protein}) + K\mathbf{F}_c(\text{solvent}) \exp(-B_{sol}/4d^2)$$

where K may be treated as a refineable scale factor that can be used to adjust the average solvent-scattering density. This type of calculation may be easily carried out, for example, using the X-PLOR program.[5] The optimal

[3] L. F. Ten Eyck, *Acta Crystallogr.* **A33,** 486 (1977).
[4] R. C. Agarwal, *Acta Crystallogr.* **A34,** 791 (1978).
[5] A. T. Brünger, "X-PLOR 3.1 Manual." Yale University Press, New Haven and London. 1992.

value for the average solvent density may be obtained by finding the minimum value of

$$\Sigma\ (|\mathbf{F}_{obs}| - |\mathbf{F}_c(\text{total})|)^2$$

where the summation is over the low-resolution reflection data.[6] Although there is an analytical solution to finding the best values for K and B_{sol} for this function, trial-and-error calculations are more commonly used. The value of the solvent density that is obtained should be close to the value expected for the composition of the crystal mother liquor.

Although this model appears to be a correct physical description of bulk solvent, the R factors that are obtained for the low-resolution data often remain surprisingly high. For example, calculations on two different crystal forms of lysozyme gave R values of 0.63 and 0.44 for the data for which $d > 10$ Å.[6] This is clearly a significant improvement on the R values of 1.63 and 2.85 that were obtained without the solvent correction, but is much worse than one might hope when one has accurate atomic models and good data. A minor practical difficulty with the application of this method for protein structure refinement is that one must either update $\mathbf{F}_c$(solvent) on each refinement cycle, at some significant cost in computer time, or fix the values of $\mathbf{F}_c$(solvent) and assume that any change in the protein shape during the refinement does not significantly change the solvent-scattering contribution.

Reciprocal-Space Approximations to Effect of Solvent Scattering

An alternative approximation that may be used to compensate for the effect of bulk solvent scattering involves a reciprocal space rescaling of the calculated structure factor amplitudes (see Ref. 7, and [15] in this volume[7a]). The underlying physical assumption of this method is that the density distribution of solvent atoms is the complement of the distribution of protein atoms. If a scale factor, K, is applied to account for the difference in mean protein and solvent-scattering densities, and an artificial temperature factor, B_{sol}, is used to eliminate structural features from the protein, then Babinet's principle gives the solvent scattering as

$$\mathbf{F}_c(\text{solvent}) = -K\mathbf{F}_c(\text{protein}) \exp(-B_{sol}/4d^2)$$

[6] C. C. F. Blake, W. C. A. Pulford, and P. J. Artymiuk, *J. Mol. Biol.* **167,** 693 (1983).

[7] R. Langridge, D. A. Marvin, W. E. Seeds, H. R. Wilson, C. W. Hooper, M. H. F. Wilkins, and L. D. Hamilton, *J. Mol. Biol.* **2,** 38 (1960).

[7a] D. E. Tronrud, *Methods Enzymol.* **277,** [15], (1997) this volume.

Because

$$\mathbf{F}_c(\text{total}) = \mathbf{F}_c(\text{protein}) + \mathbf{F}_c(\text{solvent})$$

the complete calculated structure factor is simply

$$\mathbf{F}_c(\text{total}) = [1 - K \exp(-B_{\text{sol}}/4d^2)]\mathbf{F}_c(\text{protein})$$

The scale factor K usually is equal to the average solvent density divided by the average protein density (typically $0.8 < K < 1.0$) and the large artificial temperature factor, B_{sol}, is usually in the range of 200–400 Å^2. So, like the real-space formulation, this model has two free parameters that may be optimized for a particular problem, or simply set to expected values. The computer time needed to correct structure factors for solvent scattering with this approximation is negligible because all that is required is a rescaling of the calculated structure factors from the protein model. One aspect of this approximation that may be noted is that the amplitudes of the calculated structure factors are changed, but the phases are not altered.

Results using this method are often quite good. For example, an R value of ~0.30 is obtained for the cubic insulin crystal (Brookhaven Protein Data Bank entry, 9INS) for data extending from 32- to 10-Å resolution. If one simply wants to apply a quick and simple method for compensating for bulk solvent in refinement, this method may be the best choice. This method for compensating for bulk solvent scattering is incorporated in the structure refinement program TNT,[8] some versions of ProLSQ,[9] and could be incorporated in X-PLOR[5] through its scripting language. A related procedure is included in the SHELXL program.

Atomistic Modeling of Ordered Solvent

Difficulties in Modeling Solvent Structure in Protein Crystals

Many water molecules in protein crystal structures are sufficiently localized to be represented by atomic parameters, in much the same way as the protein atomic model. These water molecules are identifiable as peaks in electron-density difference maps, almost invariably at sites where the water molecule may form hydrogen bonds to the protein. The crystallographer must use caution when introducing water molecules into the structure because incorrect sites will be difficult to eliminate later.

Reliable model building and refinement of solvent molecules are more difficult than for the protein atoms because the generally higher mobility

[8] D. E. Tronrud, L. F. Ten Eyck, and B. W. Matthews, *Acta Crystallogr.* **A43,** 489 (1987).
[9] W. A. Hendrickson, *Methods Enzymol.* **115,** 252 (1988).

of much of the solvent leads to less distinct peak positions. Furthermore, although there has been some useful work toward identifying common geometries for protein–solvent interactions (for example, see Ref. 10), the underlying forces are too weak to provide a strong set of restraints for use in the solvent structure determination. Comparisons of solvent positions in independent determinations of identical crystal structures[11,12] do not give strong encouragement regarding the past reproducibility of solvent model building. In general, the best ordered water molecules are found to be similarly placed but the correspondence of the more disordered solvent sites is much poorer. Inspection of the Brookhaven Protein Data Bank also indicates that there are vast differences in the numbers of solvent molecules included in the coordinate files for comparable structure determinations. However, the emergence of validation methods such as the free R value (see Ref. 13, and [19] in this volume[13a]) should provide a consensus on what is the correct number of water molecules to include in an atomic coordinate file. In addition, the increased availability of automated solvent-modeling techniques should lead to more reliable treatments of the crystal solvent. The application of these methods is discussed in the following sections.

Incorporating Ordered Solvent Molecules into Models

Ordered water molecules should not be included in the coordinate set until the refinement of the protein model is almost complete (i.e., almost all atoms have been located and are considered to be placed in essentially the correct positions). Although it is difficult to set hard-and-fast rules, this usually means that the R factor for the protein model alone should be ~25% with tightly restrained stereochemistry and temperature factors. The number of ordered water molecules that it will be possible for one to find in a given crystal structure will depend on the resolution of the data. Although some crystallographers seem to avoid modeling any ordered water molecules for data sets when the maximum resolution of the reflection data is less than ~2.8 Å it may be noted that the superlative 3.0-Å resolution maps of human rhinovirus 14 obtained by density averaging clearly showed many well-defined water molecules.[14] At high resolution ($d < 2.0$ Å) it

[10] N. Thanki, J. M. Thornton, and J. M. Goodfellow, *J. Mol. Biol.* **202,** 637 (1988).

[11] A. Wlodawer, N. Borkakoti, D. S. Moss, and B. Howlin, *Acta Crystallogr.* **B42,** 379 (1986).

[12] D. H. Ohlendorf, *Acta Crystallogr.* **D50,** 808 (1994).

[13] A. T. Brünger, *Nature* (*London*) **355,** 472 (1992).

[13a] A. T. Brünger, *Methods Enzymol.* **277,** [19], 1997 (this volume).

[14] E. Arnold, G. Vriend, M. Luo, J. P. Griffith, G. Kramer, J. W. Erickson, J. E. Johnson, and M. G. Rossmann, *Acta Crystallogr.* **A43,** 346 (1987).

should certainly be possible to identify many ordered water molecules in the crystal.

Automated Identification and Inclusion of Ordered Solvent

Locating sites for ordered water molecules in a crystal begins by examining electron-density difference maps for significant peaks. Conventional Fourier difference maps computed from Fourier coefficients $(F_{obs} - F_c)$ $\exp(i\alpha_c)$ are usually the most useful for this purpose. As a supplementary check many crystallographers also like to examine maps computed from Fourier coefficients $(2F_{obs} - F_c)$ $\exp(i\alpha_c)$, which show the complete structure.

To ensure that all significant peaks in the difference map are accounted for in a systematic fashion, and to speed up the otherwise tedious job of placing water molecules in the map, most crystallographers now use semiautomated methods for identifying ordered solvent. The model-building system O[15] and the two commercial crystallography packages InsightII/Xsight and Quanta/X-ray (both from Molecular Simulations, San Diego, CA) contain software to assist with this task. These systems offer the advantage that the software for locating the ordered waters is integrated into interactive graphics systems for visualizing and checking the results. Brief mentions of programs written in individual academic laboratories for carrying out automated water placement may also be found in some structure-determination articles.

The basic scheme that is used to locate ordered waters molecules is usually as follows.

1. Scan the Fourier difference map for significant peaks (i.e., peaks that are at least three to four times larger than the root-mean-square density fluctuation in the difference map).
2. Go through the peak list and save only those peaks that are neither too close (less than ~2.4 Å) nor too far (>4.2 Å) from the protein for it to be likely that the peak is an ordered water molecule.
3. Optionally, check that the putative water molecule is within hydrogen-bonding distance of the protein or preexisting water.
4. Make sure that the water list is unique (i.e., does not contain copies of water molecules related by crystallographic symmetry).
5. Optionally, do a visual check on the water molecules that meet criteria 1–4. This visual check may be useful to avoid placing water mole-

[15] T. A. Jones, J.-Y. Zou, S. W. Cowman, and M. Kjeldgard, *Acta Crystallogr.* **A47,** 110 (1991).

cules in density that really is due to a missing ligand molecule or a discretely disordered protein group.

This scheme results in a list of water sites that meet a crystallographic test of significance and that appear reasonable on stereochemical grounds. Once the set of water sites has been added to the protein model, the crystallographer should re-refine the positional and thermal parameters of the complete structure, and then repeat these calculations to find further solvent molecules. Obviously at some point all water molecules that are ordered well enough to give significant peaks will have been fitted and the procedure must stop to avoid fitting noise.

How to Avoid Overfitting Data

After adding a batch of water molecules to the protein model and running a few cycles of least-squares refinement, the free *R* value for the model should be lower than before the waters were added. If the free *R* value has increased then the additional refinement parameters that were introduced by including the waters were not warranted by the diffraction data and their validity must be questioned. This could mean that the criteria that were used to select the water molecules were too lenient and more stringent criteria should be applied. The free *R* value is a useful global (reciprocal space) check on the progress of the refinement but cannot be usefully employed to check the validity of individual solvent sites.

The most rigorous method for checking individual solvent sites is through annealed omit maps, in which the questionable solvent molecule(s) are omitted and the atomic model structure is deliberately disturbed to remove any bias toward reproducing the water molecules in question. In practice this has usually meant running a few additional refinement cycles on the model with the questionable waters deleted, although it is probable that applying small random shifts to the atomic model would achieve the same result in negligible computer time. If an electron-density peak reappears in the omit map at the site from which it was removed, then one has good grounds for accepting this particular water molecule as part of the model. The final number of ordered water molecules that is included in the model will depend on the resolution of the data and on whether the crystal was frozen. As a rough guide, it may be possible to place about one water molecule per amino acid for a small protein if good data are measured to 2-Å resolution.

Modeling and Analysis of Solvent Density Distributions

Limitations of Current Solvent-Modeling Methods

So far, this chapter has described methods for the modeling of crystal solvent in terms of either a structureless continuum or by a set of atomic coordinates. The first method is a simple approximation to the total scattering from the solvent whereas the second method is usually only appropriate for accounting for a small fraction of the solvent atoms in the crystal. Techniques for analyzing and describing solvent in a way that falls between these two extreme representations have been the subject of a small number of recent studies. None of these methods can be described as routinely applicable, but a new approach is clearly required if we are to extend our understanding of protein hydration beyond studies that merely catalog the small number of highly ordered molecules that interact strongly with the protein surface.

Direct Analysis of Electron-Density Maps of Solvent

Because the probability distribution of the most poorly ordered solvent molecules in a crystal cannot be captured in terms of a unique atomic model, questions regarding overall solvent distributions may be approached through the direct analysis of a properly calculated electron-density map of the solvent.[6,16–19] In principle, one would like to calculate a complete density map of the crystal solvent, with the protein removed. This map is given by

$$\begin{aligned}\rho(\text{solvent}) &= \rho(\text{crystal}) - \rho(\text{protein}) \\ &= \text{FT}\{F_{\text{obs}} \exp[i\alpha(\text{crystal})] - F_c(\text{protein}) \exp[i\alpha(\text{protein})]\}\end{aligned}$$

where FT represents the Fourier transform operation. The values of F_c(protein) and α(protein) may be calculated from an atomic model of the protein, F_{obs} is the obesrved structure factor data, and α(crystal) is the phase angle for the scattering from all protein and solvent in the crystal. In a conventional "scalar" approximation to this difference map the value of α(crystal) is approximated by α(protein). A computational method for deriving values for α(crystal), which aims to account for density variations in the unmodeled solvent, has been described.[16,17] The analysis of the solvent electron-density distribution, ρ(solvent), as a function of distance from the protein revealed multiple hydration layers with spacings consistent with predictions from molecular simulation studies on other proteins.[20] This was a new result because the solvent density in these layers contains a significant contribution from unmodeled mobile solvent molecules (e.g., at hydrophobic surfaces). The method of analyzing solvent density distributions directly also has been taken a stage further

[16] J. Badger and D. L. D. Caspar, *Proc. Natl. Acad. Sci. U.S.A.* **88,** 622 (1991).
[17] J. Badger, *Biophys. J.* **65,** 1656 (1993).
[18] J.-S. Jiang and A. T. Brünger, *J. Mol. Biol.* **243,** 100 (1994).
[19] F. T. Burling, W. I. Weis, K. M. Flaherty, and A. T. Brünger, *Science* **271,** 72 (1996).
[20] M. Levitt and R. Sharon, *Proc. Natl. Acad. Sci. U.S.A.* **85,** 7557 (1988).

by calculating radially averaged solvent densities as a function of protein atom type.[18] Rather than try to derive improved estimates for α(crystal) by computational methods, one study has used anomalous scattering data to obtain highly accurate experimental phases for mannose-binding protein.[19] This direct method for computing an unbiased map is of great biophysical interest for investigations of protein hydration but, of course, relies on unusually good anomalous scattering data and will not be applicable to most crystallographic problems.

Analytical Methods for Modeling Solvent Density Distributions

Because the tendency of the crystal solvent to reside in distinct hydration shells around the protein was revealed in the analysis of electron-density maps described above,[16–19] the application of explicit shell models to interpret neutron diffraction data from myoglobin crystals[21] appears well motivated. In this scheme the protein is embedded in a lattice of pseudoatoms that are grouped into radial shells around the protein. The occupancies and mobility factors for the pseudo-atoms then are optimized on a shell-by-shell basis. An advantage of this method is that well-established refinement machinery (e.g., the X-PLOR program) may be adapted to carry out the necessary calculations. The main issues to be addressed when using this method for modeling radial variations in the solvent density are the uniqueness of the solution and the physical interpretation of the derived parameters.

An interesting approach to the problem of describing solvent density distributions is to employ analytical representations for solvent probability distributions around the various protein atom types.[22] The parameters employed in these models describe the expected shapes of the solvent probability distributions around atoms of each type. For a given protein structure it is then possible to compute the joint solvent probability distribution over all atoms, which leads directly to a density model of the solvent. The Fourier transform of this density model of the solvent may then be compared to the experimental diffraction data and the model parameters for the solvent probability distribution could be refined for that structure. This type of parameterization has been shown to capture solvent density distributions derived from molecular dynamics simulations reasonably well, but published comparisons with experimental diffraction data are still tentative.[22]

[21] X. Cheng and B. P. Schoenborn, *Acta Crystallogr.* **B46,** 195 (1990).
[22] V. Lounnas, B. M. Pettitt, and G. N. Phillips, Jr., *Biophys. J.* **66,** 601 (1994).

[18] Refinement and Reliability of Macromolecular Models Based on X-Ray Diffraction Data

By LYLE H. JENSEN

Introduction

When the 6-Å resolution myoglobin structure was reported in 1958 by Kendrew and associates,[1] the techniques for refining small molecular models were well advanced. Indeed, the decade of the 1950s saw the development of the $(F_o - F_c)$ synthesis for refinement,[2] the introduction of individual atom anisotropic thermal parameters to model atomic mobilities, the increasing use of three-dimensional data,[3,4] and the widespread application of relatively fast electronic computers, all important factors leading to improved models based on X-ray diffraction data. By the mid-1960s, organic structures based on data extending essentially to the limit of Cu $K\alpha$ radiation (d_{min} = 0.77 Å) were routinely being reported with R values $(\Sigma \| F_o| - |F_c\| / \Sigma |F_o|)$ less than 0.05 and positional precisions for carbon, nitrogen, and oxygen atoms better than 0.01 Å.

For most macromolecular crystals, however, the extent of the X-ray data that can be collected is severely limited. The underlying reason can be traced to the flexibility of these large molecules and to the solvent that is an essential constituent of crystals grown from aqueous media. It is the large atomic mobilities of the weakly bound, flexible macromolecules and the highly disordered solvent that lead to the steep falloff of the X-ray intensities with increasing diffraction angle.

Because of the limited data and the disordered solvent structure, most macromolecular models are simplified in two important ways. First, the atomic parameters are limited to x, y, z, and a single isotropic mobility factor B, four parameters per atom instead of the nine required to model both position and general anisotropic motion. Second, the solvent continuum in the typical macromolecular crystal is usually omitted from the model because it mainly affects only the low-order reflections, i.e., those with d spacings in the range 5 Å to ∞ $(0.10 > \sin \theta/\lambda > 0)$, and these are weighted

[1] J. C. Kendrew, G. Bodo, H. M. Dintzis, R. C. Parrish, H. Wyckoff, and D. C. Phillips, *Nature (London)* **181,** 662 (1958).

[2] W. Cochran, *Acta Crystallogr.* **4,** 408 (1951).

[3] D. P. Shoemaker, J. Donohue, V. Schomaker, and R. B. Corey, *J. Am. Chem. Soc.* **72,** 2328 (1950).

[4] H. Mendel and D. C. Hodgkin, *Acta Crystallogr.* **7,** 443 (1954).

0076-6879/97 $25.00

by zero in the refinement. Because the simplified models do not represent the structures accurately and the data are limited, both the *R* index and the precision of macromolecular structures fall far short of achieving the low values noted above for small organic structures.

The matter of mistakes in macromolecular models is a serious concern.[5] High atomic mobilities not only limit the resolution of the data that can be collected but also decrease the peak densities in the Fourier maps (see Fig. 1a and b) to such an extent that some regions of maps based on experimental phases may be extremely difficult to fit, and serious errors in the initial model are troublesome to correct in refinement. In the following sections, we consider the process of refinement, the detecting and correcting of errors in macromolecular models, the matter of the solvent, atomic mobilities and disorder, and the reliability of macromolecular models.

Refinement Process

The initial model of a macromolecular structure is derived either from an experimentally phased electron-density map or from the model of a closely related structure by molecular replacement. In either case the initial model is often seriously in error, and it becomes the problem of refinement to correct the errors and improve the model until the calculated structure amplitudes match the observed values within what is regarded as acceptable limits. Before initiating the refinement, it is worthwhile to check carefully the crystallographic data, and in particular the space group. Eliminating error at this point can subsequently save a great deal of time and frustration.

The traditional index for monitoring the progress of refinement is the crystallographic *R* factor. Although it is not a sensitive indicator of error in large models, it is nevertheless the most widely quoted index expressing the agreement between the calculated and observed structure amplitudes. To ignore it would amount to neglecting a vast body of accumulated experience.

A "free" *R* factor (designated R_{free}) has been proposed by Brünger (see Ref. 6 and [19] in this volume[7]). It is defined in the same way as *R* but is calculated only for a subset of the X-ray data that is excluded from the refinement process and therefore cannot act in driving $|F_c|$ toward $|F_o|$. Accordingly, if the subset truly represents the data as a whole, R_{free} is a more reliable indicator of refinement progress than the conventional *R*. If

[5] C.-I. Bränden and T. A. Jones, *Nature* (*London*) **343,** 687 (1990).

[6] A. T. Brünger, *Nature* (*London*) **355,** 472 (1992).

[7] A. T. Brünger, *Methods Enzymol.* **277,** [19], 1997 (this volume).

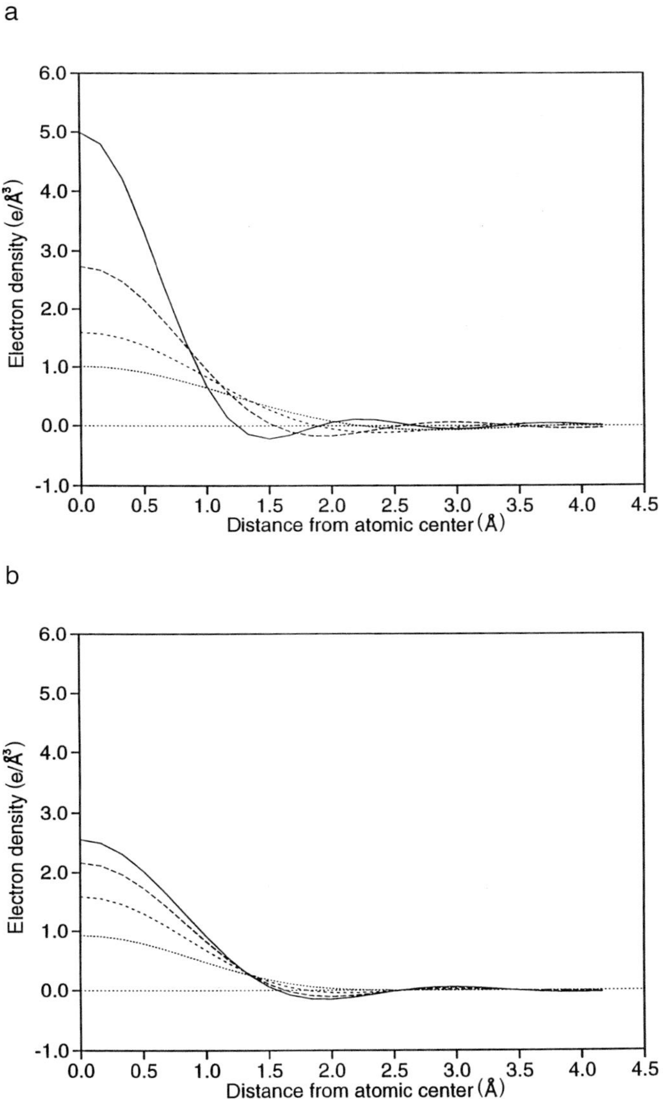

FIG. 1. Plots of electron density ρ versus r, distance from atomic center (in angstroms). (a) Oxygen atom with $B = 3.0$ Å^2 at resolutions of 1.5 Å (—), 2.0 Å (- - -), 2.5 Å (– – –), 3.0 Å (· · ·). (b) Oxygen atom at 2.0-Å resolution with $B = 5.0$ Å^2 (—), 10 Å^2 (- - -), 20 Å^2 (– – –), 40 Å^2 (· · ·). [Error-free amplitudes and phases assumed, calculated by R. E. Stenkamp (University of Washington, Seattle, Washington).]

a refinement is genuine, R_{free} should more or less track R, but at a somewhat higher value.

For a random distribution of the requisite kind and number of atoms in an acentric structure, R (and R_{free}) are expected to have a value of 0.59, but somewhat greater than this if the structure has any centric projections.[8] The approximate models used to initiate macromolecular refinements are based on maps of the Fourier series defined by the equation

$$\rho(x, y, z) = \frac{1}{V} \sum_h \sum_k \sum_l |\mathbf{F}_o| \exp i\alpha_c \exp 2\pi i(hx + ky + lz)$$

where $\rho(x, y, z)$ is the electron density, F_o values are the observed structure factors, and α_c values are the calculated phase angles in radians. The initial models are usually poor approximations and may be incomplete, leading to relatively high R values, often in the vicinity of 0.50 or even higher. Refining such models will almost certainly be difficult and require many rebuilding sessions based on what are initially poorly phased maps. To improve the interpretability of the maps, Read[9] has addressed the problem of weighting the coefficients in a Fourier series and modifying them to minimize the bias in a map toward the current model. Hodel *et al.* have tested for the reduction of model bias by several different σ_A-weighted omit-map techniques.[10]

When R (and R_{free}) no longer decrease from one cycle (or session) to the next, the refinement is said to have *converged.* To visualize what the term implies, consider R as a function in n-dimensional space where n is the number of model parameters. Over much of this space, R (and R_{free}) will fluctuate about the expected high value. In the region of the global minimum, i.e., where most of the atoms in the model are within $d_{min}/2$ of their true positions, R (and R_{free}) should decrease sharply. But for a large linked assemblage of atoms, parts of the model, such as large side chains in a protein (or even parts of the main chain), may lie outside the convergence range and therefore in a local minimum within the global minimum. For example, a relatively modest rotation about the C_α–C_β bond of a correctly positioned arginine side chain on the surface of a protein molecule will almost certainly displace its terminal guanidinium group beyond the convergence range of the usual minimization program (but see below). If in the course of refinement the model had lodged in this minimum, R would be only slightly higher than if it were in the true minimum.

[8] A. J. C. Wilson, *Acta Crystallogr.* **3,** 397 (1950).
[9] R. J. Read, *Acta Crystallogr.* **A42,** 140 (1986).
[10] A. Hodel, S.-H. Kim, and A. T. Brünger, *Acta Crystallogr.* **A48,** 851 (1992).

The preceding example suggests that for a macromolecular model the global minimum in R space encompasses numerous local minima. This view is consistent with the results reported by Ohlendorf,[11] who compared four independently determined models for the structure of interleukin 1β. After refining three of the models against one of the data sets, R values ranged from 0.172 to 0.191, comparable to the values originally reported, and differences in the coordinates among the models were little changed from the initial differences. Thus the models had not converged to a single structure in refining against the same data set. Nevertheless, a certain linear combination of the models resulted in one having R 0.159, which is substantially less than any one of the component models. Presumably this low R value could have been reduced even further by refinement.

One should note that the sharp decrease in R when the model is in the vicinity of the global minimum may not be evident in the course of refinement. Poor initial models that require numerous rebuilding sessions to correct the most flagrant errors often show a more gradual decrease in R as various parts of the model shift toward their true positions. Nor does a sharp decrease in R necessarily mean the model has locked into the true global minimum, because substantial minima elsewhere may mimic the true one, particularly if certain kinds of near symmetry are present in the structure.

Although the X-ray data from macromolecular crystals are limited, we have a great deal of stereochemical information in the form of bond lengths, bond angles, planar groups, contact distances, etc., as well as other information. Including this supplemental data in a refinement program will improve its conditioning. In practice, standard values of known features are imposed as constraints (rigid models) or incorporated as additional observations (restrained models) in the refinement program. Five such programs are listed in Table I.

The program X-PLOR listed in Table I is unique in the sense that it includes a molecular dynamics algorithm (see [13] in this volume[12]). The crystallographic term $\sum w(|F_o| - k|F_c|)^2$ is treated as a pseudo-energy term and is added to the energy of the molecular dynamics simulation. Owing to the kinetic energy in the simulation, modest energy barriers in the model may be surmounted, increasing the convergence range of the refinement. That the convergence range can be substantially increased in practice has been amply demonstrated by Fujinaga *et al.*,[13] who used GROMOS (the Gröningen molecular simulation program) in test refinements of bovine

[11] D. H. Ohlendorf, *Acta Crystallogr.* **D50,** 808 (1994).

[12] A. T. Brünger and L. M. Rice, *Methods Enzymol.* **277,** [13], 1997 (this volume).

[13] M. Fujinaga, P. Gros, and W. F. van Gunsteren, *J. Appl. Crystallogr.* **22,** 1 (1989).

TABLE I
MACROMOLECULAR REFINEMENT PROGRAMS

Program	Comments	Ref.
CORELS	COnstrained (rigid groups)–REstrained Least-Squares	*a*
PROLSQ	PROtein Least-SQuares	*b, c*
EREF	Energy and $\Sigma\, w(\lvert F_o\rvert - \lvert F_c\rvert)^2$ minimization	*d, e*
TNT	Conjugate gradient/conjugate direction	*f, g*
X-PLOR	Molecular dynamics simulation linked to crystallographic energy minimization program	*h, i*

[a] J. L. Sussman, *Methods Enzymol.* **115,** 271 (1985).
[b] J. Waser, *Acta Crystallogr.* **16,** 1091 (1963).
[c] W. A. Hendrickson, *Methods Enzymol.* **115,** 252 (1985).
[d] A. Jack and M. Levitt, *Acta Crystallogr.* **A34,** 931 (1978).
[e] J. Deisenhofer, S. J. Remington, and W. Steigemann, *Methods Enzymol.* **115,** 313 (1985).
[f] D. E. Tronrud, L. F. Ten Eyck, and B. W. Matthews, *Acta Crystallogr.* **A43,** 489 (1987).
[g] D. E. Tronrud, *Acta Crystallogr.* **A48,** 912 (1992).
[h] A. T. Brünger, J. Kuriyan, and M. Karplus, *Science* **235,** 458 (1987).
[i] A. T. Brünger and L. M. Rice, *Methods Enzymol.* **277,** [13], 1997 (this volume).

pancreatic phospholipase A_2 (123 amino acids, 957 nonhydrogen atoms, 1.7-Å data). From an R of 0.485 for the model based on the multiple isomorphous replacement (MIR) phased map, GROMOS corrected automatically four or five of the seven misoriented peptide planes and several other serious errors. Nevertheless, certain types of errors remained that still required manual correction.[13]

Macromolecular Model

The general behavior during the refining of a macromolecular model provides an overall indication of its reliability. If the refinement has behaved reasonably in the sense that the decrease in R and R_{free} is commensurate with the corrections in the model and if the most prominent solvent sites have been included, convergence can be expected at an R in the range 0.20–0.15 or less. If the data are of good quality and extend to at least 2.0-Å resolution, the model should be free of the most serious errors although some errors may remain.[11]

At this point the overall geometric indicators, such as average deviations of bond lengths and angles from standard values, should fall within acceptable ranges. However, satisfactory overall values may cover a few outliers, and it is these that point to possible trouble spots in a model. Any deviations that exceed 2.6 σ should be checked.

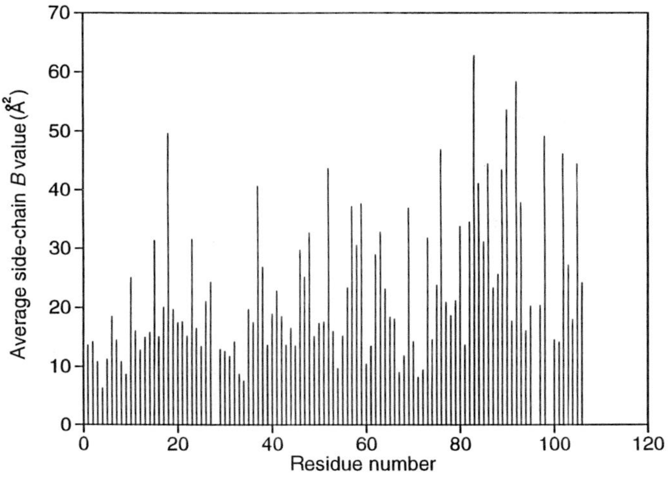

FIG. 2. Average side-chain B values versus residue number for the pH 6.5 structure of ferredoxin I from *Azotobacter vinelandii*.[15]

Excellent computer programs exist for checking not only geometric details of the model but more general features as well.[14] These programs should be exploited to the fullest extent possible, but the investigator must assess the reliability of any program used by checking it with a test model having known errors to ensure that it is working properly.

Atomic B values are useful indicators of atomic mobilities. High values (e.g., exceeding ~40–50 Å^2) indicate low electron densities, suggesting possible disorder or error in position (see Fig. 1b). In the case of proteins, a plot of the average B value for main-chain and side-chain atoms vs residue number is useful for comparing atomic mobilities throughout the molecule. Indeed, it is a rewarding exercise to compare the atomic B values of the model with the structural features of the molecule and its packing. Atoms toward the molecular surface tend to have higher B values, as do those in main-chain loops connecting secondary structural features of the molecule. Side-chain B values often spike at high B values as is evident in Fig. 2.[15]

The most sensitive crystallographic indicator of error is a map of the Fourier series with coefficients ($F_o - F_c$), i.e., a difference map. When a refinement is regarded as complete, such a map should be calculated rou-

[14] R. A. Laskowski, M. W. MacArthur, and J. M. Thornton, *in* "Proceedings of the CCP4 Study Weekend," p. 149. Daresbury Laboratory, Daresbury, Warrington, England, 1994.

[15] E. A. Merritt, G. H. Stout, S. Turley, L. C. Sieker, and L. H. Jensen, *Acta Crystallogr.* **D49,** 272 (1993).

tinely. To be most easily interpretable, the map should be on an absolute scale, i.e., the F_o values should be properly scaled and a ΔF_{000} term must be included. The map should be contoured at reasonable intervals, often in the range of 1.0–2.0 σ. All positive or negative peaks exceeding $\pm 2\ \sigma$ should be accounted for by the model. Any indicated corrections can be estimated directly from the map, and one or more additional refinement cycles minimizing $\sum w(|F_o| - |F_c|)^2$ should be calculated before another $(F_o - F_c)$ checking map is calculated.

Solvent Model

A detailed view of the solvent in a protein crystal first emerged in the refinement of the small iron–sulfur protein rubredoxin from *Clostridium pasteurianum* (RdCp, 54 amino acids, ~43% solvent[16]). Prominent peaks occurred in the solvent space within hydrogen-bonding distance of nitrogen and oxygen atoms at the surface of the protein molecules. Progressively weaker peaks occurred in a transition region, which merged into a more or less structureless continuum beyond 5 to 7 Å from the protein surface. (A few of the smaller proteins crystallize with solvent spaces sufficiently restricted in size that essentially all water molecules are hydrogen bonded in networks that span the spaces, linking the protein molecules in a more rigid structure. These crystals are atypical in the sense that they have little or no solvent continuum, and they diffract X-rays more intensely, providing data to higher resolution.)

The contribution to the Bragg reflections from the solvent continuum of typical macromolecular crystals decreases rapidly with increasing diffraction angle, becoming negligible for $\sin \theta/\lambda > 0.10$ ($d < 5$ Å). Figure 3 is a plot of R vs $\sin \theta/\lambda$ for ferredoxin from *Azotobacter vinelandii* (106 amino acids, ~60% solvent), where omitting the solvent continuum accounts for the sharp increase in R for $\sin \theta/\lambda < 0.10$.[15]

Despite the fact that a simple model for the solvent continuum has been included in the refinement of oxymyoglobin[17] and in the study of two different lysozymes,[18] the practice of neglecting the disordered solvent has persisted because only the low-order reflections are strongly affected. Omitting the corresponding terms in calculating the Fourier maps, however, raises or lowers the electron density over broad regions of the unit cell, degrading the maps and complicating their interpretation. Furthermore,

[16] K. D. Watenpaugh, L. C. Sieker, and L. H. Jensen, *Acta Crystallogr.* **B29,** 943 (1973).
[17] S. E. V. Phillips, *J. Mol. Biol.* **142,** 531 (1980).
[18] C. C. F. Blake, W. C. A. Pulford, and P. J. Artymiuk, *J. Mol. Biol.* **167,** 693 (1983).

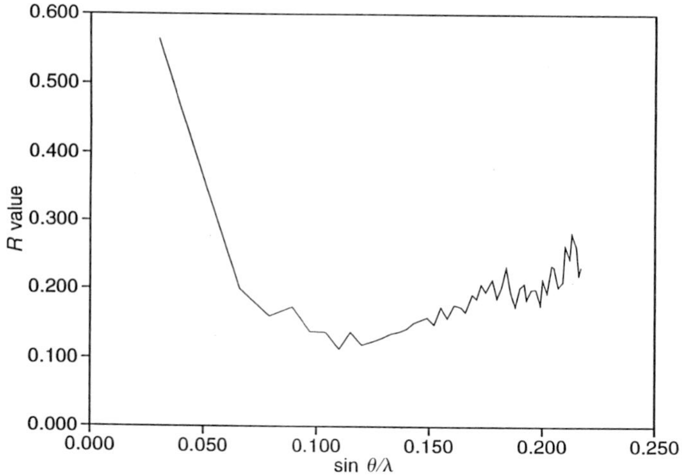

FIG. 3. R as a function of sin θ/λ for the pH 6.5 structure of ferredoxin I from *A. vinelandii*, $R = 0.170$ for 10- to 2.3-Å data.[15]

the low-order reflections are important in scaling the data by minimizing the correlation between the thermal parameters and scale factor.

In a careful study, Jiang and Brünger[19] tested four different models of the solvent continuum in penicillopepsin [323 amino acids, 38% (v/v) solvent] and in neuraminidase [380 amino acids, ~60% (v/v) solvent], treating as part of the protein all water sites with occupancies greater than 0.75. In one of the tests, a simple flat model was assumed for the bulk solvent (including water sites with occupancies < 0.75), multiplied by the factor $\exp(-B \sin^2\theta/\lambda^2)$ to smooth what would otherwise have been a sharp boundary between the protein and solvent. This simple model greatly improved the agreement between the calculated and observed reflections with $d > 6$ Å and differed only marginally from two of the more complex models tested.

The $F_o - F_c$ map has been used to model disordered regions of small structures and has been applied to the solvent structure of insulin.[20] Jiang and Brünger also tested this model, but for reasons that remain obscure, the calculated structure factors with $d > 5$ Å were little improved. The method has the potential, however, to fit the solvent structure realistically, in particular the transition region within 3–7 Å of the protein surface. The

[19] J.-S. Jiang and A. T. Brünger, *J. Mol. Biol.* **243,** 100 (1994).

[20] J. Badger and D. L. D. Caspar, *Proc. Natl. Acad. Sci. U.S.A.* **88,** 622 (1991).

$F_o - F_c$ solvent model can be refined within limits, but the process must be checked carefully to make certain the model does not overfit the data.

In experimentally phased electron-density maps based on data to at least 2.0-Å resolution, some water peaks usually appear in the solvent spaces. If these peaks are within hydrogen-bonding distance of the surface of the macromolecule, it is reasonable to include these sites in the model, but if the reflections are poorly phased or if the structure is solved by molecular replacement, no solvent sites should be included initially.

Current practice often includes only the most prominent solvent peaks at unit occupancy and absorbs any departure from full occupancy in the B parameters. If the data extend to at least 2.0-Å resolution, however, it is preferable to include an occupancy parameter, Q, for each solvent site and to assign occupancies proportional to peak densities in a $(2F_o - F_c)$ map, calibrating the density against a fully occupied site taken as unity.[21] Because of correlation between the B and Q parameters of a given site, it is not feasible to refine them simultaneously, but it is reasonable in alternate refinement cycles to fix one of the parameters (Q or B) while refining the other (B or Q).

It is at the point of adding partially occupied water sites that including the solvent continuum becomes important in clarifying structural details. The peak electron densities at the solvent sites will be affected by adding the continuum, as will details of the surface features of the macromolecule. Adding partially occupied solvent sites will have little effect on R and R_{free}, which are insensitive to minor structural features. Nevertheless, attention to details of the solvent structure can improve the interpretability of surface features of the macromolecule, as was found in the *A. vinelandii* ferredoxin structure.[15]

In choosing an electron-density level at which to include a partially occupied solvent site, it is helpful to consider two kinds of error. Figure 4 shows a partially occupied water site W and an error peak E along with two threshold levels *a* and *b*. If level *a* is chosen, error peak E is rejected, as it should be, but the true peak W is also rejected, which is an error. This is designated as an *error of the first kind.* If level *b* is chosen as a threshold, the true peak W is accepted as it should be, but the error peak is also accepted. This is designated as an *error of the second kind.*

Both X-ray[16] and neutron diffraction results[22] support the concept of partially occupied hydration sites beyond the water molecules hydrogen bonded to the macromolecule. Any reasonable level at which a peak will be accepted as a partially occupied site will lead in general to both kinds

[21] L. H. Jensen, *Acta Crystallogr.* **B46,** 650 (1990).

[22] X. D. Cheng and B. P. Schoenborn, *Acta Crystallogr.* **B46,** 195 (1990).

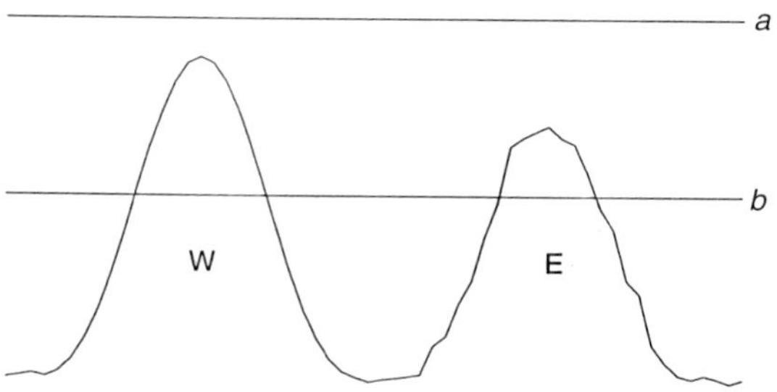

FIG. 4. Hypothetical water peak W and error peak E with acceptance/rejection thresholds *a* and *b*.

of error, but whatever the level chosen, it should be based on the σ of the map. Thus if errors in electron densities in the region of the continuum are normally distributed, no more than one error peak out of 100 will exceed 2.3 σ if the standard deviations are accurate.

When the refinement has converged, the discrete solvent sites should be omitted from the model and completely redetermined from a $(2F_o - F_c)$ map based on data scaled to the content of the unit cell and should include the F_{000} term for both the protein and solvent.

Atomic Mobilities and Disorder

The complexity of large biomolecules leads one to expect structural disorder, which means that atoms are distributed in multiple configurations among a multiplicity of energy minima. If the barriers separating the minima are high with respect to kT, the disorder is *static* in the sense that the distribution of configurations does not change with time. A single B value for each atom will cover its atomic motion and also its multiplicity of sites, leading to excessively high apparent mobilities. If the barriers are low with respect to kT, the distribution of configurations will vary with time. This type of disorder is termed *dynamic* and can be classed as thermal motion, but it certainly is not harmonic and in general not isotropic. A single B parameter for each atom will again cover mobility and multiplicity of sites and lead to excessively high B values.

The problem of representing the atomic mobilities and disorder in macromolecules can be dealt with by a molecular dynamics simulation as described by Gros *et al.*[23] in an exploratory study of bovine pancreatic phospholipase A_2. The calculated structure amplitudes in the simulation

[23] P. Gros, W. F. van Gunsteren, and W. G. J. Hol, *Science* **249,** 1149 (1990).

were restrained by the observed X-ray data, and the distribution of the various configurations over time represents not only the anisotropy of the atomic mobility but also any possible anharmonicity and disorder. For the structure averaged over a 40- to 80-psec interval (20,000 configurations), R decreased from 0.171 for the traditional refinement to 0.104 for data from 8 to 1.7 Å. As expected, R for the high-resolution data decreased more than it did for data at intermediate resolution, but contrary to expectation, R for the lowest order data also decreased more than at intermediate resolutions, suggesting that one should exercise caution in considering the validity of the results.

In a similar study of a myoglobin mutant, Clarage and Phillips[24] reported two different approaches. In the first, B values for the atomic mobilities were gradually decreased to zero in the simulation. At that point R continued to decrease as the simulation progressed, but R_{free} increased, indicating overfitting of the data. In the second approach, B values from the traditional refinement were retained in the simulation. In this case, R decreased to 0.144 while R_{free} decreased to 0.205. C_α plots of seven structures equally spaced in time from the last 7 psec of the simulations for both approaches indicate totally unreasonable mobilities throughout the model in the first approach but reasonable ones in the second. The reason for the different results from the two approaches is not evident at this time. The simulations reported in Refs. 23 and 24 did not include the solvent continuum, i.e., the simulations were done *in vacuo*.

Reliability of Macromolecular Models

In the early refinement of the protein RdCp, four $(F_o - F_c)$ cycles that reduced R from 0.372 to 0.224 were followed by four block-diagonal least-squares cycles that reduced R to 0.126 for the 5005 reflections with $I > 2\,\sigma(I)$ in the range 10–1.54 Å.[16] Because the refinement was unconstrained, the standard deviations in the atomic positions could be estimated from the inverse matrix, and an independent estimate of the average value could be derived from the distribution of bond lengths of a given bond type. The 48 C_α–C_β bonds were selected for this purpose, but the N-terminal methionine and two C-terminal glutamate residues were excluded because of high B values. On the basis of the remaining 45 C_α–C_β bonds, the estimated standard deviations (ESDs) in the bond length, σ_{bond}, equaled 0.20 Å. Assuming equal ESDs in the positions of the C_α and C_β atoms leads to $\sigma_{\text{position}} = 0.20/2^{1/2} = 0.14$ Å for the carbon atoms. This is to be compared

[24] J. B. Clarage and G. N. Phillips, Jr., *Acta Crystallogr.* **D50,** 24 (1994).

with the value from least squares, $\sigma_{\text{position,L.S.}}$, of 0.13 Å for the corresponding atoms.

Subsequently, data for RdCp extending to 1.2-Å resolution were collected and the model further refined, again by unconstrained block-diagonal least squares. R was reduced in nine cycles to 0.128 for the 10,936 reflections with $I > 2\sigma(I)$ in the range 10 to 1.2 Å.[25] The ESDs improved by a factor of ~2 as estimated in the same way as the 1.5-Å resolution refinement.

Although the RdCp molecule is relatively small, the atomic B values vary widely depending on the atomic location, and accordingly a wide range in ESDs of the atomic positions was observed. The C_α and C_β atoms in RdCp are among the best determined in the molecule, and as a result their ESDs are among the lowest. Atoms at or near the surface of the molecule, however, will have much higher ESDs, at least two to three times greater than the values derived above.

Modern refinement programs that impose geometric restraints preclude assessing the accuracy of positional parameters in a model by either of the methods used for RdCp, but an overall estimate of the accuracy can be determined from a plot of R as a function of sin θ/λ, as shown in Fig. 3. On the basis of error theory, Luzzati has calculated R as a function of sin θ/λ for positional errors of various magnitudes.[26] Superimposing a plot of the error lines on Fig. 3 enables one to estimate the overall error simply by comparing the more or less linear part of the plot above sin $\theta/\lambda = 0.1$ with the theoretical lines.[27] In principle, one should use R_{free} in a Luzzati plot rather than R because R_{free} is a more reliable indicator of error in refined models. This would raise the estimated error somewhat, but it would be more realistic. In any case, the overall estimate from a Luzzati plot is probably more representative of the better determined parts of the model with ESDs of other parts ranging upward from the overall value by factors of at least two or three (see summary in Ref. 11).

For a given atom type in a structure, B values can be used as a measure of accuracy in atomic positions, although high values suggest either disorder or error in the atomic positions. If restraints have been imposed on the B parameters in the refinement, their value as indicators of accuracy will be impaired. Nevertheless, the range in B values provides an idea of the variation in accuracy within the model that is not evident from an overall estimate.

[25] K. D. Watenpaugh, L. C. Sieker, and L. H. Jensen, *J. Mol. Biol.* **138,** 615 (1980).
[26] V. Luzzati, *Acta Crystallogr.* **5,** 802 (1952).
[27] L. H. Jensen, *Methods Enzymol.* **115,** 233 (1985).

Conclusion

Despite the fact that X-ray diffraction data sets from crystals of large biomolecules are limited and the models have been simplified accordingly, X-ray crystallographic methods have been extraordinarily successful in establishing the structures of these intricate molecules. But at the cost of only a modest effort, the solvent continuum could be added at the electron density of the bulk solution from which the crystals were grown. This would improve the interpretability of the electron density in the vicinity of the macromolecular surface and at the same time reduce the correlation between the scale factor and the mobility parameters.

The method of time-averaged molecular dynamics simulation can account for both anisotropic and anharmonic atomic mobilities. At present, however, the method must be regarded as exploratory and in need of further development and testing.

When the refinement of a macromolecular model has converged, we should consider the possibility that the data and the methods used may not be of sufficient power to differentiate among what may be numerous local minima within the global minimum. In reporting crystallographic results for macromolecular structures, we have an obligation to transmit a realistic assessment of the reliability of the model and to point out any questionable features of the model or regions of uncertainty.

Acknowledgments

I thank my colleagues at the University of Washington, particularly Ron Stenkamp, Ethan Merritt, Wim Hol, and Dave Teller, for technical support and valuable discussions, and for keeping me in touch with current progress.

[19] Free *R* Value: Cross-Validation in Crystallography

By AXEL T. BRÜNGER

Introduction

In 1990 Brändén and Jones[1] published a communication in *Nature,* entitled "Between Objectivity and Subjectivity," demonstrating the need for better validation methods in macromolecular crystallography. Their

[1] C. I. Brändén and T. A. Jones, *Nature* (*London*) **343,** 687 (1990).

0076-6879/97 $25.00

report was written in response to the publication of several partially or completely incorrect structures that appeared to be acceptable by standard validation criteria.

Existence of errors in crystal structures is somewhat surprising given the rigorous physical principles underlying X-ray structure determination. Fortunately, gross errors represent an exception. They can be recognized by independent structure solution or by comparison to known structures of homologous macromolecules. However, diffraction data can be misinterpreted in more subtle ways. For example, it is possible to overfit the diffraction data by introduction of too many adjustable parameters. A typical problem arises when too many water molecules are fitted to the diffraction data, thus compensating for errors in the model or the data. A related problem is overinterpretation of models by putting too much faith in the accuracy of atomic positions at the particular resolution of the crystal structure. This problem is caused by the difficulty of obtaining realistic estimates for the coordinate error of macromolecular crystal structures.

Models of conformational variability and solvation represent another potential source of overfitting. X-Ray diffraction data provide limited information about these phenomena. Because they can be of biological importance, it is tempting to push the interpretation of the X-ray diffraction data to the limit or beyond.

Powerful methods have been developed to lower the chances of misinterpreting or overinterpreting diffraction data. Some of the more important methods measure the agreement of the structure with empirical rules about protein folds,[2–4] comprehensive conformational analyses,[5,6] the real-space correlation coefficient,[7] and the free *R* value.[8] The protein-folding rules and conformational analyses depend on empirical knowledge of protein structure. They validate the model regardless of the fit to the diffraction data. In contrast, the real-space correlation coefficient and the free *R* value are entirely diffraction data based and are applicable to any macromolecule. They validate the extent to which the model explains the diffraction data. This chapter focuses on the free *R* value and other applications of cross-validation in crystallography.

[2] M. J. Sippl, *J. Mol. Biol.* **213,** 859 (1990).

[3] R. Lüthy, J. U. Bowie, and D. Eisenberg, *Nature* (*London*) **356,** 83 (1992).

[4] D. Jones and J. Thornton, *J. Comput. Aided Mol. Design* **7,** 439 (1993).

[5] R. A. Laskowski, M. W. MacArthur, D. S. Moss, and J. M. Thornton, *J. Appl. Crystallogr.* **26,** 283 (1993).

[6] G. Vriend, *J. Mol. Graph.* **8,** 52 (1990).

[7] T. A. Jones, J.-Y. Zou, S. W. Cowan, and M. Kjeldgaard, *Acta Crystallogr.* **A47,** 110 (1991).

[8] A. T. Brünger, *Nature* (*London*) **355,** 472 (1992).

Structure determination of macromolecules by crystallography involves fitting atomic models to the observed diffraction data. The traditional measure of the quality of this fit is the R value, defined as

$$R = \frac{\sum_{\mathbf{h}} w_{\mathbf{h}} \| \mathbf{F}_{\text{obs}}(\mathbf{h})| - k|\mathbf{F}_{\text{calc}}(\mathbf{h})\|}{\sum_{\mathbf{h}} |\mathbf{F}_{\text{obs}}(\mathbf{h})|} \quad (1)$$

where $\mathbf{h} = (h, k, l)$ are the indices of the reciprocal lattice points of the crystal, $w_{\mathbf{h}}$ are weights, k is a scale factor, and $|\mathbf{F}_{\text{obs}}(\mathbf{h})|$ and $|\mathbf{F}_{\text{calc}}(\mathbf{h})|$ are the observed and calculated structure factor amplitudes, respectively. Despite stereochemical restraints,[9] it is possible to overfit the diffraction data: an incorrect model can be refined to low R values, as several examples have shown.[1] The underlying reason can be found in the close relationship between the R value and the target function for the refinement that is aimed at minimizing the crystallographic residual

$$E_{\text{xray}} = \sum_{\mathbf{h}} w_{\mathbf{h}} (|\mathbf{F}_{\text{obs}}(\mathbf{h})| - k|\mathbf{F}_{\text{calc}}(\mathbf{h})|)^2 \quad (2)$$

If one assumes that all observations are independent and normally distributed, it can be shown that E_{xray} is a linear function of the negative logarithm of the likelihood of the atomic model.[10] Crystallographic refinement consists of minimizing E_{xray} and thereby maximizing the likelihood of the atomic model. However, E_{xray}, and thus R, can be made arbitrarily small simply by increasing the number of model parameters used during refinement regardless of the correctness of the model.

The theory of linear hypothesis tests has been employed to decide whether the addition of parameters or the imposition of fixed relationships between parameters results in a significant improvement or a significant decline in the agreement between atomic model and diffraction data.[11] This theory strictly applies to normally distributed errors and linear chemical restraints, and thus may not be applicable to macromolecular refinement (see [13] in this volume[12]).

[9] W. A. Hendrickson, *Methods Enzymol.* **11,** 252 (1985).

[10] W. H. Press, B. P. Flannery, S. A. Teukolosky, and W. T. Vetterling, "Numerical Recipes." Cambridge University Press, Cambridge, 1986.

[11] W. C. Hamilton, *Acta Crystallogr.* **18,** 502 (1965).

[12] A. T. Brünger and L. M. Rice, *Methods Enzymol.* **277,** [13], 1997 (this volume).

The statistical method of cross-validation[13–15] is not subject to these limitations. As an example of cross-validation, the free *R* value is a better statistic with which to gauge the progress of structure determination and refinement. It measures the degree to which an atomic model predicts a subset of the observed diffraction data (the test set) that has been omitted from the refinement. In all test calculations to date, the free *R* value has been highly correlated with the phase accuracy of the atomic model. In practice, about 5–10% of the observed diffraction data (chosen at random from the unique reflections) become sequestered in the test set. The size of the test set is a compromise between the desire to minimize statistical fluctuations of the free *R* value and the need to avoid a deleterious effect on the atomic model by omission of too much experimental data. In some cases, fluctuations in the free *R* value cannot be avoided because the test set is too small; this is especially true at low resolution. The dependence of the free *R* value on a particular test set can be avoided by repeating the same refinement with different test sets and combining the results.[16] The advantage of cross-validation is that it does not make any assumptions about the distribution of errors and about the form of the chemical restraints.

Cross-validation is not restricted to computing the free *R* value. It can be applied to any statistic that is expressed as a function of the diffraction data and model parameters. For example, it can be applied to estimators of coordinate error[17] that are derived from the *R* value, or to the σ_a statistic between observed and computed data.[18] As we show below, cross-validation produces a much more realistic estimation of the coordinate error. Cross-validation can also be applied when an atomic model is not yet available and the electron density is described by a crude model. Solvent flattening uses a simple partition of the unit cell into solvent and macromolecule and applies the constraint of flat solvent density to modify the initial phases. A cross-validated *R* value can be used to decide if the modification of the phases actually represents an improvement.

In a first approximation, the free *R* value is related to likelihood estimation in which the predictability of subsets of diffraction data is tested using maximum-entropy theory.[19] Although the free *R* value is a reciprocal-space

[13] M. Stone, *J. R. Stat. Soc. Ser. B* **36,** 111 (1974).

[14] B. Efron, "The Jackknife, the Bootstrap, and Other Resampling Plans," Vol. 38. CBMS-NSF Regional Conference Series in Applied Mathematics. Society for Industrial and Applied Mathematics, Philadelphia, Pennsylvania, 1982.

[15] B. Efron, *Society for Industrial and Applied Mathematics Rev.* **30,** 421 (1988).

[16] J.-S. Jiang and A. T. Brünger, *J. Mol. Biol.* **243,** 100 (1994).

[17] V. Luzzati, *Acta Crystallogr.* **5,** 802 (1952).

[18] R. J. Read, *Acta Crystallogr.* **A42,** 140 (1986).

[19] G. Bricogne, *Acta Crystallogr.* **A40,** 410 (1984).

measure, it is obtained from computer simulation in real space; this allows the inclusion of arbitrarily complex restraints such as geometric energy functions. The likelihood estimation is formulated in reciprocal space, which makes it nearly impossible to include detailed chemical real-space information. In a different context, partitioning of observed reflections into test and working sets has been used by Karle[20] to aid the convergence behavior of direct computation of structure factors from intensities for small molecules.

There are two issues that this chapter addresses: (1) that free *R* values are generally higher than expected (typically 20% and sometimes above 30%) and (2) that there should be a uniform standard to define the free *R* value. We show that the free *R* value is empirically related to coordinate error, and thus high free *R* values may be caused by a relatively high coordinate error of the model. We also propose a standardization of the free *R* value. The statistical theory of cross-validation is first described, and it is then applied to crystallographic refinement, solvent flattening, and the modeling of solvent and conformational variability. Several issues are addressed pertaining to the practical use of the free *R* value and several open questions are posed that will require future research.

Theoretical Background

In this section we review classic statistical regression theory, and the methods of cross-validation and bootstrap.

Classic Statistical Regression Theory

A data set consists of *n* pairs of points

$$(x_i, y_i); \quad i = 1, i, \ldots, n \tag{3}$$

where $y = (y_1, \ldots, y_n)$ are the observations referenced to a basis set of indices x_i. An example are the intensities $I(\mathbf{h})$ and indices $(\mathbf{h})$ of a crystallographic diffraction experiment.

Least-squares theory assumes that a set of model functions $M(x_i, \mathbf{b})$, $i = 1, \ldots, n$ are given as a function of *m* parameters $\mathbf{b} = (b_1, \ldots, b_m)$. Ideally, one wants to find parameters $\mathbf{b}^o = (b_1^o, \ldots, b_m^o)$ such that

$$y_i = M(x_i, \mathbf{b}^o) + e_i \tag{4}$$

where e_i is an experimental error drawn at random from a population with zero mean. In X-ray crystallography, the model parameters are typically

[20] J. Karle, *Proc. Natl. Acad. Sci. U.S.A.* **88,** 10099 (1991).

chosen as the atomic coordinates, temperature factors, and occupancies of the crystal structure. The functions $M(x_i, \mathbf{b})$ refer to the computed structure-factor amplitudes.

It can be shown that the best estimator of $\mathbf{b}^o$ is given by the minimum $\hat{\mathbf{b}}^o$ of the residual $L(\mathbf{b})$,[10]

$$\mathbf{b}^o \quad \text{minimizes} \quad L(\mathbf{b}) = \sum_{i=1}^{n} [y_i - M(x_i, \mathbf{b})]^2 \tag{5}$$

The estimator $\hat{\mathbf{b}}^o$ is best in the sense that it represents the maximum-likelihood estimation of $\mathbf{b}^o$.

Linear regression analysis is a special case of least-squares theory in which the model functions are linear functions of the parameters $\mathbf{b}$,

$$M(x_i, \mathbf{b}) = \sum_{j=1}^{m} b_j F_{ji} = (\mathbf{F}^* \mathbf{b})_i \tag{6}$$

The star denotes the transposition operation. The solutions to this linear least-squares problem are the well-known normal equations

$$\mathbf{b} = (\mathbf{FF}^*)^{-1}\mathbf{Fy} \tag{7}$$

If the model functions M are not linear, the problem can be solved approximately by linearizing M,

$$M(x_i, \mathbf{b}') \approx M(x_i, \mathbf{b}) + \sum_{j=1}^{m} \frac{\partial M(x_j, \mathbf{b})}{\partial b_j} (b_j' - b_j) \tag{8}$$

and iterative application of the normal equation [Eq. (7)].

A fundamental issue in regression analysis is the determination of the optimal number of parameters m required to maximize the information content of the model. Too few parameters will not fit the data satisfactorily, whereas too many parameters will fit noise. At a first glance, one could use the residual $L(\hat{\mathbf{b}}^o)$ [Eq. (5)] as a measure of the information content. However, the problem with this choice is that the same data are used for the least-squares estimator $\hat{\mathbf{b}}^o$ as to assess the information content. $L(\hat{\mathbf{b}}^o)$ always decreases monotonically as the number of parameters m increases. The classic approach to circumvent this problem is to compute the significance F of a change in $L(\hat{\mathbf{b}}^o)$ on changing the number of parameters from m_1 to m_2. The change is considered insignificant if the F value is close to one. The decision as to whether F is close to one can be derived from a theoretical distribution of F. Hamilton[11] expressed the F test in terms of the more widely used R values in crystallography. Although the F test enjoys great popularity in small-molecule crystallography it is strictly limited to linear models and normally distributed experimental errors e_i.

Cross-Validation Theory

Cross-validation[13–15] is a general "model-free" method that overcomes the limitations of the classic F test. It is based on the idea that the information content of the model should be related to the residual L computed for new data that were not involved in the determination of the least-squares estimator $\hat{\mathbf{b}}^o$ in the first place. However, instead of measuring entirely new data, it is more useful to omit a single data point i from the original data set and to determine the least-squares estimator $\hat{\mathbf{b}}^o(i)$ using Eq. (5) with i omitted. The expression $\{y_i - M[x_i, \hat{\mathbf{b}}^o(i)]\}^2$ then becomes an estimate of the quality of the fit. However, because a single value is prone to large statistical fluctuations, it may be necessary to determine the least-squares estimators $\hat{\mathbf{b}}^o(i)$ for each data point i in the original data set. The "completely" cross-validated residual is then defined as

$$L_{\mathrm{CV}} = \sum_{i=1}^{n} \{y_i - M[x_i, \hat{\mathbf{b}}^o(i)]\}^2 \tag{9}$$

In the case of linear least squares, L_{CV} can be computed analytically from a single least-squares solution of another residual.[21] However, in the general case, n least-squares evaluations are required to compute L_{CV}. In particular, most crystallographic optimization problems are nonlinear, which means that cross-validation must be performed numerically.

Figure 1 illustrates this procedure with a simple example. Consider a hypothetical, two-dimensional data set

$$[x, y(x) = x + N(0, 20)]; \qquad x = 1, \ldots, 200 \tag{10}$$

where N is a normal (Gaussian) distribution with zero mean and standard deviation of 20 (Fig. 1a). The Gaussian noise is supposed to simulate measurement errors of the observations y. Let a set of functions

$$f_m(x) = a_1 + a_2 x + \sum_{i=3, i \le m}^{m} a_i \sin[2^{(i-2)}x], \qquad m = 2, \ldots, 22 \tag{11}$$

be fitted to this data set by linear least squares [Eq. (6)]. The least-squares residual $L(\hat{\mathbf{b}}^o)$ is plotted in Fig. 1b as a function of the number of parameters m for $2 \le m \le 22$. As expected, the residual can be made arbitrarily small by increasing the number of parameters m. Yet, by definition, the only reasonable fit is the linear function f_2.

Figure 1c shows the completely cross-validated residual L_{CV}, using Eq. (9). Note that complete cross-validation involved n least-squares evaluations for each fitted function, i.e., this required 21×200 evaluations of Eq.

[21] W. Härdle, P. Hall, and J. S. Marron, *J. Am. Stat. Assoc.* **83,** 86 (1988).

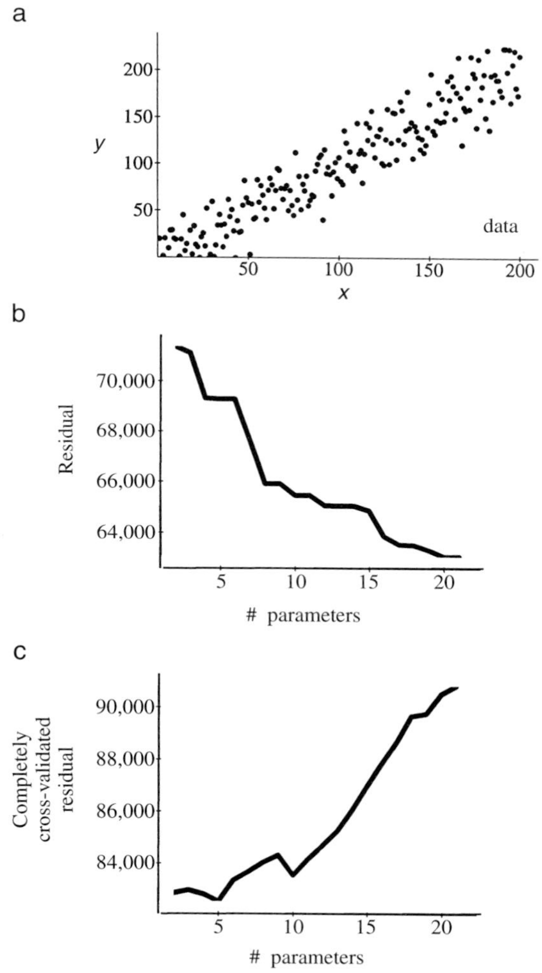

FIG. 1. Application of cross-validation to the least-squares fit of the noisy data set described in Eq. (10). (a) Actual data set used in the example. (b) Residuals $L(\hat{\mathbf{b}}^o)$ obtained by linear least-squares fits [Eq. (7)] using the functions described in Eq. (11) for $2 \leq m \leq 22$; the residuals are plotted as a function of the parameters m used in the least-squares fit. (c) Completely cross-validated residual as defined in Eq. (9).

(6). The completely cross-validated residual tends to increase as the number of parameters increases. Thus, complete cross-validation predicts that the functions f_m for m larger than five significantly overfit the data. The statistical significance of L_{CV} can be improved by the bootstrap method.[15] Artificial, bootstrap data sets are simulated by randomly drawing n (not necessarily distinct) data pairs from the original data set. Complete cross-validation is then carried out for the bootstrap data sets. The distribution of the resulting cross-validated residuals can be used to compute the mean value of L_{CV} and the standard deviation.

Cross-validation and bootstrap are examples of modern statistics wherein computer experiments produce probability distributions[15,22,23] Raw computing power has replaced tedious and often impossible analytical calculations. The beauty of this approach is that it can be applied to any statistical modeling procedure, not just least-squares fitting.

Complete cross-validation as described in Eq. (9) is usually impractical for large data sets. For many crystallographic applications it is often sufficient to consider a single test set T and to compute the residual for the test set[24]

$$L_{\mathrm{TCV}}(T) = \sum_{i \in T} \{y_i - M[x_i, \hat{\mathbf{b}}_o(T)]\}^2 \tag{12}$$

However, there are applications (discussed below) in which complete cross-validation is required.[16,24,25] Instead of using Eq. (9), it is computationally less demanding, but often sufficient, to partition the original data set D into t disjoint test sets T_r of equal size, which are obtained by randomly drawing them from the original data set, i.e.,

$$\begin{aligned} D &= \cup_{r=1}^{t} T_r \\ T_i \cap T_j &= \varnothing \end{aligned} \tag{13}$$

Least-squares evaluations are then carried out with the test sets T_r omitted and the corresponding solutions are referred to as $\hat{\mathbf{b}}^o(T_r)$. In analogy to Eq. (9) one can define the residual

$$L_{\mathrm{TCV}} = \sum_{r=1}^{t} \sum_{i \in T_r} \{y_i - M[x_i, \hat{\mathbf{b}}^o(T_r)]\}^2 \tag{14}$$

Complete cross-validation requires as many least-squares evaluations as there are test sets. Statistical significance can be checked by repeating the procedure with different test set partitionings.

[22] D. V. Hinkley, *J. R. Stat. Soc. Ser. B* **50,** 321 (1988).

[23] B. Efron and R. Tibshirani, *Science* **253,** 390 (1991).

[24] A. T. Brünger, *Acta Crystallogr.* **D49,** 24 (1993).

[25] A. T. Brünger, G. M. Clore, A. M. Gronenborn, R. Saffrich, and M. Nilges, *Science* **261,** 328 (1993).

Figure 2 illustrates cross-validation and complete cross-validation for the example in Eq. (10) using 10 test sets, each containing 20 data points. Because of the small size of the test set, the individual test residuals $L_{\mathrm{TCV}}(T)$ [Eq. (12)] show large statistical fluctuations (Fig. 2a). Therefore, complete cross-validation [Eq. (14)] must be applied to obtain statistically significant results (Fig. 2b). The mean and standard deviation of the completely cross-validated residual L_{TCV} was estimated by using 20 different partitionings of the data as defined in Eq. (13) (Fig. 2c and d).

Cross-Validated R Values

Crystallographic diffraction data typically are overdetermined. In most cases it is thus possible to set aside a certain fraction of the diffraction data for cross-validation. The influence of the size of the test set on the standard deviation or precision of the free R value is discussed below.

For cross-validation with a single test set T the free R value is defined as

$$R_{\mathrm{free}} = \frac{\sum_{\mathbf{h}\in T} w_{\mathbf{h}} \left| |F_{\mathrm{obs}}(\mathbf{h})| - k|\mathbf{F}_{\mathrm{calc}}(\mathbf{h})| \right|}{\sum_{\mathbf{h}\in T} w_{\mathbf{h}} |\mathbf{F}_{\mathrm{obs}}(\mathbf{h})|} \tag{15}$$

where the calculated structure factors $[\mathbf{F}_{\mathrm{calc}}(\mathbf{h})]$ are obtained from a model that was constructed and refined against the working set, i.e., without knowledge of the test set. In general, R_{free} will be higher than R because the test set has been omitted in the refinement process.

The choice of test set usually has little influence on the behavior of the free R value, provided the selection is purely random and the test set contains a sufficient number of data points (see below). However, considerable variation can occur in the free R value distribution as a function of resolution.[16] This is a concern when bulk solvent is to be modeled, because the fit to relatively few reflections at low resolution is critical for these models. Similarly, density modification procedures, such as solvent flattening, depend on low-resolution reflections, which are relatively few in number.[26]

To reduce fluctuations of the free R value or the free R distribution, one can use complete cross-validation [Eqs. (13) and (14)]. The observed diffraction data set is partitioned into 10 nonoverlapping test sets ($T_1, \ldots, T_{10}$), where each set contains 10% of the data. For each test set T_i, a corresponding working set A_i is defined consisting of all data excluding T_i. Refinement, density modification, or any other crystallographic procedure

[26] A. L. U. Roberts and A. T. Brünger, *Acta Crystallogr.* **D51,** 990–1002 (1997).

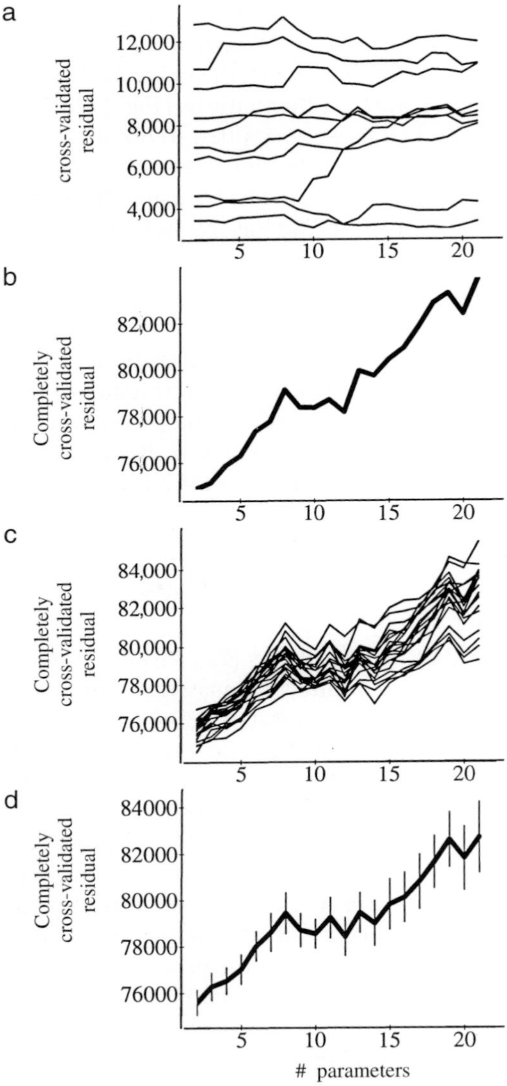

FIG. 2. Illustration of cross-validation with test sets that partition the data set [Eq. (13)]. (a) Residuals $L_{\mathrm{TCV}}(T)$ [Eq. (12)] computed over the individual test sets after least-squares optimization with the corresponding working sets. (b) Completely cross-validated residual $L_{\mathrm{TCV}} = \sum_{r=1}^{t} L_{\mathrm{TCV}}(T_r)$ [Eq. (14)]. (c) Twenty complete cross-validations using different test sets for the data partitioning described in Eq. (13). (d) Mean and standard deviation (σ) for the 20 complete cross-validations.

is carried out 10 times, once for each of the working sets A_i. Structure factors are calculated for the test set T_i and then merged to produce the cross-validated structure factor $\mathbf{F}_{cv}$, i.e, $\mathbf{F}_{cv}(\mathbf{h}) = \mathbf{F}_{calc}(\mathbf{h})$, where $\mathbf{F}_{calc}$ is the calculated structure factor obtained after the crystallographic procedure using the working set A_i with $\mathbf{h} \notin A_i$. The free R value obtained by complete cross-validation is defined as

$$R_{free} = \frac{\sum_{\mathbf{h}} w_{\mathbf{h}} \big| |\mathbf{F}_{obs}(\mathbf{h})| - k|\mathbf{F}_{cv}(\mathbf{h})| \big|}{\sum_{\mathbf{h}} w_{\mathbf{h}} |\mathbf{F}_{obs}(\mathbf{h})|} \tag{16}$$

Restrained Refinement

Crystallographic refinement consists of searching for the global minimum of the target[12,27]

$$E = E_{chem} + w_{xray} E_{xray} \tag{17}$$

where E_{chem} is composed of empirical information about chemical interactions, and w_{xray} is a weight chosen to balance the gradients arising from each term. If w_{xray} is too small, too much emphasis is put on the geometry, resulting in a poor fit to the diffraction data. If w_{xray} is too large the conventional R value becomes very small, but the structure will fit the diffraction data. Thus, the normal R value cannot be used to determine the optimal weight.

Jack and Levitt[27] proposed that w_{xray} be chosen so that the gradients of E_{chem} and E_{xray} have the same magnitude for the current structure. Hendrickson[9] suggested that w_{xray} be adjusted so that the deviations of bond lengths and bond angles from ideal geometry are close to the root-mean-square (r.m.s.) distributions found in small molecules. Both approaches require readjustment of w_{xray} during the course of the refinement. Brünger *et al.*[28,29] suggested an empirical procedure for obtaining a value for w_{xray} that can be kept constant throughout the refinement. It consists of performing a short molecular dynamics simulation at 300 K with w_{xray} set to zero, then calculating the final r.m.s. gradient due to the empirical energy term E_{chem} alone. Next, one calculates the gradient due to the experimental restraints E_{xray} alone, and chooses w_{xray} to balance the two. The underlying idea is that a molecular dynamics simulation without X-ray

[27] A. Jack and M. Levitt, *Acta Crystallogr.* **A34,** 931 (1978).

[28] A. T. Brünger, M. Karplus, and G. A. Petsko, *Acta Crystallogr.* **A45,** 50 (1989).

[29] A. T. Brünger, "X-PLOR: A System for X-Ray Crystallography and NMR," version 3.1. Yale University Press, New Haven, Connecticut, 1992.

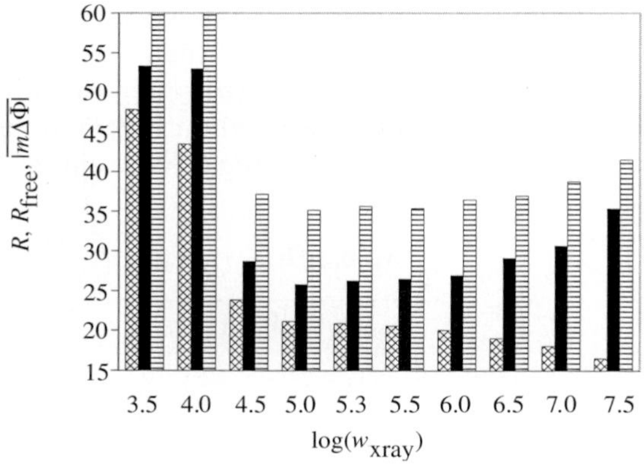

FIG. 3. Optimization of w_{xray} [Eq (17)] for refinements of the 1.8-Å crystal structure of penicillopepsin[31] at 6- to 1.8-Å resolution. Refinements began with the crystal structure of penicillopepsin with water molecules omitted and uniform *B* factors. Each refinement consisted of simulated annealing using a Cartesian-space slow-cooling protocol[32] starting at 2000 K, overall *B*-factor refinement, and individual restrained *B*-factor refinement.[9] Simulated annealing the *B*-factor refinements were carried at 6- to 1.8-Å resolution with 10% of the data randomly omitted for evaluating the free *R* value. Phase accuracy of the model is assessed by $\overline{m|\Delta\Phi|}$, which is the figure-of-merit (m) weighted mean phase difference between model phases and the most probable multiple isomorphous replacement phases[33] at 6- to 2.8-Å resolution. *R* (%; cross-hatched bars), R_{free} (%; black bars), and $\overline{m|\Delta\Phi|}$ (degrees; striped bars) are plotted vs log(w_{xray}).

restraints will estimate the average gradient for the chemical restraints. None of the three methods is optimal because they depend on the quality and correctness of the structure that is being refined. Furthermore, subjective adjustment of the final weight is commonplace, resulting in a rather broad distribution of deviations from ideal geometry in the Brookhaven Data Bank.[30]

A more objective way to optimize w_{xray} makes use of the free *R* value.[8] This is illustrated in Fig. 3,[31–33] which shows that model phases of the penicillopepsin crystal structure are most accurate near the minimum of the free *R* value. It should be noted that the most optimal weight is a factor two to three smaller than that obtained by balancing the gradients between E_{chem} and E_{xray}. Weights for restrained thermal *B*-factor refinement[8] or

[30] F. C. Bernstein, T. F. Koetzle, G. J. B. Williams, E. F. Meyer, Jr., M. D. Brice, J. R. Rodgers, O. Kennard, T. Shimanouchi, and M. Tasumi, *J. Mol. Biol.* **112,** 535 (1977).

[31] M. N. G. James and A. R. Sielecki, *J. Mol. Biol.* **163,** 299 (1983).

[32] A. T. Brünger, A. Krukowski, and J. Erickson, *Acta Crystallogr.* **A46,** 585 (1990).

[33] I.-N. Hsu, L. T. J. Delbaere, M. N. G. James, and T. Hofmann, *Nature* (*London*) **266,** 140 (1977).

noncrystallographic symmetry (NCS) restraints (see [11] in this volume[34]) can be optimized in the same way.

In addition to the overall weight factor w_{xray}, relative weighting between individual terms E_{chem} may be required. Cross-validation and phase accuracy suggested that the relative distribution of bond lengths and bond angles as it is found in the Cambridge Structural Database,[35,36] although derived purely from small molecules, is also optimal for protein structures.[24]

The absolute distribution of bond lengths and bond angles (deviations of the geometry from ideality) for a 1.8-Å crystal structure is fairly small for the optimal weight factor (0.008 Å and 1° for bond lengths and bond angles, respectively, using the protein parameter database of Engh and Huber[35]). This is a surprising result because the observed average deviations in small-molecule crystal structures are nearly twice as large. Apparently, at 1.8-Å resolution a tighter geometry is required than is actually present in the molecule to avoid overfitting.

Cross-Validated Coordinate Error

The quality of a crystal structure is related to the high-resolution limit of the observable diffraction data. A low coordinate error is a prerequisite for an understanding of macromolecular structure and function. Figure 4a illustrates this relationship between resolution and coordinate error. The penicillopepsin diffraction data were artificially truncated to the specified high-resolution limits and the crystal structure was extensively refined against the truncated data set. The resulting coordinate errors probably represent a lower bound because, in a real situation, the quality of the diffraction data would gradually deteriorate at the high-resolution limit in contrast to the truncation employed here. Moreover, in realistic cases, refinement would not start from a high-resolution initial model, and model building may have introduced additional errors that cannot be mimicked by resolution truncation alone.

At 1.8-Å resolution, the crystal structure has a coordinate error of approximately 0.2 Å as measured by the r.m.s. difference between independently refined structures and the crystal structure (Fig. 4a). The r.m.s. coordinate error $[\overline{\Delta(\mathbf{r})^2}]^{1/2}$ is equal to the difference error $[\overline{\Delta(\mathbf{r})}]$ used by Luzzati[17] if one assumes that the errors are normally distributed. We chose the r.m.s. coordinate error because it is related to the commonly used r.m.s. difference statistic between coordinate sets. The r.m.s. coordinate error

[34] G. J. Kleywegt and T. A. Jones, *Methods Enzymol.* **277,** [11], 1997 (this volume).

[35] R. A. Engh and R. Huber, *Acta Crystallogr.* **A47,** 392 (1991).

[36] F. H. Allen, O. Kennard, and R. Taylor, *Acc. Chem. Res.* **16,** 146 (1983).

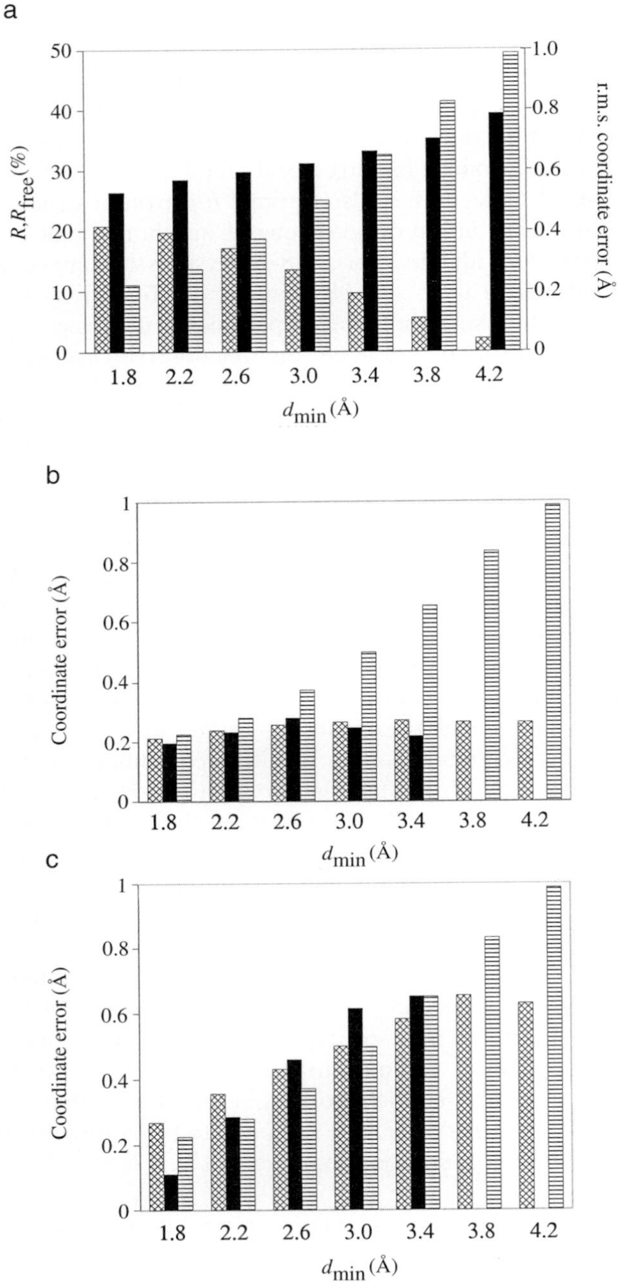
a
50
40
30
20
10
0
R,Rfree(%)
1.0
0.8
0.6
0.4
0.2
0
r.m.s. coordinate error (Å)
1.8 2.2 2.6 3.0 3.4 3.8 4.2
dmin (Å)
b
1
0.8
0.6
0.4
0.2
0
Coordinate error (Å)
1.8 2.2 2.6 3.0 3.4 3.8 4.2
dmin (Å)
c
1
0.8
0.6
0.4
0.2
0
Coordinate error (Å)
1.8 2.2 2.6 3.0 3.4 3.8 4.2
dmin (Å)

increases monotonically as the resolution is lowered. At 3-Å resolution, the r.m.s. difference is 0.5 Å, and at 4.2-Å resolution, the r.m.s. difference is around 1 Å. The reason for the deterioration of the quality of the crystal structure can be found in the increasingly adverse parameter-to-observable ratio, which makes a lower resolution structure intrinsically less accurate and that also lowers the chance of refinement to reach the global minimum (see [13] in this volume[12]). Although the latter problem can be addressed by using torsion-angle refinement,[37] one can never achieve the quality of a high-resolution structure at lower resolution without including additional information in the refinement process.

The normal *R* value improves as the quality of the refined structure deteriorates (Fig. 4a). This paradoxical behavior again clearly illustrates that the normal *R* value can be a rather useless statistical quantity. In contrast, the free *R* value shows the correct behavior by being correlated with the r.m.s. difference and, consequently, the phase error of the model. The failure of the normal *R* value to assess model quality becomes even more striking when estimating the coordinate error of the refined model using the methods of Luzzati[17] and Read[18] (Fig. 4b). The estimated coordinate errors stay approximately constant regardless of the resolution of the refined model in contrast to actual coordinate error as estimated by the r.m.s. difference to the crystal structure.

[37] L. M. Rice and A. T. Brünger, *Proteins* **19,** 277 (1994).

FIG. 4. Effect of resolution on *R* values and coordinate error: accuracy as a function of resolution. Refinements were begun with the crystal structure of penicillopepsin with water molecules omitted and uniform *B* factors. The low-resolution limit was set to 6 Å. The penicillopepsin diffraction data were artificially truncated to the specified high-resolution limit. Each refinement consisted of simulated annealing using a Cartesian-space slow-cooling protocol[32] starting at 2000 K, overall *B*-factor refinement, and individual restrained *B*-factor refinement.[9] Refinements were carried out with 10% of the data randomly omitted for cross-validation. Averages are shown over 10 refinements with different test sets. (a) The coordinate error is measured by the atomic r.m.s. difference between the refined structure and the original crystal structure.[31] [*R* (cross-hatched bars); R_{free} (black bars); r.m.s. difference (striped bars).] (b) Coordinate error estimators of the refined structures using the methods of Luzzati[17] and of Read.[18] All observed diffraction data were used, i.e., no cross-validation was performed. The coordinate error is the same as in (a) and is included for purposes of comparison. No σ_a estimates are shown below 3.4-Å resolution because the method became numerically unstable. [Luzzati (cross-hatched bars); σ_A (black bars); r.m.s. (striped bars).] (c) Cross-validated coordinate error estimators [Luzzati (cross-hatched bars); σ_A (black bars); r.m.s. (striped bars)]. The test set was used to compute the coordinate error estimators. Complete cross-validation was used to compute σ_a, whereas simple averaging over individual estimates of Luzzati's method was used.

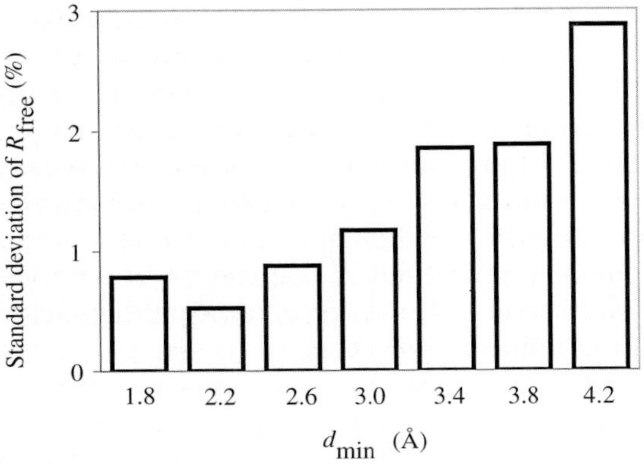

FIG. 5. Standard deviation of the free R value as a function of resolution. Refinements were started with the crystal structure of penicillopepsin with water molecules omitted and uniform B factors. Each refinement consisted of simulated annealing using a Cartesian-space slow-cooling protocol[32] starting at 2000 K, overall B-factor refinement, and individual restrained B-factor refinement.[9]

It was suggested that cross-validation can address this problem, i.e., the test set is used to compute the Luzzati plot or the σ_a coordinate error.[38] Figure 4c shows for one specific case that the cross-validated estimators of coordinate error using the methods of Luzzati and Read are fairly close to the atomic r.m.s. difference to the crystal structure. This suggests that the free R value is related to the coordinate error of the model. In the penicillopepsin case, free R values of 26, 30, and 32% correspond to r.m.s. coordinate errors of about 0.2, 0.5, and 0.6 Å, respectively. Clearly, systematic studies on other systems are needed to establish the generality of this observation. However, it is safe to conclude that estimators of coordinate error without cross-validation must be viewed with caution, whereas their cross-validated counterparts appear to be more useful.

Test Set Size

The size of the test set should be kept small to minimize the impact on the structure, but it must be large enough to produce a statistically well-defined average for the free R value. On the basis of model calculations it was suggested that 10% of the diffraction data should be used for cross-validation.[8,24] The standard deviation of the free R value using cross-validation with 10% of the data is shown in Fig. 5 as a function of resolution and

[38] G. J. Kleywegt, T. Bergfors, H. Senn, P. Le Motte, B. Gsell, K. Shudo, and T. A. Jones, *Structure* **2,** 1241 (1994).

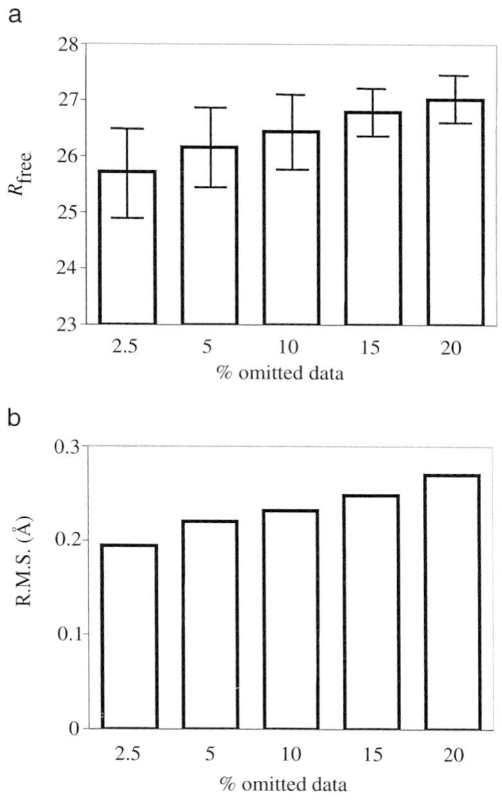

FIG. 6. Standard deviation [$\sigma(R_{free})$] of the free *R* value as a function of percentage of omitted data. Each refinement consisted of simulated annealing using a Cartesian-space slow-cooling protocol[32] starting at 2000 K, overall *B*-factor refinement, and individual restrained *B*-factor refinement.[9] Refinements were started with the crystal structure of penicillopepsin with water molecules omitted and uniform *B* factors. (a) Average free *R* value. Error bars indicate $\pm\sigma(R_{free})$. (b) The coordinate error is measured by the atomic r.m.s. difference between the refined structure and the crystal structure.

in Fig. 6a as a function of the percentage of omitted data. The standard deviation was obtained by performing 10 refinements with different test sets. The standard deviation generally increases with decreasing resolution and with decreasing test set size (Figs. 5 and 6a).

The influence of test set size on the model coordinates at 1.8-Å resolution is shown in Fig. 6b. There is a slight decrease in model accuracy as the test set size increases, a trend reflected by the increase in free *R* value (Fig. 6a). Clearly, one must compromise between reducing the standard deviation of the free R value and minimizing the effect on the refined model. Omission of 10% of the data seems like a good compromise. It should be noted that the difference in free R values for different test set sizes is close to the standard deviation of the free R value at that resolution (~0.7%, Fig. 6a).

One can argue that the fluctuation of the free R value reflects the variation of a normal R value computed over a small set of numbers and, thus, is a consequence of the size of the test set. For example, the normal R value computed over 10% test sets increases linearly from 0.5% at 1.8-Å resolution to 2% at 4.2-Å resolution for the penicillopepsin diffraction data (not shown). Furthermore, a 2.5% test set at 1.8-Å resolution produces approximately the same standard deviation as a 10% test set at 3-Å resolution that has about the same number of data points.

The relationship between test set size and standard deviation of the free R value (Fig. 5) appears to be general. For example, in the case of the amylase inhibitor, the standard deviation of the free R value is 1.5% at 2.5-Å resolution using a test set of 166 reflections (see [13] in this volume[12]). Interestingly, the standard deviation of the free R value is approximately given by $R_{free}/(n)^{1/2}$, where n is the number of reflections in the test set.

Given that the standard deviation of the free R value is simply related to the size of the test set, it is important to specify a lower bound for the test set size. If one makes the reasonable requirement that the standard deviation of the free R value be better than 1%, then the test set size must be greater than 500. If this is not possible because of limited resolution, or if one wants to compute a free R value distribution, averaging over multiple cross-validations or complete cross-validation may be necessary. If in doubt, the standard deviation of the free R value should be assessed empirically by performing multiple, independent refinements.

Conformational Variability

Macromolecular motions occur over a large range of amplitudes (up to tens of angstroms) and a remarkable range of time (from subpicoseconds to seconds). Examples in which such motions play an important biological role include protein folding, macromolecular association, and ligand binding. Understanding these motions is a long-standing problem because detailed experimental studies are often difficult and macromolecular simulation techniques are limited by computing power and inaccuracies of empirical energy functions.

During macromolecular refinement, thermal motion of atoms is typically described by isotropic Gaussian distributions around each atom in a single molecular conformation.[9] This model is an oversimplification of a macromolecule in the crystalline state. Experiments and calculations show that macromolecules exhibit anisotropic atomic distributions, which include thermal motion and crystal lattice disorder.[39–43] In addition, anharmonic motions are present that cannot be described by Gaussian distributions. Thermal parameters used in crystallographic refinement cannot distinguish between static disorder in the crystal lattice and thermal motion of a macromolecule in the crystalline environment.[43] Furthermore, the data-to-parameter ratio obtained from a typical macromolecular diffraction data set is not sufficient to allow refinement of anisotropic thermal factors.

Various modifications of conventional macromolecular refinement have been proposed that attempt to describe thermal motion and conformational variability more accurately. These include molecular dynamics time-averaging refinement,[44,45] twin model refinement,[46] and normal mode refinement.[47,48]

The quality of models arising from the former two techniques were assessed in the case of the penicillopepsin crystal structure.[49] The results are summarized in Fig. 7, which shows another example of a correlation between the free *R* value and the phase accuracy of the model. The multiconformer method is an extension of the twin method using multiple copies of the macromolecule. The multiconformer technique yields a lower free *R* value and better agreement with the multiple isomorphous replacement (MIR) phases than does time-averaging or single-conformer refinement. This improvement is significant for up to eight copies of the protein. Using 16 copies of the molecule led to an increase in the free *R* value and a decrease in the *R* value, indicating overfitting. This suggests that the anisotropy of the protein in the crystalline environment can be accounted for by multiconformer refinement when the number of conformers is adjusted by cross-validation.

[39] H. Hartmann, F. Parak, W. Steigemann, G. A. Petsko, D. Ringe, and H. Frauenfelder, *Proc. Natl. Acad. Sci. U.S.A.* **79,** 4967 (1982).

[40] H. Frauenfelder, G. A. Petsko, and D. Tsernoglou, *Nature* (*London*) **280,** 558 (1979).

[41] T. Ichiye and M. Karplus, *Proteins* **2,** 236 (1987).

[42] T. Ichiye and M. Karplus, *Biochemistry* **27,** 3487 (1988).

[43] J. Kuriyan, G. A. Petsko, R. M. Levy, and M. Karplus, *J. Mol. Biol.* **190,** 227 (1986).

[44] P. Gros, W. F. van Gunsteren, and W. G. J. Hol, *Science* **249,** 1149 (1990).

[45] J. B. Clarage and G. N. Phillips, Jr., *Acta Crystallogr.* **D50,** 24 (1994).

[46] J. Kuriyan, K. Osapay, S. K. Burley, A. T. Brünger, W. A. Hendrickson, and M. Karplus, *Proteins* **10,** 340 (1991).

[47] A. Kidera and N. Gō, *Proc. Natl. Acad. Sci. U.S.A.* **87,** 3718 (1990).

[48] R. Diamond, *Acta Crystallogr.* **A46,** 425 (1990).

[49] F. T. Burling and A. T. Brünger, *Israel J. Chem.* **34,** 165 (1994).

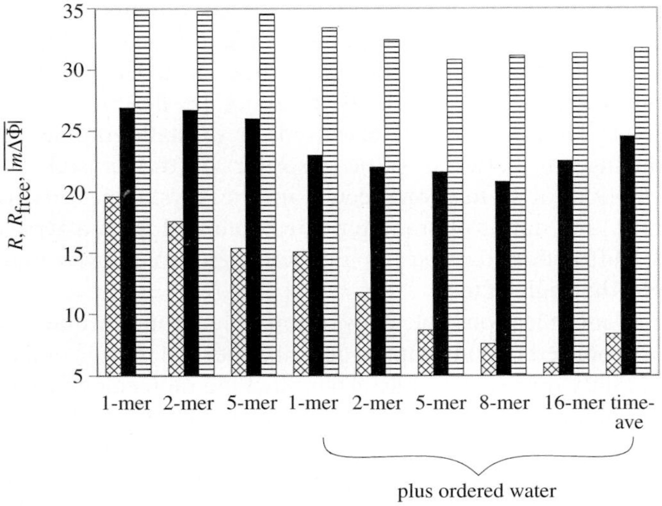

FIG. 7. *R* value (cross-hatched bars), free *R* value (black bars), and figure-of-merit weighted phase differences (striped bars) for the ensemble multiconformer and time-averaging refinements of the penicillopepsin crystal structure at 1.8-Å resolution with and without ordered water molecules.[49] Shown for comparison are the corresponding values from the conventional single-conformer model of penicillopepsin. The single-conformer and multiconformer refinements consisted of simulated annealing using a Cartesian-space slow-cooling protocol[32] starting at 2000 K, overall *B*-factor refinement, and individual restrained *B*-factor refinement.[9] Details of the time-averaging refinement that retained individual *B* values are described in Ref. 49. Phase differences were computed using the experimental MIR phases from 6.0- to 2.8-Å resolution. The weight w_{xray} [Eq. (17)] was set to $10^{5.3}$, which approximately optimizes the free *R* value and phase accuracy (Fig. 3).

Bulk Solvent Modeling

Water associated with the solvation of macromolecules plays an important role in biological processes such as enzymatic reactions, folding, specific and nonspecific macromolecular association, and ligand binding. Solvent constitutes a large portion of the volume in macromolecular crystals.[50] Fully occupied hydration sites—bound or ordered water—represent only a small fraction of the solvent. The remaining solvent is disordered but not completely featureless.

One approach to modeling solvent is atomistic, using individual scattering atoms for each solvent molecule. Although this approach is reasonable for modeling fully occupied hydration sites, it is less appropriate for the

[50] B. W. Matthews, *J. Mol. Biol.* **33,** 491 (1968).

bulk solvent regions, which are largely disordered. Furthermore, atomistic modeling of solvent introduces a large number of adjustable parameters, increasing the danger of overfitting the data.[8] It is therefore more appropriate to use continuous models of electron density to describe disordered solvent.

The simplest possible model comes from the simplest possible assumptions—that scatterers are positioned with equal likelihood everywhere in the bulk solvent regions, giving rise to constant, or flat, electron density outside the macromolecule. Implementations of this "flat" model use Babinet's principle[51,52] or the molecular surface to define a "solvent mask."[53] In the vicinity of the solvated macromolecule one expects deviations from this flat model. A more detailed "radial shell" model divides the solvent volume into shells of constant (but possibly different) electron density extending outward from the macromolecular surface.[54] The radial shell model was devised to minimize the number of adjustable parameters by radial averaging. However, the surface of a macromolecule is normally anisotropic both in terms of shape and chemical composition. Badger and Caspar[55] attempted to model the resulting anisotropic solvent distribution through an iterative difference map procedure ("difference map model"). An extension of the difference map model was aimed at damping large density fluctuations in the solvent regions; it is referred to as the "density modification model."[16]

The quality of these bulk solvent models was assessed by complete cross-validation.[16] Figure 8 summarizes the results. Without any solvent model the free *R* value is 26.2% and the phase error of the model is 34.4°. On inclusion of 134 ordered water molecules, the free *R* value drops by 1.5%, accompanied by a 1.5° decrease in the phase error of the model. If the flat solvent model is added both quantities drop further (1.7% and 2.3°, respectively). The radial shell model and the density modification model produce a slight further improvement (around 0.5% and 0.2°, respectively). The difference map model, however, overfits the diffraction data at high resolution, indicated by a lower *R* value but a higher free *R* value compared to the radial shell model.

Although the flat solvent model fits the diffraction data quite well, the electron density in the solvent region is not completely homogeneous. Fluctuations of solvent density are observed giving rise to characteristic

[51] R. D. B. Fraser, T. P. MacRae, and E. Suzuki, *J. Appl. Crystallogr.* **11,** 693 (1978).
[52] P. C. Moews and R. H. Kretsinger, *J. Mol. Biol.* **91,** 201 (1975).
[53] S. E. V. Phillips, *J. Mol. Biol.* **142,** 531 (1980).
[54] X. D. Cheng and B. P. Schoenborn, *Acta Crystallogr.* **B46,** 195 (1990).
[55] J. Badger and D. L. D. Caspar, *Proc. Natl. Acad. Sci. U.S.A.* **88,** 622 (1991).

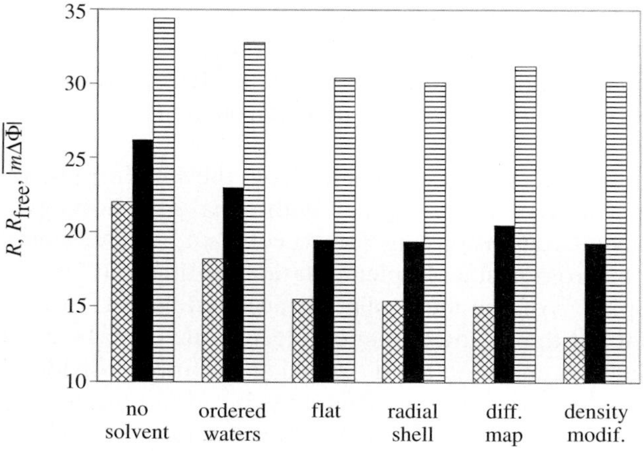

FIG. 8. Assessment of the quality of solvent models for the penicillopepsin crystal structure at 1.8- to 23-Å resolution. The R value (%; cross-hatched bars), free R value (%; black bars), and the model's difference from the MIR phases $(\overline{m|\Delta\Phi|})$ (degrees; striped bars) are shown for a number of solvent models. Complete cross-validation was used to compute the free R value. The weight w_{xray} [Eq. (17)] was set to $10^{5.3}$, which approximately optimizes the free R value and phase accuracy (Fig. 3). The following cases are shown: no solvent model, inclusion of ordered water molecules, additional modeling of disordered solvent by the flat model, radial shell model, difference-map model, and density-modification model. For details, consult Ref. 16.

solvent distributions around polar and nonpolar atoms. Detailed analysis of the solvent features in electron density as obtained by the difference map and density modification models must be viewed with caution because of the low level of solvent electron density that is observed. An alternative approach consists of using radial averaging over multiple sites on the surface of the protein, which will reduce the amount of noise present in the distribution functions.[16,56] The resulting solvent distribution functions show a well-defined solvation shell at hydrogen-bonding distance (2.8 Å) between water and nitrogen or oxygen atoms, and a roughly van der Waals distance (4 Å) between water and carbon atoms.[16] No statistically significant higher order solvation shells emerged in the cases studied. The observation of a largely flat average electron-density distribution in the bulk solvent region justifies the use of solvent-flattening procedures for initial phase improvement.

[56] J. Badger, *Biophys. J.* **65,** 1656 (1993).

Data Quality

Although multiconformer and bulk solvent refinement produce significant improvements, the resulting free *R* value does not approach the low noise level (R_{merge} = 3.4%) of the diffraction data of penicillopepsin. What would happen if the noise level of the diffraction data were actually much higher than estimated? To investigate this question a synthetic data set was calculated from the penicillopepsin crystal structure at 1.8-Å resolution. Noise derived from a Gaussian distribution was added to the calculated amplitudes, and the crystal structure was refined against these synthetic data. For noise-free data, the refinement converges almost perfectly to the crystal structure with a phase error of 3° and an r.m.s. coordinate error below 0.02 Å. This convergence is achieved despite an aggressive refinement protocol involving simulated annealing at 2000 K and the application of cross-validation.

Figure 9 shows that the *R* values and the phase error of the model increase as more noise is added to the data. To reach a free *R* value greater

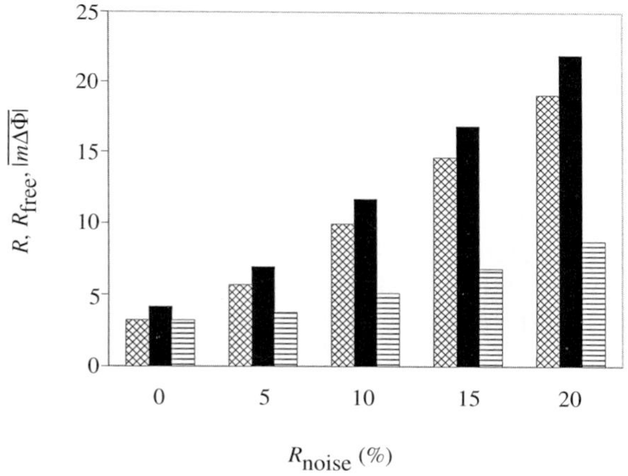

FIG. 9. Influence of normally distributed noise on the *R* (%; cross-hatched bars) and free *R* values (%; black bars), and on $\overline{m|\Delta\Phi|}$ (degrees; striped bars) for the penicillopepsin crystal structure, using synthetic data computed from the crystal structure and noise added to it. The level of noise is indicated by the *R* value (R_{noise}) between the noise-free synthetic data set and the data set with noise added. The *R* and free *R* values were computed at 6- to 1.8-Å resolution whereas $\overline{m|\Delta\Phi|}$ was computed at 6- to 2.8-Å resolution. A single-conformer model without water molecules was used in the computation of the synthetic data and the refinement against the noisy data. Refinements were started with the crystal structure of penicillopepsin with water molecules omitted and uniform *B* factors. Refinement consisted of a Cartesian-space slow-cooling protocol[32] starting at 2000 K, overall *B*-factor refinement, and individual restrained *B*-factor refinement.[9] The weight w_{xray} [Eq. (17)] was set to $10^{5.3}$. Averages over 10 refinements with different test sets are shown.

than 20% (as observed in Figs. 3, 7, and 8) one must add about 20% noise to the data. It seems unlikely that the real uncertainty of the measurements is this high, considering the high statistical quality of the penicillopepsin diffraction data. Furthermore, the difference between *R* value and free *R* value is much smaller than that observed in Figs. 3, 7, and 8. Thus, one can conclude safely that one cannot account for the free *R* value of the penicillopepsin crystal structure exclusively by normally distributed noise.

Incorrect Structures

The free *R* value shows enhanced sensitivity to global model errors.[8] For example, the *R*-value difference between the correct and incorrect models of plant ribulose-1,5-bisphosphate carboxylase/oxygenase (RuBisCO, Ec 4.1.1.39)[57,58] is only 4% for comparable geometry, whereas the free *R* value difference is 13%, suggesting that the incorrect model had been overfit.

Perhaps even more dramatic is a study showing that an inverted chain tracing of the cellular retinoic acid-binding protein type II (i.e., the polypeptide chain runs in the opposite direction) can be refined to a reasonably low *R* value of around 21% with good geometry and satisfaction of many criteria of correct protein folds.[38,59] However, the free *R* value for this incorrect model is 62%, which exceeds slightly the expected value for a random distribution of atoms in an acentric space group.

It should be noted that the free *R* value is mostly sensitive to global errors in the chain tracing. Detection of more subtle errors, such as misplaced side chains, will be more difficult unless the change in free *R* value exceeds the estimated standard deviation (Fig. 5). Annealed omit maps[60] and the real-space *R* value[7] between them and the model are probably more useful in detecting local errors.

Density Modification

Solvent flattening is a useful constraint for the early stages of crystallographic structure determination. However, sometimes it fails to produce significant improvement of poor experimental or molecular replacement

[57] P. M. Curmi, D. Cascio, R. Sweet, D. Eisenberg, and H. Schreuder, *J. Biol. Chem.* **267,** 16980 (1992).

[58] M. S. Chapman, S. W. Suh, P. M. G. Curmi, D. Cascio, W. W. Smith, and D. Eisenberg, *Science* **241,** 71 (1988).

[59] G. J. Kleywegt and T. A. Jones, *Structure* **3,** 535–540 (1995).

[60] A. Hodel, S.-H. Kim, and A. T. Brünger, *Acta Crystallogr.* **48,** 851 (1992).

phases. This may result from incorrect parameterization. In addition, as with any powerful method there is the potential to overfit or misinterpret the data. Cross-validation avoids this problem.

In the context of solvent flattening, cross-validation is implemented by omitting a test set of the observed data, modifying electron-density maps computed using the remaining data, updating calculated structure factors by inverse Fourier transformation, and evaluating the free *R* value.[26] Only the working reflections are used to compute the electron-density map, but updated solvent-flattened calculated structure factors are obtained for all the reflections, including the test set, by inverse Fourier transformation. Because of the sensitivity of the free *R* value to the test set selection at low resolution complete cross-validation is required.

Figure 10 illustrates the use of cross-validation to optimize the smoothing radius used to compute the molecular boundary from an initial electron density map.[61] Too small a radius is likely to produce cavities within the protein and assign external protein loops to the solvent region, whereas too large a value will remove detail from the solvent–macromolecule boundary. This is reflected in the poor behavior of the free *R* value and the phase error for small (2Å) and large (15 Å) values of the smoothing radius. The free *R* value can thus be used to optimize parameters of density modification procedures when independent phase error estimates are unavailable.

Practical Issues

A Posteriori Free R Value

It may be useful to obtain the free *R* value for crystal structures that were modeled and refined against all observed data. Because all data were used to determine the structure, a procedure is required to remove the "memory" of the test set before the free *R* value can be computed. To accomplish this, one needs a refinement method capable of escaping local minima, such as simulated annealing (see [13] in this volume[12]). A few cycles of conjugate gradient minimization are insufficient to remove the memory of the test set (Fig. 11). It would take hundreds more cycles to achieve convergence of the free *R* value.[24] Thus, it is clearly more efficient to use simulated annealing to remove the memory of the test set. Figure 11 shows that even a fairly modest initial simulated annealing temperature

[61] B.-C. Wang, *Methods Enzymol.* **115,** 90 (1985).

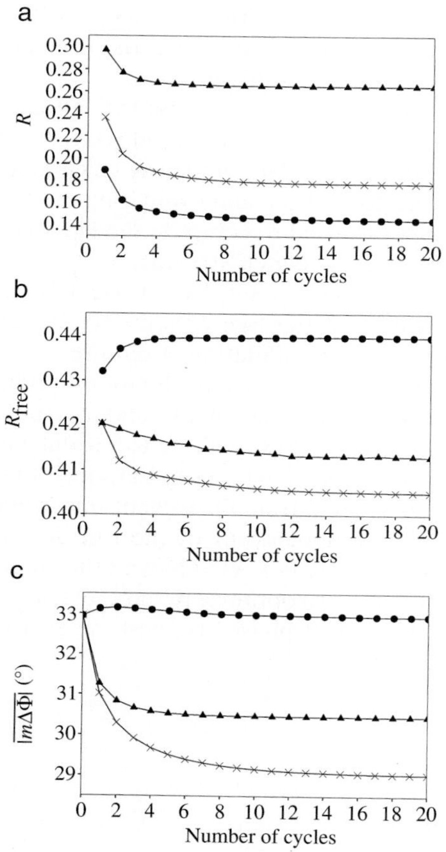

FIG. 10. Comparison of different smoothing radii r_s (●, 2 Å; ×, 8 Å; ▲, 15 Å) for solvent flattening starting with the multiple isomorphous replacement (MIR) phases of penicillopepsin at 2.8-Å resolution.[26,62] Shown are (a) the R value, (b) the free R value, and (c) the difference between the solvent-flattened MIR phases and the phase accuracy of the crystal structure ($\overline{m|\Delta\Phi|}$) as a function of density modification cycles. Complete cross-validation was used.[26]

is sufficient to achieve convergence of the free R value (within the standard deviation of the free R value in this case).

Even the best refinement method does not guarantee the complete removal of knowledge of the test set. Ideally, the test set would be removed right from the start and kept fixed throughout the modeling and refinement process. If higher resolution data become available, it is a good idea to extend the test set to higher resolution instead of simply redefining it.

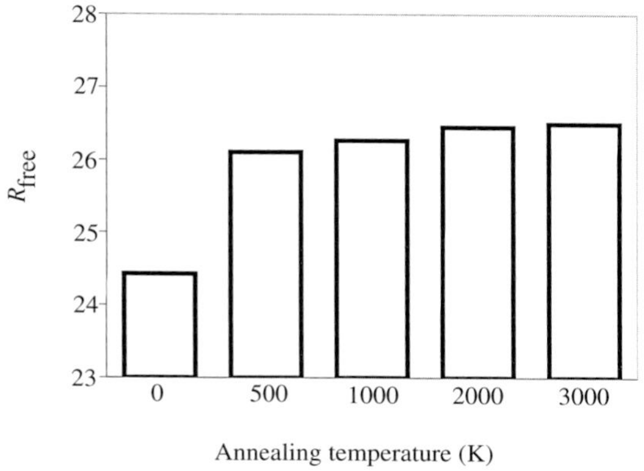

FIG. 11. *A posteriori* determination of the free R value. The penicillopepsin crystal structure was first refined against all observed data at 1.8-Å resolution. The water molecules were removed and the structure refined against all data using simulated annealing and conjugate gradient minimization. This refined structure was the starting point for all calculations presented here. Refinement against the working set only was used to reduce the memory toward the test set. Refinement at 0 K consisted of 40 steps of conjugate gradient minimization followed by 40 steps of restrained conjugate gradient B-factor refinement. Refinement at nonzero temperatures used a Cartesian-space slow-cooling protocol[32] starting at the specified temperature, followed by the same refinement protocol as at 0 K. The weight w_{xray} [Eq. (17)] was set to $10^{5.3}$. Averages over 10 refinements with different test sets are shown.

To remain completely faithful to the idea of cross-validation, electron-density maps should be computed from the working set of diffraction data. However, one expects the effect of omitting the test set on the map to be small.

Water Placement

The free R value can be used to gauge if the inclusion of water molecules improves the model.[8] It should be monitored as more and more water molecules are added. If the free R value stays constant or increases, one is likely overfitting the diffraction data. However, the placement of a single water molecule has probably an insignificant effect on the free R value.

It is important to refine the model and water molecules before computing the free R value. Assuming that the model was refined without the test set, a conjugate gradient minimization will probably suffice to compute the free R value after additional water molecules were placed. Deciding about water placement by direct inspection of the free R value without prior refinement is inappropriate because it would be equivalent to a direct fitting of the test set.

Data Manipulations

Data transformations and manipulations that affect both the working and test sets may disrupt the cross-validation protocol. Examples are resolution-dependent weighting procedures. Because they will influence all data, a change in the free R value can be caused directly by the weighting procedure or indirectly by the change in the model on refinement against the weighted working set of reflections. Cross-validation may still be useful, but care must be taken in the interpretation of the results.

Monitoring Free R Value

It is good practice to monitor the free R value during the whole course of structure determination and refinement. If the computational resources are available, it is useful to perform multiple refinements with different initial velocities for simulated annealing.[37] The refinement protocol that produces the lowest free R value would then be used for electron-density map calculations or for further refinement.

Free R Value Standard

The outcome of the free R calculation depends to some degree on the choice of the test set. It is thus important to set a standard. Once the standard has been accepted, comparison of free R values among different crystal structures becomes possible.

We propose the following standard for evaluation of the free R value. Ten percent of the diffraction data should be selected for the test set. According to our empirical rule discussed above, the standard deviation of the free R value is given by $R/(n)^{1/2}$. Thus if $n < 500$ and $R_{\text{free}} = 20\%$ one obtains $\sigma(R_{\text{free}}) > 1\%$, which is too large for most purposes. If the test set size is below 500, if free R value distributions are desired, or if the particular phasing or refinement algorithm depends on particular small subsets of the diffraction data, one should perform complete cross-validation. Ideally, cross-validation should be used right from the start of the structure determination and refinement process. The definition of the test set or test sets in the case of complete cross-validation should be kept fixed. If cross-validation was not used from the start, the memory toward the test set(s) needs to be removed by a simulated-annealing protocol, but it is good practice to indicate in publications that the quoted free R value is *a posteriori.*

We chose a definition of the test set based on relative rather than absolute size. This allows one to extend the test set to higher resolution if additional data become available while maintaining consistency of the free *R* definition.

If a single test set is used the free *R* calculation does not require any additional computational resources. However, for complete cross-validation, the phasing or refinement procedure needs to be carried out 10 times, once for each test set.

Open Issues

Noncrystallographic Symmetry

Inclusion of noncrystallographic symmetry through restraints or constraints[62] can produce fairly low free *R* values. In principle, this behavior is correct because model phases will improve by averaging if the individual molecules are sufficiently close. However, the noncrystallographic symmetry introduces correlations between reflections that may affect the independence of the test set from the working set of reflections. An extreme example would be the incorrect description of a crystallographic symmetry as noncrystallographic symmetry. In this case, there are sets of reflections that are exactly related to each other. If one were to choose at random a test set from the asymmetric unit of the lower symmetry space group, some elements in the test set would probably be equal to some in the working set. In this case, the free *R* value would always be close to the normal *R* value regardless of the quality of the model.

The situation in the case of true noncrystallographic symmetry is more complicated. Nonlocal relationships between reflections are introduced that make it difficult to choose a perfectly independent test set. One can imagine a number of approximate schemes to make the test set more independent of the working set (e.g., using randomly selected resolution shells), but these alternative schemes do not offer any advantages.[59] Clearly, when free *R* values are quoted it is necessary to indicate the presence of noncrystallographic symmetry.

Acceptable Free R Values

Brändén and Jones[1] suggested 25% as a critical threshold for the normal *R* value. However, it has since been shown that incorrect models can be refined to lower *R* values.[8,59] Because the free *R* value is much more sensitive to model errors, there may indeed be a meaningful threshold for the free *R* value above which the crystal structure is likely to be incorrect or the

[62] W. I. Weis, A. T. Brünger, J. J. Skehel, and D. C. Wiley, *J. Mol. Biol.* **212,** 737 (1989).

data quality so poor that it must be viewed with caution. If one makes the reasonable assumption that an overall r.m.s. coordinate error of more than 1 Å could indicate a potential error in model, this threshold could be defined at around 40% (Fig. 4a).[59]

Conclusion

Even the best bulk solvent and multiconformer models result in free *R* values of around or slightly above 20% for the penicillopepsin crystal structure (Figs. 7 and 8). A free *R* value of 20% is significantly higher than one might expect from the estimated high statistical quality of the diffraction data. Thus, it is conceivable that the present models for solvation and thermal motion are incomplete. Alternatively, the intensity data might be affected by systematic errors of unknown origin. These facts point to the need for the solution of benchmark macromolecular structures at high resolution and with accurate experimental phases.[63] With such benchmark structures improved refinement strategies can be developed in an attempt to provide a more accurate model of macromolecules in the crystalline state. More accurate models will provide new insights into flexibility and solvation and their role in biological processes.

Acknowledgments

I am grateful to F. T. Burling, P. Gros, J.-S. Jiang, G. J. Kleywegt, R. J. Read, L. M. Rice, and A. L. U. Roberts for many fruitful discussions, and Paul Adams and Sapan Shah for critical reading of the manuscript. This work was funded in part by grants awarded to the author from the National Science Foundation (DIR 9021975 and BIR 9317832). Use of the Cray supercomputer center at the National Cancer Institute is also acknowledged.

[63] F. T. Burling, W. I. Weis, K. M. Flaherty, and A. T. Brünger, *Science* **271,** 72 (1996).

[20] VERIFY3D: Assessment of Protein Models with Three-Dimensional Profiles

By DAVID EISENBERG, ROLAND LÜTHY, and JAMES U. BOWIE

Introduction

As methods for experimental and computational determination of protein three-dimensional structure develop, a continuing problem is how to

0076-6879/97 $25.00

verify that the final protein model is correct.[1] One effective test of the correctness of a 3D protein model is the compatibility of the model to its own amino acid sequence as measured by a 3D profile. The 3D profile[2] of a protein structure is a table, computed from the atomic coordinates of the structure, that can be used to score the compatibility of the three-dimensional structure model with any amino acid sequence. Three-dimensional profiles computed from correct protein structures match their own sequences with high scores. In contrast, 3D profiles computed from protein models known to be wrong score poorly. An incorrectly modeled segment in an otherwise correct structure can be identified by examining the profile score in a moving-window scan. Thus the correctness of a protein model can be verified by its 3D profile, regardless of whether the model has been derived by X-ray, nuclear magnetic resonance (NMR), or computational procedures. For this reason, 3D profiles are useful in the evaluation of undetermined protein models, based on low-resolution electron-density maps, on NMR spectra with inadequate distance constraints, or on computational procedures.

Three-dimensional models of proteins are derived today by a variety of means, and sometimes the validity of the model is in doubt. This can occur in X-ray analysis if the crystals studied diffract only moderately well, or if the phases are poorly determined. The challenge is to distinguish between a mistraced or wrongly folded molecule and one that is basically correct, but not adequately refined. Several published protein structures have had to be revised, correcting errors ranging from local misregistration of the model in the electron density to interchanged β strands to wholesale mistracings. This situation has prompted the development of new criteria for judging X-ray models to supplement the traditional R-factor and Ramachandran plots.[2–7] In energetic analysis of protein structure, the difficulty in evaluating models is a topic of ongoing discussion.[8–10] To help in these situations, we describe a method for verification of protein models in which the test of correctness is the compatibility of the model with its own amino acid sequence.

[1] R. Lüthy, J. U. Bowie, and D. Eisenberg, *Nature (London)* **256,** 83 (1992); this chapter follows closely this original report of the method.
[2] J. U. Bowie, R Lüthy, and D. Eisenberg, *Science* **253,** 164 (1991).
[3] T. A. Jones, J.-Y. Zou, S. W. Cousan, and M. Kjeldgaad, *Acta Crystallogr.* **A47,** 110 (1991).
[4] C.-I. Bränden and T. A. Jones, *Nature (London)* **343,** 687 (1990).
[5] A. T. Brünger, *Nature (London)* **355,** 472 (1992).
[6] C. Colovos and T. O. Yeates, *Protein Sci.* **2,** 1511 (1993).
[7] M. J. Sippl, *Proteins Struct. Funct. Genet.* **17,** 355 (1993).
[8] J. Novotny, A. A. Rahin, and R. Bruccoleri, *Proteins Struct. Funct. Genet.* **4,** 19 (1988).
[9] D. Eisenberg and A. D. McLachlan, *Nature (London)* **319,** 199 (1986).
[10] J. Novotny, R. Bruccoleri, and M. Karplus, *J. Mol. Biol.* **177,** 787 (1984).

Three-Dimensional Profiles

The method (see Fig. 1[1,2,10,11] for a description) measures the compatibility of a protein model with its sequence, using a 3D profile. Each residue position in the 3D model is characterized by its environment[2] and is represented by a row of 20 numbers in the profile. These numbers are the statistical preferences (called 3D–1D scores) of each of the 20 amino acids for this environment. Environments of residues are defined by three parameters: the area of the residue that is buried, the fraction of side-chain area that is covered by polar atoms (oxygen and nitrogen), and the local secondary structure. The 3D profile score S for the compatibility of the sequence with the model is the sum, over all residue positions, of the 3D–1D scores for the amino acid sequence of the protein. As described below, the compatibility of segments of the sequence with their 3D structures can be assessed by plotting, against sequence number, the average 3D–1D score in a window of 21 residues.

For 3D protein models known to be correct, the 3D profile score S for the amino acid sequence of the model is high. This is illustrated in Fig. 2,[12–14] where the scores of well-determined structures are indicated by dots. In contrast, the profile score S for the compatibility of a wrong 3D protein model with its sequence is generally low, as shown by the squares in Fig. 2 and discussed below.

The profile score of a model depends on its length and its validity. Profile scores of correct models increase with the length of the protein, simply because more positive residue preferences are added into the sum. The scores for computationally based models vary, depending on their correctness. The deliberately misfolded models of Novotny *et al.*[10] receive poor scores, because the environments of residues in the incorrect 3D structures are not compatible with the residues in the corresponding positions of the sequence. In contrast, models based on structures having closely related amino acid sequences, such as those for cyclic-AMP protein kinase (Brookhaven Protein Data Bank model 2APK),[15] insulin-like growth factor (1GF1, 2GF1),[16] and apolipoprotein D (1APD),[17] receive high scores, comparable to scores received by many X-ray and NMR models. This contrast

[11] R. E. Stenkamp, L. C. Sieker, and L. H. Jensen, *Acta Crystallogr.* **B38,** 784 (1978).

[12] F. C. Bernstein, T. F. Koetzle, G. J. Williams, E. F. Meyer, M. D. Brice, J. R. Rodgers, O. Kennard, T. Shimanouchi, and M. Tasumi, *J. Mol. Biol.* **112,** 535 (1977).

[13] D. Eisenberg, J. U. Bowie, R. Lüthy, and S. Chloe, *Faraday Discuss.* **93,** 25 (1992).

[14] C. P. Hill, D. H. Anderson, L. Wesson, W. F. DeGrado, and D. Eisenberg, *Science* **249,** 543 (1990).

[15] I. T. Weber, T. A. Steitz, J. Bubis, and S. S. Taylor, *Biochemistry* **26,** 343 (1987).

[16] T. A. Blundell, S. Bedarkar, and R. E. Humbel, *FASEB J.* **42,** 2592 (1983).

[17] M. C. Peitsch and M. S. Boguski, *New Biol.* **2,** 197 (1990).

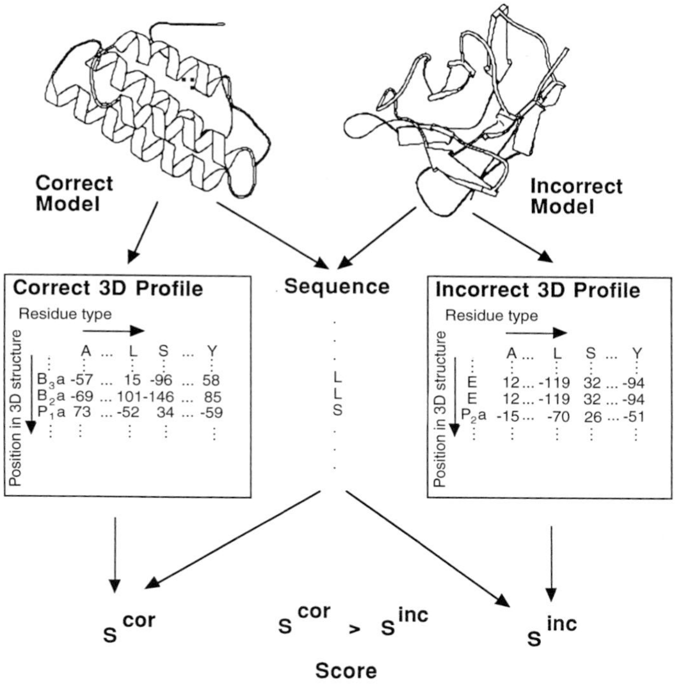

FIG. 1. The use of 3D profiles to verify protein models, illustrated with the correct and misfolded structures of hemerythrin,[10] both having 113 residues. The model on the left is the X-ray-derived structure.[11] A 3D profile calculated from its coordinates matches the sequence of hemerythrin with a score $S^{\text{cor}} = 38$. The model on the right is the misfolded hemerythrin model of Novotny *et al.*[10] A 3D profile calculated from it matches its sequence poorly, with score $S^{\text{inc}} = 15$.

The actual profile consists of 113 rows (1 for each position of the folded protein). In this schematic example only three rows are shown, those for positions 33 (where the residue is L), 34 (L), and 35 (S). The first column of the profile gives the environmental class of the position, computed from the coordinates of the model,[2] and the next 20 columns give the amino acid preferences (called 3D–1D scores) in that position. In this schematic example, there are only four columns of 3D–1D scores shown, those for residues, A, L, S, and Y. In the correct profile, on the left, position 33 is computed to be in the buried polar α class ($B_3\alpha$); positions 34 and 35 are computed to be in the buried moderately polar α class ($B_2\alpha$) and the partially buried α polar class, $P_1\alpha$, respectively. The scores for the residues L, L, and S for these three positions are 15, 101, and 34. The profile for the midfolded model assigns positions 33, 34, and 35 to the environmental classes E, E, and $P_2\alpha$, giving 3D–1D scores for residues L, L, and S of -119, -119, and 26, respectively. That is, in the incorrect structure, the leucine residues are exposed, giving low 3D–1D scores and leading to a summed total score S^{inc} that is much smaller than S^{cor} when all other 3D–1D scores are summed along with the three shown here. (Reprinted from Lüthy *et al.*[1])

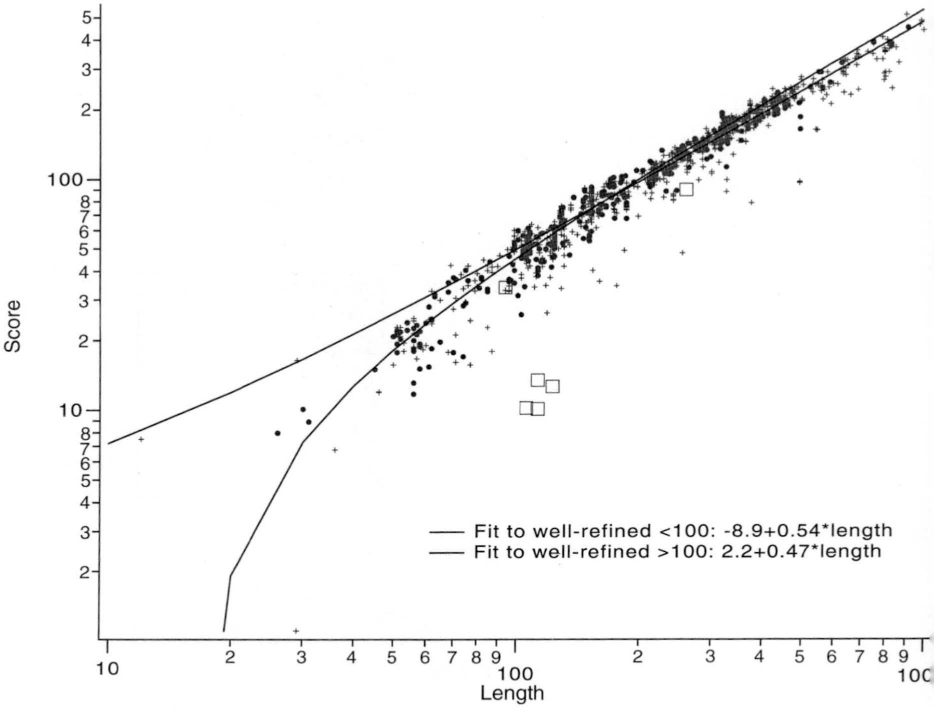

FIG. 2. Three-dimensional profile scores (indicated by plus symbols, +) for X ray-determined protein coordinate sets in the Brookhaven Protein Data Bank,[12] as a function of sequence length, on a log–log scale. Scores for highly refined X-ray determinations are indicated by dots (•). These are structures determined at resolutions of at least 2 Å and with R factors less than 20%. Two lines are fit by least squares for well-refined structures: the upper for structures of greater than 100 residues; the lower for structures of fewer than 100 residues. Misfolded structures are indicated by squares (□): These correspond to the entries of Table I and also to the lower curves in Fig. 3. Environmental classes for 3D profiles of oligomeric proteins were generally computed from oligomeric structures, rather than protomers. The difference is that the accessible surface areas of residues positioned at interfaces are greater for the protomers, producing a poorer fit of the profile to the sequence. This matters little for large structures, but very much for small structures.[13] As an extreme case, the profile for the 12-residue designed protein $\alpha 1$[14] (the leftmost plus symbol) matches its sequence only when the molecule is surrounded by its neighbors as in the crystal structure. The application of VERIFY3D to small proteins is discussed by Eisenberg *et al.*[13]

TABLE I
THREE-DIMENSIONAL PROFILE SCORES FOR COMPATIBILITY OF CORRECT AND FLAWED PROTEIN MODELS WITH THEIR OWN AMINO ACID SEQUENCES[a]

Model	Correct	Flawed	Length (residues)
RuBisCO small subunit	55.0 (3RUB)[b,c]	15.1[d]	123
Ferredoxin of *Azotobacter vinelandii*	45.9 (4FD1)[b]	10.8 (2FD1)[b]	106
Ig κ V region	47.0[e]	6.9[e]	113
Hemerythrin	37.9[f]	14.9[e]	113
p21^{ras}	79.8	50.5	177

[a] Notice that correct models match their own sequences with high scores, but that flawed models match with low scores. Scores of some flawed models are shown in Fig. 2 by squares.
[b] Brookhaven Protein Data Bank codes.[21]
[c] From Ref. 19.
[d] From Ref. 18.
[e] Energy-minimized models, described in Novotny *et al.*[8]
[f] From Ref. 20.

suggests that 3D profiles can distinguish between correct and misfolded models, however developed, and are useful in assessing computational models.

Examples of Model Verification with VERIFY3D Program

Several examples of profiles from models having problems are given in Table I[8,18] and illustrated in Fig. 3. One example is that of the small subunit of ribulose-1,5-bisphosphate carboxylase/oxygenase (RuBisCO), which was traced essentially backward from a poor electron-density map.[18] The profile calculated from this mistraced model gives a score of only 15 when matched to the sequence of the small subunit of RuBisCO. This score is well below the value of about 58 expected for a correct structure of this length (123 residues). In fact, the profile for the correct model the small subunit of tobacco RuBisCO[19] matches its sequence with a score of 55. Also, when the score for the mistraced small subunit of RuBisCO is plotted as a function of the sequence, as in Fig. 3, the average score is often below the value of

[18] M. S. Chapman, S. W. Suh, P. M. G. Curmi, D. Cascio, W. W. Smith, and D. Eisenberg, *Science* **241,** 71 (1988).
[19] P. M. G. Curmi, D. Cascio, R. M. Sweet, D. Eisenberg, and H. Schreuder, *J. Biol. Chem.* **267,** 16980 (1991).

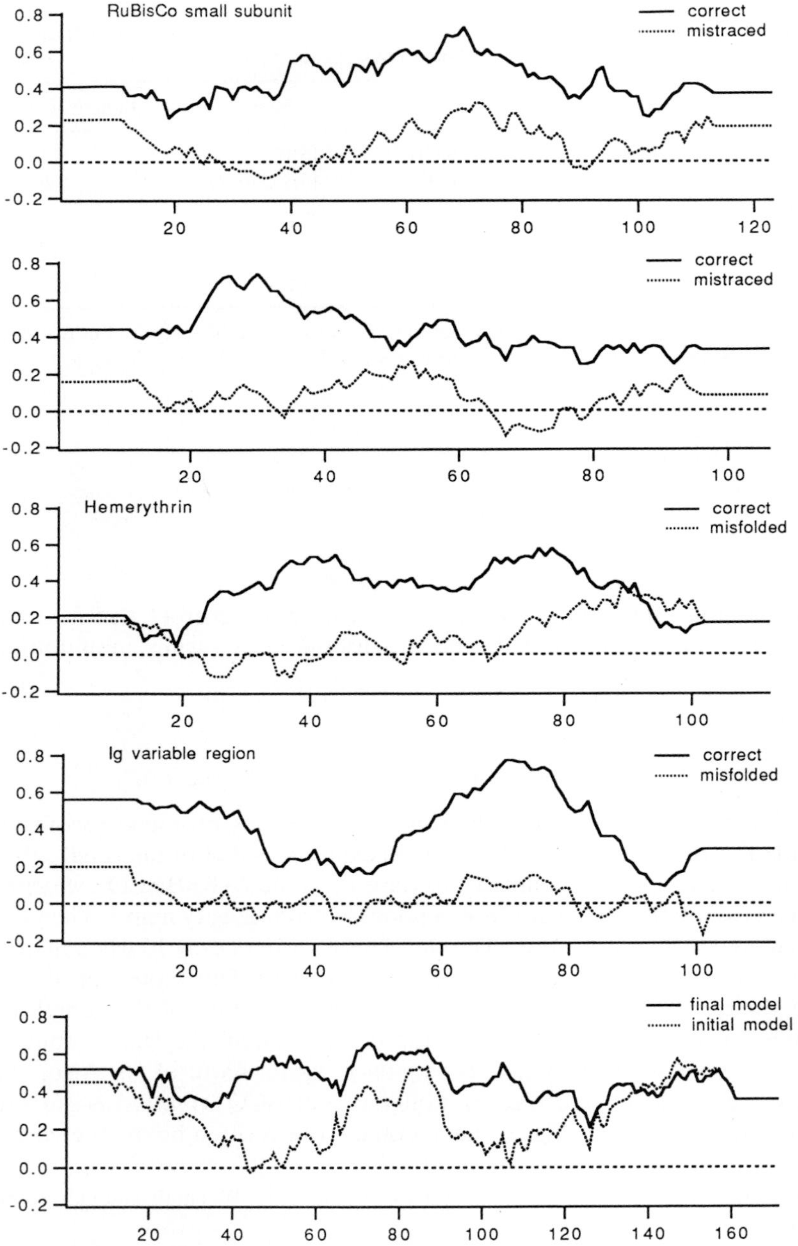
RuBisCo small subunit
correct
mistraced
0.8
0.6
0.4
0.2
0.0
-0.2
20
40
60
80
100
120
correct
mistraced
20
40
60
80
100
Hemerythrin
correct
misfolded
Ig variable region
correct
misfolded
final model
initial model
140
160

0.1 and dips below 0 at two points. When scores for well-refined structures, such as the correct model for the RuBisCO small subunit, are displayed in this way, in a running window of 21 residues, the score rarely dips below 0.1 and virtually never below 0. Similar results are found for other mistraced models and for deliberately misfolded protein models.[1] Taken together, the examples of Table I and Fig. 3 suggest that a correct protein model can be verified by compatibility of its 3D profile with its sequence, and a wrong model can be detected by incompatibility.

If a model is flawed but largely correct, a 3D profile plotted in a 21-residue window reveals the segments of sequence that are improperly constructed. An initial model for one protein illustrates this power: in an initial model, β strands 1 (residues 1–9) and 3 (residues 51–57) were interchanged, but the electron-density map had otherwise been properly interpreted. The overall profile score for the initial model is toward the lower range of profile scores for proteins of this length (171 residues), but perhaps not low enough to signal an error in the chain tracing. However, when the profile score is plotted in a 21-residue window, as in the bottom portion of Fig. 3, the dip of the score below 0 is a more serious signal of a problem in the region of β strand 3. Thus profile window plots, such as those in Fig. 3, may be effective in revealing local problems in initial structures.

Conclusion

An advantage of using 3D profiles for testing models is that profiles have not themselves been used in the determination of the structure. Traditional *R*-factor tests in X-ray analysis depend on comparison of observed properties (i.e., the X-ray structure factor magnitudes) with the same property calculated from the final protein model. In X-ray analysis, the final model is generally obtained by systematic variation of the model, forced by minimization of a function closely related to the *R* factor. Thus in essence, the model is judged by a test that is inherent in its formulation. The

---

FIG. 3. Profile scores for various incorrect and flawed protein models compared to scores for their correct counterparts, shown as a function of position along the sequence. The vertical axis gives the average 3D–1D scores for residues within a 21-residue sliding window, the center of which is at the sequence position indicated by the horizontal axis. Scores for the first nine and the final nine sequence positions have no meaning. Notice that the scores for incorrect models dip below 0 in some windows. A window length of 21 residues strikes a useful balance between smoothing fluctuations and localizing errors (Adapted from Lüthy *et al.*[1])

free R value[5] is a device to avoid this circularity. Similarly, in construction of models by molecular mechanics and molecular dynamics, the final model is arrived at by minimization of energy. If no experimental structure is available for comparison, energetic analysis enters into both the determination and the evaluation of the structure. In contrast, the 3D profile method for model evaluation proposed here relies only on a comparison of the model to its own amino acid sequence. Thus the present method not only seems effective in assessing protein models, but it also avoids the circularity that is to some extent inherent in some common methods of evaluation of 3D protein models.

VERIFY3D on World Wide Web

The program described here is available on the World Wide Web for noncommercial applications. The URL is http://www.doe-mbi.ucla.edu/verify3d.html. VERIFY3D expects a coordinate file in PDB format. The program returns a profile window plot of the type shown in Fig. 3.

Acknowledgments

We thank the NIH and DOE for support of this research.

Section III

Dynamic Properties

A. From Static Diffraction Data
Article 21

B. From Time-Resolved Studies: Laue Diffraction
Articles 22 through 24

[21] Analysis of Diffuse Scattering and Relation to Molecular Motion

By JAMES B. CLARAGE and GEORGE N. PHILLIPS, JR.

Introduction

Structural biology is on the verge of a major transformation. We are moving beyond static pictures of macromolecules to a point where dynamic descriptions are necessary for defining mechanisms of function. Traditional X-ray crystallography has provided a wealth of protein and nucleic acid structures, revealing intricate details for many biological activities. In many cases, however, the dynamic behavior of these structures contains the essence of their function. Allostery, induced fit, conformational changes: these sometimes subtle alterations in the atomic positions allow proteins to work. Thus, function is not determined by form, but rather by the way in which form changes in time.

Although most X-ray crystallographic scattering experiments are insensitive to time, it does not follow that the technique cannot sense conformational fluctuations. We employ atomic B values to describe a sort of motion, but these quantities embody only the isolated displacements of individual atoms and not how the displacements of different atoms couple with one another.

Statistically, Bragg experiments are sensitive to the mean Fourier components in the electron density for a crystal,

$$I_{\text{Bragg}} = |\langle \text{FT}\rho \rangle|^2 = |\text{FT}\langle \rho \rangle|^2 \tag{1}$$

from which we obtain the mean atomic positions, $\langle r_i \rangle$, and displacements

$$I_{\text{Bragg}} \rightarrow \langle \delta_i^2 \rangle \tag{2}$$

where δ_i is the displacement in the atomic coordinate r_i relative to its average value $\langle r_i \rangle$.

Note that these individual atomic displacements (often expressed as B values, $B_i = 8\pi^2\delta_i^2$) still do not distinguish correlations, e.g., whether the displacements in a molecule are due to a global hinge motion or to some random rearrangement of atoms. Many different dynamic models give rise to the same mean electron density.

To learn about higher order properties in the atomic displacements one should measure, in addition to the mean Fourier spectrum of the electron density, the variance of the Fourier components of the electron density,

0076-6879/97 $25.00

$\Sigma_{\text{time}}(\text{FT}\rho - \langle\text{FT}\rho\rangle)^2$. This can be done. By rewriting this variance in terms of other experimentally measured intensities, we find

$$I_{\text{variance}} = \sum_{\text{time}} (\text{FT}\rho - \langle\text{FT}\rho\rangle)^2 \tag{3}$$

$$= \langle|\text{FT}\rho|^2\rangle - |\text{FT}\langle\rho\rangle|^2 \tag{4}$$

$$= I_{\text{Total}} - I_{\text{Bragg}} \tag{5}$$

$$\equiv I_{\text{Diffuse}} \tag{6}$$

which is simply all X-ray intensity on a detector besides the sharp Bragg peaks. This intensity, referred to as *diffuse scattering* because often it is not concentrated at sharp Bragg positions, contains valuable information about variations from the average structure.

In a sense, then, study of diffuse scattering from a crystal is study of a higher moment in the statistical distribution of matter in a macromolecule. Indeed, whereas standard Bragg analysis yields the individual mean-square displacements, diffuse intensity contains information on all of the cross-correlations among displacements in atomic coordinates i and j,

$$I_{\text{Diffuse}} \rightarrow \langle\delta_i\delta_j\rangle \tag{7}$$

A particular pair-correlation coefficient, $\langle\delta_i\delta_j\rangle$, will be large if a pair of atoms is moving concertedly as part of some correlated unit in the molecule (e.g., a rigid domain); conversely, if two atoms are uncoupled from each other and move independently, their correlation coefficient will vanish.

Diffuse studies are important because they give us the only way to distinguish between different models for correlated movement.[1,2] In this chapter we often use interchangeably the terms *movement* and *displacement,* reflecting the mixing of space and time averages in an X-ray scattering experiment. Despite this ambiguity, much of the disorder in macromolecular crystals is obviously dynamic.[3] Even if the interconversion of conformations within crystallized molecules does not occur on fast time scales, diffuse scattering studies still reveal the conformational flexibility of the molecules, which is crucial to understanding their biological function.

Note that any variation in the periodicity of the electron density of a crystal, not just the disorder introduced by atomic displacements, gives rise

[1] For a visually striking example of how neither diffuse nor Bragg, and therefore none of the intensity data, can provide information on moments higher than the second, see Welberry and Butler.[2] By implication, even though diffuse scattering does not suffer the same degree of model degeneracy that Bragg scattering does, it still cannot guarantee the uniqueness of a model.

[2] T. R. Welberry and B. D. Butler, *J. Appl. Crystallogr.* **27,** 205 (1994).

[3] G. Pesko and D. Ringe, *Annu. Rev. Biophys. Bioeng.* **13,** 331 (1984).

to diffuse scattering. Figure 1 presents a survey of what the diffuse scattering looks like from various kinds of crystal disorder. For example, both mosaicity[4,5] and substitution disorder[6,7] produce diffuse scattering. Mosaicity involves distortions on mesoscopic scales much larger than a molecule; these scattering effects usually can be separated from those due to intramolecular processes. The functional form of diffuse scattering from substitution imperfections is distinct from that due to displacements, and for macromolecular systems with large-amplitude movements is only an issue in cases[6] with a high numbers of vacancies.

Although all imperfections are interesting from a condensed-matter perspective, displacement disorder is most relevant to the biological behavior of molecules. This is the focus of this chapter.

Theory of Diffuse Scattering

Atomic displacements damp the intensity of Bragg reflections from the ideal values that could be assumed were there no disorder. Structure-factor amplitudes are reduced by mean atomic displacements[8] δ through the Debye–Waller factor, which is usually modeled assuming Gaussian statistics (or, dynamically speaking, harmonic potentials) so that Bragg intensities are damped by $e^{-(2\pi s\delta)^2} = e^{-(1/2)Bs^2}$, where $\mathbf{s}$ is the scattering vector,[9] with magnitude $|\mathbf{s}| = s = 2 \sin \theta/\lambda$.

The energy lost from the Bragg peaks does not vanish, however. Rather, because thermal agitation of molecules in the lattice has broken the translational symmetry of the space group, the Fourier components of the crystal need no longer be restricted to a lattice. This resulting diffuse scattering component therefore can assume nonzero values between Bragg nodes, and is spread continuously throughout reciprocal space. The functional form of the Debye–Waller factor implies that this diffuse intensity will rise from the origin as $[1 - e^{-(2\pi s\delta)^2}]$, to complement the Bragg intensity, becoming more intense with resolution. We invite the reader to verify that for a typical protein crystal with $\delta = 0.5$ Å ($B \approx 20$), more photons are scattered diffusely than into Bragg reflections beyond resolutions of $1/s =$ 3.8 Å.

[4] C. G. Darwin, *Phil. Mag.* **27,** 315, 675 (1914).
[5] J. M. Cowley, "Diffraction Physics." North-Holland, Amsterdam, 1984.
[6] A. Guinier, "X-Ray Diffraction." W. H. Freeman & Company, New York, 1963.
[7] J. Doucet, J. P. Benoit, W. B. T. Cruse, T. Prange, and O. Kennard, *Nature* (*London*) **337,** 190 (1989).
[8] Along the direction of the scattering vector.
[9] Note that because of the subject matter, this vector will in general be defined on all of reciprocal space, unless we explicitly restrict it to the reciprocal lattice by writing $\mathbf{s}_{hkl}$.

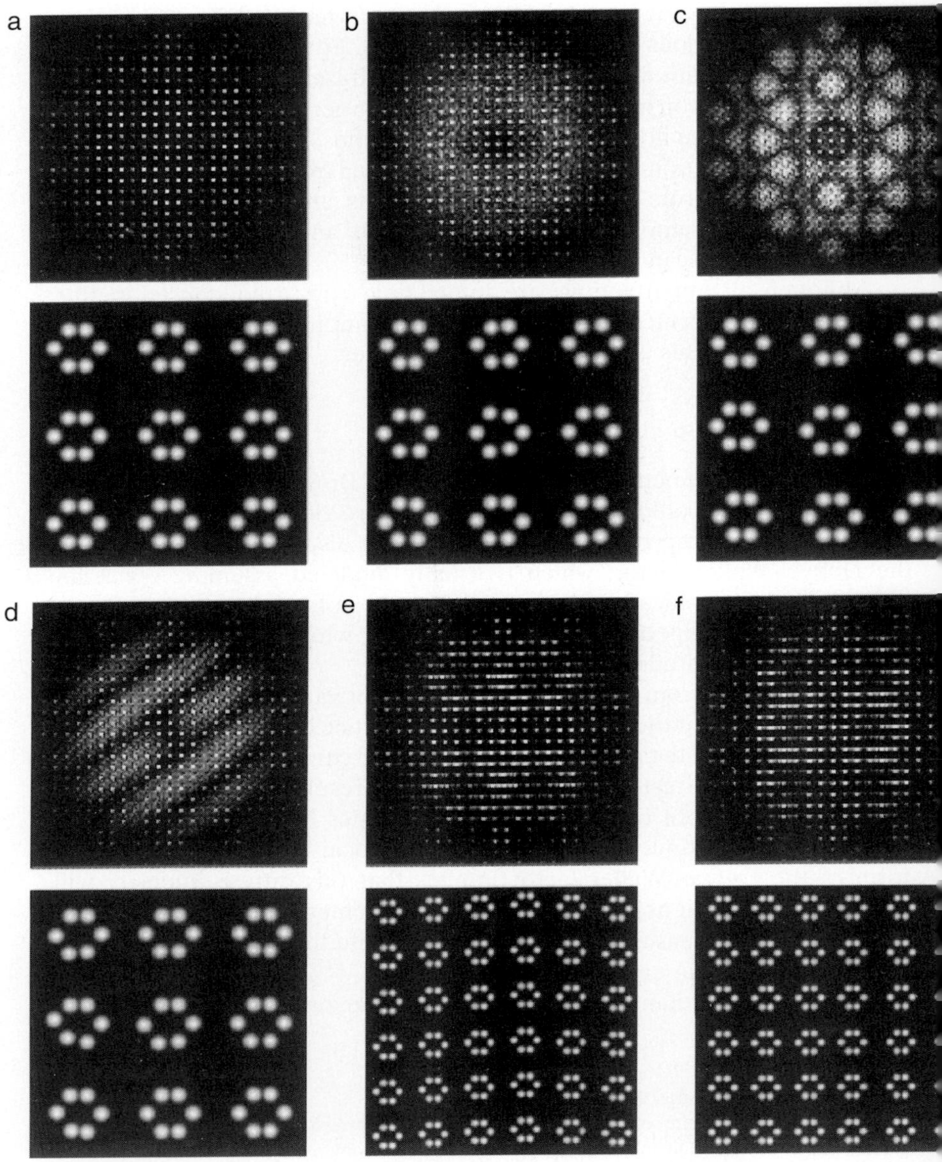

FIG. 1. The effect of different kinds of disorder on the diffraction patterns from crystals. (a–g) Generated by building a two-dimensional crystal with perturbations in the positions of "atoms," followed by calculation of the intensities that would result. Only part of the crystal used in the calculation is shown. (a) A perfect crystal comprising quasihexagonal "molecules," each made up of six "atoms." (b) An Einstein crystal, in which each atom is randomly

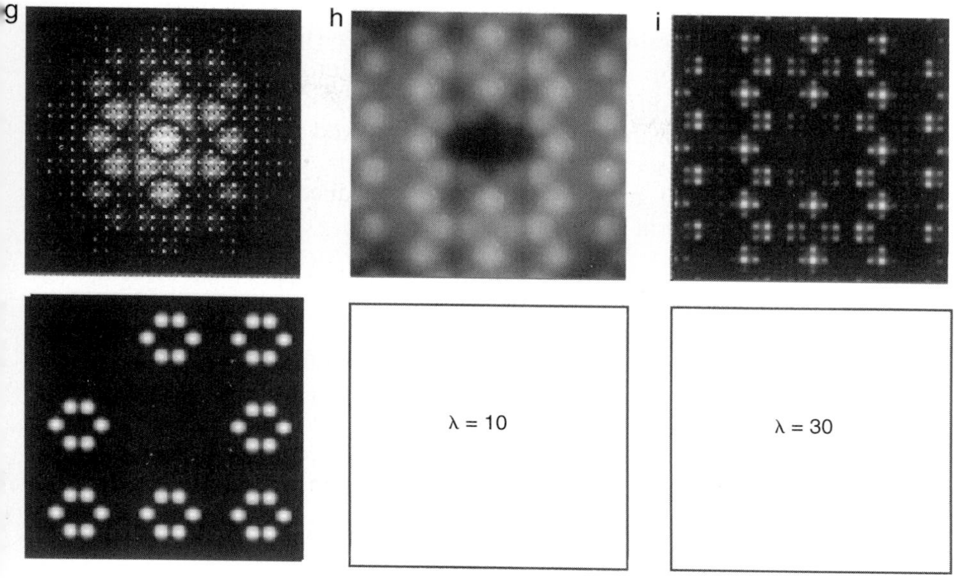

displaced by a small amount. The resulting diffuse scatter is then a ring, representing the scattering from a single atom modulated by the complement of the Debye–Waller factor [see Eq. (9)]. (c) Each molecule has been displaced by a small random vector. In this case, because each molecule moves as a unit, the constructive and destructive interference is maintained in the diffuse scattering, and the transform of an intact molecule is seen in the background. (d) In this case, representing "domain" movement within molecules, a single pair of atoms, the lower left two of each molecule, are displaced. The fringes in the diffuse scattering thus arise from the pairs of moving atoms. (e) Lattice coupled modes. Here, columns of molecules are moved up or down in "waves" that travel in the horizontal direction, hence the term *transverse* waves. These are quite common in macromolecular crystals. (f) Another type of wave is the "compression" wave, with columns of molecules moved to the left or to the right in a sinusoidal pattern that extends horizontally. Note that in both (e) and (f) the diffuse scattering consists of halos close to the Bragg reflections, their shapes smeared in the direction of propagation of the displacement waves, which is horizontal for both these disturbances. However, the amplitude of the diffuse scattering is larger in the vertical directions for (e) and in the horizontal directions for (f), reflecting the fact that the magnitude of the diffuse scattering is strongest along the direction of atomic displacement generated by a disturbance [due to the $\mathbf{s} \cdot \delta$ term in eq. (11)]. (g) Substitution disorder also produces diffuse scattering patterns that follow the transform of the missing motifs. The difference between the substitution case and the displacement case, is that at low resolution the diffuse intensity is still strong with substitution disorder. (h and i) The liquid-like approximation of Caspar and Clarage [Eq. (9)] can be generated for the same crystal. Although it does not correspond to a specific set of atomic correlations in real space, it uses a global correlation length, γ, to characterize the length scale over which atomic movements are correlated. (h) The result of spherically symmetric, short-range correlations ($\gamma = 10$ pixels, or about one-sixth of a unit cell repeat). (i) The result of longer range correlations ($\gamma = 30$ pixels, or roughly one-half of a cell constant). In the latter case, the effect is clear "haloes" around Bragg peaks. [Reproduced from *Acta Cryst.* **D50,** 210 (1994). Copyright © International Union of Crystallography.]

Besides the increase in diffuse intensity with resolution, another phenomenological restraint is that because all diffraction phenomena sample the Fourier transform of the coherently scattering unit in one way or another, we expect diffuse scattering from atomic displacements to depend on the transforms of the groups of atoms involved in concerted movement in the crystal (e.g., see Fig. 1 and Ref. 10).

Thus on general grounds alone we can deduce that diffuse scattering should be of the form

$$[1 - e^{-(2\pi s\delta)^2}]\mathrm{FT}(\text{Concerted unit}) \tag{8}$$

which is our rule of thumb for these studies.

The quantities diffuse scattering lets one obtain, then, are the magnitudes of displacement and dimension of the concertedly moving unit. Because Bragg scattering also gives displacement magnitudes (via B values), diffuse scattering studies usually focus on displacement correlations. We proceed from here with two classic models for how atomic displacements are correlated in condensed matter.

Einstein and Debye Models

The two paradigms for the dynamics of a thermally fluctuating crystalline solid were developed early this century, motivated by the need for explaining the observed low-temperature dependence of heat capacities. In 1907 Einstein[11] assumed individual atoms in a crystal move independently in harmonic wells. Debye,[12] in 1912, devised a model that could fit the data even better by postulating nearest neighbor harmonic coupling between atoms; the motion of the system is then decomposed into the normal modes of the elastic solid, and the displacement correlations fall off algebraically as $1/r$, where r is the separation between two atoms. Thus, in the two paradigms for crystalline solids there is either no correlation, or quasi-long-range correlation in displacements.

Before discussing what the scattering looks like from these two models, let us first give their mathematical form. The total scattering from a monatomic Einstein crystal is[13]

$$I(\mathbf{s}) = e^{-(2\pi s\delta)^2}F_0^2(\mathbf{s}) + [1 - e^{-(2\pi s\delta)^2}]f_0^2(\mathbf{s}) \tag{9}$$

[10] J. L. Amorós and M. Amorós, "Molecular Crystals: Their Transforms and Diffuse Scattering." John Wiley & Sons, New York, 1968.

[11] A. Einstein, *Jb. Radioakt.* **4,** 411 (1907).

[12] P. Debye, *Ann. Phys.* **39,** 789 (1912).

[13] P. Debye, *Ann. Phys.* **43,** 49 (1914).

where $F_0^2(\mathbf{s})$ is the ideal (no motion) structure factor intensity and $f_0(\mathbf{s})$ is the individual atomic form factor. The first term is the Bragg scattering, consisting of structure factors damped by a Debye–Waller factor. The second term, the diffuse scattering, is a featureless ring because atomic form factors are spherical with little modulation. For an isotropic Debye crystal[14] the scattering is[15]

$$I(\mathbf{s}) = e^{-(2\pi s\delta)^2} F_0^2(\mathbf{s}) + e^{-(2\pi s\delta)^2} \frac{(2\pi s)^2 k_B T}{1/2mv^2} \left[F_0^2(\mathbf{s}) * \frac{1}{(\Delta\mathbf{s})^2} \right] \tag{10}$$

where $\Delta\mathbf{s}$ is a vector measured from each reciprocal lattice point, and the asterisk (*) represents convolution. Again, the first term is Bragg scattering. The diffuse scattering pattern, from the second term, consists of ideal Bragg peaks convolved with a sharp function, $1/(\Delta\mathbf{s})^2$, so that the diffuse scattering is concentrated into dense halos surrounding each Bragg reflection.

In both the Einstein and Debye models, the intensity is composed of two parts. Note that, despite the difference in correlations embodied by the two models, their Bragg scattering is identical, demonstrating a result whose importance cannot be overemphasized: *Bragg experiments are insensitive to the underlying dynamic model.*

But the diffuse scattering components are completely different, reflecting the distinct atomic correlations postulated by the two models. Because $[1 - e^{-(2\pi s\delta)^2}] \approx (2\pi s\delta)^2$, to first order both models obey the rule of thumb [Eq. (6)] that diffuse intensity reflects the transform of the correlated unit of movement, modulated by the complement of the Debye–Waller factor. In the Einstein crystal, the featureless diffuse scattering is consistent with the correlated unit being an isolated atom. For the Debye case, the array of halos in the diffuse scattering reflects the fact that the atomic displacements are correlated throughout the crystal lattice.

How well do these two models for the crystalline solid apply to macromolecular crystals? We note first that there must be an Einstein-like component to any disordered crystal; for if there were no random decorrelation between nearest neighbor atomic pairs, only global translation of the entire crystal would be possible. Second, on experimental grounds a Debye component also has been demonstrated for protein crystals, at least in the long-wavelength limit. By carefully measuring the intensity surrounding a Bragg reflection in ribonuclease, Glover *et al.*[16] showed that the observed diffuse intensity in the halos falls off with the inverse square of the distance from

[14] m is the mass of each atom, T is the temperature, $k_B T$ is Boltzmann's constant, and v is the velocity at which the normal modes propagate, i.e., the speed of sound for the solid.

[15] I. Waller, *Z. Phys.* **17,** 398 (1923).

[16] I. D. Glover, G. W. Harris, J. R. Helliwell, and D. S. Moss, *Acta Crystallogr.* **B47,** 960.

the reciprocal lattice point, out to about 1/15 lattice constants. Thus for fluctuations in atomic positions owing to sound vibrations with wavelengths of 15 lattice constants or larger, the crystal behaves harmonically. Farther from the Bragg reflection, however, the intensities of the halos deviate from the predictions of the Debye model [Eq. (10)]. This means that for fluctuations approaching length scales on the order of the protein, and protein–protein interaction, atomic displacements are not well accounted for by the elastic transmission of acoustic waves. This deviation indicates that other modes contribute to the intensity in this regime. Indeed, besides the propagating Debye waves in a crystal, a polyatomic crystal will have a multitude of internal degrees of freedom, termed *optical* modes.[17] The discrepancy at short length scales may also indicate nonharmonic, diffusive modes; for example, macromolecules are known to have a glass transition at 200 K, which leads to departures from harmonicity.[18,19]

Obviously, then, macromolecules and the crystals they form are states of matter that cannot be readily categorized at present. There are more formal treatments of scattering theory (e.g., see Ref. 17) that we have not addressed, but they are presently difficult to apply to biological systems with large numbers of parameters. Nevertheless, two relatively successful means for analyzing diffuse scattering from macromolecules have been found: the analytic and the multicell (or supercell) treatments.

Analytic Method: Use of Global Correlation Function

Although the information content of diffuse scattering observations is unknown, it is safe to assume one cannot use these data to fit all $3N \times 3N$ elements of the atomic displacement correlation matrix, $\Gamma(i, j) = \langle \delta_i \delta_j \rangle$, the cross-correlation between displacements in atomic coordinates i and j.

A key approximation, then, for macromolecular diffuse scattering studies was the simplification of $\Gamma(i, j)$.[20,21] We assume that the displacement correlation function depends only on the radial distance, r, between the pair of atoms, so that every atom effectively shares the same mean-square displacement δ^2 and correlation function $\Gamma(r)$. This drastically reduces the complexity of the problem to determining a handful of parameters that best characterize the global function $\Gamma(r)$. The reduced correlation function $\Gamma(r)$ determines how the displacements of two atoms, one at r' and the

[17] M. Born and K. Huang, "Dynamical Theory of Crystal Lattices." Clarendon Press, Oxford, 1954.

[18] F. Parak, E. W. Knapp, and D. Kucheida, *J. Mol. Biol.* **161,** 177 (1982).

[19] W. Doster, S. Cusack, and W. Petry, *Nature* (*London*) **337,** 754 (1989).

[20] D. L. D. Caspar, J. Clarage, D. M. Salunke, and M. Clarage, *Nature* (*London*) **332,** 659 (1988).

[21] J. Clarage, M. Clarage, and D. L. D. Caspar, *Proteins* **12,** 145 (1992).

other at $r' + r$, are correlated, namely $\Gamma(r) \equiv \langle \delta(r')\delta(r' + r)\rangle_{r'}/\delta^2$. This homogeneous function is unity at $r = 0$, where there is total correlation, and vanishes for large separations where displacement correlations eventually disappear.

The diffuse scattering resulting from this simplifying assumption was originally derived in terms of a generalized Patterson function[21] defined by Fourier inverting, not only the Bragg reflections, but the entire scattering pattern. Restricting ourselves to isotropic (one-dimensional) disorder for now, the diffuse intensity is given by[22–24]

$$I_D(s) = e^{-(2\pi s\delta)^2}(2\pi s\delta)^2\{F_0^2(s_{hkl}) * \tilde{\Gamma}(\Delta s)\} \tag{11}$$

$F_0(s_{hkl})$ represents the set of ideal structure factors for the regularly ordered structure. $\tilde{\Gamma}(\Delta s) \equiv \mathrm{FT}[\Gamma(r)]$ is the Fourier transform of the displacement correlation function; as in the Debye model above [Eq. (10)], Δs is a vector measured from each reciprocal lattice point. Equation (11) represents the diffuse scattering as a convolution of the ideal structure factors with a halo function, $\tilde{\Gamma}(\Delta s)$, modulated by a factor that increases with increasing scattering angle (see Fig. 1h and i).

In a sense this is a quantification of the rule of thumb [in Eq. (8) and Fig. 1] that diffuse intensity follows the transform of the correlated unit. Because the Fourier transform of the halo function describes how displacements are correlated, modulations of $I_D(s)$ as a function of s in reciprocal space are inversely proportional to the distance in real space over which displacements are correlated. Simple examination of an X-ray image can provide an estimate of the correlation scales by measuring the halo widths, which are inversely related to the distances over which movements are correlated.

For example, in many macromolecular diffuse scattering patterns (such as tRNA shown in Figs. 2–5) the sharp halos around Bragg reflections have widths roughly 0.5–1 reciprocal lattice constants; inverting implies a component of disorder in the crystal where displacement correlations extend over 1–2 unit cells. Another diffuse feature seen in macromolecular diffraction patterns is a more continuous ring that is not obviously associated

[22] The total intensity is actually a power series, whose zeroth-order term is the Bragg, and whose first-order term is the diffuse scattering expression given here. Higher order diffuse scattering terms also contribute when the mean square displacements are large relative to the resolution of the data.[21,23,24] This expression also can be derived in real space following Waller's derivation[15] of the intensity from the Debye crystal, but replace $\langle \delta_i \delta_j \rangle$ with the homogeneous correlation function $\Gamma(r)$ instead of with the expectation value predicted by normal modes.

[23] A. Kolatkar, J. B. Clarage, and G. N. Phillips, Jr., *Acta Crystallogr.* **D50,** 210 (1994).

[24] B. D. Butler and T. R. Welberry, *Acta Crystallogr.* **A49,** 736 (1993).

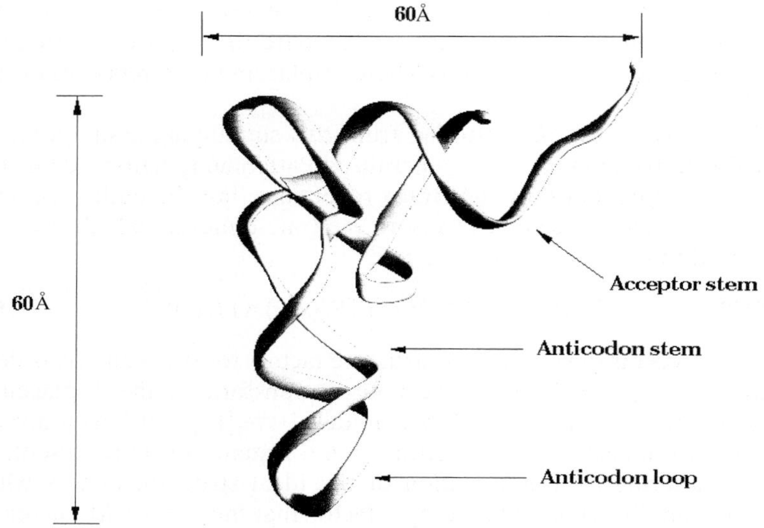

FIG. 2. Ribbon drawing of yeast initiator tRNA. Analysis of diffuse scattering data from crystals (see Figs. 3–5) implies a motion strongly coupled along the helical axis of the acceptor stem, but relatively uncorrelated with the other end of the structure (the anticodon stem and loop). This results in a flexing action. Information derived from simulation of the diffuse clouds of diffraction shows that the anticodon loop also exhibits liquid-like motion, with atomic displacement locally correlated over $\gamma \approx 3$ Å, the approximate base-pair stacking for tRNA.

with the reciprocal lattice positions. This arises from the underlying halo function $\tilde{\Gamma}(s)$ being so broad that its convolutions at the different Bragg nodes blur all lattice sampling; this implies that the correlations giving rise to this component of the scattering extend only over distances smaller than the unit cell dimension, i.e., intramolecular motion.

This analytic method is a phenomenological useful way of looking at diffuse scattering. It introduces few parameters: the average mean-square displacement of the atoms and the correlation length, γ, over which atomic displacements are correlated for atoms separated by a distance r. The parameter values can be refined beyond visual estimates using computer simulation (outlined in Comparing Models with Observations, below). Also the method is model independent in the sense that different correlation functions can be tested. In fact, the intensity formulas given earlier [Eqs. (9) and (10)] for the Einstein and Debye models of crystal disorder can be derived using the analytic method, simply by using their effective correlation functions: $\Gamma(r) = \Delta(r)$ for the Einstein crystal[25] and $\Gamma(r) = 1/r$ for the Debye case.[21,26]

Extension to Anisotropic Correlations. The discussion thus far has been restricted to disorder that is both isotropic (correlations depend only on the radial coordinate between atoms) and homogeneous (there is one global correlation function governing the behavior of all atoms). The clearest example of anisotropic correlations manifests itself as the oblong streaks (often observed around Bragg reflections, e.g., see Figs. 3 and 5) due to lattice movements being more strongly correlated along some directions, usually along lattice contacts. Keeping to the program of not refining the entire $\Gamma(i, j)$ matrix, one can define a simple extension of the exponential correlation function:

$$\Gamma(\mathbf{r}) = \exp - \left(\frac{r_x^2}{\gamma_x^2} + \frac{r_y^2}{\gamma_y^2} + \frac{r_z^2}{\gamma_z^2}\right)^{1/2} \tag{12}$$

written here in diagonal form. The values of γ_x, γ_y, and γ_z, which define a correlation ellipsoid, now can be adjusted separately until $\mathrm{FT}[\Gamma(\mathbf{r})]$ matches the observed halo shapes.[23]

It must be kept in mind that because we are now working in three dimensions, atomic displacements can be factored into statistically independent components, δ_i, so that the total mean-square displacement of an atom is the Pythagorean sum $\delta_{\mathrm{Total}}^2 = \sum_i \delta_i^2$. Here a component may be the contribution from displacements along the different Cartesian or crystal directions, or the contribution from different dynamic modes, e.g., lattice-coupled or intramolecular movements. The preceding three-dimensional (3D) $\Gamma(\mathbf{r})$ refers to the correlations along one of these displacement components; each component has its associated correlation function $\Gamma_i(\mathbf{r})$ (just as, in the Bragg scattering analogy, each spatial component has its own contribution, B_i, to the overall anisotropic B value).

The generalization of Eq. (11) is[21]

$$I_{\mathrm{D}}(\mathbf{s}) = e^{-\sum_i (2\pi\mathbf{s}\cdot\delta_i)^2} \sum_i (2\pi\mathbf{s}\cdot\delta_i)^2 \{F_0^2(\mathbf{s}) * \mathrm{FT}[\Gamma_i(\mathbf{r})]\} \tag{13}$$

which implies that the total diffuse scattering is the direct sum from each of the components, adding in proportional to their relative contribution to the overall mean-square displacement.[27]

[25] $\Delta(r)$ is the delta function, which is unity at the origin and zero elsewhere.

[26] J. Clarage, "Disorder in Protein Crystals." Ph.D. thesis, Brandeis University, Waltham, Massachusetts, 1989.

[27] Again, like the isotopic expression in Eq. (9), this is the lowest order in a power series.

a

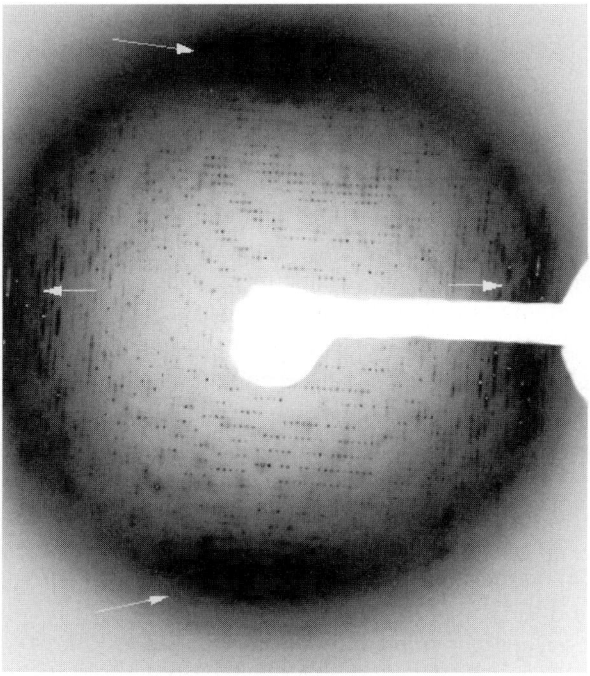

b

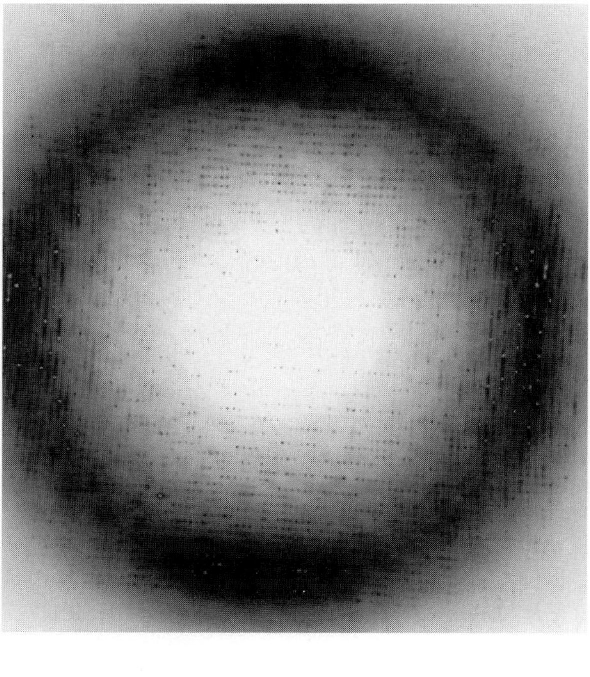

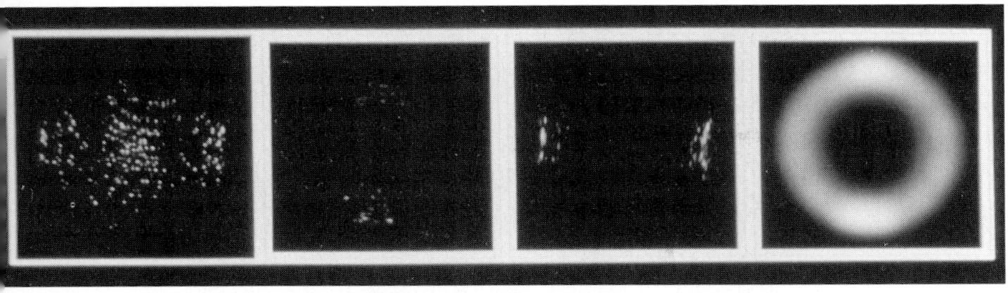

FIG. 4. The individual components used in the calculated total intensity from crystals of tRNA (Fig. 3 and Ref. 23). From left to right are the Bragg scattering, long-range lattice-coupled motion perpendicular to the *c* axis, long-range lattice-coupled motion along the *c* axis, and short-range motion local to the anticodon loop.

Extension to Inhomogeneous Correlations. The homogeneity assumption allows us to think of diffuse scattering as consisting of a halo surrounding Bragg reflections, and that the transform of this halo is the displacement correlation function.[28–30] It also allowed us to reduce the complexity of the diffuse intensity formula from an $O(N^2)$ sum to a simple $O(N)$ sum as in structure-factor calculations. The limitation of homogeneity, of course, is that every pair of atomic correlations is assumed to be governed by the same function; there is no allowance for some rigid part in a structure having longer range correlations than another, more floppy part.

The simplest was to relax homogeneity without dealing with the full displacement correlation matrix is to partition the molecule into pieces, each with a single correlation function, $\Gamma_{\text{piece}}(r)$, applying to all atoms in the piece. If there is negligible correlation between different pieces, the diffuse scattering from each piece is then given by convolving the halo function $\tilde{\Gamma}_{\text{piece}}(\Delta s)$ with structure factors computed using only those atoms

[28] In this sense it is an X ray-scattering application of the Wiener–Kintchine theorem[29,30] from classic harmonic analysis, as pointed out in Ref. 21.

[29] J. M. Ziman, "Models of Disorder." Cambridge University Press, Cambridge, 1979.

[30] M. Weiner, *Acta Math.* **55,** 117 (1930).

FIG. 3. (a) Observed and (b) calculated diffuse scattering from yeast initiator tRNA crystals.[23,38,49] Arrows in the experimentally observed total intensity (a) point to the major features in the data. The vertical streaks elongated perpendicular to the *c* axis have been modeled as lattice-coupled motions along the pseudohelix axis, to give the calculated intensity in (b). The diffuse clouds were modeled as local intramolecular motion in the anticodon arm. The diffraction patterns are on a gray scale, where darker features are more intense. [Reproduced from *Acta Cryst.* **D50,** 210 (1994). Copyright © International Union of Crystallography.]

a

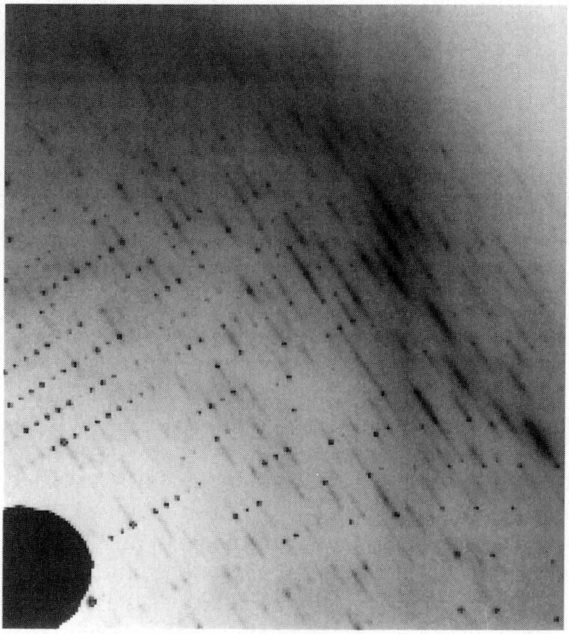

b

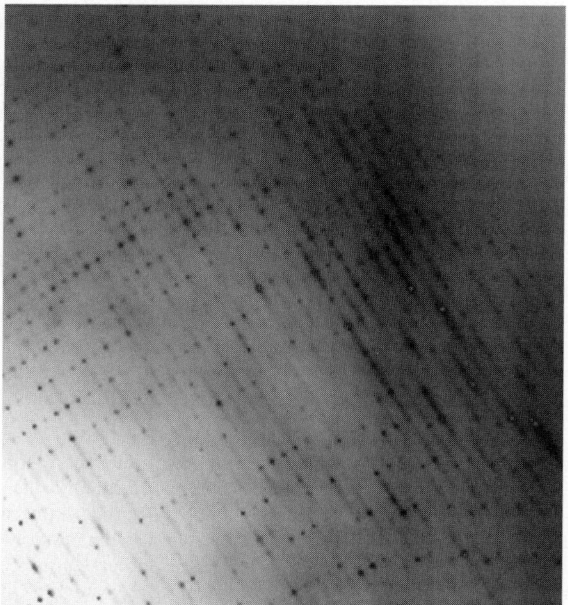

within the piece.[23] There has also been some success in lifting the homogeneity approximation from the magnitude of the displacements,[23] allowing for individual B values to be used.

Multicell Method: Using Explicit Correlation Models

The preceding analytic methods are still limited by a certain degree of ideality, as are all analytic models in science. In this regard the computer becomes useful; aside from shear speed, it allows us to study the consequences of any model that can be encoded as an algorithm—not just those whose homogeneity or isotropy make them easier to treat. If some model for the atomic disorder is constructed in the computer, then the diffuse scattering can be generated from the defining formula,

$$I_{\text{Diffuse}} = \langle |\text{FT}\rho|^2 \rangle - |\text{FT}\langle \rho \rangle|^2 \qquad (14)$$

without any further mathematical steps than a series of Fourier transforms. The averaging here is over multiple independent copies of a model cell. What to use for the cell depends on the disorder one is modeling. In general, to model correlations with relaxation length γ, the cell must be no smaller than γ. For instance, to model Debye-type lattice waves, one needs a superlattice with many connected unit cells.[31] If one is interested only in intramolecular correlations, then the cell need consist only of the unit cell of the crystal, and the averaging in Eq. (14) taken over the multiple conformations of the molecule.

Besides allowing a wide class of models for disorder, this method is exact, in the sense that it not a series expansion like the analytic method. A caveat to this exactness is that the results do not automatically extrapolate to the much larger ensembles to which experiments have access; thus one must be certain that enough cells are used to obtain good statistics. For example, simple hinge motion in an isolated protein would require only a few "snapshots" of the hinge mode inside a single unit cell. Lattice modes, however, would demand sampling of all wavelength disturbances from the shortest to the longest mode measured on the X-ray detector, each with proper weighting.

[31] D. Boylan and G. N. Phillips, *Biophys. J.* **49,** 76 (1986).

FIG. 5. Magnified quadrant from another orientation of the (a) experimental and (b) calculated diffraction patterns from tRNA crystals.[23] The diffuse R value between these images is less than 20%.

Whether to employ analytic or multicell methods depends on the system being studied and the questions being asked. For instance, lattice disorder has a straightforward analytical treatment, so the multicell method may not be the optimal approach. A multicell calculation is ideally suited, however, for comparing molecular dynamics simulations with experimental diffuse scattering data, which we discuss below.

Measuring Diffuse Scattering Data

All crystals have diffuse scattering. Although intensity is lost from Bragg reflections owing to atomic disorder, this intensity does not disappear—it scatters continuously throughout reciprocal space. Because the diffuse signal is not confined to discrete scattering directions, it tends to appear weak compared to Bragg peaks. For instance, with good collimation, the dimension of a Bragg reflection can be made less than 1/10 the distance between reciprocal lattice points, so that Bragg scattering occupies less than 10^{-3} times the volume of reciprocal space. Therefore, at a resolution of about 3–4 Å, where there is comparable integrated intensity in Bragg and diffuse scattering (see Theory of Diffuse Scattering, above), the Bragg reflections will be approximately 1000 times brighter than the diffuse scattering.

Nevertheless, given the dynamic range of modern area detectors and the brilliance of synchrotron sources, this "weak" signal can be measured today—although it is not always clear exactly what to collect. Measurement of Bragg scattering is straightforward. For the continuous diffuse intensity, however, there are in principle as many data points as there are detector pixels. Physically speaking, whereas the average electron density, by definition, has Fourier components that must be commensurate with the crystal lattice, variations in the electron density can have any wavelength, from macroscopic sound waves to atomic vibrations.

In practice, the size of the Bragg reflections sets an upper limit on the complexity of a diffuse scattering pattern. For example, in an X-ray experiment with an order-to-order resolution of 10 pixels, the reciprocal space will be coarse grained into cells 1/10 the length of a reciprocal lattice constant. This means density components greater than 10 unit cells are integrated out, and the effective ensemble being measured consists of crystal blocks, or superlattices, 10 unit cells to a side.

We now discuss ways in which diffuse data are collected and processed.

Data Collection

X-Ray Detectors and Sources. Diffuse scattering can be seen on almost any X-ray detector, even with ordinary rotating anode sources.[20] An advantage of film is the high "pixel resolution" allowing intricate diffuse features to be separated and mapped. Imaging plates and the newer charge-coupled device (CCD) detectors have the added advantage of high signal-to-noise ratios. For example, we have recorded the diffuse intensity values for myoglobin using an imaging plate and a rotating anode, obtaining signal-to-noise ratios between 10 : 1 and 100 : 1 for the diffuse intensities. Furthermore, both components of the diffraction can be contained in the same data set without overexposure; for instance, at a resolution of 4 Å the diffuse scattering values between reciprocal lattice nodes are 500–5000 times smaller per scanned imaging plate pixel than a typical Bragg peak intensity. Synchrotron sources can dramatically reduce the exposure time. Interpretable patterns can usually be recorded in minutes[21] from these facilities, as opposed to hours using a rotating anode.[20]

Point-Spread Function. A monochromatic, well-collimated beam is important for diffuse scattering studies. As part of our imaging plate measurements of the diffuse scattering from P6 myoglobin ($a = b = 91.20$ Å, $c = 45.87$ Å, $\alpha = \beta = 90°$, $\gamma = 120°$), we took an image of the direct beam. The full width at half-maximum (FWHM) of the direct beam profile was 0.0021 $Å^{-1}$ = 1/480 Å, or about 1/5 the P6 reciprocal lattice constant, and 1/10 the reciprocal dimensions of a myoglobin molecule. The point spread will convolve, or blur, the diffuse intensity that arises from atomic disorder. One notable effect is that any estimate of the diffuse halo widths in reciprocal space will be uncertain by the FWHM. This will not severely affect the corresponding correlation lengths in real space, γ, as long as these lengths are significantly smaller than the reciprocal of the FWHM, which is the effective "grain size" of the analysis.[32–34]

Even if the wings of the point spread function are relatively weak, the fact that this function smears each Bragg reflection means that, in principle, additional intensity features will arise that are indistinguishable from the

[32] Such instrumental grain size implies that diffuse scattering experiments are not sensitive to fluctuations correlated over distances greater than the inverse of the point spread function (in the case of the myoglobin measurements, 480 Å). This fact also has implications for Bragg studies of atomic motion. Because all diffuse scattering closer to a reciprocal lattice node than the beam width is integrated into the structure factor, temperature factors derived from the Bragg amplitudes include only the contribution from movements correlated over distances shorter than the inverse of the effective beam width, typically 10 unit cells or less. Misunderstanding on this point has spread confusion. In particular, comparisons between B values and the mean square displacements obtained from analysis of the inelastic properties of a crystal[33,34] are specious; the values of δ^2 derived from bulk properties include the entire spectrum of thermal fluctuations in the crystal, including all of the longer wavelength modes whose effects do not manifest themselves in crystallographically measured B values.

[33] C. Edwards, S. B. Palmer, P. Emsley, J. R. Helliwell, I. D. Glover, G. W. Harris, and D. S. Moss, *Acta Crystallogr.* **A46,** 315 (1990).

[34] V. N. Morozov and T. Y. A. Morozova, *J. Theor. Biol.,* **121,** 73 (1986).

genuine diffuse scattering halos that result from displacement disorder. Therefore, as part of our measurements on myoglobin, in addition to the FWHM we determined the intensity response at a distance from the beam center corresponding to half a P6 myoglobin reciprocal lattice constant, $0.0055\ \text{Å}^{-1} = 1/(2 \times 91.2\ \text{Å})$; it was at least 100 times smaller than the signal measured between nearest neighbor Bragg peaks when the crystal was present, and at least 10 times smaller than the intensity that remained after processing the crystal data to remove the circular symmetric background from Compton and water scattering. Therefore, true diffuse intensity is larger than the spurious diffuse intensity due to the point spread in the beam.

Absorption effects. Although crystal and capillary absorption has not been treated rigorously in diffuse scattering work, it has been possible in some studies[20] to circumvent capillaries by suspending the crystal in a moist air stream. Collection of data in cryoloops also should help reduce absorption.

Extraneous Background Scattering. Unlike in Bragg work, which need only distinguish sharp features on the detector, any weak background scattering can interfere severely with the measurement of diffuse intensity. A clean beam with guard slits will minimize spurious diffraction. Mother liquor around the crystal should be kept to a minimum to reduce the liquid scattering ring. Air scattering should be minimized by using a helium tunnel.

Sources of background that cannot be attenuated include scattering from the crystal solvent and Compton scattering due to incoherent (Einstein-like) diffraction from electrons in the sample; these must be subtracted as described below.

Data Processing

Polarization. Crude polarization factors can be applied as for Bragg data. Still, this effect is usually ignored because it is small compared with other factors.

Background Corrections. There have been two successful means of eliminating background scattering. Thüne and Cusack[35] have taken "control" data without the crystal, thereby collecting the background scattering from capillary, mother liquor, air or helium, and the camera apparatus. To subtract this background, radial averages of it and the crystal data are scaled to one another, using the profile at low resolution where diffuse scattering from displacement disorder should vanish.

[35] Thüne, T. (1993). *The structure of seryl-tRNA synthetase from E. coli at 100K and the analysis of thermal diffuse scattering of its crystals.* Ph.D. Thesis, Technical University of Munich.

Another method[21,26] comes from the more mature field of small-molecule diffuse scattering: to remove all circularly symmetric scattering from the data. To define this circular component, the minimum intensity value is located at each radius in reciprocal space[36] (or in practice the average of the few lowest values). The radial curve that interpolates these minima is subtracted from the data.[37]

Although successful in producing data sets from which important information has been obtained, a drawback to both these schemes is the assumption of a spherically symmetric background (which asymmetric absorption or ordered crystal solvent, for example, will violate). Methods have been devised to accommodate asymmetries.[26,38]

Separating Bragg and Diffuse Components. There are two schools of thought on interpreting diffuse scattering: to separate and not to separate. Superimposed Bragg peaks can obscure the accompanying diffuse distribution, especially in macromolecular diffraction where reciprocal lattice constants are small. Because biologically motivated researchers are interested in intramolecular motions, they must measure that diffuse scattering whose modulations vary no more rapidly than the reciprocal lattice spacing. To distinguish Bragg and diffuse components, techniques have been developed[39] similar to "mode filtering" used by astronomers, who face a similar data-processing problem in separating sharp starlike features from diffuse nebulas. A box somewhat larger than the measured width of a Bragg reflection is slid over the image. Then the average background intensity value within the box is defined as the diffuse signal at this point in the image. The technique is robust in that it is insensitive to details of the routine (i.e., box shape, or other background measures such as the mode or median of the box values). Numerical tests show that this method does give a good representation of the intramolecular diffuse scattering. In any case, it is always wise to check the result by subtracting the isolated intramolecular diffuse pattern from the original data, to make sure the residual contains nothing but Bragg peaks and sharp halos.[20] Note that by changing

[36] T. R. Welberry, *J. Appl. Crystallogr.* **16,** 192 (1983).

[37] Note that in addition to instrumental background, this technique also removes the diffraction effects arising from bulk crystal solvent, Compton scattering, and the completely uncorrelated Einstein component in the atomic displacements. An Einstein component, although a manifestation of atomic displacements in the macromolecule, is often not missed because one is concerned with correlated displacements.

[38] A. Kolatkar, "Correlated Atomic Displacements in Crystals of Yeast Initiator tRNA and Aspartate Aminotransferase Analyzed by X-Ray Diffuse Scattering." Ph.D. Thesis, Rice University, Houston, Texas, 1994.

[39] M. E. Wall, "Diffuse Features in X-Ray Diffraction from Protein Crystals." Ph.D. thesis. Princeton University, Princeton, New Jersey, 1996.

the box size to, say, a little larger than the breadth of a Bragg reflection, the algorithm will isolate both the intra- and intermolecular components from the Bragg peaks. More sophisticated methods such as wavelet decomposition,[40] whose fractal-like basis is expert at recognizing and separating sharp features in an image, may also prove useful.

Another school of thought does not bother separating Bragg from diffuse scattering. Rather, at the stage of constructing models for the disorder, a model for the Bragg peaks is also included. This comprehensive model is then used to simulate the same total X-ray scattering collected on the detector: Bragg peaks from the average, sharp halos from lattice movements, and slowly varying diffuse scattering from internal motions. This method has the advantage that it can test models for movement correlated on many length scales because all information is retained in the data. The data are also kept in a relatively unprocessed form, leaving the extra work to the modeling side of the analysis.

Symmetry Averaging. Symmetry-related Bragg intensities often are averaged to increase signal to noise (S/N) for the measurement. This same principle is extremely useful for diffuse scattering, where the strength per detector element is weak. With photographs taken down high-symmetry zones, averaging of diffuse data also can alleviate many absorption problems, for the data in the different symmetric sections of a 2D image effectively are rescaled to a common average. Averaging of all symmetry related pixels from a still taken down the sixfold axis of P6 myoglobin gave much better images of the intramolecular diffuse scattering.[41]

Crystallographers often avoid shooting down high-symmetry directions when collecting structure factors; however, exploiting symmetry axes could prove to be the best strategy for accurate diffuse scattering data sets.

Displaying Data. A feature of articles on diffuse X-ray scattering that distinguishes them from standard crystallographic articles is that the data are always shown. This is partly because diffuse scattering analysis has not yet been automated[42]; thus the data are not taken for granted. Also, unlike structure factors, diffuse intensities do not have a natural representation as integer-indexed values. As with other continuous data sets (galactic X-ray sources, weather maps, etc.) diffuse scattering can be visualized using

[40] J. E. Odegard, H. Guo, M. Lang, C. S. Burrus, R. O. Wells, Jr., L. M. Novak, and M. Hiett, Wavelet based SAR speckle reduction and image compression. *In* "SPIE Symposium on OE/Aerospace Sensing and Dual Use Photonics, Algorithm for Synthetic Aperture Radar Imagery II." April 17–21, 1995, Orlando, Florida (http://jazz.rice.edu/publications/).

[41] J. B. Clarage, T. Romo, B. K. Andrews, B. M. Pettitt, and G. N. Phillips, Jr., *Proc. Natl. Acad. Sci. U.S.A.* **92,** 3288–3292 (1995).

[42] Although we have developed a suite of programs, XCADS, used to process and analyze diffuse data.

2D color graphics. When displaying diffuse data, and the corresponding model calculations, one must communicate to the audience what the color values stand for, as well as use the same color table when comparing two or more images.

Three-Dimensional Diffuse Scattering Maps. In a sense diffuse scattering studies are in the same nascent stage as very early Bragg studies, when scientists gathered a handful of photographs down different directions to derive a few geometric parameters related to the crystal lattice. Just as Bragg crystallography has since evolved to the collection and processing of entire 3D data sets, making possible the refinement of many structural parameters, there is promise that diffuse studies will burgeon as soon as complete 3D data sets can be collected with precision. Progress has been made on such a 3D diffuse scattering data set. Using a complete set of 2D diffraction measurements from crystals of staphylococcal nuclease (each separated by 1°) Wall and Gruner[39,43] were able to obtain a map of the full 3D reciprocal-space intensity.

Comparing Models with Observations

Given the experimental geometry corresponding to a 2D frame of diffuse data, and some initial model for the atomic disorder in the crystal, the diffuse intensity predicted by the model can be simulated at each point on the detector. The observed and calculated patterns then can be compared to determine whether the correlations in the model are reasonable, or need to be modified. Usually trial-and-error visual refinement is used to arrive at the best overall results. However, in a promising study of the diffuse scattering from tRNA crystals, Kolatkar[38] used a diffuse R value (mean-square difference between observed and calculated pixel values) to refine the best values for the anisotropic correlation lengths due to displacement of whole tRNA molecules in the lattice (see Figs. 3–5).

To simulate diffuse intensities, the crystal orientation corresponding to the detector image must be determined accurately. This can be done with the same standard methods used to determine orientations of X-ray films.

Generating the entire 3D intensity, $I_D(\mathbf{s})$, on a reciprocal space grid fine enough to match the pixel size with which the detector samples the Ewald sphere, would be a massive investment of computer time and space, involving the calculation and storage of many intensities neither collected on the film nor independent of each other. It is best to begin simulations in "film space." A point on the detector array is mapped back to the corresponding

[43] M. E. Wall, S. E. Ealick, and S. M. Gruner, *Proc. Natl. Acad. Sci., U.S.A.* **94,** 6180–6184 (1997).

coordinate $\mathbf{s}_E$ on the Ewald sphere, and the diffuse scattering $I_D(\mathbf{s}_E)$ is computed at this point to give the simulated detector intensity.

Analytic Method

For the analytic method [Eq. (11)], simulation of the diffuse scattering for a still photograph proceeds as follows. Given a trial correlation function $\Gamma(r)$ with characteristic decay length γ, consider all reciprocal lattice nodes $\mathbf{s}_{hkl}$ within a distance of $\approx 1/\gamma$ of the sphere point $\mathbf{s}_E$, that is, all nodes close enough to affect the intensity at this point on the sphere of reflection. The contribution from each of these lattice nodes is then summed at $\mathbf{s}_E$, the weights being determined by the values of the halo functions $\tilde{\Gamma}(\mathbf{s}_E - \mathbf{s}_{hkl})$, and the ideal structure factors $F_0^2(\mathbf{s}_{hkl})$. Structure factors for the ideally ordered crystal are computed from atomic coordinates for the known average structure. Last, the convolution products $F_0^2(\mathbf{s}_{hkl}) * \tilde{\Gamma}(\mathbf{s}_E - \mathbf{s}_{hkl})$ are multiplied by $e^{-(2\pi s_E \delta)^2}(2\pi s_E \delta)^2$, where δ is the estimate for the root-mean-square (r.m.s.) displacement.[44] Oscillation photographs are handled by superposing several still simulations contained within the oscillation range.

Because the simulation method just outlined does not rely on actually constructing the 3D distribution of $I_D(\mathbf{s})$ in the computer, but rather on performing operations on precise points in reciprocal space, one can explore arbitrarily fine reciprocal spacings on the film, i.e., arbitrarily long length scales in the crystal lattice. For example, to simulate all 20 pixels recorded between 2 Bragg peaks in some data set, it is not necessary to construct, store, and perform calculations on an actual 3D model consisting of 20 unit cells on a side, which would entail thousands of molecules and be computationally prohibitive. Instead, the only arrays involved are the 2D array of simulated data and the list of ideal structure factors.

Note that this same algorithm can be used to simulate the Bragg scattering, $e^{-(2\pi s_{hkl}\delta)^2}F_0^2(\mathbf{s}_{hkl})$, for those reciprocal lattice points that lie within the beam divergence of the Ewald sphere, by treating the X-ray beam profile, estimated from the low-angle Bragg reflections, as a sharp Gaussian halo function centered at each reciprocal lattice point.

Multicell Method

If, instead of analytic correlation functions, one is working from an ensemble of structures (generated from a molecular dynamics trajectory, a spectrum of normal modes, etc.), then serial computation of the multicell diffuse scattering, $I_D = \langle |\mathrm{FT}\rho|^2 \rangle - |\mathrm{FT}\langle\rho\rangle|^2$, at each point on the detector

[44] Higher order terms in the diffuse scattering expansion can be simulated with an entirely similar algorithm.[21]

would involve a large number of Fourier transforms: the product of the number of detector pixels and the number of structures in the ensemble. To reduce this to something that scales only with the size of the ensemble, the fast Fourier transform (FFT)[45] should be used. If the FFT is employed, then I_D will be defined only on a Gibbs lattice[46] reciprocal to the real-space lattice used to hold the atomic structures. For example, if each molecular configuration in the ensemble fits in a single unit cell of the crystal, then the diffuse intensity will be defined only at the reciprocal lattice nodes (the same points on which the structure factors are defined); if models involving two unit cells are used, then the diffuse intensity will be sampled on a Gibbs lattice twice as dense as the standard reciprocal lattice.

Whatever the case, determining I_D at the points collected on the detector is now entirely similar to the procedure above for the analytic method; one need store and operate only on the 2D simulated intensity array and the list of diffuse intensities generated by the FFT at Gibbs lattice nodes. After a detector point is mapped back to reciprocal space, the diffuse intensity at this point is given by interpolating the surrounding intensity values from the nearest Gibbs lattice nodes.

Note that although the interpolation used in this procedure can yield a simulated pattern with arbitrary pixel fineness, this simulated pattern does not actually contain any information on correlation lengths greater than the size of the array holding the original atomic models. This must be kept in mind when one compares simulations with actual data, which can contain information from many length scales. That is, if the multicell method is used to compute the diffuse diffraction from a molecular dynamics trajectory done on a single unit cell, then the resulting intensity can not contain reciprocal-space features (harmonics) corresponding to halos from lattice-correlated movements. Also, if the starting atomic model does not possess the same symmetry as the system on which the data were collected, then the space group operators must be applied to the diffuse intensities so that a meaningful comparison can be made with data. Such averaging at the level of intensities assumes there are no displacement correlations between symmetry-related atoms; this approximation is unlikely to be strictly true in the actual crystal, but is the only assumption that can be made if the original model did not incorporate crystal symmetry.

Current Results from Diffuse Scattering Studies

Several proteins and nucleic acids have now been studied by diffuse scattering methods. The first was tropomyosin, whose cable-like structure

[45] L. F. Ten Eyck, *Methods Enzymol.* **115,** 324 (1985).

[46] A generalization of the reciprocal lattice, defined as the Fourier dual of any real-space lattice, including those not generated by the primitive unit cell.

allows large-scale motions in regions of the crystal that are not closely packed. Both lattice-coupled modes and random displacements of the tropomyosin filaments have been described and the results related to the function of the protein in muscle regulation.[31,47]

The streaks of diffraction surrounding Bragg peaks in orthorhombic lysozyme were shown to arise from rigid-body displacement of rods composed of protein molecules connected along principal lattice directions.[48] The quality of the data and simulations allowed numerical values to be refined for the average number of contiguous proteins responsible for this lattice disorder, as well as the r.m.s. amplitude of movement about the ideal lattice sites.

Analysis of the slowly varying diffuse ring due to intramolecular atomic displacements has been carried out on insulin[20] and lysozyme[21] crystals. In both cases it was found that correlations in displacements of neighboring atoms in the protein fall off roughly exponentially with the distance between the atoms, with a characteristic length of 4–8 Å. Thus, in analogy with liquid structures, where the correlations in atomic positions decay approximately exponentially with pair separation, the motion observed here is termed liquid-like, in the sense that correlations in atomic displacements fall off rapidly with distance.

The first nucleic acid whose flexibility was studied rigorously using diffuse scattering was yeast initiator tRNA.[23,38,49] Elucidation of both intra- and intermolecule displacements, as well as treatment of anisotropy and inhomogeneity in the disorder, yielded a good match between experimental and calculated diffuse scattering patterns (Figs. 3 and 5). Some of the types of motions postulated to occur during tRNA action were found to occur in the crystal environment, a discovery that could not be made by standard crystallographic analysis of the average structure.

A study of the anisotropic diffuse scattering from seryl-tRNA synthetase, at room temperature and 80 K,[35] has provided good evidence for a degree of freedom that might be necessary for the enzyme to bind to its cognate tRNA. Such identification of a correlated unit in the macromolecule would be impossible with Bragg data alone.

Each of these studies has provided particular, often biochemically relevant, insight into the displacement correlations in these systems. At a more general level we can conclude from all the systems currently in the litera-

[47] S. Chacko and G. N. Phillips, Jr., *Biophys. J.* **61,** 1256 (1992).

[48] J. Doucet and J. P. Benoit, *Nature* (*London*) **325,** 643 (1987).

[49] A. Kolatkar, J. Clarage, and G. N. Phillips Jr., *Rigaku J.* **9**(2), 4 (1992).

ture[50,51] that intramolecular movements have the greatest contribution to the total mean-square displacement measured in X-ray diffraction studies, and encompassed in the *B* values. These intramolecular motions are coupled over relatively short length scales, be they the filament size in tropomyosin, a few side chains in globular proteins, or neighboring base pairs in the case of nucleic acids, such as tRNA.

Testing Molecular Dynamics and Normal Mode Calculations

In light of modern molecular dynamics and normal-mode simulations, one might be wondering why we should bother to study this diffuse component in first place, since atomic displacements and their correlations can be computed. Indeed, presumably once such simulations can predict the average structure of macromolecules, then not even the Bragg component will need to be measured anymore.

In testing whether the so-called hinge-bending motion in lysozyme was consistent with the diffuse scattering data measured from these crystals, Clarage *et al.*[21] found that the lowest frequency normal mode for the molecule did not fit the data as well as the liquid-like model.

Mizuguchi *et al.*[52] calculated the diffuse scattering from a more complete spectrum of normal modes on lysozyme. Although no data were used for comparison, the theoretical patterns seem to show modulations in reciprocal-space characteristic of correlations greater than 10 Å, indicating that normal modes predict longer correlation lengths than observed in globular proteins.

Faure *et al.*[53] tested both a 600-psec molecular dynamics and a spectrum of the lowest 15 normal modes against the observed diffuse scattering from lysozyme. Although neither model for the displacement correlations fits the data well, the normal-mode model did appear to fit better than molecular dynamics, giving at least the correct reciprocal-space intensity modulation, and hence the correct correlation length for pair displacements.

A reason for the poor quality of fit attained thus far by molecular dynamics has been traced to a sampling problem in molecular dynamics simulations on macromolecules. Clarage *et al.*[41] showed that the diffuse scattering predicted by current nanosecond time scale dynamics simulations is not unique; two different simulations give two different predictions for

[50] Including novel time-resolved diffuse scattering studies on myoglobin,[51] using Mössbauer radiation instead of the X radiation that is the focus of this book.

[51] G. U. Nienhaus, J. Heinzl, E. Huenges, and F. Parak, *Nature* (*London*) **338,** 665 (1989).

[52] K. Mizuguchi, A. Kidera, and N. Go, *Proteins* **18,** 34 (1994).

[53] P. Faure, A. Micu, D. Perahia, J. Doucet, J. C. Smith, and J. P. Benoit, *Nat. Struct. Biol.* **1,** 124 (1994).

what the diffuse scattering should be. Whereas the average molecular transform (Bragg scattering) converges for these dynamics simulations, the variance in the molecular transform (diffuse scattering) does not.

This sampling dilemma may explain why normal modes gave a somewhat better fit to the measured diffuse scattering data. Such methods are analytic and thus suffer no sampling problem. That is, even though the harmonic potential function may be a poor approximation to reality, perhaps this error would be made up for by the complete sampling that is possible.

These latest results show what a powerful tool diffuse scattering is in empirically testing and guiding theoretical molecular simulation studies. This sampling problem points out why analytic formalisms are still important in studying disorder; there is not always a complete, explicit real-space model for atomic motion and thus more model-independent schemes (as discussed in Theory of Diffuse Scattering, above) that can be used to refine parameters from experimental data are still quite invaluable.

Future: Complete Refinement of Structure and Dynamics

As this chapter has shown, the features of diffuse scattering from macromolecules are now fairly well understood. This long neglected component present in every diffraction pattern now can be accounted for at least semiquantitatively to provide a physical picture of atomic movements, which complements traditional Bragg scattering studies, and that can be useful in discovering the modes of motion exploited by functioning biological molecules.

In the future it is hoped that enough motivated researchers take up the challenge of placing diffuse studies on the same rigorous level that current Bragg crystallography enjoys. This will lead to a complete refinement of macromolecular structure and dynamics, wherein a global target function

$$E = E_{\text{chem}} + E_{\text{Bragg}} + E_{\text{Diffuse}} \tag{15}$$

can be used to refine the extended parameter set: r_i, $B_i = \langle \delta_i^2 \rangle$ and $\Gamma_{i,j} = \langle \delta_i \delta_j \rangle$. In addition to giving information on movement, such a target function should also improve refinement of the structure itself because diffuse scattering also depends on the average atomic positions. This function also could be used as a pseudo-energy to generate molecular dynamics ensembles, $r_i(t)$, that are consistent with all X-ray scattering observations, i.e., a family of structures having the correct atomic structure and displacement correlations.

Acknowledgments

Work supported by NSF MCB-9315840, NIH AR40252, the Robert A. Welch Foundation, and the W. M. Keck Center for Computational Biology.

[22] Laue Diffraction

By KEITH MOFFAT

Introduction

Although Laue diffraction, in which a stationary crystal is illuminated by a polychromatic beam of X rays, was the original crystallographic technique, it was largely superseded by rotating crystal, monochromatic techniques in the 1930s. With the advent of naturally polychromatic synchrotron X-ray sources, the Laue technique has undergone a renaissance. As with any experimental technique, accurate results are obtained only if its principles, advantages, and the sources of its limitations are well understood and applied in practice. The thrust of this chapter is to address three questions. What is the best way to conduct a Laue experiment? What limitations remain to be overcome? What classes of structural problems are attacked best by the Laue technique? I begin by reviewing the principles briefly, then show how the advantages and disadvantages are derived from these principles, and how these are deployed in real experiments.

Chapters [23] and [24] in this volume[1,2] address the key topic of the quantitation and evaluation of Laue diffraction patterns, and the related topic of reaction initiation in time-resolved crystallography.

Principles of the Laue Technique

Many articles describe the principles of the Laue technique,[3–7] and only the main points are reviewed here.

A Laue diffraction pattern is obtained when a stationary crystal is illuminated by a polychromatic X-ray beam spanning the wavelength range

[1] I. J. Clifton, S. Wakatsuki, E. Duke, and Z. Ren, *Methods Enzymol.* **277,** [23], 1997 (this volume).

[2] I. Schlichting and R. S. Goody, *Methods Enzymol.* **277,** [24], 1997 (this volume).

[3] J. L. Amorós, M. J. Buerger, and M. C. Amorós, "The Laue Method." Academic Press, New York, 1975.

[4] D. W. J. Cruickshank, J. R. Helliwell, and K. Moffat, *Acta Crystallogr.* **A43,** 656 (1987).

[5] D. W. J. Cruickshank, J. R. Helliwell, and K. Moffat, *Acta Crystallogr.* **A47,** 352 (1991).

[6] J. R. Helliwell, J. Habash, D. W. J. Cruickshank, M. M. Harding, T. J. Greenhough, J. W. Campbell, I. J. Clifton, M. Elder, P. D. Machin, M. Z. Papiz, and S, Zurek, *J. Appl. Crystallogr.* **22,** 483 (1989).

[7] A. Cassetta, A. Deacon, C. Emmerich, J. Habash, J. R. Helliwell, S. McSweeney, E. Snell, A. W. Thompson, and S. Weisgerber, *Proc. R. Soc. Lond.* **A442,** 177 (1993).

0076-6879/97 $25.00

from λ_{min} to λ_{max}. A reciprocal lattice point that lies between the two limiting Ewald spheres of radii $1/\lambda_{min}$ and $1/\lambda_{max}$, and within a radius $\mathbf{d}^*_{max}$ of the origin where $\mathbf{d}^*_{max} = 1/\mathbf{d}_{min}$, the limiting resolution of the crystal, is in diffracting position for the wavelength λ and will contribute to a spot on the Laue diffraction pattern (Fig. 1). All such points, for which $\lambda_{min} \leq \lambda \leq \lambda_{max}$, diffract simultaneously. A Laue pattern may be thought of as the superposition of a series of monochromatic still diffraction patterns, each taken with a different wavelength of X rays. Laue spots arise from a mapping of rays (lines emanating from the origin in reciprocal space) onto

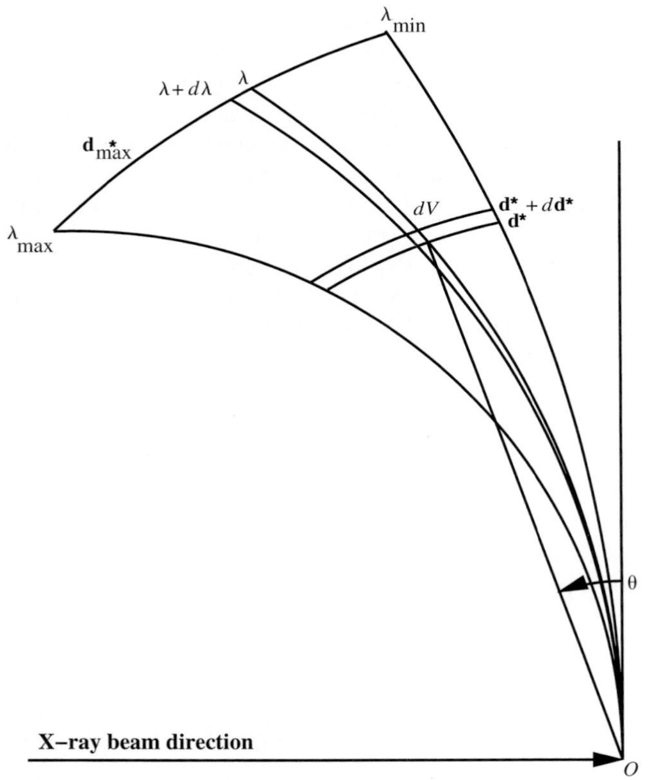

FIG. 1. Laue diffraction geometry. The volume element dV stimulated in a Laue experiment lies between $\mathbf{d}^*$ and $\mathbf{d}^* + d\mathbf{d}^*$, between the Ewald spheres corresponding to λ and $\lambda + d\lambda$, and between ϕ and $\phi + d\phi$, where ϕ denotes rotation about the incident X-ray beam direction. The entire volume stimulated in a single Laue exposure lies between 0 and $\mathbf{d}^*_{max}$, between the Ewald spheres corresponding to λ_{min} and λ_{max}, and between values of ϕ ranging from 0 to 2π.

the detector, in contrast to spots in a monochromatic pattern (oscillation, precession, or Weissenberg) that arise from a mapping of individual reciprocal lattice points. Because a ray may contain only one reciprocal lattice point ($h\,k\,l$) or several ($h\,k\,l$, $2h\,2k\,2l$, ... , $nh\,nk\,nl$) between the origin and $\mathbf{d}^*_{max}$, a Laue spot may be single, arising from only one reciprocal lattice point, wavelength, and structure factor, or multiple, arising from several. If a spot is multiple its integrated intensity is the sum of the integrated intensities of each of the components. The components also are known as orders or harmonics. The spots lie on conic sections, each corresponding to a zone $[UVW]$. Prominent nodal spots, surrounded by clear areas devoid of spots, lie at the intersection of well-populated zones and correspond to rays whose inner point (the reciprocal lattice point closest to the origin along the ray) is of low, coprime, indices ($h\,k\,l$). The nodal spots always are multiple. Each single spot is characterized by a unique wavelength λ, associated with the Ewald sphere on which the reflection lies. Extraction of structure amplitudes from single Laue spots requires the derivation and application of a wavelength-dependent correction factor known as the wavelength-normalization curve or λ curve, the value of which varies from reflection to reflection.

The volume element dV stimulated in a Laue experiment lying between $\mathbf{d}^*$ and $\mathbf{d}^* + d\mathbf{d}^*$, θ and $\theta + d\theta$, and ϕ and $\phi + d\phi$, where θ is the Bragg angle and ϕ denotes rotation about the incident X-ray beam direction, is given by (Fig. 1)

$$\begin{aligned} dV &= \mathbf{d}^* \cos\theta\, d\phi \cdot \mathbf{d}^*\, d\theta \cdot d\mathbf{d}^* \\ &= \mathbf{d}^{*2} \cos\theta\, d\theta \cdot d\phi \cdot d\mathbf{d}^* \end{aligned}$$

The volume of the shell dV_s in reciprocal space lying between $\mathbf{d}^*$ and $\mathbf{d}^* + d\mathbf{d}^*$ is obtained by integrating over ϕ between 0 and 2π, and over θ between θ_{min} and θ_{max}, where $\sin\theta_{min} = 1/2\lambda_{min}\mathbf{d}^*$ and $\sin\theta_{max} = 1/2\lambda_{max}\mathbf{d}^*$, and is

$$\begin{aligned} dV_s &= 2\,\pi\mathbf{d}^{*2} d\mathbf{d}^*(\sin\theta_{max} - \sin\theta_{min}) \\ &= \pi\mathbf{d}^{*3} d\mathbf{d}^*(\lambda_{max} - \lambda_{min}) \end{aligned}$$

The number of reciprocal lattice points within dV_s is

$$N_s = dV_s/V^*$$

where V^* is the volume of the reciprocal cell. The total number of reciprocal lattice points lying within the hemispherical shell between $\mathbf{d}^*$ and $\mathbf{d}^* + d\mathbf{d}^*$ is

$$N_{s,total} = 2\,\pi\mathbf{d}^{*2} d\mathbf{d}^*/V^*$$

The fraction of reciprocal lattice points at a resolution **d*** stimulated in a Laue experiment is therefore

$$N_s/N_{s,total} = 1/2\mathbf{d}^*(\lambda_{max} - \lambda_{min})$$

which is proportional both to the wavelength range and to the resolution. Laue experiments therefore afford more complete coverage of reciprocal space at higher resolution (larger **d***) than at lower resolution (smaller **d***), and coverage at all resolutions depends on the wavelength range.

The complete volume stimulated in a single Laue image is given by integrating the volume element dV over ϕ, θ, and **d***, and is

$$V_v = \tfrac{1}{4}\mathbf{d}^{*4}_{max}(\lambda_{max} - \lambda_{min}) \qquad (1)$$

This volume contains N_v reciprocal lattice points, where $N_v = V_v/V^*$, and N_v can be large, particularly for crystals that diffract to high resolution. Hence, Laue diffraction patterns often are crowded and spatial overlaps between adjacent spots may be common.[5,8]

The complete volume stimulated in a single Laue exposure usually is much larger than in a typical monochromatic rotation exposure. The complete volume stimulated in a Laue experiment is (approximately) equal to that in a monochromatic rotation experiment covering an angle α when

$$\alpha = 33.75\mathbf{d}^*_{max}(\lambda_{max} - \lambda_{min}) \quad \text{degrees} \qquad (2)$$

For example, if $(\lambda_{max} - \lambda_{min}) = 1$ Å, $\mathbf{d}^*_{max} = 0.5$ Å^{-1}, then $\alpha \approx 17°$. A Laue data set may therefore contain far fewer images than a monochromatic data set. Indeed, if the crystal is of high symmetry, a single Laue image of a carefully oriented crystal may contain a large fraction of the unique data.[9] However, one image (or a set of a few images) may not offer the significant redundancy that is essential to the derivation of an accurate wavelength-normalization curve from the Laue intensities alone,[6,8,10] and to accurate estimates of the mean and variance of the structure amplitudes.

The Laue experiment affords a significantly lower exposure time than the monochromatic rotation experiment, as may be shown readily. Following Kalman,[11] the integrated reflected energy E_L for the Laue experiment is

$$E_L = j_e|F|^2 I'(k)[V/(2k_L^2 \sin^2\theta_L V_0^2)]\,\Delta t_L \qquad (3)$$

[8] Z. Ren and K. Moffat, *J. Appl. Crystallogr.* **28,** 461 (1995).

[9] I. J. Clifton, M. Elder, and J. Hajdu, *J. Appl. Crystallogr.* **24,** 267 (1991).

[10] J. W. Campbell, J. Habash, J. R. Helliwell, and K. Moffat, *Inf. Q. Protein Crystallogr.* **18,** 23 (1986).

[11] Z. H. Kalman, *Acta Crystallogr.* **A35,** 634 (1979).

and the corresponding quantity E_M for the monochromatic rotation experiment is

$$E_M = j_e|F|^2\{I_0/\omega\}[V/(k_M^3 \sin 2\theta_M V_0^2)] \tag{4}$$

where j_e is the radius squared of the electron times the polarization factor; F is the structure factor of the reflection h; θ is the Bragg angle of the reflection h; k_L is the wave vector $1/\lambda_L$ in the Laue case; k_M is the wave vector $1/\lambda_M$ in the monochromatic case; $I'(k)$ is the spectral intensity of the polychromatic Laue beam, dI/dk; I_0 is the intensity of the monochromatic beam; V is the volume of crystal illuminated; V_0 is the volume of the direct unit cell; ω is the angular velocity of the rotating crystal, and Δt_L is the Laue exposure time. Require that the integrated reflected energies be equal in the two experiments, i.e., $E_L = E_M$. Further, assume that in the two experiments the polarization factors are identical (not strictly true, but a close approximation); that the reflection h is stimulated by the same wavelength and hence $k_L = k_M$ and $\theta_L = \theta_M$; and that $\omega = \alpha/\Delta t_M$, where Δt_M is the time to rotate through the angle α. Then by rearranging Eqs. (3) and (4), we obtain

$$\Delta t_L/\Delta t_M = \frac{\tan\theta}{\alpha} \cdot \frac{I_0}{kI'(k)} \tag{5}$$

which expresses the ratio between the Laue and monochromatic exposure times for the reflection h. The monochromatic intensity I_0 typically is obtained by passing a polychromatic synchrotron beam over a silicon or germanium monochromator, which transmits a bandpass $\Delta k/k$. Hence

$$\begin{aligned} I_0 &\approx dI/dk \cdot \Delta k \\ &= I'(k)\,\Delta k \end{aligned}$$

and

$$\Delta t_L/\Delta t_M = \frac{\tan\theta}{\alpha} \cdot \frac{\Delta k}{k} \tag{6}$$

The bandpass $\Delta k/k$ typically has a value $\leq 2 \times 10^{-4}$. If we assume realistic values of $\tan\theta = 0.1$ and $\alpha = 2^\circ = 0.035$ rad, then $\Delta t_L/\Delta t_M \approx 6 \times 10^{-4}$. Under these stated assumptions, the Laue exposure time is thus between three and four orders of magnitude less than the corresponding monochromatic exposure time.

For example, a carboxymyoglobin crystal required a Laue exposure time Δt_L of 100 msec on beamline X26C at the National Synchrotron Light Source (NSLS) (Brookhaven National Laboratory, NY). When a channel-cut silicon (1 1 1) monochromator was inserted in the beamline, a monochromatic rotation exposure through 5° on the same crystal required an exposure time Δt_M of 3 min (Ref. 12 and T. Y. Teng, unpublished results, 1996). In this experiment, $\Delta t_L/\Delta t_M = 5.5 \times 10^{-4}$.

However, this is not the whole story. The ability to measure integrated intensities (integrated reflected energies) accurately depends also on the background underlying each reflection, and this differs in the Laue and monochromatic experiments. However, Moffat *et al.*[13] show that the signal-to-noise (S/N) ratio, namely the ratio of the integrated intensity to the background intensity, is comparable to within an order of magnitude in the two classes of experiment.

Crucial to quantitative application of the Laue method is the fact that each reflection is measured at a different wavelength. Equation (3) shows that the integrated intensity for a Laue spot depends on crystal factors (V and V_0), on X-ray beam factors [j_e, $I'(k)$, Δt] and on geometric factors (k, θ), where we now drop the subscript L denoting the Laue experiment. Thus for the reflection h,

$$E = KG(k, \theta)|F|^2 \tag{7}$$

where K is a known constant. The recording of this integrated intensity on the detector involves additional factors that depend on wavelength, on the Bragg angle, and possibly also on the position x on the detector on which this spot falls. The normal assumption (see, e.g., Ref. 7) is that all quantities that depend on more than one variable are factorable. The recorded, integrated intensity I for the reflection h may therefore be written as

$$I = Kg(\theta)j(x)f(\lambda)|F|^2 \tag{8}$$

hence

$$|F|^2 = I[Kg(\theta)j(x)f(\lambda)]^{-1} \tag{9}$$

Now I is the measured intensity, and the values of K and the functions $g(\theta)$ and $j(x)$ are generally known. Hence, the extraction of integrated intensities $|F|^2$ from the measured intensities depends on knowledge of the function $f(\lambda)$, known as the wavelength-normalization curve. This curve may be derived by one of several methods: comparison with an external, reference, monochromatic data set,[14] explicit measurement of the source

[12] T. Y. Teng, V. Srajer, and K. Moffat, *Nat. Struct. Biol.* **1,** 701 (1994).

[13] K. Moffat, D. Bilderback, W. Schildkamp, D. Szebenyi, and T.-Y. Teng, *In* "Synchrotron Radiation in Structural Biology" (R. M. Sweet and A. D. Woodhead, eds.), pp. 325–330. Plenum Press, New York and London, 1989.

[14] I. G. Wood, P. Thompson, and J. C. Mathewman, *Acta Crystallogr.* **B39,** 543 (1983).

spectrum[15] and detector sensitivity, and as is now usual, purely internally,[6,8,10] by examination of redundant measurements of the same reflection, or its symmetry mates, that are stimulated by different wavelengths at different crystal orientations.

With these principles in mind, we can identify the advantages and disadvantages of the Laue technique. We caution, however, that the relevance of particular advantages or disadvantages may differ from experiment to experiment and experimental conditions must be chosen carefully to match the goals of the investigation.[16]

Advantages and Disadvantages

Because all spots are in reflecting position simultaneously and throughout the exposure, three main advantages (A1–A3) follow:

A1. Shortest possible exposure time, therefore well suited to rapid, time-resolved studies that require high time resolution
A2. Unlike all monochromatic methods, insensitive to fluctuations in the source, in the optics that deliver the synchrotron beam to the crystal, and in the exact shutter timing
A3. No (geometrically) partial reflections; all spots in a local region of detector space have an identical profile

The use of a wide wavelength range ($\lambda_{max} - \lambda_{min}$) offers both advantages (A4 and A5) and disadvantages (D1–D6).

A4. A large volume of reciprocal space is stimulated in a single image, hence only a few images may yield a data set of the required completeness
A5. Greater coverage of reciprocal space, hence greater redundancy and completeness, is obtained automatically at higher resolution where the intensities are naturally weaker

However:

D1. Multiple Laue spots, stimulated by several wavelengths and containing several superimposed structure amplitudes (energy overlaps), must be resolved into their component single spots if com-

[15] D. M. Szebenyi, D. H. Bilderback, A. LeGrand, K. Moffat, W. Schildkamp, B. Smith Temple, and T.-Y. Teng, *J. Appl. Crystallogr.* **25,** 414 (1992).
[16] J. Hajdu and I. Andersson, *Annu. Rev. Biophys. Biomol. Struct.* **22,** 467 (1993).

plete Laue data sets are to be obtained, particularly at lower resolution[17–19]

D2. Lesser coverage is obtained at lower resolution, which may contribute to a "low-resolution hole" in the data set. Completeness is therefore systematically lower at low resolution, which may introduce series termination errors in Laue-derived electron-density maps[16,20,21]

D3. A wide wavelength range may lead to dense diffraction patterns in which numerous spatial overlaps are present that must be resolved[8]

D4. The rate of X-ray absorption and energy deposition in a crystal exposed to an intense polychromatic Laue beam can be very high,[22,23] yet crystals of macromolecules rarely will withstand a temperature jump of more than a few degrees without disorder

D5. A wide wavelength range stimulates more spots, but at the expense of an increased background underlying those spots.[13] There is therefore a tradeoff between coverage of reciprocal space and accuracy

D6. The Laue technique is inherently sensitive to crystal disorder, which introduces elongated spots whose profile varies smoothly, but substantially, across the complete image. Crystal disorder may be present naturally, or introduced by the processes of reaction initiation in a time-resolved experiment, or by inadvertent temperature jump. Whatever its origin, it increases spatial overlaps and diminishes the peak intensity value of the spot, and must be accounted for if accurate integrated intensities are to be obtained[8]

More generally, an advantage (A6) is that

A6. The Laue technique requires a stationary crystal and no (or relatively simple) X-ray optics, and Laue images are therefore simpler to acquire

[17] Q. Hao, J. W. Campbell, M. M. Harding, and J. R. Helliwell, *Acta Crystallogr.* **A49,** 528 (1993).

[18] J. W. Campbell, A. Deacon, J. Habash, J. R. Helliwell, S. McSweeney, Q. Hao, J. Raftery, and E. Snell, *Bull. Mater. Sci.* **17**(1), 1 (1994).

[19] Z. Ren and K. Moffat, *J. Appl. Crystallogr.* **28,** 482 (1995).

[20] E. M. H. Duke, A. Hadfield, S. Walters, S. Wakatsuki, R. K. Bryan, and L. N. Johnson, *Phil. Trans. R. Soc. London Ser. A* **340,** 245 (1992).

[21] E. M. H. Duke, S. Wakatsuki, A. Hadfield and L. N. Johnson, *Protein Sci.* **3,** 1178 (1994).

[22] Y. Chen, Ph.D. thesis. Cornell University, Ithaca, New York, 1994.

[23] K. Moffat and R. Henderson, *Curr. Opin. Struct. Biol.* **5,** 656 (1995).

However, a disadvantage (D7) also exists:

D7. The Laue images require less familiar and somewhat more complicated analysis to account for the wavelength-dependent correction factors, the energy overlaps, the spatial overlaps, and the effects of crystal disorder. If this analysis is not performed carefully, then inaccurate structure amplitudes and incomplete data sets will result, which in turn yield distorted electron-density maps and poor structural results

Practical Details: Acquisition of an Excellent Laue Data Set

The Laue experiment must be planned with the preceding principles in mind. Prior Laue studies have not necessarily paid close attention to all critical experimental details, and it is only recently that advanced software for Laue image analysis has been widely available.[1,6,8,19,24,25] The limitations of the Laue approach therefore have received more attention (for example, see Ref. 16) than its strengths. It is all too easy to collect inadequate Laue data, quickly; it is more difficult, but now entirely possible, to collect excellent Laue data even more quickly.

Source and Optics

Almost all Laue patterns have been obtained using a bending magnet or wiggler synchrotron source, which provides a broad, smooth spectrum extending from the beryllium window cutoff around 6 keV (2 Å) up to roughly three times the critical energy of the source. An undulator source provides a much peakier spectrum in which relatively narrow lines, the harmonics, are superimposed on a more or less continuous background.[15] The exact shape of the spectrum and its polarization depend strongly on where the limiting aperture in the beamline is located with respect to the axis of the undulator, and the shape is therefore sensitive to drift or jitter in the particle orbit or optics. Undulator sources should be avoided unless experimental reasons compel the highest source brilliance, and unless software is readily available that can fit a highly structured spectrum (see, for example, Refs. 15 and 26).

[24] S. Wakatsuki, *in* "Data Collection and Processing" (L. Sawyer, N. W. Isaacs, and S. Bailey, eds.), Publication DL/SCI/R34. SERC Daresbury Laboratory, Warrington, England, 1993.

[25] Z. Ren, K. Ng, G. E. O. Borgstahl, E. D. Getzoff, and K. Moffat, *J. Appl. Crystallogr.* **29,** 246 (1996).

[26] B. R. Smith Temple, Ph.D. thesis. Cornell University, Ithaca, New York, 1989.

It is usual to increase λ_{min} (decrease E_{max}) by total external reflection from a mirror, which also may be bent to allow focusing. Even a simple plane mirror set to deflect in the vertical plane offers the major advantage of moving the crystal and the experiment out of the plane of the particle orbit and thereby decreasing the *bremsstrahlung* background, which can be substantial. The desired value of λ_{min} can be set by adjusting the grazing angle on the mirror. For float glass, the product of the grazing angle and the cutoff energy E_{max} is 32 mrad keV; for platinum, 80 mrad keV; and for rhodium, 68 mrad keV. Metal coating of the mirror has the disadvantage of introducing sharply varying features into the spectrum at the absorption edges of the metal. For example, the platinum L_{III} edge lies at 11.57 keV (1.072 Å) and its effect can be discerned readily in an accurately determined wavelength-normalization curve (see, for example, Ref. 8). Again, software must be available to fit such spectra, or alternatively, spots stimulated by energies around the absorption edge may be rejected or weighted down. There appears to be no appropriate minimum value of λ_{min}; excellent data were obtained at the European Synchrotron Radiation Facility (ESRF, Grenoble, France) with λ_{min} of 0.4 Å.[27]

Adjustment of λ_{max} is not so convenient to achieve. Because spots stimulated by longer wavelengths are subject to much larger absorption errors, which are difficult to correct fully, reducing λ_{max} from (say) 2.0 to 1.4 Å may not reduce substantially the amount of excellent data that can be quantitated. Initial attempts have been made that employ a thin-film transmission mirror to set λ_{max}[7,28] but this device is not yet fully developed. The typical practice is to attenuate the longer wavelengths and reduce λ_{max} by insertion of thin carbon or aluminum foils. An aluminum foil of 100-μm thickness will attenuate 1.5-Å X rays by 63 %, and 1.9-Å X rays by 88%.

We recommend adjusting λ_{min} to 0.5 Å or to the wavelength corresponding to three times the critical energy of the source, whichever is lower; and adjusting λ_{max} to two to three times λ_{min} or to 1.5 Å, whichever is higher. A value for λ_{max} of $\leq 2\ \lambda_{min}$ eliminates those double, low-order reflections with $0 < \mathbf{d}^* < 0.5\ \mathbf{d}^*_{max}$[4] and has been found effective in practice.[29]

Shutters and Other Beamline Components

Laue exposures on a focused bending magnet source such as beamline X26C at the NSLS[30] and other second-generation sources may lie easily in

[27] V. Srajer, T.-Y. Teng, T. Ursby, C. Pradervand, Z. Ren, S. Adachi, D. Bourgeois, M. Wulff, and K. Moffat, *Science* **274,** 1726 (1996).

[28] B. M. Lairson and D. H. Bilderback, *Nucl. Instr. Methods* **195,** 79 (1986).

[29] R. M. Sweet, P. T. Singer, and A. Smalås, *Acta Crystallogr.* **D49,** 305 (1993).

[30] E. D. Getzoff, K. W. Jones, D. McRee, K. Moffat, K. Ng, M. L. Rivers, W. Schildkamp, P. T. Singer, P. Spanne, R. M. Sweet, T.-Y. Teng, and E. M. Westbrook, *Nucl. Instrum. Methods Phys. Res.* **B79,** 249 (1993).

the 500-μsec to 50-msec range. Thus, a fast shutter is essential. It must open and close reproducibly in the desired time, possess a smooth opening function such as a boxcar with short triangular ramps at either end, attenuate hard X rays satisfactorily when closed without distorting or melting, and be robust. Most such shutters are based on a rotating slot milled in lead, tungsten, or titanium, or on a tungsten blade that can be flicked in and out of the X-ray beam. If the beam at the shutter is elliptical in cross-section rather than round, as is normally the case, the axis of rotation of the shutter should coincide with the long dimension of the beam.

Laue exposures on a focused insertion device source such as beamline BL3 at the ESRF or the Structural Biology Sector of the Consortium for Advanced Radiation Sources (BioCARS) at the Advanced Photon Source (APS; Chicago, IL) can be only a few microseconds, a time that is comparable with the revolution time of a single particle bunch in the storage ring. A fast shutter train has been devised[31] that can isolate the individual X-ray pulse of around 100-psec duration emitted by the particle bunch, and that occurs once per revolution. Hence, single X-ray pulse diffraction experiments are possible.

Circuitry and software have been devised and applied successfully[27,31,32] that enable the shutter opening and closing to be linked to the master accelerator clock and to the triggering of an external device that initiates a structural change in the crystal, such as a pulsed laser. This is essential for rapid time-resolved Laue experiments.

There is substantial scattering from an intense polychromatic beam. All beamline components therefore must be carefully shielded and air paths eliminated or minimized, particularly in the experimental station itself and around the crystal. Excellent Laue images depend on minimizing the background as well as on enhancing the signal!

Detectors

Most Laue data have been collected on film or imaging plates, although charge-coupled device (CCD) detectors are beginning to be employed. The wavelength-normalization curve also accounts for the properties of the detector and the presence of, for example, bromine in film and in imaging plates, which may introduce a significant further absorption edge. The *K* edge of bromine lies at 0.92 Å, in the middle of the wavelength range typically used.

[31] D. Bourgeois, T. Ursby, M. Wulff, C. Pradervand, A. Legrand, W. Schildkamp, S. Labouré, V. Srajer, T.-Y. Teng, M. Roth, and K. Moffat, *J. Synch. Rad.* **3,** 65 (1996).

[32] U. Genick, G. E. O. Borgstahl, K. Ng, Z. Ren, C. Pradervand, P. M. Burke, V. Srajer, T.-Y. Teng, W. Schildkamp, D. E. McRee, K. Moffat, and E. D. Getzoff, *Science* **275,** 1471 (1997).

The detective quantum efficiency (DQE) of most detectors diminishes at shorter wavelengths. The value of λ_{min} set by the source and optics must correspond to a detector DQE of no less than (roughly) 70% of its maximum value. It is pointless to illuminate the crystal with X rays of a wavelength that cannot be detected efficiently. A problem that has not yet been addressed is that the DQE also may depend significantly on the obliquity (the angle at which the diffracted beam falls on the detector), particularly for harder X rays.

The detector must be located sufficiently close to the crystal that reflections at $\mathbf{d}^*_{max}$ stimulated by λ_{max} are intercepted by the detector. That is,

$$\sin\,\theta_{max} \leq 1/2\lambda_{max}\mathbf{d}^*_{max}$$

where $2\theta_{max}$ is the maximum scattering angle intercepted by the detector. It often has been tempting to locate the detector further away from the crystal, which both minimizes the component of the background that falls off as the inverse square of the crystal-to-detector distance and provides more space between spots. However, this may both eliminate single reflections that otherwise could be recorded,[4] and more important, compromise the accurate determination of the wavelength-normalization curve, particularly at longer wavelengths where fewer reflections are now recorded. Data at all wavelengths are adversely affected by an inaccurate wavelength-normalization curve.

Because Laue exposure times are so short, the time taken to exchange detectors between images usually controls the elapsed data-collection time. This requires a speedy detector carousel for film or imaging plates, or a rapid readout time for an electronic detector.

There is another approach to the recording of the time course of a reaction in a crystal. For time-resolved experiments, several diffraction patterns may be laid down on a single image, each slightly offset from the others, in which each pattern corresponds to a different experimental condition, such as time after initiation of a structural reaction. This approach minimizes the total number of images and scaling errors, but unavoidably enhances the background and spot spatial overlaps. A careful analysis of this experimental approach[25] demonstrated that the best data (at least in the system studied) are obtained with only a single pattern per image. Unless there are compelling reasons that point in the direction of recording multiple patterns per image (such as the necessity to replace the entire crystal after each reaction initiation because the reaction is irreversible), a single pattern per image should be acquired.

Analysis Software

The available Laue analysis software and its strengths and limitations are described in detail elsewhere (see [23] in this volume,[1] and references therein). We emphasize here that the properties of the software (e.g., its ability to resolve adjacent spots at high scattering angle from a mosaic, disordered crystal) may influence the experimental design greatly, and must therefore be borne in mind in the planning and conduct of the experiment itself.

Laue Experiment: Exposure Time

The exposure time of an individual exposure must not be so long that unacceptable disorder is introduced by crystal heating arising from X-ray absorption. For example, with the focused bending magnet beamline X26C at the NSLS, the adiabatic heating rate of a typical protein crystal is 200–300 K sec^{-1}.[22,23] If a crystal will withstand a temperature jump of 5 K without significant disorder, a maximum single exposure time of 15–25 msec is therefore indicated. If an exposure time longer than this is required because, for example, the crystal is small or weakly diffracting, the total exposure may be accumulated as a series of subexposures. Each subexposure is followed by a cooling period of a few seconds to allow the crystal to return to its initial temperature and the crystal lattice to regain its initial values, before delivery of a further subexposure. For example, Genick *et al.*[32] found it necessary to deliver a total exposure of 150 msec in 15 subexposures of 10 msec each, separated by several seconds. The adiabatic heating rate depends on the form of the incident X-ray spectrum and is of course further enhanced at insertion device sources such as X25 at the NSLS.[30] The peak temperature rise depends also on the balance between the rate of heating and of cooling. The latter varies roughly as $1/L^2$, where L is a typical crystal dimension.[33] Because the required exposure time varies as $1/L^3$, small crystals are liable to suffer more from heating effects.

If a heating artifact is suspected, deliver the same total exposure as 1, 10, and 50 subexposures. If all three images are identical, no heating artifact is present; if the first image exhibits disorder that is less prominent or absent in the latter two, the artifact is present. Heating artifacts may be minimized by reducing $\lambda_{max} - \lambda_{min}$, or preferably by decreasing λ_{max}. It seems quite likely that this artifact, whose magnitude is insufficiently appreciated, has reduced the quality of many Laue data sets or even led to the rejection of crystals as too disordered when in fact they are well ordered. However, this artifact is believed to be essentially absent in even the most intense monochromatic beam lines.[34] With the availability of intense focused undulator beams, this requires further study.

[33] K. Moffat, Y. Chen, K. Ng, D. McRee, and E. D. Getzoff, *Phil. Trans. R. Soc. London Ser. A* **340,** 175 (1992).

[34] J. R. Helliwell, "Macromolecular Crystallography with Synchrotron Radiation." Cambridge University Press, Cambridge, 1992.

The total exposure time must not be so large as to saturate the detector or to exceed its calibrated range on a significant number of spots. The form of the Laue Lorentz factor and the existence of nodal spots mean that Laue patterns may exhibit a particularly large range of intensities. In one example,[35,36] a 1.7-Å resolution Laue data set was collected twice using an imaging plate detector, once with a short exposure time of 0.5 msec to record the most intense, predominantly low-angle reflections, and again with a 20-fold longer exposure time of 10 msec to record the weaker, predominantly high-angle reflections. Despite the fact that imaging plates have a large dynamic range, the entire pattern could not be recorded accurately with a single exposure time and data quality was significantly enhanced by combining data from the two exposure times.

Laue Experiment: Angular Settings of the Crystal

As with any diffraction experiment, there is a tradeoff between speed and accuracy. A single Laue image from a suitably oriented crystal will yield a substantial fraction of the unique data,[9] but with low redundancy and accuracy, and poor coverage of reciprocal space at low resolution. The wavelength-normalization curve cannot be determined accurately from such a single image. More images, recorded at different angular settings, increase the coverage of reciprocal space, the redundancy, the accuracy of the wavelength-normalization curve, and hence the quality and quantity of the observations in the data set. A quantitative assessment of the required number of images is provided by Ren and Moffat[8] for a particular example. It appears that a good Laue data set requires at a minimum between 5 and 20 images, with the smaller number appropriate to cases in which the crystal is of high symmetry, the wavelength range is larger, the value of $\mathbf{d}_{min}$ is greater than 2.0 Å, and the data-collection strategy is carefully designed by prior simulation to afford good coverage of reciprocal space. A strategy that combines several angular settings with brief exposures (to record the strong, lower-resolution data) and fewer angular settings at long exposures (to record the higher resolution data) may be effective.

Laue Experiment: Time-Resolved Data Collection

Additional considerations come into play in a time-resolved experiment, the chief of which is whether the structural reaction is irreversible or reversible. If the former holds (e.g., the liberation of GTP from caged GTP and

[35] X.-J. Yang, Ph.D. thesis. The University of Chicago, Chicago, Illinois, 1995.
[36] X.-J. Yang, Z. Ren, and K. Moffat, *Acta Crystallogr.* **D,** submitted (1997).

its turnover by the p21^{ras} GTPase[37]) then it is highly advantageous to collect a series of time points on a single crystal after one reaction initiation, if necessary at a single crystal orientation. The entire experiment then is repeated with fresh crystals, each at a new orientation, until a sufficiently complete and redundant data set is acquired. If the latter holds (e.g., the photolysis of carboxymyoglobin[27]) then more flexibility exists in experimental design. In both cases, it is highly desirable to collect a reference data set on the same crystal, immediately prior to reaction initiation. This minimizes crystal-to-crystal variability and enhances the accuracy of measurement of the small structure amplitude differences on which the experiment depends.

Conclusion

A well-designed and executed Laue experiment, whose data are analyzed by appropriate algorithms and software, yields structure amplitudes that are at least as accurate and complete as those of conventional monochromatic data sets.[8,9] The very brief exposure times characteristic of the Laue technique clearly fit it for the fastest time-resolved experiments, as has long been realized (see, e.g., Refs. 14 and 38). Slow time-resolved experiments may of course still use the tried-and-true monochromatic techniques. "Slow" here presently means reactions with half-times in excess of a few seconds, although these half-times will drop to a few tens of milliseconds as new insertion device, monochromatic beamlines become available at the ESRF and the APS. The Laue technique is also well suited to data collection at high resolution, where its more efficient coverage of reciprocal space promotes high data redundancy. However, care must be taken to minimize the background, because this governs the accuracy of measurement of weak reflections.

The Laue technique may be used for essentially any experiment to which monochromatic techniques are applied; it should be used only if it offers significant advantages, and these must be carefully evaluated for each experiment.

Acknowledgments

I have benefited greatly over many years from discussions of the Laue method with Don Bilderback, Ying Chen, Durward Cruickshank, John Helliwell, Zhong Ren, Wilfried Schildkamp, Vukica Srajer, Marian Szebenyi, Tsu-Yi Teng, and Brenda Smith Temple: I thank them all. Supported by NIH Grants GM36452 and RR07707.

[37] I. Schlichting, S. C. Almo, G. Rapp, K. Wilson, K. Petratos, A. Lentfer, A. Wittinghofer, W. Kabsch, E. F. Pai, G. A. Petsko, and R. S. Goody, *Nature* (*London*) **345,** 309 (1990).

[38] K. Moffat, D. Szebenyi, and D. Bilderback, *Science* **223,** 1423 (1984).

[23] Evaluation of Laue Diffraction Patterns

By I. J. CLIFTON*, E. M. H. DUKE, S. WAKATSUKI, and Z. REN

Introduction

The Laue method is the most senior technique in the armory of X-ray crystallographers. The first experiments of von Laue and assistants[1,2] were performed with unfiltered radiation that contained a range of wavelengths. The beautiful patterns produced have intrigued crystallographers ever since. The 1920s and 1930s were a boom period for small-molecule Laue crystallography,[3] but the method became eclipsed by more popular techniques such as the Weissenberg and precession methods, which allowed more direct interpretation of the patterns. However, renewed interest in the technique has resulted in its resurrection.The spectrum of X radiation from bending magnets and insertion devices at synchrotrons provides an excellent source of X rays for the Laue method and this has opened up new applications of the Laue method in both the small-molecule and biological macromolecule spheres. The extreme simplicity of the method means it can be very fast—subnanosecond exposure times with current synchrotrons are quite feasible.[4,5] The quality of data obtained from the method can be, given certain inherent limitations discussed below, as good as that obtained using monochromatic methods.

In this chapter we discuss a number of important aspects in the data analysis of Laue diffraction experiments: autoindexing and prediction, integration, deconvolution of spatial overlaps, wavelength normalization, absorption correction, and deconvolution of energy overlaps.

Overview of Laue Data Processing

Daresbury Laue Data-Processing Software Program

The development of the Daresbury suite of programs was a large collaborative project that started around 1985. The Daresbury project co-workers

* Present address: Dyson Perrins Laboratory, South Parks Road, Oxford OX1 3Q4, United Kingdom.

[1] W. Friedrich, P. Knipping, and M. Laue, "Interferenzerscheinungen bei Röntgenstrahlen," pp. 303–322. Sitzungsberichte der (Kgl.) Bayerische Akademie der Wissenschaften, 1912.

[2] W. Friedrich, P. Knipping, and M. Laue, *Ann. Phys.* **41**(10), 971 (1913).

[3] D. W. J. Cruickshank, *Phil. Trans. R. Soc. Lond.* **340,** 169 (1992).

[4] D. M. E. Szebenyi, D. Bilderback, A. LeGrand, K. Moffat, W. Schildkamp, and T.-Y. Teng, *Trans. Am. Crystallogr. Assoc.* **24,** 167 (1988).

[5] D. M. E. Szebenyi, D. H. Bilderback, A. LeGrand, K. Moffat, W. Schildkamp, B. Smith Temple, and T.-Y. Teng, *J. Appl. Crystallogr.* **25,** 414 (1992).

METHODS IN ENZYMOLOGY, VOL. 277 0076-6879/97 $25.00

were P. A. Machin, I. J. Clifton, S. Zurek, J. W. Campbell, and M. Elder. T. J. Greenhough and A. K. Shrive developed the integration method at the University of Keele, UK. M. Z. Papiz of Daresbury Laboratory (Warrington, UK) devised and programmed the basic algorithm for unscrambling the intensities of wavelength overlap spots. The first programs to be developed were the prediction program LGEN (M. Elder) and the film-pack scaling program [ABSCALE (I. J. Clifton), the ancestor of AFSCALE]. Soon afterward I. J. Clifton and S. Zurek created GENLAUE, AFSCALE, and UNSCRAM on the basis of these early programs. Together with MOSLAUE (adapted by T. J. Greenhough from his version of the monochromatic integration program MOSFLM) and the interface program to the preexisting CCP4 suite for monochromatic data LAUENORM (written by J. W. Campbell), these programs formed the first usable suite and made possible the early results from Daresbury.[6,7] This "release 1" of the software was based on film as a detector.

The current "release 2" of the Daresbury software accommodates both film and imaging plate data, and is composed of LAUEGEN, INTLAUE, AFSCALE/UNSCRAM, LAUENORM, and several ancillary programs.

LCHK, SCHK Explores efficient orientations. (Authors: M. Elder, I. J. Clifton)

INTANAL Calculates intensity statistics. (Authors: I. J. Clifton, T. J. Greenhough)

AFSCALE Analyzes the resulting measurements for a film pack, scales together the six films of the pack, applies Lorentz, polarization, and obliquity corrections. (Authors: I. J. Clifton, S. Zurek)

UNSCRAM Uses multipack measurements to unscramble the intensity components of wavelength multiplets. (Authors: I. J. Clifton, S. Zurek, M. Z. Papiz)

DIFFLAUE, LAUEDIFF Processes data from "difference Laue" experiments. (Author: J. W. Campbell)

LAUENORM, LAUESCALE Normalizes wavelength. (Author: J. W. Campbell)

INTLAUE Makes intensity measurements on digitized film at positions in the Generate file, with overlap deconvolution, and handling of elliptical spots. (Authors: T. J. Greenhough, A. K. Shrive)

[6] J. Hajdu, P. Machin, J. W. Campbell, T. J. Greenhough, I. Clifton, S. Zurek, S. Gover, L. N. Johnson, and M. Elder, *Nature (London)* **329,** 178 (1987).

[7] J. R. Helliwell, J. Habash, D. W. J. Cruickshank, M. M. Harding, T. J. Greenhough, J. W. Campbell, I. J. Clifton, M. Elder, P. A. Machin, M. Z. Papiz, and S. Zurek, *J. Appl. Crystallogr.* **22,** 483 (1989).

LAUEGEN Represents a much improved combination of GENLAUE plus NEWLAUE with X-Windows interface. (Author: J. W. Campbell)

Independent Laue Data-Processing Software Efforts

About the same time that the Daresbury suite was being written, a CHESS Laue data reduction package was developed at Cornell University (Ithaca, NY) by M. Szebenyi, B. Smith Temple, and Y. Chen.[8] Some new programs have been developed in several other laboratories.

LEAP Intended for use in conjunction with the Daresbury suite. It takes predictions from LAUEGEN and makes intensity measurements at the positions predicted in the Generate file. It also can scale data, deconvolve multiplets, and carry out extensive analyses, aided by graphics. (Author: Soichi Wakatsuki, developed at Oxford)

Bourgeois program Similar to LEAP, developed at the European Synchrotron Radiation Facility (ESRF, Grenoble, France)

Laue View[9–12] An independent Laue diffraction data-processing software system developed by Z. Ren at the University of Chicago (Chicago, IL). This work was initiated in late 1991 as a subproject of the time-resolved studies on photoactive yellow protein[13,14] in the laboratory of K. Moffat to extract accurately the weak time-resolved signal. LaueView, finished at the beginning of 1994, is now a complete software system with the main program LaueView, the utility program LauePlot and a set of data interfaces. LaueView features analytical profile fitting for integration, Chebyshev polynomials for λ-curve modeling, and a systematic approach to harmonic deconvolution, and is able accurately to reduce structure-factor amplitudes from mosaic crystals and high-resolution diffraction patterns. LaueView allows the extraction of sharp features in λ curves and is therefore more suitable for the processing of data from undulator sources. In addition, LaueView is able to recover most of the spatial and energy overlaps in the data.

[8] B. Smith Temple, "Extraction of Structure Factor Amplitudes from Laue X-Ray Diffraction." Ph.D. thesis. Cornell, Ithaca, New York, 1989.

[9] Z. Ren and K. Moffat, *Synchrotron Radiat.* **1**(1), 78 (1994).

[10] Z. Ren and K. Moffat, *J. Appl. Crystallogr.* **28**(5), 482 (1995).

[11] Z. Ren and K. Moffat, *J. Appl. Crystallogr.* **28**(5), 461 (1995).

[12] Z. Ren, K. Ng, G. Borgstahl, E. Getzoff, and K. Moffatt, *J. Appl. Crystallogr.* **29(3),** 246 (1996).

[13] E. Blum, Y. Chen, K. Moffat, and T. Teng, *Biophys. J.* **57**(2), A422 (1990).

[14] K. Ng, Z. Ren, K. Moffat, G. Borgstahl, D. Mcree, and E. Getzoff, *Biophys. J.* **64**(212), A373 (1993).

Prediction

Prediction programs for Laue patterns are not much more complicated than their monochromatic cousins. The principles are well established[15]; the main additional task in the Laue case is the identification of the various types of overlap. The assignment of wavelength arises naturally from the Laue geometry and is trivial to code.

Planning Laue Experiments

Programs LCHK and SCHK. Whenever possible, it is important to plan the experiment carefully to collect as much of the symmetry-unique reflection set as one can in a minimal number of exposures. Clifton and co-workers[16] have demonstrated how much can be gained from this for favorable space groups. In current practice, the initial orientation of the crystal usually is estimated from its morphology, which of course can be unclear, and the procedure as a whole is rather inaccurate. If a suitable electronic detector is available, however, the orientation could be solved accurately from an X-ray image taken with a tiny dose of very hard radiation.

The program LCHK simulates Laue experiments and generates statistics to help find crystal orientations that minimize the number of exposures needed and/or maximize the overall coverage of reciprocal space achieved. A variant of LCHK, called SCHK, was written to generate completeness tables given a list of orientations expressed as crystallographic directions.[16]

LaueView. Because a Laue diffraction simulator is built into the LaueView system for the purposes of autoindexing, geometry refinement, and prediction, it is relatively easy to extend the simulation program to optimize data-collection strategy. Multiple optimizers can be built depending on the application. For example, one optimizer of the LaueView system identifies the best ϕ coverage under the circumstances of fixed crystal orientation on a single axis (ϕ) camera. The program suggests how many images of what ϕ spacing are necessary and where to locate these ϕ settings around the spindle to cover most of reciprocal space.

Predicting Laue Pattern to Match an Image

The inputs to the programs used to generate Laue patterns comprise (1) the cell parameters a, b, c, α, β, γ; (2) crystal-to-film distance D; (3) wavelength range λ_{min}–λ_{max}; (4) minimum d spacing d_{min}; and (5) crystal orientation ϕ_x, ϕ_y, ϕ_z.

[15] D. Taupin, *J. Appl. Crystallogr.* **18,** 253 (1985).

[16] I. J. Clifton, M. Elder, and J. Hajdu, *J. Appl. Crystallogr.* **24,** 267 (1991).

In many applications of the Laue method most of these parameters are known (e.g., from the characteristics of the white radiation used) or can be reasonably estimated beforehand. The only parameters usually determined by autoindexing from each Laue pattern are therefore those in item 5, the crystal orientation.

It is possible (but not as easy as in monochromatic methods) to estimate the cell indirectly from the Laue pattern or its gnomonic projection.[17,18]

Soft Limits. The wavelength range and minimum spacing parameters are rather imprecise values and are called the "soft limits." The soft limits often are known approximately beforehand, for example from the characteristics of the X-ray source. However, there is a danger that incorrect estimates will cause reflections to be assigned to the wrong overlap class. This can affect the categorization of both wavelength and spatial overlaps. Therefore it is important to make as good an estimate of the soft limits as possible. The soft limits can be determined from the Laue pattern itself via the gnomonic projection.[19] Alternatively, using a prediction program with graphic output (e.g., LAUEGEN), one can closely inspect the superposition of the predicted on the actual pattern while the soft limits are adjusted. It also is a good idea to scrutinize the final statistics produced by processing [e.g., tables of $I/\sigma(I)$ as functions of $d_{\min}$, $\lambda_{\min}$, $\lambda_{\max}$] and perhaps to readjust the soft limits as a postrefinement step. LaueView can perform a preliminary stage of processing to facilitate soft-limit estimation. If the soft limits are overpredicted at the first cycle of data processing, better estimates can be obtained by visual inspection of statistical histograms output by the first cycle.[11]

Automatic Indexing of Diffraction Patterns. In the monochromatic oscillation method, the projection of the regular reciprocal lattice onto the detector is itself made up of uniform two-dimensional lattices. Automatic searches are commonly made to generate a list of peaks, and then, given the crystal–detector distance, λ, and the position of the direct beam, robust algorithms can determine both the cell parameters and the crystal orientation. In the Laue geometry, the projection of reciprocal space is no longer uniform (see Fig. 1). However, the elliptical "lunes," with "nodal spots"[20] at their intersections, stand out clearly. For this reason, autoindexing of Laue patterns usually begins with the manual picking of lunes or nodal spots from an image display rather than with an automatic search.

[17] P. D. Carr, D. W. J. Cruickshank, and M. M. Harding, *J. Appl. Crystallogr.* **25,** 294 (1992).
[18] P. Carr, I. Dodd, and M. Harding, *J. Appl. Crystallogr.* **26**(3), 384 (1993).
[19] D. W. J. Cruickshank, P. D. Carr, and M. M. Harding, *J. Appl. Crystallogr.* **25,** 285 (1992).
[20] D. W. J. Cruickshank, J. R. Helliwell, and K. Moffat, *Acta Crystallogr.* **A43,** 656 (1987).

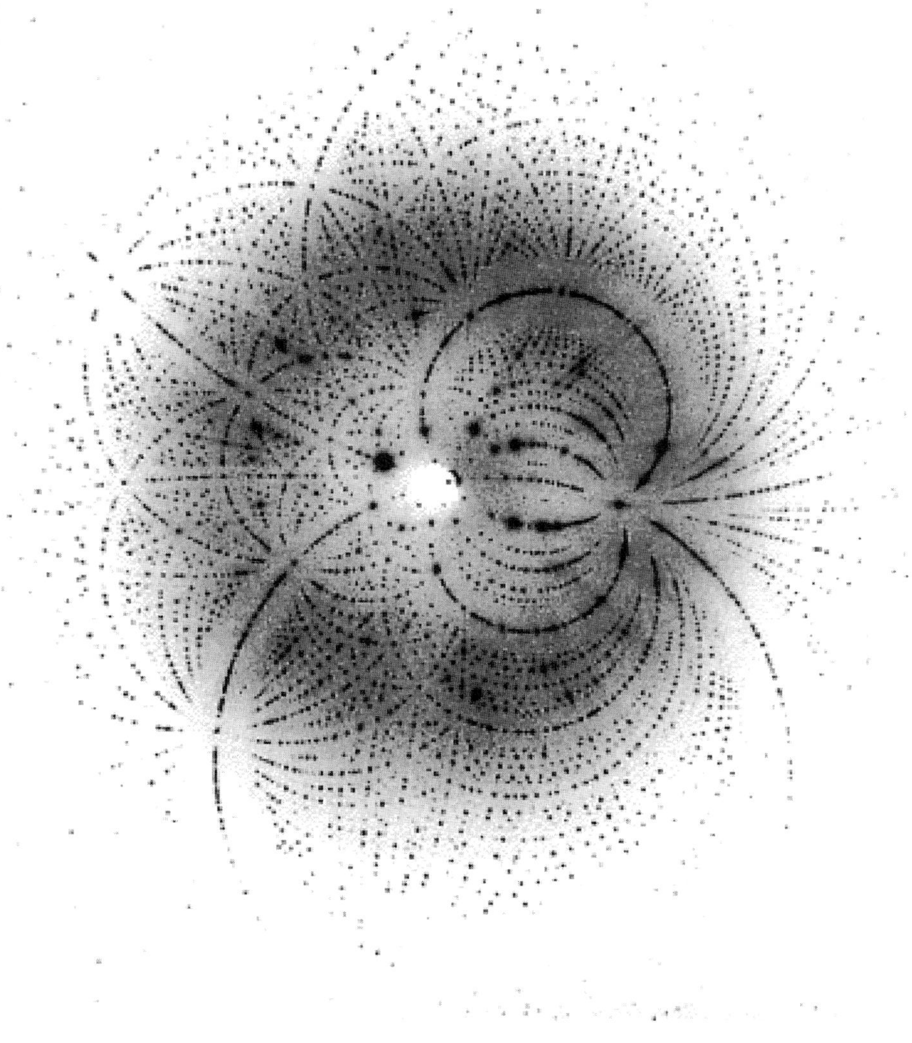

FIG. 1. A typical Laue image.

Each nodal spot corresponds to a low-index (low-resolution) reflection $(h\,k\,l)$ and its superimposed harmonics $(nh\,nk\,nl)(n > 0)$. The lunes are formed from belts of reflections belonging to a low-index zone $[UVW]$. There are thus two possible methods of autoindexing of Laue patterns: indexing the nodals, or indexing the zones that correspond to lunes. The location of a nodal spot on a Laue pattern indicates the direction of the central ray or axis on which that low-index reflection lies. The size and orientation of an ellipse indicate the direction of a zone axis. By matching a set (four or more) of major axes or zones reduced from an observed Laue pattern with a set of theoretical axes or zones, the orientation of the crystal can be obtained (given the cell parameters).

Autoindexing Method Implemented in LAUEGEN. This first method was designed and principally coded by M. Elder[21] in LGEN. The same algorithm has been used in later programs such as NEWLAUE and LAUEGEN.

Cell parameters and the crystal–film distance, as well as an accurate estimate of the position of the beam center, must be known. These are required because the algorithm works with the angular separation between diffracted rays. Using an auxiliary program (e.g., SPOTIN) or otherwise, the user prepares a list of coordinates, relative to the center, of prominent nodals. These should have small reduced indices, hence the angles between them are drawn from a relatively small subset that can be generated easily. The program calculates reciprocal lattice vectors for low-index spots and the angles between them. These angles are then matched with those derived from the entered spot positions to index these spots, hence to extract the crystal orientation.

Centering and Autoindexing from Ellipses. Laue diffraction patterns show prominent ellipses composed of a high density of reflection spots. (Notice that here we restrict our discussion to X-ray Laue diffraction of macromolecular crystals and forward reflections of small-molecule crystals. Back reflections from small-molecule crystals and from neutron diffraction of macromolecular crystals are not considered.) Every pair of ellipses intersects exactly twice with one intersection at the direct beam center. The reflections forming the ellipse belong to a common major zone in reciprocal space. The elliptical arrangement of Laue diffraction spots can be used in at least two ways: to find the direct beam center and to determine the crystal orientation by autoindexing. The success of the former is necessary for that of the latter. A visually easy approach to centering the direct beam position is the use of the gnomonic projection,[19,22] which converts the ellipses into straight lines if the center is correct. Bent lines indicate an error in the center position. In the CHESS LaueView package, one can simply adjust the center until the lines in the gnomonic projection become straight. However, a method based on gnomonic projection rather than

[21] M. Elder, *Inf. Q. Protein Crystallogr.* **19,** 31 (1986).

[22] J. L. Amorós, M. J. Buerger, and M. Canut de Amorós, "The Laue Method." Academic Press, New York, 1975.

visual adjustment confers no particular advantage in the automatic refinement of the center position. Fitting several (three is sufficient) ellipses that have distinct noncentral intersections can be used easily to determine the position of the direct beam center. This is another method utilized by LaueView.

Two Autoindexing Methods Compared. The nodal and ellipse methods differ in several respects. First, it is easier to find several nodals than ellipses on one Laue image with order less than or equal to three. Matching nodals with axes or ellipses with zones at low orders (≤ 3) has a much greater chance of indexing a pattern correctly than matching at higher orders. This is one reason why the nodal method is most commonly used. However, the directions of major axes deduced from nodals have larger uncertainties than those of zone directions deduced from ellipses because an ellipse can be fitted against many (say, 20) spots located on it, resulting in a more accurate mean. For streaky Laue patterns, the ellipse method would seem to have an advantage because the nodal positions are harder to eliminate accurately than the least-squares fitted ellipses, which are considerably more accurate. Both methods have been incorporated into the LaueView system.

In addition, the principle of the ellipse method also has been used in indexing and identification of multiple-crystal Laue diffraction patterns (C. Wilkinson, personal communication, 1996). Laue patterns from multiple crystals composed of several similar-size single crystals of the same molecule and space group, but in random orientations, show many ellipses intersecting each other. In such cases visual inspection cannot identify the pattern from any one crystal. However, many axes can be obtained from the ellipses and an algorithm based on a statistical approach has been developed to group these zones to match a combination of several orientations.

Integration

The intensity of each reflection is obtained by the integration of the individual pixel intensities within the spot. A correction also must be made for the background. The background has two significant sources: noise in the detector, and scattered X rays from the experimental apparatus, air, and the sample itself. The second component varies considerably across the detector, so it is important to take a local background sample for each spot or group of spots.

Various methods exist for carrying out the integration process, but broadly speaking they fall into two categories: those based on a simple box summation, and the "profile-fitting" methods that apply profiles determined empirically from suitable prototype spots. The profile-fitting methods can be subdivided further according to whether they use numerical or analytical profiles. The numerical profile methods are tied closely to the pixelization of the image; this generally means that they are computationally faster but interpolation becomes necessary. Generating an average numerical profile from spots whose centers are not exactly at the same position, relative to the grid points, flattens the resulting profiles and gives rise to up to 1% error in integrated intensities.[23] The analytical methods are usually free from interpolation issues.

Many of these integration methods use the idea of an "integration box," larger than the size of the spot, which is centered on its previously predicted center. The interior of the box identifies the pixels to be integrated, its rim provides a local sample of background pixels, and often marginal pixels in the box can be flagged to be ignored. It is usually possible to vary the size of the box depending on the position of the spot on the detector. This definition of an integration box, widely applied in monochromatic integration programs, must be modified to cope with Laue patterns where the "background" and even the "peak" pixels are often contaminated with intensity from a close neighbor.

Box-Summation Method

The integrated intensity is obtained by the direct summation of each pixel value within the box. To correct for background, the average intensity of the pixels on the frame of the box is subtracted from each pixel within the box.

We consider the integration of a spot with detector coordinates (x_0, y_0) in pixels. It has neighbors at $(x_1, y_1), \ldots, (x_n, y_n)$. The intensity of each pixel is given by $\rho(x, y)$. Box summation, the simplest form of integration, is represented by

$$I_{\text{box}} = \sum^{\text{peak}(x_0,y_0)} \{\rho(x, y) - [p_x(x - x_0) + p_y(y - y_0) + p_{\text{b}}]\} \tag{1}$$

where $\sum^{\text{peak}(x_0,y_0)}$ represents summation over the subset of peak pixels $\rho(x, y)$ in the box centered at (x_0, y_0). p_x, p_y, and p_{b} model a "background plane" and are determined by least-squares fitting of the pixels around the frame of the integration box, that is, by minimizing Q in

$$Q = \sum^{\text{frame}(x_0,y_0)} \{\rho(x, y) - [p_x(x - x_0) + p_y(y - y_0) + p_{\text{b}}]\}^2 \tag{2}$$

where $\sum^{\text{frame}(x_0,y_0)}$ represents summation over the subset of rim pixels in the box centered at (x_0, y_0).

[23] A. G. W. Leslie, P. Brick, and A. J. Wonacott, *Inf. Q. Protein Crystallogr.* **18,** 33 (1986).

Profile Fitting

In the profile-fitting methods, a least-squares fit is determined between a profile function $P(x, y)$ and the area of detector around (x_0, y_0), $\rho(x, y)$.

$$\rho(x, y) = pP(x, y) \tag{3}$$

The integrated intensity is then given by

$$I_{\text{profile}} = p \int^{\text{peak}} P(x, y) \tag{4}$$

This approach has the advantage that it can be generalized for the simultaneous integration and deconvolution of overlapping spots: For n spots in a section of detector:

$$\rho(x, y) = \sum_{i=0}^{n} p_i P_i(x, y) \tag{5}$$

where $P_i(x, y)$ is the profile function centered at (x_i, y_i) and p_i are weights determined simultaneously by least-squares fitting.

Numerical Profiles. The numerial profiles method[24] has several variations; however, the basic idea is that a set of standard spot profiles is obtained, usually from the strong, singlet, nonoverlapping spots on the image. A weighting function then can be used to produce a smoothly varying profile suitable for every part of the image. For each individual reflection spot a background plane is determined using least-squares fitting between observed and calculated background levels. At this stage, pixels deviating significantly from the proposed background level (e.g., because they include contributions from an adjacent spot) are discarded. The final integrated intensity is determined by the summation of the background plane corrected pixel intensities and the weighted profiles. Integration based on this method has a significant advantage over the box method in that the background calculation is not skewed by the close proximity of neighboring spots.

Analytical Profiles Determined by Empirical Parameters

Rabinovich and Lourie[25] suggested a two-dimensional Gaussian function as an analytical profile for Laue diffraction spots.

LaueView uses extended two-dimensional Gaussian functions[11] to fit unimodal peaks. This approach resolves simultaneously the spot-streaking and spatial overlap problems.

[24] M. G. Rossmann, *J. Appl. Crystallogr.* **12,** 225 (1979).

[25] D. Rabinovich and B. Lourie, *Acta Crystallogr.* **A43,** 774 (1987).

The following is an expression for the model using the analytical profiles.

$$P_{\text{analytical}}(x, y) = \sum_{i=0}^{n} p_i \exp(-A_i^{g_a} - B_i^{g_b}) + p_x(x - x_0) + p_y(y - y_0) + p_b \quad (6)$$

where x, y are the coordinates of a pixel on the image; x_0, y_0 are the predicted coordinates of the desired spot; and x_i, y_i for $i = 1, \ldots, n$ are the predicted coordinates of adjacent, overlapping spots. The final three terms represent the slopes in two directions, p_x, p_y, and the level, p_b, of the background plane, respectively, in that region of the pattern. p_0 and p_i ($i = 1, 2, \ldots, n$) are the coefficients to be fitted, which represent the intensities of the desired spot and its adjacent spot, respectively. g_a and g_b are parameters for non-Gaussian kurtosis of the diffraction spots and

$$A_i = \frac{[(x - x_i + d_x)\cos(\phi_i + \varepsilon) + (y - y_i + d_y)\sin(\phi_i + \varepsilon)]^2}{[a + s_a(x - x_i) + t_a(y - y_i)]^2} \quad (7)$$

$$B_i = \frac{[-(x - x_i + d_x)\sin(\phi_i + \varepsilon) + (y - y_i + d_y)\cos(\phi_i + \varepsilon)]^2}{[b + s_b(x - x_i) + t_b(y - y_i)]^2} \quad (8)$$

where a, b are the half-long axis and half-short axis of an elliptical peak; s_a, t_a, s_b, t_b are nonelliptical corrections. ϕ_0 and ϕ_i ($i = 1, 2, \ldots, n$) are the radial polar angles of the desired spot and its adjacent spots, respectively; ε is an additional nonradial correction angle to ϕ_i. d_x, d_y are local corrections for prediction errors.

Thus the profile of a spot is defined by $15 + n$ parameters. They are divided into two groups. Group 1 has $2 + n$ parameters [p_b, p_0, p_i ($i = 1, 2, \ldots, n$)] and group 2 has 13 parameters (a, b, s_a, t_a, s_b, t_b, ε, d_x, d_y, g_a, g_b, and p_x, p_y). The parameters in group 2 are identical for the desired spot ($i = 0$), and for its adjacent overlapping neighbors ($i = 1, 2, \ldots, n$), but those in group 1 are not. Group 2 parameters are determined initially by profile fitting of some strong reflections that are ideally not spatially overlapped, and then the parameters are modified to form a smooth surface for each across the detector frame. LaueView utilizes a detector-binning method that divides a detector frame into many very small bins (a few millimeters squared; see Ren and Moffat[11]) Standard profiles are allowed to vary smoothly from one bin to another. Every shape-related parameter of the analytical profile forms a continuous function across the detector frame. The final group 2 parameters that form the standard profiles are fixed within a single detector bin (a small portion of the detector frame) and vary slowly across the entire detector space.

A unique feature of this approach is the use of the shape parameters to fit streaky spots. If for some reason spots are nonunimodal, i.e., consisting of more than one maximum, this method needs to be modified.[11] Numerical profiles, when applied correctly, do not exhibit this kind of problem. An additional advantage of the analytical profiles is that they do not require interpolation.

Integration of Streaked and Spatially Overlapped Spots

Integration of Streaked Spots. A characteristic of Laue diffraction patterns is a tendency toward radial streaking, attributable to disorder within the crystal. A crystal of moderate inherent mosaicity, which would give well-shaped diffraction spots with monochromatic data collection, typically would give significant streaking in Laue mode. It also is possible for streaking to arise owing to a transient disorder within the crystal, such as one that might occur if a reaction were taking place in the crystal.[26–28] The common occurrence of radial streaking and the fact that the streaking is often worst at precisely the time point at which data collection is desirable has meant that effort has been put into developing procedures to analyze streaked diffraction patterns.

In some cases the Laue diffraction pattern is only slightly streaked and the individual spots still remain discrete entities. In these cases it is relatively easy for the spots to be integrated. However, care must be taken over the criteria used for the flagging of spatial overlaps when the spot shape is asymmetric.

INTLAUE Modifications to Deal with Streakiness. A method has been developed to perform the integration of severely radially streaked Laue spots.[29] In general the width of a streak (perpendicular to the radius) remains fairly constant over a Laue pattern, whereas the length of the streak can vary. The length of the elliptical mask in the radial direction can be written as

$$a = 2\eta D(1 + k \tan^2 2\theta) \tag{9}$$

where η is the mosaic block width, D is the crystal-to-detector distance, and k is a constant. For $k = 0$ the streaks have a constant length. For $k > 0$ the streaks will increase with the radius on the detector whereas with $k < 0$ the streaks will decrease.

[26] L. N. Johnson and J. Hajdu, Synchrotron studies on enzyme catalysis in crystals. *In* "Biophysics and Synchrotron Radiation" (S. Hasnain, ed.), pp. 142–155. Horwood, Chichester, 1989.

[27] C. Reynolds, B. Stowell, K. Joshi, M. Harding, S. Maginn, and G. Dodson, *Acta Crystallogr.* **B44,** 512 (1988).

[28] K. Moffat, Y. Chen, K. Ng, D. McRee, and E. Getzoff, *Phil. Trans. R. Soc. Lond.* **340,** 175 (1992).

[29] T. Greenhough and A. Shrive, *J. Appl. Crystallogr.* **27,** 111 (1994).

Once a mask appropriate for the observed streak length variation has been established, identification of spatial overlap spots is carried out and integration by either box or profile methods then can take place.

Integration of Spatially Overlapped Spots. The simplest approach to integration of spatial overlaps is to exclude all reflections that have a neighbor closer than some minimum spacing δ. In the earliest Daresbury software, this could be done at the prediction stage. However, if anything much more sophisticated than a crude δ cutoff is to be performed, it is important to predict all the spots observable on the detector, otherwise subsequent software could be misled. As we describe below, programs such as LEAP and LaueView attempt to solve whole systems of mutually overlapping spots, and they need to know the predicted positions of all spots.

A better approach is to exclude pixels, not spots, that is, suppress in the integration of a spot those pixels that are overlapping from neighboring spots.[30]

The next level of sophistication is to separate overlapping spots by simultaneous fitting of multiple profiles. In general, a minimization function for deconvolution of n diffraction spots is defined as follows:

$$Q = \sum_{j=0}^{m} \left[\sum_{i=0}^{n} p_i P_i(x_j, y_j) + ax_j + by_j + c - \rho(x_j, y_j) \right]^2 \qquad (10)$$

or

$$Q = \sum_{j=0}^{m} \left[\sum_{i=0}^{n} p_i P_i(x_j, y_j) - \rho_{BG}(x_j, y_j) \right]^2 \qquad (11)$$

where $\rho(x, y)$ is the pixel value, $\rho_{BG}(x, y)$ is the background subtracted pixel intensity, and j runs through all the pixels to be considered.

The summation over j can cover different areas (see Fig. 2). The simplest approach to obtain an integrated intensity for one spot out of a set of spatially overlapped spots is to use pixels that belong only to that spot but not to the neighbors (Fig. 2; and Shrive *et al.*[30]). LaueView takes one further step and uses all the pixels in a given area that contains the spot of interest and a few other overlapped neighboring spots. It then solves the least-squares equations for the spots within the window, but uses the result only for the spot of interest. It then moves to the neighboring spot and repeats the procedure. Whether the use of this rather small box is free of "edge effects" needs to be verified.

[30] A. K. Shrive, J. Hajdu, I. J. Clifton, and T. J. Greenhough, *J. Appl. Crystallogr.* **23,** 169 (1990).

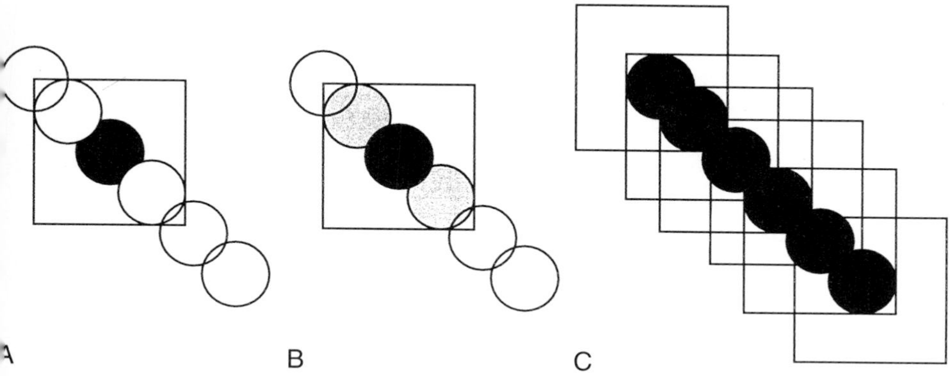

FIG. 2. Three approaches to the spatial overlap problem in Laue integration. (A) INTLAUE excludes pixels from encroaching reflections (white circles) from the spot of interest (black circle). (B) LaueView profile fits to the spot of interest (black circle) simultaneously with nearby spots (gray circles). (C) LEAP simultaneously profile fits a whole chain of mutually overlapping reflections (black spots).

LEAP and Bourgeois's deconvolution programs form more extensive linear equations to solve all the spots in one step. They use all the pixels in a whole set of spots connected by mutual spatial overlap (Fig. 2C).

Bourgeois's deconvolution method differs from that of LEAP in that it uses fewer pixels for the normal equation. The selection of the pixels is based on the σ/I plot.[31] It selects pixels according to their height compared to the highest pixel value for each spot so as to give the best ratio, σ/I.

Treatment of Background. Commonly, background planes are estimated for each spot and used to subtract the background. Fitting the background separately or together with profiles could be a problem when fitting very weak diffraction spots.

LaueView treats the local background in the same fashion as the diffraction spots (see Analytical Profiles Determined by Empirical Parameters, above). The slopes of the background, p_x and p_y, are shape-related parameters and belong to group 2; the background level p_b belongs to group 1.

Postintegration Processing

Lorentz Correction for Monochromatic and Laue Photography

A further difference between Laue and monochromatic methods is the way in which reflections are integrated, that is, the nature of the Lorentz

[31] C. Wilkinson, H. W. Khamis, R. F. D. Stansfield, and G. J. McIntyre, *J. Appl. Crystallogr.* **21,** 471 (1988).

correction. For monochromatic rotation methods this reflects the relative speed at which each reflection passes through the Ewald sphere (more properly, the time each reflection is in diffracting position) and is given by an expression such as[32]

$$L = \frac{1}{\sin 2\theta} \tag{12}$$

For Laue diffraction the Lorentz correction accounts for how much wavelength range an infinitesimally small reflection cuts through as a function of θ. This is[25,33]

$$L = \frac{1}{2 \sin^2\theta} \tag{13}$$

Interpack Scaling and Wavelength Normalization

AFSCALE. The earlier approach (AFSCALE/UNSCRAM in the Daresbury package) based on X-ray film as a detector uses a stack of up to six X-ray films in a cassette. The decay of the integrated intensities of singlets as a function of film number in the cassette can be used to derive the absorption coefficient at the energy of the diffraction spot. Combining the absorption coefficients determined from the singlets at different energies, one can establish a general absorption curve using the Victoreen model.[34] This is then used in two ways: (1) to increase the effective dynamic range of the detector, and to make it more efficient at higher energies, and (2) to solve the energy overlaps as a linear least-squares problem. The function of AFSCALE is to scale intensity measurements on the successive films in a pack (up to six) into a single intensity for each reflection. AFSCALE itself deals with the nonwavelength overlapped spots (singlets); a companion program UNSCRAM uses the interfilm scaling coefficients found by AFSCALE to "unscramble" the intensities of wavelength overlapped reflections (multiplets).

Once Victoreen coefficients a have been established, for each spot the six intensity and σ values are scaled up to the A film by applying interfilm scaling factors of the form

$$K \exp(a\lambda^3/\cos 2\theta) \tag{14}$$

[32] G. H. Stout and L. H. Jensen, "X-Ray Structure Determination." John Wiley & Sons, New York, 1989.

[33] B. Buras and L. Gerward, *Acta Crystallogr.* **A31,** 372 (1975).

[34] C. H. MacGillavry and G. D. Rieck (eds.), "International Tables for X-Ray Crystallography," Vol. III, pp. 157–161. Kynoch Press, Birmingham, England, 1962.

The mean intensity

$$I_{\text{mean}} = \frac{\sum_i I_i/\sigma_i^2}{\sum_i 1/\sigma_i^2} \tag{15}$$

and scaled-together σ

$$\sigma_{\text{mean}} = \max(\sigma', \sigma'') \tag{16}$$

where

$$\sigma' = \left[\left(\sum_i 1/\sigma_i^2\right)^{1/2}\right]^{-1} \tag{17}$$

and

$$\sigma'' = \left\{\frac{\sum_i [(I_i - I_{\text{mean}})/\sigma_i^2]^2}{\sum_i 1/\sigma_i^4}\right\}^{1/2} \tag{18}$$

are then evaluated.

LAUENORM and LAUESCALE. The programs LAUENORM and LAUESCALE calculate and apply a wavelength normalization to the Laue data from AFSCALE. The wavelength normalization curve is based on the idea of splitting the data into wavelength bins. In LAUENORM, an internal scaling procedure is carried out to minimize the difference between the scaled measurements for symmetry-equivalent reflections. In LAUESCALE, an external scaling against another reference data set is performed. A scaling factor for each bin is found so that the mean Laue intensities in each bin match the mean reference intensity in that bin. Both programs use polynomial curves (applying to disjoint ranges of bins) to calculate interpolated scale factors for individual reflections. LAUENORM also calculates interpack scale and temperature factors using the Fox and Holmes method.[35]

On output, the data are merged using $[1/\sigma(I)]^2$ weighting, and $\sigma(F)$ is evaluated as follows, following the program TRUNCATE[36]:

$$\sigma(F) = [I + \sigma(I)]^{1/2} - (I)^{1/2} \tag{19}$$

[35] G. C. Fox and K. C. Holmes, *Acta Crystallogr.* **20,** 886 (1966).

[36] CCP4 (Collaborative Computational Project, Number 4), *Acta Crystallogr.* **D50,** 760 (1994).

LEAP and LaueView. X-ray polarization correction, interimage scaling, temperature factor correction, wavelength normalization, and absorption correction are done by LaueView in an integrated manner.[11]

The use of Chebyshev polynomials, as opposed to Taylor polynomials, to model the wavelength normalization curve has some advantages: (1) the error is equally distributed over the range in which the data are fitted, and (2) changing the order of the polynomials requires minimum readjustment of already calculated coefficients. Both LaueView and LEAP use individual observations for calculating Chebyshev coefficients. This allows proper weighting of the observations and therefore can cope with normalization curves containing high-frequency components. Fitting to rapidly changing parts of the wavelength normalization curve can be avoided, however, by excluding the wavelength ranges corresponding to absorption edges (LAUENORM), and in this case, fourth-order polynomials are normally sufficient to describe typical bending magnet or wiggler sources.

Deconvolution of Harmonic Overlaps

The sampling of reciprocal space achieved in a single Laue exposure tails off to zero at low resolution because of the narrowing of the volume between the limiting Ewald spheres (Fig. 3). This insufficient sampling gives

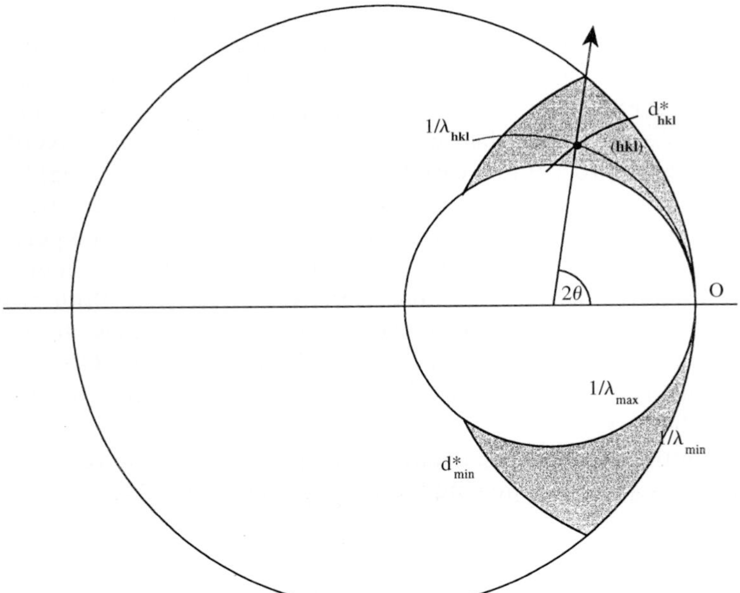

FIG. 3. The Ewald construction for the Laue geometry.

rise to the "low-resolution hole." The problem is exacerbated if harmonic overlaps are omitted in the processing of data: Many reflections with $d < 2d_{\min}$ are missing from the final data set. The low-resolution hole not only causes ragged electron density but also produces artifacts at a distance of half the unit cell dimensions from a strong peak.[37] The use of a narrower band width, for example $\lambda_{\max} \leq 2\lambda_{\min}$, enhances the proportion of singlets.[38] This is not always easy because the short wavelength cutoff is not sharp if a focusing mirror is used.

UNSCRAM. The role of UNSCRAM in the Daresbury suite is to process the wavelength-overlapped spots. For each overlapped spot, it uses the different film–film ratios between intensity measurements on successive films at different wavelengths to calculate ("unscramble") the individual single-wavelength components.

Given Victoreen coefficients from AFSCALE, and a wavelength-overlap spot of order n, scaling factors $s_{2,\lambda_1}, s_{3,\lambda_1}, \ldots, s_{6,\lambda_1}, s_{2,\lambda_2}, \ldots, s_{6,\lambda_n}$ can be evaluated for the individual components (whose wavelengths are λ_1, $\lambda_2, \ldots, \lambda_n$). On each film, each measured intensity is the sum of its single-wavelength components, that is,

$$\begin{aligned} I_A &= I_{\lambda_1} + \cdots + I_{\lambda_n} \\ I_B &= s_{2,\lambda_1} I_{\lambda_1} + \cdots + s_{2,\lambda_n} I_{\lambda_n} \\ \vdots &= \ddots \\ I_F &= s_{6,\lambda_1} I_{\lambda_1} + \cdots + s_{6,\lambda_n} I_{\lambda_n} \end{aligned}$$

UNSCRAM calculates a least-squares solution for this set of linear simultaneous equations in the n unknowns I_{λ_i}. In practice, there are only m good measurements, where $m \leq 6$, and unscrambling is usually worthwhile only for doublets and perhaps triplets. The drawback of this reflection-by-reflection approach is that the number of good observations of intensities that are decaying exponentially through the film pack often is not enough to obtain a sound estimate of the individual components.

LEAP and LaueView. LEAP and LaueView take a different approach[39]: They delay unscrambling of multiplets until wavelength normalization has been established. This allows the combination of all the intensities of symmetry equivalents to solve the harmonic overlaps. LEAP uses all the observations of symmetry equivalents recorded on the X-ray films or image plates in a single cassette. The use of intensities from only one frame

[37] E. M. H. Duke, A. Hadfield, S. Walters, S. Wakatsuki, R. K. Bryan, and L. N. Johnson, *Phil. Trans. R. Soc. Lond.* **340,** 245 (1992).

[38] R. Sweet, P. Singer, and A. Smalås, *Acta Crystallogr.* **D49**(2), 305 (1993).

[39] J. W. Campbell and Q. Hao, *Acta Crystallogr.* **A49**(6), 889 (1993).

often is insufficient to deconvolute higher harmonics. LaueView takes this approach one step further and combines the observations from all the frames, and thus greatly increases the ratio of the number of observations to that of parameters to fit. Therefore, LaueView is able to deconvolute multiples of multiplicity as high as 15.[10]

A further difference between the two program packages is that LEAP treats the problem as a "linear" least-squares problem whereas LaueView tackles it as a "nonlinear" least-squares problem, that is, solving using structure amplitudes F and thus automatically avoiding negative solutions.

In either case, it is important to note that wavelength normalization curves must be as accurate as possible for successful deconvolution.

Another approach to solving the energy harmonics problem is to use the positiveness of Patterson maps.[40] In addition, it would also be possible to apply maximum likelihood methods to the solution of general overlap problems in Laue crystallography.

Time-Resolved Applications

In the reaction system of the crystal is fully reversible, thus allowing multiple experiments on the same crystal as is possible for the photoreactive yellow protein,[14] one can collect dozens of frames at different orientations to obtain complete data sets on a single crystal. It is not so clear how to achieve this for many other systems that are not repeatable for various reasons such as radiation damage or irreversible release of caged compounds. In this case many frames from different crystals must be scaled together to make the final data set as complete as one obtained using monochromatic techniques.

DIFFLAUE and LAUEDIFF

The programs DIFFLAUE and LAUEDIFF are used to process data from "difference Laue" experiments in which (in the simplest case) two Laue exposures are taken before and after some event in the crystal but otherwise under identical conditions. The main difference between the programs is that LAUEDIFF takes data on which interpack scaling has been done by AFSCALE or another equivalent program, whereas DIFFLAUE reads intensities from individual images and combines them itself (the very nature of the difference method means that λ-dependent corrections such as interpack scaling can be bypassed[41]). Both programs scale the two Laue

[40] Q. Hao, J. W. Campbell, M. M. Harding, and J. Helliwell, *Acta Crystallogr.* **A49**(3), 528 (1993).

[41] D. Bilderback, K. Moffat, and D. M. E. Szebenyi, *Nucl. Instr. Methods* **222,** 245 (1984).

data sets using an anisotropic temperature factor initially, and then the differences are put on a λ-independent basis using a reference data set.

LaueView

The LaueView system includes some special facilities for time-resolved applications. Spatial overlap deconvolution has been extended to handle multiple diffraction patterns on a single detector frame, because the spatial overlap problem can be severe if users choose such a data collection mode in time-resolved studies.[12] LaueView supports both those methods that bypass wavelength normalization and harmonic deconvolution and those that do not.

Concluding Remarks

Laue diffraction patterns can be measured accurately to give structure amplitudes nearly as good as those from standard monochromatic oscillation data. In protein crystallography, unless the crystal has high symmetry, frames collected at only a few different angles usually do not yield a data set of sufficient completeness.[10] The problem arises from the systematic nature of the data incompleteness inherent in Laue experiments.

[24] Triggering Methods in Crystallographic Enzyme Kinetics

By ILME SCHLICHTING and ROGER S. GOODY

Introduction

Knowledge of three-dimensional protein structures, determined by conventional monochromatic X-ray diffraction techniques, and also by multidimensional nuclear magnetic resonance (NMR) methods, has provided a solid base for understanding their architectures. However, relatively little direct information has been extracted from these studies about the workings of the enzymes. This arises from the essentially static nature of the information obtained.[1] Knowledge about the dynamics of important events in biological macromolecules has originated mainly from spectroscopic and other

[1] Or as H. Frauenfelder has formulated, "...it is difficult to assess from a sitting skier how well he skis..."; H. Frauenfelder, *in* "Protein Structure: Molecular and Electronic Reactivity" (R. Austin, E. Buhks, B. Chance, D. De Vault, P. L. Dulton, H. Frayenfelder, and V. I. Gol'danskii, eds.), pp. 245–263. Springer-Verlag, New York, 1987.

0076-6879/97 $25.00

rapid measurements in solution. This is because there is a large gap between the time scale of enzymatic reaction rates on the one hand (typically milliseconds to seconds) and the long X-ray data acquisition times on the other hand (typically hours to days). Despite the static character of crystallographic data collection, which averages both over time and space, the study of unstable species by X-ray diffraction, known as time-resolved crystallography, is nevertheless made feasible by slowing the observed reaction (e.g., by cooling or/and using poor substrates, suboptimal pH, and slow mutants) or by decreasing the data acquisition time (by using the high-intensity X-rays provided by synchrotrons, where a further reduction in data collection time can be achieved with fast data collection strategies employing the Weissenberg or Laue geometries).[2–8] Another approach to the study of unstable species is to work under steady state conditions if the kinetics of the system allows significant accumulation of the intermediate of interest. Common to these approaches is the need to find a way to initiate the reaction in the crystal efficiently, rapidly, uniformly, and gently. In the following sections we discuss some triggering methods and considerations one has to bear in mind when choosing one to initiate a reaction in a crystal.

Kinetic Considerations

Unfortunately there is no general recipe for performing time-resolved X-ray diffraction experiments. The overriding consideration, and potential limitation, is determined by the detailed kinetics of the reaction to be studied. Thus, of vital importance are not only the general time scale of events to be followed in relation to the rapidity of the trigger and data collection method used, but also the ratio of the effective rate coefficients for buildup and decay of the species (intermediates) of interest. The latter factor decides whether an intermediate accumulates sufficiently to be observable by X-ray diffraction (or indeed any other direct method) as shown

[2] J. Hajdu, K. R. Acharya, D. I. Stuart, D. Barford, and L. N. Johnson, *Trends Biochem. Sci.* **13,** 104 (1988).

[3] K. Moffat, *Annu. Rev. Biophys. Biophys. Chem.* **18,** 309 (1989).

[4] J. Hajdu and L. N. Johnson, *Biochemistry* **29,** 1669 (1990).

[5] E. F. Pai, *Curr. Opin. Struct. Biol.* **2,** 821 (1992).

[6] L. N. Johnson, *Protein Sci.* **1,** 1237 (1992).

[7] D. W. J. Cruickshank, J. R. Helliwell, and L. N. Johnson (eds.), *Phil. Trans. R. Soc. Lond. A* **340** (1992).

[8] J. Hajdu and I. Andersson, *Annu. Rev. Biophys. Biomol. Struct.* **22,** 467 (1993).

in Fig. 1. Thus, before embarking on the structural aspect of the experiment, the kinetics both in solution and in the crystal should be studied thoroughly.

Protein crystals may be regarded as concentrated solutions (typically 5–50 m*M* protein) because of their large solvent contents of, usually, 30–80%. The molecules are held in the crystal lattice by relatively few weak interactions, allowing for some flexibility and motion. Crystalline proteins or enzymes are therefore often biochemically active.[9–11] A large difference from working with concentrated solutions, however, stems from the arrangement of the solvent in ordered arrays of solvent channels. Thus convection is impossible, and diffusion is restricted in space. Reaction initiation by a simple mixing of reactants, as done in the stopped- or quenched-flow methods for rapid kinetic experiments in solution, generally will take too long. Also, enzymatic inactivity or reduced activity in the crystalline state may be caused by steric restrictions, such as blocked active sites or impaired motions of the molecules, or by inhibition of substrate binding or catalysis caused by the crystallization conditions, such as high salt concentration or unfavorable pH in the crystals. Because of such effects, the kinetics of the process being investigated must be measured in the crystal. This information is also essential for knowing when to collect the diffraction data of an intermediate after the reaction has been initiated in the crystal.

Figure 1a–c shows simulations of the accumulation of intermediates and products for a simple three-state reaction. Knowledge of the rate constants can help in devising a strategy to capture an intermediate at a moment when it has accumulated in the crystal. One can see from Fig. 1a–c that if the rates in solution and the crystal differ, one might easily collect diffraction data at the wrong time. In the ideal case, kinetics in the crystal should be analyzed spectroscopically and noninvasively *in situ,* adding the advantage that this analysis may also be done simultaneously with the collection of X-ray diffraction data. Thus, unpleasant surprises, such as faster-than-expected reactions caused by heating by the X-ray beam,[12] other-than-expected reactions from production of electrons produced by the X-ray beam (I. Schlichting *et al.*, unpublished, 1996), and incomplete photolysis due to misalignment of the apparatus,[13] can be detected while performing the diffraction experiment. This offers the possibility of correction or abandonment of the experiment, rather than recognition of the problem after extensive analysis of the diffraction data. In addition,

[9] M. Makinen and A. L. Fink, *Annu. Rev. Biophys. Bioeng.* **6,** 301 (1977).

[10] G. L. Rossi, *Curr. Opin. Struct. Biol.* **2,** 816 (1992).

[11] G. L. Rossi, A. Mozzarelli, A. Peracchi, and C. Rivetti, *Phil. Trans. R. Soc. A* **340,** 191 (1992).

[12] I. Schlichting, S. C. Almo, G. Rapp, K. Wilson, K. Petratos, A. Lentfer, A. Wittinghofer, W. Kabsch, E. F. Pai, G. A. Petsko, and R. S. Goody, *Nature* (*London*) **345,** 309 (1990).

[13] E. M. H. Duke, S. Wakatsuki, A. Hadfield, and L. N. Johnson, *Protein Sci.* **3,** 1178 (1994).

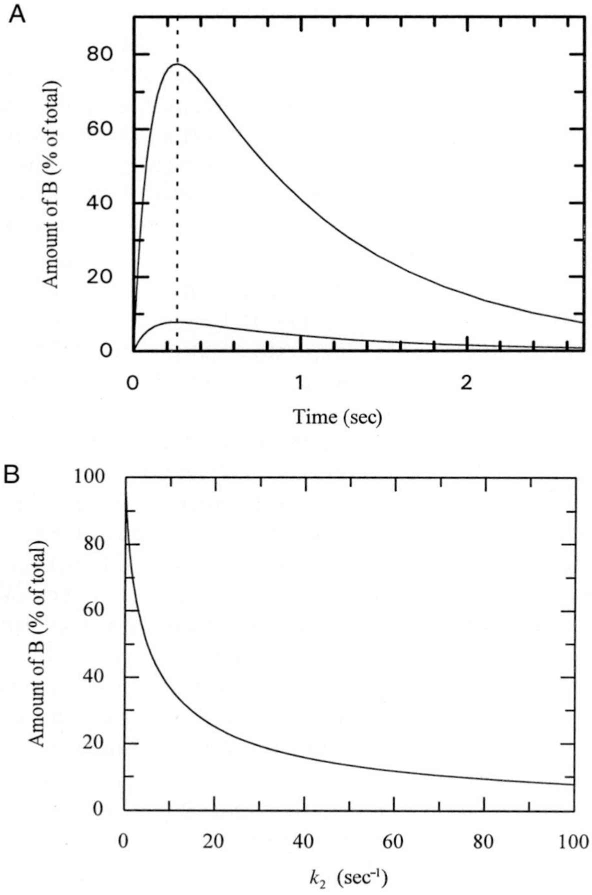

FIG. 1. (A) Simulation of the reaction

$$A \xrightarrow{k_1} B \xrightarrow{k_2} C$$

with $k_1 = 10$ sec^{-1}, $k_2 = 1$ sec^{-1} (upper curve) and with $k_1 = 1$ sec^{-1}, $k_2 = 10$ sec^{-1} (lower curve), showing the appearance and disappearance of B. B accumulates only to high occupancy in the first case ($k_1 \gg k_2$), although the maximum is reached at the same time after reaction initiation in both cases in this special situation. (B) Dependence of the maximum amplitude of formation of B in (A) on the magnitude of k_2 (with k_1 held constant at 10 sec^{-1}). Note that it is the ratio of the two rate constants that determines the maximal amount of B formed, independent of their exact magnitude, but that the time at which the maximum is reached depends on the absolute values.

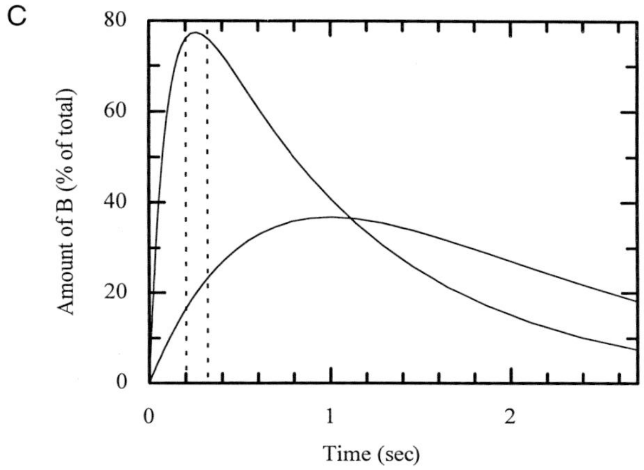

(C) Simulation of the reaction

$$A \xrightarrow{k_1} B \xrightarrow{k_2} C$$

with $k_1 = 10\ \text{sec}^{-1}$, $k_2 = 1\ \text{sec}^{-1}$ (upper curve) and with $k_1 = 1\ \text{sec}^{-1}$, $k_2 = 1\ \text{sec}^{-1}$ (bottom curve). Collection of the diffraction data of state B in the time window defined by the dashed vertical lines inferred from the rate constants $k_1 = 10\ \text{sec}^{-1}$, $k_2 = 1\ \text{sec}^{-1}$ obtained from kinetic studies performed in solution results in significantly reduced relative occupancy of B when the rate constants for the reaction are different in the crystal.

having an "objective" independent identification of the species is often crucial for the interpretation of electron-density maps of mixed states that are structurally similar, as mentioned in several of the following examples.

For crystallographic studies on enzymes or proteins in action, it is necessary not only that the crystal lattice does not inhibit the activity of the protein, but also that the crystalline order should not be affected by the reaction or its initiation through loss of crystal contacts caused by changes in the protein structure. These requirements will not be met by all (protein) crystal systems; some examples of transient disorder during reaction initiation are discussed in the section on light-induced reactions. Further prerequisites for time-resolved studies are that the reaction to be studied can be initiated uniformly in space and time throughout the crystal and that the time necessary for triggering the reaction and collecting the data is much shorter than typical lifetimes of the reaction intermediates to be studied. Intermediate states can be observed only if they are sufficiently occupied and live longer than it takes to collect the data. It pays, therefore, to explore the possibilities of observing intermediates by working under steady-state conditions, because this circumvents the difficulties of finding a fast trigger to start a single turnover experiment and the technical problems of having to work on a fast data collection time scale (see Ref. 14 and examples cited therein).

Triggering Methods

Reactions can be initiated by changes in thermodynamic parameters such as temperature or pressure; by changes in the concentration of substrates, cofactors, protons, or electrons; or by light or other radiation. The choice of the trigger depends largely on the physicochemical properties of the system and the reaction studied. An ideal starting point for time-resolved protein crystallography is provided by the presence of a "built-in" trigger. This may be the case for proteins involved in the conversion of light into other energy forms (e.g., photosynthetic reaction centers), in the transduction of light signals [e.g., (bacterio) rhodopsin], or that have light-sensitive bonds (e.g., carbonmonoxy complexes of heme proteins). A somewhat more general approach to fast reaction initiation involves the use of components of the system under investigation that have been rendered inactive by chemical modification, so that the normally reactive species can be generated by photolysis. Such a compound could be a substrate, cofactor, or the enzyme itself, as discussed below.

Light

The photochemical release of biologically active compounds from photosensitive biochemically inert precursors, so-called caged compounds,[15] can be used to trigger reactions.[16–18] To exploit fully the fast time scale of photolysis, it is desirable to avoid the potentially slow diffusion of the deprotected compound to the active site by using caged compounds that bind to the site of action of the photoreleased product, and to cocrystallize or diffuse them into the crystal prior to photolysis. It is important to know whether and how strongly a caged compound binds to the active site of the crystalline macromolecule (also for knowing what concentrations to use), and whether it is a substrate, but also whether side products of photolysis react with the macromolecule, thereby possibly inactivating it.

Compounds as diverse as nucleotides, metals, neurotransmitters, protons, amino acids, and sugars have been caged and used in kinetic experi-

[14] V. N. Malashkevich, M. D. Toney, and J. N. Jansonius, *Biochemistry* **32,** 13451 (1993).

[15] J. H. Kaplan, B. Forbush III, and J. F. Hoffman, *Biochemistry* **17,** 1929 (1978).

[16] J. A. McCray and D. R. Trentham, *Annu. Rev. Biophys. Chem.* **18,** 239 (1989).

[17] J. E. T. Corrie, Y. Katayama, G. P. Reid, M. Anson, and D. R. Trentham, *Phil. Trans. R. Soc. Lond. A* **340,** 233 (1992).

[18] J. E. T. Corrie and D. R. Trentham, *in* "Bioorganic Photochemistry" (H. Morrison, ed.), Vol. 2, pp. 203–305. John Wiley & Sons, New York, 1993.

ments in solution or on intact or semiintact biological entities such as muscle fibers or cells.[19] In crystallographic studies, most difficulties associated with photolysis of caged compounds are caused by their high concentration in the crystal, which is required to form a 1:1 complex between enzyme and caged compound. Even in favorable cases, in which the ratio between the dissociation constant and the enzyme concentration is small compared to one ($K_d/[E] \ll 1$), so that an excess of caged substance over enzyme need not be used, the concentration range is in the tens of millimolar. This leads to certain requirements for caging groups that are useful for time-resolved structural studies. For complete and rapid photolysis and minimization of heating of the crystal by the photolysis light, the quantum yield should be high and the absorption of the photolyzing light by the protein or photoproducts should be minimal. This, together with handling considerations,[19a] limits the useful photolytic wavelength range to 300–400 nm.

The photolysis yield under a given set of external conditions is determined by the absorption coefficient at the photolytic wavelength, the intensity of the exciting light source, and the quantum yield. In practice, other factors such as the geometry of the sample are also important. Ideally, the size of the photolysis beam should be larger than the crystal, illuminating it completely and uniformly, thus minimizing potential problems with the X-ray beam probing an incompletely photolyzed region. To reduce the drop in intensity owing to absorption in the direction of the light beam, it is advisable to rotate the crystal while photolyzing it or to illuminate it simultaneously from opposite directions. Taking a realistic extinction coefficient of 10^4 cm^{-1} M^{-1} for the photolytic excitation and a concentration of 10 mM (which is lower than in most crystals), the optical density (OD) of a 100-μm-thick layer will be 1.0, meaning that the light intensity drops exponentially across the 100-μm layer from 100 to 10%. To help reduce the severity of the problem of nonuniform illumination, several factors can be varied. Moving away from the absorption maximum decreases the drop in intensity across the illuminated width of the crystal, but more intensity will be needed to compensate for the lower absorption. Reducing the size of the crystal (some structure determinations have been performed with crystals with a smallest dimension of 5–10 μm) also will help to alleviate this problem, but having less material to scatter the X-ray beam will lead to longer exposure times and poorer signal-to-noise (S/N) ratios. In practice, these factors must be traded against each other for the specific problem at hand.

[19] A. M. Gurney and H. A. Lester, *Physiol. Rev.* **67,** 583 (1987).

[19a] To avoid unintended photolysis it is necessary to work in red or orange lighting, avoiding in particular fluorescent light, which has significant intensity in this wavelength range, and to keep exposures short to all but the most subdued lighting.

Light sources typically used for photolysis are xenon flash lamps,[20] and dye, frequency-doubled ruby, or Nd:YAG (neodymium–yttrium-aluminum garnet) lasers (using the third harmonic of the 1065-nm line). Argon-pumped sapphire and other laser systems can be used for two-photon molecular excitation (reviewed, e.g., by Williams *et al.*[21]). In this case absorption is less of a problem because of the long wavelength used (double the excitation wavelength). To increase the energy density, the laser beam is focused through a microscope. This makes two-photon laser photolysis a powerful tool for time-resolved microscopy,[22] but less attractive for crystallography because of the extreme spatial localization.

Commonly used cage groups are substituted 2-nitrobenzyls such as the 2-nitrophenylethyl[16] group, which can be cleaved with light of around 350-nm wavelength with concomitant production of a nitroso ketone. Because the photochemical characteristics of a cage group also depend on the compound to which it is attached, care must be taken to choose an appropriate group. Also, other considerations may be important for the choice of a certain cage group, e.g., sensitivity of the system toward the addition of thiols to scavenge the 2-nitrosoacetophenone photolysis by-product of the 2-nitrophenylethyl cage group, as exemplified in the case of phosphorylase[13] (see below). Properties of caged compounds have been reviewed.[17,18] Depending on the particular caged compound used, the photolysis rate is in the range of 10^1–10^5 sec^{-1}. However, many flashes may be necessary to liberate enough of the compound for crystallographic purposes (many millimolar; see above), bringing the effective photolysis time into the minute range.[12,13] The intensity and frequency of the flashes must be balanced against heating of the crystal and may also be limited by the light sourced used.

Examples. Caged GTP [1-(2-nitrophenylethyl)GTP] was used in a study of the GTP complex of the *ras* oncogene product p21. This complex is the active form of the protein in signal transduction and, because of the intrinsic GTPase activity of the enzyme, it has a half-life of ca. 40 min at room temperature.[12] Because the affinity of p21 for caged GTP is ca. 10^{10} M^{-1} (an order of magnitude lower than for GDP), p21 could be cocrystallized with caged GTP in a 1:1 complex.[23] About 10 pulses of a xenon flash lamp, which took about 1 min, were used to photolyze the crystals containing ca. 40 m*M* p21·caged GTP. To minimize heating effects by the white photolysis

[20] G. Rapp and K. Güth, *Eur. J. Physiol.* **411,** 200 (1988).
[21] R. M. Williams, D. W. Piston, and W. W. Webb, *FASEB J.* **8,** 804 (1994).
[22] W. Denk, J. H. Strickler, and W. W. Webb, *Science* **248,** 73 (1990).
[23] I. Schlichting, J. John, G. Rapp, A. Wittinghofer, E. F. Pai, and R. S. Goody, *Proc. Natl. Acad. Sci. U.S.A.* **86,** 7687 (1989).

light, a filter was used, which was transparent only in the 300- to 400-nm range. In addition, the flashes were separated by 5 sec to allow heat dissipation, and the crystals were mounted wet (i.e., bathed in mother liquor). The generated crystalline p21·GTP complex hydrolyzes GTP at the same rate as in solution, as shown by high-performance liquid chromatography (HPLC) analysis of dissolved crystals. Thus, the crystal lattice did not affect the activity of the protein, or vice versa. The initial studies were done using a mixture of the *R*- and *S*-diastereomers of (2-nitrophenylethyl)GTP (the protecting group has a chiral center),[13] whereas a follow-up study used the pure diastereomers.[24] Interestingly, in the crystal form of the racemic mixture, caged GTP was bound to p21 in a fashion unlike the binding of GTP or GDP, and was stable in the crystals that diffracted mono- or polychromatic X rays to 2.8 Å.[23] However, the pure caged GTP diastereomers were bound to crystalline p21 in the same way as was GTP or GDP, were not very stable in the crystals (particularly the *S*-isomer), and diffracted monochromatic X rays to 2.2 Å (*S*) and 1.8 Å (*R*), but polychromatic X rays only to 2.8 Å.[24] After photolysis of the p21·caged GTP crystals, diffraction data of the p21·GTP complex were collected either at room temperature with the Laue method for the wild-type protein[23,24] (or a nononcogenic mutant, G12P[24,25]); or at 4° by conventional monochromatic methods using X rays generated by a rotating anode for an oncogenic p21 mutant (G12V)[12] that has a 10-fold reduced GTPase rate[23] (see Fig. 2).

The experiment produced a clear result, regardless of whether a pure isomer or the racemic mixture of caged GTP was in the starting crystals. The structure of the p21·GTP·Mg^{2+} Michaelis–Menten complex is similar to the structure of the stable complex of p21 with the nonhydrolyzable GTP analog GppNHp (β,γ-imido-GTP) determined to a resolution of 1.35 Å,[26] thereby confirming its biological significance. The structures of mixed GTP–GDP complexes of p21 obtained during the reaction (at 3, 12, 15, 31, and 56 min after photolysis) were well defined in regions that did not undergo structural change on hydrolysis of GTP, but the density could not be interpreted in places (i.e., the "effector region," or loop 2, of the protein) known from the earlier work to undergo significant conformational changes. It could be seen, however, that the electron density of the position at the γ-phosphate of GTP decreased smoothly with time. These results are in agreement with those of solution kinetic studies, which suggest that an

[24] A. Scheidig, E. F. Pai, J. Corrie, G. Reid, I. Schlichting, A. Wittinghofer, and R. S. Goody, *Phil. Trans. R. Soc. A* **340,** 263 (1992).

[25] A. Scheidig, A. Sanchez-Llorente, A. Lautwein, E. F. Pai, J. E. T. Corrie, G. Reid, A. Wittinghofer, and R. S. Goody, *Acta Crystallogr.* **D50,** 512 (1994).

[26] E. F. Pai, U. Krengel, G. A. Petsko, R. S. Goody, W. Kabsch, and A. Wittinghofer, *EMBO J.* **9,** 2351 (1990).

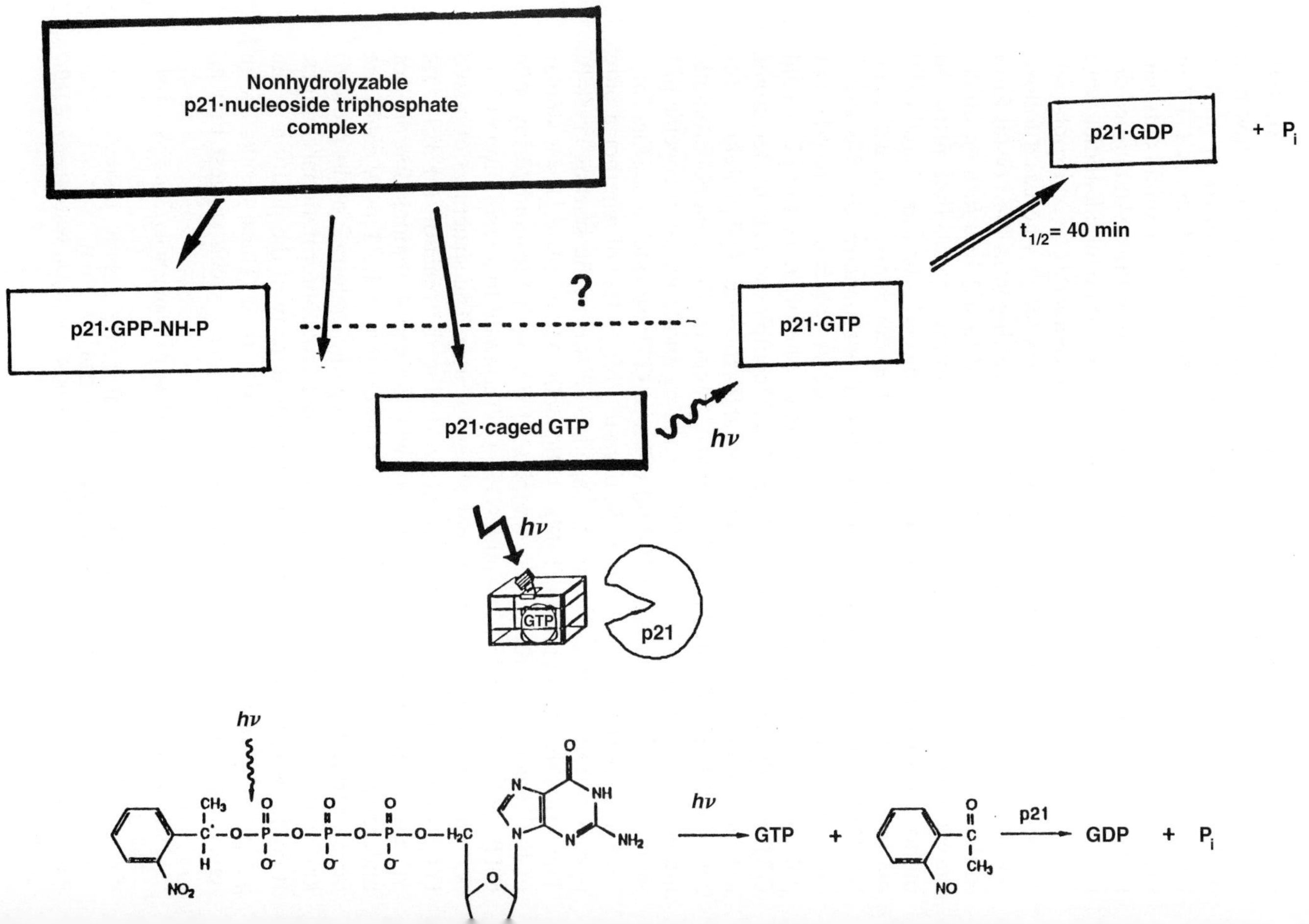
Nonhydrolyzable
p21·nucleoside triphosphate
complex
p21·GPP-NH-P
?
p21·GTP
p21·GDP
+ Pi
t1/2= 40 min
p21·caged GTP
hν
GTP
p21
CH3
NO2
NH
NH2
hν
GTP
+
NO
CH3
p21
GDP
+ Pi

intermediate between the GTP and GDP states (e.g., the enzyme–GDP–P_i complex) does not accumulate on the kinetic pathway. In both studies, the structure of the p21·GDP·Mg^{2+} complex obtained at the end of the reaction differs significantly from the starting p21·GTP·Mg^{2+} complex in the effector region and the loop around amino acid 60. Interestingly, the coordination of the Mg^{2+} ion seems to be different in the structures of the p21·GDP·Mg^{2+} complexes obtained after hydrolysis of photogenerated crystalline p21·GTP·Mg^{2+} or of cocrystallized p21·GDP·Mg^{2+}. In the latter, as in the p21·GTP or p21·GppNHp complexes, an indirect interaction of the highly conserved Asp-57 (via a water molecule) with the metal ion is observed whereas in p21·GDP arising from p21·GTP produced from p21·caged GTP the aspartic acid side chain and the metal ion interact directly.

Another example is glycogen phosphorylase, which catalyzes the reversible phosphorylation of glycogen to glucose 1-phosphate. The enzyme is active in the crystalline state, with similar K_m values and a 30-fold reduced rate compared to solution studies.[27] Initial studies of the phosphorylation reaction were attempted by diffusing phosphate compounds into the crystals. Despite the rapid diffusion of phosphate ($t_{1/2} < 1$ min) or glucose 1-phosphate ($t_{1/2} \approx 1.8$ min) into the crystals, these experiments were hampered by a concomitant transient disorder of the crystal lattice that takes about 7 min to resolve.[28] To avoid this dead time, which limits the first time point after which the reaction can be followed by X-ray diffraction, a different approach was taken to initiate the reaction, namely photolysis of caged phosphate (DNPP in Fig. 3). Because phosphorylase crystals crack in the presence of high concentrations of reducing agents such as dithiothreitol (DTT), which is necessary to avoid modification of the enzyme by the nitroso ketone photoproduct of the commonly used caged phosphate 1-(2-nitrophenyl) ethyl phosphate,[4] 3,5-dinitrophenyl phosphate (DNPP) was used instead. DNPP is converted (at 10^4 sec^{-1}) to 3,5-dinitrophenol and inorganic phosphate by irradiation at 300–360 nm with a quantum yield of 0.67.[18] For determination of the amount of phosphate released from DNPP under true experimental conditions, a crystal ($0.32 \times 0.2 \times 2$ mm^3) was soaked in a solution containing 21 m*M* DNPP. A diode array spectrophotometer was used to measure the increase in absorbance at 400 nm *in situ*

[27] P. J. Kasvinsky and N. B. Madsen, *J. Biol. Chem.* **251,** 6852 (1976).

[28] L. N. Johnson and J. Hajdu "Biophysics and Synchrotron Radiation" (S. Hasnain, ed.), pp. 142–155. Ellis Horwood, Chichester, UK, 1989.

FIG. 2. Schematic diagram for the study of p21·GTP complexes obtained after photolysis of p21·caged GTP crystals (see text). [Modified from J. Schlichting, *Chemie in Unserer Zeit* **29,** 230 (1995).]

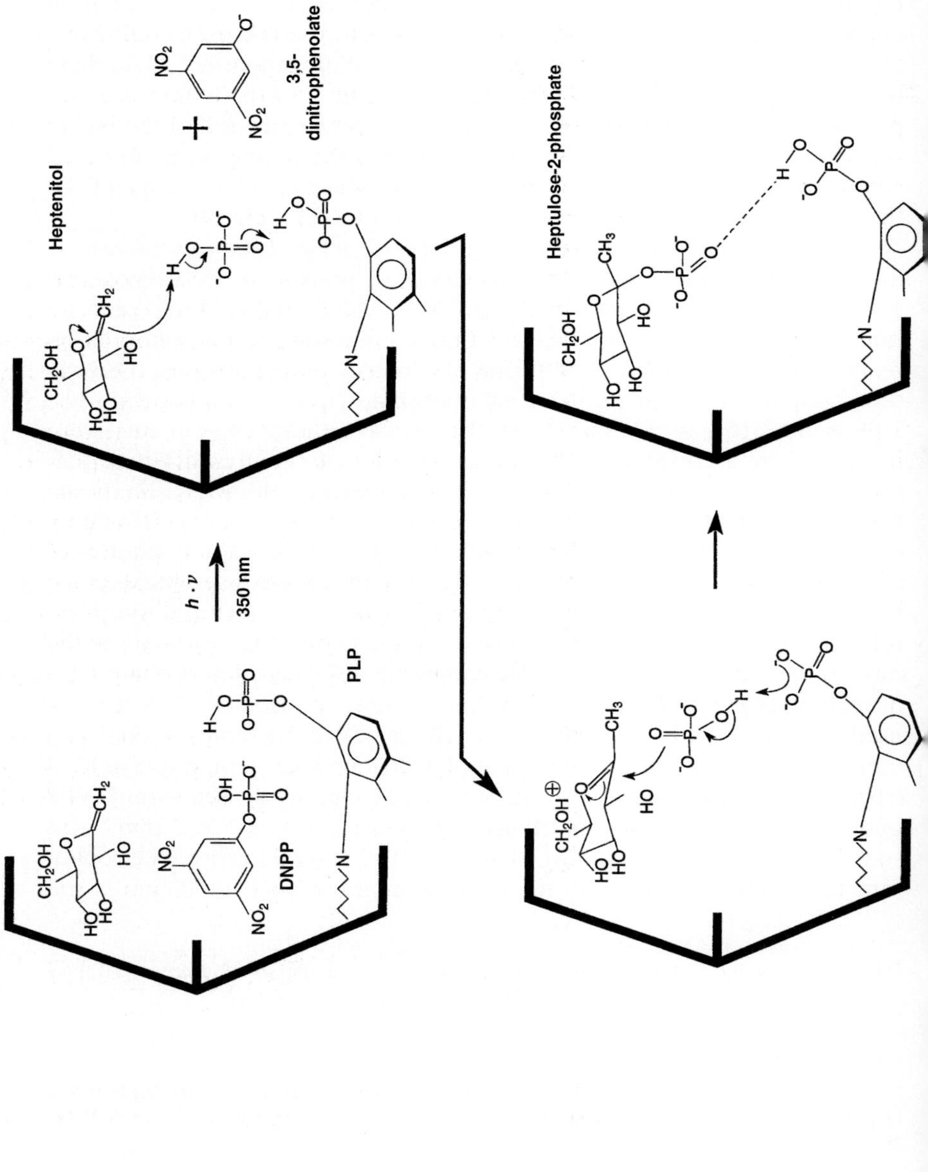
DNPP
PLP
h·ν
350 nm
Heptenitol
3,5-dinitrophenolate
Heptulose-2-phosphate

to determine the concentration of the released 3,5-dinitrophenolate.[29,30] It was found that about 12.3% of the total concentration of DNPP was released after the first pulse of a xenon flash lamp and 5–6% after each subsequent pulse, i.e., 12–14 m*M* phosphate is generated from 45 m*M* DNPP after 5 pulses.

To study the phosphorylation of heptenitol by glycogen phosphorylase *b* crystallographically (see Fig. 3), crystals of the enzyme were soaked in a solution containing 50 m*M* heptenitol and 30 m*M* DNPP. The reaction was initiated by photolysis using five pulses of a xenon flash lamp (approximately 100 mJ each), which took about 30 sec. Laue data sets were collected at 3 min, 15 min, and 1 hr after photolysis. No catalysis had occurred in this time, as shown by the clear electron density for heptenitol (and not for the product heptulose 2-phosphate). Monochromatic X-ray diffraction data from a similarly prepared crystal show that sufficient phosphate was released to promote catalysis within the 24 hr of data collection. The authors discuss the difficulty of distinguishing between a ternary enzyme–substrate complex and a product–enzyme complex when there is no other diagnostic available to provide estimates of the different states of the enzyme in the crystal, demonstrating again the necessity of following the kinetics *in situ.*

A different approach, caging of an enzyme, was taken in a study on the protease chymotrypsin. The active site serine (of the catalytic triad) was reacted with a light-sensitive mechanism-based inhibitor, a cinnamate derivative (*trans-p*-diethylamino-*o*-hydroxymethylcinnamate).[31,32] In the dark, a suicide substrate (3-benzyl-6-chloro-2-pyrone) was diffused into caged chymotrypsin crystals. Single-exposure Laue data sets were taken before and during photolysis using a single 1-msec flash of a xenon lamp, and 1 min, 5 min, 10 min, 20 min, 3 hr, and 24 hr afterward.[33] A transient increase in the mosaicity of the crystals was observed for the data collected concomitant with photolysis. The enzyme–cinnamate complex was observed in the data collected prior to photolysis, the photolysis product (a coumarin),

[29] A. T. Hadfield and J. Hajdu, *J. Appl. Crystallogr.* **26,** 839 (1993).

[30] A. Hadfield and J. Hajdu, *J. Mol. Biol.* **236,** 995 (1994).

[31] B. L. Stoddard, J. Bruhnke, N. A. Porter, D. Ringe, and G. A. Petsko, *Biochemistry* **29,** 4871 (1990).

[32] B. L. Stoddard, P. Koenigs, N. Porter, D. Ringe, and G. A. Petsko, *Biochemistry* **29,** 8042 (1990).

[33] B. L. Stoddard, P. Koenigs, N. Porter, K. Petratos, G. A. Petsko, and D. Ringe, *Proc. Natl. Acad. Sci. U.S.A.* **88,** 5503 (1991).

FIG. 3. Caged phosphate (3,5-dinitrophenyl phosphate, DNPP) was used in a study of glycogen phosphorylase *b*. The E-shaped structure denotes the enzyme; the proposed mechanism is shown.

concomitant with photolysis, and the free enzyme 1 min after photolysis. Electron density to represent the bound pyrone starts to build up in the data set collected 3 hr after photolysis; after 24 hr the pyrone is fully bound in the active site. In solution, using 5 m*M* pyrone, a steady state is achieved 4–5 hr after photolysis, which seems to correspond to the crystal structure obtained 24 hr after photolysis. The authors speculate that the lack of observation of pyrone-bound intermediates (if any) prior to the formation of the final complex may be due to the nonoptimal time of the X-ray data collection. This problem could be dealt with by a careful examination of the kinetics of the reaction under the conditions used for the X-ray experiment, preferably by spectroscopic monitoring of the reaction during the structural experiment.

The next two examples belong to the group of systems with built-in triggers. Myoglobin is a heme protein that binds reversibly ligands such as O_2 and CO and serves as an oxygen storage protein. Single photons of visible light can break the covalent bond between CO and the heme iron in carbonmonoxy myoglobin (MbCO); subsequently an unstable intermediate (Mb*CO) is formed, with CO still inside the protein. The ensuing rebinding process has been studied extensively as a model for the interplay of dynamics, structure, and function in protein reactions. In studies on the Mb*CO complex, the dynamics of light and heat transfer in the crystals, together with the strong temperature dependence of rebinding rate coefficients, place fundamental limitations on crystal thickness (ca. 150 μm), photolysis power, measurement temperature, and time scale. To make CO rebinding slow enough for monochromatic data collection time scales, the crystals must be kept at liquid helium temperatures during photolysis and X-ray data collection.

The structure of Mb*CO was determined by two groups using different experimental protocols and leading to distinct results.[34,35] In one case[34] recombinant sperm whale myoglobin was used with a mutation of a surface residue causing crystallization in a hexagonal space group. The crystals were reduced and reacted with CO within 2 hr; reduction and production of the CO complex were monitored visually by color changes. The crystal was flash-frozen and kept at 20 K (at which rebinding is negligible on the time scale of diffraction data collection) and photolyzed with white light from a fiber-optic illuminator run continuously at low power for the duration of the X-ray data collection. The size of the photolysis beam was much larger than the crystal, thus avoiding potential problems with the X-ray

[34] I. Schlichting, J. Berendzen, G. N. Phillips, Jr., and R. M. Sweet, *Nature (London)* **371,** 808 (1994).

[35] T.-Y. Teng, V. Šrajer, and K. Moffat, *Nat. Struct. Biol.* **1,** 702 (1994).

beam probing an incompletely photolyzed region. The absorbed power density of the visible and X-ray beams were low enough to ensure that heating by the beam was negligible. Photolysis for 20 min before the start of the data collection ensured that the photolysis was complete. The diffraction data were collected at 20 K to a resolution of 1.5 Å. The Mb*CO structure was refined to 1.5 Å and shows the photolyzed CO atop heme pyrrole CD, 2.3 Å from its ligated position. In parallel with this movement, the heme "domes," the heme iron moves by 0.22 Å toward the proximal histidine, the iron-proximal histidine bond is "compressed," the F helix is strained, and the distal histidine swings toward the outside of the ligand-binding pocket.

In the other case,[35] monoclinic wild-type sperm whale myoglobin crystals were reduced and reacted with CO overnight. A crystal was flash-frozen and photolyzed for 90 sec and then exposed to X rays for 3 min. The crystal orientation was then changed and the photolysis/X-ray exposure cycle repeated until a complete data set was acquired at 40 K to a resolution of 1.7 Å. Microspectrophotometry of the frozen crystal showed that it contained 30% metmyoglobin and that 15–20% ligand rebinding occurred during collection of the diffraction data. The Mb*CO structure was refined to 2.0 Å and shows the photolyzed CO 0.8 Å from its ligated position, the heme "domes" somewhat, the heme iron moved by 0.04 Å toward the proximal histidine, and no shifts larger than 0.15 Å of any residue surrounding the heme.

The photoactive yellow protein (PYP) is a 14-kDa photoreceptor protein with a fully reversible photocycle similar to rhodopsin, and is thus also a system with a built-in trigger. The chromophore is a 4-hydroxycinnamyl group covalently bound to the sole cysteine in the protein. It has been proposed that the photobleached intermediate in the photocycle corresponds to a protonated form of the chromophore.[36] During trial photolytic studies on PYP, it became apparent that there is only a narrow range accessible for photolysis in terms of power and wavelength (where the extinction coefficient is low) minimizing gradients of bleaching and temperature as judged from an increase in mosaicity under nonoptimal photolysis conditions.[37]

Diffusion

Concentration jumps can be achieved by diffusion of substrates, cofactors, protons, etc., into a crystal mounted in a flow cell[38] or, as we have

[36] G. E. O. Borgstahl, R. W. DeWight, and E. D. Getzoff, *Biochemistry* **34,** 6278 (1995).
[37] K. Ng, E. D. Getzoff, and K. Moffat, *Biochemistry* **34,** 879 (1995).
[38] G. A. Petsko, *Methods Enzymol.* **114,** 141 (1985).

described, by photolysis of "caged" biologically inert precursors such as caged nucleotides, amino acids, or metals.[17,18] Both methods have advantages and disadvantages. Although diffusion is experimentally straightforward, it is slow; typical diffusion times for small molecules across a 200-μm-thick crystal are one to many minutes. Because of the intrinsic generation of temporal and spatial concentration gradients and the competing effects of diffusion and enzymatic reaction (which can be modeled and therefore estimated[9]), reaction initiation by diffusion is suitable only for slow enzymatic processes. Because diffusion in crystals is not free, but is restricted to solvent channels that may be small or blocked, it may be necessary to prediffuse a large, hence slowly diffusing, ligand into the crystal under nonreactive conditions and then to start the reaction rapidly on completion of diffusion, e.g., by a diffusive change in pH, a light flash, or by a temperature jump.

Changes in pH constitute a powerful tool as long as they are tolerated by the crystal lattice, because they can be performed quite rapidly and efficiently. pH changes can be used in a dynamic way to initiate a reaction whose time course will be followed, or in a static way by a trapping of the reaction at certain stages that are more stable at a certain pH. This is particularly useful if only a certain protonation state of an amino acid, such as a histidine, is catalytically active.

Examples. Catalysis by crystalline glycogen phosphorylase *b* (see also the section on light as an initiating factor) has been followed crystallographically by use of the diffusion of substrates to start the reaction and by collection of the diffraction data (within as little as 25 min) with monochromatic synchrotron radiation.[39] The allosteric control properties of the enzyme were used to slow the reactions studied: the conversion of heptenitol to heptulose 2-phosphate, the phosphorolysis of maltoheptaose to yield glucose 1-phosphate, and oligosaccharide synthesis involving maltotriose and glucose 1-phosphate. Changes in difference Fourier electron-density maps were observed at the catalytic, allosteric, and glycogen storage sites as the reactions proceeded.

Cytochrome-*c* peroxidase (ccp) from yeast is a heme protein found in the mitochondrial electron transport chain, where it probably acts as a peroxide scavenger. The peroxidase reaction involves a doubly oxidized intermediate, compound I, which contains a semistable free radical on Trp-191 and an oxyferryl heme group. Crystalline compound I was generated by flowing 10 m*M* peroxide across a ccp crystal mounted in a flow

[39] J. Hajdu, K. R. Acharya, D. I. Stuart, P. J. McLaughlin, D. Barford, H. Klein, N. G. Oikonomakos, and L. N. Johnson, *EMBO J.* **6,** 539 (1987).

cell.[40] *In situ* microspectrophotometry was used to show that 90–95% conversion to compound I had occurred after 2 min, and that the conversion was nearly complete after 10–15 min at 6°. The relative concentration of compound I remained above 90% for an additional 20–30 min. Laue diffraction data of both native and compound I states of ccp were collected at 6° before and 10–15 min after the addition of peroxide to the crystal. Clear structural changes were observed at the peroxide-binding site but no significant differences were seen at the radical site.

The serine protease trypsin has a two-step mechanism: The peptide bond is broken by nucleophilic attack of the active site serine, resulting in an esterification of its hydroxyl group. This acyl–enzyme intermediate is subsequently hydrolyzed by a water molecule, regenerating the hydroxyl group and releasing the N-terminal segment of the cleaved protein. It has been assumed that this water molecule is activated by a "catalytic triad" (formed by the active site histidine, aspartate, and serine), analogous to the activation of the first half of the reaction. In a study designed to follow the hydrolysis of the ester intermediate, a metastable *p*-guanidinobenzoyltrypsin (GB) complex was formed by reaction of the enzyme with *p*-nitrophenylguanidinobenzoate.[41] Release of the GB group can be controlled by pH: it is stable for days at low pH, but only for hours at a pH slightly higher than neutral. To initiate hydrolysis, a GB–trypsin crystal was mounted in a flow cell at pH 5.5. After collection of a Laue data set, the pH was raised to pH 8.5 and Laue data sets were collected 3 and 90 min after reaction initiation. Subsequent analysis of the crystal showed that it was 58% deacylated. A comparison of the structures (solved to 1.8-Å resolution) before and after the pH jump revealed a new water molecule positioned to attack the acyl group of the GB–trypsin complex. In addition, in the course of the experiment the imidazole group of the active site histidine gradually shifts away from this catalytic water molecule. It appears to be stabilized by the change in charge of the imidazole side chain on the pH jump.

The catalytic mechanism of haloalkane dehalogenase (HD) has also been investigated by the use of changes in pH to stabilize the reaction at different steps.[42] Haloalkane dehalogenase converts 1-haloalkanes into primary alcohols and a halide ion by hydrolytic cleavage of the carbon–halogen bond, with water as a cosubstrate, at an optimal pH of 8.2. Asp-124,

[40] V. Fülöp, R. P. Phizackerley, S. M. Soltis, I. J. Clifton, S. Wakatsuki, J. Erman, J. Hajdu, and S. L. Edwards, *Structure* **2,** 201 (1994).

[41] P. T. Singer, A. Smalas, R. P. Carty, W. F. Mangel, and R. M. Sweet, *Science* **259,** 669 (1993).

[42] K. H. G. Verschueren, F. Seljée, H. J. Rozeboom, K. H. Kalk, and B. W. Diikstra, *Nature (London)* **363,** 693 (1993).

Asp-260, and His-289 have been proposed to form a catalytic triad. The structure of the active site of HD suggests two possible reaction mechanisms: either a nucleophilic substitution by the carboxylate of Asp-124, resulting in a covalently bound ester intermediate that might be hydrolyzed subsequently by a water molecule activated by His-260, or by general base catalysis with a water molecule activated by the histidine. To discriminate between these two possibilities, the catalytic pathway of HD was dissected using pH and temperature as shown in Fig. 4. At low pH the imidazole side chain of His-260 is protonated and no longer can act as a base. Therefore, the covalently bound intermediate should accumulate in the crystal according to the first mechanism, whereas the substrate should not be degraded according to the second mechanism. Haloalkane dehalogenase crystals were soaked in mother liquor containing a 10 m*M* concentration of the substrate 1,2-dichloroethane (1) at pH 5.0 and 4° for 3 hr, (2) at pH 5.0 and room temperature for 24 hr, and (3) at pH 6.2 and room temperature for 96 hr. Monochromatic X-ray diffraction data of these complexes were collected (data collection time, 2 days). The structures obtained were as follows: (1) dehalogenase with substrate bound in its active site, (2) HD with the covalently bound intermediate, and (3) HD with chloride as the product of the reaction bound in the active site. From these studies, it was concluded

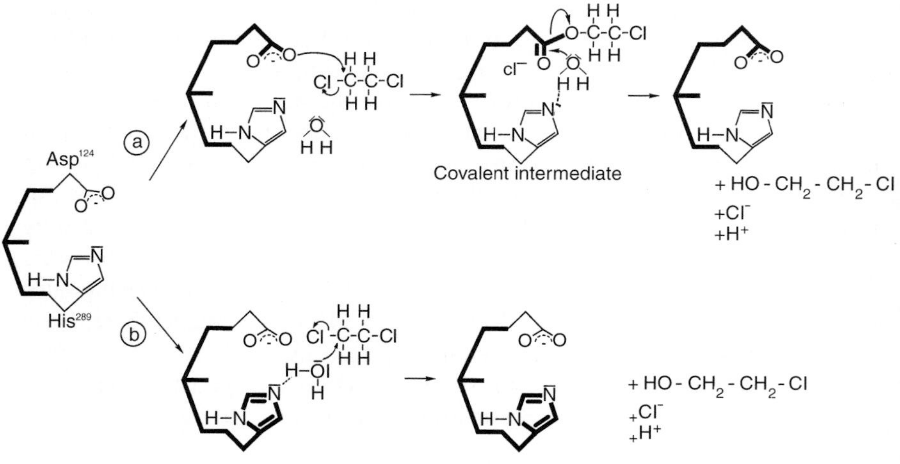

FIG. 4. Possible reaction mechanisms for the hydroytic dehalogenation by haloalkane dehalogenase, denoted by the E-shaped structure: (a) nucleophilic substitution by the carboxylate anion of Asp-124 resulting in a covalent bound intermediate ester that is subsequently hydrolyzed by a water molecule activated by His-289; (b) general base catalysis with a water molecule activated by His-289.

that the dehalogenation catalyzed by HD proceeds through a two-step mechanism involving a covalently bound intermediate.

Temperature

Lowering the temperature at which a reaction takes place not only may slow the reaction to an instrumentally convenient time scale, but also may allow the separation of sequential kinetic processes that would otherwise be convoluted. In any given mechanism, individual steps are likely to have nonidentical temperature dependencies of rate and equilibrium coefficients; the relative populations of various intermediates in a reaction mechanism may change dramatically with temperature.

Crystallographic cryoenzymology can be performed either on flash-cooled crystals mounted on a loop[43] or on crystals transferred to solutions that do not freeze at the cryogenic temperature desired,[44] so that conventional crystal mounting techniques, including flow cells, can be used.[45] In both cases care must be taken that the added cryoprotectant does not affect the crystal quality or the enzymatic reaction.[46] The latter is especially critical for the "nonfrozen" case employing mixed organic–aqueous solvents, for which the dielectric constant and proton activity of the cryoprotectant should be kept as close as possible to the value for the "normal" mother liquor.[47]

Temperature jumps can be used to initiate reactions. Depending on the time scale, temperature jumps (typically 10–15 K) can be made conductively or by laser flashes (see, e.g., Ref. 48). Laser T-jumps are fast and require data acquisition within seconds because of cooling of the specimen. This makes fast data collection techniques such as the Laue method necessary. Owing to the problems associated with working on fast time scales and the inherent limitations of the Laue geometry, laser T-jumps are mostly attractive for fast reactions, where they are really needed. A relatively straightforward way to measure the change in temperature in a crystal *in situ* is to diffuse a chromophore with a temperature-dependent absorption spectrum into the crystal, to measure the absorption spectrum of the crystal, and to compare it to a known calibration curve. The laser T-jump method has been used successfully in some applications with noncrystalline samples such as muscle fibers or lipid phase transitions; specific problems envisaged

[43] T.-Y. Teng, *J. Appl. Crystallogr.* **23,** 387 (1990).
[44] P. Douzou and G. A. Petsko, *Adv. Protein Chem.* **36,** 245 (1984).
[45] B. F. Rasmussen, A. M. Stock, D. Ringe, and G. A. Petsko, *Nature (London)* **357,** 423 (1992).
[46] P. Douzou, *Q. Rev. Biophys.* **12,** 521 (1979).
[47] A. L. Fink and G. A. Petsko, *Adv. Enzymol. Relat. Areas Mol. Biol.* **52,** 177 (1981).
[48] G. Rapp and R. S. Goody, *J. Appl. Crystallogr.* **24,** 857 (1991).

for crystalline samples concern the change in population of states that may be achieved with realistic T-jumps. It is only in exceptional cases, for example, when the starting temperature is poised strategically with respect to the critical temperature for a phase transition, that a temperature change can be used to take a system from essentially 100% state A to 100% state B. In general, the conditions must be chosen so that the starting situation represents a mixture of the starting and ending states, with only a small change in these populations being achievable by a T-jump. This is not likely to lead to easily interpretable X-ray diffraction data, even if all the other associated problems can be solved.

In principle, temperature cycling can be used to follow a reaction by a "stop-and-go" approach. The reaction is initiated by an increase in temperature and allowed to proceed until a reaction intermediate is formed, which then can be stabilized by a lowering of the temperature. Many processes appear to be frozen out when the system is cooled below what is referred to as its glass transition temperature, which is often around 220 K (e.g., Ref. 45). After collection of the X-ray diffraction data, the temperature is raised again so that the reaction can continue until another intermediate is formed, which again can be stabilized by a lowering of the temperature. In a favorable case, this process could be repeated until the product is formed. It should also be noted that while this approach is quite promising for certain types of study, it must be realized that cooling to a temperature at which a quasistable species can be examined at leisure can work only if the combined processes of diffusion, binding, and subsequent covalent bond formation (as described in the following example) are rapid compared to the lifetime of the generated intermediate at the temperature used for the soaking process.

Although this may pertain for some enzymatic reactions, for most enzymatic reactions chemistry is likely to be much faster than diffusion into the crystal. In such cases, a more promising approach is the combination of caged compounds and cryocrystallography: A caged substrate is positioned at the active site of a crystalline enzyme, and the temperature is then lowered to a value at which photolysis can still be achieved but at which the enzymatic reaction is slowed to such an extent that intermediates of interest can be characterized structurally without resorting to high-speed data collection methods. In appropriate cases, temperature changes could then be used to take the system stepwise through the enzymatic mechanism. An advantage of this approach would be that the temperature could be chosen to allow time for careful and complete data recording on a time scale commensurate with monochromatic data collection, avoiding the problems often encountered with Laue data collection (e.g., incomplete data, especially lack of low- to medium-resolution reflections, sensitivity to crystal imperfections, radiation damage). Another advantage might be that photolysis can be achieved in a relatively leisurely manner, i.e., without the use of short, high-intensity pulses and the accompanying problems of temperature effects and nonuniform release of the reacting species if a single pulse is used. A variant of this procedure, which might allow structural characterization of a relatively short-lived state, would be to conduct the photolysis at ambient temperature and subsequently to cool the crystal rapidly to a temperature at which the state induced is stable. This approach may obviate the need for rapid data collection, but will be applicable only to reactions that are slow compared to flash-cooling the crystal.

Examples. As already mentioned, catalysis by serine proteases proceeds via formation of a covalent acyl–enzyme complex. Crystal structures of this intermediate species have been determined by observing complexes rendered nonproductive in terms of substrate turnover by working at low pH values, at which the active site histidine is protonated (see trypsin example under the section Diffusion, above), or by using substrates that bind incorrectly (carboxyl oxygen points away from the oxyanion hole; see chymotrypsin example under the section Light, above). The structure of the reactive acyl–enzyme has been obtained for elastase by forming the intermediate at 247 K and stabilizing it at 218 K during data collection, taking advantage of the glass transition that occurs around 223 K.[49] For this experiment, elastase crystals were transferred to a cryoprotectant solution containing 70% methanol by increasing the alcohol concentration stepwise. The crystals were mounted in the cryoprotectant solution in a flow cell, cooled to 247 K, and substrate was added to the solution. A few medium-intensity reflections at high resolution were recorded at regular intervals to observe binding. After leveling off of the intensity changes, the crystal was cooled to 218 K to stabilize the expected covalent acyl–enzyme against further breakdown and a complete 2.3-Å data set was collected with a diffractometer. A native data set was collected under similar conditions. Comparison of the structures showed that the acyl–enzyme had indeed formed.

A similar experiment on elastase using another substrate was performed with a different strategy[50]): Elastase crystals were mounted in a flow cell and the mother liquor was gradually replaced with 70% methanol, they were cooled to 220 K, and substrate was flowed over the crystal for several hours. The temperature was then increased in steps of 5–10 K and scanning Laue diffraction patterns were recorded. A strong increase in mosaicity at 250 K was interpreted as arising from formation of the Michaelis complex.

[49] X. Ding, B. F. Rasmussen, G. A. Petsko, and D. Ringe, *Biochemistry* **33,** 9285 (1994).

[50] H. D. Bartunik, L. J. Bartunik, and H. Viehmann, *Phil. Trans. R. Soc. A* **340,** 209 (1992).

Data sets were collected at 244 and 264 K, using monochromatic synchrotron radiation. They revealed a mixture of the Michaelis complex and the acyl–enzyme. The authors addressed the problem of interpreting electron densities of mixed states and their chemical identification.

Radiolysis

Radicals, such as hydrated electrons, can be generated by X rays or by pulse radiolysis using bursts of high-energy electrons (typically 2–15 MeV) provided by radioactive sources or electron accelerators. There is an analogy between pulse radiolysis and flash photolysis methods for the study of short-lived species; the difference is that in radiolysis energy transfer is primarily to the solvent, whereas in flash photolysis the solvent is generally chosen to be transparent and energy transfer is to the solute. Although radiolysis has rarely been used in crystallography, it should be kept in mind as a potential trigger in studying redox reactions, in particular in heme-containing proteins.[51]

Assessment and Prospects of Kinetic Crystallographic Experiments

The revolutionary short data collection times possible with synchrotron radiation and the revived Laue method have led to high expectations for kinetic crystallographic studies on enzymes or proteins "at work." However, only a few such experiments have been performed so far. In each case, the success relied heavily on ingenious "tricks" specific for the particular system under investigation—thus there is no general recipe for time-resolved crystallographic experiments. Common to most of the studies, however, is that the reactions studied are either slow or have been slowed to experimentally convenient time scales, thus taking at least minutes at room temperature. Fast and uniform reaction initiation throughout the crystal, using photolysis or diffusion as trigger, was therefore no problem. Data collection times were also short compared to the lifetimes of the intermediates to be observed. One of the problems encountered during these studies was the quality of the crystals. This is an especially severe limitation when using the Laue method. Because all of the reciprocal lattice points that are lying between the spheres $1/\lambda_{max}$, $1/\lambda_{min}$, and within $1/d_{resolution}$ are in diffraction condition, all irregularities in the crystal lattice such as high mosaicity will result in streaked or otherwise misshapen reflections. Even if an initially "perfect" crystal is used in a dynamic experiment, the crystal can disorder (temporarily) when the reaction is initiated by, for instance, diffusion or photolysis, or when structural changes occur. This means that only poor

[51] M. H. Klapper and M. Faraggi, *Q. Rev. Biophys.* **12,** 465 (1979).

or no diffraction data can be obtained at the potentially most interesting time point of the reaction. Transient disorder during reaction initiation was indeed observed by Hajdu *et al.*,[39] Reynolds *et al.*,[52] H. Bartunik (personal communication, 1995), and Stoddard *et al.*.[33] In the latter case the slight elongation of the reflections collected concurrently with photolysis of a cinnamate-inhibited chymotrypsin crystal resulted in a significant deterioration in the processing and refinement statistics of this Laue data set compared to data sets collected before or after photolysis. Using monochromatic radiation, the effect of the transient increase in mosaicity of the crystal on the data quality would have been much less, and perhaps insignificant. Many crystals that give nonanalyzable Laue data yield good monochromatic data. In these cases it is important to seek factors (a change in temperature or pH, the use of mutants or substrate analogs, etc.) that slow the reaction to a time scale at which fast monochromatic data collection techniques are applicable.

For a large number of biochemical processes, the available X-ray intensity is not a real issue. Exposure times for collecting Laue data are now down to 10^{-5} to 10^{-3} sec. However, there are indications that radiation damage[53] and heating of the crystal become a serious problem for intense polychromatic X-ray beams, which takes away some of the attraction of short exposure times. In addition, the limiting step in the data collection is not the exposure time itself (down to picoseconds[54]), but rather the mechanical rotation or translation of the crystal required to obtain a full data set. Thus, experiments in which a substance or enzyme is activated by a single flash of light at ambient temperature, and the data collected on a time scale that allows characterization of intermediates under these conditions, require either reactions that can be cycled, or many crystals, and will probably remain the exception.

Ideal biochemical systems for kinetic crystallographic studies bind their substrates and cofactors tightly and catalyze the reaction slowly. Their intermediates accumulate to high occupancy because they are formed much faster than they decay. In addition, initial or final states should differ in a manner that can be detected crystallographically (e.g., more than the movement of a hydrogen or electron). Unfortunately, there are only a few biochemical systems that fulfill these conditions. Finding them and a way to initiate the reaction is the most challenging aspect of time-resolved

[52] C. D. Reynolds, B. Stowell, K. K. Joshi, M. M. Harding, S. J. Maginn, and G. G. Dodson, *Acta Crystallogr.* **B44,** 512 (1988).

[53] P. T. Singer, R. P. Carty, L. E. Berman, I. Schlichting, A. M. Stock, A. Smalas, Z. Cai, W. F. Mangel, K. W. Jones, and R. M. Sweet, *Phil. Trans. R. Soc. A* **340,** 285 (1992).

[54] D. M. E. Szebenyi, D. H. Bilderback, A. LeGrand, K. Moffat, W. Schildkamp, B. Smith Temple, and T.-Y. Teng, *J. Appl. Crystallogr.* **25,** 414 (1992).

crystallographic studies. The trigger must be faster than both the reaction and data collection. In many cases, only photoactivation will be rapid enough, but at present caged compounds exist only for a limited number of reactions. Although additional caged compounds need to be synthesized, we can expect to see more examples with built-in triggers or using caged compounds and caged enzymes, particularly in combination with other approaches such as low-temperature crystallography.

Helpful for the interpretation of the electron density of an intermediate occurring during a reaction is the availability of good starting phases from a model of a "stationary" state, and a single, high-occupancy conformation of the species of interest. Methods must be developed for the deconvolution and refinement of multiple conformations. Because the concentration of the intermediate may be low, or its presence may be difficult to deduce crystallographically (e.g., oxidation states), it is advisable to follow the reaction (e.g., spectroscopically) as it proceeds in the crystal. This information is useful not only for interpretation of the structure, but also helps to detect problems such as heating effects due to absorption of the X rays or photolyzing light during the experiment.

Kinetic crystallography is still a young field. In addition to the major technical breakthrough represented by development of synchrotron radiation sources, there have been a large number of ingenious individually developed tricks and devices, none of which can be generalized. Each case to be investigated therefore presents a new challenge and can be undertaken only when an intimate knowledge of structure and basic mechanistic properties has been accumulated for the system in hand. Kinetic crystallography does, however, have the potential of providing detailed and accurate dynamic structural information at the atomic level on the mode of action of important biological molecules.

Acknowledgments

We thank Jochen Reinstein for suggestions concerning Fig. 1, and the Alexander von Humboldt Stiftung for financial support.

Section IV

Presentation and Analysis

A. Illustrating Structures
Articles 25 and 26

B. Modeling Structures
Article 27

C. Databases
Articles 28 through 30

D. Program Packages
Articles 31 and 32

[25] Ribbons

By MIKE CARSON

The *Ribbons* software interactively displays molecular models, analyzes crystallographic results, and creates publication-quality images.

The ribbon drawing popularized by Richardson[1] is featured. Space-filling and ball-and-stick representations, dot and triangular surfaces, density map contours, and text are also supported. Atomic coordinates in Protein Data Bank[2] (PDB) format are input.

Output may be produced in the Inventor/VRML format. The VRML (virtual reality modeling language) format has become the standard for three-dimensional interaction on the World Wide Web. The on-line manual is presented in HTML (hypertext markup language) suitable for viewing with a standard Web browser.

The examples give the flavor of the software system. Nearly 100 commands are available to create primitives and output. Examples include creating spheres colored by residue type, fitting a cylinder to a helix, and making a Ramachandran plot. The user essentially creates a small database of ASCII files. This provides extreme flexibility in customizing the display.

Examples

Ribbons is typically used on a terminal in a darkened room. The default settings give a dark screen background and vivid colors for display. This is recommended to produce slides for presentations. A white background with pastel colors is recommended for publication.

Example 1

Given any PDB entry or file in PDB format *entry.pdb,* create default files and view all macromolecular chains as ribbons and all other atoms as balls and sticks:

```
ribbons  -e entry.pdb
```

[1] J. S. Richardson, *Adv. Protein Chem.* **34,** 167 (1981).

[2] F. C. Bernstein, T. F. Koetzle, G. J. B. Williams, E. F. Meyer, Jr., M. D. Brice, J. R. Rogers, O. Kennard, T. Shimanouchi, and M. Tasumi, *J. Mol. Biol.* **112,** 535 (1977).

0076-6879/97 $25.00

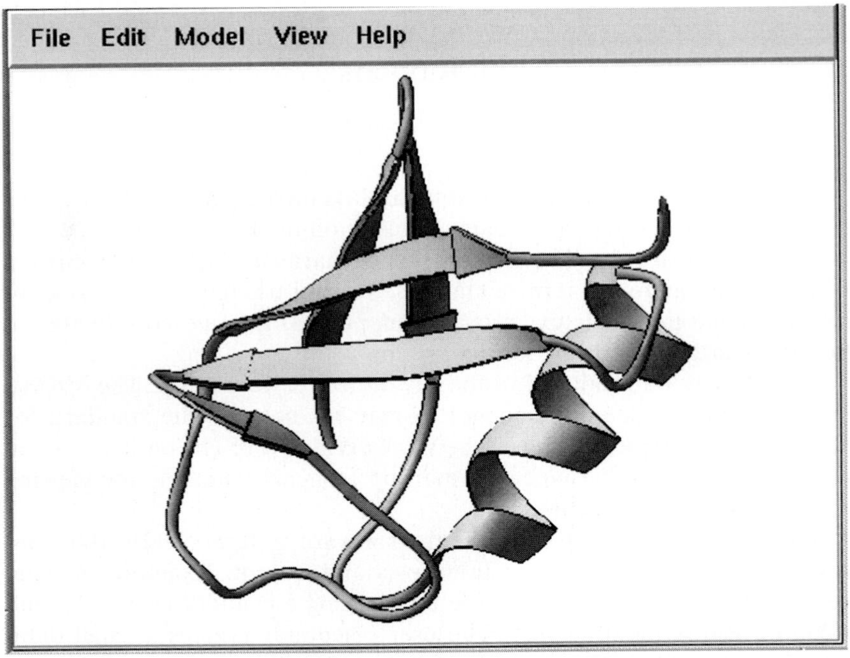

FIG. 1. The *Ribbons* graphics window.

Example 2

Given the single-chain protein coordinates in file *ubiquitin.pdb*,[3] create the required files for a ribbon drawing, then display it:

```
ribbon-model  ubiquitin.pdb  ubiq.model
ribbons  -n ubiq
```

Figure 1 shows the display screen, presenting an X/Motif interface. The user interacts by pointing and clicking with the mouse. Choosing "Help" from the menu bar accesses the hypertext manual.

Example 3

Given the crystal structure of the influenza virus neuraminidase complexed with the inhibitor DANA (PDB entry *1nnb.ent*[4]), create the required

[3] S. Vijay-Kumar, C. E. Bugg, K. D. Wilkinson, and W. J. Cook, *Proc. Natl. Acad. Sci. U.S.A.* **82,** 3582 (1985).

[4] P. Bossart-Whitaker, M. Carson, Y. S. Babu, C. D. Smith, W. G. Laver, and G. M. Air, *J. Mol. Biol.* **232,** 1069 (1993).

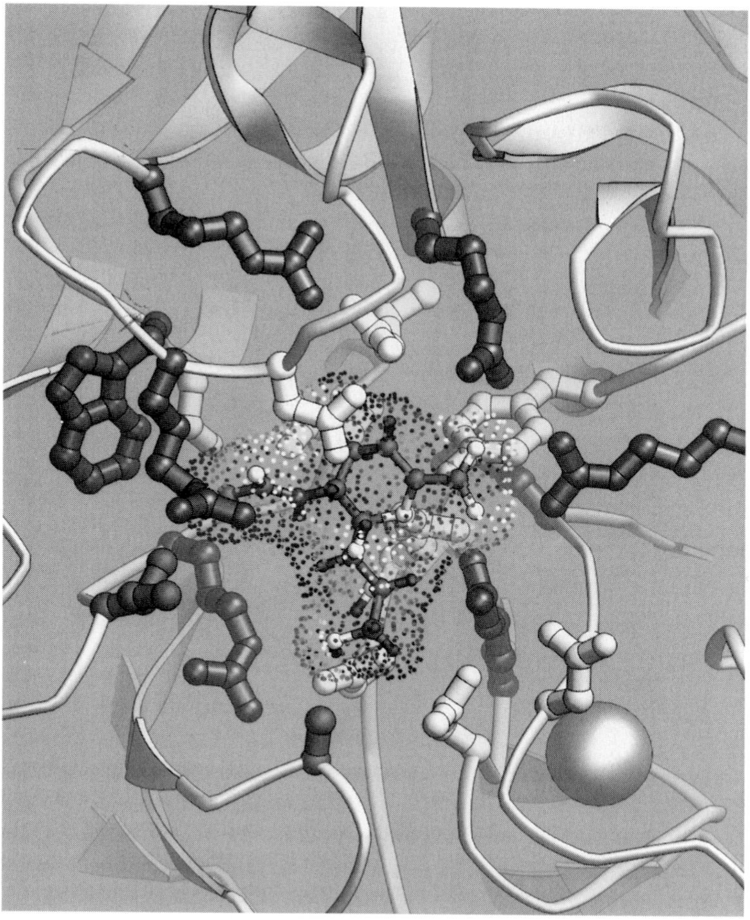

FIG. 2. Active site of neuraminidase.

files for a more complicated model, then display and interactively adjust to the desired visual result. The image saved is shown in Fig. 2.

The PDB entry is edited into three coordinate files for the protein (*na.pdb*), the bound calcium ion (*cal.pdb*), and the inhibitor (*dana.pdb*).

```
ribbon-model  na.pdb  na.model
```

creates the files necessary for the display of the protein as a ribbon, as in the previous example.

```
pdb-sele-pdb na.pdb site.pdb
  sele = ( not ( name N or name C or name 0 or hydro )
  and byres point (27.0 18.5 63.5) cut 8.0 )
```

selects all nonhydrogen side-chain atoms of *na.pdb* belonging to residues having any atom within 8 Å of the inhibitor center and writes coordinates as *site.pdb.* The X-PLOR[5] atom selection syntax is used.

```
pdb-atom-sph cal.pdb cal.sph
pdb-atm-sph dana.pdb dana.sph
pdb-res-sph site.pdb site.sph
sph-bond dana.sph dana.cyl
sph-bond site.sph site.cyl
sph-ms dana.sph dana.dot
```

creates spheres colored by atom type for the calcium and inhibitor, and colored by residue type for the active site; creates cylinders for bonds colored as the spheres for the inhibitor and the active site; and creates the dot surface[6] of the inhibitor colored as the nearest sphere.

```
ls dana.sph cal.sph > na.atoms
ls dana.cyl site.cyl > na.bonds
ls dana.dot > na.ndots
```

creates the required files to link the display of the primitives generated in the previous step to the model "na." The latest version of *Ribbons* provides a point-and-click graphic interface as an alternative to the command line mode to manage primitives.

```
ribbons -n na
```

opens the graphics window for viewing and interactive adjustment. Mouse motions rotate, scale, and translate to focus on the active site. Primitives are adjusted with collections of widgets called Control panels. Selecting Edit from the menu bar (see Fig. 1) presents the options. The Atom Panel is shown in Fig. 3.

The Atom Panel scales the radii of the three groups of spheres to different values (see Fig. 2). The calcium is set fairly high and the inhibitor set very low. The complexity of the calcium sphere is set high to ensure smoothness. The Bond Panel scales the radii of the two groups of cylinders in line with their respective atoms. The N-Dot Panel scales the dots. The

[5] A. T. Brünger, "X-PLOR: A System for X-Ray Crystallography and NMR," version 3.1. Yale University Press, New Haven, Connecticut, 1992.

[6] M. L. Connolly, *Science* **221,** 709 (1983).

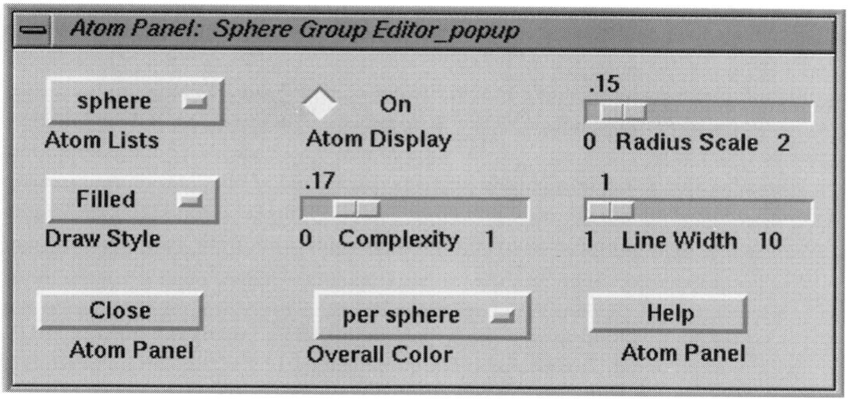

FIG. 3. The Atom Panel editor.

Ribbon Style Panel forces the ribbon through the C_α atoms in the coils, otherwise defaults are used. (There are nearly 40 widgets in three panels to set aspects of the style, dimensions, and complexity of ribbon drawings.) The Light Panel adjusts the lighting and depth cueing. The Image Panel sets full-screen antialiasing. Selecting the Save Image option of File from the menu bar completes the process.

Example 4

Given a partially refined crystal structure, determine and tabulate the quality of the results, and display either interactively or through static plots. Preliminary data are from the aldehyde reductase/NADPH complex.[7] The full details are given in the *Ribbons* manual. Five input files must be present: the final coordinates with individual *B* factors (here *alr1.pdb*), the corresponding coordinates of the previous refinement, the best "observed" map in FRODO[8] format, a purely calculated map at the same scale, and a short list of X-PLOR commands to set parameters.

```
ribbon-errors
```

executes the analysis protocol. The protocol[9] claims crystallographic data are required to assess reliably the quality of a coordinate file. Statistical

[7] O. El-Kabbani, K. Judge, S. L. Ginell, D.. A. A. Myles, L. J. DeLucas, and T. G. Flynn, *Nature Struct. Biol.* **2,** 687 (1995).

[8] T. A. Jones, *J. Appl. Crystallogr.* **11,** 268 (1978).

[9] M. Carson, T. W. Buckner, Z. Yang, S. V. L. Narayana, and C. E. Bugg, *Acta Crystallogr.* **D50,** 900 (1994).

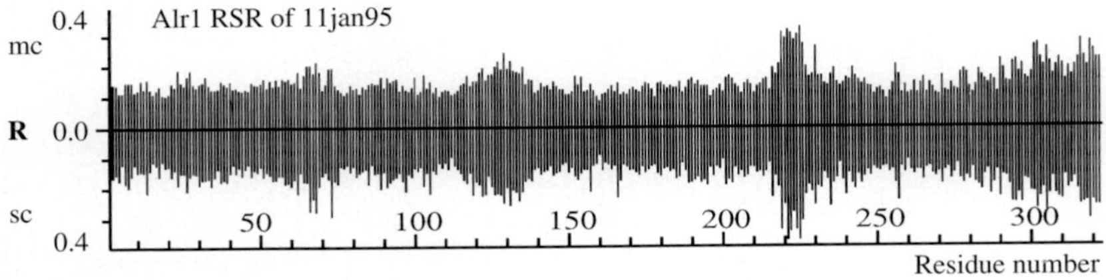

FIG. 4. Bar graph from *ribbon-errors* protocol.

analysis implies a linear model of five independent variables fits the error function. The temperature factors (*B*), real-space fit residuals (*R*), geometric strains (*G*), dihedral angles (*D*), and shifts from the previous refinement cycle (*S*) for main-chain (mc) and side-chain (sc) atoms determine an overall error factor (*E*). Output files summarize the results for each protocol criteria.

```
rsr-ps  -t ''Alr1 RSR of 11jan95''  alr1_rsr.list > alr1_rsr.ps
```

creates the PostScript file *alr_rsr.ps* from an output list of *ribbon-errors.* The *rsr* refers to the real-space residual advocated by Jones *et al.*,[10] which is the best single error-detection criterion. This plot is shown in Fig. 4.

A *.*ss* (secondary structure) file is required to display a ribbon drawing. The file is created automatically by the *ribbon-model* command in the previous examples. The user adds columns to this file for custom color coding. The *ribbon-errors* protocol produces an extended *_*xa.ss* file with a letter grade assigned for each error analysis criterion, in addition to the standard sequence and secondary structure information. A few lines of *alr_xa.ss* follows:

res#	seq	ss	sshb	hb	Eres	Emc	Esc	Rmc	Bmc	Smc	Gmc	Dmc	Rsc	Bsc	Ssc	Gsc	Dsc
288	Q	H	H	H	B	A	B	A	C	A	A	A	B	C	A	A	A
289	L	H	H	B	A	A	A	A	B	C	A	A	A	A	A	A	A
290	D	c	T	H	C	C	C	C	C	E	A	B	C	C	F	A	A
291	A	c	T	x	B	B	A	B	B	C	A	A	A	A	B	A	A
292	L	c	c	H	A	A	A	A	B	A	A	A	A	A	A	A	A

```
pdb-ss-model alr1.pdb alr1_xa.ss
ribbons  -n alr1
```

creates required files then displays the ribbon. The Ribbon Style Panel selects among *seq*, *ss*, etc., for per-residue color codings.

[10] T. A. Jones, J.-Y. Zou, S. W. Cowan, and M. Kjeldgaard, *Acta Crystallogr.* **A47,** 110 (1991).

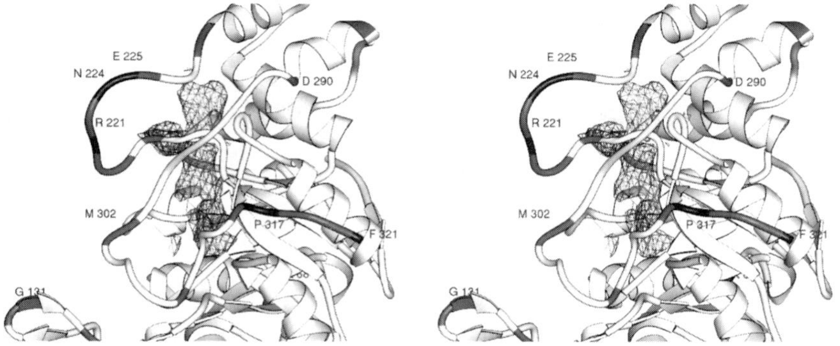

FIG. 5. Residue error in preliminary Alr1 refinement.

Figure 5, a stereo drawing, has the Eres key selected to color the ribbon based on residue error. Hot spots in the structure should be obvious. Dark colors indicate problems here. A residue may be picked to display information. The contour surface is created from a cofactor difference map in FRODO format.

Summary

Each user-created file is eligible for display and editing as a separate graphics object. All files have a simple format. For example, each sphere file line has coordinates (*xyz*), radius, color code, and label. The file *cal.sph* from example 2 is shown here:

```
28.73 30.05 62.86 2.30 11 cal_471_X
```

Ribbons can write/read files to save/restore the current orientation, styles and scale factors, colors, and lighting. Data from other programs may be reformatted to generate the lists of spheres, cylinders, or triangles for display. An example is the visualization of ribosomal models.[11] *Ribbons* can output primitives suitable for input into several ray-tracing packages.

Ribbons saves raster images as Silicon Graphics (SGI) **.rgb* files. Many utilities support this file format. Figure 6 is a final example compositing several *Ribbons* images of the DNA/Trp repressor complex (PDB entry *1tro.ent*[12]). All figures herein were converted to PostScript for printing.

[11] A. Malhotra and S. C. Harvey, *J. Mol. Biol.* **240,** 308 (1994).

[12] Z. Otwinowski, R. W. Schevitz, R.-G. Zhang, C. L. Lawson, A. J. Joachimiak, R. Marmorstein, B. F. Luisi, and P. B. Sigler, *Nature* (*London*) **335,** 321 (1988).

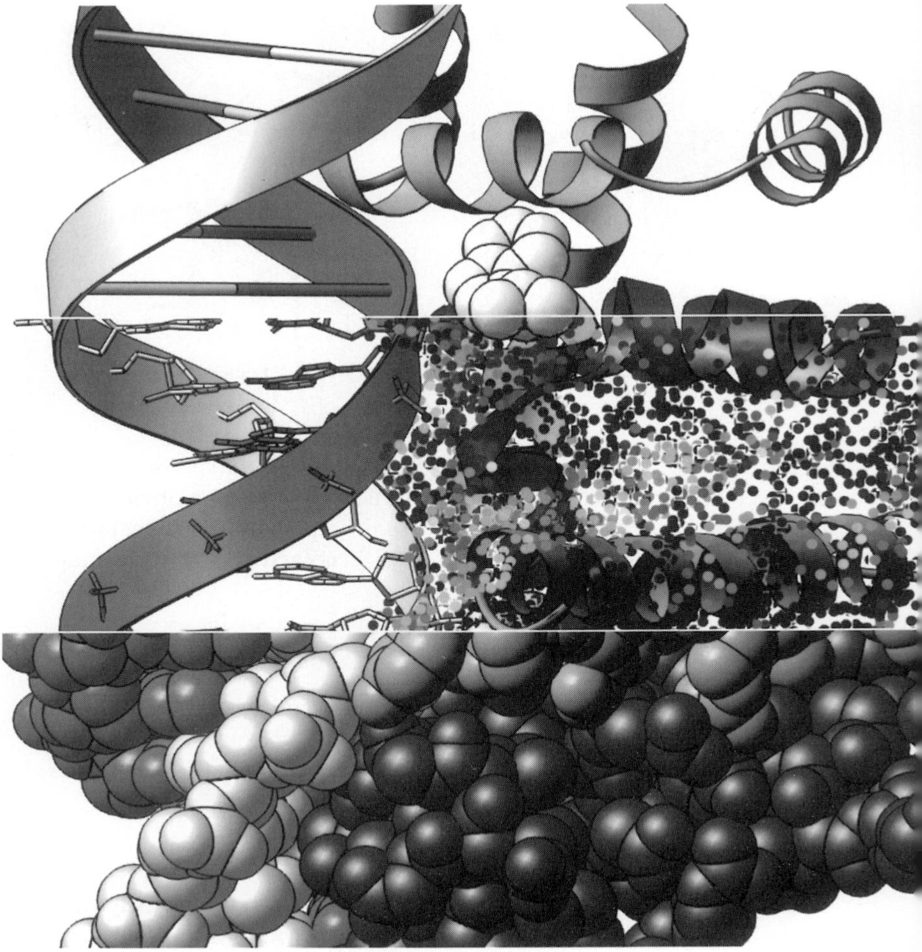

FIG. 6. DNA/protein rendering styles.

Methods

The first computer graphics software created by the author was intended to enhance understanding of molecular dynamics simulations.[13] The C language programs developed under UNIX created ASCII files suitable for display on the Evans and Sutherland (E&S) PS300, the state-of-the-art

[13] M. Carson and J. Hermans, "Molecular Dynamics and Protien Structure" (J. Hermans and W. van Gunsteren, eds.), pp. 119 and 165. Polycrystal Book Service, Dayton, Ohio, 1984.

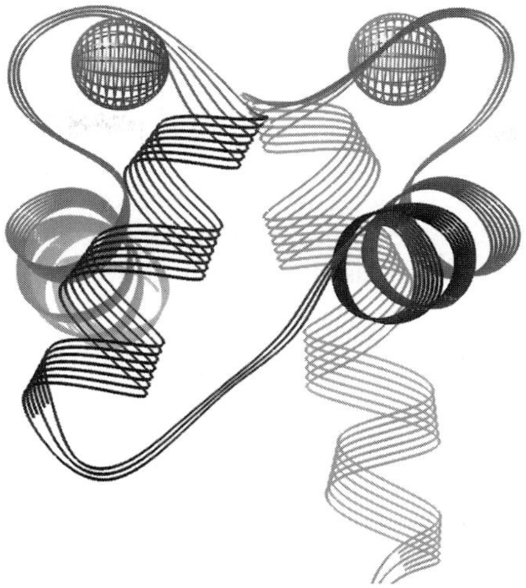

FIG. 7. B-spline ribbon.

graphics device of its day (the current programs still carry some of this historical baggage). The interaction with the molecular graphics group at North Carolina, including Jan Hermans, Mike Pique, Fred Brooks, and Dave and Jane Richardson, proved most stimulating.

Lesk and Hardman[14] described computer programs to plot ribbon drawings. Pique[15] and Burridge[16] created interactive versions. We modeled a protein ribbon in the manner of a ribbon used to wrap presents: composed of many roughly parallel smooth threads running the length of the ribbon.[17] This conceptually simple algorithm fit cubic polynomial "B-spline" curves to the peptide planes. (Most modern graphics systems provide B-splines as a basic drawing primitive.) The algorithm quickly became incorporated into several commercial molecular modeling packages.

The program *BSribbon* was developed on a PS300 in 1985 at the University of Alabama at Birmingham (UAB). Figure 7 is an example presenting the protein calmodulin.[18] The acquisition of an SGI IRIS 2400 workstation in 1986 led to the program *Ribbon*. Figures 8 and 9 (see color insert) are

[14] A. M. Lesk and K. D. Hardman, *Science* **216,** 539 (1982).

[15] M. Pique, Video tape. "What does a protein look like?" University of North Carolina, Chapel Hill, North Carolina, 1982.

[16] R. F. Doolittle, *Sci. Am.* **253,** 88 (1985).

[17] M. Carson and C. E. Bugg, *J. Mol. Graphics* **4,** 121 (1986).

[18] Y. S. Babu, J. S. Sack, T. J. Greenhough, C. E. Bugg, A. R. Means, and W. J. Cook, *Nature (London)* **315,** 37 (1985).

sample images. Computation of the individual splines, sectioning of the ribbon into residues, extension of the method to nucleic acids, and production of solid, shaded, and textured ribbon drawings on raster devices were described.[19] This style has been widely copied, and the geometry put into the data flow visualization environments AVS[20] and Explorer.[21]

The article by Carson[19] defines a "ribbon space curve." Figure 9 (see color insert) shows how a mapping of the curvature/torsion values of the ribbon space curve effectively highlights the classes of secondary structure, offering an alternative parameterization to the ϕ/ψ values.

The best curve is debatable: Should it pass through all C_α atoms, should it have a pleat in the sheets, should it be a different order or type of polynomial? The B-spline method[17] sought to capture the feel of Richardson's drawings.[1] A cubic polynomial was the lowest order that gave good results; higher order splines showed little difference. Several alternative formulations of ribbon construction have been given.[22–24] These other types of curves may be given by alternative spline formulations.[19]

We presented *Ribbons* as a visual sanity check of a structure by mapping properties of crystallographic interest to the ribbon drawing.[25] Residues were color coded by main-chain and side-chain dihedral sensibility, agreement with the electron-density map, and potential energy. The program *Ribbons 2.0* exploited the expanding capabilities of graphics hardware and software for hidden surface removal and lighting models. Real-time manipulation of solid models became possible. *Ribbons 2.0* has been widely distributed and a detailed methodology published.[26]

The *Ribbons*$^{++}$ prototype[27] used the SGI Inventor, a 3D object-oriented toolkit. The toolkit provided advanced texturing and transparency, a direct manipulation interface, and a good introduction to the C^{++} language. Figure 10 (see color insert) is a rendering of the protein PNP[28] in this environment. Figure 11 (see color insert) is a *DNurbs*[29] representation of DNA.

[19] M. Carson, *J. Mol. Graphics* **5,** 103 (1987).

[20] A. Shah, personal communication (1993).

[21] O. Casher, S. M. Green, and H. S. Rzepa, *J. Mol. Graphics* **12,** 226 (1994).

[22] J. M. Burridge and S. J. P. Todd, *J. Mol. Graphics* **4,** 220 (1986).

[23] P. Evans, "Daresbury Laboratory Information Quarterly for Protein Crystallography," p. 19. Daresbury Laboratory, Daresbury, Warrington, UK, 1987.

[24] J. Priestle, *J. Appl. Crystallogr.* **21,** 572 (1988).

[25] M. Carson and C. E. Bugg, Abstract G2. American Crystallographic Association Annual Meeting, 1988.

[26] M. Carson, *J. Appl. Crystallogr.* **24,** 958 (1991).

[27] M. Carson, *J. Mol. Graphics* **12,** 67 (1994).

[28] S. E. Ealick, S. A. Rule, D. C. Carter, T. J. Greenhough, Y. S. Babu, W. J. Cook, J. Habash, J. R. Helliwell, J. D. Stoeckler, R. E. Parks, Jr., S. Chen, and C. E. Bugg, *J. Biol. Chem.* **265,** 1812 (1990).

[29] M. Carson and Z. Yang, *J. Mol. Graphics* **12,** 116 (1994).

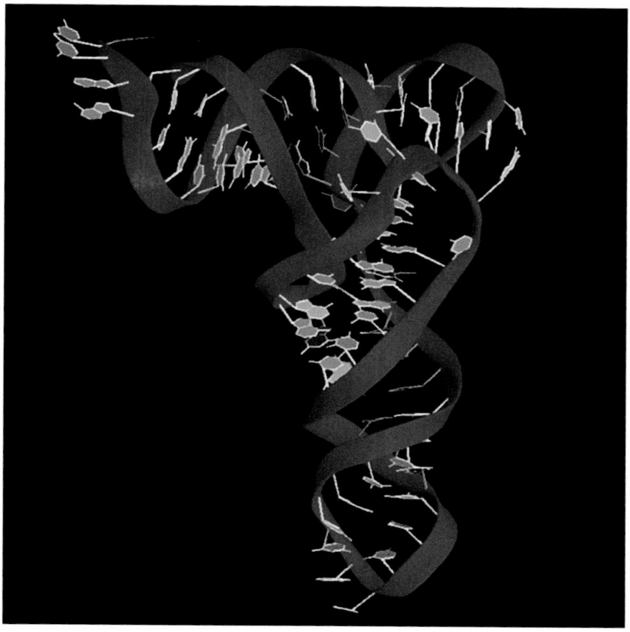

FIG. 8. tRNA model (*Ribbon '87*).

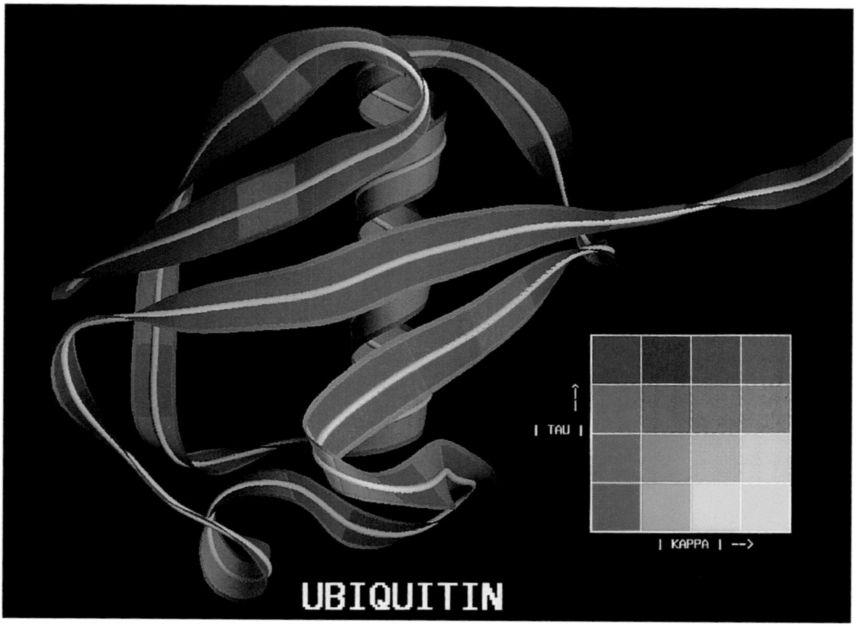

FIG. 9. Ribbon space curve of ubiquitin (*Ribbon '88*).

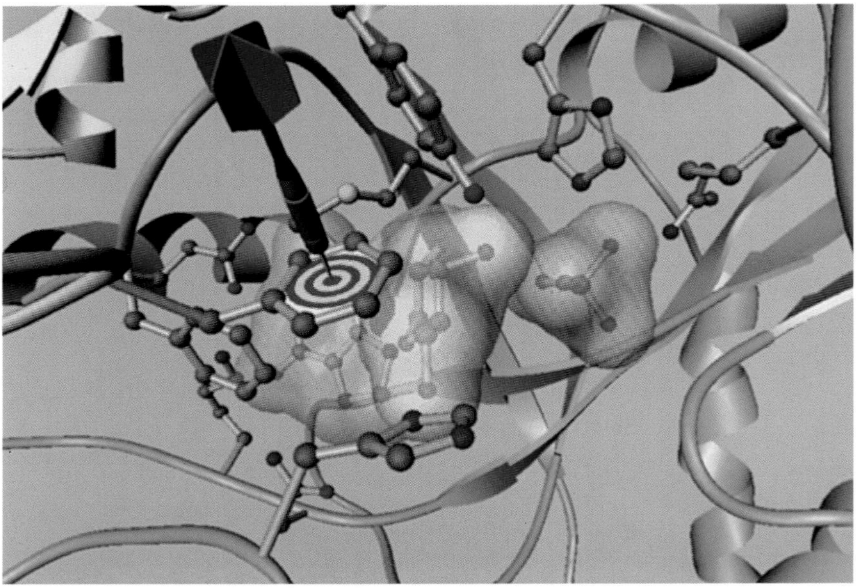

FIG. 10. PNP: Target for drug design (*ribbons*$^{++}$ *'93*).

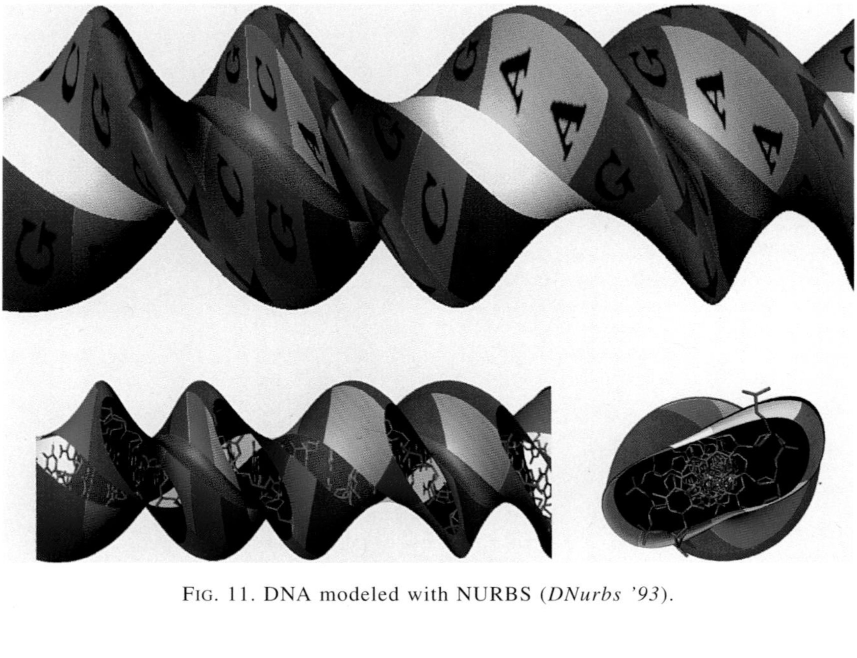

FIG. 11. DNA modeled with NURBS (*DNurbs '93*).

Ribbons 3.0 is the current Unix/X/Motif/OpenGL/C++ code. Elements of the previous programs are combined, along with an XPLOR-like command language and a graphic interface to create and manage data. Image-processing techniques[30] and stenciled outlines have been added, allowing black-and-white graphics that approach the beautiful line-drawing plots of the program Molscript.[31] There are still no methods to correct the residues in error. XtalView or O (see [10] in this volume[32]) is recommended. Further development will visualize aspects of structure-based drug design.

The changes in computers, graphics, and networking since the publication of Volume 115 in this series[33] is astounding. The author (having computed on slide rules, mainframes, workstations, and now PCs) intends *Ribbons* shall keep pace with changing "standards."

The molecular modeling and drug design program WHAT IF[34] uses *Ribbons* as its artwork module. Some favorite examples of *Ribbons* usage by others are journal covers,[35–37] display of protein–DNA interaction,[38] and for a few seconds in the movie "Jurassic Park."

Beauty Is Truth?

Ribbon drawings are an elegant way to visualize the folding and secondary structure of proteins. They have been observed to become "smoother" during the course of structural refinement. The ribbons present a global view of the molecule: color cues and irregularities are immediately obvious to the eye.

Ribbon drawings make striking illustrations. Is this the only reason they grace so many journal covers? Keats's line that truth is beauty may be applicable here. The geometry of a ribbon curve may offer clues to the protein-folding problem.

[30] D. S. Goodsell and A. J. Olson, *J. Mol. Graphics* **10,** 235 (1992).

[31] P. J. Kraulis, *J. Appl. Crystallogr.* **24,** 946 (1991).

[32] T. A. Jones, *Methods Enzymol.* **277,** [10], 1997 (this volume).

[33] H. W. Wyckoff, C. H. W. Hirs, and S. N. Timasheff (eds.), *Methods Enzymol.* **115,** (1985).

[34] G. Vriend, *J. Mol. Graphics* **8,** 52 (1990).

[35] A. M. de Vos, L. Tong, M. V. Milburn, P. M. Matias, J. Jancarik, S. Noguchi, S. Nishimura, K. Miura, E. Ohtsuka, and S.-H. Kim, *Science* **239,** 888 (1988).

[36] B. Shaanan, H. Lis, and N. Sharon, *Science* **254,** 769 (1991).

[37] R. L. Rusting, *Sci. Am.* **267**(6), 130 (1992).

[38] C. W. Muller, F. A. Rey, M. Sodeoka, G. L. Verdine, and S. C. Harrison, *Nature (London)* **373,** 311 (1995).

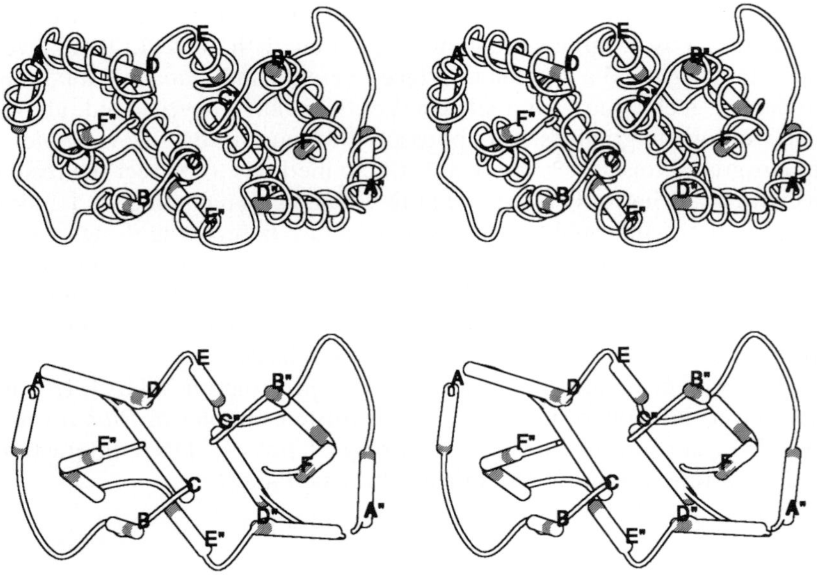

FIG. 12. Multiresolution curve analysis of γ-interferon.

Multiresolution curve analysis of end-point interpolating B-spline curves via the wavelet transform[39] generates the best approximating curve from reduced sets of control points. Such a B-spline models a protein fold by points from the peptide planes and termini of the chain.[17] This offers possibilities for display, editing, and topological comparison at lower resolutions.

Figure 12 is a multiresolution curve analysis with *Ribbons 3.0.* Each monomer of dimeric γ-interferon[40] is labeled per helix (e.g., A and A′). Cylinders fit to the helical residues have their C-terminal end shaded. At the top of Fig. 12 is the *BSribbon* curve based on the 123 residues. At the bottom of Fig. 12 is the B-spline curve defined by 32 points. This curve generally passes through the cylinders. Preliminary results suggest a recognizable fold is generated from a number of points equal to one-sixth the number of residues.[41]

[39] A. Finkelstein and D. H. Salesin, "SIGGGRAPH Proceedings," p. 261. 1994.

[40] S. E. Ealick, W. J. Cook, S. Vijay-Kumar, M. Carson, T. L. Nagabhushan, P. P. Trotta, and C. E. Bugg, *Science* **252,** 698 (1991).

[41] M. Carson, *J. Comput. Aided Mol. Design* **10,** 273 (1996).

Software Available

The *Ribbons* software package is copyrighted by the author and licensed through the UAB Center for Macromolecular Crystallography (CMC). The CMC NASA grant encourages the commercialization of our work. The executable *Ribbons* version 2.6, used in the Example sections, is free to the academic crystallographer. Please consult our web site at http://www.cmc.uab.edu/ribbons.

Versions now exist for several UNIX workstations. We plan to port *Ribbons* to the PC, Microsoft having chosen OpenGL as its 3D graphics standard. (RASMOL[42] is a free program that already supports raster graphics on workstations and PCs.)

Acknowledgments

We thank Susan Baum for editorial help, Yang Zi for technical assistance, and NASA for support.

[42] R. Sayle, personal communication (1995).

[26] Raster3D: Photorealistic Molecular Graphics

By ETHAN A. MERRITT and DAVID J. BACON

Raster3D Program Suite

Raster3D is a suite of programs for molecular graphics (Fig. 1). It is intended to fill a specific niche in a world where sophisticated graphics workstations are proliferating rapidly: that of preparing figures for publication and for presentation in situations where it is impractical to have the viewing audience gather around a workstation screen. At the heart of the package is a rendering program that is not dependent on any special graphics hardware and that can render images of any desired size and resolution. The algorithms used by this rendering program have been chosen to maximize both the speed of execution and the quality of the resulting image when applied to scenes typical of molecular graphics. For these scenes it is faster, usually much faster, than general-purpose ray-tracing programs.

Crystallographers were among the first and most avid consumers of graphics workstations. Rapid advances in computer hardware, and particularly in the power of specialized computer graphics boards, have led to

0076-6879/97 $25.00

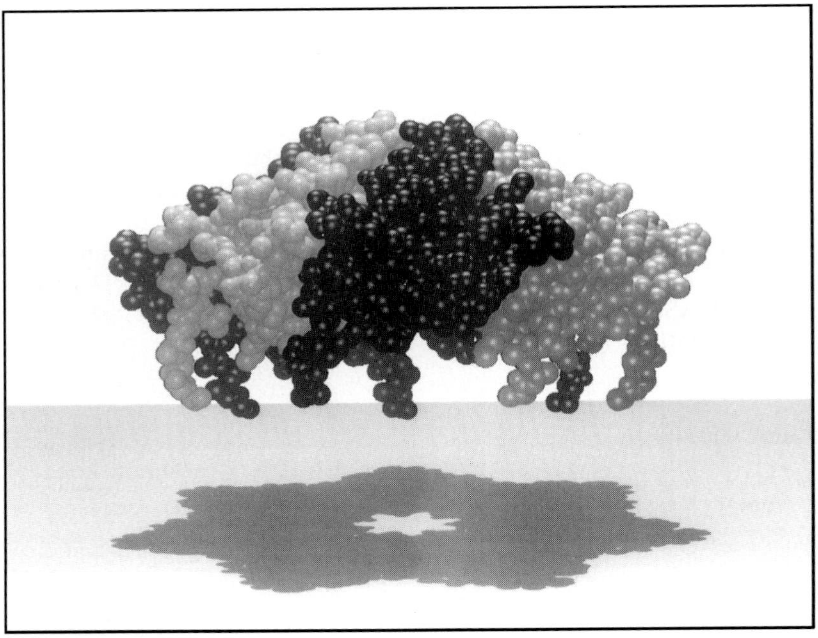

FIG. 1. "*No substance can be comprehended without light and shade.*" (the notebooks of Leonardo da Vinci)

successive generations of personal workstations with ever more impressive capabilities for interactive molecular graphics. For many years, it was standard practice in crystallography laboratories to prepare figures by photographing directly from the workstation screen. No matter how beautiful the image on the screen, however, this approach suffers from several intrinsic limitations. Among these is the inherent limitation imposed by the effective resolution of the screen. Virtually all current workstations employ raster displays. That is, the image on the screen is composed of a finite array of colored pixels. A typical large workstation screen can display a 1280 × 1024 pixel raster. Although this resolution is impressive (it is, for instance, roughly twice that of commercial television in the United States) it pales before that used in high-quality printed images. To gain maximum benefit from the much higher pixel resolution of a film recorder or dye sublimation printer, the image must be rendered directly for that resolution rather than captured photographically from a screen.

Use of the graphics hardware in a workstation to generate images for later presentation can also impose other limitations. Designers of workstation hardware must compromise the quality of rendered images to achieve rendering speeds high enough for useful interactive manipulation of three-dimensional objects. Specular highlighting and surface-shading algorithms, for example, are chosen with an eye to speed rather than purely for fidelity or realism. Similarly, few hardware-based graphics libraries support shadowing and thus few molecular visualization programs built on top of these libraries offer the option of displaying shadows. A further consideration is the proliferation of graphics standards and platform-dependent graphics libraries. This has resulted in many visualization programs that are tied to a single graphics architecture or can be ported to others only with difficulty. The Raster3D programs sidestep this problem by providing platform-independent tools for composition and rendering. Elements of a complex scene can be imported from interactive viewers running on different platforms and merged into a composite description for Raster3D rendering.

Programs in the Raster3D suite are under continuous development.[1,2] The description given here applies to Raster3D version 2.3. Source code and full documentation for the programs are freely available from the authors via anonymous FTP or via the World Wide Web.[3]

Programs in Raster3D Suite

Four categories of programs are distributed with the Raster3D suite: composition tools, input conversion utilities, the central rendering program, and output conversion filters. The Raster3D composition tools *balls, ribbon,* and *rods* generate simple representations of atomic structure directly from an input file in Brookhaven Protein Data Bank (PDB) format. The three programs, respectively, generate a van der Waals surface representation, a peptide backbone trace, and a ball-and-stick model of bonded atoms. The output in each case is a stream of object descriptions formatted for input to the *render* program. These object descriptions are plain text files that may be mixed and matched using a text editor prior to rendering.

The Raster3D utilities themselves do not use any interactive graphics or screen display. They are designed for use either at the command line level during a keyboard session or inside a script run in the background. Quite complex scenes may be built up using these utilities alone. It is important to note, however, that a growing number of interactive molecular

[1] D J. Bacon and W. F. Anderson, *J. Mol. Graph.* **6,** 219 (1988).

[2] E. A. Merritt and M. E. P. Murphy, *Acta Crystallogr.* **D50,** 869 (1994).

[3] Raster 3D source and documentation are available from the authors via anonymous FTP from ftp.bmsc.washington.edu/pub/raster3d, or through the Raster3D Web page at URL http://www.bmsc.washington.edu/raster3d/raster3d.html.

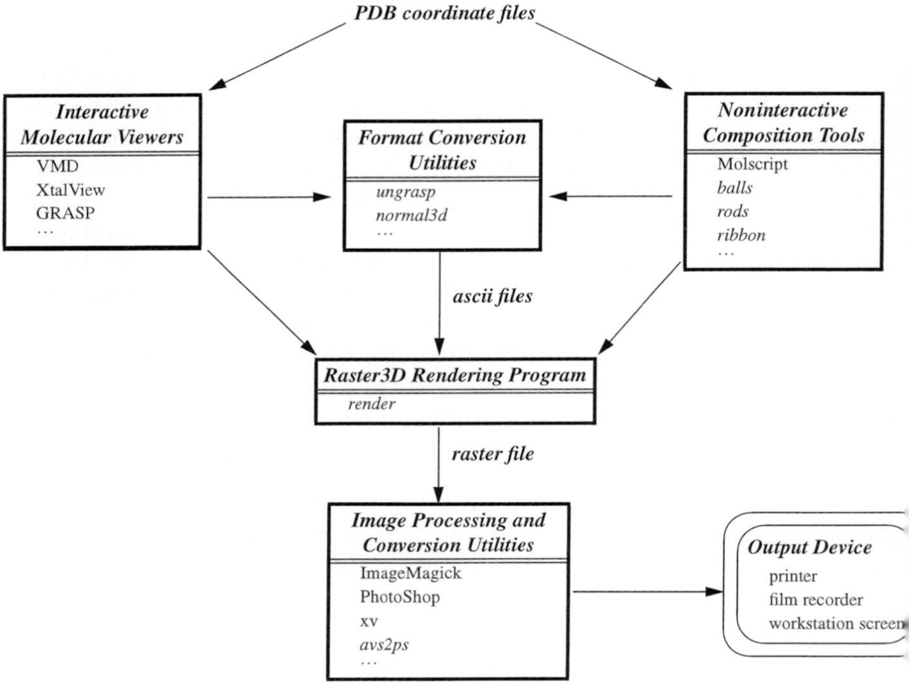

FIG. 2. The flow of data through the Raster3D rendering program. Both interactive and noninteractive composition tools can be used to translate the atomic coordinates found in a Protein Data Bank (PDB) file into object descriptions that can be used by the rendering program. The *render* program itself produces a raster image in one of three standard formats (AVS, TIFF, or SGI libimage). If desired, the rendered scene can be enhanced by further image processing before being printed in its final form. Programs included in the Raster3D suite itself are listed in italic.

graphics tools developed by other groups have a built-in capability to produce scene description files for Raster3D directly. Most notable among these at the time of writing are Molscript,[4] VMD,[5] and XtalView.[6] The Raster3D suite also contains two conversion utilities *ungrasp* and *normal3d* to assist in importing objects from other commonly used molecular visualization tools, e.g., GRASP.[7] A schematic of the various pathways from initial atomic coordinates to finished image is given in Fig. 2.

[4] P. Kraulis, *J. Appl. Crystallogr.* **24,** 946 (1991).

[5] W. F. Humphrey, A. Dalke, and K. Schulten, *J. Mol. Graph.* **14,** 33 (1996).

[6] D. E. McRee, "Practical Protein Crystallography." Academic Press, San Diego, California (1993).

[7] A Nicholls, K. Sharp, and B. Honig, *Proteins* **11,** 281 (1991).

The *render* program produces raster images in one of three standard formats: AVS, TIFF, and SGI. The AVS image format is useful when it is desirable to pipe the rendered image directly from *render* to another application, in particular to display the resulting image directly on the screen. The TIFF image format offers the advantages of machine independence and widespread compatibility with other image-processing programs. The SGI (libimage) format is used by a set of freeware image-processing tools distributed with the SGI workstations commonly used in crystallography and molecular modeling. Rendered images in any of these formats may be processed further or modified if necessary prior to printing. For example, standard image-processing programs may be used to add labels, to apply color correction, to form composite images, or to convert to other raster image formats for specific printer compatibility.

Raster3D *render* Program

The heart of the Raster3D suite is the *render* program. To create figures, it is not strictly necessary for the user to understand the format of the input and output files used by *render,* but a general understanding will be helpful in composing complex scenes and in choosing rendering options. The *render* program works from a single formatted input stream. This consists of an initial series of header lines specifying global rendering options, followed by a series of individual object descriptors. The required header records of a *render* input file are described below. Except where noted, each set of parameters listed below is given in free format on a single input line. The parameter names given below are those of variables in the program source code. Maximum values listed below for certain parameters may be increased by recompiling the *render* program.

TITLE

Anything, up to 80 characters

NTX, NTY

Number of tiles in each direction (see Algorithms Used for Rendering, below). Maximum = 192

NPX, NPY

Number of pixels per tile in each direction. Maximum = 36

SCHEME

Pixel averaging (anti-aliasing) scheme

0—No anti-aliasing; calculate and output alpha channel in rendered image
1—No anti-aliasing; no alpha channel
2—Anti-aliasing using 2 × 2 computing pixels for one output pixel

3—Anti-aliasing using 3 × 3 computing pixels for 2 × 2 output pixels (NPX, NPY must be divisible by three)
4—Anti-aliasing as in scheme 3, but program computes required raster size and tiling

Anti-aliasing options are discussed in more detail below

BKGR, BKGG, BKGB
Background color (red, green, and blue components, each in the range 0 to 1)

SHADOW
T to calculate shadowing within the scene, *F* to omit shadows

IPHONG
Control parameter for specular highlights (reflections of the light sources).[8] A smaller value results in a larger spot. IPHONG = 0 disables specular highlighting and estimation of surface normals for ribbon triangles[2]

STRAIT
Straight-on (secondary) light source contribution (typical value, 0.15). The primary light source contribution (see also SOURCE below) is given by PRIMAR = 1 − STRAIT

AMBIEN
Ambient illumination contribution (typical value, 0.05). Increasing the ambient light will reduce the contrast between shadowed and nonshadowed regions

SPECLR
Specular reflection contribution (typical value, 0.25). The diffuse reflection quantity is given by DIFFUS = 1 − (AMBIEN + SPECLR). Ambient and diffuse reflections are chromatic, taking on the specified color of each object, whereas specular reflections are white by default. This behavior can be modified through the inclusion of explicit material descriptors within the object list, as discussed below

EYEPOS
One can think of the image produced by *render* as corresponding to a photograph taken by a camera placed a certain distance away from the objects making up the scene. This distance is controlled by the EYEPOS parameter. EYEPOS = 4 describes a perspective corresponding to a viewing distance four times the narrow dimension of the described scene. A very large value of EYEPOS will yield a scene with essentially no perspective. EYEPOS = 0 disables perspective altogether.

SOURCE
Direction of the primary light source (typically 1.0 1.0 1.0). This is a white light point source at infinite distance in the direction of the vector given (see note on coordinate convention, below). The secondary light source is always head-on. Only the primary light source casts shadows

TMAT
Homogeneous global transformation[9] for input objects, given as a 4 × 4 matrix on four lines just as it would be written if it were intended to be a postfix (suffix) operator

[8] B.-T. Phong, (1975). *CACM* **18,** 311 (1975).

[9] L. G. Roberts, "Homogeneous Matrix Representations and Manipulation of *N*-Dimensional Constructs." Document MS 1405. Lincoln Laboratory, Massachusetts Institute of Technology, Cambridge, Massachusetts, 1965.

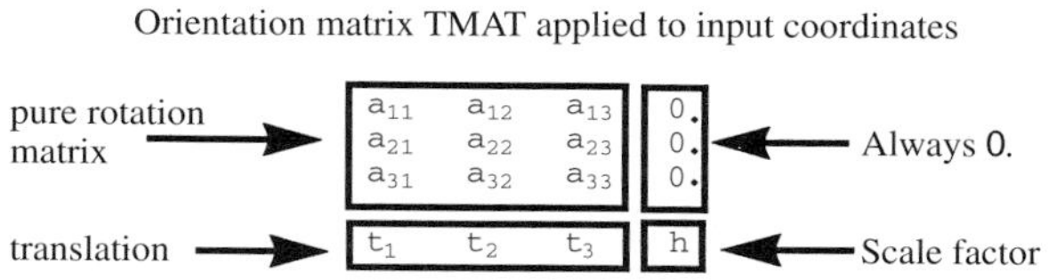

FIG. 3. The TMAT matrix describing the transformation of input coordinates.

(Fig. 3). The upper left 3×3 submatrix expresses a pure rotation, the lower left 1×3 submatrix gives a translation, the upper right 3×1 submatrix should be zero (otherwise extra perspective is introduced), and the lower right scalar (h) produces global scaling. Note that the scale factor h ends up being applied as an inverse; i.e., a larger value of h will result in shrinking the objects in the picture. Input coordinate vectors $[x\ y\ z]$ are extended with a 1 to make them homogeneous, and then postmultiplied by the entire matrix; i.e., $[x'\ y'\ z'\ h'] = [x\ y\ z\ 1]$[TMAT]. The coordinates ultimately used for rendering are then $[x''\ y''\ z''] = (1/h')[x'\ y'\ z']$. In the rendered image, the positive x axis points to the right; the y axis points upward, and the positive z axis points toward the viewer

INMODE
Object input mode (1, 2, or 3). Mode 1 means that all objects are triangles, mode 2 means that all objects are spheres, and mode 3 means that each object will be preceded by a record containing a single number indicating its type. The Raster3D composition tools always use mode 3

INFMT or INFMTS
Object input format specifier(s). For object input modes 1 and 2, there is just one format specifier INFMT for the corresponding object type, whereas for mode 3, there are three format specifiers INFMTS on three lines. The first describes the format for a triangle, the second for a sphere, and the third for a cylinder. Each format specifier is either a Fortran format enclosed in parentheses, or a single asterisk to indicate free-format input. The free-format option is normally preferred, particularly when merging object descriptors produced by different programs

An example of a set of header records input to *render* is given in Fig. 4.

Object Types

Immediately following the header records in the input stream to *render* is a series of object descriptors. Each object descriptor consists of one or more input lines. When INMODE = 3, the first line of the descriptor consists of a single integer specifying the object type. Subsequent lines of the descriptor contain numerical values in the format specified by one of the INFMTS records in the header. The object types currently supported are listed as follows.

type 1 = Triangle, specified by three vertices $[x_1\ y_1\ z_1][x_2\ y_2\ z_2][x_3\ y_3\ z_3]$ and an $[R\ G\ B]$ color triplet

type 2 = Sphere, specified by a center $[x\ y\ z]$, a radius r, and an $[R\ G\ B]$ color triplet

A ***render*** input file

```
Picture title for a 1280 x 1024 pixel anti-aliased image
80 64          NTX,NTY tiles in x,y
24 24          NPX,NPY pixels in x,y per tile
3              SCHEME  anti-aliasing level 3; 3x3->2x2
0 0 0          BKGND(3) black background
F              SHADOW  no shadows cast
25             IPHONG Phong power for specular highlights
0.25           STRAIT secondary light contribution
0.05           AMBIEN ambient light contribution
0.25           SPECLR specular reflection component
4.0            EYEPOS eye position
1. 1. 1.       SOURCE(3) primary light source (from upper right)
1 0 0 0    ]
0 1 0 0    ]-TMAT   4x4 view matrix describing
0 0 1 0    ]        input coordinate transformation
0 0 0 0.6  ]   no translation; enlarge objects in picture by 66% (1/.
3              INMODE mixed objects
*              (free format triangle and plane descriptors)
*              (free format sphere descriptors)
*              (free format cylinder descriptors)
2
0.1 0.1 0.1      0.05     1.0 0.0 0.0
9 This line is purely a comment, begin material description
8
17.0     0.6     -1.0 -1.0 -1.0     0.9         0 0 0 0
@bunch_of_objects.r3d
@another_bunch.r3d
9 End of previous material description
...
[many more object descriptors]
...
0
```

header records
first object descriptor
comment
material descriptor
file indirection

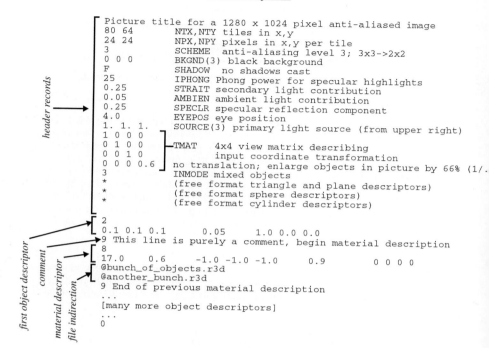

FIG. 4. An sample input file for the *render* program, illustrating the required header records, a single object descriptor, a material descriptor, and the use of file indirection. The initial object descriptor specifies a red sphere of radius 0.05 centered at [0.1 0.1 0.1]. The material descriptor specifies a transparent surface (CLRITY = 0.9) that will have specular highlights matching the base color of all objects rendered using this material. Two files containing additional object descriptors are included via file indirection; the objects in these files will be rendered according to the properties specified in the previous material descriptor. The files specified as indirect input must not, however, contain any material termination records (object type 9) if the transparent material description is to apply across both instances of file indirection.

type 3 = Round-ended cylinder, specified by a center $[x_1\ y_1\ z_1]$ and radius r_1 for one end, $[x_2\ y_2\ z_2]$ and radius r_2 for the other end, and a single $[R\ G\ B]$ triplet. The second radius is ignored in the current implementation. The object is rendered as two spheres of radius r_1 connected by a constant radius cylinder

type 4 = Not used

type 5 = Flat-ended cylinder, specified identically to the round-ended cylinder

type 6 = Plane, specified by three points $[x_1\ y_1\ z_1][x_2\ y_2\ z_2][x_3\ y_3\ z_3]$ and an $[R\ G\ B]$ color triplet. The plane is rendered as a triangle with infinite extent whose color fades to half intensity at infinite z

type 7 = Explicit surface normals at the vertices of the previous triangle (object type 1), specified as three unit vectors $[u_1\ v_1\ w_1][u_2\ v_2\ w_2][u_3\ v_3\ w_3]$. Pairs of objects of types 1 and 7, triangles with corresponding explicit surface normals, may be used to describe an arbitrary surface in terms of a triangular mesh

type 8 = Explicit material properties applied to subsequent objects until terminated by a trailing record of type 9. Material properties are specified by 10 parameters that supersede global parameters in the header records: MPHONG, MSPEC, SPECRGB(3), CLRITY, OPT(4). MPHONG sets a local control parameter for specular highlights, MSPEC sets the relative contribution of specular highlighting to the shading model, an $[R\ G\ B]$ triplet specifies the color of reflected light from this material (by default all specular highlights are white), and the CLRITY parameter indicates the degree of transparency (0.0 = fully opaque; 1.0 = fully transparent). A negative value for any component of the $[R\ G\ B]$ color triplet causes specular highlights to take on the base color of the object being rendered. This is useful for adding color to a mostly transparent material. The four required OPT parameters are reserved for future additions to the material descriptor

type 9 = Terminate previous explicit material properties, no following data lines. This descriptor type may also be used as a comment. Any data on the same line as the initial "9" are ignored

type 0 = End of input file, return to previous level of file indirection, if any (see below). This record is optional; processing of object descriptors terminates when the end of the input stream is detected

File Indirection

At any point in the input stream to *render* where an object descriptor would be legal, it is also legal to insert a line beginning with "@" (Fig. 4). In this case the remainder of the line is interpreted as the name of a file from which further input is taken. This mechanism makes it possible to reuse standard objects in multiple rendered scenes, e.g., a set of bounding planes or standard definitions of material properties. When input from this level of file indirection ends, control returns to the previous input stream. Multiple levels of file indirection are possible.

General Guidelines for Producing Raster3D Pictures

The Raster3D distribution includes sample shell scripts and sample input files in addition to full documentation on the use of the programs. These illustrate the composition and description of several typical molecular graphics scenes. Recommended procedures for integration of Raster3D rendering with specific external programs may change with time, and are thus beyond the scope of this chapter. Several general considerations apply to producing optimal rendered images of a wide variety of scenes, however, and some of these are discussed below.

Shadowing

The primary purpose of shadowing is to convey an impression of depth in the rendered image. This can be particularly effective when the scene contains a relatively small number of elements, or when there is a grouping of foreground objects that cast shadows onto a separate grouping of background objects. However, if the scene contains a large number of elements, for example, a ribbon-and-arrow representation of a large molecular assembly, the use of shadows may complicate the image to the point where the visual impact is reduced. In this case try rendering the image both with and without shadows, or experiment with the location of the primary light source (the SOURCE parameter in the twelfth header record input to *render*). Finally, the shadows can be deemphasized by increasing the secondary (straight-on) light source contribution parameter or the ambient light level. A separate use of shadowing is to convey information that otherwise would not be apparent in the figure. Figure 1 uses shadowing to indicate a central pore, for example, avoiding the need for a second orthogonal view.

Adapting Figures to Output Devices

The quality of pictures generated by Raster3D ultimately is limited by the output device. Although figures will probably be composed and previewed on a workstation screen, generally the final version should be rerendered with a larger number of pixels before being sent to a film recorder or high-performance color printer. For example, a typical 35-mm film recorder can produce slides with a resolution of roughly 4000×3000 pixels (much larger than can be displayed on a workstation screen). The number of pixels in the rendered image is controlled by the parameters (NTX, NTY) and (NPX, NPY) in the second and third header records input to the *render* program.

Color balance and particularly the appropriate "gamma correction" vary from one output or display device to another.[10] Raster3D itself applies no gamma correction; if correction is needed, apply it to the generated image files afterward. This is a standard image-processing procedure and may even be a selectable print option for the output device. If a particular output device is used regularly, it is worth an initial round of experimentation to determine the best gamma value for future runs. The appropriate gamma correction can then be applied to each rendered picture before sending it for printing.

[10] C Poynton, "A Technical Introduction to Digital Video." John Wiley & Sons, New York, 1996.

Side-by-Side Figures

The EYEPOS parameter input to the *render* program specifies a viewing distance for the resulting image; this is equivalent to the distance between a camera and the object being photographed. Generally the sense of depth conveyed by the rendered image is increased slightly by positioning the virtual camera reasonably close to the object; the default (and recommended) value for EYEPOS is 4, which means that the distance from the camera to the center of the object is four times the width of the narrow dimension of the field of view. However, if the figure being composed contains two or more similar objects that are next to each other, e.g., a comparison of two variants of the same protein structure, the resulting parallax may be more of a hindrance than a help. Because the virtual camera is centered, it will "see" the right-hand object slightly from the left, and the left-hand object slightly from the right. This results in different effective viewpoints for paired objects that would otherwise be identical. To overcome this effect set EYEPOS to some large value, which will generate an image with very little parallax.

Stereo Pairs

The Raster3D rendering options, particularly shadowing, were intended to obviate the need for stereo pairs to convey a sense of depth. Nevertheless there may be times when a stereo pair is appropriate. To prepare a stereo pair using Raster3D, render the same set of object descriptors twice, using a different set of header records to *render* for the left and right members of the pair. It does not work to describe two copies of the scene in a single input stream to *render,* even though some composition tools are capable of generating such descriptions. This single-pass approach would fail because the light sources are fixed, and thus the copy of the scene presented to one eye effectively would be illuminated from a different angle than the one presented to the other eye. At the same time, unfortunately, one cannot trust all composition tools to create two equivalent input files to *render* corresponding to left-eye and right-eye viewpoints. This is because the recalculation of clipping in the second view may cause some objects to appear in one half of the pair but not in the other.

The recommended procedure for producing a stereo pair is therefore to start with a single scene description, and modify the header records twice to describe a left-eye view and a right-eye view separately. After rendering the two views, either print the two images separately and mount them side by side, or merge them electronically before printing. The details of this process are somewhat dependent on the way the initial scene description was generated, and the set of image-processing tools that is available. The first step is to convert the object descriptions in the *render* input file to a normalized coordinate system. Some composition tools, including Molscript and VMD, do this automatically. Otherwise the Raster3D utility program *normal3d* can be used. The normalized input file to *render* will now contain the identity matrix in the TMAT field of the header records (Fig. 3).

The second step is to edit the header records to alter the transformation matrix described by TMAT, and possibly the direction of the primary light source (SOURCE). The new transformation matrix may describe either a rotation operation or a shear operation. If a rotation operation is used, shadowing differences can be minimized by applying a corresponding rotation to the vector defining the direction of the primary light source. Secondary specular highlights will be incorrectly rendered in this case, however, because the secondary light source is always treated as being coincident with the view direction. A further problem is that some composition tools (e.g., Molscript version 1.4) describe only the front surface of three-dimensional objects. If the scene contents are rotated, the hollow back is partially revealed. Both of these problems can be ameliorated by altering TMAT to describe a shear operation, as shown here, rather than a rotation.

1	0	0	0
0	1	0	0
.03	0	1	0
0.	0.	0.	1.0

Because of the off-diagonal term added to the matrix, objects will be shifted to the left or right as a function of their z coordinate. The shear operation will result in correct shading and specular highlighting, but may still result in slight errors in shadow calculation.

The entire sequence of operations to convert a single Raster3D scene description into a stereo pair may be automated through the use of a shell script. Further discussions of stereo pairs derived from specific composition tools, along with worked examples, are provided in the Raster3D user documentation. A shell script, *stereo3d,* is provided to illustrate the automatic generation of side-by-side stereo pairs from a single *render* input file.

Algorithms Used for Rendering

Although the images produced by Raster3D rendering give the appearance of having been ray traced, this is achieved through the use of Z buffer algorithms rather than through true ray tracing. This compromise produces a remarkably photorealistic appearance for the rendered objects while requiring much less computation than full ray tracing of the same scene. The hidden surface removal method used by Raster3D is general enough to render any opaque surface that has an analytic description. It is especially efficient for van der Waals molecular surfaces and ball-and-stick representations, because it capitalizes on the rather uniform spatial distribution of atoms and bonds. It handles shadow calculation for a single shadowing light source, and encompasses a simple model of transparency. It does not, however, allow calculation of reflected objects, refraction, multiple shadowing light sources, or arbitrarily nested transparent surfaces. Texture mapping of arbitrary patterns onto rendered surfaces would be compatible with the algorithms in *render,* but this capability is not implemented in the current version of Raster3D.

Hidden Surface Removal

The internal coordinate system used by *render* is an orthogonal, right-handed three-space with the origin at the center of the field of view and z increasing toward the viewer. The color or brightness at each pixel center $[x\ y]$ of the output image is determined by a two-stage process:

1. Find the object with the highest z at $[x\ y]$.
2. Calculate shading parameters for the point $[x\ y\ z]$ on the surface of the object.

The first stage is called hidden surface removal, because it removes from view all but the surface closest to the viewer. The second stage, shading, depends on whether $[x\ y\ z]$ is in shadow from the point of view of the primary light source. We have found that a vivid three-dimensional effect can be achieved with the use of a single strong point light source located, say, behind and over one shoulder of the viewer, provided this is supplemented by a weak straight-on light source and an "ambient" term to keep shadowed curved surfaces from appearing flat or disappearing completely. The default relative contributions of these three lighting effects are set in the header record parameters STRAIT and AMBIEN. These defaults may be superseded by material descriptors specified individually for selected objects.

A simple and correct algorithm for hidden surface removal would be to search through all objects in the scene at each pixel position $[x\ y]$, calculate the elevation (z) at this point for every object that occludes $[x\ y]$, and pick the object with the greatest elevation (highest z) at that point. For a medium-size protein of 2500 atoms represented as a van der Waals space-filling model, and an output image of 1032×768 pixels, this would require nearly 2 billion ($2500 \times 1032 \times 768$) boundary checks and possible z calculations. The size of the problem easily can become orders of magnitude bigger in the case of larger output rasters, large proteins or protein assemblies, and triangular mesh representations of molecular surfaces.

If, however, the problem could be broken up into smaller pieces encompassing, say, 25 atoms each, the time performance of this simple algorithm on the 2500-atom protein would improve by a factor of about 100. As it happens, there is an easy way to divide the image so as to preserve a reasonably uniform distribution of atoms across the picture: split the whole scene into nonoverlapping rectangular "tiles," each containing perhaps 12 pixels in each direction (giving $86 \times 64 = 5504$ tiles in our example). A fairly well-balanced distribution of atoms onto tiles in projection is then guaranteed by the naturally uniform distribution of atoms in space. Of course, the projection of each atom can intersect more than one tile, but the number of such overlaps will still be far less than the total number of atoms. By forming a list for each tile of the atoms that might impinge on that tile, we can reduce our hidden surface removal problem to a set of much smaller problems and still use a simple hidden surface removal algorithm to solve each of them. The same argument applies to rendering representations of secondary structure built up from primitive objects other than spheres. For example, a "solid" ribbon-like helix or sheet can be built from two triangular mesh faces and cylindrical edges. If the ribbon is broken into 6 segments per residue, this comes to approximately 24 objects per residue. The spatial distribution of these objects is somewhat less uniform than the atomic centers in a van der Waals representation of the same residues, but not drastically so.

This subdivision by tiling is exactly how hidden surface removal is implemented in *render*. For each tile, a list is made of objects that could overlap the tile in projection; then, for each pixel position $[x\ y]$ within the tile, the list is searched to find out which object is closest to the viewer at $[x\ y]$. Note that in constructing these lists it is possible to use quite a crude overlap test, so long as all errors consist of estimating that an overlap exists where there is really none. Errors of this kind have a subsequent cost in efficiency, but not in correctness of the algorithm. The current implementation simply checks whether a tile overlaps a rectangle bounding the extreme x and y values of the object in projection.

Choosing the optimal size for tiles is not a completely straightforward task. Very large tiles can lead to excessively long searches at each pixel, but very small tiles can require large amounts of memory for per-tile object arrays, as each object then overlaps many tiles. In general, if ample virtual memory is available, it is best to choose rather small tiles (Fig. 5). The algorithm is quite well behaved in its memory access pattern, keeping page faults to a minimum even when the total amount of virtual memory used is large.

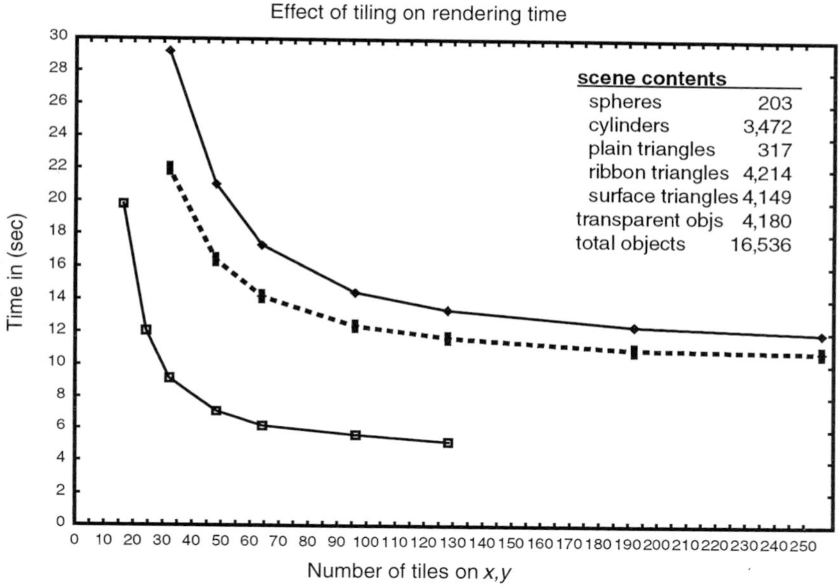

FIG. 5. The speed increase obtained by subdividing the image into tiles is substantial. Shown here is the effect of increasing the number of tiles used to render a scene given a fixed raster size. Timing runs were performed on a DEC AlphaServer 2100 5/250, using as a test image the scene described in example 6 and distributed with Raster3D version 2.2. No antialiasing was performed. Note that this scene includes a transparent molecular surface, and that as discussed later the presence of transparent objects adds considerable overhead to the rendering process. Top curve: Rendering times for a 1536 × 1536 pixel raster. The fastest rendering shown is obtained when the image is treated as an array of 256 × 256 tiles each made up of a 6 × 6 pixel array. Changing the surface to opaque leads to an approximately 10% decrease in rendering times for the same set of objects (middle curve). Bottom curve: Rendering times for the same scene rendered at lower resolution on a smaller (768 × 768) pixel raster. When increasing the resolution of a rendered image it is clearly preferable to increase the number of tiles before increasing the number of pixels per tile.

There are further refinements to this procedure that enhance its speed significantly. One is to sort the lists initially such that the object with the highest overall z is considered first. This tends to maximize the probability of finding the correct object early in the list search, and usually allows the search to terminate well before the end of the list is reached, because the search stops when it encounters an object that lies entirely below the highest z recorded so far. This approach puts the hidden surface removal into a class of algorithms known as Z buffer algorithms. Rendering speed is also affected by the precise sequence of tests made on each object to determine, with no error, whether it occludes the current pixel. The optimal ordering of tests is different for each primitive object type (sphere, cylinder, triangle).

Shading

The hidden surface removal algorithm just described is applied for each pixel $[x\ y]$ to find the z coordinate of the object closest to the viewer at that point. At this point the shading computation begins. This calculation requires the nominal color of the object, the surface normal of the object at $[x\ y\ z]$, and an indication as to whether the point is in shadow. The object color is taken from the explicit red/green/blue components given in the object descriptor record, possibly modified by information in a previous material descriptor. The surface normal for the three basic object types (spheres, cylinders, and flat triangles) is calculated analytically. The surface normal for interior points of a triangle with associated explicit vertex normals (object type 7) is interpolated using the algorithm proposed by Phong.[8] Such triangles are usually part of a triangular mesh surface imported from a surface generation program such as GRASP.[7] Other triangles are identified by the program as belonging to a linear mesh, or ribbon, and assigned approximate vertex normals accordingly.[2]

Shadowing is calculated by a Z buffer algorithm exactly like that used for hidden surface removal, an approach that is analogous to one originally suggested by Williams.[11] A point in shadow is one that is hidden from the primary light source by some other intervening object. To find out if a point $[x\ y\ z]$ is so hidden, we rotate it to $[x'\ y'\ z']$ using a transformation that would rotate the primary light source vector to lie along the positive z axis. Then we apply the hidden surface removal algorithm at $[x'\ y']$ to obtain an elevation z'' for the object closest to the primary light source in the new reference frame. If z'' is greater than z', i.e., closer to the light source, then the original point $[x\ y\ z]$ is in shadow; otherwise it is fully lit. This shadowing calculation is implemented by maintaining a set of "shadow tiles" with associated sorted shadow object lists exactly analogous to those described above. In the event of a near-tie between z'' and z', the point is considered to be strongly lit, so that roundoff errors will not put objects in their own shadows. The converse problem, of light reaching places it should not, does not occur in practice for two reasons. First, if a component of a surface normal in the direction of the strong light source is opposite to its component in the direction of the viewer, the surface is automatically depicted as being in shadow at $[x\ y\ z]$. Second, because surfaces are always

[11] L. Williams, *Comput. Graph.* **12,** 270 (1978).

used to enclose volumes that are not thinner than floating-point roundoff tolerance, light does not "leak through" at surface boundaries as it otherwise might where triangles interpenetrate.

The shading and highlighting procedures performed after hidden surface removal and shadow determination are quite standard. There is one strong and one weak point light source, and one white "ambient source," as mentioned previously. Diffuse reflections of the light sources are simulated using the Lambert cosine law, and specular reflections are treated using the exponential model of Phong.[8] The reader is referred to the textbook of Foley *et al.*[12] for an excellent treatment of this subject.

Transparency

The Raster3D rendering program supports a simple model of transparency, primarily for the purpose of including transparent molecular surfaces in a scene containing other, opaque, objects. Transparency is specified as part of an explicit set of material properties (object type 8). The CLRITY parameter in a material descriptor sets the degree to which an object passes light at normal incidence on its surface; i.e., CLRITY = 1.0 specifies a fully transparent object. The program models incomplete opacity as arising from finite thickness of a translucent material. The effective transparency of a surface at a given point may then be calculated from the clarity and the thickness of the surface in projection at that point relative to the thickness at normal incidence. The thickness in projection may be estimated from the surface normal and an approximation of the local surface curvature. Thus even a highly transparent object appears opaque when viewed obliquely.

The Z buffer algorithm used for hidden surface removal would become computationally expensive if forced to support a fully general treatment of transparency. The compromise used in *render* is to search through the per-tile list of objects to find the three objects with the highest z values at each pixel $[x\ y]$. This means that the presence of any transparent objects in the scene potentially slows the rendering of every pixel because of an increased overhead in object list sorting. If the object with the highest z coordinate is opaque, however, no computational price is paid in the shading computation. On the other hand, if this object is transparent, its partial contribution to the rendered pixel is stored and the shading calculation is repeated to find the contribution of the object with the second highest

[12] J. D. Foley, A. van Dam, S. K. Feiner, and J. F. Hughes, "Computer Graphics: Principles and Practice." Addison-Wesley, Reading, Massachusetts, 1990.

z coordinate. This second object is similarly tested for transparency, its contribution stored, and if necessary the third and final object is shaded to complete the rendered value for this pixel. Thus two interpenetrating transparent objects will be rendered "correctly," but any further transparent objects hidden by these surfaces will appear opaque. This simple model suffices for any single convex transparent surface, and in practice is adequate for a single convoluted surface, although imperfections in the rendered image may become noticeable in the case of a very irregular surface viewed edge on.

Alpha Channel

A concept closely tied to the transparency of an object is the notion that the rendered appearance of that object contributes only partially to the net color assigned to pixels it occludes. This effective transparency is conventionally denoted by the symbol α. As described above, the *render* program calculates α values in the process of modeling transparent objects. Another use of α is to define an algebra for compositing digital images.[13] Each pixel in the image is represented by a quadruple $[\alpha R\ \alpha B\ \alpha G\ \alpha]$. In the case that $\alpha = 1$ everywhere, this may be reduced to the familiar $[R\ G\ B]$ color triplet without loss of information. If α is not uniform over the image, however, its value may be encoded in a separate "alpha channel" stored in the digital image. A common use for such an alpha channel is to distinguish between foreground and background pixels, so that a complex scene rendered against an arbitrary background may later be superimposed unambiguously onto a different background altogether. The *render* program will include an alpha channel in the output image if SCHEME = 0. Each pixel is assigned an associated α value: $\alpha = 0$ if the pixel was rendered as background; $\alpha = 1$ if the pixel was rendered as part of an opaque object; $0 < \alpha < 1$ if the pixel corresponds to a superposition of one or more transparent objects through which the background is partially visible. If transparent objects in the image produced by *render* are in fact to be matted against an externally supplied background, then the background used during rendering must be black ($[BKGR\ BKGG\ BKGB] = [0\ 0\ 0]$).

Antialiasing

The term *antialiasing* is applied to techniques that reduce the jaggedness of lines and edges in a raster image. The *render* program allows antialiasing of the output image by performing an initial calculation using a larger, internal, raster. It then reduces the final raster size by assigning each pixel

[13] T. Porter and T. Duff, *Comput. Graph.* **18,** 253 (1984).

to be a weighted average of several pixels from the initial raster. This operation noticeably improves the appearance of the rendered image at a cost of additional computational time needed to calculate and average the larger raster. The antialiasing mode is selected by the SCHEME header parameter. Schemes 0 and 1 perform no antialiasing. Scheme 2 uses the requested raster size (NTX*NPX, NTY*NPY) internally, but outputs a final image that is only half this size in each dimension. That is, four pixels are calculated and averaged to generate each output pixel. Schemes 3 and 4 both calculate an initial raster that is three-halves that of the output raster in each dimension. In this case each 2×2 block of pixels in the final raster is derived from a corresponding 3×3 block in the initial raster. Again, each output pixel is formed by averaging four initially computed pixels, but now the average is weighted to reflect the areas of overlap generated by overlaying the 2×2 block on top of the 3×3 block. The central pixel of the initial 3×3 block contributes to four output pixels, each corner pixel contributes to a single output pixel, and so on. Scheme 3 interprets the parameters (NTX, NPX, NTY, NPY) as defining the initial raster size, whereas scheme 4 interprets the same parameters as defining the final raster size. Images may thus be previewed by rendering with scheme 0 or 1 for greater speed, and then rendered again to produce a final antialiased image with the same size simply by changing to scheme 4.

Future Directions

The Raster3D suite is continually evolving. We are currently working to achieve better integration with a wide variety of interactive visualization tools and to provide better support for describing and rendering molecular surfaces. In keeping with our goal of platform independence, we favor the use of generic graphic interfaces such as the MIT X-windows protocol. Two such X-windows tools under active development are a graphic user interface for the Raster3D suite, called TkRaster3D,[14] and a Raster3D output mode for the XralView crystallographic modeling program Xfit.[6]

We also expect to see increased use of Raster3D to prepare both static and dynamic images served over the World Wide Web on demand. Earlier versions of Raster3D have already been used to render MPEG animations of molecular motion.[15,16] Version 2.3 has added several features that should

[14] TkRaster3D was developed by Hillary Gilson at the National Institute of Standards and Technology, and is available from URL http://indigo15.carb.nist.gov/TkRaster3D.

[15] C. Vonrhein, G. J. Schlauderer, and G. E. Schulz, *Structure* **3,** 483 (1995).

[16] H. Grubmüller, B. Heymann, and P. Tavan, *Science* **271,** 997 (1996). The animation itself may be found at URL http://attila.imo.physik.uni-muenchen.de/~grubi/streptavidin.html.

facilitate its use for animation. The mechanism for file indirection is ideal for describing successive frames in which only a subset of the total scene is changed. The support for a separate alpha blend channel[13] in the output image can be used to isolate the rendered foreground objects for subsequent matting against an externally supplied background image. The fact that the Raster3D programs can be invoked inside a shell script to compose, render, and pipe images on demand make them well suited for use in a distributed environment such as the Web.

Raster3D is able to provide rapid rendering of photorealistic molecular graphics scenes largely because it is able to capitalize on prior knowledge of the type and distribution of objects likely to be present in such scenes. However, the support for molecular surface representations in Raster3D (including the simple model of transparency) is a more recent addition and still rather minimal. There is considerable scope for optimizing the choice of rendering algorithms as we gain more experience with figures that use surface representations of molecular structure. Future versions of *render* will probably allow the user to guide the program by specifying additional information about individual surfaces. A prime example of this would be the ability to associate texture maps with the objects making up a rendered surface.[17] Texture mapping provides a general mechanism for portraying additional properties of the surface by varying color or other visible attributes. It also allows superposition of patterns, contour lines, and graphic or textual data onto the rendered surface. This approach has been demonstrated elsewhere with great success.[7,18,19]

Acknowledgments

We thank our colleagues for contributing encouragement, suggestions, and code to Raster3D, with special thanks to Wayne Anderson, Mark Israel, Albert Berghuis, and Michael Murphy.

[17] P. S. Heckbert, "Fundamentals of Texture Mapping and Image Warping." M.Sc. thesis. Department of Electrical Engineering and Computer Science, University of California, Berkeley, California, 1989.

[18] M. Teschner, C. Henn, H. Vollhardt, S. Reiling, and J. Brickmann, *J. Mol. Graph.* **12,** 98 (1994).

[19] B. S. Duncan and A. J. Olson, *J. Mol. Graph.* **13,** 258 (1995).

[27] Detecting Folding Motifs and Similarities in Protein Structures

By GERARD J. KLEYWEGT and T. ALWYN JONES

Introduction

As the number of solved protein structures increases, it becomes more and more difficult to keep track of all these structures and their particularities, motifs, folds, etc. This implies that it becomes ever more difficult to recognize tertiary structural features in new protein structures, and to answer such questions as "in which proteins does this motif occur as well (if any)?" or "does this protein have a new fold?" That this is actually a problem, even for seasoned protein scientists, was amplified by Murzin,[1] who found that "All three protein structures reported in a recent issue of *Nature* display folding motifs previously observed in other protein structures. In only one case did the authors concerned notice this similarity." This was not an isolated incident. For example, when the structure of protein G was first solved, it was believed to have a novel fold,[2] but it was soon pointed out that, in fact, it was similar to ubiquitin.[3] Also, there are many anecdotes about how structural similarities were detected during seminars, conference lectures, and poster sessions, sometimes in time to prevent the publication of an article with the phrase "novel fold" in the title, sometimes not. The reason for this is that, until recently, one had to rely fully on human memory for the recognition of common motifs and similar folds. Only in the past few years has software been developed to aid in this process.[4]

Detecting similarities at the level of tertiary structure is of interest for at least three reasons: (1) It may provide insight into the *modus operandi* of proteins that share a common structural and functional trait (e.g., if a set of proteins uses a similar motif to bind a substrate or cofactor); (2) it may reveal evolutionary pathways (either divergent or convergent); and (3) it may provide insight into protein folding and stability by revealing that a certain arrangement of helices and strands occurs in unrelated proteins. In all cases, if similarities at the tertiary structure level exist, one is also

[1] A. G. Murzin, *Nature (London)* **360,** 635 (1992).

[2] A. M. Gronenborn, D. R. Filpula, N. Z. Essig, A. Achari, M. Whitlow, P. T. Wingfield, and G. M. Clore, *Science* **253,** 657 (1991).

[3] P. J. Kraulis, *Science* **254,** 581 (1991).

[4] L. Holm and C. Sander, *Proteins Struct. Funct. Genet.* **19,** 165 (1994).

0076-6879/97 $25.00

interested in sequence alignments that are based on these similarities. Such structure-based sequence alignments are expected to correlate with functional similarities. They may reveal, for example, that particular residues are structurally conserved in different proteins, and from that information hypotheses regarding the functional reasons underlying this conservation may be constructed. Another application, hitherto largely unexplored, lies in the recognition of motifs even before a protein structure has been completely built or refined.[5] As one traces the chain of a protein in an electron-density map, secondary structure elements (α helices and β strands) are often recognized early on, even though their direction and connectivity may still be unknown. Clearly, comparing a set of identified structural elements to a database of known structures may provide hints as to the direction and connectivity, and may even point to the presence of other structural elements.

Because it is easily foreseen that the problem of recognizing structural similarities is going to be more and more serious as the number of solved protein structures increases, we set out to develop software to facilitate this process. In our view, such software should contain at least two types of tool: one for rapid screening of the richest source of information regarding protein structure, the Brookhaven Protein Data Bank[6] (PDB), to single out a few structures ("hits") that appear to be structurally similar to the protein one is interested in (either because they share a particular, user-defined motif, or because they display overall similarity), and another that finds the optimal structural superposition of the protein in question and each of the hits and, as a result of that, a structure-based partial sequence alignment. In addition, such software should allow for visual inspection and interactive manipulation of the query protein and all of the hits, and it should be fast (real-time response), highly automatic, and practical (i.e., easy to use and employing search criteria to which protein scientists can easily relate). Moreover, we wanted to be able to look for structural similarities even before a structure has been completely built or refined, and when there still may be ambiguities regarding the direction and/or connectivity of structural elements. This requires that the software be closely coupled to O, our program of choice for building protein structures in electron-density maps. Although there have been many reports of programs that

[5] G. J. Kleywegt and T. A. Jones, *in* "From First Map to Final Model" (S. Bailey, R. Hubbard, and D. A. Waller, eds.), p. 59. SERC Daresbury Laboratory, Daresbury, Warrington, UK, 1994.

[6] F. C. Bernstein, T. F. Koetzle, G. J. B. Williams, E. F. Meyer, M. D. Brice, J. R. Rodgers, O. Kennard, T. Shimanouchi, and M. Tasumi, *J. Mol. Biol.* **112,** 535 (1977).

provide a subset of this functionality,[7–18] none of them satisfied all our conditions (for a review, see Ref. 4). Some of them appear to be slow, others cumbersome and nonintuitive to use. Moreover, all of them require atomic coordinates of at least the C_α atoms, which are often unavailable when one is still tracing the density.

Here we describe a collection of software that satisfies all previously mentioned criteria (rapid database screening; flexible, extensive superposition and alignment analysis of a few selected "hits"; fast, automatic, and intuitive). We have built a database of secondary structure elements (SSEs), based on known structures from the PDB. A cluster analysis using several known, high-resolution structures was carried out to find "typical" geometries of consecutive stretches of five C_α atoms in α helices and β strands, respectively (T. A. Jones, unpublished results, 1987). The central structures of the α and β clusters are used as templates. To classify a residue as being α helical, its C_α atom and those of the two neighboring residues on both sides are superimposed onto the α-helical template structure. If the root-mean-square distance (RMSD) between the corresponding atoms is less than a cutoff value (typically, 0.5 Å), the central residue is classified as α helical. The same approach is used to classify residues as being in a β strand, albeit that a slightly higher cutoff value is used (usually 0.8 Å) to account for the fact that β strands show larger deviations from the central cluster. The algorithm, called YASSPA, has been incorporated into the macromolecular modeling program O.[19] To investigate a newly solved structure, the user must delineate the SSEs in the protein (either manually or employing YASSPA). A program called DEJAVU uses this information to compare the set of SSEs to the whole database, using search criteria pertaining to the primary, secondary, and tertiary structure. DEJAVU produces a macrofile for O that contains instructions for displaying the protein being examined, and for carrying out the superposition, structure-based sequence alignment, and display of each of the proteins retrieved

[7] M. R. N. Murthy, *FEBS Lett.* **168,** 97 (1984).
[8] R. A. Abagyan and V. N. Maiorov, *J. Biomol. Struct. Dynam.* **5,** 1267 (1988).
[9] E. M. Mitchell, P. J. Artymiuk, D. W. Rice, and P. Willet, *J. Mol. Biol.* **212,** 151 (1989).
[10] G. Vriend and C. Sander, *Proteins Struct. Funct. Genet.* **11,** 52 (1991).
[11] C. A. Orengo, N. P. Brown, and W. R. Taylor, *Proteins Struct. Funct. Genet.* **14,** 139 (1992).
[12] R. B. Russell and G. J. Barton, *Proteins Struct. Funct. Genet.* **14,** 309 (1992).
[13] Z. Y. Zhu, A. Sali, and T. L. Blundell, *Protein Eng.* **5,** 43 (1992).
[14] M. S. Johnson, J. P. Overington, and T. L. Blundell, *J. Mol. Biol.* **231,** 735 (1993).
[15] L. Holm and C. Sander, *J. Mol. Biol.* **233,** 123 (1993).
[16] C. A. Orengo and W. R. Taylor, *J. Mol. Biol.* **233,** 488 (1993).
[17] Y. Matsuo and M. Kanehisa, *CABIOS* **9,** 153 (1993).
[18] S. D. Ruffino and T. L. Blundell, *J. Comput. Aided Mol. Design* **8,** 5 (1994).
[19] T. A. Jones, J. Y. Zou, S. W. Cowan, and M. Kjeldgaard, *Acta Crystallogr.* **A47,** 110 (1991).

from the database. The superposition and structure-based sequence alignment are carried out by an algorithm that has been implemented in O (T. A. Jones, unpublished results, 1988). In addition, DEJAVU can produce an input file for a separate least-squares superpositioning program called LSQMAN. This program is a faster and more versatile implementation of the algorithm that is used in O. LSQMAN, in turn, can produce a macrofile for O to read, align, and display the structures of the proteins found in the database.

In the case of so-called "bones searches," i.e., using approximately delineated SSEs from traced electron density, there are of course no atomic coordinates available. At present, the user must generate an SSE file with an editor. It is identical to normal SSE files, except that the residue names are irrelevant, the number of residues in an SSE must be estimated, and the coordinates of the C- and N-terminal C_α atoms are replaced by the coordinates of the skeleton atoms that are judged to be close to the C and N termini of the SSE. In the future, new commands will be added to O to simplify this process.

Databases

We use a simple representation of protein structure, in which SSEs have only the following attributes: a type ("alpha" or "beta"), the number of residues, their length, and a direction vector. The latter two properties are derived on the fly when DEJAVU searches the database from the coordinates of the N-terminal and C-terminal C_α atoms (which are stored in the database): The length of the SSE is the distance between these two atoms, the direction vector of the SSE is the vector from the N- to the C-terminal C_α atom. All secondary structure designations for the database proteins were carried out by the YASSPA algorithm as implemented in O. The results of YASSPA may differ somewhat from those obtained with other programs (such as DSSP[20]), because YASSPA classifies residues on the basis of structural criteria rather than on an analysis of hydrogen-bonding patterns. However, the results obtained with YASSPA are self-consistent so that, if one processes the query structure with YASSPA as well, one can be reasonably assured that the hits that DEJAVU comes up with are really similar to the structure of interest. In addition, YASSPA is fast and it does not require human interpretation of its output.

To generate the SSE database, we run an auxiliary program (called PRO1) that creates a macro for O. This program, in turn, reads the appropriate PDB files, executes YASSPA, and stores the results of YASSPA

[20] W. Kabsch and C. Sander, *Biopolymers* **22,** 2577 (1983).

(the names of the protein residues as well as their C_α coordinates) in files. These files are then processed by a second auxiliary program (called PRO2) that creates an ASCII SSE file suitable for use with DEJAVU. Users may process their own structures in the same way to create input SSE files for DEJAVU.

The database (and the user input SSE files) contain, for each protein, a record with the name of the molecule (usually, its PDB code), a record with remarks, and a record with the name of the corresponding PDB file. In addition, every SSE is represented by one record containing its type (ALPHA or BETA), its name (e.g., "A1," "B15"), the names of the first and last residues, the number of residues in the SSE, and the coordinates of the N-terminal and C-terminal C_α atoms; an example of such a file is shown in Fig. 1. It should be noted that DEJAVU is not limited to using only

```
!
! ===    0CRB
!
MOL      0CRB
NOTE     cellular retinol binding protein - model M26
PDB      /nfs/public/dombo/0crb.pdb
!
! type, name, first res, last res, #res, CA-first, CA-last
!
BETA  'B1'    '7'   '9'   3  2.060   4.567  -6.874   3.422   4.578  -0.194
BETA  'B2'   '12'  '14'   3  7.004  -0.256   7.085  13.083  -2.626   8.893
ALPHA 'A1'   '16'  '23'   8 17.376   0.123   7.184  27.353   0.903   2.554
ALPHA 'A2'   '27'  '35'   9 25.356   8.526  10.690  12.795   8.450   9.571
BETA  'B3'   '36'  '45'  10 11.734   9.291   5.980   6.034  -1.338 -19.834
BETA  'B4'   '48'  '54'   7 11.259  -1.686 -19.722  10.724   9.515  -4.087
BETA  'B5'   '58'  '63'   6 16.117  13.026  -1.904  15.462   2.945 -15.419
BETA  'B6'   '68'  '75'   8 20.056  -6.614 -19.327  24.768   8.802  -6.565
BETA  'B7'   '80'  '87'   8 29.082   3.916  -6.181  15.433 -10.159 -13.700
BETA  'B8'   '92'  '98'   7  9.823 -12.245  -7.698  28.057  -6.330  -8.726
BETA  'B9'  '105' '111'   7 21.190  -6.791  -4.255   1.189  -9.800  -1.821
BETA  'B10' '114' '121'   8  1.397  -4.545  -1.028  24.520  -5.704   1.303
BETA  'B11' '124' '131'   8 22.748  -7.396   6.205   0.872  -0.114   0.019
ENDMOL
```

FIG. 1. Example of an ASCII SSE file that can be used as input to DEJAVU (or as part of the database file), in this case for cellular retinol-binding protein, crbp (PDB code 1CRB). The lines that start with an exclamation mark are comment lines. The header contains three records: MOL, which defines the name or code of the protein; NOTE, which contains textual information; and PDB, which defines the name of the appropriate PDB file. The SSE records begin with their type (ALPHA or BETA), followed by the name of the SEE, the names of the first and last residues, the number of residues and, finally, the coordinates of the N-terminal and C-terminal C_α atoms, respectively.

α helices and β strands, but the underlying assumption that a representative direction vector can be derived from the terminal C_α coordinates will not be true, in particular, for turns. However, if a different database was used, one could easily include other types of helix, for instance.

Multiple databases can be chained together. Typically, one would use a small database containing the structures of interest, which is chained to a database derived from local structures that are not yet in the PDB, which in turn is chained to the PDB-derived database. At present, we have two different databases. The first was created using the structures available from the PDB in October 1994. It contains more than 70,000 SSEs from almost 2400 protein structures, made up of ~28,000 α helices and ~42,000 β strands. Mostly crystallographically determined structures were included (this will be changed in future), and no effort was made to remove identical or very similar proteins (e.g., mutants). Also, we have opted not to remove multiple copies of the same molecule inside the asymmetric unit of the unit cell. Because such molecules sometimes make up a functional dimer or trimer, etc., it is conceivable that one might actually be interested in a motif that is made up of SSEs from two different monomers (for instance, when a ligand is bound at the interface of a dimer). In addition, multiple copies of the same molecule within the asymmetric unit may have slightly different structures (in particular if the molecules were refined without noncrystallographic symmetry constraints or restraints). Indeed, occasionally one finds a hit with only one copy of the molecule but not with another one of the same molecule. DEJAVU, however, contains an option to include or skip multiple copies of a protein. To further reduce the problem of very similar structures, we have also created a separate database using the 95% homology list of Hobohm and Sander.[21] This database contains ~27,000 SSEs (~11,000 helices and ~16,000 strands) from 1381 protein structures (X-ray and NMR) that were available from the PDB in December, 1996.

Input

The input to DEJAVU consists of three parts: a database file with SSEs, a file containing information about the SSEs in the query structure, and the query and search parameters. A query may be defined in two different ways. One may define an explicit structural motif by selecting a subset of the protein SSEs and instructing the program to find proteins that contain

[21] U. Hobohm and C. Sander, *Protein Sci.* **3**, 522 (1994).

SSEs in a similar orientation (motif search). Alternatively, one may select all SSEs that contain a sufficient (user-definable) number of residues and instruct the program to find proteins that have as many similar SSEs as it can find in a similar spatial arrangement (similarity search).

The search parameters and criteria are the following:

The maximum allowed difference in the number of residues that potentially comparable SSEs comprise (typically, two to five residues)

The maximum allowed difference in their lengths (typically, 5 to 20 Å)

The maximum allowed difference in inter-SSE distances (typically, 5 to 10 Å)

The maximum allowed difference in the cosines of inter-SSE angles (typically, 0.1 to 0.5)

The type of distance that is to be used: either the center-to-center distances, or the minimum of the head-to-tail distances (for completely antiparallel motifs), or the minimum of the head-to-head and the tail-to-tail distances (for completely parallel motifs) can be used, where the "head" is the N-terminal C_α atom, the "tail" is the C-terminal C_α atom, and the center is the point midway between these two

Application or not of a directionality constraint; if this is used, the SSEs of the query are sorted from N terminus to C terminus and the restriction is imposed that when a query SSE is matched to an SSE in a database structure, then any SSEs following it in the sequence can be matched only to SSEs in the database structure that do not precede the matched SSE

Conservation or not of the absolute motif; if this is used then the complete distance and cosine matrices are compared and all entries must differ by less than the corresponding maximum allowed difference; otherwise only those matrix elements that define the distance and cosine of the angle between subsequent SSEs need to satisfy that condition

Conservation or not of sequential neighbors; if this is used, then any pair of SSEs that are neighbors in the query structure (i.e., there are no other SSEs between them in the sequence) must also be neighbors in the database structures

Attempt, or not, to avoid multiple-chain hits; if this is used, then the program attempts to avoid hits that contain SSEs from different molecules in the asymmetric unit (this works only if alphabetical chain identifiers were used in the original PDB file)

Attempt, or not, to avoid hits with multiple copies of essentially the same protein; if this is used, then if, for example, database protein

1LYZ has yielded a hit, then all proteins whose identifier ends in LYZ are not scrutinized (note that this is not 100% foolproof)

Weight factors for the mismatches in the number of residues, SSE lengths, distances, and cosines that are used to compute a score (see below) for every hit (typical values are 1, 1, 10, and 5, respectively; these are scaled by DEJAVU so that their sum equals one)

In our experience, in most cases where structurally similar proteins occur in the database, using the program default values suffices to retrieve these proteins. If a search with the default parameters yields no satisfactory hits, one may relax the cutoff values and/or release some of the constraints.

In the case of a bones search, we tend to use fairly strict cutoff values for the distance and cosine mismatches (5 Å and 0.2, respectively). Also, if the direction and connectivity of the SSEs are not known reliably, the neighbor and directionality constraints should be switched off.

Algorithm

In the following, the algorithm for motif searches is discussed. The algorithm for similarity searches is similar. The difference is that there is not a fixed set of SSEs that must be matched with other proteins, but rather a large set of which as many as possible SSEs must be matched.

The first things DEJAVU does in the motif-matching process is to put the selected SSEs in a list, to count the number of SSEs of each type, and to set up the appropriate distance and cosine matrices for the query motif. Subsequently, a loop over all the proteins in the database starts (optionally, one may limit the set of proteins that is to be used in the comparison). For each protein, the following is done.

1. DEJAVU checks if the protein contains a sufficient number of SSEs of the appropriate types (for instance, if the query structure contains four helices, there is no point in considering proteins that contain fewer than four helices).

2. For each SSE in the query, DEJAVU finds all the SSEs in the database structure that match it in terms of the number of residues they contain and their lengths; if any of the SSEs in the query cannot be matched in this fashion, this protein is not considered further.

3. Now DEJAVU embarks on an exhaustive, depth-first tree search with backtracking to find sets of SSEs in the database structure that match the query. This is done by generating combinations of SSEs found in the previous step that satisfy all the user-imposed criteria (distances, cosines, conservation of neighbors, etc.). Rather than generating all possible combinations of SSEs and subsequently testing them, DEJAVU applies all tests

as soon as an SSE of the database structure is matched to a query SSE. In this fashion, the search tree is pruned as early as possible and combinations of SSEs that cannot possibly yield hits are eliminated from further consideration. This algorithm is basically the same as that which was used to generate sequence-specific assignments for protein residues using two- and three-dimensional proton nuclear magnetic resonance (NMR) spectra.[22]

An example may help to clarify the algorithm. Suppose that the query contained three SSEs, Q1, Q2, and Q3, of which Q1 and Q2 are sequential neighbors. Suppose further that Q1 turns out to have three possible counterparts T1a, T1b, and T1c in a database structure of the same type and similar number of residues and length, that Q2 has counterparts T2a and T2b, and that Q3 has counterparts T3a, T3b, T3c, and T3d. There are $3 \times 2 \times 4 = 24$ ways of matching these SSEs of the database structure with those in the query. DEJAVU will start at the top and initially match Q1 to T1a. The next step is to try and match Q2 to T2a and T2b and testing that, given the fact that Q1 is matched to T1a, this match still satisfies all criteria. Suppose we want to conserve directionality and neighbors. Now if T2a is not the C-terminal neighbor of T1a, we can eliminate this match and we do not even have to try any of the four possible matches of Q3; in other words, we may skip searching this particular branch of the tree. If T2b is actually the neighbor of T1a, then we investigate the possibilities for matching Q3. When this is done, irrespective of the outcome (because we want to find all hits), the procedure backtracks and subsequently matches Q1 to T1b. Subsequently, the tree is traversed again, Q2 is matched to T2a and T2b and, if successful, the search will continue to the level of Q3, etc.

When all suitable combinations (if any) have been generated, DEJAVU prints the corresponding matching of SSEs and computes a score. To this end, the root-mean-square differences of the numbers of residues, the lengths of the matched SSEs and the elements of the distance and cosine matrices are computed, each is multiplied by the corresponding weight factor, and the four terms are added together. This means that a "perfect hit" has a score of zero and that the worse a hit is, the higher its score will be. For each database structure, the match that gave the lowest score is selected and appropriate instructions to further investigate this SSE alignment are written to a macro file for O and/or an input file for LSQMAN.

The complete database search procedure is extremely fast (usually limited by the rate at which output can be written to the terminal screen); it is conveniently executed interactively and typically takes between 30 sec and a couple of minutes of real time per query.

[22] G. J. Kleywegt, G. W. Vuister, A. Padilla, R. M. A. Knegtel, R. Boelens, and R. Kaptein, *J. Magn. Reson. Ser. B* **102,** 166 (1993).

Alignment

The O macro, when executed, will read the PDB file containing the protein of interest and draw a C_α trace of it; the SSEs that were part of the query motif are colored green, and the rest of the molecule is displayed in yellow. For each of the "best hits" the following is done.

1. The appropriate structure is read into O.
2. An explicit least-squares superposition is carried out.[23,24] In this procedure, the residues in the query SSEs are matched with equal numbers of residues from the corresponding SSEs of the database protein. The coordinate transformation thus obtained is usually nonoptimal, because:
   a. The SSEs are aligned on their first residues (even though they may contain different numbers of residues and, for example, matching the first residue in the query SSE with the second or third residue in the other structure may be much better).
   b. There may be parts in both proteins that were not used in the query (other helices or strands or even loops and turns) that superimpose much better than some of the SSEs that were used in the query.
   c. The explicit least-squares procedure will find a coordinate transformation matrix between explicitly defined corresponding atoms, but there may be one or two SSEs that superimpose badly (e.g., because relaxed search parameters were used), whereas four or five others may fit snugly on top of each other. In this case, the best superposition of all SSEs will be considerably worse than the optimal superposition of the smaller subset of SSEs.

For these reasons, an iterative least-squares operator improvement is also carried out. In this procedure, the initial coordinate transformation matrix is gradually improved. The procedure begins by finding the longest consecutive stretches (of, say, at least four residues) whose paired C_α atoms are closer than a certain distance cutoff (e.g., 6 Å) after application of the original transformation matrix. These groups of matched C_α atoms are then used in another round of least-squares superpositioning, which yields a new operator. This procedure is repeated until the results no longer improve.

The result of this procedure is an improved transformation matrix and the corresponding structure-based sequence alignment. The macro then instructs O to apply the transformation to the database structure and to

[23] W. Kabsch, *Acta Crystallogr.* **A32,** 922 (1976).
[24] W. Kabsch, *Acta Crystallogr.* **A34,** 827 (1978).

draw a C_α trace of its matched SSEs in red (optionally, the rest of the database protein can be drawn in blue). From the display of the query structure and the optimally aligned SSEs of the various database structures it is fairly simple to pick out those proteins that contain a motif that is similar to the query motif in the structure of interest. Of course, poor matches sometimes occur but they tend to be a result of using search parameters that are a trifle too relaxed. There is one pitfall, and that is that the program cannot discern an arrangement of SSEs and its mirror image. Such cases, however, are easily identified on closer study of the successful matches with O, which should always follow in any case.

An alternative is to use a separate program, called LSQMAN, to do the superpositioning and alignment improvement. This program, which is much faster than O, uses a similar algorithm but with some extensions. By default, the input file for this program as created by DEJAVU will first do the explicit superpositioning of the residues in the matched SSEs. Subsequently, two operator improvement macrocycles are carried out. In the first macrocycle, a distance cutoff of 6 Å is used, a minimum fragment length of four residues, and the optimization criterion is the total number of matched residues. In the second cycle, the initial distance cutoff is set to 4 Å, with a minimum fragment length of five residues. A maximum of 10 improvement cycles is carried out, and in each cycle the distance cutoff is multiplied by a factor of 0.975. This optimization criterion used is the so-called similarity index (SI), which is defined as shown in Eq. (1).

$$\mathrm{SI} = \mathrm{RMSD} \times \min(N_1, N_2)/N_\mathrm{m} \tag{1}$$

where RMSD is the root-mean-square distance between matched residues, N_1 and N_2 are the number of residues in the two molecules, and N_m is the number of matched residues. It is therefore a combined measure of RMSD (which should be small) and N_m (which should be large); for identical proteins, SI has a value of zero; the less similar two proteins are, the higher the value of SI becomes. In our experience, this operator-improvement scheme gives good results. However, it can easily be modified by the user if necessary. In addition to the number of matched residues and the value of SI, one may also use the RMSD as the optimization criterion, or another combined criterion called the match index (MI), which is defined as shown in Eq. (2).

$$\mathrm{MI} = (1 + N_\mathrm{m})/[(1 + w \times \mathrm{RMSD})(1 + \min(N_1, N_2)] \tag{2}$$

where w is a weight factor (default 0.5 Å^{-1}). The value of MI lies between zero (no similarity) and one (identical protein structures).

LSQMAN, in turn, produces a macro for O that will read, superimpose, and display the hits. The advantage of using LSQMAN is that if functions as a rapid filter of good, reasonable, and poor solutions. The O macro can then be edited (with a separate utility program, DEJANA, or with a text editor) to keep only the interesting hits, which saves time at the display.

Examples

Cro Repressor

As an example of looking for a tiny motif, we have searched our database with the two helices in the structure of cro repressor[25] (PDB code 1CRO), which, together with the turn in between them, make up the DNA-binding helix–turn–helix motif. We searched for this "motif," using maximum mismatches of two residues, 6 Å in the lengths of the SSEs, 4 Å in their head-to-tail distance (because the two helices are antiparallel) and 0.250 in the cosines of their angle, conserving directionality, the absolute motif and, of course, neighbors. In less than 3 CPU sec, DEJAVU found 63 matching proteins. LSQMAN was used to improve the alignments (and thereby to sort out the real hits from the false positives). Proteins of which 15 or more residues could be aligned to the structure of cro repressor are listed in Table I (some duplicates have been omitted for brevity). The best hits are all helix–turn–helix DNA-binding proteins, but there appear to be more proteins that contain a similar motif, in particular some calcium-binding proteins. Naturally, a structural similarity does not necessarily imply a functional similarity, but database searches such as these may yield interesting and unexpected results. The conserved residues for some of the hits are shown in Fig. 2. Note that all the DNA-binding proteins have a conserved Q*xxx*A*xxx*G sequence motif.

Glutathione Transferase A1-1

As a further example of a small-motif search, we used the structure of human α-class glutathione transferase A1-1[26] (GTA; PDB code 1GUH). One of the striking features of this protein is a tandem of two long, kinked α helices (26 and 29 residues, respectively). We searched for this motif, using maximum mismatches of four residues, 13 Å in the lengths of the SSEs, 5 Å in their center-to-center distance and 0.3 in the cosines of their angle, conserving directionality and the absolute motif but not neighbors.

[25] Y. Takeda, J. G. Kim, C. G. Caday, E. Steers, D. H. Ohlendorf, W. F. Anderson, and B. W. Matthews, *J. Biol. Chem.* **261,** 8608 (1986).

[26] I. Sinning, G. J. Kleywegt, S. W. Cowan, P. Reinemer, H. W. Dirr, R. Huber, G. L. Gilliland, R. N. Armstrong, X. Ji, P. G. Board, B. Olin, B. Mannervik, and T. A. Jones, *J. Mol. Biol.* **232,** 192 (1993).

TABLE I
RESULTS OF DEJAVU SEARCH FOR PROTEINS CONTAINING HELIX–(TURN)–HELIX MOTIF SIMILAR TO THAT OF cro REPRESSOR, SORTED BY SIMILARITY INDEX[a]

PDB	N_m	RMSD (Å)	SI (Å)	MI	N_c	Protein
1cro	264	0.000	0.0	1.000	264	Cro repressor
3fis	23	0.51	3.2	0.130	5	Fis-protein
1fia	23	0.50	3.4	0.122	5	Fis-protein
1lmb	23	0.62	4.8	0.101	5	λ repressor
1lrd	23	0.66	5.1	0.100	5	λ repressor
2or1	31	1.36	5.5	0.150	6	434-repressor
1tnc	17	1.29	10.9	0.075	0	Troponin
1trc	22	1.86	11.5	0.087	3	Calmodulin
1phs	31	1.76	15.0	0.064	3	Phaseolin
1gal	26	1.77	18.0	0.054	0	Glucose oxidase
2cyp	19	1.39	19.3	0.045	1	Cytochrome *c* peroxidase
1baa	23	1.85	19.5	0.051	3	Endochitinase
1ccp	19	1.41	19.6	0.044	1	Cytochrome *c* peroxidase
1csc	15	1.12	19.6	0.039	1	Citrate synthase
1aco	19	1.52	21.1	0.043	1	Aconitase
5acn	19	1.53	21.3	0.043	1	Aconitase
1gsg	20	1.77	23.3	0.042	1	Glutaminyl-tRNA synthetase
3cp4	23	2.07	23.8	0.044	1	Cytochrome P-450_{cam}
1vsg	24	2.28	25.1	0.044	4	Variant surface glycoprotein

[a] PDB, PDB identifier of the protein; N_m, number of matched residues; RMSD, root-mean-square distance of the matched C_α atoms; SI, Similarity Index (see text); MI, match index (see text); N_c, number of conserved residue types among those that were matched.

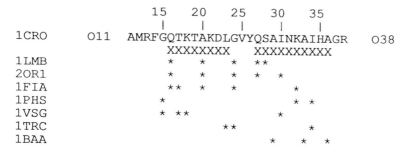

```
               15   20   25   30   35
               |    |    |    |    |
1CRO       011 AMRFGQTKTAKDLGVYQSAINKAIHAGR  038
                    XXXXXXXX   XXXXXXXXXX
1LMB                *   *   *  **
2OR1                *   *   *  *  *
1FIA                **  *   *       *
1PHS               *                * *
1VSG               * **           *
1TRC                       **         *
1BAA                             *   *  *
```

FIG. 2. Conserved residues in the structurally aligned sequences of some proteins that contain a helix–turn–helix motif similar to that observed in cro repressor. The two helices are marked with X's; asterisks indicate conserved residue types.

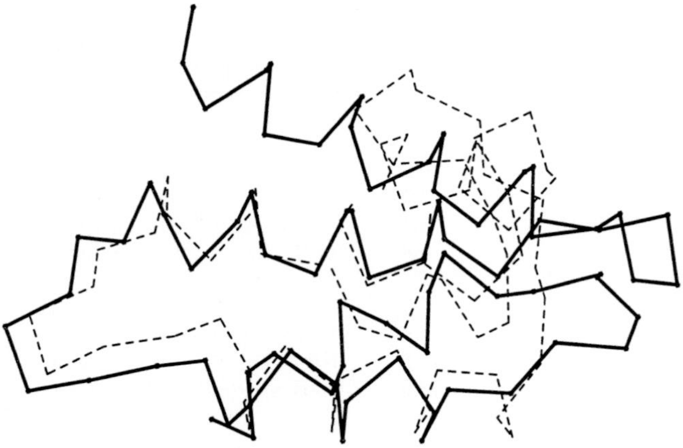

FIG. 3. The structure of the N-terminal four-helix bundle of RNR1 is shown (solid lines) with the equivalent part of the structure of thermolysin after alignment with DEJAVU and LSQMAN (dashed lines).

The first and best hit we found (other than the hit of GTA with itself) was with RNR2, ribonucleotide reductase R2[27] (PDB code 1MRR), which gave a score of 0.748. In this case, explicit least-squares superposition in O matched 55 atoms with an RMS distance of 5.24 Å; improvement with the zone composing the two helices aligned 52 atoms, whereas improvement with both complete molecules aligned 86 atoms (3.47 Å). If one were interested only in structure-based sequence alignment of the two helices, one would not use the latter alignment. Other hits occurred with 11 citrate synthase structures and with the photosynthetic reaction center.[28] The latter gave a DEJAVU score of 0.290, but the superposition of the 2 helices aligned only 48 atoms with an RMS distance of 3.27 Å. This demonstrates that there is no simple correlation between the DEJAVU scores and the results of the least-squares superposition. However, on the whole (unless extremely relaxed search parameters are used) DEJAVU usually produces hits that cluster well around the query motif.

Ribonucleotide Reductase Protein R1

As an example of looking for a larger motif, we screened the database using the four-helix bundle as found in the structure of ribonucleotide reductase protein R1[29] (RNR1). This protein, the only enzyme to catalyze

[27] P. Nordlund, B. M. Sjöberg, and H. Eklund, *Nature* (*London*) **345,** 593 (1990).

[28] J. Deisenhofer and H. Michel, *EMBO J.* **8,** 2149 (1988).

[29] U. Uhlin and H. Eklund, *Nature* (*London*) **370,** 533 (1994).

TABLE II
RESULTS OF DEJAVU SEARCH FOR PROTEINS WITH SAME 10-STRANDED β BARREL AS CRABP II[a]

PDB	N_m	RMSD (Å)	SI (Å)	MI	N_c	Protein
1opa	118	1.19	1.4	0.542	45	CRBP II
1opb	123	1.35	1.5	0.537	48	CRBP II
1alb	119	1.39	1.5	0.537	44	ALBP
1ifb	123	1.60	1.7	0.521	39	I-FABP
2hmb	122	1.56	1.7	0.524	51	M-FABP
1mdc	105	2.06	2.6	0.386	26	*Manduca sexta* FABP

[a] PDB, PDB identifier of the protein; N_m, number of matched residues; RMSD, root-mean-square distance of the matched C_α atoms; SI, Similarity Index (see text); MI, match index (see text); N_c, number of conserved residue types among those that were matched. Note that the current database contains more examples of proteins with this fold.

the *de novo* formation of deoxyribonucleotides, contains a 10-stranded α/β barrel as the major motif. Extensive DEJAVU searches yielded no other proteins that contain a similar barrel or overall fold. The N-terminal domain of RNR1 contains a four-helix bundle. DEJAVU was used to search for similar arrangements, using maximum mismatches of five residues, 15 Å in the lengths of the SSEs, 8 Å in their center-to-center distances and 0.4 in the cosines of their angles, conserving directionality and the absolute motif but not neighbors. The program came up with 5 hits, but after improvement of the alignments with LSQMAN only 2 thermolysin structures (PDB codes 3TMN and 4TLN) turned out to have parts of all four helices matched (in both cases, 52 residues were matched with an RMSD of 2.00 and 2.02 Å, respectively). The relevant parts of the structure of RNR1 (residues 19–93) and 3TMN (residues E240–E309) after alignment are shown in Fig. 3.

Cellular Retinoic Acid-Binding Protein Type II

To demonstrate how DEJAVU can be used to rapidly retrieve and align all structures of a certain fold, we used it with the structure of holo-cellular retinoic acid-binding protein type II (CRABP II).[30] This protein has the 10-stranded β-barrel structure typical of cellular lipid-binding proteins.[31,32] With the default input parameters, DEJAVU finds all known

[30] G. J. Kleywegt, T. Bergfors, H. Senn, P. Le Motte, B. Gsell, K. Shudo, and T. A. Jones, *Structure* **2,** 1241 (1994).

[31] T. A. Jones, T. Bergfors, J. Sedzik, and T. Unge, *EMBO J.* **7,** 1597 (1988).

[32] L. Banaszak, N. Winter, Z. Xu, D. A. Bernlohr, S. W. Cowan, and T. A. Jones, *Adv. Protein Chem.* **45,** 89 (1994).

```
[...]
s_a_i /nfs/taj/gerard/progs/secs/cbh2/native_riga.pdb 0cb2
mol 0cb2 obj c0cb2
pai_zo 0cb2 ; yellow
pai_zo 0cb2 102 109 green
[...]
pai_zo 0cb2 433 441 green
ca ; end
cent_id term_id 0cb2 102    CA ;
!
db_set_dat .lsq_integer 1 1 60
db_set_dat .lsq_integer 2 2 4
db_set_dat .lsq_integer 3 3 16999999
[...]
print ... comparing 9xia
print d-*xylose isomerase (e.c.5.3.1.5) complex
print ... score = 2.339508
print ... nr of matched SSEs = 5
s_a_i /nfs/public/pdb/xia9.pdb 9xia pdb
!
lsq_expl 0cb2 9xia
162 168 CA
50
191 209 CA
65
212 217 CA
84
233 253 CA
109
344 350 CA
176
; 9xia_to_0cb2
!
lsq_impr 9xia_to_0cb2 0cb2 ; 9xia ; CA 9xia_to_0cb2
9xia_to_0cb2
lsq_mol 9xia_to_0cb2 9xia ;
mol 9xia obj c9xia
pai_zo 9xia ; blue
pai_zo 9xia 50 54 red
[...]
pai_zo 9xia 176 183 red
ca ; end_obj
[...]
```

proteins in the database that have the same fold, plus a few false positives that are easily recognized after structural alignment with LSQMAN. The results are shown in Table II.

Cellobiohydrolase II

As an example of a straightforward similarity search, we have used cellobiohydrolase II[33] (CBH II; PDB code 3CBH). CBH II is an enzyme with exoglucanase activity, i.e., it is able to break down cellulose by sequentially cleaving off terminal cellobiose disaccharides. We used the 13 SSEs in CBH II, which contain at least 5 residues and required hits to have at least 5 similar SSEs. We ran DEJAVU with the default values for the parameters, which gave 28 hits. After superpositioning with O, we found the best hit to be D-xylose isomerase[34] (PDB code 9XIA), for which 199 atoms could be matched with an RMS distance of 3.71 Å. This result was more or less expected, because both proteins contain a so-called TIM barrel or variant. Figure 4 shows parts of the O macro generated by DEJAVU to align and display 9XIA and CBH II. Figure 5 shows some edited output from O when it executes the macro, in particular the optimal least-squares operator and the fragments that were used in the alignment of CBH II and D-xylose isomerase.

Cellobiohydrolase I

As an example of a nontrivial similarity search, we have used the structure of the catalytic core of cellobiohydrolase I[35] (CBH I; PDB code 1CEL). The structure of CBH I consists of a large β sandwich that encloses a substrate-binding tunnel that runs the length of the protein and is open at both ends. We used all SSEs containing at least four residues and required a minimum of six matched SSEs for the hits. We searched the entire database, using maximum mismatches of three residues, 10 Å in the lengths of the SSEs, 7 Å in their center-to-center distances and 0.4 in the cosines of their angles, conserving directionality and the absolute motif but not

[33] J. Rouvinen, T. Bergfors, T. Teeri, J. Knowles, and T. A. Jones, *Science* **249,** 380 (1990).
[34] H. L. Carrell, J. P. Glusker, V. Burger, F. Manfre, D. Tritsch, and J. F. Biellmann, *Proc. Natl. Acad. Sci. U.S.A.* **86,** 4440 (1989).
[35] C. Divine, J. Ståhlberg, T. Reinikainen, L. Ruohonen, G. Pettersson, J. K. C. Knowles, T. T. Teeri, and T. A. Jones, *Science* **265,** 524 (1994).

FIG. 4. Part of the O macro generated by DEJAVU for comparing CBH II and the hits resulting from a similarity search. The macro has been edited for the sake of brevity.

```
Lsq >  Loop =   11 ,r.m.s. fit =     3.708 with   199 atoms
Lsq >  x(1) =     0.7746*x+    0.3732*y+   -0.5106*z+   15.9403
Lsq >  x(2) =    -0.4485*x+    0.8934*y+   -0.0273*z+   28.1045
Lsq >  x(3) =     0.4460*x+    0.2501*y+    0.8594*z+  -31.3210
Lsq > Here are the fragments used in the alignment
Lsq > 0   133 FMWLD     137
Lsq >      12 TFGLW      16
Lsq > 0   140 DKTPLMEQTLADIR    153
Lsq >      34 LDPVESVQRLAELG      47
Lsq > 0   163 AGQFVVYDL     171
Lsq >      48 AHGVTFHDD      56
Lsq > 0   175 DCAAL     179
Lsq >      96 HPVFK     100
Lsq > 0   183 GEYSI     187
Lsq >      60 PFGSS      64
Lsq > 0   190 GGVAKYKNYIDTIRQIVVEYS     210
Lsq >      63 SSDSEREEHVKRFRQALDDTG      83
Lsq > 0   211 DIRTLLVIEPD     221
Lsq >      83 GMKVPMATTNL      93
Lsq > 0   225 NLVTNLGTP     233
Lsq >     103 GFTANDRDV     111
Lsq > 0   239 QSAYLECINYAVTQLN     254
Lsq >     116 LRKTIRNIDLAVELGA     131
Lsq > 0   258 VAMYLDAGH     266
Lsq >     130 GAETYVAWG     138
Lsq > 0   272 WPANQDPAAQLFANVYKNASSPR     294
Lsq >     154 ALDRMKEAFDLLGEYVTSQGYDI     176
Lsq > 0   295 ALRGLATN     302
Lsq >     175 DIRFAIEP     182
Lsq > 0   324 YNEKLYIHAIGPLLAN     339
Lsq >     193 LPTVGHALAFIERLER     208
Lsq > 0   343 SNAFFITDQGRSGKQP     358
Lsq >     210 ELYGVNPEVGHEQMAG     225
Lsq > 0   371 IGTGF     375
Lsq >     253 KYDQD     257
Lsq > 0   383 TGDSLLDSFVW     393
Lsq >     236 LWAGKLFHIDL     246
Lsq > 0   394 VKPGG     398
Lsq >     286 FDFKP     290
Lsq > 0   399 ECDGTS     404
Lsq >      22 GRDPFG      27
Lsq > 0   422 QPAPQ     426
Lsq >      25 PFGDA      29
```

TABLE III
RESULTS OF DEJAVU SEARCH FOR PROTEINS THAT DISPLAY STRUCTURAL SIMILARITIES TO CBH I[a]

PDB	N_m	RMSD (Å)	SI (Å)	MI	N_c	Protein
1ayh	126	1.64	2.8	0.325	17	β-Glucanase
1byh	125	1.73	3.0	0.315	15	β-Glucanase
1lte	108	2.11	4.7	0.221	12	Lectin
1lec	103	2.09	8.7	0.208	8	Lectin
2nn9	30	1.68	21.8	0.043	3	Neuraminidase
1bbk	44	2.24	22.1	0.049	0	Methylamine dehydrogenase
1ncc	39	2.12	23.6	0.045	3	Neuraminidase
1phs	32	2.10	23.9	0.044	0	Phaseolin
1nca	37	2.08	24.4	0.043	3	Neuraminidase
1mcp	34	2.06	26.3	0.040	3	Fab fragment
1mad	40	2.45	26.6	0.042	3	Methylamine dehydrogenase
1nsb	41	2.54	26.9	0.043	3	Neuraminidase sialidase
2bat	34	2.37	27.0	0.041	2	Neuraminidase
1ncb	37	2.39	28.0	0.040	2	Neuraminidase
1cn1	29	2.20	32.9	0.033	1	Concanavalin A
1ncd	29	2.38	35.5	0.032	2	Neuraminidase
1aoz	25	2.48	43.0	0.027	1	Ascorbate oxidase

[a] PDB, PDB identifier of the protein; N_m, number of matched residues; RMSD, root-mean-square distance of the matched C_α atoms; SI, Similarity Index (see text); MI, match index (see text); N_c, number of conserved residue types among those that were matched.

neighbors. DEJAVU found 40 hits, the best of which (after improvement of the alignment with LSQMAN) are listed in Table III. The top two hits belong to an endoglucanase, and the similarity between CBH I and this protein had been noted previously. Nontrivial, however, were the similarities with the next-best hits, two plant lectins. Lower down the list the structure of concanavalin A shows up. In fact, this protein has a β sandwich similar to the one found in CBH I, but the participating strands have been rearranged in the sequence. This means that it is an "accidental" hit, i.e., the right protein but with an incorrect alignment operator. Nevertheless, the similarity was obvious when the hits were inspected on the graphics display. The lesson to be learned from this is that when there are no very similar proteins in the database (i.e., perhaps a novel fold), it is all the

FIG. 5. Edited output from O for the superposition of CBH II and D-xylose isomerase. The final improved least-squares operator is shown, followed by the structure-based sequence alignment of the matched parts of both proteins. Residues that are identical in both proteins have been indicated with boldface type. This demonstrates how completely different sequences may fold into similar three-dimensional structures.

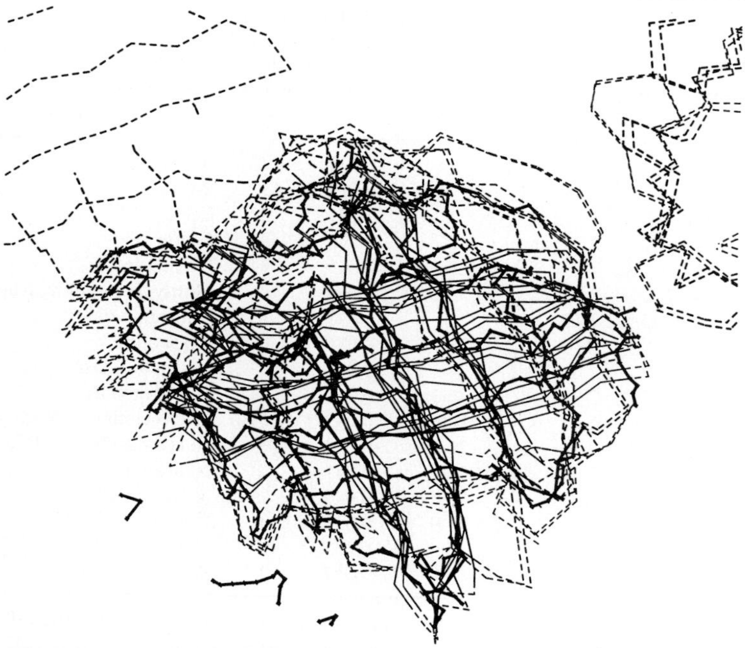

FIG. 6. Skeletonized MIR density for P2 myelin protein (thick solid lines) with the structures of other small lipid-binding proteins superimposed. In this case, the skeletonized density was used to delineate seven approximate SSEs with unknown directionality and connectivity, which were used in a DEJAVU "bones search." The seven hits are shown with thin lines; the solid lines represent SSEs that were matched; and dashed lines are used to draw the rest of their structures.

more important to screen all the hits interactively on the display, even if they show up with low scores.

P2 Myelin Protein

P2 myelin protein (PDB code 1PMP) is a member of a family of hydrophobic ligand-binding proteins. It occurs in peripheral nervous system myelin and probably functions as a part of a lipid transport and storage system in myelinating Schwann cells. Its structure was solved and refined in our laboratory.[31,36] We used this protein as a test case for the option to use SSEs found in a partly edited skeleton as input to DEJAVU. The structure of P2 myelin protein is similar to that of CRABP II (see above). Using a skeleton derived from a threefold-averaged multiple isomorphous replacement (MIR) map, it was fairly simple to delineate many of the SSEs.

[36] S. W. Cowan, M. E. Newcomer, and T. A. Jones, *J. Mol. Biol.* **230,** 1225 (1993).

We used only six strands and one helix. They were not put in any particular (i.e., sequential) order in the SSE file, and the direction of some of them was intentionally reversed to test the program. DEJAVU was run, requiring at least six matching SSEs, using maximum mismatches of three residues, 10 Å in the lengths of the SSEs, 5 Å in their center-to-center distances and 0.2 in the cosines of their angles. Directionality and neighbors were not constrained to be conserved. This reflects the "real-life" situation, in which one often does not know in which direction SSEs run, nor how they are connected. DEJAVU finds seven hits, all cellular lipid-binding proteins known to have the same fold as P2 myelin protein. The RMSDs of the centroid positions of the SSEs after superpositioning vary between 1.23 and 2.40 Å. Figure 6 shows the P2 skeleton and the aligned proteins. Note that the results are not always as good as in this example[5]; further work is needed to optimize the program in this respect. In addition, for smooth operation, some new commands will have to be implemented in O. Also, in the case of P2 myelin protein, this procedure would not have simplified model building because none of the structures found by DEJAVU were available at the time. In fact, most of them can be traced to P2 myelin protein in molecular replacement ancestry.

Software

DEJAVU and LSQMAN have been written in Fortran-77; they run on Silicon Graphics, Evans & Sutherland ESV, and DEC ALPHA/OSF1 workstations. YASSPA and LSQ have been incorporated into O,[19] which runs on a variety of workstations. For details regarding the availability of DEJAVU, LSQMAN, and O, contact T. A. Jones (e-mail: alwyn@xray.bmc.uu.se). A description of the latest version of the programs and databases can be found on the World Wide Web at URL: http://alpha2.bmc.uu.se/~gerard/manuals/. Some other applications of DEJAVU can be found in Refs. 37–39. The program LSQMAN is described in Ref. 40.

Acknowledgments

This work was supported by the Netherlands Organization for Scientific Research (NWO) through a postdoctoral fellowship to G.J.K. Additional funding was provided by the Swedish Natural Science Research Council and Uppsala University. RNR1 and RNR2 coordinates were kindly provided by Professor H. Eklund prior to their general release.

[37] C. Fan, P. C. Moews, C. T. Walsh, and J. R. Knox, *Science* **266,** 439 (1994).

[38] C. Fan, P. C. Moews, Y. Shi, C. T. Walsh, and J. R. Knox, *Proc. Natl. Acad. Sci. U.S.A.* **92,** 1172 (1995).

[39] X. Qiu, C. L. M. J. Verlinde, S. Zhang, M. P. Schmitt, R. K. Holmes, and W. G. J. Hol, *Structure* **3,** 87 (1995).

[40] G. J. Kleywegt, *Acta Crystallogr.* **D52,** 842 (1996).

[28] Biological Macromolecule Crystallization Database

By GARY L. GILLILAND

Introduction

The ability of a biological macromolecule to crystallize is related directly to its solution properties, determined by factors such as size, shape, surface complexity, and conformational stability. Recognition of these facts has prompted studies of the biophysical properties of biological macromolecules aimed at discovery of simple methods that would have predictive capabilities (i.e., George and Wilson[1] and Veesler *et al.*[2]). Such studies have met with mixed success. Traditionally, to circumvent this lack of understanding needed to predict crystallization behavior, one performs an empirical procedure consisting of a series of experiments that vary several different parameters such as pH, temperature, ionic strength, and macromolecule concentration (for a review, see McPherson[3]). The number of experiments required for success depends on the macromolecule and the choices made by the investigator. Occasionally the search ends quickly, either because crystallization occurs over a broad range of conditions or because the right choice was made early. More often many experiments are required to discover crystallization conditions. Sometimes no crystallization conditions are found, no matter how many experiments are done.

Currently no universal strategy for searching for the crystal growth parameters for a biological macromolecule has been accepted by experimentalists even after more than 50 years of experience in the production of diffraction-quality crystals. However, several systematic procedures and strategy suggestions have been put forth.[4–11] Many laboratories have begun

[1] A. George and W. W. Wilson, *Acta Crystallogr.* **D50,** 361 (1994).
[2] S. Veesler, S. Marcq, S. Lafont, J. P. Astier, and R. Boistelle, *Acta Crystallogr.* **D50,** 355 (1994).
[3] A. McPherson, *Eur. J. Biochem.* **189,** 1 (1990).
[4] A. McPherson, Jr., *Methods Biochem. Anal.* **23,** 249 (1976).
[5] T. L. Blundell and L. N. Johnson, "Protein Crystallography." Academic Press, New York. 1976.
[6] C. W. Carter, Jr., and C. W. Carter, *J. Biol. Chem.* **254,** 12219 (1979).
[7] A. McPherson, "Preparation and Analysis of Protein Crystals." John Wiley & Sons, New York. 1982.
[8] G. L. Gilliland and D. R. Davies, *Methods Enzymol.* **104,** 370 (1984).
[9] G. L. Gilliland, *J. Crystal Growth* **90,** 51 (1988).
[10] G. L. Gilliland and D. Bickham, "Methods: A Companion to Methods in Enzymology," Vol. 1, p. 6. Academic Press, San Diego, California, 1990.

0076-6879/97 $25.00

using experimental procedures called "fast screens"[12,13] that use sparse-matrix sampling techniques.[6] These techniques scan a wide pH range and a variety of precipitants, buffers, and additives that have proved successful in other crystal growth studies. These strategies are based on the successes of many scientists who have produced suitable crystals for diffraction studies for a variety of macromolecules.

The Biological Macromolecule Crystallization Database (BMCD) catalogs and summarizes the information concerning crystallization that is available in the literature as an aid in the development of crystallization strategies to produce large single crystals suitable for X-ray structural investigations. The BMCD also has become the National Aeronautics and Space Administration (NASA) Protein Crystal Growth (PCG) Archive, providing access to details and results of crystallization experiments undertaken in microgravity (outer space). We present a brief history of the development of the BMCD, a description of the data abstracted from the literature, a description of the PCG Archive, and examples for using the BMCD in the development of crystallization strategies.

History

The BMCD has its roots in work initiated in the laboratory of D. Davies at the National Institutes of Health in the late 1970s and early 1980s.[8] Working on a variety of difficult protein crystallization problems, the Davies group abstracted a large body of crystallization information from the literature. This led to a systematic search of the literature and a compilation of data that included most of the crystallization reports of biological macromolecules at that time (the end of 1982). In 1983 the data, consisting of 1025 crystal forms of 616 biological macromolecules, were deposited in the Brookhaven PDB.[14] This provided the first public access to the data.

In 1987, with assistance from the National Institute of Standards and Technology (NIST) Standard Reference Data Program, the data contained in the file previously submitted to the PDB were incorporated into a true database. The menu-driven software used to access the data was written

---

[11] E. A. Stura, A. C. Satterthwait, J. C. Calvo, D. C. Kaslow, and I. A. Wilson, *Acta Crystallogr.* **D50,** 448 (1994).

[12] J. Jancarik and S.-H. Kim, *J. Appl. Crystallogr.* **24,** 409 (1991).

[13] B. Cudney, S. Patel, K. Weisgraber, Y. Newhouse, and A. McPherson, *Acta Crystallogr.* **D50,** 414 (1994).

[14] F. C. Bernstein, T. F. Koetzle, G. J. B. Williams, E. F. Meyer, Jr., M. D. Brice, J. R. Rogers, O. Kennard, T. Shimanouchi, and M. Tasumi, *J. Mol. Biol.* **112,** 535 (1977).

and compiled with Clipper[15,16] and ran as an independent program on personal computers (PCs). All database files were in dBase III Plus[17] format, and all of the index files were in the Clipper indexing format. The database was released to the public in 1989 as the NIST/CARB (Center for Advanced Research in Biotechnology) Biological Macromolecule Crystallization Database, version 1.0.[9] In 1991 a second version of the software and data for the PC database was released.[10] The data from 1465 crystal forms of 924 biological macromolecules were included. In 1994 the BMCD began including data from crystal growth studies supported by the NASA to fulfill its new function as the NASA PCG Archive.[18] The software was expanded and released as the NIST/NASA/CARB BMCD version 3.0, including data for 2218 crystal forms of 1465 biological macromolecules.

The BMCD has now been ported to a UNIX platform to take advantage of the development of network capabilities that employ client–server tools.[19] This implementation of the BMCD uses the POSTGRES database management system[20] and World Wide Web/NCSA Mosaic client–server protocols.[21] This not only provides all of the features of the earlier PC versions of the BMCD, but also gives the user community access to the most recent updates and allows rapid implementation of new features and capabilities of the software.

Biological Macromolecule Crystallization Database

The BMCD includes data for all classes of biological macromolecules for which diffraction-quality crystals have been obtained and reported in the literature. These include peptides, proteins, protein–protein complexes, nucleic acids, nucleic acid–nucleic acid complexes, protein–nucleic acid complexes, and viruses. In addition, the BMCD, serving as the NASA PCG Archive, contains the crystallization data generated from ground-based and microgravity crystallization studies supported by NASA. Crystallization

[15] Certain commercial equipment, instruments, and materials are identified in this chapter to specify the experimental procedure. Such identification does not imply recommendation or endorsement by the National Institute of Standards and Technology, nor does it imply that the materials and equipment identified are necessarily the best available for the purpose.

[16] Clipper is a registered trademark of Nantucket Corp., 12555 West Jefferson Boulevard, Los Angeles, California 90066.

[17] dBASE III Plus is a registered trademark of Borland International, Inc.

[18] G. L. Gilliland, M. Tung, D. M. Blakeslee, and J. Ladner, *Acta Crystallogr.* **D50,** 408 (1994).

[19] D. Comer and D. Stevens, "Internetworking with TCP/IP," Vol. III: Client Server Programming and Applications. Prentice-Hall, Englewood Cliffs, New Jersey, 1994.

[20] M. Stonebraker, E. Hanson, and C.-H. Hong, "Proceedings of the 1986 ACM-SIGMOD Conference on Management of Data." Washington, DC, 1986.

[21] B. R. Schatz and J. B. Hardin, *Science* **265,** 895 (1994).

data from microgravity experiments sponsored by other international space agencies are also included.

The information contained in the BMCD can be divided into three major categories: biological macromolecule, crystallization, and summary and reference data. The data are organized into separate entries for biological macromolecules and for crystal forms. Each biological macromolecule in the database has a unique biopolymer sequence. For example, a site-directed mutant protein with a single amino acid change from the wild-type protein is considered a separate macromolecule entry. Analogously, a crystal entry extracted from the literature must have unique unit cell constants. Crystal entries that are nearly isomorphous with previously reported crystals are included if the author(s) describes significant differences in the intensity distribution of the diffraction pattern from that previously reported. Thus, each macromolecule entry has at least one, but may have more than one, crystal entry associated with it. The data elements for each type of entry, summary information, and reference data contained in the BMCD are described below.

Biological Macromolecule Entry

The data in a macromolecule entry include the preferred name of the macromolecule, the systematic name, and/or other aliases. Each entry includes biological source information that consists of the common name, genus and species name, tissue, cell, and organelle from which the macromolecule was isolated. Attempts have also been made to include this information for recombinant proteins expressed in a foreign host. The subunit composition and molecular weight also are provided. This information consists of the total number of subunits, the number of each type of distinct subunit, and the corresponding molecular weights. (A subunit of a biological macromolecule entity is defined as a part of the assembly associated with another part by noncovalent interactions. For example, the two oligomeric nucleic acid strands of a double-stranded nucleic acid fragment are considered as two subunits.) The name of a prosthetic group associated with a biological macromolecule is included if it is reported in the crystallographic studies. If the macromolecule is an enzyme, the EC number[22] and catalytic reaction are present. Finally, general remarks are included if any special features of the macromolecule were relevant to, or might influence, its crystallization behavior. Each biological macromolecule entry is assigned

[22] E. C. Webb, "Enzyme Nomenclature." Academic Press, San Diego, California, 1992.

Macromolecule Entry ME#: M050
Macromolecule Name: chymosin
Aliases: acid protease
proteinase, carboxyl
rennin
Common Name: bovine, calf
EC: 3.4.23.4
specificity resembles that of pepsin A; clotting of milk
Total Molecular Weight: 35,000
Total No. Subunits: 1
Remarks: Structurally homologous with pepsin. Formed from prorennin (prochymosin). Formerly EC 3.4.4.3, under the name rennin.
Total Number of Crystal Entries: 1

FIG. 1. A representative example of a biological macromolecule entry (M050) in the BMCD.

a four-character alphanumeric identifier beginning with the letter M. A typical macromolecule entry is illustrated in Fig. 1.

Crystal Entry

For each biological macromolecule entry there is at least one crystal entry. The data in each crystal entry include the crystallization procedure, the crystal data, crystal morphology, and complete references. The crystallization procedure includes the crystallization method, the macromolecule concentration, the temperature, the pH, the chemical additives to the growth medium (buffer, precipitant, and/or stabilizers), and the length of time required to produce crystals of a size suitable for diffraction experiments. If the crystallization deviates from standard protocols,[7] a description of the procedure is provided in the Comments section. As described in the References section, the crystal size and shape are given along with the resolution limit of X-ray diffraction. If crystal photographs or diffraction pictures are published, the appropriate references are indicated. The crystal data include the unit cell dimensions (a, b, c, α, β, γ), the number of molecules in the unit cell (Z), the space group, and the crystal density. The data also include cross-references to other structural biology databases such as the Brookhaven PDB.[14] Each of these crystallization entries is also given a four-character alphanumeric identifier beginning with C. A crystal entry for the macromolecule entry illustrated in Fig. 1 is shown in Fig. 2.

Summary and Reference Information

The summary information gives the user a convenient mechanism for browsing the BMCD data. The user has access to a complete list of macromolecule names, tabulations of the number of macromolecules and crystal

Crystal Entry CE#: C08I

a	b	c	alpha	beta	gamma	Z	Space Group*	CS
72.80	80.30	114.80	90.00	90.00	90.00	8	I222	Orthorhombic

Method(s) used for the crystallization: vapor diffusion in hanging drops
Macromolecule concentration: 8.600 mg/ml
pH: 5.000 to 6.200
Chemical additions to the crystal growth medium:

sodium chloride	2.0000	M
MES	0.0500	M

Crystal density: 1.243 gm/cm^3
Diffraction limit: 2.000 Angstroms
Type of x-ray photograph: 3.0 Ang hk0-R07I
Reference's description of the crystal habit: rectangular equidimensional blocks, sometimes elongated
Cross reference to Protein Data Bank: 1CMS
Comments: Cross-linked crystals were used in the heavy atom search. Large-scale crystals of the recombinant protein were grown by repeated macroseeding -R0YS.
References:

R07I: Bunn, CW; Camerman, N; T'sai, LT; Moews, PC; Baumber, ME (1970) *Phil Trans Roy Soc Lond*, **257**, 153-158.

R0YS: Gilliland, GL; Winborne, EL; Nachman, J; Wlodawer, A (1990) *Proteins*, **8(1)**, 82-101. "The three-dimensional structure of recombinant bovine chymosin at 2.3 Å resolution."

R07J: Jenkins, J; Tickle, I; Sewell, T; Ungaretti, L; Wollmer, A; Blundell, T (1977) *Adv Exp Med Biol*, **95**, 43-60. "X-ray analysis and circular dichroism of the acid protease from Endothia parasitica and chymosin."

FIG. 2. A representative example of a crystal entry (C08I for macromolecule entry M050) in the BMCD. It should be noted that the concentrations of chemical additives use four decimal places to accommodate the range of values of concentrations of the database, not to indicate the precision of the data.

forms for each source, prosthetic group, space group, chemical addition, crystallization method, and complete references. References can be queried for matches with a particular author or for a key word or phrase. The BMCD also provides a listing of general references concerning all aspects of crystal growth. For convenience, these references have been divided into categories that include reviews and books, articles concerning procedures, and references concerning nomenclature. All references include complete titles, and remarks are often added to emphasize important aspects of a reference that may not be evident from the title.

National Aeronautics and Space Administration Protein Crystal Growth Archive

As briefly mentioned above, a new function of the BMCD is to serve as the NASA PCG Archive. The interest on the part of NASA in protein crystal growth has been sparked by the results of several crystallization experiments indicating that sometimes protein crystals grown in microgravity diffract to higher resolution than those grown in laboratories on earth.[23] A complete description of the biological macromolecule crystallization experiments performed in microgravity in the NASA space shuttle and/or in laboratory control experiments are included in the BMCD. In addition, the BMCD includes data and results from microgravity crystallization experiments sponsored by other international space agencies. The data are organized in the same general manner as other entries in the BMCD. The data composing the PCG Archive contain biological macromolecule and crystal entries, as described above, with additional data items that relate to the microgravity experiment(s). In the PCG Archive, a single crystal form of a macromolecule may have multiple entries, one for each microgravity experiment. Each of these crystallization entries is also given a four-character alphanumeric identifier beginning with the letter C.

Also, added to the BMCD are new summary displays of information specific to the microgravity crystallization experiments. Included are lists of the microgravity missions, of the acronyms and names of sponsors, of the apparatuses used in microgravity experiments, and of references discussing theory or experimental results of crystallization of biological macromolecules in microgravity. Detailed descriptions of the crystallization apparatuses that have grown crystals and/or are in use for microgravity crystallization experiments are also available.

Querying Biological Macromolecule Crystallization Database

The BMCD provides a user interface for querying the database. Database searches may be carried out for data elements of any of the categories listed in Table I. Depending on the type of query, a parameter value (numerical or text) or a range of values (numerical) is requested. The Boolean logic AND and OR functions can be used to generate quite complex searches. The results of searches are displayed at the monitor in a variety of ways and can be written to an ASCII file or printed.

Crystallization Strategies

From its beginnings, the primary purpose of the BMCD has been to develop crystallization strategies for biological macromolecules. These

[23] L. J. DeLucas, C. D. Smith, W. Smith, S. Vijay-Kumar, S. E. Senadhi, S. E. Ealick, D. C. Carter, R. S. Snyder, P. C. Weber, F. R. Salemme, D. H. Ohlendorf, H. M. Einspahr, L. L. Clancy, M. A. Navia, B. M. McKeever, T. L. Nagabhushan, G. Nelson, A. McPherson, S. Koszelak, G. Taylor, D. Stammers, K. Powell, G. Darby, and C. E. Buff, *J. Crystal Growth* **110,** 302 (1991).

TABLE I
DATABASE SEARCH PARAMETERS

Macromolecule	**Crystallization conditions**
1. Macromolecule name	11. Crystallization method
2. Biological source	12. Macromolecule concentration
3. Molecular weight	13. Crystallization temperature
4. Subunit composition	14. pH of crystallization
5. Prosthetic group	15. Crystal growth time
6. Multiple crystal forms	16. Chemical additions to crystallization solution
Crystal data	**Reference**
7. Space group	17. Author
8. Unit cell dimensions	18. Year
9. *Z*, molecules/unit cell	19. Journal
10. Crystal density	
	Database
	20. Database cross-reference

range from reproducing published protocols to crystallizing a biological macromolecule that is unrelated to any that previously has been crystallized.[10] Discussed here are four general categories of problems that arise. These include strategies for (1) the crystallization of a previously crystallized biological macromolecule, (2) crystallization of a modified or mutant biological macromolecule for which the unmodified or wild-type biological macromolecule has been reported, (3) crystallization of a biological macromolecule that is homologous to previously crystallized macromolecule(s), and (4) *de novo* crystallization of a biological macromolecule that was not previously crystallized.

Previously Crystallized Macromolecules

The BMCD provides the information to reproduce the conditions required for crystallization of a biological macromolecule reported in the literature. At present many laboratories initiate crystallographic studies for structures that have been determined previously to assist in protein engineering, rational drug design, protein stability, and other studies. Usually the reported crystallization conditions are the starting points to initiate the crystallization trials. The crystallization of the biological macromolecule may be quite routine, but differences in the isolation and purification procedures, reagents, and crystallization methodology of laboratories can influence the outcome dramatically. The crystallization conditions in the database should be considered as a good starting point that may require experiments that vary pH, macromolecule and reagent concentrations, temperature, along with the crystallization method.

Variant Macromolecules or Macromolecule–Ligand Complexes

Determination of the structures of sequence variants, chemically modified macromolecules, or macromolecule–ligand complexes is becoming routine for crystallographic laboratories. Often the altered biopolymer sequence, chemical modification, or presence of the ligand does not interfere with crystal-packing interactions, significantly alter the conformation, or influence the solution properties of the macromolecule. If this is the case, the crystallization of the biological macromolecule may be similar to that of the wild-type, unmodified, or "free" biological macromolecule. This may be true even for quite radically modified variants such as the nine-residue, high-affinity calcium site deletion mutant of subtilisin BPN′.[24] This variant crystallized under conditions virtually identical to those required for crystallization of the wild-type macromolecule.

Homologous Biological Macromolecules

Frequently, crystallographic studies are initiated with a biological macromolecule related through sequence (and presumably structural) homology to a set of macromolecules that have been crystallized and that have had structures determined. The BMCD can be used to tabulate the crystallization conditions for the members of the family; one would then use this information to set up a series of experiments to initiate crystallization trials with the desired macromolecule. For example, an examination of crystallization conditions for Fab fragments of immunoglobulins shows that most crystals are grown using a protein concentration of 10 mg/ml, ammonium sulfate or polyethylene glycol 4000 to 6000 as the precipitant, at 20°, between pH 4.0 and 7.0.[10] These reagents and parameters would then represent a logical starting point for initiating a limited set of crystallization trials. If no crystals were obtained using this information, a more general procedure (outlined in the next section) would be warranted.

General Crystallization Procedure

The use of the BMCD has been incorporated into a more general procedure required for the crystallization of a unique biological macromolecule that has never been crystallized.[9,10] Briefly, in this procedure the purified biological macromolecule is concentrated (if possible) to at least 10

[24] T. Gallagher, P. Bryan, and G. L. Gilliland, *Protein Struct. Funct. Genet.* **16,** 205 (1993).

mg/ml and dialyzed into 0.005 to 0.025 *M* buffer at neutral pH or at a pH required to maintain solubility of the biopolymer. Other stabilizing agents such as EDTA and/or dithiothreitol may be included at low concentrations to stabilize the biological macromolecule during the crystallization trials. Using the BMCD, 3 to 10 of the reagents that have been most successful at inducing crystallization for a particular class of macromolecule (soluble proteins, DNA, viruses, membrane proteins, etc.) are selected to begin crystallization attempts. The data in the BMCD are also used to set the limits for parameters such as pH and temperature.

To illustrate the procedure, it will be assumed that the crystallization of a soluble protein is wanted. After one examines the data in the BMCD, one might select the precipitating agents, ammonium sulfate, polyethylene glycol 8000, 2-methyl-2,4-pentanediol, and sodium-potassium phosphate for the initial crystallization attempts, and experiments might be restricted to a pH range of 3.0 to 9.0 and temperatures ranging from 6 to 35°. The protein is then titrated with each of the selected reagents[4] at pH 4.0, 6.0, and 8.0 and at both cold room and room temperatures (~6 and 20°, respectively). This establishes the concentration ranges for the reagents for setting up hanging drop (or any other commonly used technique) experiments.[25] Next, separate sets of experiments that would sample the pH range in steps of 1.0 pH unit and reagent concentrations near, at, and above what might induce precipitation of the protein would be set up at temperatures of 6, 20, and 35°. The assessment of the results of experiments after periodic observations might show (with, for example, an abrupt precipitation at a particular reagent concentration, pH, and temperature) a need for finer sampling of any or all of the parameters near the observed discontinuity. If these crystallization trials are unsuccessful, the addition of small quantities of ligands, products, substrate, substrate analogs, monovalent or divalent cations, organic reagents, etc., to the crystallization mixtures of otherwise identical experiments may induce crystallization. If this does not prove fruitful, additional reagents may be selected with the aid of the BMCD, and new experiments initiated.

Besides the reagents mentioned above, the BMCD shows that about 10% of the soluble proteins crystallize at low ionic strength ($<0.2\ M$). Thus, microdialysis experiments that equilibrate the protein solutions against low ionic strength over time in a stepwise manner, over a pH range of pH 3.0 to 9.0 in steps of 0.5 to 1.0 should also be undertaken. It is also worthwhile to do microdialysis experiments at or near the isoelectric point of the protein, a point at which a protein is often the least soluble. As with the vapor diffusion experiments mentioned above, if crystallization does not

[25] A. Wlodawer and K. O. Hodgson, *Proc. Natl. Acad. Sci. U.S.A.* **72,** 398 (1975).

occur, the introduction of small quantities of ligands, products, substrate, substrate analogs, monovalent or divalent cations, organic reagents, etc., to the crystallization mixtures may facilitate crystal growth. Also, analogous to the vapor diffusion experiments, the search may be expanded to finer increments of pH if results warrant.

The example above illustrates how the data in the BMCD can be used to develop a general procedure for soluble proteins. The BMCD can be used to develop analogous procedures for other classes of biological macromolecules.

[29] Protein Data Bank Archives of Three-Dimensional Macromolecular Structures

By ENRIQUE E. ABOLA, JOEL L. SUSSMAN, JAIME PRILUSKY, and NANCY O. MANNING

Introduction

The growing reliance on the availability of structural data on macromolecules to help one understand biological processes highlights the importance of information resources such as the Protein Data Bank (PDB). Established at Brookhaven National Laboratory (BNL, Upton, NY) in 1971, the PDB has a 25-year history of service to a global community of researchers, educators, and students in a variety of scientific disciplines.[1–3] The common interest shared by this community is a need to access information that can relate the biological functions of macromolecules to their three-dimensional structures.

The PDB was started at the urging of members of the scientific community who in the late 1960s anticipated the growth in structural biology. The scientific and commercial importance of the resource was recognized even then by this community, who urged the adoption of a policy to encourage deposition, archiving, and distribution of data. Furthermore, it was deemed important to seek ways of having these data available free of charge to the community.

[1] F. C. Bernstein, T. F. Koetzle, G. J. B. Williams, E. F. Meyer, Jr., M. D. Brice, J. R. Rodgers, O. Kennard, T. Shimanouchi, and M. Tasumi, *J. Mol. Biol.* **112,** 535 (1977).

[2] E. E. Abola, F. C. Bernstein, S. H. Bryant, T. F. Koetzle, and J. Weng, Protein Data Bank. *In* "Crystallographic Databases—Information Content, Software Systems, Scientific Applications" (F. H. Allen, G. Bergerhoff, and R. Sievers, eds.), pp. 107–132. Data Commission of the International Union of Crystallography, Bonn, 1987.

[3] *Nature New Biol.* **233,** 223 (1971). [Crystallography, Protein Data Bank: Announcement].

0076-6879/97 $25.00

The PDB was established as an international collaboration between the groups headed by W. Hamilton at BNL and O. Kennard at the University of Cambridge (Cambridge, UK). This collaboration was expanded later to include centers at the Commonwealth Scientific and Industrial Research Organization (CSIRO, Clayton, Australia) and at the University of Osaka (Osaka, Japan). Today, this international collaboration is being strengthened by the establishment of data collection and distribution centers at the European Bioinformatics Institute (EBI; Hinxton Hall, UK), at the University of Osaka, and at the Weizmann Institute of Science (Rehovot, Israel). In addition to these centers, a number of official "PDB mirror" sites are operating Web and FTP sites, facilitating access to data within regional areas (see PDB Web home page for a list of current mirror sites). Funds to operate the PDB initially were provided by the U.S. National Science Foundation (NSF, Arlington, VA). More recently, the U.S. Department of Energy (Washington, DC), the U.S. National Institutes of Health (Bethesda, MD), and the U.S. National Library of Medicine (Bethesda, MD) along with the NSF provide funds to operate the resource.

Today, the challenge facing the PDB is to keep abreast of the increasing flow of data, to maintain the archive as error free as possible, and to organize and present the stored information in ways that facilitate data retrieval, knowledge exploration, and hypothesis testing, without interrupting current services. We discuss in this chapter various facets of our activities with the hope that this will help users understand the nature and scope of the data in the PDB archives and help them access these data. We also provide practical guidelines to depositors that will help them in depositing data with the PDB.

Contents of Protein Data Bank

Several pieces of information related to an entry are archived by the PDB (see Table I). In addition to the coordinate entry file, the PDB stores files related to the experiment such as structure factors, nuclear Overhaüser effect (NOE) restraints, and lists of chemical shifts. Also archived are

TABLE I
CONTENTS OF PROTEIN DATA BANK ARCHIVES

Coordinate entries (released and obsolete) with correction history
Raw coordinate data files
Structure factors
NMR–NOE and chemical shifts data files
Topology and parameter file used in refinement

auxiliary files used in structure analysis and refinement such as X-PLOR parameter and topology files. Currently, the archives are managed as a set of individual files, and each entry may have several associated files. The PDB is in the process of building a relational database, 3DBase, that will replace the current data management and access system. A description of 3DBase, including an outline of how users can access its contents is provided below.

A summary of the contents of the November 1996 PDB release is given in Table II. More than 5000 coordinate entries are available, and another 1000 currently either are being processed or are not to be released for up to 1 year at the request of the depositors. In 1996, the PDB receives 5 new entries per day and the rate of deposition is growing exponentially as shown in Fig. 1. It is expected that by the year 2000, the PDB will contain between 20,000 and 30,000 entries.

Coordinate entries in the PDB are stored in separate files, each of which reports the results of an experiment or analysis that elucidates the structure of proteins, necleic acids, polysaccharides, and other biological macromolecules. Although most of the data are generated from single crystal X-ray diffraction studies, a growing number of PDB entries are from nuclear magnetic resonance (NMR) studies (see Table II).

Files are distributed using the PDB data interchange format that was introduced in 1976 and has been used since then without significant changes. Entries are distributed as flat files consisting of fixed length records. Each of these records is identified by a tag word such as HEADER, COMPND and ATOM. The PDB records are divided into fields, most of which are

TABLE II
PROTEIN DATA BANK HOLDINGS LIST: NOVEMBER 15, 1996[a]

Molecule type
4500 proteins, peptides, and viruses
202 protein–nucleic acid complexes
361 nucleic acids
12 carbohydrates
Experimental technique
142 theoretical modeling
714 NMR
4219 diffraction and other
576 Structure Factor files
169 NMR Restraint files

[a] A total of 5075 released atomic coordinate entries, by molecule type, were available as of November 1996.

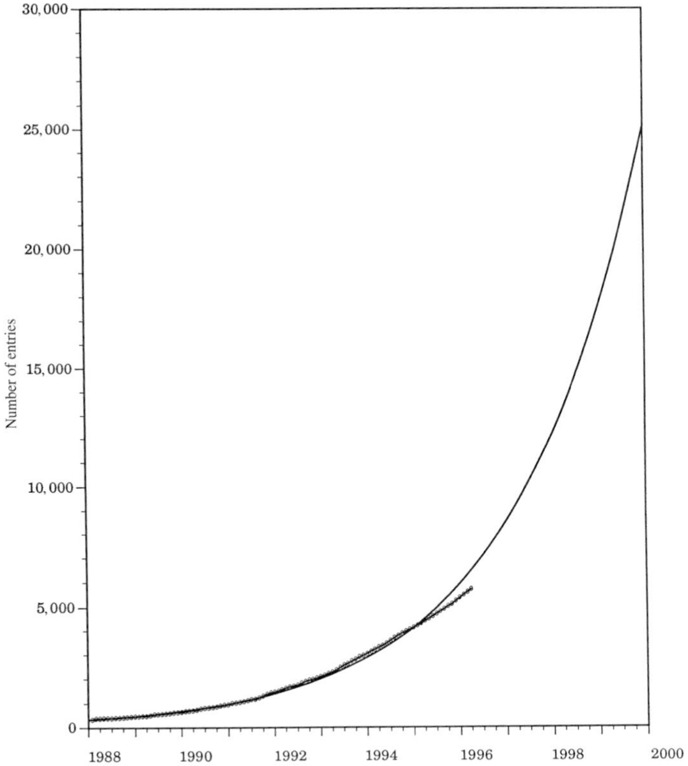

FIG. 1. Total number of atomic coordinate entries in the PDB extrapolated to year 2000 (exponential fit to 1988–1996 data).

fixed in length. More recently a number of records with varying field lengths and terminated by a character token (e.g., semicolon ";") have been introduced. The contents of the PDB file along with a full description of the format and general rules used by the PDB in representing biological and chemical data are given in the *Protein Data Bank Contents Guide*.[4] Table III lists all PDB record types in use in 1996. The REMARK section of PDB entries contains not only free text comments but also a number of standard tables such as those containing matrices needed to generate complete viral particles from the icosahedral asymmetric unit supplied by the depositor.

[4] "Protein Data Bank Contents Guide: Atomic Coordinate Entry Format Description," version 2.1, October 25, 1996. Available anonymous FTP: ftp.pdb.bnl.gov, and at URL http://www.pdb.bnl.gov.

TABLE III
PROTEIN DATA BANK RECORD TYPES[a]

Section	Description	Record type
Title	Summary descriptive remarks	HEADER, OBSLTE, TITLE, CAVEAT, COMPND, SOURCE, KEYWDS, EXPDTA, AUTHOR, REVDAT, SPRSDE, JRNL
Remark	Bibliography, refinement, annotations	REMARKs 1, 2, 3 and others
Primary structure	Peptide and/or nucleotide sequence and relationship between PDB sequence and that found in sequence database(s)	DBREF, SEQADV, SEQRES, MODRES
Heterogen	Description of nonstandard groups	HET, HETNAM, HETSYN, FORMUL
Secondary structure	Description of secondary structure	HELIX, SHEET, TURN
Connectivity annotation	Chemical connectivity	SSBOND, LINK, HYDBND, SLTBRG, CISPEP
Miscellaneous features	Features within macromolecule	SITE
Crystallographic	Description of crystallographic cell	CRYST1
Coordinate transformation	Coordinate transformation operators	ORIGXn, SCALEn, MTRIXn, TVECT
Coordinate	Atomic coordinate data	MODEL, ATOM, SIGATM, ANISOU, SIGUIJ, TER, HETATM, ENDMDL
Connectivity	Chemical connectivity	CONECT
Bookkeeping	Summary information, end-of-file marker	MASTER, END

[a] Various sections of a PDB coordinate entry and the records comprising them.

3DBase: Relational Database Management System for Protein Data Bank

In 1994, the PDB started work on building a new relational database for managing and accessing the contents of the data bank. The new database, 3DBase, is constructed with the SYBASE[5] Relational Database Management System (RDBMS), the Object-Protocol Model (OPM), and the OPM data management tools[6] developed by the Markowitz group at Lawrence Berkeley National Laboratory (Berkeley, CA). SYBASE provides a powerful and robust environment for data management, the OPM tools allow rapid development of SYBASE databases, and the object-oriented view

[5] "SYBASE SQL Server," Unix version 10.0. Sybase, Inc., Emeryville, California, 1994.

[6] I. A. Chen and V. M. Markowitz, *Inf. Syst.* **20**(5), 393 (1995); article and related information available at URL:http://gizmo.lbl.gov/DM_TOOLS/OPM/opm.html.

of the OPM provides a scientifically intuitive representation of data. For example, a purely relational view of the data requires the construction of several tables to store all the information related to the coordinates of a residue. Queries that require access to these data will then have to be constructed by naming and joining data found in each of these tables. In contrast, only one object definition is required to store and access the same data in an object-oriented view.

This development effort attempts to address the needs of the diverse user community served by the PDB. A conceptual view (also referred to as a *conceptual schema*) of the data was developed that supports queries by those interested in answering both crystallographic and molecular biology questions. The system is designed to federate 3DBase easily with other biological databases. It is expected that federation will permit complex queries to be submitted to the database, returning a composite answer built from a set of diverse databases. Interoperability is addressed through the use of a schema that shares with other OPM-based databases and supports a variety of data interchange formats in the query results. In addition to providing users with a powerful environment to do complex *ad hoc* queries, 3DBase also will facilitate management of the growing archive, which is expected to contain more than 20,000 structural reports by the year 2000.

This work is being done as a collaboration among the following groups: The Protein Data Bank (Brookhaven National Laboratory), Bioinformatics Unit (Weizmann Institute of Science), OPM Data Management Tools Project (Lawrence Berkeley National Laboratory), and Genome Database (GDB, Johns Hopkins University, Baltimore, MD).

Schema Development

The OPM is a semantic data model that includes constructs that are powerful enough to represent the diversity and complexity of data found in PDB entries. The OPM has constructs such as object class, object attribute, class hierarchy and inheritance, and derived attribute. In addition, multiple views of the data are constructed in the OPM by using derived classes. This mechanism allows others to develop their own conceptual view of the data without having to alter the underlying database. The schema for 3DBase has been developed using the OPM, and is available for perusal through the PDB WWW (World Wide Web) home page. Among its notable features is a description of the coordinate data set from two perspectives. The object class *Experiment* provides users with the classic view of a PDB entry, which is a report of crystallographic or NMR analysis. An alternative view is presented in the class *Aggregate,* which describes the macromolecule and its complexes with ligands, or other macromolecules. A clear example that demonstrates the differences between these classes is the case of the hemoglobin molecule. The *Experiment* object contains the coordinates for the crystallographic asymmetric unit, which is usually a dimer of α and β subunits. The full tetramer generated by using a crystallographic twofold symmetry operator will, however, be presented in the *Aggregate* object. The latter case is normally what molecular biologists are interested in when accessing PDB entries. Those wanting to do crystal-packing studies or further crystallographic refinement will need access to the *Experiment* object to obtain the asymmetric unit.

In 3DBase, literature citation data are being loaded into the citation database (CitDB) of references that was developed by the GDB.[7] A pointer to the appropriate entry in the CitDB is loaded in the *Experiment* object of 3DBase. This is an example of the strategy that the PDB is following in linking to external databases. The CitDB will be managed as a federation run by a number of database centers that include the GDB and PDB. There are several advantages to this scenario. By sharing the schema and management of the citation databases, access to information stored in each of the databases via the bibliographic citation becomes straightforward. Duplication of effort is also minimized. Today, it is still common to have several public databases build and maintain their own bibliographic databases. This will no longer be economically feasible with the expected rapid growth in database size.

A notable feature of the 3DBase schema is the inclusion of an object class that allows users, depositors, and other editors to add their own annotation to objects. This can, for example, be used to attach to an *Experiment* object the output of a program that describes error estimates using a novel data-checking technique. The PDB and its advisers, along with the rest of the community, will have to arrive at a workable editorial policy before general use of this annotation mechanism is permitted.

Building Semantic Links to External Data Sources

Links to contents of sequence databases are provided in 3DBase via the *PrimarySeq* and *SeqAdv* classes. These classes form another set of objects that link 3DBase objects to external databases. Representing, building, and maintaining these links will be one of the primary tasks of the PDB in the coming years. There are several issues that must be addressed for this effort to succeed. Data representation issues are foremost. Each database uses different data models to represent and store information. Semantic contents are rarely the same; for example, the primary sequence data stored in

[7] K. H. Fasman, A. J. Cuticchia, and D. T. Kingsbury, *Nucleic Acids Res.* **22,** 3462 (1994).

sequence databases such as SWISS-PROT[8] (Medical Biochemistry Department, University of Geneva, Switzerland and European Bioinformatics Institute, Hinxton, England) or Protein Information Resource[9] (PIR, National Biomedical Research Foundation, Washington, DC) are presented using a view that differs significantly from that used by the PDB.

In general, PIR and SWISS-PROT entries contain information on the naturally occurring wild-type molecules. Each entry normally contains the sequence of one gene product, and some entries include the complete precursor sequence. Annotation is provided to describe residue modifications. In both databases, the residue names used are limited to the 20 standard amino acids.

In contrast, PDB entries contain multichain molecules with sequences that may be wild type, variant, or synthetic. Sequences also may have been modified through protein-engineering experiments. A number of PDB entries report structures of domains cleaved from larger molecules.

The *PrimarySeq* object class was designed to account for these differences by providing explicit correlations between contiguous segments of sequences as given in PDB ATOM records and PIR or SWISS-PROT entries. Several cases are easily represented using this class. Molecules containing heteropolymers may be linked to different sequence database entries. In some cases, such as those PDB entries containing immunoglobulin Fab fragments, each PDB chain may be linked to several different SWISS-PROT entries.

This facility is needed because these databases represent sequences for the various immunoglobulin domains as separate entries. *PrimarySeq* also should be able to represent molecules engineered by altering the gene (fusing genes, altering sequences, creating chimeras, or circularly permuting sequences). In addition, it will be possible to link segments of the structure to entries in motif databases (e.g., PROSITE,[10] BLOCKS[11]).

Initial building of these links is straightforward, and requires analysis of a few entries coming out of a sequence comparison search using the program FASTA[12] or BLAST[13] against the sequence databases. An issue

[8] A. Bairoch and B. Boeckmann, *Nucleic Acids Res.* **22,** 3578 (1994); available at URL: http://expasy.hcuge.ch/sprot/top.html.

[9] D. G. George, W. C. Barker, H. W. Mewes, F. Pfeiffer, and A. Tsugita, *Nucleic Acids Res.* **22,** 3569 (1994); available at URL: http://www-nbrf.georgetown.edu/pir.

[10] A. Bairoch and P. Bucher, *Nucleic Acids Res.* **22,** 3583 (1994); available at URL: http://expasy.hcuge.ch/.

[11] S. Henikoff and J. G. Henikoff, *Genomics* **19,** 97 (1994); available at URL: http://blocks.fhcrc.org/.

[12] W. R. Pearson and D. J. Lipman, *Proc. Natl. Acad. Sci. U.S.A.* **85,** 2444 (1988).

[13] S. F. Altschul, W. Gish, W. Miller, E. W. Myers, and D. J. Lipman, *J. Mol. Biol.* **215,** 403 (1990).

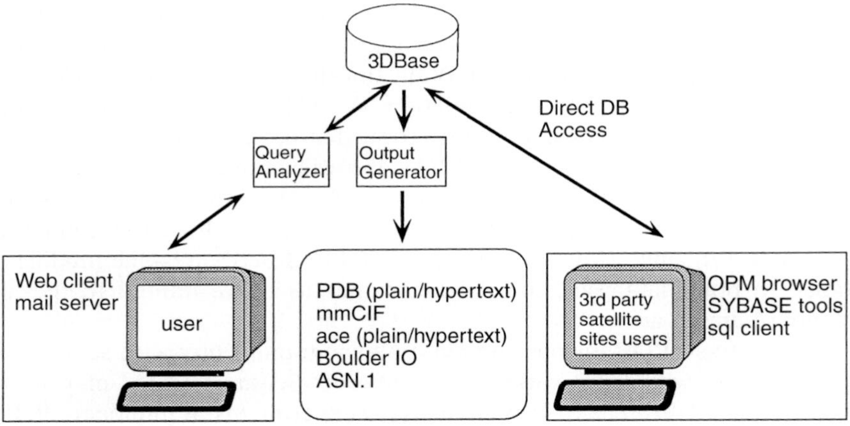

FIG. 2. Access to 3DBase.

that must be resolved in the long run will be the updating of these links as new experimental evidence is encountered, leading to a correction in either database. Both PIR and SWISS-PROT have similar problems as they build pointers to PDB entries. To help obviate these difficulties the PDB has agreed to establish a closer interaction among the databases. A protocol is being set up that will broadcast to each database changes that occur that could, in turn, affect specific entries.

Accessing Data in 3DBase

User queries to 3DBase will be via the Internet, using general-purpose graphical user interfaces such as Mosaic and Netscape. Access also will be possible through the use of software developed by third parties (commercial developers). As diagrammed in Fig. 2, user queries will be addressed to the Query Analyzer (PDB-QA), a program module running at the server site that will parse queries and pass them on to 3DBase. Query results will be returned through the Output Generator (PDB-OG) in the format requested by the user. Queries placed over the network generally will be in the form of uniform resource locators (URLs), which are easily generated from HyperText links, HyperText markup language (HTML)-based forms, or by use of programs or scripts employing the National Center for Supercomputing Applications libraries[14] for more sophisticated applications. As part of the query, the user may specify the format of the response, as we

[14] National Center for Supercomputing Applications, University of Illinois at Urbana-Champaign, Illinois. Available at URL: http://www.ncsa.uiuc.edu.

do at the present time in the PDB 3DB Browser. The response will be frequently in the form of an HTML document, but it can also be a PDB- or Crystallographic Information File (CIF)-formatted file.[15] The information returned may be either a complete or partial entry, and may include information from linked databases or external programs.

A 3DBase browser has been built using the Letovky Genera system.[16] Users specify search criteria by filling out an HTML form. Software at BNL processes this form and generates the required structured query language (SQL). System performance is improved by using stored SYBASE SQL procedures that access each predefined object. The fields available are similar to those in our PDBBrowse program, and answer most of the questions that users have been asking.

For those familiar with (or willing to learn about) the OPM protocol, access to the object layer will be provided using a high-level OPM-based query language. As part of the PDB open database policy, direct access to the underlying RDBMS will be allowed and actively supported. These queries are not parsed by the PDB-QA module, so better response time can be expected. This provides third-party developers with the opportunity either to incorporate SQL clients in their products or to learn more of the OPM protocol and, thereby, gain access to all of the benefits that the Object model affords (e.g., active external links, programs). As depicted in Fig. 2, the output generator will return query results using a variety of data interchange formats. The PDB will continue to support its current format for the foreseeable future. We plan also to extend this format to allow us to represent objects being stored in 3DBase. In addition, a "raw format" is being provided that returns an attribute/value pair. This form is easily parsed and is more compact than the PDB format.

Submitting Data to Protein Data Bank

The PDB is evolving to operate as a direct-deposition archive, providing mechanisms allowing depositors to load data with minimal staff intervention. This strategy is essential if the PDB is to meet present projections of exponential growth in depositions against a fixed staff size. This is particularly challenging owing to the complexity of the data being handled, the need for a common viewpoint of the entry description, and the community requirement that these data be accessible immediately on receipt.

[15] P. M. D. Fitzgerald, H. M. Berman, P. E. Bourne, and K. Watenpaugh, *American Crystallographic Association Annual Meeting* [*program and abstracts*], Ser.2, **21,** 33 (1993).

[16] S. I. Letovsky, "Genera" (computer program, 1994). URL: http://gdbdoc.gdb.org/letovsky/genera.

With direct deposition, there will be a concomitant need to increase the power of data validation procedures. These procedures must reflect current models for identifying errors and must be as complete as possible. Quality control issues assume a more central and difficult role in direct deposition strategies. Distributed data must be of the highest quality; otherwise users will lose their trust in the archived data and will have to revalidate data received from the PDB before using them, clearly an unproductive scenario.

Current Data Deposition Procedures

Since its inception in 1971, the method followed by the PDB for entering and distributing information has paralleled the review-and-edit mode used by scientific journals. The author submits information that is converted into a PDB entry and is run against PDB validation programs by a PDB processor. The entry and the output of the validation suite are then evaluated by a PDB scientific staff member, who completes the annotations and returns the entry to the author for comment and approval. Table IV summarizes checks included in the current data validation suite. Corrections from the author are incorporated into the entry, which is reanalyzed and validated before being archived and released.

Originally data flow was a manual system, designed for a staff of 1–2 scientists, and a deposition rate of about 25–50 entries per year. One person processed an entry from submission through its release. By the late 1980s, when the first steps at automation were being introduced, running the

TABLE IV
DATA VALIDATION WITH CURRENT SYSTEM

Class	What is checked
Stereochemistry	Bond distances and angles, Ramachandran plot (dihedral angles), planarity of groups, chirality
Bonded/nonbonded interactions	Crystal packing, unspecified inter- and intraresidue links
Crystallographic information	Matthews coefficient, Z value, cell transformation matrices
Noncrystallographic transformation	Validity of noncrystallographic symmetry
Primary sequence data	Correlations with sequence databases
Secondary structure	Generated automatically or visually checked
Heterogen groups	Identification, geometry, and nomenclature
Miscellaneous checks	Solvent molecules outside hydration sphere, syntax checks, internal data consistency checks

validation programs took about 4 hr per entry. Today, the same step, which includes a vastly improved set of validation programs, takes about 1 min.

The current deposition load of ~100 entries a month is handled by about 10 staff members who annotate and validate entries. The process is a production line in which checking is repeated at various steps to ensure that errors and inconsistencies in data representation are minimized. Prior to June 1994, a significant number of depositions required that administrative staff manually input information provided in a deposition form. Introduction of the current Electronic Deposition Form, together with a new parsing program, has greatly reduced hand entry of information.

Today, most of the processing time is spent resolving data representation issues and ensuring that outliers are identified and annotated. The most troublesome areas are consistently those involving handling of heterogens, resolving crystal-packing issues, representing molecules with noncrystallographic symmetry, and resolving conflicts between the submitted amino acid sequence and that found in the sequence databases. Publications and other references are sometimes consulted to verify factual information such as crystal data, biological details, and reference information. Although much improved over those used in 1991, processing programs still allow errors to pass undetected through the system, requiring a visual check of all entries. The PDB continually improves these programs, and also acquires software from collaborators to address deficiencies that both we and our users have identified. In addition, the PDB now has formed a quality control group that will be identifying sources of errors and recommending steps to improve data quality.

Development of Automatic Deposition and Validation

The PDB must overcome many challenges for direct deposition to work. In a workshop held to assess the needs of PDB users, crystallographers and NMR spectroscopists were unanimous in their desire for a system that did not require additional work on their part when depositing data. On the other hand, consumers (who included these same depositors) were vocal in their desire for entries to contain more information than is currently available within the PDB. The PDB is striving to develop a suite of deposition and validation programs that accommodates these somewhat conflicting desires while ensuring that the archives maintain the highest standard of accuracy.

A considerable variety of information is archived about each structure, which must be supplied by the authors. The new PDB Web-based data deposition program, AutoDep, simplifies the process (see Table V). It includes a convenient and interactive electronic deposition protocol that

TABLE V
IMPORTANT HIGHLIGHTS OF AutoDep

Program allows author to fill in form automatically from existing PDR entry or from previous deposition. AutoDep enters data from designated file to appropriate fields in new form. Author need only update fields to reflect new structure
X-Ray structural refinement software is available to write PDB records that can be merged automatically into the deposition form. For example, new releases of X-PLOR and SHELXL write refinement details as PDB records that will be read by program and entered in relevant sections. PDB is continuing to work with authors of various programs and anticipates that increasing numbers of programs will be integrated with PDB
Each session has Help files, examples, and links to related documentation and useful URLs to support author during AutoDep session
At any time during AutoDep session, Deposition Form or resultant header portion of PDB file can be viewed to check progress
AutoDep session can be interrupted at any time and resumed later. Session ID number and password must be recorded to continue with same deposition
When author is satisfied with completed Deposition Form, Submit button is provided to initiate following: 1. Coordinates are passed through syntax checker 2. If they fail, depositor is asked to correct problem and resubmit coordinates 3. If they pass, depositor immediately is sent acknowledgment letter containing PDB ID code 4. Entry enters PDB processing flow

guides the author in providing information. It also contains tools for data verification and validation, and is able to flag errors in syntax or spelling. The form requests approximately 1000 items, including a description of the experiment and the molecule under study. Steps are being taken to help ease the burden of filling out this form. For example, the program can fill in fields using data from an existing PDB entry or from any file containing PDB format-compliant records. These data then can be modified to reflect the contents of the new deposition. Checks against other databases are an important and evolving part of this process. Thus, names of organisms are checked against the taxonomy database of the National Center for Biotechnology Information (NCBI, Bethesda, MD),[17] chemical names against IUPAC (International Union of Pure and Applied Chemistry[18])

[17] National Center for Biotechnology Information, National Library of Medicine, National Institutes of Health, Bethesda, Maryland (producer). Available URL: http://www.ncbi.nih.gov. Available anonymous FTP: ncbi.nlm.nih.gov. Directory: /repository/taxonomies.

[18] C. Liebecq (ed.), "Biochemical Nomenclature and Related Documents: A compendium," 2nd Ed. International Union of Biochemistry and Molecular Biology, Portland Press, London, 1992.

nomenclature tables, and author names and citations against MEDLINE[19] (CitDB when it becomes available). FASTA/BLAST programs are run against the SWISS-PROT and PIR databases to verify protein sequences. Links between the PDB entry and these databases are established in the process. To handle the increasing number of entries with nonstandard residues (*heterogens*), a standard residue and heterogen dictionary has been developed for use in the data entry and checking process. The PDB is also adopting programs developed by the Cambridge Crystallographic Data Center[20] (CCDC, Cambridge, UK), and elsewhere, to handle heterogens automatically for use in AutoDep.

In addition to the deposition form that is filled out by AutoDep, authors are requested to submit the coordinate data entry and other experimental data files for processing and archiving. Facilities are provided by AutoDep that simplify this process. An FTP script is provided that uploads author-specified local filenames to the PDB server site.

The completed form is then converted automatically into a file in PDB format and, along with the coordinate data, is submitted to a set of validation programs for checking and further annotation. These programs are designed to check (1) the quality, consistency, and completeness of the experimental data; (2) possible violations of physical or stereochemical constraints (e.g., no two atoms in the same place, appropriate bond angles); (3) compliance with our data dictionary (syntax checks); and (4) in the near future, the correspondence of the experimental data to the derived structure. Development of the validation suite will continue to evolve with advice from the community and encompass programs currently in use, written both within and outside the PDB.

The validation software automatically generates, and includes in the entry, measures of data quality and consistency, as well as annotations giving details of apparent inconsistencies and outliers from normal values. This output is returned to the depositor for review. Entries whose data quality and consistency meet appropriate standards may then be sent by the depositor directly for final review by the PDB staff and entry into the database. Entries that do not pass the quality and consistency checks may be revised by the depositor to correct inadvertent errors; alternatively, more experimental work may be needed to resolve problems uncovered.

Apparent inconsistencies or outliers may remain in a submitted entry, provided these are explained by the depositor in an annotation. In the most

[19] MEDLINE (on-line and CD-ROM). National Library of Medicine, National Institutes of Health, Bethesda, Maryland (producer). Available: NLM, DIALOG, BRS, SilverPlatter.

[20] F. H. Allen, J. E. Davies, J. J. Galloy, O. Johnson, O. Kennard, C. F. Macrae, E. M. Mitchell, G. F. Mitchell, J. M. Smith, and D. G. Watson, *J. Chem. Inf. Comput. Sci.* **31,** 187 (1991).

interesting cases, unusual features are a valid and important part of the structure. However, all such entries will be reviewed for possible errors by PDB staff, who may discuss any important issues with the depositor. The PDB staff will then forward acceptable entries to the database.

To make automatic deposition as easy as possible, the PDB is working with developers of software commonly used by our depositors. By modifying these programs to produce compliant data files and performing validation and consistency checks before submission, it may be possible to bypass most of the tedious steps in deposition. We are already working with A. Brünger (Yale University, New Haven, CT) to use procedures available through X-PLOR[21] to replace part of the validation suite for structures produced by X-ray crystallography and NMR. Diagnostic output will be included automatically as annotations in the entry. A limited version of X-PLOR will be available from BNL to all depositors for validation purposes only.

Validation of coordinate data against experimental X-ray crystallographic data requires access to structure-factor data, which are requested by the PDB, the International Union of Crystallography (IUCr), and some journals, but are not always supplied by the depositor. We are working toward building consensus in the community that structure-factor data are a necessary component of deposits of structures derived by X-ray crystallography. Statistics such as number of F and R values vs sin θ/λ will be calculated and included in the PDB entry as annotation for the experiment.

To make it easier for depositors to submit structure factors (as well as to exchange these data between laboratories), the PDB, in close collaboration with a number of macromolecular crystallographers, has developed a standard interchange format for these data. This standard is in CIF (Crystallographic Information File)[15,22] and was chosen both for simplicity of design and for being clearly self-defining, i.e., the file contains sufficient information for the file to be read and understood by either a program or a person. Details of this format are available through the PDB WWW server.

A consensus is still developing in the NMR community as to what types of experimental data should be deposited and what kinds of validation and consistency checks should be performed. Structural data produced by other methods also may have special features that should be archived or checked, for example, the sequence alignment used for modeling studies. Require-

[21] A. T. Brünger, "X-PLOR, Version 3.1: A System for X-Ray Crystallography and NMR." Yale University Press, New Haven, Connecticut, 1992.

[22] S. R. Hall, F. H. Allen, and I. D. Brown, *Acta Crystallogr.* **A47,** 655 (1991); related information available at URL: http://www.iucr.ac.uk/cif/home.html.

ments for the types of data to be deposited and proper ways of checking the validity and consistency of the data will be developed in cooperation with the experimental community for each category of structure data archived by the PDB.

[30] Macromolecular Crystallographic Information File

By PHILIP E. BOURNE, HELEN M. BERMAN, BRIAN MCMAHON, KEITH D. WATENPAUGH, JOHN D. WESTBROOK, and PAULA M. D. FITZGERALD

Introduction

The Protein Data Bank (PDB) format provides a standard representation for macromolecular structure data derived from X-ray diffraction and nuclear magnetic resonance (NMR) studies. This representation has served the community well since its inception in the 1970s[1] and a large amount of software that uses this representation has been written. However, it is widely recognized that the current PDB format cannot express adequately the large amount of data (content) associated with a single macromolecular structure and the experiment from which it was derived in a way (context) that is consistent and permits direct comparison with other structure entries. Structure comparison, for such purposes as better understanding biological function, assisting in the solution of new structures, drug design, and structure prediction, becomes increasingly valuable as the number of macromolecular structures continues to grow at a near exponential rate. It could be argued that the description of the required content of a structure submission could be met by additional PDB record types. However, this format does not permit the maintenance of the automated level of consistency, accuracy, and reproducibility required for such a large body of data.

A variety of approaches for improved scientific data representation is being explored.[2] The approach described here, which has been developed under the auspices of the International Union of Crystallography (IUCr), is to extend the Crystallographic Information File (CIF) data representation used for describing small-molecule structures and associated diffraction experiments. This extension is referred to as the macromolecular Crystallo-

[1] F. C. Bernstein, T. F. Koetzle, G. J. B. Williams, E. F. Meyer, Jr., M. D. Brice, J. R. Rogers, O. Kennard, T. Shimanouchi, and M. Tasumi, *J. Mol. Biol.* **112,** 535 (1977).
[2] IEEE Metadata: *http://www.llnl.gov/liv_comp/metadata/* (1996).

0076-6879/97 $25.00

graphic Information File (mmCIF) and is the subject of this chapter. We briefly cover the history of mmCIF, similarities to and differences from the PDB format, contents of the mmCIF dictionary, and how to represent structures using mmCIF. The mmCIF home page (mmCIF[3]) contains a historic description of the development of the dictionary, current versions of the dictionary in text and HyperText markup language (HTML) formats, software tools, archives of the mmCIF discussion list, and a detailed online tutorial.[4]

Background

The CIF was developed to describe small-molecule organic structures and the crystallographic experiment by the IUCr Working Party on Crystallographic Information at the behest of the IUCr Commission on Crystallographic Data and the IUCr Commission on Journals. The result of this effort was a core dictionary of data items[4a] sufficient for archiving the small-molecule crystallographic experiment and its results.[5,6] This core dictionary was adopted by the IUCr at its 1990 Congress in Bordeaux. The format of the small-molecule CIF dictionary and the data files based on that dictionary conform to a restricted version of the Self-Defining Text Archive and Retrieval (STAR) representation developed by Hall and others.[7,8] STAR permits a data organization that may be understood by analogy with a spoken language (Fig. 1). STAR defines a set of encoding rules similar to saying the English language is composed of 26 letters. A dictionary definition language (DDL) is defined that uses those rules and that provides a framework from which to define a dictionary of the terms needed by the discipline. The DDL is a computer-readable way of declaring that words are made up of arbitrary groups of letters and that words are organized into sentences and paragraphs. The DDL provides a convention for naming and defining data items within the dictionary, declaring specific attributes of those data items (e.g., a range of values and the data type), and for declaring relationships between data items. In other words, the DDL defines the format of the dictionary and any new words that are added must conform to that format. Just as words are constantly being added to a language, data items will be added to the dictionaries as the discipline evolves. The

[3] mmCIF: *http://ndbserver.rutgers.edu/mmcif/*(1996).

[4] P. E. Bourne, *http://www.sdsc.edu/pb/cif/overview.html* (1996).

[4a] A *data item* refers to a data name and its associated value, as discussed subsequently.

[5] S. R. Hall, F. H. Allen, and I. D. Brown, *Acta Crystallogr.* **A47,** 655 (1991).

[6] IUCr: *ftp://ftp.iucr.ac.uk/pub/cifdic.c91* (1996).

[7] A. Cook and S. R. Hall, *J. Chem. Inf. Comput. Sci.* **31,** 326 (1992).

[8] S. R. Hall and N. Spadaccini, *J. Chem. Inf. Comput. Sci.* **34,** 505 (1994).

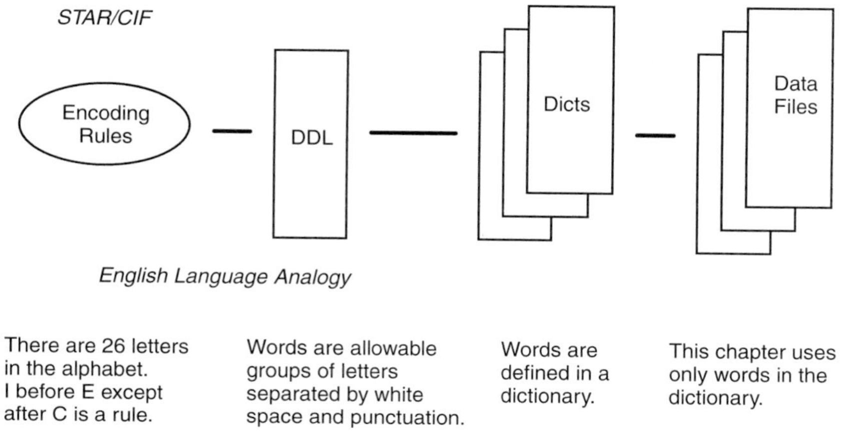

FIG. 1. Components of the STAR/CIF data representation and their analogy to a natural language.

STAR encoding rules and the DDL are being used to develop a variety of dictionaries and reference files, for example, the powder diffraction dictionary, the modulated structures dictionary, a file of ideal geometry for amino acids, and an NMR dictionary. This extensibility is attractive because the same basic reading and browsing software (context-based tools) can be used irrespective of the data content. Data files (this chapter is an example in our language analogy) are composed of data items found in the dictionaries.

In 1990, the IUCr formed a working group to expand the core dictionary to include data items relevant to the macromolecular crystallographic experiment. Version 1.0 of the mmCIF dictionary (Fitzgerald *et al.*[9] mmCIF[3]), which encompasses many data items from the current core dictionary (IUCr[10]), is in the final stage of review by COMCIFs, the IUCr appointed committee overseeing CIF developments. This dictionary has been written using DDL version 2.1.1 (Westbrook and Hall[11]), which is significantly enhanced, yet upwardly compatible with DDL version 1.4 (IUCr[12]) currently used for the small-molecule dictionary.

[9] P. M. D. Fitzgerald, H. M. Berman, P. E. Bourne, B.McMahon, K. Watenpaugh, and J. D. Westbrook, *Acta Crystallogr.* (*Suppl.*) **A52,** C575 (1996).
[10] IUCr: *ftp://ftp.iucr.ac.uk/pub/cifdic.c96* (1996).
[11] J. D. Westbrook and S. R. Hall, *http://ndbserver.rutgers.edu/mmcif/ddl/* (1995).
[12] IUCr: *ftp://ftp.iucr.ac.uk/pub/ddldic.c95* (1995).

Considerations in Development of mmCIF Dictionary

In developing version 1.0 of the mmCIF dictionary we made the following decisions.

1. Every field of every PDB record type should be represented by a data item if that PDB field is important for describing the structure, the experiment that was conducted in determining the structure, or the revision history of the entry. It is important to note that it is straightforward to convert a mmCIF data file to a PDB file without loss of information because all information is parsable. It is not possible, however, to automate completely the conversion of a PDB file to an mmCIF, because many mmCIF data items either are not present in the PDB file or are present in PDB REMARK records that in some instances cannot be parsed. The content of PDB REMARK records are maintained as separate data items within mmCIF so as to preserve all information, even if that information is not parsable.
2. Data items should be defined such that all the information described in the Materials and Methods section of a structure paper could be referenced. This includes major features of the crystal, the diffraction experiment, phasing methodology, and refinement.
3. Data items should be defined such that the biologically active molecule could be described as well as any structural subcomponents deemed important by the crystallographer.
4. Atomic coordinates should be representable as either orthogonal (in Ångströms) or fractional.
5. Data items should be provided to describe final *h*,*k*,*l* values, including those collected at different wavelengths.
6. For the most part data items specific to an NMR experiment or modeling study would not be included in version 1.0. Exceptions are the data items that summarize the features of an ensemble of structures and permit the description of each member of the ensemble.
7. Crystallographic and noncrystallographic symmetry should be defined.
8. A comprehensive set of data items for providing a higher order structure description, for example, to cover supersecondary structure and functional classification, was considered beyond the scope of version 1.0.
9. Data items should be present for describing the characteristics and geometry of canonical and noncanonical amino acids, nucleotides, and heterogen groups.
10. Data items should be present that permit a detailed description of the chemistry of the component parts of the macromolecule, including the provision for two-dimensional (2D) projections.

11. Data items should be present that provide specific pointers from elements of the structure (e.g., the sequence, bound inhibitors) to the appropriate entries in publicly available databases.

12. Data items should be present that provide meaningful 3D views of the structure so as to highlight functional and structural aspects of the macromolecule.

On the basis of the preceding decisions, a mmCIF dictionary with approximately 1500 data items (including those data items taken from the small-molecule dictionary) was developed. It is not expected that all relevant data items will be present in each mmCIF data file. What data items are mandatory to describe the structure and experiment adequately need to be decided by community consensus.

Comparing mmCIF Data Files with PDB Files

The format of an mmCIF containing structural data can best be introduced through analogy with the existing PDB format. A PDB file consists of a series of records each identified by a keyword (e.g., HEADER, COMPND) of up to six characters. The format and content of fields within a record are dependent on the keyword. An mmCIF, on the other hand, always consists of a series of *name-value* pairs (a data item) defined by STAR, where the data name is preceded by a leading underscore (_) to distinguish it from the data value. Thus, every field in a PDB record is represented in mmCIF by a specific data name. The PDB HEADER record,

```
HEADER    PLANT SEED PROTEIN                          11-OCT-91   1CBN
```

becomes

```
_struct.entry_id                          '1CBN'
_struct.title                             'PLANT SEED PROTEIN'

_struct_keywords.entry_id                 '1CBN'
_struct_keywords.text                     'plant seed protein'

_database_2.database_id                   'PDB'
_database_2.database_code                 '1CBN'

_database_PDB_rev.rev_num                  1
_database_PDB_rev.date_original           '1991-10-11'
```

The *name-value* pairing represents a major departure from the PDB file format and has the advantage of providing an explicit reference to each item of data within the data file, rather than having the interpretation left to the software reading the file. The *name* matches an entry in the mmCIF dictionary where characteristics of that data item are defined explicitly. When multiple values for the same data item exist, the name of the data item or items concerned is declared in a header and the associated values follow in strict rotation. This is a STAR rule referred to as a *loop_* construct. This *loop_* construct is illustrated in the representation of atomic coordinates.

```
     loop_
    _atom_site.group_PDB
    _atom_site.type_symbol
    _atom_site.label_atom_id
    _atom_site.label_comp_id
    _atom_site.label_asym_id
    _atom_site.label_seq_id
    _atom_site.label_alt_id
    _atom_site.cartn_x
    _atom_site.cartn_y
    _atom_site.cartn_z
    _atom_site.occupancy
    _atom_site.B_iso_or_equiv
    _atom_site.footnote_id
    _atom_site.auth_seq_id
    _atom_site.id
ATOM N  N   VAL  A  11 . 25.369  30.691  11.795  1.00  17.93 . 11 1
ATOM C  CA  VAL  A  11 . 25.970  31.965  12.332  1.00  17.75 . 11 2
ATOM C  C   VAL  A  11 . 25.569  32.010  13.881  1.00  17.83 . 11 3
#           [data omitted]
```

Note that the *name* construct is of the form *_category.extension.* The category explicitly defines a natural grouping of data items such that all data items of a single category are contained within a single *loop_*. There is no restriction on the length of *name,* beyond the record length limit of 80 characters mentioned below, and while there is no formal syntax within *name* beyond the category and extension separated by a period, by convention the category and extension are represented as an informal hierarchy of parts, with each part separated by an underscore (_). The *names _atom_site.label_atom_id* and *_atom_site.label_comp_id* are examples.

Questions that arise concerning the separation of data names and data values are solved with some additional syntax. For example, what if the data value contains white space, an underscore, or runs over several lines? Similarly, what if a value in a *loop_* is undefined or has no meaning in the context in which it is defined? The following syntax rules, which are a more restricted set of rules than permitted by STAR, complete the mmCIF description.

Comments are preceded by a hash (#) and terminated by a new line.
Data values on a single line may be delimited by pairs of single (′) or double (″) quotes.
Data values that extend beyond a single line are enclosed within semicolons (;) as the first character of the line that begins the text block and the first character of the line following the last line of text.
Data values that are unknown are represented by a question mark (?).
Data values that are undefined are represented by a period (.).
The length of a record in mmCIF is restricted to 80 characters.
Only printable ASCII characters are permitted.
Only a single level of *loop_* is permissible.

To complete the introductory picture of the appearance of an mmCIF data file consider the notion of scope. A PDB file has essentially one form of scope—the complete file. Thus, a single structure or an ensemble of structures is represented by a single file with each member of the ensemble separated by a PDB MODEL keyword record. There is no computer-readable mechanism for associating components of, say, the REMARK records with a particular member of the ensemble. The mmCIF representation deals with this issue by using the STAR data block concept. Data blocks begin with *data_* and have a scope that extends until the next *data_* or an end-of-file is reached. A *name* may appear only once in a data block, but data items may appear in any order. A consequence of these STAR rules is that the combination of data block name and data name is always unique.

Contents of mmCIF Dictionary

Table I summarizes the category groups, their associated individual categories, and their definitions as found in the mmCIF dictionary version 0.9.01 dated Jan. 31, 1997. This comprehensive hierarchy of categories follows closely the progress of the experiment and the subsequent structure description.

Structure Representation Using mmCIF

The categories describing the crystallographic experiment are relatively self-explanatory and are not detailed here. We do, however, outline the data model used to describe the resulting structure and its description.

The structural data model can be described most simply as containing three interrelated groups of categories: *ATOM_SITE* categories, which give coordinates and related information for the structure; *ENTITY* categories, which describe the chemistry of the components of the structure; and *STRUCT* categories, which analyze and describe the structure.

TABLE I
mmCIF CATEGORY GROUPS AND ASSOCIATED CATEGORIES[a]

Category groups and members	Definition
INCLUSIVE GROUP	All category groups
ATOM GROUP	
ATOM_SITE	Details of each atomic position
ATOM_SITE_ANISOTROP	Anisotropic thermal displacement
ATOM_SITES	Details pertaining to all atom sites
ATOM_SITES_ALT	Details pertaining to alternative atom sites as found in disord etc.
ATOM_SITES_ALT_ENS	Details pertaining to alternative atom sites as found in ensembles, e.g., from NMR and modeling experiments
ATOM_SITES_ALT_GEN	Generation of ensembles from multiple conformations
ATOM_SITES_FOOTNOTE	Comments concerning one or more atom sites
ATOM_TYPE	Properties of an atom at a particular atom site
AUDIT GROUP	
AUDIT	Detail on the creation and updating of the mmCIF
AUDIT_AUTHOR	Author(s) of the mmCIF including address information
AUDIT_CONTACT_AUTHOR	Author(s) to be contacted
CELL GROUP	
CELL	Unit cell parameters
CELL_MEASUREMENT	How the cell parameters were measured
CELL_MEASUREMENT_REFLN	Details of the reflections used to determine the unit cell parameters
CHEM_COMP GROUP	
CHEM_COMP	Details of the chemical components
CHEM_COMP_ANGLE	Bond angles in a chemical component
CHEM_COMP_ATOM	Atoms defining a chemical component
CHEM_COMP_BOND	Characteristics of bonds in a chemical component
CHEM_COMP_CHIR	Details of the chiral centers in a chemical component
CHEM_COMP_CHIR_ATOM	Atoms comprising a chiral center in a chemical component
CHEM_COMP_LINK	Linkages between chemical groups
CHEM_COMP_PLANE	Planes found in a chemical component
CHEM_COMP_PLANE_ATOM	Atoms comprising a plane in a chemical component
CHEM_COMP_TOR	Details of the torsion angles in a chemical component
CHEM_COMP_TOR_VALUE	Target values for the torsion angles in a chemical component
CHEM_LINK GROUP	
CHEM_LINK	Details of the linkages between chemical components
CHEM_LINK_ANGLE	Details of the angles in the chemical component linkage
CHEM_LINK_BOND	Details of the bonds in the chemical component linkage
CHEM_LINK_CHIR	Chiral centers in a link between two chemical components
CHEM_LINK_CHIR_ATOM	Atoms bonded to a chiral atom in a linkage between two chemical components
CHEM_LINK_PLANE	Planes in a linkage between two chemical components
CHEM_LINK_PLANE_ATOM	Atoms in the plane forming a linkage between two chemical components
CHEM_LINK_TOR	Torsion angles in a linkage between two chemical componen
CHEM_LINK_TOR_VALUE	Target values for torsion angles enumerated in a linkage between two chemical components

TABLE I (*continued*)

ategory groups and members	Definition
HEMICAL GROUP	
HEMICAL	Composition and chemical properties
HEMICAL_CONN_ATOM	Atom position for 2-D chemical diagrams
HEMICAL_CONN_BOND	Bond specifications for 2-D chemical diagrams
HEMICAL_FORMULA	Chemical formula
ITATION GROUP	
ITATION	Literature cited in reference to the data block
ITATION_AUTHOR	Author(s) of the citations
ITATION_EDITOR	Editor(s) of citations where applicable
OMPUTING GROUP	
OMPUTING	Computer programs used in the structure analysis
OFTWARE	More detailed description of the software used in the structure analysis
ATABASE GROUP	
ATABASE	Superseded by DATABASE_2
ATABASE_2	Codes assigned to mmCIFs by maintainers of recognized databases
ATABASE_PDB_CAVEAT	CAVEAT records originally found in the PDB version of the mmCIF data file
ATABASE_PDB_MATRIX	MATRIX records originally found in the PDB version of the mmCIF data file
ATABASE_PDB_REMARK	REMARK records originally found in the PDB version of the mmCIF data file
ATABASE_PDB_REV	Taken from the PDB REVDAT records
ATABASE_PDB_REV_RECORD	Taken from the PDB REVDAT records
ATABASE_PDB_TVECT	TVECT records originally found in the PDB version of the mmCIF data file
IFFRN GROUP	
IFFRN	Details of diffraction data and the diffraction experiment
IFFRN_ATTENUATOR	Diffraction attenuator scales
IFFRN_DETECTOR	Describes the detector used to measure scattered radiation
IFFRN_MEASUREMENT	Details on how the diffraction data were measured
IFFRN_ORIENT_MATRIX	Orientation matrices used when measuring data
IFFRN_ORIENT_REFLN	Reflections that define the orientation matrix
IFFRN_RADIATION	Details on the radiation and detector used to collect data
IFFRN_SOURCE	Details pertaining to the radiation source
IFFRN_REFLN	Unprocessed reflection data
IFFRN_REFLNS	Details pertaining to all reflection data
IFFRN_SCALE_GROUP	Details of reflections used in scaling
IFFRN_RADIATION_ VAVELENGTH	Wavelength of radiation used to measure diffraction intensities
IFFRN_STANDARD_REFLN	Details of the standard reflections used during data collection
IFFRN_STANDARDS	Details pertaining to all standard reflections
NTITY GROUP	
NTITY	Details pertaining to each unique chemical component of the structure
NTITY_KEYWORDS	Keywords describing each entity
NTITY_LINK	Details of the links between entities

continued

TABLE I (*continued*)

Category groups and members	Definition
ENTITY_NAME_COM	Common name for the entity
ENTITY_NAME_SYS	Systematic name for the entity
ENTITY_POLY	Characteristics of a polymer
ENTITY_POLY_SEQ	Sequence of monomers in a polymer
ENTITY_SRC_GEN	Source of the entity
ENTITY_SRC_NAT	Details of the natural source of the entity
ENTRY GROUP	
ENTRY	Identifier for the data block
EXPTL GROUP	
EXPTL	Experimental details relating to the physical properties of the material, particularly absorption
EXPTL_CRYSTAL	Physical properties of the crystal
EXPTL_CRYSTAL_FACE	Details pertaining to the crystal faces
EXPTL_CRYSTAL_GROW	Conditions and methods used to grow the crystals
EXPTL_CRYSTAL_GROW_COMP	Components of the solution from which the crystals were grow
GEOM GROUP	
GEOM	Derived geometry information
GEOM_ANGLE	Derived bond angles
GEOM_BOND	Derived bonds
GEOM_CONTACT	Derived intermolecular contacts
GEOM_TORSION	Derived torsion angles
JOURNAL GROUP	
JOURNAL	Used by journals and not the mmCIF preparer
PHASING GROUP	
PHASING	General phasing information
PHASING_AVERAGING	Phase averaging of multiple observations
PHASING_ISOMORPHOUS	Phasing information from an isomorphous model
PHASING_MAD	Phasing via multiwavelength anomalous dispersion (MAD)
PHASING_MAD_CLUST	Details of a cluster of MAD experiments
PHASING_MAD_EXPT	Overall features of the MAD experiment
PHASING_MAD_RATIO	Ratios between pairs of MAD datasets
PHASING_MAD_SET	Details of individual MAD datasets
PHASING_MIR	Phasing via single and multiple isomorphous replacement
PHASING_MIR_DER	Details of individual derivatives used in MIR
PHASING_MIR_DER_REFLN	Details of calculated structure factors
PHASING_MIR_DER_SHELL	As above but for shells of resolution
PHASING_MIR_DER_SITE	Details of heavy atom sites
PHASING_MIR_SHELL	Details of each shell used in MIR
PHASING_SET	Details of data sets used in phasing
PHASING_SET_REFLN	Values of structure factors used in phasing
PUBL GROUP	
PUBL	Used when submitting a publication as a mmCIF
PUBL_AUTHOR	Authors of the publication
PUBL_MANUSCRIPT_INCL	To include special data names in the processing of the manuscript

TABLE I (*continued*)

Category groups and members	Definition
REFINE GROUP	
REFINE	Details of the structure refinement
REFINE_ANALYZE	Items used to assess the quality of the refined structure
REFINE_B_ISO	Details pertaining to the refinement of isotropic B values
REFINE_HIST	History of the refinement
REFINE_LS_RESTR	Details pertaining to the least squares restraints used in refinement
REFINE_LS_RESTR_NCS	Details of the restraints applied to atomic positions related by noncrystallographic symmetry
REFINE_LS_SHELL	Results of refinement broken down by resolution
REFINE_OCCUPANCY	Details pertaining to the refinement of occupancy factors
REFLN GROUP	
REFLN	Details pertaining to the reflections used to derive the atom sites
REFLNS	Details pertaining to all reflections
REFLNS_SCALE	Details pertaining to scaling factors used with respect to the structure factors
REFLNS_SHELL	As REFLNS, but by shells of resolution
STRUCT GROUP	
STRUCT	Details pertaining to a description of the structure
STRUCT_ASYM	Details pertaining to structure components within the asymmetric unit
STRUCT_BIOL	Details pertaining to components of the structure that have biological significance
STRUCT_BIOL_GEN	Details pertaining to generating biological components
STRUCT_BIOL_KEYWORDS	Keywords for describing biological components
STRUCT_BIOL_VIEW	Description of views of the structure with biological significance
STRUCT_CONF	Conformations of the backbone
STRUCT_CONF_TYPE	Details of each backbone conformation
STRUCT_CONN	Details pertaining to intermolecular contacts
STRUCT_CONN_TYPE	Details of each type of intermolecular contact
STRUCT_KEYWORDS	Description of the chemical structure
STRUCT_MON_DETAILS	Calculation summaries at the monomer level
STRUCT_MON_NUCL	Calculation summaries specific to nucleic acid monomers
STRUCT_MON_PROT	Calculation summaries specific to protein monomers
STRUCT_MON_PROT_CIS	Calculation summaries specific to cis peptides
STRUCT_NCS_DOM	Details of domains within an ensemble of domains
STRUCT_NCS_DOM_LIM	Beginning and end points within polypeptide chains forming a specific domain
STRUCT_NCS_ENS	Description of ensembles
STRUCT_NCS_ENS_GEN	Description of domains related by noncrystallographic symmetry
STRUCT_NCS_OPER	Operations required to superimpose individual members of an ensemble
STRUCT_REF	External database references to biological units within the structure

continued

TABLE I (*continued*)

Category groups and members	Definition
STRUCT_REF_SEQ	Describes the alignment of the external database sequence with that found in the structure
STRUCT_REF_SEQ_DIF	Describes differences in the external database sequence with that found in the structure
STRUCT_SHEET	Beta sheet description
STRUCT_SHEET_HBOND	Hydrogen bond description in beta sheets
STRUCT_SHEET_ORDER	Order of residue ranges in beta sheets
STRUCT_SHEET_RANGE	Residue ranges in beta sheets
STRUCT_SHEET_TOPOLOGY	Topology of residue ranges in beta sheets
STRUCT_SITE	Details pertaining to specific sites within the structure
STRUCT_SITE_GEN	Details pertaining to how the site is generated
STRUCT_SITE_KEYWORDS	Keywords describing the site
STRUCT_SITE_VIEW	Description of views of the specified site
SYMMETRY GROUP	
SYMMETRY	Details pertaining to space group symmetry
SYMMETRY_EQUIV	Equivalent positions for the specified space group

[a] Taken from *http://ndbserver.rutgers.edu/mmcif/dictionary/dict-html/cif-mm.dic/Index/.*

The data items in the *ATOM_SITE* category record details about the atom sites including the coordinates, the thermal displacement parameters, the errors in the parameters, and a specification of the component of the asymmetric unit to which an atom belongs.

The *ENTITY* category categorizes the unique chemical components of the asymmetric unit as to whether they are polymer, nonpolymer, or water. The characteristics of a polymer are described by the *ENTITY_POLY* category and the sequence of the chemical components composing the polymer by the *ENTITY_POLY_SEQ* category. The *CHEM_COMP* categories describe the standard geometries of the monomer units such as the amino acids and nucleotides as well as that of the ligands and solvent groups.

The *STRUCT_BIOL* category allows the author to describe the biologically relevant features of a structure and its component parts. The *STRUCT_BIOL_GEN* category provides the information about how to generate the biological unit from the components of the asymmetric unit, which are in turn specified by the *STRUCT_ASYM* category. Various features of the structure such as intermolecular hydrogen bonds, special sites, and secondary structure are specified in *STRUCT_CONN, STRUCT_SITE,* and *STRUCT_CONF,* respectively. Figure 2 illustrates the interrelationships among these categories.

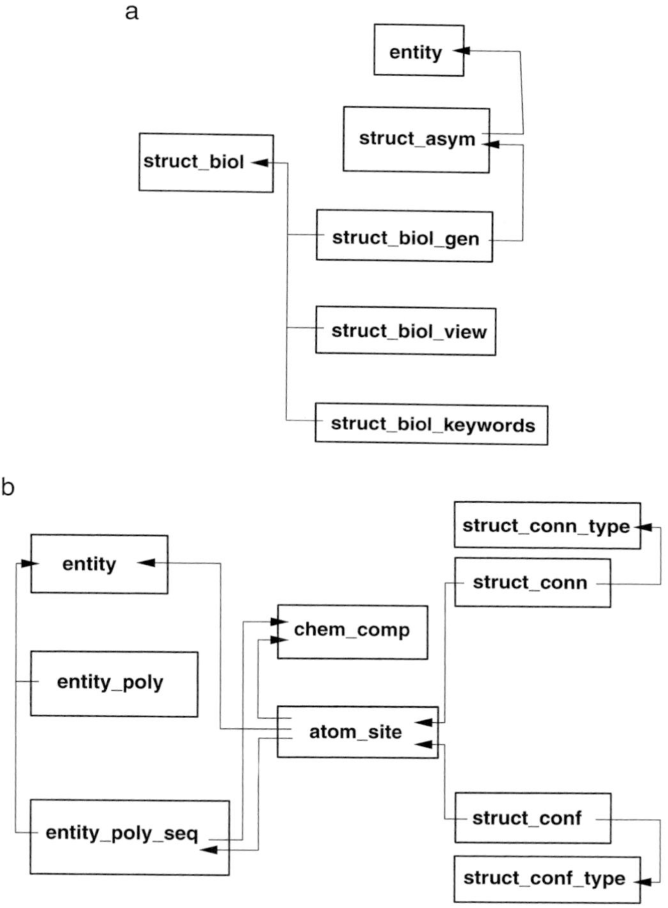

FIG. 2. (a) The relationships between categories that describe biologically relevant structure. (b) The relationships between categories describing polymer structure, the atomic coordinates, and those categories that describe structural features such as hydrogen bonding and secondary structure.

These and other major descriptive features of the mmCIF dictionary are best explored by example. A browsable dictionary can be found at the mmCIF World Wide Web (WWW) site (mmCIF[3]) as well as some complete examples. Complete examples for all nucleic acids can be found at the Nucleic Acid Database WWW site (NDB[13]). Partial mmCIFs for every

[13] NDB: *http://ndbserver.rutgers.edu/* (1996).

structure in the PDB are available at two WWW sites [PDB[14], San Diego Supercomputer Center (SDSC, San Diego, CA)[15]] having been generated with the program *pdb2cif* (Bernstein *et al.*[16]).

Example 1

Starting simply, consider the protein crambin, which is a single polypeptide chain of 48 residues, and in the low-temperature form at 0.83 Å resolution (Teeter *et al.*[17]; PDB code 1CBN) has nearly all the protein-bound solvent resolved as well as an ethanol molecule cocrystallized. The protein shows recognizable sequence microheterogeneity at positions 22 (Pro/Ser) and 25 (Leu/Ile) and 24% of residues show discrete disorder. Although microheterogeneity and disorder are described using data items in the mmCIF dictionary, they are not detailed here for the sake of simplicity.

Because the biological function of this molecule is unknown, no biologically relevant structural components are justified. A single identifier (*crambin_1*) is used to identify the unknown biological function of this molecule.

```
_struct_biol.id                        crambin_1
_struct_biol.details
;     The function of this protein is unknown and therefore the
      biological unit is assumed to be the single polypeptide
      chain without co-crystallization factors i.e. ethanol.
;
```

The single biological descriptor, *crambin_1,* is generated from the single polypeptide chain found in the asymmetric unit without any symmetry transformations applied. The polypeptide chain is designated *chain_a.*

```
_struct_biol_gen.biol_id               crambin_1
_struct_biol_gen.asym.id               chain_a
_struct_biol_gen.symmetry              1_555
```

The chemical components of the asymmetric unit are three entities: a single polypeptide chain characterized as a polymer, ethanol characterized as nonpolymer, and water. Whether the source of the entity is a natural product, or has been synthesized, is also indicated.

[14] PDB: *http://www.pdb.bnl.gov/cgi-bin/pdbmain* (1996).

[15] SDSC: *http://www.sdsc.edu/moose* (1996).

[16] H. J. Bernstein, F. C. Bernstein, and P. E. Bourne, *J. Appl. Cryst.* accepted (1997).

[17] M. M. Teeter, S. M. Roe, and N. Ho Heo, *J. Mol. Biol.* **230,** 292 (1993).

```
 loop_
_entity.id
_entity.type
_entity.formula_weight
_entity.src_method
      A           polymer          4716         'NATURAL'
      ethanol     non-polymer      52           'SYNTHETIC'
      H2O         water            18           .
```

It is then possible to expand on this basic description of each entity using the *entity.id* as a reference. So, for example, the common and systematic names are specified as

```
_entity_name_com.entity_id           A
_entity_name_com.name                crambin

_entity_name_sys.entity_id           A
_entity_name_sys.name                'Crambe Abyssinica'
```

Similarly, the natural and synthetic description can be given in more detail, so for the natural product we have

```
_entity_src_nat.entity_id            A
_entity_src_nat.common_name          'Abyssinian cabbage seed'
_entity_src_nat.genus                Crambe
_entity_src_nat.species              Abyssinica
_entity_src_nat.details              ?
```

By using the entities as building blocks, the contents of the asymmetric unit are specified. Crambin is straightforward because each entity appears only once in the asymmetric unit.

```
 loop_
_struct_asym.id
_struct_asym.entity_id
_struct_asym.details
      chain_a     A           'Single polypeptide chain'
      ethanol     ethanol     'Cocrystallized ethanol molecule'
      H2O         H2O         .
```

Entities classified as polymer, in this instance only that entity identified as *A*, are further described. First, the overall features of the polypeptide chain:

```
_entity_poly.entity_id           A
_entity_poly.type                polypeptide(L)
_entity_poly.nstd_chirality      no
_entity_poly.nstd_linkage        no
_entity_poly.nstd_monomers       no
_entity_poly.type_details        'Microheterogeneity at 22 and 25'
```

and then the component parts,

```
        loop_
       _entity_poly_seq.entity_id
       _entity_poly_seq.num
       _entity_poly_seq.mon_id
             A      1      THR    A      2      THR
#      [data omitted]
             A      22     PRO    A      23     GLU
             A      24     ALA    A      25     LEU
#      [data omitted]
             A      47     ALA    A      48     ASN
```

The entity also may exist in other databases and these references may be cited and described. For the entity designated *A*, which is defined in GenBank but without sequence microheterogeneity, we have

```
 loop_
_struct_ref.id
_struct_ref.entity_id
_struct_ref.biol_id
_struct_ref.db_name
_struct_ref.db_code
_struct_ref.seq_align
_struct_ref.seq_dif
_struct_ref.details
1     A     crambin_1    'Genbank'    '493916'    'entire'    'no' .
2     A     crambin_1    'PDB'        '1CBN'      'entire'    'no' .
```

Once each polymer entity is defined, the details of the secondary structure are defined using the *STRUCT_CONF* category.

```
 loop_
_struct_conf.id
_struct_conf.conf_type.id
_struct_conf.beg_label_comp_id
_struct_conf.beg_label_asym_id
_struct_conf.beg_label_seq_id
_struct_conf.end_label_comp_id
_struct_conf.end_label_asym_id
_struct_conf.end_label_seq_id
_struct_conf.details
H1  HELX_RH_AL_P ILE chain_a 7  PRO  chain_a  19 'HELX-RH3T 17-19'
H2  HELX_RH_AL_P GLU chain_a 23 THR  chain_a  30 'Alpha-N start'
S1  STRN_P       CYS chain_a 32 ILE  chain_a  35 .
S2  STRN_P       THR chain_a  1 CYS  chain_a   4 .
S3  STRN_P       ASN chain_a 46 ASN  chain_a  46 .
S4  STRN_P       THR chain_a 39 PRO  chain_a  41 .
T1  TURN-TY1_P   ARG chain_a 17 GLY  chain_a  20 .
T2  TURN-TY1_P   PRO chain_a 41 TYR  chain_a  44 .
```

These assignments are further enumerated over those made in a PDB file for the record types *HELIX, TURN,* and *SHEET.* Moreover, the *STRUCT_CONF_TYPE* category (Table I) specifies the method of assignment that could, for example, be deduced by the crystallographer from the electron-density maps or defined algorithmically.

```
 loop_
_struct_conf_type.id
_struct_conf_type.criteria
_struct_conf_type.reference
HELX_RH_AL_P      'author judgement'     .
STRN_P            'author judgement'     .
TURN_TY1_P        'author judgement'     .
HELX_RH_P         'Kabsch and Sander' 'Biopolymers (1983) 22:2577'
```

The commented entry at the end is a hypothetical example of a calculated assignment. Data items also exist (Table I) for the description of β sheets, but are not shown in this introductory example.

Interactions between various portions of the structure are described by the *STRUCT_CONN* and associated *STRUCT_CONN_TYPE* categories.

```
 loop_
_struct_conn.id
_struct_conn.conn_type_id
_struct_conn.ptnr1_label_comp_id
_struct_conn.ptnr1_label_asym_id
_struct_conn.ptnr1_label_seq_id
_struct_conn.ptnr1_label_atom_id
_struct_conn.ptnr1_role
  _struct_conn.ptnr1_symmetry
  _struct_conn.ptnr2_label_comp_id
  _struct_conn.ptnr2_label_asym_id
  _struct_conn.ptnr2_label_seq_id
  _struct_conn.ptnr2_label_atom_id
  _struct_conn.ptnr2_role
  _struct_conn.ptnr2_symmetry
  _struct_conn.details
  SS1 disulf CYS  chain_a  3  S . 1_555  CYS  chain_a 40  S . 1_555  .
  SS2 disulf CYS  chain_a  4  S . 1_555  CYS  chain_a 32  S . 1_555  .
    [data omitted]
  HB1  hydrog  SER  chain_a  6  OG positive  1_555
               LEU  chain_a  8  O  negative  1_556  .
  HB2  hydrog  ARG  chain_a 17  N  positive  1_555
               ASP  chain_a 43  O  negative  1_554  .
    [data omitted]
```

These intermolecular interactions are partially specified on PDB *CONNECT* records. However, mmCIF provides an additional level of detail such that the criteria used to define an interaction may be given using the *STRUCT_CONN_TYPE* category. Here is a hypothetical example used to describe a salt bridge and a hydrogen bond:

```
 loop_
_struct_conn_type.id
_struct_conn_type.criteria
_struct_conn_type.reference
saltbr  'negative to positive distance > 2.5 \%A and < 3.2 \%A ' .
hydrog  'N to O distance > 2.5 \%A, < 3.2 \%A, NOC angle < 120\%' .
```

Example 2

Consider an mmCIF representation of a more complex structure: the gene regulatory protein 434 CRO complexed with a 20-base pair DNA segment-containing operator (Mondragon and Harrison[18]; PDB code 3CRO).

```
      loop_
     _struct_biol.id
     _struct_biol.details
                complex
;          The complex consists of 2 protein domains bound to a
           20 base pair DNA segment.
;
                protein
;          Each of the 2 protein domains is a single homologous
           polypeptide chain of 71 residues designated L and R.
;
                DNA
;          The two strands (A and B) are complementary given a one
           base offset.
;
```

The protein–DNA complex, the protein, and the DNA are considered as three separate biological components each generated from the contents of the asymmetric unit. No crystallographic symmetry need be applied to generate the biologically relevant components.

[18] A. Mondragon and S. C. Harrison, *J. Mol. Biol.* **219,** 321 (1991).

```
 loop_
_struct_biol_gen.biol_id
_struct_biol_gen.asym.id
_struct_biol_gen.symmetry
      complex     L     1_555
      complex     R     1_555
      complex     A     1_555
      complex     B     1_555
      protein     L     1_555
      protein     R     1_555
      DNA         A     1_555
      DNA         B     1_555

 loop_
_entity.id
_entity.type
      dimer        polymer
      DNA_A        polymer
      DNA_B        polymer
      water        water
```

Because each protein domain is chemically identical they constitute a single entity that has been designated *dimer*. The complementary DNA strands are not chemically identical and therefore constitute two separate entities:

```
 loop_
_struct_asym.id
_struct_asym.entity_id
_struct_asym.details
      L     dimer       '71 residue polypeptide chain'
      R     dimer       '71 residue polypeptide chain'
      A     DNA_A       '20 base strand'
      B     DNA_B       '20 base strand'
      H2O   water       'solvent'
```

Features of the CRO 434 secondary structure and intermolecular contacts can be described in the same way in which crambin was represented and are not repeated.

Conclusion

In preparing these examples of representing macromolecular structure using mmCIF it was necessary to return to the original articles because not all the relevant information could be retrieved from the PDB entry. This is evidence that mmCIF provides additional information that also has the advantage of being in a computer-readable form. The consequence is that it places additional emphasis on the person preparing the mmCIF. It is anticipated that full use of the expressive power of mmCIF will be made only when existing structure solution and refinement programs are modified to maintain mmCIF data items and software tools are developed to help prepare and use an mmCIF effectively. A variety of software tools have been developed for mmCIF (Bernstein *et al.*[16]; Westbrook *et al.*[19]). A description of a variety of other efforts can be found elsewhere (Bourne[20]). Code and documentation are available at the mmCIF WWW site (mmCIF[3]). A long-term goal might be to maintain all aspects of the structure determination in an electronic laboratory notebook that uses mmCIF as its underlying data representation.

Acknowledgments

The development of the mmCIF dictionary has been a community effort. The Background and Introduction sections of the mmCIF WWW site describe the contributions of the many people who have participated in this project (mmCIF[3]).

[19] J. D. Westbrook, S. H. Hsieh, and P. M. D. Fitzgerald, *J. Appl. Crystallogr.* **30,** 79–83 (1997).
[20] P. E. Bourne (ed.), "Proceedings of the First Macromolecular CIF Tools Workshop." Tarrytown, New York, 1993.

[31] PHASES-95: A Program Package for Processing and Analyzing Diffraction Data from Macromolecules

By W. Furey and S. Swaminathan

Background

The determination of macromolecular crystal structures has always been a computationally demanding task, often further complicated by the need for frequent algorithm modifications, by lack of general, compatible software for the many different procedures required, and by difficulties in exchanging data with the (usually incompatible) hardware needed for visual feedback. In PHASES the idea was to overcome these problems by creating a processing package that is efficient, general, reasonably thorough, easily upgradeable, nearly hardware independent, compatible with other popular software, and (most of all) easy to use. Several events took place in the late 1980s and early 1990s that facilitated such a package. First, increases in CPU speed aided efficiency and justified use of general algorithms for

0076-6879/97 $25.00

most applications. Second, although technically a software issue, the general acceptance of UNIX[1] as an operating system standard nevertheless enabled a significant degree of hardware independence to be obtained. Third, the field of macromolecular crystallography had matured to such a point that general procedures for determining structures were well established. Fourth, the proliferation and general acceptance of high-speed workstations capable of locally generating high-quality graphics displays for map fitting with familiar and accepted software provided an ideal platform. Finally, the emergence of the X Window system as an accepted standard for two-dimensional graphics displays enabled simple hardware-independent graphics code to be produced. By exploiting these developments the goal has essentially been realized as the PHASES package[1a] has been distributed to roughly 150 laboratories worldwide, where it has been instrumental in the determination of a large number of protein structures. The description that follows is strictly valid only for versions of PHASES distributed after February 1996, although many of the features were present in earlier releases. The standard distribution contains equivalent versions for Silicon Graphics, Sun, IBM R6000, ESV, and DEC Alpha AXP (both OSF and OpenVMS) workstations.

Overview

The basic philosophy employed while developing PHASES was to create software that is reasonably thorough (able to deal with most common phasing situations), general (able to deal with all space groups, even nonstandard ones), simple to use and understand (requiring minimal input, often interactive), flexible, easy to interface with other software, and efficiently implemented on most popular hardware, particularly workstations. No attempt was made to be completely comprehensive for several reasons. Very good software addressing certain areas of macromolecular crystallography such as molecular replacement, refinement of protein structures, and three dimensional (3D) graphics for chain tracing and display already existed, and significant improvements were not obvious. Some areas such as primary data reduction may be strongly tied to the particular data collection

[1] *Trade marks:* UNIX is a registered trademark of UNIX System Laboratories, Inc. Silicon Graphics is a registered trademark of Silicon Graphics, Inc. (Mountain View, CA). IBM is a registered trademark of International Business Machines, Inc. Alpha AXP, DEC, VMS, and OpenVMS are registered trademarks of Digital Equipment Corporation. Sun is a registered trademark of Sun Microsystems, Inc. OSF is a registered trademark of Open Software Foundation, Inc. X Window is a registered trademark of the Massachusetts Institute of Technology (Boston, MA). PostScript is a registered trademark of Adobe Systems, Inc.

[1a] W. Furey and S. Swaminathan, *Am. Crystallogr. Assoc. Meet. Abstr.* **18,** PA33, 73 (1990).

equipment, and are perhaps best dealt with by the producers of the hardware. Accordingly, PHASES (as its name implies) focuses on solution of the phase problem in macromolecular crystallography, primarily by the methods of isomorphous replacement, anomalous scattering, solvent flattening, negative density truncation, phase extension, noncrystallographic symmetry averaging, and partial structure phase combination. Software tools are provided allowing one to start with unique reflections for native and/or derivative data sets and ultimately produce from them electron-density maps and skeletons that can be directly displayed in popular graphics programs for chain tracing.

The package currently consists of 44 individual Fortran programs, about a dozen of which are "workhorse" programs used in most applications, and a single C interface subroutine used by some of the graphics programs. Most of the PHASES programs are interactive and will prompt for any needed input; however, some of the programs (in general, those that either take considerable time to run, require more than about five pieces of input information, or are usually run repeatedly as part of an iterative procedure) are geared for submission as a batch process. All communication between programs is by files, with a particularly simple, common format chosen to facilitate flexibility, future extension, and interfacing with other software. The programs can be combined in many ways, and template scripts for frequently needed iterative tasks are provided. Apart from atomic coordinate records, all user-supplied information can be read by the programs in free format. Memory allocation for the major programs is generally through a large, single array that is partitioned as needed for the individual problem at execution time. Thus if faced with an exceedingly large problem one may have to change, at most, one or two lines of code in the MAIN routine and recompile.

Parameter Files

The user starts work on a new problem by creating a simple "parameter file" that contains the lattice type, cell constants, space group symmetry operations, and (optionally) the name of a "running" log file. Most of the programs will require this file, thus its creation ensures uniformity in cell and symmetry throughout all computations, and eliminates the need for redundant input and the associated increased likelihood of typing errors. It also allows whole strings of calculations to be easily repeated with a different symmetry, just by changing a single file. Space group symmetry operations are specified by explicitly providing the equivalent positions, and all symmetry-related information such as systematic absences, multiplicity, which reflections are centric and their allowed phase values, are derived directly from the positions so that one can utilize nonstandard settings if desired. If requested, a "running" log file will be opened in append mode, so that in addition to the normal output from each program, a copy of each output will be entered in the running log preceded by the program name and a time stamp. Thus one can maintain a complete history of all calculations and printed results in a single file.

Major Capabilities

The three main tasks addressed by PHASES are as follows: (1) the computation of phase angles by heavy atom-based methods such as isomorphous replacement[2] or anomalous scattering,[3] (2) the improvement/extension of phases by solvent flattening with negative density truncation,[4] and (3) phase improvement/extension by noncrystallographic symmetry averaging.[5,6] In addition to the major programs, numerous auxiliary programs are included to accomplish these tasks, to examine and assess results, and to interface data or maps with other popular software.

Heavy atom-based phasing is initiated when one or more "scaled" files is input to the program PHASIT. Each file can contain either isomorphous replacement, derivative anomalous scattering, or native anomalous scattering data. On reading each data set and applying user-specified rejection criteria, a subset of reflections (all centric reflections, plus the 25% largest differences if there are insufficient centric data) is automatically selected and used to scale structure-factor amplitudes calculated from input heavy-atom parameters to the observed isomorphous or anomalous differences. If needed, user-supplied scattering factors may be input to facilitate phasing with MAD (multiple wavelength anomalous diffraction) data. Initial estimates of the "standard error" E (expected lack of closure) are then determined from this subset as a function of F magnitude, treating centric and acentric data separately. SIR (single isomorphous replacement) or SAS (single wavelength anomalous scattering) phase probability distributions are then computed for each set, and the distributions are cast in the A, B, C, D form suggested by Hendrickson and Lattman.[7] R factors are given and phasing power estimates are then computed on the basis of the SIR or SAS phase information for each set. After all sets are processed, the individual probability distributions for common reflections are combined

[2] D. W. Green, U. M. Ingram, and M. F. Perutz, *Proc. R. Soc. Lond.* **A225,** 287 (1954).
[3] R. Pepinsky and Y. Okaya, *Proc. Natl. Acad. Sci. U.S.A.* **42,** 286 (1956).
[4] B. C. Wang, *Methods Enzymol.* **115,** 90 (1985).
[5] M. G. Rossmann and D. M. Blow, *Acta Crystallogr.* **16,** 39 (1963).
[6] G. Bricogne, *Acta Crystallogr.* **A30,** 395 (1974).
[7] W. A. Hendrickson and E. E. Lattman, *Acta Crystallogr.* **B26,** 136 (1970).

and all distributions are then integrated to yield a centroid phase and figure of merit for each reflection. On the basis of the new phases, figure of merit statistics are given along with new phasing power estimates. The standard error estimates as a function of F magnitude are then updated by including all reflections for each set, this time by using a probability-weighted average over all possible phase values for the contribution from each reflection.[8] Using the updated standard error estimates, the individual distributions are then recomputed. The new distributions are then combined and processed as before, and final phases, updated phasing power, and figure of merit statistics are output. A similar strategy is employed during refinement, which is discussed in more detail below. The probability-weighted averaging in standard error estimation (always done), when coupled with the maximum likelihood option, enable the key features of maximum likelihood refinement to be carried out in a manner similar to that in MLPHARE.[9] However, as in practically all refinement programs distributed at this time, combination of phase information is still by simple multiplication of contributing SIR or SAS distributions, leading to a slight overweighting of the native data set.[10]

Solvent flattening is carried out following the strategy suggested by Wang,[4] including a reciprocal space equivalent[11] of the automated solvent-masking procedure. During solvent mask construction, however, density near input heavy-atom sites is automatically ignored, leading to more accurate masks. The entire process of creating 3 solvent masks with at least 16 cycles of phase combination (4 with the first mask, 4 with the second, 8 with the third, and then, optionally, an arbitrary number of additional cycles with the third mask for phase extension) is carried out by executing a single

[8] T. C. Terwilliger and D. Eisenberg, *Acta Crystallogr.* **A43,** (1987).

[9] Z. Otwinowski, *in* "Isomorphous Replacement and Anomalous Scattering: Proceedings of the CCP4 Study Weekend 25–26 January 1991" (W. Wolf, P. R. Evans, and A. G. W. Leslie, eds.), pp. 80–86. Science and Engineering Research Council, Daresbury Laboratory, Daresbury, Warrington, UK, 1991.

[10] R. J. Read, *in* "Isomorphous Replacement and Anomalous Scattering: Proceedings of the CCP4 Study Weekend 25–26 January 1991" (W. Wolf, P. R. Evans, and A. G. W. Leslie, eds.), pp. 67–79. Science and Engineering Research Council, Daresbury Laboratory, Daresbury, Warrington, UK, 1991.

[11] The reciprocal space equivalent of the Wang direct space boundary determination method as coded in PHASES was independently derived, both in our laboratory and in the Westbrook laboratory in 1983 almost immediately after the direct space formalism was proposed. A joint paper from the two laboratories was submitted to *Acta Crystallographica* in 1984, but was *rejected* as a reviewer *did not believe solvent flattening was useful anyway,* and therefore *saw no reason to speed up the calculation!* Three years later a similar derivation, again independently developed, finally was published by A. G. W. Leslie, *Acta Crystallogr.* **A43,** 134 (1987). The algorithm employed in PHASES differs from that of Leslie only by a constant multiplicative factor equal to the sphere volume; thus the results are identical apart from scale.

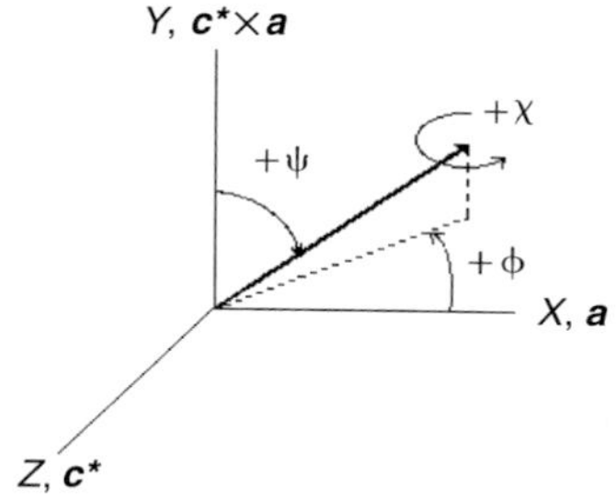

FIG. 1. Relationships between noncrystallographic symmetry rotation axis direction, orthogonal reference system axes X, Y, Z, and crystallographic axes. The X axis is aligned with crystal **a**. The Y axis is parallel to crystal **a** × **c***. The Z axis is parallel to $X \times Y$, i.e., crystal **c***. ψ is the angle between the NC rotation and $+Y$ axes. ϕ is the angle between the projection of the NC rotation axis in the XZ plane and the $+X$ axis, with $+\phi$ counterclockwise when viewed from $+Y$ toward the origin. χ is the amount of rotation about the directed axis, with $+\chi$ clockwise when viewed from the axis toward the origin.

doall shell script or command procedure, making it easy to try different phasing/flattening strategies or parameters. A program is provided to examine/edit solvent masks interactively, or even create them from scratch.

Noncrystallographic (NC) symmetry averaging problems are treated in direct space by a series of programs designed to operate on submaps, i.e., electron-density map regions that can span less than a full cell or asymmetric unit volume, but that encompass all molecules to be averaged that are unique by crystal symmetry. Many of these programs were derived from routines originally developed by Hendrickson and Smith, and described by Bolin *et al.*,[12] but were modified and in most cases substantially rewritten for incorporation into PHASES. Initial estimates of the NC symmetry operators (at least the orientations) must be provided from external software (usually self-rotation functions), while estimates of the location of the operators can usually be obtained within the package either directly from initial electron-density maps or from heavy-atom coordinates. In all cases NC symmetry operators are specified in terms of the parameters ϕ, ψ, χ, O_x, O_y, O_z, and T, which refer to a Cartesian coordinate system (in angstroms) obtained by orthogonalization of the unit cell as in the Protein Data Bank. The angles ϕ and ψ define the orientation of the NC rotation axis while χ defines the amount of rotation about it. The relationships between the angles, orthogonal reference axes X, Y, Z and unit cell are given in Fig. 1. O_x, O_y, and O_z are coordinates of a point on the rotation axis and thus define its absolute location, while T is a postrotation transla-

[12] J. T. Bolin, J. Smith, and S. W. Muchmore, *Am. Crystallogr. Assoc. Meet. Abstr.* **21,** V001, 51 (1993).

tion parallel to the rotation axis, allowing for arbitrary screwlike translations. Coordinates of two points related by noncrystallographic symmetry are then expressed in the orthogonal system by

$$\mathbf{P}_2 = \mathbf{R}_{\phi,\varphi,\chi}(\mathbf{P}_1 - \mathbf{O}) + \mathbf{O} + T \cdot \mathbf{D}_{\phi,\varphi} \tag{1}$$

where $\mathbf{P}_1$ and $\mathbf{P}_2$ are three-element column vectors containing coordinates for the related points, $\mathbf{R}_{\phi,\varphi,\chi}$ is a 3×3 rotation matrix derived from the angles, $\mathbf{O}$ is a three-element column vector containing coordinates for a point on the axis, T is the postrotation translation scalar in angstroms and $\mathbf{D}_{\phi,\varphi}$ is a three-element column vector containing direction cosines for the rotation axis. Programs are provided to generate and examine the required map regions, refine the operators, interactively create envelope masks, average electron density within the envelopes, convert submaps back to full cell maps, invert the modified maps, and combine the phases with those from another source. A shell script *extndavg.sh* is provided, which is an extension to the *doall* script that enables an arbitrary number of averaging/phase combination cycles to be performed in addition to solvent flattening/negative density truncation. The script also allows for gradual phase extension to be carried out, extending by one reciprocal lattice point at a time for a given number of cycles.

The major paths to carry out these procedures are illustrated in flow chart form in Fig. 2. Native and derivative data enter the package and are scaled/merged in programs CMBISO and/or CMBANO. Initial phases are then computed in PHASIT. These phases are then used both to seed the solvent-flattening process by creating the first map, and as "anchor" phases for combination with information obtained from map inversion after each cycle. Solvent flattening/negative density truncation/phase combination iterations are then carried out within the FSFOUR–BNDRY–MAPINV loop. Optionally, noncrystallographic symmetry averaging can be carried out by following the path to the right of the dashed line in Fig. 2, where the submap region to be averaged is first extracted from the map by EXTRMAP, averaged according to the local symmetry in MAPAVG, and then converted back to a normal full cell map in BLDCEL. As a further option, phase extension can be included during phase combination by selecting appropriate reflections with program MISSNG. The reflections used for phase extension may contain only amplitude data, amplitude plus phase probability distribution data, or simply Miller indices in any combination.

All of the programs contained in PHASES can be loosely divided according to their functionality, although some of the programs could easily be included in more than one category. All programs dealing with initial (not yet scaled/merged or phased) reflection data accept the data in either MULIST, SCALEPACK, or "free" format, with the format type deduced

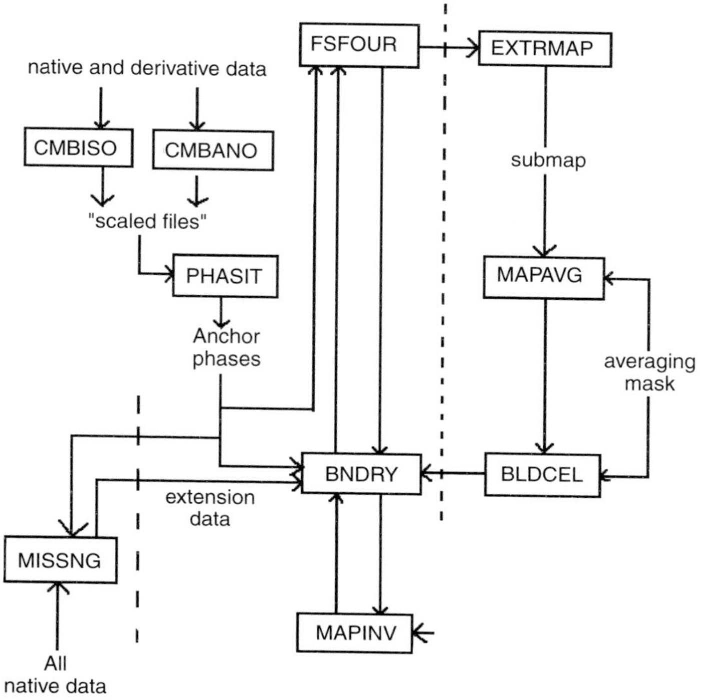

FIG. 2. Flow chart for the major phasing path encompassing native/derivative scaling, heavy-atom-based phasing, solvent flattening, negative density truncation, and phase combination. Boxed entries represent programs; lines represent files. Optional paths for noncrystallographic symmetry averaging and phase extension are included by considering the additional programs offset from the main path by dashed lines.

from the file name. The general categories and programs within each category are now described.

Initial Processing: Native/Derivative Scaling, Preparation for Difference Patterson Maps and Difference Fouriers

Programs CMBISO and CMBANO

The interactive programs CMBISO and CMBANO combine derivative isomorphous and anomalous scattering data sets, respectively, with a native data set and scale the derivative data to that for the native. All common reflections or those related by symmetry are paired up, scaled, and output as a single record to an ASCII file, which is then referred to as a "scaled" file. Scaling is initially carried out by means of a relative Wilson[13] plot in which the native and derivative scattering is set to be equal. Following initial scaling the user may optionally select additional non-Wilson scaling, either anisotropic by fitting an ellipsoid to the entire data set, or by local scaling. If anisotropic scaling is chosen the unique parameters of a symmetric 3×3 tensor S are determined by two cycles of least-squares minimizing

$$\sum_{hkl} W_{hkl}(|F_{\mathrm{nat}}|_{hkl} - S|F_{\mathrm{der}}|_{hkl})^2 \qquad (2)$$

where

$$S = S_{11}O_x^2 + S_{22}O_y^2 + S_{33}O_z^2 + 2(S_{12}O_xO_y + S_{13}O_xO_z + S_{23}O_yO_z) \qquad (3)$$

and O_x, O_y, O_z are direction cosines of the reciprocal lattice vector expressed in an orthogonal system. If local scaling is selected a scale factor for each reflection is still determined by minimizing Eq. (2), but S is then a scalar and the sum is taken only over neighboring reflections within a sphere centered on the reflection being scaled. The sphere size is initially set for each reflection so that about 125 neighbors should be contained, and the scale factor is accepted if at least 80 are actually found. If needed, the sphere size is automatically increased in increments, up to a preset maximum. If the maximum is reached, the acceptance criterion is then reduced to 40 reflections for 1 final attempt, after which the program stops and informs the user that the data set is too sparse for local scaling.

Following pairing up and scaling, the relative Wilson plot-derived scale and thermal factors are printed and if additional scaling was selected the mean, minimum, and maximum scale factors found are reported. The merging R factors on F values and intensities are then given, both overall and broken down as a function of resolution, F magnitude, and F/σ. The mean absolute isomorphous difference is then given independently for centric and acentric data in a resolution-dependent table to facilitate assessment of the derivative. The isomorphous merging R factor reported is defined as

$$R_{\mathrm{iso}} = \frac{\sum_{hkl} \| F_{\mathrm{der}}|_{hkl} - |F_{\mathrm{nat}}|_{hkl}|}{\sum_{hkl} |F_{\mathrm{der}}|_{hkl} + |F_{\mathrm{nat}}|_{hkl}} \qquad (4)$$

with a corresponding definition used for the R factor when reported on intensities. These definitions are used because it can be shown that when there are only two contributing reflections (as we have here), the definition commonly employed for R_{sym}, i.e., the R factor for internal agreement

[13] A. J. C. Wilson, *Acta Crystallogr.* **2,** 318 (1949).

between symmetry related reflections within the same data set, is equivalent to this expression. For this reason the merging *R* factors from CMBISO are lower (by roughly 50%) than those reported by some other programs, but the advantage is that they are directly comparable to the internal *R* factors obtained during primary data reduction and thus are useful to determine whether heavy-atom signals exceed internal noise within a data set. CMBISO and CMBANO function identically except that in CMBANO it is the mean $|F|$ for the Bijvoet pair used in determining the scale factor, only reflections where both members of the Bijvoet pair were explicitly input are used, and both measurements propagate to the output file.

Programs PRECESS and PRECESS_X

These interactive programs are used to visually examine and evaluate data sets displayed in the form of precession photographs. Reflection data can be read in a variety of formats, such as simple sets containing only h, k, l, F and $\sigma(F)$ in free format, XENGEN[14]-style MULISTS or UREFLS formats, SCALEPACK files, or the "scaled" files output from CMBISO or CMBANO. If the latter types are used, one has the option of displaying either native or "difference" precession photographs, with the intensities then being either isomorphous or anomalous differences. One specifies an initial zone such as $hk0$, along with a resolution cutoff and whether a pseudo-background is to be generated. All desired reflections, including those related by symmetry, are then either directly found or generated, and the intensities are scaled with the range mapped to colors or to gray-scale values between 0 and 128 (or 0 and 100 for the X Window version of the program). One of four possible sizes is then chosen for each reflection spot based on the relative intensity. If a pseudo-background is requested, resolution-dependent background values are generated from the average standard deviations in the intensities, and a small random value is added to minimize quantization errors. A reasonable initial display intensity is chosen and the pattern appears on the screen along with a menu. As the mouse cursor moves across the screen the intensity and d spacing are continuously updated and displayed, and when the cursor is near a diffraction maximum (within 0.2 of integral Miller indices) the Miller indices, intensity, and standard deviation are also shown. Mapping of the reflection intensities to the color or gray scale can be finely adjusted from the menu, and one can step up or down to generate the pseudo-photograph corresponding to an adjacent (and parallel) zone. One can also select a new

[14] A. J. Howard, G. L. Gilliland, B. C. Finzel, T. L. Poulos, D. H. Ohlendorf, and F. R. Salemme, *J. Appl. Crystallogr.* **20,** 383 (1987).

zone or direction, and a new resolution cutoff. If a gray scale is chosen along with a good intensity mapping (one in which there is significant contrast between the strongest and weakest reflections), the pseudo-photograph looks very much like a real, properly exposed precession photograph taken with Polaroid film. A sample pseudo-precession photograph is given in Fig. 3.

The programs are useful to aid or confirm space group assignments (if the lowest likely symmetry was used during data reduction), to check diffraction quality and completeness, and to check the quality and distribution of isomorphous or anomalous differences. Good derivatives have differences fairly evenly distributed throughout the film, and good data sets still show variation in intensities near the resolution limit even when proper contrast over the entire film is obtained. Systematically missing regions (uncollected data) are blatantly obvious.

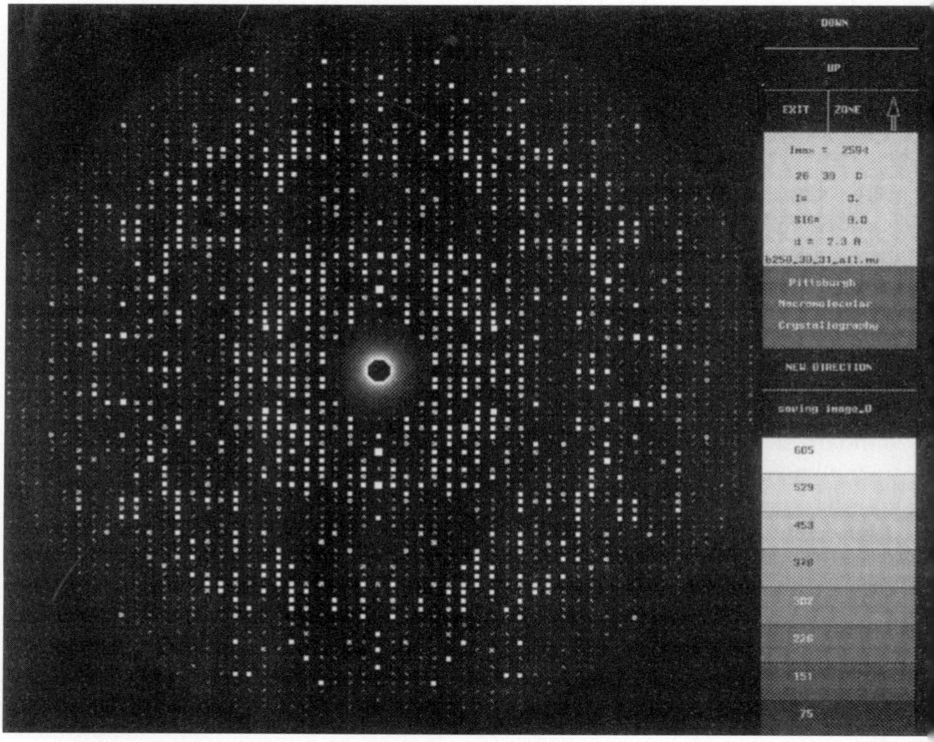

FIG. 3. Native pseudo-precession photograph made by PRECESS, captured from a workstation screen and converted from color to a gray-scale PostScript file for printing. The *hk*0 zone for pyruvate decarboxylase to 3-Å resolution is shown.

Program TOPDEL

The interactive program TOPDEL accepts a "scaled" file and is used both to identify and reject outliers and to produce a structure-factor file suitable for difference Patterson calculations. It applies minimum and maximum resolution cutoffs, F/σ cutoffs, and sorts the reflections in descending order of absolute ΔF (either isomorphous or anomalous differences). It then displays the top (largest) ΔF values and prompts the user to decide if any are deemed outliers and are to be rejected. Selected data are then written to a structure-factor file with coefficients suitable for isomorphous or anomalous difference Patterson calculations to identify heavy-atom sites.

Programs MRGDF and MRGBDF

The interactive programs MRGDF and MRGBDF are used to merge amplitude differences from a "scaled" file with phases and weights from a "structure-factor" file for the purpose of computing difference Fouriers. They accept phases and possibly figures of merit from one file, and corresponding amplitude differences from another to produce a hybrid coefficient structure-factor file suitable for cross-difference or cross-Bijvoet difference Fouriers. Thus one can use phases from one derivative to solve for heavy atoms with properly correlated origin and hand in a new derivative. Both *d* spacing and F/σ cutoffs can be applied. The Bijvoet difference Fourier version may also be used to check the hand of the heavy-atom constellation.

Structure-Factor Calculations and Parameter Refinement

Program PHASIT

The program PHASIT, invoked from a script or command procedure, can be used in one of two modes: protein phasing mode or structure-factor calculation mode. In protein phasing mode structure factors are computed only from heavy-atom positions associated with derivative or anomalously scattering atoms, and the derived information is used only for SIR, multiple isomorphous replacement (MIR), etc., calculations, possibly during parameter refinement. The program expects one or more "scaled" files to be input, and phases are initially computed as described earlier. The heavy-atom and scaling parameters may then be refined for any data set in two basic ways: either by classic "phase refinement" or by maximum likelihood phase refinement. Numerous options are available in both cases. Refinable parameters include all positional values, temperature factors (either isotropic or anisotropic), occupancy factors, native-to-derivative scaling factors, and the scale factor relating the calculated heavy-atom structure factor amplitudes to the observed data. In all situations, parameters are refined by least squares to minimize the quantity

$$\sum_{hkl} W_{hkl} \sum_{\phi_P} P_{\phi_P}[|FPH_{obs}|_{hkl} - |FPH_{calc}(\phi_P)|_{hkl}]^2 \quad (5)$$

or

$$\sum_{hkl} W_{hkl} \sum_{\phi_P} P_{\phi_P}\{(|FPH_{obs}|^{+}_{hkl} - |FPH_{obs}|^{-}_{\bar{h}\bar{k}\bar{l}}) - [|FPH_{calc}(\phi_P)|^{+}_{hkl} - |FPH_{calc}(\phi_P)|^{-}_{\bar{h}\bar{k}\bar{l}}]\}^2 \quad (6)$$

for isomorphous or anomalous scattering data sets, respectively, where

$$|FPH_{calc}(\phi_P)|^2_{hkl} = |FP_{obs}|^2_{hkl} + |FH_{calc}|^2_{hkl} + 2|FP_{obs}|_{hkl}|FH_{calc}|_{hkl}\cos(\phi_P - \phi_H)_{hkl} \quad (7)$$

FP, FH, and FPH are the native, heavy-atom, and derivative structure factors, ϕ_P is the native protein phase, ϕ_H the heavy-atom phase, and the plus and minus superscripts indicate measurements for reflections with indices hkl and $\bar{h}\bar{k}\bar{l}$, respectively. Native anomalous scatterers can be refined with a definition analogous to Eq. (6), but using native amplitudes. Invoking various options, including switching to maximum likelihood mode, involves setting a figure of merit cutoff for reflections in the outer summation, setting the weights W_{hkl}, deciding on the source for ϕ_P, and deciding what phases are to be included in the inner summation. The external weights W_{hkl} can be either $1/E^2$, $1/\langle E^2 \rangle$ where the average is over all data sets contributing to the protein phase distribution, or unity, and E is the expected lack of closure. The inner summation (e.g., the assumed protein phase) can either be restricted to a single value, or can be stepped over all allowed phase values for each reflection. For conventional phase refinement W_{hkl} is typically $1/E^2$, ϕ_P is restricted to the current "best" (centroid) phase, P_{ϕ_P} is set to unity, and a fairly high (0.4–0.6) figure of merit cutoff is used. One has the additional option of temporarily removing the contribution to the centroid phase arising from the derivative being refined to remove bias. In maximum likelihood refinement mode typically the inner summation ranges over all possible phase values for each reflection, P_{ϕ_P} is the corresponding phase probability, the external weight W_{hkl} is either $1/E^2$ or unity, and a low (0.10–0.20) figure of merit cutoff is used. Regardless of the mode employed, after each pass of parameter refinement over all selected data sets, protein phases are computed on the basis of the new parameters, the standard lack of closure estimates are updated (again by probability weighted averaging over all allowed phases) and the protein phases are then recomputed this time using the updated lack of closure estimates.

After all such passes the final protein phases and distribution information are written to a file and a statistical breakdown is given.

In either conventional or maximum likelihood refinement mode one has the option of using protein phase information within the inner summation obtained from either the current data via MIR-type calculations, or from an external file. Temporarily fixing the phase information to that from an external file essentially allows refinement against solvent flattened, negative density truncated and/or NC symmetry averaged maps. This option has been found to be important in several structure determinations, and is particularly useful when refining native-to-derivative scaling parameters and/or when common sites are present in multiple derivatives. Although the heavy-atom parameters and statistical indicators such as mean figure of merit and phasing power do not change much, phases based on the new parameters, when used to initiate another round of density modification, often generate superior maps.

In structure-factor mode PHASIT is usually employed to compute structure factors from atomic coordinates for a complete, or nearly complete, protein model. Files containing atomic parameters and observed structure factor data are input, and the resulting calculated amplitudes and phases are then output along with an R factor and correlation coefficient. Prior to output the scale factor relating F_{calc} to F_{obs} is determined by least squares. One can optionally specify a series of residues to be neglected, so that coefficients appropriate for a residue deleted "omit map" will be output. Any of several output coefficient types may be selected. In one case h, k, l, F_{obs}, F_{calc}, and ϕ_{calc} are written, and the file is thus appropriate for straight electron density, $F_o - F_c$ maps, etc. With another option the calculated and scaled structure factors are sorted into bins based on resolution, and a three-term polynomial is fit by least squares to the mean abs$(|F_o|^2 - |F_c|^2)$ as a function of d spacing. For each reflection a unimodal phase probability distribution based on Bricogne's modification[15] of the Sim procedure[16] is then computed from the observed and calculated amplitudes, calculated phases, and polynomial result. The distributions are then evaluated to obtain centroid phases and figures of merit (fom). An output file containing h, k, l, fom $*F_o$, F_o, ϕ(centroid), and the distribution coefficients is then written. One may also request that the distributions be computed using the Read SIGMAA[17] algorithm instead of the modified Sim procedure. These options allow one to create an MIR-like representation of phase information obtained from conventional structure-factor calculations, and

[15] G. Bricogne, *Acta Crystallogr.* **A32,** 832 (1976).
[16] G. A. Sim, *Acta Crystallogr.* **12,** 813 (1959).
[17] R. J. Read, *Acta Crystallogr.* **A42,** 140 (1986).

thus compute figure of merit-weighted model-based maps or use molecular replacement-derived phase information in subsequent solvent flattening and/or averaging computations. One can also request that the Read SIGMAA procedures be used to output coefficients appropriate for either reduced bias native or reduced bias difference maps.

Program GREF

The program GREF, invoked from a script or command procedure, can be used to compute structure factors from atomic coordinates for an input heavy-atom or protein model. It can also be used to refine heavy-atom parameters against isomorphous or anomalous difference amplitudes, although entire protein domains can be refined as rigid bodies against native amplitudes as well. It accepts "scaled" files as well as free format native files. The user specifies input and output files, resolution and F/σ cutoffs, as well as a series of option flags indicating what data, scattering factors, weights, and variables are to be used. Because it is a GROUP refinement program only rigid bodies can be refined, although the groups can be defined in any arbitrary manner. Indeed, if dealing with individual heavy atoms, one simply defines each "group" to contain only a single atom and then does not select any rotational parameters for refinement. One benefit of this treatment is that if the heavy-atom reagents are known to be rigid groups, the entire group can be properly dealt with, and the program can also be used for conventional rigid body refinement of entire protein molecules, domains, sheets, helices, etc. In general, one can select either structure-factor difference amplitudes or straight amplitudes for the target value; either centric, acentric, or all data to be used; either normal (real) or anomalous dispersion ($\Delta f''$) scattering factors to be used; any of overall scale, overall thermal factor, group centroid x, y, z positions, group rotations about x, y, z axes, group thermal factors, or group occupancies for refinement. If dealing with differences and refinement against centric data is requested, then if an insufficient number of reflections is found the 25% largest acentric differences can automatically be included. This simplifies heavy-atom refinement of all parameters when multiple sites and certain polar space groups are used. When dealing with anomalous dispersion data, the 25% largest differences are automatically selected. Refinement is by least-squares minimization, either with unit weights or weights based on the input standard deviations. The standard R factor and shifts for each group are output and the new coordinates are written to a file. Optionally, a structure-factor file based on the most recent coordinates can be output for normal or "double-difference"-type Fourier map calculations.

Program MAPINV

The program MAPINV, invoked from a script or command procedure, is used to compute structure factors by Fourier inversion of a continuous density function sampled on a finite grid. It is used to compute structure factors from a map that has been modified in some way, usually solvent flattened, negative density truncated, and/or NC symmetry averaged. It accepts only maps created by FSFOUR and can produce from them all structure factors (amplitudes and phases) within a hemisphere. The user specifies input and output files along with the range of indices for which structure factors are desired. One can optionally truncate density below an input value or square the densities prior to inversion.

Maps, Masks, and Related Calculations

Program FSFOUR

The program FSFOUR, invoked from a script or command procedure, is responsible for all map calculations. It is a space group-general, variable radix 3D fast Fourier transform program to produce maps from an input set of unique structure factors. The input reflections are first expanded to a hemisphere and the calculation then proceeds in P1. The user can control the grid spacing, but the number of grid points on each axis must be even and be a composite of the factors 2, 3, 4, or 5, although the program will automatically adjust if the user requests unsuitable values. A variety of coefficients may be chosen to generate Patterson maps, electron-density maps, $2F_o - F_c$ maps, Bijvoet difference maps, etc. The output map always covers one full cell. The map can be searched for peaks, contoured either interactively or offline, skewed, converted to graphics map format, and further manipulated by other programs described below.

Program GMAP

The interactive program GMAP serves as an interface between FSFOUR maps and graphics maps employed by the powerful external programs TOM (SGI version of FRODO[18]), O,[19] and CHAIN.[20] It first prompts for the input FSFOUR map file and a region to be extracted. Any region, including one spanning multiple cell edges, may be requested. It then requests an output file name and desired output format. If a CHAIN

[18] T. A. Jones, *J. Appl. Crystallogr.* **11,** 268 (1978).
[19] T. A. Jones, J. Y. Zou, S. W. Cowan, and M. Kjeldgaard, *Acta Crystallogr.* **A47,** 110 (1991).
[20] J. S. Sack, *J. Mol. Graphics* **6,** 224 (1988).

format map is requested, the program inquires as to whether CHAIN will be run on an SGI or ESV workstation, and the appropriate file is generated. In all cases standard deviations in the input and output maps (which differ due to scaling) are reported. It then asks if any skeleton[21] files are to be generated. If so, either TOM-style skeleton files, O-style skeleton data blocks, or both can be created corresponding to the output region. One useful feature is that although the output (map and TOM skeleton) files are binary, the programs are written such that the created files can be used directly by the target graphics program on SGI or ESV graphics workstations, regardless of the particular hardware on which GMAP was run. Thus even if GMAP is run on a DEC VMS, IBM R6000, or other system, the binary files can simply be transferred to the SGI or ESV workstation (say, for example, by FTP with type binary set) and be used immediately within the target programs, without the need for any additional byte swapping, floating point representation, or format conversion programs.

Programs MAPVIEW and MAPVIEW_X

The interactive programs MAPVIEW and MAPVIEW_X are used to contour and display electron-density or Patterson maps and masks. They can work with any FSFOUR map or with submaps. If an FSFOUR map is input one can select for display any region, including one spanning multiple cell edges, and view it as contoured *xz*, *yz*, or *xy* sections. One can change the region, orientation, and contour levels at any time from a menu. If a submap is used, the region and orientation are fixed as input. The menu always allows one to scroll up or down instantly through the sections, or to add (superimpose) the previous or next section to create projections. As the mouse cursor moves over the display, the corresponding fractional coordinates are continuously updated and shown. If desired, masks (either solvent or averaging) can be displayed superimposed on the contoured density. When masks are being used (either created "from scratch" or read in from a file) additional menu items become functional, allowing one to display and "trace out" (i.e., create or edit) the mask for any section. Because the masks are typically slowly varying functions, at least when near molecular centers, one can also "copy" the next or previous sections mask, to see rapidly if it is still applicable or needs editing. These features enable one to view/edit all masks, and to create envelope masks rapidly for averaging over NC symmetry. The options are particularly powerful when the NC symmetry involves only an *N*-fold rotation axis, as a skewed map can be input and the averaging mask can be created while

[21] J. Greer, *J. Mol. Biol.* **82,** 270 (1974).

looking directly down the NC rotation axis, where the breakdown of the local symmetry is usually obvious. Creating masks is simplified by first forming a projection over about four map sections as contrast in the boundary is then enhanced. From the menu one can toggle between 12 different mask values, and each is color coded in the display. This allows up to 12 independent envelopes to be created and later distinguished, facilitating averaging when translational components and/or arbitrary rotations are involved. Once the mask sections encompassing the molecules to be averaged are created, a *make asymmetric unit* menu item can be selected to verify that all points within mask envelopes are unique (not related to other points within the envelopes by crystal symmetry). If the user, for example, strayed into a neighboring asymmetric unit when creating the mask, the redundant points will be flagged in red to facilitate further editing. Once this symmetry check within the masked regions is completed, an outward expansion by crystal symmetry is made so that all points outside the masked regions, but related to points within them by crystal symmetry are flagged in green. This is useful to determine packing contacts and to ensure completeness of the structure. After this step, if any significant density remains that is not in either a primary envelope or against a green background, a part of the structure must have been omitted as it should have been assigned to some molecular envelope. Finally, on exiting the map region and/or mask can be saved to a file. MAPVIEW and MAPVIEW_X are powerful programs that are useful for examining any map or mask, and are also used, for example, to determine the region isolating a molecule that will be needed later by a graphics program for chain tracing. They are particularly important for NC symmetry averaging applications. A sample MAPVIEW display is shown in Fig. 4 (see color insert).

Program CTOUR

The program CTOUR, invoked from a script or command procedure, is used to create generic contoured plot files from FSFOUR maps. Any region, including those spanning multiple cell edges, can be extracted. Individual sections, monoprojections, or stereoprojections can be created. Multiple plot files corresponding to any combination of different types, regions, and directions may all be produced in a single run. The generic plot files created can be displayed on workstation monitors, X-terminals, dumb terminals in Tektronix 4010 mode, or converted to PostScript by the appropriate driver programs (which are described below). An example of an output plot is shown in Fig. 5.

Program PSRCH

The program PSRCH, invoked from a script or command procedure, can be used to search the FSFOUR map and produce a list of unique peaks.

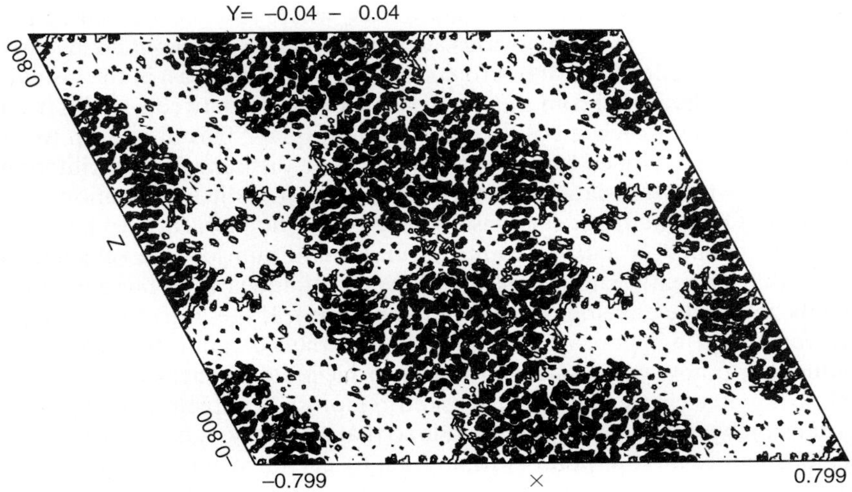

FIG. 5. Example of a CTOUR plot, converted to PostScript by MKPOST for printing. A projection over seven sections in a 3-Å solvent-flattened, NC symmetry-averaged MIRAS map for a pyruvate decarboxylase tetramer (space group *C*2) is shown contoured at the 1 σ level. The map is viewed down the *b* axis and illustrates both the general packing and dimer–dimer interactions forming the tetramer.

It is most useful for examination of difference Fouriers to identify minor sites, or for examination of cross-difference Fouriers to find heavy atoms in a new derivative based on phases obtained from other derivatives. Interpolated heights and positions are provided for a set of unique peaks in the map. One can select either positive or negative peaks to be sought, with the latter option useful in certain MAD phasing applications when the anomalous scatterer peaks are expected to be negative.

Programs MDLMSK and MRGMSK

The interactive programs MDLMSK and MRGMSK can be used to create masks either for solvent flattening or averaging from input atomic models, and to combine multiple mask files into a single one. MDLMSK prompts for a map region and grid, an atomic coordinate file, mask number, and a masking radius. A 3D mask is then constructed and output such that all grid points within the specified radius of any atom in the input model are assigned the specified mask value. MRGMSK combines two mask files created by MDLMSK into a single mask file. The two mask files must cover identical regions with identical grids, although they may (and usually do) have different mask values. If a point is found to be assigned to both input

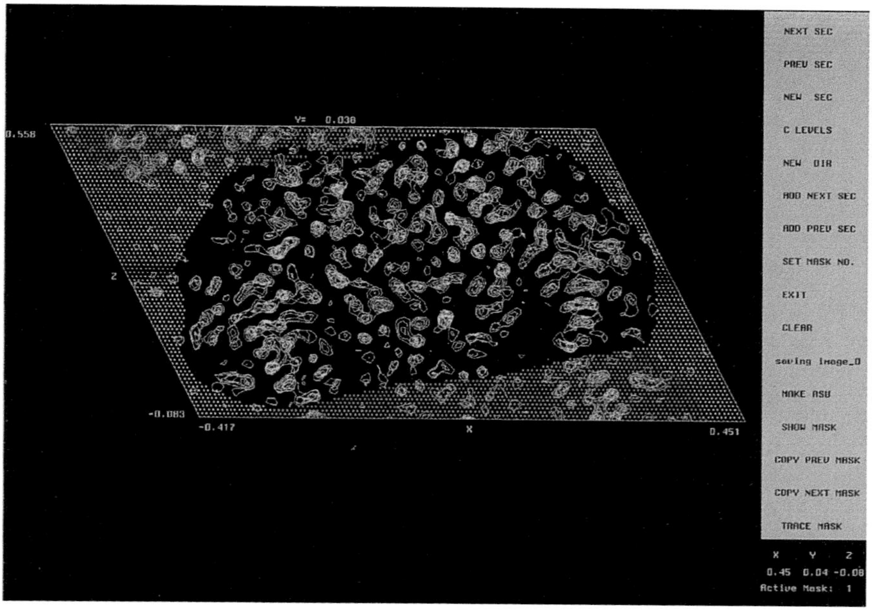

FIG. 4. Image displayed by MAPVIEW, captured from a workstation screen and converted to a PostScript file for printing. A projection over four electron-density sections (in a 3-Å map) is shown for part of a pyruvate decarboxylase dimer, with contours at the 1 σ level. Part of an averaging mask envelope traced seconds earlier via the mouse cursor is displayed simply as the black background, whereas grid points outside the envelope appear as dots, either blue for the solvent region or green if related to the primary protein envelope by crystallographic symmetry. A noncrystallographic twofold axis lies within the projected volume roughly parallel to the xz plane and is inclined by 102° to the x axis. The mask used for averaging in the structure determination was created in a similar way, but while using a skewed map viewed directly down the local twofold axis.

masks, it is redefined as a nonmasked point on output. The total number of masked points and number of overlapping points are then listed. Ideally, one creates a model-based mask using van der Waals radii, and then quickly edits the mask in MAPVIEW to maintain the outer boundary, but to fill in "holes" within the molecular interior. The resulting mask can be used for NC symmetry averaging, operator refinement, etc., or it can be expanded to a full cell mask by BLDCEL for later use as a solvent mask.

Solvent Flattening, Phase Combination, and Phase Extension

Program BNDRY

The program BNDRY, invoked from a script or command procedure, is used for automatic masking, solvent flattening, negative density truncation, phase combination, and phase extension. Four options are possible, two related to automatic construction of solvent masks, one to density modification, and one to phase combination (possibly with phase extension). For the first option an input reflection file obtained (from MAPINV) by inverting a map in which all density with values below zero have been set to zero is supplied, along with a sphere radius for weighted averaging during solvent mask construction. For this map the density scale is arbitrary and the F_{000} term is neglected in the summation. Following Wang,[4] the radius R is generally taken to be 2.5–3 times the data resolution. The Fourier transform $F(s)$ of the weighting function $W(r) = 1 - r/R$ with $W(r) = 0$ for $r > R$ is then computed and each unique structure factor is multiplied by $F(s)$, where

$$F(s) = 4\pi R^3\{2[1 - \cos(A)] - A\sin(A)\}/A^4 \tag{8}$$

and

$$A = 4\pi R\sin(\theta)/\lambda \tag{9}$$

The modified structure factors are then output, and can be used (in FSFOUR) to create a "smeared" map for boundary determination that corresponds to convolution of all nonnegative density in the original map with the weighting function $W(r)$.

For the second option, the "smeared" map is read in along with an estimate of the solvent fraction. The solvent fractional volume is converted to the corresponding number of map grid points and a histogram is constructed that identifies the number of grid points having each density value. Starting with the lowest observed density a threshold value is increased in increments, and a running sum is maintained that identifies the current number of grid points with density values below the threshold. When the number of points reaches that expected in the solvent volume, the corresponding threshold indicates the cutoff density for the solvent region. A mask having a one-to-one correspondence with the map grid is then created such that if the density is below the threshold, the grid point is deemed to be in the solvent region; otherwise it is in the protein region. The mask is then written to a file, which can be viewed/edited in MAPVIEW.

For the map modification option, electron-density map and mask files are read in along with an empirical constant S that is used to estimate the value of F_{000}/V on the scale of the input map. The estimation, again following Wang,[4] is based on the assumption that for typical solvent conditions and proteins without heavy metals the ratio of mean solvent electron density to maximum protein electron density is relatively constant, although it tends to be resolution dependent. Thus by coupling values of S from known systems with density values extracted from the arbitrarily scaled map, an estimate of F_{000}/V on the appropriate (but unknown) scale can be obtained by solving the equation

$$\frac{\langle \rho \rangle_{\text{solvent}} + F_{000}/V}{\rho_{\text{max,protein}} + F_{000}/V} = S \tag{10}$$

Once the estimate of F_{000}/V is determined, solvent flattening and negative density truncation are carried out simply by resetting all map values according to the relationships

$$\begin{aligned} \rho &= \langle \rho \rangle_{\text{solvent}} + F_{000}/V \qquad \text{if in solvent region} \\ \rho &= \text{maximum}(\rho_{\text{input}} + F_{000}/V, 0) \qquad \text{if in protein region} \end{aligned} \tag{11}$$

After reporting the estimate of F_{000}/V, the modified map is written to a file suitable for inversion by MAPINV.

For the final phase combination option two structure factor files are supplied along with a flag indicating whether any phase extension is desired. One of the files contains an "anchor" phase set (usually SIR, MIR, etc., phases from PHASIT) and includes phase probability distribution information in the form of A, B, C, D coefficients.[7] The second file contains new phase information (obtained by inversion of modified maps or by structure-factor calculations for a partial structure), and includes only indices, amplitudes, and phases. All unique reflections in the new file are matched up with their counterparts in the anchor set, and the calculated amplitudes are scaled to the anchor set by least squares. The R factor and correlation coefficient are then reported and the data are sorted into bins according to d spacing. A three-term polynomial is then fit to the mean values of $|F_o^2 - F_c^2|$ as a function of resolution. For each matched reflection in the new phase set a unimodal phase probability distribution is constructed following Bricogne's modification[15] of the Sim weighting scheme,[16] with the appropriate distribution coefficients given by

$$\begin{aligned} W &= 2F_o F_c / \langle |F_o^2 - F_c^2| \rangle_d \\ A &= W \cos(\phi_c) \\ B &= W \sin(\phi_c) \\ C &= 0 \\ D &= 0 \end{aligned} \tag{12}$$

where the denominator in W is determined from the polynomial. Phase combination is then carried out by simple addition of the distribution coefficients, and the resulting combined distributions are evaluated and reduced to centroids and associated figures of merit. Optionally, one can request that the unimodal distributions are constructed using the Read SIGMAA[17] method instead of the modified Sim procedure. Figure of merit statistics are then listed and the combined phases and distribution coefficients are written to a file for use in map calculations or additional processing. A damping factor in the range 0–1 is also supplied, which can be used to damp the contributions of the anchor set. Although normally set to 1 (no damping), if less than 1 more emphasis will be placed on the map inverted/partial structure phases. If set to zero the new phases are essentially accepted as is, because there is effectively no phase combination.

If phase extension is requested, then an additional file (prepared by MISSNG) is supplied along with a d spacing cutoff. This file contains all unique reflections that are absent from the "anchor" file (due to application of cutoffs, rejection criteria, etc.) but for which observed native amplitudes (and possibly phase probability distribution coefficients) are available. Pairing of the observed and calculated amplitudes, scaling (with the previously determined scale factor), phase distribution coefficients, centroids, and figures of merit are then computed exactly as for the original data, and the results are included in the output file. If phase probability distribution coefficients were supplied on the extension list then phase combination is carried out; otherwise the calculated phase is output exactly as input. On the first cycle the calculated F values are weak and the figures of merit low for these reflections, but after a few iterations they usually improve significantly. As a final option, phase and amplitude extension may be requested, in which case any remaining unpaired unique reflections input are also included in the output. In that case $F_c/2$ simply replaces F_o in the numerator of W, with the remaining processing identical. These options enable one to extend phasing, either to reflections within the initial resolution range or to higher resolution, and if desired even to generate reasonable values for unmeasured data. Phase extension is essential when phasing purely by single-wavelength anomalous scattering, as it is the only way to ensure that centric reflections are phased. Amplitude extension, although rarely done, may be useful to include low-resolution data typically masked by a beamstop.

Program SLOEXT

The program SLOEXT is automatically invoked from any script or command procedure dealing with phase extension, either with or without NC symmetry averaging. It is used to control the rate and range of phase extension. The user specifies the initial and final *d* spacings as well as the number of map modification/phase combination cycles to be performed at each *d* increment. The *d* spacing cutoff is then incremented in steps corresponding to the smallest reciprocal lattice spacing, allowing slow phase extension in a series of shells expanding by essentially one reciprocal lattice point.

Noncrystallographic Symmetry Averaging and Related Calculations

Programs EXTRMAP and EXTRMSK

The programs EXTRMAP and EXTRMSK, invoked from a script or command procedure, are used mainly in NC symmetry-averaging applications to extract regions encompassing only the dimer, trimer, etc., to be averaged from full cell FSFOUR maps or solvent masks, respectively. Any region can be extracted, including one that spans multiple cell edges. The resulting region is written to a new submap file, which can be contoured and examined or utilized by other programs in the package. For EXTRMSK the corresponding mask region is written to a file.

Program SKEW

The program SKEW, invoked from a script or command procedure, is used mainly in NC symmetry-averaging applications in which one wants to create an electron-density map in sections orthogonal to any arbitrarily oriented vector, for the purpose of constructing an averaging envelope mask or to verify the local symmetry and averaging. A submap file (obtained from an FSFOUR map by MAPVIEW or EXTRMAP) and a direction (specified in terms of spherical polar angles) along with a desired range and grid spacing are to be supplied. A 64-point cubic spline interpolation algorithm is used to obtain the appropriate density values, and create a new submap with its *b* axis aligned along the specified direction and passing through a specified point. The skewed submap can be contoured, viewed, and used for mask creation in MAPVIEW. Optionally, a mask file adhering to the input submap region can also be skewed so as to maintain correspondence with the skewed submap.

Program MAPAVG

The program MAPAVG, invoked from a script or command procedure, is used to average electron density over NC symmetry elements. Noncrystallographic symmetry averaging can be done either within a single crystal, between multiple nonisomorphous crystals (up to six), or both simultaneously. One submap file encompassing the monomer, dimer, trimer, etc., to be averaged in each crystal is input along with the corresponding averaging mask file, and the NC symmetry operations are specified along with the associated mask numbers. A copy of each input submap is first made, and the copy is then modified by stepping over each grid point in turn and examining the associated mask value. If the grid point is within an envelope to be averaged, all points related to it by the local symmetry are generated and examined to see if they also lie within the appropriate averaging envelope. If so, the density associated with each related point is determined from the input submaps by interpolation via a 64-point cubic spline algorithm, the average over all values is computed, and the result is deposited at the original point in the new map copy. When multiple crystals are used, scale factors relating the density within each submap are automatically refined by least squares prior to the averaging. After processing all grid points in this manner, each new, averaged submap is output to a file, and the correlation coefficient associated with each operator is output.

Program BLDCEL

The program BLDCEL, invoked from a script or command procedure, is used to expand an NC symmetry averaged submap created by MAPAVG back to a full cell FSFOUR-type map that obeys space group symmetry. Input files include the unaveraged full cell FSFOUR map from which the region to be averaged was originally extracted, the averaged submap created by MAPAVG, and the averaging mask. First the new map is initialized as an exact copy of the unaveraged FSFOUR map to ensure that all unaveraged points will retain their original values. Then the averaged submap and associated averaging mask are examined point by point. If a grid point lies within an averaging envelope, its value in the averaged map is inserted into the new map both directly and at all points in the unit cell related by crystallographic symmetry. Following this procedure the new map spans a complete cell and contains NC symmetry-averaged density within all averaging envelopes and at all points related to them by crystal symmetry, and the original, unaveraged density everywhere else. The new map is then written to a file and is consistent in format with any FSFOUR map. Thus it is suitable for subsequent solvent flattening and inversion by BNDRY and MAPINV, respectively.

As an option, the averaging mask can also be expanded to its symmetrized, full cell version and output, in which case it could later be used for solvent flattening everywhere outside the averaging envelopes. In practice, however, it is safer to use an independent, objectively determined mask for solvent flattening as this is much more tolerant of errors near the averaging mask edges, and allows one to reap most of the benefits of averaging early on, often before the outer averaging envelope boundary can be precisely determined.

Program MAPORTH

The program MAPORTH, invoked from a script or command procedure, is used only as a precursor to refinement of NC symmetry operators, and even then it is needed only if the unit cell is not orthogonal. The computations are similar to that in SKEW, but the new submap is simply an orthogonalized version of the input submap. Optionally, an averaging mask can also be orthogonalized in a corresponding manner so that it can be used later to identify masked grid points for operator refinement.

Programs LSQROT and LSQROTGEN

The programs LSQROT and LSQROTGEN, invoked from a script or command procedure, are used to refine NC symmetry operators against electron density. LSQROT is used when the operator is purely rotational with the rotation given by $360/n$ where n is a small integer, i.e., pure twofolds, threefolds, etc. LSQROTGEN is used for general transformations that may involve arbitrary rotation angles (including zero for pure translations) and/or postrotation translations (including screwlike operations). An input submap (which must be on an orthogonal grid; see Program MAPORTH) is supplied along with starting values for the operator and, optionally, an averaging mask. Sphere centers and radii can be supplied in lieu of (or in addition to) the averaging mask. In general, any of the transformation defining parameters ϕ, ψ, χ, O_x, O_y, O_z, and T can be refined or held fixed, although χ (order of rotation) and T (postrotation translation parallel to the rotation axis) can be refined only with LSQROTGEN. The programs step over all grid points in the submap, selecting for refinement those points that, along with their specified NC symmetry-related mates, fall within the appropriate sphere, specified averaging mask, or both. The selected parameters are then optimized by least-squares minimization of the difference in electron density at the related points. When needed, density is interpolated by the 64-point cubic spline algorithm as in SKEW. Shifts in the parameters and the correlation coefficient between density related by the operator and within the selected volume are then output. LSQROT functions with a single submap whereas LSQROTGEN may use more than one submap, allowing refinement of operators relating density in nonisomorphous crystals. Beginning with a low-resolution map (6 Å) on a coarse grid (2-Å grid spacing) usually results in rapid convergence, even when the operator is off by up to 10° or several angstroms.The correlation coefficient is a particularly good indicator of progress, with values around 0.4 or higher (in a 3-Å MIR map, 1-Å grid) usually indicating that the position and nature of the operator have been correctly identified. Use of spheres allows operator refinement early in the phasing process, prior to determining an averaging mask. When pure N-fold rotations are involved, a single sphere can be centered on the NC symmetry axis roughly near the oligomer center of mass. For general transformations multiple spheres, each centered on a molecule within the oligomer or centered on the appropriate entity in each crystal, suffice.

Program TRNMSK

The program TRNMSK, invoked from a script or command procedure, is used to transform a mask that was created in a skewed submap to the corresponding mask in the original, unskewed submap. Although in some instances creating an averaging mask is simplified when working in a skewed cell, there is no need for skewing during the actual averaging cycles (indeed, it unnecessarily increases CPU time). Thus one can create the mask in a skewed cell, but then transform it to correspond to the original cell for repeated use in averaging cycles.

Graphics Programs and Displaying Plots

The interactive graphics programs PRECESS, PRECESS_X, MAPVIEW, and MAPVIEW_X have already been described. The multiple versions of these programs, and also VIEWPLT and VIEWPLT_X described below, function identically with the same "look" and response, but differ in two minor respects. First, the original (non-X) versions run only on Silicon Graphics workstations and make use of the proprietary graphics libraries of the vendor. The "X" versions instead are relatively hardware independent as they utilize standard X Window graphics libraries found on most workstations (including SGI). The X versions can create their displays directly on the workstation monitors or on X terminals. Second, because of the different graphics libraries in use, the menu items allowing saving of the screen to RGB files is disabled on the X versions. All of the graphics programs work best on color monitors with screens 16 inches or larger.

Programs VIEWPLT and VIEWPLT_X

The interactive programs VIEWPLT and VIEWPLT_X are used to display the generic contoured plot files created by CTOUR on Silicon Graphics workstation monitors or on any devices supporting the X Window protocol. Up to 10 plot files can be displayed simultaneously. Although not nearly as flexible as MAPVIEW, they are the fastest way to preview CTOUR plots (more than 100 times faster than PostScript previewers) and are also useful when solving difference Pattersons as all contoured Harker sections can be displayed simultaneously.

Program PLTTEK

The interactive program PLTTEK is used to display the generic contoured plot files created by CTOUR on any device supporting Tektronix 4010 emulation. Although slow, it allows the plots to be viewed on relatively "dumb" terminals, facilitating work away from the laboratory where more powerful workstations may not be available.

Program MKPOST

The interactive program MKPOST is used to convert one or more of the generic contoured plot files created by CTOUR to equivalent PostScript files. It is thus useful as a prerequisite to obtaining hard copies of the plots on PostScript printers, or to create portable, machine-readable versions of the plots for incorporation into manuscripts. It is actually a script (on UNIX systems) and a command procedure (on VMS systems) that invokes another supplied program (POSTPLOT) for the conversions.

Miscellaneous Programs

Program MISSNG

The interactive program MISSNG is used to prepare a file containing reflections for possible phase extension. It prompts for a master file containing all native reflection amplitudes, for the current (phased) structure-factor file, for an output file, and for a *d* spacing cutoff. It also inquires whether an additional file containing both amplitudes and phase probability distributions for extended reflections is to be used (for example, SIR distributions out to high resolution for one derivative). Reflections that are absent from the phased file but present in the other input files are written to the output extended reflection file, and are suitable candidates for phase extension by solvent flattening, negative density truncation, and/or NC symmetry averaging.

Program RMHEAVY

The program RMHEAVY, invoked from a script or command procedure, is used to temporarily remove density near heavy-atom sites from maps during solvent mask creation. It is automatically invoked by the *doall* procedure. Strong density near heavy-atom sites in initial MIR or SIR maps can distort solvent masks by extending the protein area into what should be the solvent region. This occurs because density near the heavy-atom sites is often strong in early maps, and because the heavy-atom sites are nearly always on the protein surface bordering the solvent region. If a tight mask is used (as it should be, for maximum effectiveness!), artificial extension of the protein envelope into the solvent region implies that the envelope must also be depleted elsewhere. To avoid this a copy of the original map is made such that all density within a specified radius of any heavy atom used in the phasing (or its symmetry mates) is zeroed out. This map is then used only for solvent mask construction.

Program HNDCHK

The interactive program HNDCHK is used to examine electron-density values at specific locations within a map, usually for the purpose of determining the absolute configuration from a Bijvoet difference map phased with SIR or MIR data. The user is prompted for the map file and a list of coordinates, which should correspond exactly to the tentative anomalous scatterer locations. The density values are then obtained precisely at those coordinates and at coordinates related by a center of symmetry, by interpolation using a 64-point cubic spline algorithm. If the input heavy-atom configuration had the correct hand large positive peaks should occur exactly at the input locations. If it was incorrect even larger negative peaks should occur at the true heavy-atom sites, i.e., those related to the input (incorrect) locations by a center of symmetry.

Program PSTATS

The interactive program PSTATS is used to compare phases in two different structure-factor files. It prompts for two input files, and prepares a list of mean phase differences for common reflections as a function of d spacing. It is useful when comparing the results from different phasing strategies, when testing new procedures, to check for convergence in iterative procedures, or to assess relative contributions of phase sets during phase combination.

Program RDHEAD

The interactive program RDHEAD can be used to read and output the header record in any mask or submap file. Thus one can examine the grid spacing and regions covered within the file and verify consistency.

Auxiliary Conversion and Interface Programs

Program IMPORT

The interactive program IMPORT allows the user to import phase information derived from programs external to the PHASES package so that it can be used for subsequent calculations within the package. Thus, for example, one can use phases and associated probability distributions obtained elsewhere to initiate solvent flattening, negative density truncation, and/or NC symmetry averaging in PHASES, or simply to generate and display maps or convert maps to graphics format. Reflection indices, F_{obs}, figure of merit, phase, and phase probability distribution coefficients must be supplied, although free format can be used.

Programs RD31 and MK31B

The interactive programs RD31 and MK31B convert PHASES structure-factor files back and forth between the internal (possibly machine-specific), binary file formats and readable ASCII versions. They are useful if one wants to export processed data to a computer of different architecture, to other software packages, or simply to examine/edit phases and probability distribution information.

Program PDB_CDS

The interactive program PDB_CDS is used to convert atomic coordinate files back and forth between standard Protein Data Bank (PDB) and PHASES formats. Recognized residue types include the standard 20 amino acids, DNA bases, and a series of selected common cofactors and solvent molecules. The program prompts for input and output file names, direction of the conversion, chain/residue ranges and whether to reset occupancies and/or thermal factors to specified values. The coordinate range spanned by the input model is also listed.

Program XPL_PHI

The interactive program XPL_PHI is used to convert a structure factor file from PHASES to a reflection file that can be used in X-PLOR.[22] It prompts only for input and output file names. The output file also contains phase and figure of merit information, and therefore is suitable for refinement in X-PLOR by restraining to MIR or solvent flattened/NC symmetry averaged, etc., phases, if desired.

Program O_to_SP

The interactive program O_to_SP is used to convert NC symmetry operators represented by a 3×3 rotation matrix and 1×3 column vector as in O[19] to the spherical polar representation used in PHASES. It prompts only for the elements of the matrix and vector. The corresponding spherical polar angles ϕ, φ defining the rotation axis orientation, the degree of rotation χ, orthogonal coordinates O_x, O_y, O_z of a point the axis passes through, and the postrotation translation along the rotation axis are then listed.

Conclusion

The PHASES software package described, including earlier releases, has been found to be both efficient and easy to use by many researchers and has been effective in the solution of a large number of protein structures.[23] References to a few representative structure determinations in which PHASES was used, some involving classic MIR or SIRAS (single isomorphous replacement with anomalous scattering) phasing, some with NC symmetry averaging and some with MAD data, include staphylococcal enterotoxin B (Swaminathan *et al.*[24]), glutathione *S*-transferase (Ji *et al.*[25]), human immunodeficiency virus (HIV) reverse transcriptase–antibody complex (Jacob-Molina *et al.*[26]), pyruvate decarboxylase (Dyda *et al.*[27]),

[22] A. Brunger, M. Karplus, and G. A. Petsko, *Science* **235,** 458 (1987).
[23] B. C. Finzel, *Curr. Opin. Struct. Biol.* **3,** 741 (1993).
[24] S. Swaminathan, W. Furey, J. Pletcher, and M. Sax, *Nature (London)* **359,** 801 (1992).
[25] X. Ji, P. Zhang, R. N. Armstrong, and G. L. Gilliland, *Biochemistry* **31,** 10169 (1992).
[26] A. Jacob-Molina, J. Ding, R. G. Nanni, A. D. Clark, Jr., X. Lu, C. Tantillo, R. L. Williams, G. Kamer, A. L. Ferris, P. Clark, A. Hizi, S. H. Hughes, and E. Arnold, *Proc. Natl. Acad. Sci. U.S.A.* **90,** 6320 (1993).
[27] F. Dyda, W. Furey, S. Swaminathan, M. Sax, B. Farrenkopf, and F. Jordan, *Biochemistry* **32,** 6165 (1993).

histone H5 globular domain (Ramakrishnan *et al.*[28]), DNA methyltransferase (Cheng *et al.*[29]), T7 RNA polymerase (Sousa *et al.*[30]), tyrosine phosphatase (Barford *et al.*[31]), HIV-1 integrase catalytic domain (Dyda *et al.*[32]), and Epstein–Barr nuclear antigen 1 (Bochkarev *et al.*[33]).

Acknowledgment

The authors are grateful to the many users who have provided feedback on their experiences with the package, and to Digital Equipment Corp. for providing hardware facilitating development of the DEC alpha AXP versions and general X Window codes.

[28] V. Ramakrishnan, J. T. Finch, V. Graziano, P. L. Lee, and R. M. Sweet, *Nature* (*London*) **362,** 219 (1993).
[29] X. Cheng, S. Kumar, J. Posfai, J. Pflugrath, and R. Roberts, *Cell* **74,** 299 (1993).
[30] R. Sousa, Y. J. Chung, J. P. Rose, and B. C. Wang, *Nature* (*London*) **364,** 593 (1993).
[31] D. Barford, A. J. Flint, and N. K. Tonks, *Science* **263,** 1397 (1994).
[32] F. Dyda, A. B. Hickman, T. M. Jenkins, A. Engelman, R. Craigie, and D. R. Davies, *Science* **266,** 1981 (1994).
[33] A. Bochkarev, J. A. Barwell, R. Pfuetzner, W. Furey, A. M. Edwards, and L. D. Frappier, *Cell* **83,** 39 (1995).

[32] Collaborative Computational Project, Number 4: Providing Programs for Protein Crystallography

By ELEANOR J. DODSON, MARTYN WINN, and ADAM RALPH

Introduction

CCP4 (Collaborative Computational Project, number 4[1]) is funded by the UK Biotechnology and Biological Sciences Research Council (BBSRC) to support the development and use of computational tools in the field of macromolecular crystallography. Its first aim is to make available to the community a suite of programs to aid the determination of macromolecular structures by crystallography. Its second, and equally important, aim is to promote discussion on state-of-the-art techniques, and to educate users in these techniques and the associated computer programs. The latter aim is pursued via an annual workshop, a twice yearly newsletter, and an active e-mail discussion list (the *Bulletin Board*).

[1] Collaborative Computational Project, No. 4, *Acta Crystallogr.* **D50,** 760 (1994).

0076-6879/97 $25.00

CCP4 was established in 1979, modestly funded by the then UK Science and Engineering Research Council (SERC). This was a time when the number of institutions doing macromolecular structure determination was increasing, while the size of the new groups remained quite small. There was a widespread awareness that without cooperation in the development of software, a great deal of effort would be duplicated, and many projects handicapped by not having used the latest techniques. So from the beginning it was agreed that code would be pooled, that standard file formats would be accepted, and that effort should be put into guaranteeing that the programs were portable. Collaboration on the development of programs was subsequently extended into Europe under the auspices of the European Science Foundation (ESF) Network of the European Association of the Crystallography of Biological Macromolecules (EACBM).

From these beginnings the present suite of about 100 programs has grown, covering all aspects of macromolecular crystallography from data reduction through initial phasing to final structure refinement. In line with the collaborative nature of the project, these programs have been donated to the suite by many people. There has been no attempt to be exclusive, and there may well be more than one program to cover a particular function. All these programs, however, use standard formats for input and output of data, and in the majority of cases standard keywords control the precise actions of each program. The suite, the associated library routines, and the standard file formats are discussed in detail below.

As part of the educational remit of CCP4, the annual Study Weekend takes place at the beginning of January. Each year, a particular topic is chosen, and a series of introductory talks and more specialized talks is given. These cover background theory, specific algorithms and programs, and instructive case studies. The resulting published proceedings are often one of the most up-to-date texts available on the chosen topic. The meetings held to date are shown in Table I.

Other sources of up-to-date information are provided by the *CCP4 Newsletter* (distributed twice yearly, and now also available from the CCP4 Web page: http://www.dl.ac.uk/CCP/CCP4/main.html) and the *Bulletin Board.* The latter is an e-mail discussion list for general queries on techniques and recommended strategies. Specific questions about the CCP4 program suite are handled by the Daresbury Laboratory (Daresbury, Warrington, UK) staff (e-mail: ccp4@dl.ac.uk).

The funding of CCP4, including important contributions from industrial companies, has grown to allow the employment of full-time staff based at Daresbury Laboratory to coordinate the activities of the project. In addition to the resources already mentioned, this has allowed the production of extensive documentation for the program suite (including both a general

TABLE I
CCP4 STUDY WEEKENDS

Year	Topic	Reference code
1980	Refinement of protein structures	DL/SCI/R16
1985	Molecular replacement	DL/SCI/R23
1987	Computational aspects of protein crystal data analysis	DL/SCI/R25
1988	Improving protein phases	DL/SCI/R26
1989	Molecular simulation and protein crystallography	DL/SCI/R27
1990	Accuracy and reliability of macromolecular crystal structures	—
1991	Isomorphous replacement and anomalous scattering	DL/SCI/R32
1992	Molecular replacement	DL/SCI/R33
1993	Data collection and processing	DL/SCI/R34
1994	From first map to final model	DL/SCI/R35
1995	Making the most of your model	DL-CONF-95-001
1996	Macromolecular refinement	DL-CONF-96-001
1997	Recent Advances in Phasing	

manual and individual program documentation), numerous example scripts and tutorials, and a simple installation procedure for both Unix and VMS machines. Furthermore, CCP4 has funded several postdoctoral positions and short-term contracts for individuals who are interested in tackling specific perceived problems. Their work is used to enhance the suite. Finally, CCP4 has also supported financially various macromolecular crystallography meetings on an occasional basis, such as the International Union of Crystallographers (IUCr) Macromolecular Crystallography Computing School (Western Washington University, Bellingham, WA, August 1996).

Overview of CCP4 Program Suite

Unlike many other packages, the CCP4 suite is designed to be loosely organized, so that it is easy for different developers to add new programs or to modify existing ones without upsetting other parts of the suite. The suite thus consists of a set of separate programs, each of which performs a specific task that may be as minor as adding free-R flags to a set of reflections (*freerflag*) or relatively complex such as performing simultaneously several kinds of density modification (*dm*). The programs communicate via standard data files, and higher level tasks are effected by chaining together several programs with the output data file of one program forming the input data file of the next. This is the approach successfully taken by Unix, and now apparently being embraced by some of the large commercial software houses.

This approach allows great flexibility in the use of the suite. However, it has the disadvantage that the use of the suite may appear complicated to a new user, compared with a more integrated approach. A complete structure solution requires the use of several command scripts (Unix shell scripts or VMS DCL scripts), and examples are distributed with the suite. A draft specification for a graphical user interface to the suite has been drawn up, and implementation of it has been started. It is hoped that this will increase the user friendliness of the suite and aid in problem solving, but it is equally important that it does not add constraints to what is a flexible system.

The programs have been provided from a variety of sources. Some were written specifically for the suite, whereas others were donated to the suite at a later date. Converting a program to a CCP4-like style, and in particular to use standard CCP4 file types, is facilitated by the subroutine library in which most of the standard operations are centralized. The suite is distributed as source code, and installation procedures are provided for VMS and most Unix platforms. To aid portability, most programs are written in standard Fortran-77, and some are in ANSI C. There are a few X Windows-based programs, but most are command line driven.

Briefly, the suite contains application programs covering all aspects of macromolecular crystallography from data processing to analysis of a refined model, for example, the reduction and analysis of intensity data, structure solution by isomorphous replacement and molecular replacement, and refinement and analysis of the structure. These programs use a standard subroutine library to access standard format files. Jiffy programs are included to convert between CCP4 file formats and other common data formats. In addition to the code itself, the distribution includes a manual, individual program documentation, and example scripts for both the VMS and Unix operating systems. We now describe the programs and the associated subroutine library in more detail.

Standard File Formats and Library Routines

The unifying feature of the suite is the use of three basic file types, namely the MTZ format for reflection files, the CCP4 map file format, and the Protein Data Bank (PDB) format for coordinates. Closely connected with these file types is a set of library routines that can be used to read and write information to and from these files. Other library routines exist to perform standard functions, such as symmetry operations. The suite includes many jiffy programs (e.g., *cad* or *maprot*), which use library routines to display or alter information in the data files. Users therefore do not need detailed information about the file formats or the routines used to access them. However, we describe the file formats and libraries here because they are central to the suite as a whole.

The three basic file types consist of the data itself together with crystallographic and bookkeeping information stored in a header section. The well-known PDB format is ASCII, whereas the MTZ and map formats are binary, principally due to considerations of size and accuracy. The binary files include a machine stamp that allows the files to be moved between systems.

Labeled Column Reflection Data Files (MTZ)

The MTZ reflection file format (renamed from LCF for three of its progenitors, S. McLaughlin, EMBL, Hamburg, H. Terry, EMBL, Hamburg, and J. Zelinka, University of York) uses fixed-length records for each reflection with a minimum of 4 columns (H K L plus at least one data column) and currently a maximum of 200 columns of data per reflection record. The columns of the reflection data records are identified by alphanumeric labels and column-type flags held as part of the file header information. The user relates the item names used by the program to the labels of the required data columns by means of assignment statements in the program control data (LABIN and LABOUT keywords). The programs check to see that the associated column type is valid for the program operation, e.g., that a phase is not being assigned to a standard deviation (this may bring to mind "tables" or "relations" in relational databases—intentionally so). Definitions of acceptable types, and a list of common program labels, are given in Table II. Additional crystallographic information (title, cell dimensions, column labels, symmetry information, resolution range, history information, and, if necessary, batch titles and orientation data) is contained in header records identified by keywords.

The model for an MTZ file is thus based on two components; one (the header) keyed on keywords such as SYMMETRY, CELL, etc., and the other (comprising the reflections) keyed on the H, K, and L attributes/columns. Library routines contained in the source file *mtzlib.f* allow one to read and write information to the header or the reflection records. Access to the MTZ files should always be via these routines. A mechanism has been put in place for handling missing data; see Changes (below) for details.

An example of the use of column labels is provided by the reflection file used in many of the CCP4 example scripts. This file (*toxd.mtz*) contains the native data set plus three derivative data sets (Hg, I, and Au, the latter with anomalous measurements) for the dendrotoxin from green mamba.[2] The labels used, with associated column types in brackets, are as follows:

[2] T. Skarzynski, *J. Mol. Biol.* **224,** 671 (1992).

TABLE II
MTZ STANDARD PROGRAM LABELS AND COLUMN TYPES

Program label	Type	Description
H, K, L	H	Miller indices
M/ISYM	Y	Partiality flag and symmetry number
BATCH	B	Batch number
I	J	Intensity I
SIGI	Q	σI (standard deviation)
FRACTIONCALC	R	Calculated partial fraction of intensity
IMEAN	J	Mean intensity
SIGIMEAN	Q	σI_{mean}
FP	F	Native F value
FC	F	Calculated F
FPHn	F	F value for derivative n
DP	D	Anomalous difference for native data $(F^+ - F^-)$
DPHn	D	Anomalous difference for derivative n
SIGFP	Q	σFP (standard deviation)
SIGDP	Q	σDP
SIGFPHn	Q	σFn
SIGDPHn	Q	σDPHn
PHIC	P	Calculated phase
PHIB	P	Phase
FOM	W	Figure of merit
WT	W	Weight
HLA, HLB, HLC, HLD	A	*ABCD* H/L coefficients
FreeR_flag	I	Free R flag (as file label)
Miscellaneous	R or I	Any attribute required

```
 K L                 (H)       - indices
TOXD3 SIGFTOXD3      (F, Q)    - native F and sd
MM11  SIGFMM11       (F, Q)    - Hg derivative F and sd
I100  SIGFI100       (F, Q)    - I derivative F and sd
AU20  SIGFAU20       (F, Q)    - Au derivative F and sd
NAU20 SIGANAU20      (D, Q)    - Au derivative (F+ - F-) and sd
reeR_flag            (I)       - free-R flags
```

The header of the MTZ file contains important information that can be viewed most easily with the program *mtzdump*. A sample of the output is

```
Cell Dimensions :
73.58  38.73  23.19  90.00  90.00  90.00
```

```
Resolution Range :

0.00074     0.18900     (    36.761 -     2.300 A )
```

This information is read in automatically by each program. *mtzdump* also lists the data columns and associated statistics, and can list a subset of the reflections. Columns of data are typically assigned as input to a program using the LABIN keyword, for example for the phasing program *mlphare*:

```
LABIN FP=FTOXD3 SIGFP=SIGFTOXD3 -
      FPH1=FAU20 SIGFPH1=SIGFAU20 -
      DPH1=ANAU20 SIGDPH1=SIGANAU20 -
      FPH2=FMM11 SIGFPH2=SIGFMM11 -
      FPH3=FI100 SIGFPH3=SIGFI100
```

The output labels required for the multiple isomorphous replacement (MIR) phase and its figure of merit could then be named using the LABOUT keyword as follows:

```
LABOUT  PHIB=PHI_Au_Hg_I FOM=FOM_Au_Hg_I
```

Maps

The electron-density map is stored in a randomly accessible binary file as a three-dimensional array preceded by a header that contains all the information needed to describe it. This includes the extent of the array, and the grid it is calculated on; the axis order; the cell and symmetry; a title; and the minimum, maximum, and mean density. Maps are structured as a number of sections each containing a (fixed) number of rows and each row containing a (fixed) number of columns. The format is also used for envelope masks and images. Library routines in the source file *maplib.f* allow one to read and write information to the header section or the map data.

Coordinates

The standard format adopted for coordinate data is that used in the Brookhaven Protein Data Bank (PDB). The programs of the suite will handle either complete files or files containing only a subset of the allowed record types. In particular, the records containing the cell (CRYST1 and SCALEx) and coordinate data (ATOM, HETATM, or TER records) are used by CCP4 programs, the others being ignored. The PDB provides a full description of the complete format. Library routines for reading and writing to PDB files are provided in the source file *rwbrook.f.*

The standard setting of the orthogonal axes relative to the crystallographic for the Brookhaven format is

$$\mathbf{x}\|\mathbf{a} \quad \mathbf{y}\|\mathbf{c}^* \times \mathbf{a} \quad \mathbf{z}\|\mathbf{c}^*$$

The suite assumes these settings if the SCALEx cards are not present in a coordinate file.

It is hoped to replace the PDB format soon by the new macromolecular Crystallographic Information File (mmCIF) format, which has many of the features incorporated in the reflection format [see [30] in this volume[3]; and Changes (below) for details].

Library Routines

One fruit of the collaborative nature of program development has been an extensive and thoroughly tested set of routines, covering most basic crystallographic applications. This is desirable both for speed in developing new software, which can utilize these, and for accuracy; defects in code are best uncovered by frequent and varied use. These library routines form a relatively stable core to the suite.

The CCP4 library subroutines perform the basic crystallographic and programming operations. In addition to the routines in *mtzlib, maplib,* and *rwbrook,* which have already been mentioned in relation to the standard data formats, the following library modules are distributed that deal with general crystallographic operations.

fftlib Crystallographic fast Fourier transform (FFT) routines[4]
modlib Various vector and matrix operations
symlib Useful routines for handling symmetry operations

Finally, there are a number of utility routines concerned with input/output and other administrative operations.

parser Processing free-format program input containing "keywords"
  keyparse A higher level interface to the parser routines
ccplib Contains various utility routines that are potentially machine dependent, such as I/O operations, file assignment, etc. It is built up from either VMS-specific code in *vms.for* and *vmsdiskio.for* or Unix-specific code in *unix.m4, diskio.f* and *library.c*
plot84lib Low-level graphics with plot84 metafiles
  plotsubs Higher level interface

[3] P. E. Bourne, H. M. Berman, B. McMahon, K. D. Watenpaugh, J. D. Westbrook, and P. M. D. Fitzgerald, *Methods Enzymol.* **277,** [30], 1997 (this volume).
[4] L. Ten Eyck, *Acta Crystallogr.* **A29,** 183 (1973).

pack_f, pack_c Image-(un)packing routines
binsortint Interface to binsort from Fortran programs

This group includes the machine-specific routines that ultimately allow CCP4 to run on a wide range of platforms.

In addition to the library routines, a data library is also distributed. This includes files of space group symmetry operators, atomic form factors, the standard dictionary used by *protin* and that used by *restrain,* and much other useful basic data.

Application Programs

A full list of programs distributed by CCP4 is given in the manual. The principal programs are contained in the directory *$CCP4/src* and its subdirectories. New programs are continually being donated to the suite, corresponding either to an improved solution to an existing computational task or to an implementation of a new technique. The suite is inclusive and some duplication does occur, but when a program has clearly been superseded it is moved to the *$CCP4/unsupported/src* directory. These are still distributed for those who have formed an attachment to them. Finally, some programs ("aggregated" software) that are not strictly part of the CCP4 suite are nevertheless distributed by CCP4.

The infrastructure of CCP4 means that new programs can be distributed to the community of users quickly. Users thus obtain state-of-the-art programs representing recent advances in the field. In return, the author obtains a large audience for the program, providing valuable feedback. This interchange between programmer and users is one of the most important features of the CCP4 project.

Details on general procedures can be found in the manual supplied with the suite, and details on individual programs are supplied as *man* files, HTML files and/or formatted text files. Here, we review the principal areas covered by the suite, and discuss the main associated programs.

Data Scaling and Merging

Reflection data may be introduced to the suite via a number of routes: *abscale* (data from *mosflm*), *absurd* (data from *madnes*), *scalepack2mtz* (merged data from *scalepack*), and *rotaprep* (most other data formats). These programs give data files in MTZ format, which in general should be sorted (*sortmtz*), the different batches scaled (*scala*) and merged (*agrovata*), and then the intensities converted to amplitudes (*truncate*). Finally, *cad* will combine different data sets (e.g., native and derivatives) into one file, ensuring that the sort order and asymmetric unit are correct.

As larger proteins are studied, and multiwavelength anomalous dispersion (MAD) phasing becomes more routine, there is a need for better experimental data. Part of this improvement must come from better scaling and merging algorithms. Also, refinement programs require a reliable estimate of the standard uncertainty of each reflection and this must be determined at this stage. P. Evans (MRC–LMB, Cambridge) is developing a new version of *scala* (to supercede the present *scala* and *agrovata*) that will allow scaling against many variables, (rotation angle, detector position, and so on). One useful option is to include a master data set in the minimization, which gives a more robust variant of local scaling. The estimates of standard uncertainty are obtained by modifying those given by the processing package to take account of agreement between symmetry equivalent reflections.

Heavy-Atom Phasing by Multiple Isomorphous Replacement or Multiwavelength Anomalous Dispersion Techniques

Z. Otwinowski (University of Texas) appreciated that many older heavy-atom refinement programs produced biased parameters. The heavy-atom sites were used to determine preliminary protein phases, which were then treated as fixed during the subsequent refinement of the sites. By the simple improvement of testing all possible phases for each reflection, and appropriately weighting these, Otwinowski obtained more reliable parameters, more accurate protein phases, and more realistic probabilities for each phase. The program *mlphare* is now widely used for heavy-atom refinement, and for both MIR and MAD phasing in conjunction with density modification.[5]

Phase Improvement and Molecular Averaging

K. Cowtan (University of York) has developed a program for phase improvement, *dm,* incorporating density modification techniques from many authors. Cowtan's own work includes an investigation of new methods of phase recombination to reduce the characteristic overweighting commonly caused by density modification calculations—this method can produce dramatically better maps.[6] An extended version of the program, *dmmulti,* developed with help from D. Schuller (University of California, Irvine), allows simultaneous density modification with averaging across several crystal forms simultaneously.

Cowtan has also written a range of support programs for the determina-

[5] Z. Otwinowski, *in* "Isomorphous Replacement and Anomalous Scattering: Proceedings of the CCP4 Study Weekend" (W. Wolf, P. R. Evans, and A. G. W. Leslie, eds.) Daresbury Laboratory, U.K., p. 80. 1991.

[6] K. D. Cowtan and P. Main, *Acta Crystallogr.* **D52,** 43 (1996).

tion and manipulation of electron-density maps and solvent and averaging masks (*maprot, mapmask,* and *ncsmask*).

The *solomon* package[7] of J.-P. Abrahams (MRC–LMB, Cambridge), developed for the solution of a difficult averaging problem beyond the scope of the automatic *dm* approach, is also a part of the suite.

Molecular Replacement Using AmoRe

The most exciting development in molecular replacement has been the successful use of poorer and poorer models, which can be positioned in the new crystal form, and that provide sufficient initial phasing information for subsequent phase improvement techniques to be successful. This is possible only when the programs can automatically search large numbers of solutions at each stage rapidly, and without excessive user intervention. J. Navaza (Châtenay Malabry, France) has incorporated this into the program *AmoRe,* and the version distributed with CCP4 has solved many structures.[8]

Macromolecular Refinement

It has been appreciated for many years that least-squares minimization is not the optimal way of refining a set of coordinates that are a long way from their target values, and that it can become trapped in false minima. G. Murshudov (University of York) has written a program, *refmac,* that has an option to use a maximum likelihood residual, where the appropriate weighting for reflections is based on the fit of F_o and F_c for the free set of reflections, and includes the experimental standard uncertainty.[9] This converges more quickly than least squares in many cases, and generates properly weighted and less biased maps for model correction. As for the older program *prolsq,* information on geometric restraints is prepared by the program *protin,* but in contrast to *prolsq* the calculation of model structure factors and the associated gradients is done internally. *prolsq* is thus effectively made obsolete by *refmac.*

Also distributed with the suite is the refinement program *restrain.* Although this uses a least-squares residual, it has many features not found in *refmac* that may be useful.[10] For example, the residual may be minimized with respect to individual or grouped (TLS) anisotropic displacement parameters (appropriate for higher resolutions).

[7] J. P. Abrahams and A. G. W. Leslie, *Acta Crystallogr.* **D52,** 30 (1996).

[8] J. Navaza, *Acta Crystallogr.* **A50,** 157 (1994).

[9] G. N. Murshudov, E. J. Dodson, and A. Vagin, *in* "Macromolecular Refinement: Proceedings of the CCP4 Study Weekend," Daresbury Laboratory, U.K., p. 93. 1996.

[10] H. Driessen, M. I. J. Haneef, G. W. Harris, B. Howlin, G. Khan, and D. S. Moss *J. Appl. Crystallogr.* **22,** 510 (1989).

Structure Validation

The suite contains many programs giving useful information on angles, *B* values, hydrogen bonding, intermolecular contacts, etc., for example *act, distang, geomcalc,* and *sortwater.* Perhaps the most widely used program is *procheck,* developed by R. Laskowski (University College, London), which does a comprehensive check of the stereochemistry of a protein, and highlights parts of the structure where conformations are unusual.[11] The latter are due either to interesting properties of the structure, or to possible errors of interpretation, and worthy of investigation during the rebuilding step.

Changes

Version 3.0 of the suite was released in April 1996, followed by version 3.1 in May. The current version 3.2 was released in November 1996. The following new programs were included in these releases: CROSSEC (scattering cross-sections), GEOMCALC (geometry calculations), HGEN (generate hydrogen positions), MAKEDIC (make dictionary entry), MAPROT (map manipulations), MATTHEWS_COEF (Matthews coefficient), MTZMNF (add missing number flags), RASMOL (molecular graphics), REFMAC (maximum likelihood structure refinement), RESTRAIN (structure refinement), SCALEPACK2MTZ (conversion jiffy), SOLOMON (density modification), SORTWATER (sort waters), XDLMAPMAN (map manipulations) and XDLDATAMAN (data manipulations). For more details on these programs, the reader is referred to the individual program documentation. The V. Lamzin (EMBL, Hamburg) program ARP is scheduled for release with the next version. The latest releases also contain numerous improvements to existing programs. The following more general changes to the suite are also taking place.

Missing Data Treatment

It is recommended practice that the set of reflection indices in an MTZ file be made complete within the desired resolution range. The MTZ file will then contain records in which there are indices but no measured data for some or all of the data columns. For version 3.0, these columns are flagged MNF (missing number flag). In the following example,

0	0	2	MNF	MNF
0	0	4	517.0	23.0
0	0	6	1567.0	57.0
			. . .	. . .

[11] R. A. Laskowski, M. W. MacArthur, D. S. Moss, and J. M. Thornton, *J. Appl. Crystallogr.* **26,** 283 (1993).

the MNFs indicate that although $hkl = (0, 0, 2)$ is not a systematic absence, there are no data available for that reflection. Alternatively, a particular reflection may be recorded for the native protein but not for a derivative, and the corresponding combined reflection data record should indicate "missing data" for the derivative. This convention means that it is easy to estimate completeness and programs such as *refmac* and *sigmaa* can "restore" data estimates where required.

The value that the MNF is actually represented by in the data file can be chosen by the user. On Unix systems, the default is NaN, which is a real number on which arithmetic operations cannot be performed, so that the MNF cannot be confused with real data. This is in contrast to older conventions in which missing data were typically indicated with a zero, a perfectly valid value for some data columns. The program *mtzdump* (version 3.2) currently represents MNFs in its listing by the symbol "?"; this is not related to the MNF value used in the data file itself. Further details about MNFs can be found on the CCP4 Web pages.

Current programs will write MNFs to the data file as appropriate. The program *mtzmnf* attempts to identify missing data entries in an old-style MTZ file and replaces them with MNFs. In addition, a script is provided with the distribution (*$CETC/uniqueify* for Unix systems and $CETC/UNIQUEIFY.COM for VMS) that completes a data set within the desired resolution range, adds free-*R* flags, and adds MNFs where appropriate. This script would typically be run on the output from *truncate.*

mmCIF format

It is intended to replace the PDB format for coordinates by the new macromolecular CIF (mmCIF) format. P. Keller (now EBI, Cambridge), funded by CCP4, has developed low-level routines for reading and writing mmCIF files and library routines, similar to those in *rwbrook.f,* to perform basic tasks. It is planned to convert the application programs in the near future to use these routines.

Graphical User Interface

In response to demand, it is planned to develop a graphical user interface (GUI) to the CCP4 suite. A draft specification has been drawn up, which aims to provide a user-friendly interface, while maintaining the flexibility of the existing suite. L. Potterton (University of York) is being funded to implement the GUI, and started at the beginning of 1997.

Program Documentation

To date, individual program documentation has been distributed as *man* page source files (in *$CCP4/man*) and formatted files (in *$CCP4/doc*).

These could be viewed directly, or via the Unix *man* facility. From version 3.2, documentation is also being distributed as HTML files. At the moment these are derived automatically from the files in *$CCP4/man,* but later they will form the primary format. It is planned to integrate the HTML files into the GUI.

Conclusion

There are disadvantages to the diverse traditions and dispersed centers where CCP4 is under development, but these have been largely overcome by centralizing the distribution and maintenance at the Daresbury Laboratory. The professional expertise provided there is essential to administer the large body of source code now deposited. This service is possible only because of the central BBSRC funding, whose recognition of the key value of this group over the years must be acknowledged. The additional contributions from industrial companies have also played an important role in allowing the project to expand.

The CCP4 tradition of organic growth is based on the interests and enthusiasms of the individuals involved. Such a development could never have a commercial basis; there is no equitable mechanism for making payments to contributors. The CCP4 practices are in the best tradition of science, another example of how scientific research is best fueled by openness in the exchange of ideas, methodology, and solutions on a generous and shared basis, in which the individuals are rewarded by the successful usage of the contributions. We are sure this is the explanation for the successful growth of the CCP4 suite over the past 17 years.

Distribution

The program suite is licensed free to academic institutions, and is available by anonymous FTP from anonymous@ccp4a.dl.ac.uk:pub/ccp4 or can be supplied on tape for a small handling charge. Mirror sites have been set up in San Diego, California (anonymous@rosebud.sdsc.edu:/pub/sdsc/xtal/CCP4) and at the Photon Factory, Japan (anonymous@pfweis.kek.jp:/mirror/ccp4/ccp4) by P. Bourne (San Diego Supercomputer Center) and A. Nakagawa (Hokkaido University), respectively. Commercial organizations should contact CCP4 (e-mail: ccp4@dl.ac.uk) to obtain current licensing conditions.

Acknowledgments

Many people have contributed to CCP4 over the years and we thank them for their time and effort. The Daresbury staff has been pivotal in directing and maintaining standards, and handling the now extensive administration. CCP4 is supported by the BBSRC.

Author Index

Numbers in parentheses are footnote reference numbers and indicate that an author's work is referred to although the name is not cited in the text.

A

B

D

E

H

I

J

K

L

N

O

P

Q

R

S

T

U

V

W

X

Y

Z

Subject Index

A

B

C

D

M

R

S

T

V

W

X

Y

ISBN 0-12-182178-1

90038

9 780121 821784